Climate Change on Diseases and Disorders of Finfish in Cage Culture, 3rd edition

Climate Change on Diseases and Disorders of Finfish in Cage Culture, 3rd edition

Edited by

Patrick T.K. Woo

Department of Integrated Biology, University of Guelph, Guelph, Canada

and

Rohana P. Subasinghe

FUTUREFISH Co. Ltd., St James's Place, London, UK

CABI

CABI is a trading name of CAB International

CABI
Nosworthy Way
Wallingford
Oxfordshire OX10 8DE
UK

Tel: +44 (0)1491 832111
E-mail: info@cabi.org
Website: www.cabi.org

CABI
200 Portland Street
Boston
MA 02114
USA

Tel: +1 (617)682-9015
E-mail: cabi-nao@cabi.org

A catalogue record for this book is available from the British Library, London, UK.

ISBN-13: 9781800621626 (hardback)
9781800621633 (ePDF)
9781800621640 (ePub)

DOI: 10.1079/9781800621640.0000

Commissioning Editor: Caroline Makepeace
Editorial Assistant: Lauren Davies
Production Editor: Shankari Wilford

Typeset by SPi, Pondicherry, India
Printed and bound in the UK by Severn, Gloucester

Contents

Contributors

Note: corresponding authors are indicated by an asterisk.

Edgar Brun, Department of Aquatic Animal Health and Welfare, Norwegian Veterinary Institute, PO Box 64, 1431 Ås, Norway. E-mail: edgar.brun@vetinst.no

Kenneth D. Cain*, Department of Fish and Wildlife Sciences, University of Idaho, 875 Perimeter Drive, Moscow, ID 83844-1136, USA. Current address: NOAA Fisheries, Northwest Fisheries Science Center, Manchester Research Station, 7305 Beach Drive East, Port Orchard, WA 98366, USA. E-mail: kcain@uidaho.edu

Siow Foong Chang, Animal and Veterinary Service, National Parks Board, 1 Cluny Road, Singapore Botanic Gardens, Singapore 259596. E-mail: chang_siow_foong@nparks.gov.sg

Benjamín Costas, CIIMAR – Interdisciplinary Centre of Marine and Environmental Research of the University of Porto, Terminal de Cruzeiros do Porto de Leixões, Av. General Norton de Matos, S/N, 4450-208 Matosinhos, Portugal; and ICBAS-UP – Instituto de Ciências Biomédicas Abel Salazar, Universidade do Porto, Rua de Jorge Viterbo Ferreira 228, 4050-313 Porto, Portugal. E-mail: bcostas@ciimar.up.pt

Ha Thanh Dong, Aquaculture and Aquatic Resources Management, Department of Food, Agriculture and Bioresources, School of Environment, Resources, and Development, Asian Institute of Technology (AIT), PO Box 4, Khlong Luang, Pathum Thani 12120, Thailand. E-mail: htdong@ait.ac.th

Oystein Evensen, Faculty of Veterinary Medicine, Norwegian University of Life Sciences, PO Box 5003, 1432 Ås, Norway. E-mail: oystein.evensen@nmbu.no

Mark D. Fast, Department of Pathology and Microbiology, Atlantic Veterinary College, University of Prince Edward Island, Charlottetown, Prince Edward Island, Canada, C1A 4P3. E-mail: mfast@upei.ca

Ignacio Fernández, Centro Oceanografico de Vigo, Instituto Español de Oceanografia (IEO), CSIC, 36390 Vigo, Spain. E-mail: ignacio.fernandez@ieo.csic.es

Thijs Frenken, Cluster Nature & Society, HAS University of Applied Sciences, Onderwijsboulevard 221, 5223 DE 's-Hertogenbosch, The Netherlands. E-mail: t.frenken@has.nl

Kyle A. Garver, Pacific Biological Station, Fisheries and Oceans Canada, Nanaimo, British Columbia, Canada, V9T 6N7. E-mail: Kyle.Garver@dfo-mpo.gc.ca

Paulo Gavaia, CCMAR – Centre of Marine Sciences, University of Algarve, Campus de Gambelas, 8005-139 Faro, Portugal; and FMCB-UALG – Faculty of Medicine and Biomedical Sciences/ University of Algarve, Campus de Gambelas, 8005-139 Faro, Portugal. E-mail: pgavaia@ualg.pt

Matt J. Griffin, Department of Pathobiology and Population Medicine, College of Veterinary Medicine, Mississippi State University, Stoneville, MS 39762, USA. E-mail: griffin@cvm.msstate.edu

Giana Bastos Gomes, Marine Biotechnology, Temasek Life Science Laboratory, 1 Research Link, National University of Singapore, Singapore 117604. E-mail: giana@tll.org.sg

Taylor I. Heckman, Aquatic Animal Health Laboratory, Department of Medicine & Epidemiology, School of Veterinary Medicine, University of California, Davis, Davis, CA 95616, USA. E-mail: tiheckman@ucdavis.edu

Felicity A. Huntingford*, School of Biodiversity, One Health and Veterinary Medicine, University of Glasgow, Glasgow G12 8QQ, UK. E-mail: Felicity.Huntingford@glasgow.ac.uk

Bastiaan W. Ibelings, Department F.A. Forel for Environmental and Aquatic Sciences and Institute for Environmental Sciences, University of Geneva, 24 rue du Général-Dufour, 1211 Geneva 4, Switzerland. E-mail: bastiaan.ibelings@unige.ch

Stewart C. Johnson, Pacific Biological Station, Fisheries and Oceans Canada, Nanaimo, British Columbia, Canada, V9T 6N7. E-mail: Stewart.Johnson@dfo-mpo.gc.ca

Sunil Kadri, Institute of Aquaculture, University of Stirling, Stirling FK9 4LA, UK. E-mail: sunil@aquainnovation.com

Jo-Ann Leong*, Hawaii Institute of Marine Biology, School of Ocean and Earth Sciences Technology, University of Hawaii at Manoa, Manoa, HI 96744, USA. E-mail: joannleo@hawaii.edu

Gilda D. Lio-Po, Southville Subd., Molo, Iloilo City, Philippines. E-mail: gdliopo@yahoo.com

Cátia Lourenço-Marques, Portuguese Institute for the Ocean and Atmosphere (IPMA)/Aquaculture Research Station of Olhão (EPPO), Av. Parque Natural da Ria Formosa s/n, 8700-194 Olhão, Portugal; and S2AQUA – Collaborative Laboratory, Association for a Sustainable and Smart Aquaculture, Av. Parque Natural da Ria Formosa s/n, 8700-194 Olhão, Portugal. E-mail: catia.marques@ipma.pt

Ana Rita Pinheiro Marques, Department of Infectious Diseases and Public Health, Jockey Club College of Veterinary Medicine and Life Sciences, City University, Hong Kong SAR, People's Republic of China. E-mail: apinheir@cityu.edu.hk

Christine A. Parker-Graham, Pacific Region Fish Health Program, US Fish and Wildlife Service, 500 Desmond Drive SE, Lacey, WA 98503, USA. E-mail: christine_parker-graham@fws.gov

Mark P. Polinski, National Coldwater Marine Aquaculture Center, US Department of Agriculture–Agricultural Research Service, 25 Salmon Farm Road, Franklin, ME 04634, USA. E-mail: Mark.Polinski1@usda.gov

Seth M. Rudman, School of Biological Sciences, 14204 NE Salmon Creek Avenue, Washington State University, Vancouver, WA 98686, USA. E-mail: seth.rudman@wsu.edu

Krishna R. Salin*, Aquaculture and Aquatic Resources Management, Department of Food, Agriculture and Bioresources, School of Environment, Resources and Development (SERD), Asian Institute of Technology (AIT), PO Box 4, Khlong Luang, Pathum Thani 12120, Thailand. E-mail: salinkr@ait.ac.th

João L. Saraiva, FishEthoGroup Association, Rua das Cegonhas D10, 8700-159 Olhão, Portugal; and CCMAR – Centre of Marine Sciences, University of Algarve, Campus de Gambelas, 8005-139 Faro, Portugal. E-mail: jsaraiva@ualg.pt

Sonja M. Saksida*, Department of Health Management, Atlantic Veterinary College – UPEI, Charlottetown, Prince Edward Island, Canada, C1A 4P3. E-mail: ssaksida@upei.ca

Heike Schmidt-Posthaus, Institute for Fish and Wildlife Health, Department of Infectious Diseases and Pathobiology, University of Bern, Länggassstrasse 122, P.O. Box, 3001 Bern, Switzerland. E-mail: heike.schmidt@unibe.ch

Helmut Segner, Institute for Fish and Wildlife Health, Department of Infectious Diseases and Pathobiology, University of Bern, Länggassstrasse 122, P.O. Box, 3001 Bern, Switzerland. E-mail: helmut.segner@unibe.ch

Darshana Senarathna, School of Environment, Resources and Development (SERD), Asian Institute of Technology, Khlong Luang, Pathum Thani 12120, Thailand. E-mail: trishansenarathna@gmail.com

René S. Shahmohamadloo*, School of Biological Sciences, Washington State University, 14204 NE Salmon Creek Avenue, Vancouver, WA 98686, USA. E-mail: rene.shahmohamadloo@wsu.edu

Andrew P. Shinn*, INVE (Thailand), 471 Bond Street, Bangpood Sub-district, Pakkred District, Nonthaburi Province 11120, Thailand. E-mail: a.shinn@inveaquaculture.com

Florbela Soares*, Portuguese Institute for the Ocean and Atmosphere (IPMA)/Aquaculture Research Station of Olhão (EPPO), Av. Parque Natural da Ria Formosa s/n, 8700-194 Olhão, Portugal; and S2AQUA – Collaborative Laboratory, Association for a Sustainable and Smart Aquaculture, Av. Parque Natural da Ria Formosa s/n, 8700-194 Olhão, Portugal. E-mail: fsoares@ipma.pt

Esteban Soto*, Aquatic Animal Health Laboratory, Department of Medicine & Epidemiology, School of Veterinary Medicine, University of California, Davis, Davis, CA 95616, USA. E-mail: sotomartinez@ucdavis.edu

Sophie N. St-Hilaire*, Department of Infectious Diseases and Public Health, Jockey Club College of Veterinary Medicine and Life Sciences, City University, Hong Kong SAR, People's Republic of China. E-mail: ssthilai@cityu.edu.hk

Rohana P. Subasinghe*, FUTUREFISH Co. Ltd., Suite LG 11, St. James's Place, London, SW1A 1NP, UK. E-mail: rohana@futurefish.org

Vera L. Trainer, Northwest Fisheries Science Center, National Oceanic and Atmospheric Administration, Seattle, WA 98112, USA. E-mail: vera.l.trainer@noaa.gov

Nguyen Tien Vinh, Aquaculture and Aquatic Resources Management, Department of Food, Agriculture and Bioresources, School of Environment, Resources, and Development, Asian Institute of Technology (AIT), PO Box 4, Khlong Luang, Pathum Thani 12120, Thailand. E-mail: vinh5k0@gmail.com

Pieter van West, International Centre for Aquaculture Research and Development and Aberdeen Oomycete Laboratory, University of Aberdeen, Institute of Medical Sciences, Foresterhill, Aberdeen AB25 2ZD, UK. E-mail: p.vanwest@abdn.ac.uk

Thomas Wahli*, Institute for Fish and Wildlife Health, Department of Infectious Diseases and Pathobiology, University of Bern, Länggassstrasse 122, P.O. Box, 3001 Bern, Switzerland. E-mail: thomas.wahli@unibe.ch

Patrick T.K. Woo*, Department of Integrative Biology, University of Guelph, Guelph, Ontario, Canada, N1G 2W1. E-mail: pwoo@uoguelph.ca

Ratchakorn Wongwaradechkul, INVE (Thailand), 471 Bond Street, Bangpood Sub-district, Pakkred District, Nonthaburi Province 11120, Thailand. E-mail: r.wongwaradechkul@inveaquaculture.com

Preface to the Third Edition

The ongoing changes to the global environment in recent decades have been well documented. Briefly, they are caused by massive emissions of carbon dioxide, methane and nitrous oxide into the atmosphere, due in part to burning fossil fuels in power plants to generate electricity and raising large numbers of domestic animals for human consumption. Effects of climate change include the rise in global temperature and acidification of water with reduction in dissolved oxygen. These abiotic changes affect the metabolism, development, growth, reproduction and lifespan of aquatic organisms, including fish.

Changes to the environment will continue even if we can significantly reduce the output of anthropogenic greenhouse gases. However, expected impacts of climate change can be reduced if the rise in global average temperature is less than 1.5°C above the pre-industrial level. This will provide national governments and international agencies with more time to develop novel strategies to mitigate ongoing environmental changes. It is now more difficult to achieve the 1.5°C target temperature as the final agreement of the Glasgow Climate Pact (2021) is a compromise. We agree with Dr Antonino Guterres, United Nations Secretary-General, who at the conclusion of the COP 26 summit lamented 'Our fragile planet is hanging by a thread ... [and] ... We are still knocking on the door of climate catastrophe.' Delegates at the COP 27 summit (Sharm El Sheikh, Egypt; 2022) tried to but were unable to modify the agreement made at COP 26 on fossil fuels.

The current world population of 7.96 billion will increase to about 9.8 billion by 2050. Consequently, not only will the demand for food grow accordingly, the cost of food production will also escalate. The increased cost is in part because it will have to compete with other increases in human activities (e.g. housing, transportation, industry) for available usable land.

Fish is considered an excellent alternative to terrestrial animal meat. In many parts of the world, fish is less expensive than meat. It is not only a good source of protein, but also a good source of many micronutrients, minerals, essential fatty acids and vitamins. Some fish species are rich in polyunsaturated fatty acids considered essential nutrients for normal brain function in humans. Other benefits may include reduction in cardiovascular diseases, mental depression and autoimmune disease. Consequently, fish is an important component of a healthy and well-balanced diet for humans.

Many marine fish stocks have been depleted significantly over the past decades due to overfishing, as well as loss of spawning grounds and habitats because of industrial development and pollution. One option for maintaining future global fish supplies is aquaculture, which in part is the breeding and raising of fish in captivity.

Aquaculture is the fastest-growing food-producing sector in the world. Briefly, the ongoing climate change has and will continue to affect fish production. Aquaculture will have to keep responding to ongoing changes which include rise in water temperatures, reduction in dissolved oxygen and salinity, and increased prevalence and severity of disease outbreaks. Successful adaptations will also rely on abilities of fish producers to respond to changes in the environment. Producers in developing countries will find it more difficult to continue responding to ongoing abiotic environmental changes. In general, small-scale fish farmers are usually in developing nations while large-scale producers are in more developed countries. Advantages of producers in large-scale operations include better and more stable financial support, having up-to-date management skills, being able to afford more advanced technologies and having reliable supply-chain systems.

Cage culture of finfish is an important component in aquaculture. It is usually practised in countries with suitable coastlines or freshwater habitats such as lakes, reservoirs, rivers and dams. Since fish are bred and/or raised in captivity, some 'negative' effects due to climate change can be readily neutralized. Also, fish are monitored regularly for infectious diseases and disorders; early detections and treatments will prevent or at least reduce the severity of outbreaks.

Environmental changes due to ongoing changes to the aquatic systems will have direct and indirect effects on fish and their health (infectious diseases and/or non-infectious disorders); hence, practical changes and recommendations are needed. The third edition of *Diseases and Disorders of Finfish in Cage Culture* has been renamed to include it in the series on climate change and fish health. Two volumes in the three-book series were published in 2020. Our current book, *Climate Change on Diseases and Disorders of Finfish in Cage Culture, 3rd edition* (CCDDFCC.3), is the third volume. It has 12 chapters written by 43 internationally recognized experts from 12 countries in North and South America, Europe and Asia. Nine chapters from *Diseases and Disorders of Finfish in Cage Culture, 2nd edition* (2014) have been either completely rewritten by new contributors (Chapters 1, 2, 5, 6 and 8) or updated by their original authors (Chapters 3, 4, 7 and 9). CCDDFCC.3 also has three new chapters, these are on harmful algal blooms, biosecurity and fish welfare (Chapters 10, 11 and 12, respectively).

All chapters in CCDDFCC.3, especially those on infectious organisms, follow a relatively similar and logical pattern which was initially used in *Fish Diseases and Disorders, Vol. 1: Protozoan and Metazoan Infections* (1995). The format has, over the years, undergone minor changes to make it more 'user friendly'.

We hope CCDDFCC.3 will be useful to colleagues currently involved in cage-culture operations or those who have plans to use the system. We suggest this volume will also be key reading for research scientists, ecologists, veterinarians, fish health consultants and policy makers interested in fish health and ongoing changes to the environment. It will serve as a reference text for workshops on fish health and for academic courses such as on aquaculture and fish health.

Patrick T.K. Woo and Rohana P. Subasinghe

Preface to the Second Edition

The world population was 7 billion in 2011, and at the current rate of increase it will be about 8 billion by 2025. Also, the demand for animal protein as a food source will continue to increase and exert additional pressures on food production which will have to compete with other human activities (e.g. housing, transportation, industry) for the limited usable land. Animal protein contains essential amino acids which are important components of a balanced diet. However, free ranging land animals are no longer a significant source of protein, and the production costs of farm animals continue to escalate. To increase efficiency and to reduce costs animal farms are large and often close to human habitations. Wastes associated with the large scale breeding of mammals and birds can pollute the environment and also increase the risks of disease outbreaks in animals with the subsequent interspecies transmission of zoonotic diseases (e.g. Nipah virus in pigs, avian influenza virus in birds, cryptosporidian parasites in cattle) to humans.

Finfish are an excellent source of protein and many marine species have beneficial PUFA (polyunsaturated fatty acids); however, the capture-fishery is either stagnant or in decline as there are no newly discovered fishing grounds. Also, natural fish stocks in many parts of the world have been significantly reduced due to more efficient fishing technologies, over and/or indiscriminate fishing, and the loss and/or destruction of spawning grounds. Industrial wastes (e.g. heavy metals, organophosphates) discharged into the aquatic environment can affect fish growth, survival and reproduction, and in some areas pollutants have accumulated in fish to the extent they are no longer suitable for human consumption. Cage culture of finfish (especially in-shore) has lower start-up and production costs and it does not have some of the problems associated with the raising of large numbers of warm blooded animals. Intensive culture of fish is one solution to producing more affordable animal protein; however, outbreaks of diseases may occur more frequently because of numerous factors, which include enhanced transmission of infectious pathogens between fish.

A tremendous volume of research has been conducted on the diseases and disorders since the publication of the first edition of 'Diseases and Disorders of Finfish in Cage Culture' in 2002. The aims, philosophy, audience, focus and format have remained unchanged. However, significant changes in the current edition include new contributors for eight of the nine chapters, the addition of a new chapter (on 'transmission of infectious agents between wild and farmed fish'), and the deletion of one chapter (on 'the history of cage culture') have resulted in a more relevant and informative text.

Our contributors are highly respected international experts from Asia, Australia, Europe, North America and South America. They have practical experience and/or research expertise on diseases/disorders and their diagnosis, and/or solutions to problems associated with cage culture. As with the first edition our primary objective is to produce an authoritative and practical volume for colleagues in the aquaculture industry, especially those associated with the cage culture of finfish. We also hope this volume will alert industry to potential and/or emerging diseases and disorders in specific regions of the world and to point out gaps in our knowledge so as to stimulate further research.

Patrick T.K. Woo and David W. Bruno

Preface to the First Edition

In many parts of the world the primary source of animal protein for humans is finfish. The intensive culture of finfish has grown significantly since the 1980s partly because of the dramatic decline in the natural fish stocks and the increase in fish consumption by the ever increasing population. For example, the worldwide consumption of fish between 1990 and 1997 increased by 30% while the capture fisheries increased only by 9%. The demand for fish is expected to continue to increase, especially as the more affluent consumers in the developed countries become more aware of the beneficial effects of fish (e.g. marine fish are an excellent source of polyunsaturated omega-3 fatty acids). Aquaculture is the only solution to the demand as it can provide consistently high quality fish protein year round. The industry is already considered the single fastest-growing food production process in the world.

The cage culture of finfish, especially mariculture, is becoming more popular because there are many economic advantages associated with this approach. However, it also has problems and one of them is disease. Disease outbreaks tend to occur more often when fish are raised under intensive culture conditions, and consequently both infectious and non-infectious diseases are important constraints to the industry.

Our primary objective is to produce an authoritative and practical volume on diseases and disorders of finfish in cage culture. We hope the book will also alert the industry to potential and/or emerging disease problems in specific regions of the world, and to point out gaps in our knowledge so as to stimulate further research. This book is designed for aquaculturalists who are using or intend to use cage culture. It will also be useful to fish health consultants (e.g. veterinarians), microbiologists, parasitologists, fish pathologists, and managers and directors of diagnostic laboratories. Each chapter is written by international experts who have personal experience or expertise on diseases and their diagnosis, and/or solutions to problems associated with the cage culture of finfish.

This book is divided into four parts – the first part is on the cage culture system, the second and third are on diseases/disorders in warmwater fish (water temperature above 15°C) and in coldwater fish, respectively. In each of these parts, there are three chapters – one on infectious diseases in fresh water (zero salinity), one on estuarine and marine diseases and one on non-infectious disorders. The final part on emerging diseases is to alert the industry to potential problems. We hope this division of the book will make it easier for the reader to access information on known diseases/disorders within a group of fish. The arrangement will also help to highlight similarities and differences in disease problems between groups of fish (e.g. between marine warmwater and marine coldwater fish). However, such divisions also create some minor problems, e.g. a few pathogens have been isolated from both seawater and freshwater fish, so our authors and editors have worked closely to avoid extensive overlaps in coverage. For example, furunculosis is in Chapter 4, with only brief reference to it in Chapter 3, because it is often seen in freshwater fish. Similarly, important infectious agents (e.g. *Piscirickettsia salmonis*) of marine fish (Chapter 3) are only briefly mentioned in Chapter 4 because of their lesser importance to freshwater fish.

There are books on infectious and on non-infectious diseases/disorders of fish (e.g. Fish Diseases and Disorders, Volumes 1–3, CAB International), but there are none devoted specifically to problems associated with cage culture of finfish. Problems encountered in cage culture are in some ways different from those using other rearing methods. In cage culture, fish may be exposed constantly to ubiquitous pathogens. Also, the stress associated with captive rearing creates opportunities for disease, and to a lesser extent non-infectious disorders, to become significant causes of morbidity and mortality. Transmissions of infectious agents are also enhanced, and fish become more susceptible to disease partly because their immune system may be compromised due to prolonged exposure to pollutants in the water and/or crowding stress. The impact and spread of new and/or emerging diseases are also important, and are influenced by factors that include international trade in eggs or fry, unauthorized transportation of fish, and contact with migratory or naive fish species. Under natural conditions these agents in their natural hosts may not be considered important pathogens, but in an expanded geographical and/or host range, under different environmental conditions or temperatures, they may lead to epizootics with serious consequential economic impact.

As the demand for animal protein increases in the new millennium, we expect a significant increase in cage culture activity in many countries. This will be true especially in countries with limited usable land mass but with relatively long coastlines and/or extensive river–lake systems. We hope this book will fill a niche and be useful to colleagues who are active in the industry.

Patrick T.K.Woo, David W. Bruno and L.H. Susan Lim

Previous titles by Patrick T.K. Woo

Climate Change and Infectious Fish Diseases
Edited by P.T.K. Woo, J-A. Leong and K. Buchmann
2020 528 pages ISBN 978 1 78924 327 7

Climate Change and Non-infectious Fish Disorders
Edited by P.T.K. Woo and G.K. Iwama
2020 256 pages ISBN 978 1 78639 398 2

Fish Viruses and Bacteria: Pathobiology and Protection
Edited by P.T.K. Woo and R.C. Cipriano
2017 376 pages ISBN 978 1 78064 778 4

Diseases and Disorders of Finfish in Cage Culture, 2nd Edition
Edited by P.T.K. Woo and D.W. Bruno
2014 354 pages ISBN 978 1 78064 207 9

Fish Parasites: Pathobiology and Protection
Edited by P.T.K. Woo and K. Buchmann
2011 400 pages ISBN 978 1 84593 806 2

Fish Diseases and Disorders, Volume 3, 2nd Edition: Viral, Bacterial and Fungal Infections
Edited by P.T.K. Woo and D.W. Bruno
2011 944 pages ISBN 978 1 84593 554 2

Fish Diseases and Disorders, Volume 2, 2nd Edition: Non-infectious Disorders
Edited by J.F. Leatherland and P.T.K. Woo
2010 416 pages ISBN 978 1 84593 553 5

Fish Diseases and Disorders, Volume 1, 2nd Edition: Protozoan and Metazoan Infections
Edited by P.T.K. Woo
2006 800 pages ISBN 978 0 85199 015 6

1 Cage Culture of Finfish: Its Importance, Distributions and Future Modifications in Ongoing Climate Change

Krishna R. Salin[1]*, Rohana P. Subasinghe[2], Darshana Senarathna[1] and Andrew P. Shinn[3]
[1]Asian Institute of Technology, Khlong Luang, Thailand; [2]FUTUREFISH Co. Ltd., London, UK; [3]INVE (Thailand), Nonthaburi, Thailand

1.1 Trends in Global Aquaculture

In 2015, the United Nations General Assembly accepted the 2030 Agenda for Sustainable Development, which establishes a shared framework for peace and prosperity for people and the planet, today and in the future. The 17 Sustainable Development Goals (SDGs), which are at its heart, serve as an urgent call to action for developed and developing countries to collaborate in a global partnership. Reducing poverty and other forms of deprivation requires concerted efforts to improve health and education, reduce inequality and stimulate economic growth while combating climate change and safeguarding our oceans and forests. Additionally, the SDGs aim to address the issue of feeding almost 10 billion people by 2050 and include targets that can be accomplished by increasing the contribution and management of fisheries and aquaculture to food security and nutrition, particularly in terms of natural resource usage. Aquaculture remains the fastest-growing food sector and is expected to meet the global demand for supply of aquatic food. The most recent biannual assessment by the Food and Agriculture Organization of the United Nations (FAO), *State of Fisheries and Aquaculture 2020*, provides an in-depth examination of global aquaculture up to 2018 (FAO, 2022).

The capture fisheries, particularly marine, have been stagnant since the late 1980s (Hannesson, 2015), and aquaculture has been instrumental in increasing the availability of fish for human consumption. According to FAO (2020), global aquaculture production reached a new all-time high of 114.5 million tonnes of live weight in 2018 (Fig. 1.1), with a farm gate value of US$263.6 billion. Aquatic animal production totalled 82.1 million tonnes (US$250.1 billion), aquatic algae production totalled 32.4 million tonnes (US$13.3 billion), and ornamental seashell and pearl production was 26,000 tonnes (US$179,000). Aquatic animal farming was dominated in 2018 by finfish (54.3 million tonnes, US$139.7 billion), which were harvested from both inland aquaculture (47 million tonnes, US$104.3 billion) and marine and coastal aquaculture (7.3 million tonnes, US$35.4 billion). Next to finfish were molluscs (17.7 million tonnes, US$34.6 billion) – primarily bivalves – crustaceans (9.4 million tonnes, US$69.3 billion), marine invertebrates (435,400 tonnes, US$2.0 billion), aquatic turtles (370,000 tonnes, US$3.5 billion) and frogs (131,300 tonnes, US$997 million) (FAO, 2020).

*salinkr@ait.ac.th

 Climate Change on Diseases and Disorders of Finfish in Cage Culture, 3rd Edition (eds P.T.K. Woo and R.P. Subasinghe)
DOI: 10.1079/9781800621640.0001

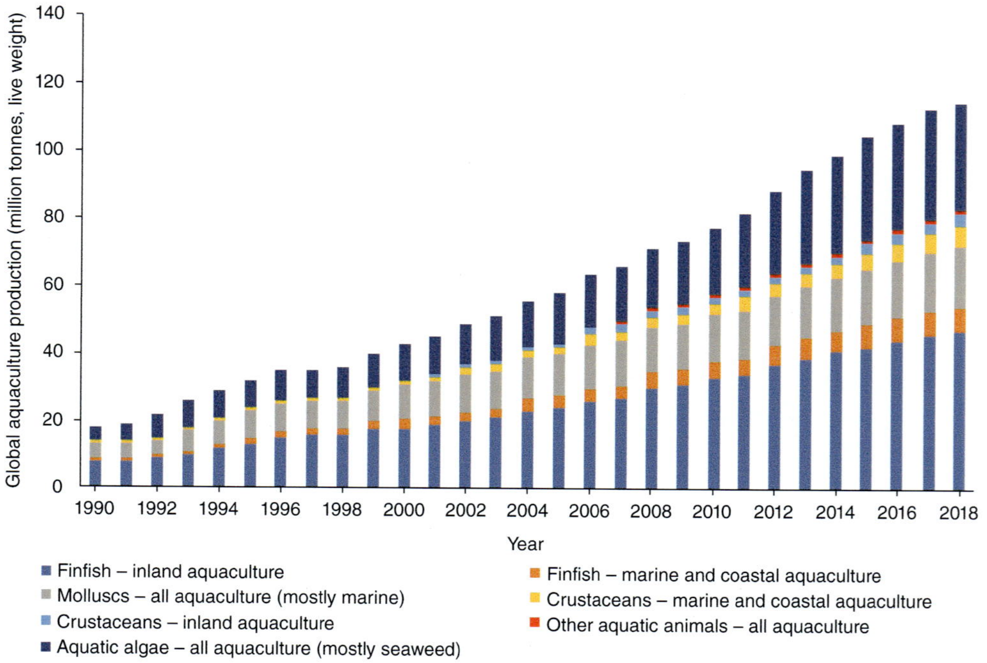

Fig. 1.1. Global aquaculture production of aquatic animals and algae (1990–2018). (From FAO, 2018.)

The contribution of aquaculture to global fish production was 46.0% in 2018. Aquaculture produced 52% of all aquatic animal food consumed in 2018 (FAO, 2020). Eighty-nine per cent of all fishery and aquaculture products (>158 million tonnes) were consumed by humans in 2019. The remaining 11% (i.e. about 20 million tonnes) comprised non-food uses including fishmeal and fish oil production (FAO, 2021). The supply of fish for human consumption has increased at a rate double that of population growth across the last five decades, increasing at an average annual rate of 3.2% from 1961 to 2013. World per capita apparent fish consumption rose from 9.9 kg in the 1960s to 14.4 kg in the 1990s and 19.7 kg in 2013, and to a level exceeding 20 kg in 2019 (FAO, 2021). Fish accounts for approximately 17.3% of the animal protein intake by the human population and 6.8% of all protein consumed. Globally, fish provides nearly 20% of the average per capita animal protein intake for approximately 3.3 billion people and 10% for approximately 5.6 billion people.

In 2019, around 37% of total fish production (equivalent live weight) was exported, with the value of exported fish and fish products surpassing US$162 billion. The populations of developed nations consume greater quantities of fish per capita than those of developing nations. Developed nations accounted for roughly 67% of total fisheries imports by value, while developing nations accounted for approximately 54% of total fishery exports by value and 61% by quantity (live weight equivalent) (FAO, 2021). The average annual fish intake per capita in industrialized nations increased progressively from 17.4 kg in 1961 to 26.4 kg in 2007 but levelled off at 24.4 kg from 2007 to 2017 (FAO, 2020). Although developing nations have a considerably lower comparable value, their yearly fish consumption rate has increased dramatically by 2.4% per annum between 1961 and 2017 (from 5.2 to 19.4 kg per capita). From 1961 to 2017, fish consumption increased by 1.5%, year-on-year (from 4.0 to 9.3 kg per capita) in low-income food-deficit countries (LIFDCs). In contrast, growth has been slow in Africa's least-developed countries (LDCs) (from 6.1 to 12.6 kg per capita), with an average annual increase of 1.3% (FAO, 2020). From 2018 to 2030, however, per capita fish consumption in Africa is anticipated to decline by 0.2% per year, from 10.0 to 9.8 kg (FAO,

2020). Despite anticipated increases in fish consumption in Asia and Latin America, the World Bank (2013) forecasted that fish consumption in Africa could decline significantly over the next two decades. In an era when the critical role of fish in global food and nutrition security has been widely acknowledged, such a forecast is concerning and merits attention (Subasinghe, 2017).

1.2 Aquaculture Systems

Fish are produced in aquaculture through a variety of farming systems that employ extensive, semi-intensive and/or intensive production techniques in fresh water, brackish water and seawater. Production systems range from small-scale, low-tech operations to sophisticated, high-tech industrial systems. Since both land and water are becoming limited because of several sectors competing for these primary resources, sustainable intensification has become the norm for aquaculture. Increased productivity (output per unit of area, water and/or energy) has become the recipe for future prosperity in aquaculture, as demonstrated by the rapid modernization and technological improvement in aquaculture production. While aquaculture is being intensified and modernized through technological innovations, over 70% of aquaculture is still produced by smallholders using ponds, cages, pens or raceways, depending on the species farmed and the availability of land and water.

In view of the increasing need for animal protein and the limited capacity of wild capture fisheries, marine aquaculture affords the potential to expand seafood production globally (Gentry *et al.*, 2017). It is obvious that feeding 10 billion people by 2050 will be a demanding task with significant challenges for securing the resource base required. Since land-based aquaculture will continue to compete with other land uses, moving aquaculture to rivers, lakes, reservoirs and the seas will become more prevalent in the coming decades.

1.3 Cage Culture

Farming of freshwater fish in earthen ponds and in cages in lakes, reservoirs and rivers is a prominent practice in Asia. Fish farming in cages originated in Southeast Asia as early as the 1800s with the traditional wooden-framed cages for freshwater fish set in the Mekong deltaic region of Cambodia (see Eng and Tech, 2002). Cage culture has expanded greatly and intensified dramatically in the last few decades with the emergence of large-scale marine fish farming in Japan and Europe since the 1950s. It is currently undergoing fast development in response to the challenges driven by globalization and increasing demand.

Due to the scarcity of suitable sites, intensive cage culture has expanded into previously unexplored open-water culture zones in inland, coastal and marine (nearshore or offshore) environments (FAO, 2020; Asmah *et al.*, 2021). There has been a drive to cluster existing cages for more intensive cage-farming techniques. Farming fish in cages has many advantages over the traditional pond culture. It is beneficial in maximizing the natural productivity of open-water resources, sharing water resources for multiple uses, developing a variety of farming practices, maintaining optimum water quality conditions, improving health management, preventing or eliminating predation, ensuring high stock survival, lowering initial investment and enabling easier harvesting. Cage-based aquaculture offers lower operational expenses per unit of production than other intensive aquaculture systems when scaled appropriately (Beveridge, 2004; Halwart *et al.*, 2007). As no energy is required for natural water exchange in cages located in areas where water current velocity is adequate, the carbon footprint from cage culture is also low because of the minimal investment cost per cubic metre of farm capacity.

1.3.1 Cage-culture systems

According to the FAO (Halwart *et al.*, 2007), 62 countries reported cage-culture production in 2005. Almost all marine finfish aquaculture is in cages. Many species of freshwater fish are also cultured in cages in Asia. In many countries, cage culture follows simple technology and locally available resources for construction and operation under the prevailing economic, social and environmental conditions. Cage-culture systems currently range from traditional family-owned and -operated small cages (typical in most Southeast Asian countries) to large-scale commercial cages used for salmon and other marine species such as sea bass, sea bream and tuna.

Cambodia has a long history of cage culture (Fig. 1.2), with aquaculture production significantly lower than that of capture fisheries (Brugère and De Young, 2015). It is dominated by small and marginal farmers in fresh water, brackish water and marine environments and continues to be the fastest-growing food production sector, with growth exceeding 18% between 2002 and 2014 (Joffre *et al.*, 2016). Despite shrinking profit margins, marine cage culture in floating net cages set in the estuaries and inshore waters is also developing in Cambodia. Although the marine cage sector contributed only 1.2% of Cambodia's total annual aquaculture production, its share in value was 6.1% at US$7 million (Joffre *et al.*, 2016). The species include finfish such as Asian sea bass, *Lates calcarifer*; shellfish such as the green mussel, *Perna viridis*, in floating rafts; blood cockles, *Tegillarca* spp., in shallow-water areas; and seaweed, *Kappaphycus* spp. and *Eucheuma* spp., on floating ropes.

Tilapia cages are one of the most popular farming systems for freshwater fish in the Mekong Delta in Vietnam (Fig. 1.3). It significantly contributes to the food security and livelihood of the rural communities. However, freshwater cage farms in Vietnam have faced challenges like water quality deterioration, diseases and fish mortalities. As cage farming for tilapia and striped catfish, *Pangasius* spp., has evolved, primarily driven by farmers' experience, the farming intensity, particularly in stocking densities, has also increased dramatically (Fig. 1.3). Many parts of the Mekong Delta have witnessed a rapid expansion in the number of cages. Cages were set up even in places with rampant industrial pollution, and in the absence of good management practices, production levels declined abruptly. The use of farm-made feeds, while reducing production costs and improving sustainability, largely contributed to environmental degradation in most cage-farming areas. However, new sustainable cage-culture practices are promising, driven mainly by Vietnam's need for

Fig. 1.2. Cages in the Mekong River, Kandal Province, Cambodia. (Photograph by K.R. Salin.)

Fig. 1.3. River cage farm for tilapia, Vietnam. (Photograph by K.R. Salin.)

certification and standards. Besides a predominance of cage installations in rivers, cage culture in freshwater reservoirs is increasingly prevalent in many parts of the world. In Vietnam, new government initiatives with strategic planning, legislation and financial support have resulted in the expansion of cage culture in several reservoirs in northern Vietnam, with up to 1750 cages contributing to an estimated production of about 8000 tonnes by 2019 (FAO, 2022). This rapid growth is also because of the perceived superior quality of cage-farmed fish that meet the export requirements with a less muddy flavour than conventional farming in earthen ponds.

Although tilapia is farmed in various production systems, cage culture is one of the most popular means of farming them in open waters. In the early days of its development (since the 1950s), there were no legal or environmental restrictions on utilizing open-water resources, and tilapia cage culture flourished in most rivers in Thailand (Fig. 1.4). Some of the suitable reservoirs in various provinces were also open to cage farming. However, the indiscriminate expansion of river cage culture and associated environmental issues coupled with legal restrictions have diminished the number of cages in most farming areas.

Cage culture is also widely practised in many African freshwater lakes, mainly the farming of Nile tilapia (Halwart and Moehl, 2006). Musinguzi *et al.* (2019) reported many cage culture operations in African lakes including Lake Volta (Ghana), Lake Victoria (Uganda and Kenya), Lake Malawi (Malawi) and Lake Kariba (Zambia). Commercial-scale tilapia cage-farming projects were reported from Lake Kariba, Zimbabwe (Blow and Leonards, 2007; Hasimuna *et al.*, 2019). In 2006, there were only 185 cages, with 17 cages on Lake Volta (Ghana), ten cages on Lake Malawi, 30 small cages on Lake Victoria (Kenya), 15 cages on Lake Victoria (Uganda), 84 cages on Lake Kariba (Zimbabwe) and 30 cages on the same lake in Zambia. There are now over 20,000 cages in the inland waters (Musinguzi *et al.*, 2019), indicating that cage culture is continuing to advance in Africa.

Fig. 1.4. Tilapia cage farms, Chao Phraya River, Thailand. (Photograph by K.R. Salin.)

Marine fish farming on an industrial scale has expanded in many countries, currently contributing to an annual production of 6.6 million tonnes, most of which come from sea cages (FAO, 2020). The earliest marine cage farms were built using nylon trawl nets and a wooden or polyethene framework for the culture of Atlantic salmon, *Salmo salar*, in the 1960s in Norway and later in Scotland, with more commercial farms established in the 1980s. More sophisticated and engineered structures have recently become popular, with various nets and mesh sizes forming net cages. These are supported by stainless steel or high-density polyethene (HDPE) platforms or rings and moorings, thereby developing the ready-to-install cage structures. Figure 1.5 depicts an HDPE-framed tilapia cage farm on Lake Victoria in Uganda, while Figs 1.6 and 1.7 depict the usage of circular HDPE cages for tuna and barramundi farming in Australia and Indonesia, respectively. Marine finfish aquaculture uses more advanced technologies and more profitable production of high-value fish than farming in inland areas, which is more oriented towards sustainable rural livelihood and nutritious food production. Marine cage culture projects do, therefore, typically attract more extensive investment than inland cage culture.

1.3.2 Species and environment

Most commercial cage-culture operations have focused on the higher-value (in marketing terms) compound-feed-fed finfish species. These include various salmonids (e.g. Atlantic salmon, coho salmon, *Oncorhynchus kisutch*; and chinook salmon, *Oncorhynchus tshawytscha*), as well as the major marine and freshwater carnivorous fish species such as Japanese amberjack, *Seriola quinqueradiata*; red sea bream, *Pagrus major*; yellow croaker, *Larimichthys polyactis*; European sea bass, *Dicentrarchus labrax*; gilthead sea bream, *Sparus aurata*; cobia, *Rachycentron canadum*; sea-raised rainbow trout, *Oncorhynchus*

Fig. 1.5. Tilapia cage farm with HDPE frame, Lake Victoria, Uganda. (Photograph by K.R. Salin.)

Fig. 1.6. Circular HDPE cage for farming tuna, Australia. (Photograph by A.P. Shinn.)

Fig. 1.7. Circular HDPE cage for farming barramundi, Bali, Indonesia. (Photograph by A.P. Shinn.)

mykiss; mandarin fish, *Siniperca chuatsi*; and snakehead, *Channa* spp. (Halwart *et al.*, 2007). Over the last two decades, however, the cage culture of omnivorous freshwater fish (including Chinese carps, tilapia, tambaqui and catfish) has also risen dramatically.

Although estimates vary, one of the earliest statistics on cage culture (Halwart *et al.*, 2007) had indicated that 40 families of fish are cultured in cages, with only five families (i.e. Salmonidae, Sparidae, Carangidae, Pangasiidae and Cichlidae) accounting for approximately 90% of total production. One family, the Salmonidae, accounted for more than 60% of total production. At the species level, around 80 species were cultured in cages. Atlantic salmon accounted for 51% of cage culture, while another four species (i.e. *O. mykiss*, *S. quinqueradiata*, *Pangasius* spp. and *O. kisutch*) made up approximately 27%. It is important to note that *Pangasius* catfish, traditionally raised in cages and pens by farmer households along the Mekong River in Vietnam, has given way to intensive farming in ponds with pellet feed. The proportion of *Pangasius* cage-farming households has decreased dramatically from 48.7% in 2012 to less than 20% in 2018, with up to 80% of the total farmland in the Mekong Delta recently being dominated by big farming corporations (Thong *et al.*, 2020).

Three genera of salmonids are farmed: (i) *Salmo* (*Salmo trutta* – brown trout; and one species of Atlantic salmon); (ii) *Oncorhynchus* (*O. mykiss* – rainbow trout; and five species of Pacific salmon: chinook, chum, coho, pink, sockeye); and (iii) *Salvelinus* (*Salvelinus fontinalis* – brook trout; *Salvelinus alpinus* – char). Atlantic salmon are bred in freshwater hatcheries. Salmon fry and fingerlings (parr) are reared in freshwater tanks, many of which are in recirculating aquaculture systems (RAS) rather than flow-through water systems. Atlantic salmon can be transferred to large sea cages or pens for grow-out at the smolt stage. Norway and Scotland are the leading producers of salmon smolts.

Smolt production capacity has also been developed significantly in Chile and Iceland.

Biological considerations for selecting cages for a particular species to be farmed are mainly related to the buoyancy regulation in fish. Many fish such as mackerel, tuna and cobia, which do not have swim bladders and are negatively buoyant, are continuous swimmers and can utilize a wide depth range in water. Several species, such as sea bream (*S. aurata*), Atlantic cod (*Gadus morhua*) and cobia, favoured in cage culture, have benthic or benthopelagic behaviour. These are best suited for growing in submerged cages that can provide lower stress to fish and optimum water quality (Dempster and Sanchez-Jerez, 2008; Sievers *et al.*, 2021).

1.3.3 Cage types and structures

Das *et al.* (2009) categorized fish cages in commercial farms into four types by the physical features of water resources: (i) fixed cages; (ii) floating cages; (iii) submerged cages; and (iv) submersible cages. The fixed cage is suitable for a water body with a depth of 1–3 m, such as streams, canals, rivers, rivulets, shallow lakes and reservoirs. It is a net bag suspended above the bottom, and is an inexpensive and simple design suitable for many types of fish. The floating cage is usually recommended for use in a water body with a depth of more than 5 m, and it is supported by a floating frame on which the net bag is fixed without touching the bottom. The submerged cage is a net bag fitted in a rugged frame and submerged underwater. The submersible cage is suspended from the surface of the water body with fixable buoyancy. The shape, size and design of cages are decided based on the requirement and condition of fish species and the geographical features of the water resources.

Freshwater cage farming in rivers and lakes is dominated by small-scale farmers and employs traditional wooden frames for net cages and plastic drums for floatation. Figure 1.8 depicts small-scale cage construction using wooden or bamboo poles and barrel floats in Thailand.

Fig. 1.8. Small-scale cage construction using wooden/bamboo poles and barrel floats, Thailand. (Photograph by A.P. Shinn.)

In contrast, marine cages involve more advanced technologies for cage installations and stock management. Rectangular, floating cages are usually used in inland areas, and batteries of individual cages are set in naturally flowing water bodies to allow for higher densities and production than is possible in ponds. In Thailand, primarily small-scale farmers are involved in cage culture, using locally made cage structures with wooden/bamboo/galvanized iron (GI) pipe framework and empty plastic barrel floats with polypropylene nets of stretched mesh size of 2.0 to 2.5 cm. Often a net bag with outer netting to avoid the entry of predators is used. A finer mesh netting stretched about half a metre above the cage bottom is also used to prevent feed loss. Additionally, each cage is covered on top with a larger mesh size (5 cm) to prevent predation by birds. Usually, the cage is designed so that it can be towed around by boat as a contingency plan.

There is an increasing tendency of moving industrial cage farms to offshore areas because of the continuing resource-use conflicts and pollution concerns in the coastal and inshore environments. Fishing, navigation, tourism, water quality issues due to domestic and industrial pollution, conservation and recreation are some of the challenges to promoting cage farming in inshore areas (Chu *et al.*, 2020). However, offshore cages need more capital investment to install cage platforms that can survive the turbulent open-ocean conditions and for the protection and husbandry of the stock in such remote locations in the sea.

The cage structures are engineered to suit the environmental conditions for their inherent stability and protection in dynamic sites. There is a great deal of interdependency between the construction, maintenance and control of cage-farming structures, stocking fish, feeding, harvesting, packaging and routine management operations in industrial cage culture. The technologies of floating, submerged and underwater (submersible) cages are also concerned with the biology of the finfish species farmed and influence the husbandry and management of fish stock (Halwart *et al.*, 2007). The HDPE floating cages that are currently used in modern industrial, marine cage culture in many countries enable easy assembly of the main structural elements (HDPE pipes) in various ways into different sizes and shapes. Such modern cages are also increasingly being used in many freshwater cage culture operations.

Both circular and square cages are used for farming Atlantic salmon. Typical marine salmon cages in Norway are 50 m in diameter and 30 m in depth. These cages can hold 200,000 salmon up to a maximum legally allowed limit of 25 kg fish per cubic metre of water. Salmon can be reared to 5 kg in 14 to 16 months. In Scotland, a marine salmon cage farm operates on a 14–20 months production cycle, excluding the hatchery and smolt production stages, which are usually carried out at other land-based sites. Salmon smolts are added to the cages to be fed to the harvest stage, which takes about 18 months, based on daily feeding and periodic disease/parasite control treatments. Cages are then left empty for about two months before restocking to eliminate the risk of diseases or parasites infecting the new stock. Cage repairs and improvements are also carried out before restocking. Salmon sea cages used in Iceland can be square or circular, from 10 to 32 m in diameter and up to 10 m deep. The cage water volume can be between 1000 and 10,000 cubic metres. Salmon stocking rates are usually about 20 kg fish per cubic metre of water (higher stocking rates lead to stress and lower feed conversion ratios, FCRs). A typical sea cage can hold up to 80,000 fish (fish weight 2–3 kg). The final harvest weight is usually 4–5 kg/fish (Ellis *et al.*, 2016; Iversen *et al.*, 2020).

Surface-based, floating cage farms are the most dominant cage system. However, they are often more prone to suboptimal water quality conditions such as fluctuating temperatures, low dissolved oxygen, hazardous algal and jellyfish blooms, and pollutants from inshore areas leading to disease outbreaks, as well as cage damage from natural calamities resulting in stock escape into the wild (Jensen *et al.*, 2010; Jackson *et al.*, 2015). One solution to these problems is to use submerged cages. This helps maintain stable water quality conditions such as temperature and salinity and provides more extensive areas suitable for cage farming in the open ocean, less affected by surface wind and waves and the impact of storms and other calamities. Deeper areas are also relatively free from parasites and have less potential for resource-use conflicts than surface waters. Their disadvantages include greater system complexity if cages are submersible and difficulties in servicing and operating the system underwater.

Newer technologies are currently applied to make cage culture more sustainable, including the introduction of closed-containment systems. 'Closed containment' refers to aquaculture technologies ranging from floating bag systems to land-based recirculating water systems. Closed containment is a barrier system that aims to limit and regulate interactions between farmed fish and the surrounding aquatic environment. Developing and implementing closed-containment systems could mitigate some of the negative environmental impacts associated with open-water cage culture. Additionally, salmon lice infestations may be prevented as the intake water is pumped from deeper water layers (CSAS, 2008; Apostle, 2012; Nilsen *et al.*, 2020). The technological focus is on minimizing potentially harmful interactions between cultured fish and the aquatic environment. While the two cage structures, namely an open net-cage system and a closed-containment tank system – floating or land-based – are suitable for nearshore or offshore cage farming, the latter is considered the next phase in the advancement of salmon production, specifically to address the perceived environmental concerns in most production areas that utilize an open net-cage farming approach.

1.3.4 Cage-culture production

Although substantial chronological data on global aquaculture production are available, especially through FAO statistics (FAO, 2020, 2021), no good official statistical information exists concerning the total production of farmed aquatic species within cage-culture systems or concerning the overall growth of the sector. The most up-to-date data are based on a statistical review published by the FAO in 2007. According to Halwart *et al.* (2007), cage-culture production from 62 reported countries and provinces/regions totalled 2,412,167 tonnes in 2005, or 3,403,722 tonnes when statistics from China are included: China (991,555 tonnes), Norway (652,306 tonnes), Chile (588,060 tonnes), Japan (272,821 tonnes), UK (135,253 tonnes), Vietnam (126,000 tonnes), Canada (98,441 tonnes), Turkey (78,924 tonnes), Greece (76,577 tonnes), Indonesia (67,672 tonnes) and the Philippines (66,249 tonnes). The same review stated that total cage-culture production from China amounted to just 2.3% of total reported aquaculture production in 2005. The review continued to indicate that cage-culture production accounted for about 70% of total aquaculture production in Canada in 2004, and cage culture in 2005 accounted for 80 to 90% of the total marine finfish production in Asia (Halwart *et al.*, 2007). According to De Silva and Phillips (2007), cage aquaculture accounts for approximately 80–90% of the 1 million tonnes of cultured marine fish in Asia.

Halwart *et al.* (2007) reported the top ten marine and brackish-water cage culture countries: Norway produced around 652,306 tonnes or 27.5% of global production, followed by Chile (588,060 tonnes or 24.8%), China (287,301 tonnes or 12.1%), Japan (286,921 tonnes or 11.3%), the UK (131,481 tonnes or 5.5%), Canada (98,441 tonnes or 4.2%), Greece (76,212 tonnes or 3.2%), Turkey (68,173 tonnes or 2.9%), Republic of Korea (31,895 tonnes or 1.3%) and Denmark (31,192 tonnes or 1.3%). In freshwater cage culture, the top ten producing countries include China, dominating with a production exceeding 700,000 tonnes or 68.4% of total fish production, followed by Vietnam (126,000 tonnes or 12.2%), Indonesia (67,700 tonnes or 6.6%), the Philippines (63,043 tonnes or 5.9%), Russian Federation (14,036 tonnes or 1.4%), Turkey (10,751 tonnes or 1%), Lao People's Democratic Republic (9900 tonnes or 1%), Thailand (7000 tonnes or 0.7%), Malaysia (6204 tonnes or 0.6%) and Japan (3900 tonnes or 0.4%).

Sea farming of Atlantic salmon started experimentally in the 1960s and became commercial in Norway in the 1980s and Chile in the 1990s. Salmon farming has expanded into Canada, Chile, Scotland, Australia, New Zealand and Iceland. Pacific salmon supports significant capture fisheries, especially the pink salmon. Atlantic salmon was the most widely cage-reared fish species by volume and value in 2005 (Halwart *et al.*, 2007), and aquaculture production of Atlantic salmon increased over 4000-fold from only 294 tonnes in 1970 to 1,235,972 tonnes in 2005 (valued at US$4,767,000 million).

Global salmon production further increased to 3.6 million tonnes in 2018, with farmed salmon accounting for about 68% at 2.43 million tonnes and the remaining (salmon, trout and smelts, combined) contributed by capture fisheries (FAO, 2021, 2022). The top farmed Atlantic salmon-producing countries were Norway (53%), Chile (27%), Scotland (6.4%), and the remaining production was spread between 12 other countries in 2018. Global demand for salmon still exceeds supply due to rapidly expanding markets in developing countries like China.

We estimated the current level of cage aquaculture production and compared it with the development of aquaculture in a few countries that primarily produce fish through cage culture. It is clear from the production figures (Tables 1.1 to 1.3) that cage-culture production of almost all marine and brackish-water fish has increased significantly during the past two decades.

Aquaculture has grown rapidly over the past few decades, with more than half of its output going towards food and nutrition security for the rising population. Intensifying aquaculture production is critical to ensuring that aquaculture continues to meet this population's growing food and nutritional needs. As an intensive aquaculture system, cage culture has great potential as an energy-efficient production system with lower operating costs per unit of fish produced than other intensive farming methods, if done at an appropriate scale, and has expanded rapidly in fresh and brackish waters and marine environments. The majority of cage culture expansion in Asia has occurred in fresh water, or brackish water, where low-cost rectangular or circular net cages suspended from a cage framework made with wooden planks, bamboo poles or GI pipes have been used for farming species of local demand. Despite the large initial investment required, marine offshore cage culture has gained popularity due to the increased demand for high-value marine fish, particularly in wealthy nations, where people consume more seafood than in the developing world. The growth of Atlantic salmon production in Norway, Chile, Scotland, etc., is an excellent example of how intensive cage culture may be used to produce seafood, and it provides a wealth of lessons to be learned. The rapid expansion, facilitated by sophisticated cage-farming structures utilizing deep and wide surface cages and improved cultured stock through selective breeding programmes and health management strategies, has also created several environmental challenges, most notably from pollution and the indiscriminate use of chemicals and therapeutic agents, drawing widespread criticism from quality-conscious consumers and regulators (Salin and Ataguba, 2018). The quick expansion of cage culture, eventually culminating in its collapse primarily due to environmental concerns, is also evident in many Asian countries, including Thailand and Vietnam. Submersible cages and floating or land-based closed-containment systems are recommended to address many of the issues associated with expanding cage culture and making it more sustainable. Cage culture will

Table 1.1. Total aquaculture production (2005 and 2018) in selected countries producing a majority of cage-cultured fish. (From FAO, 2021.)

	Total aquaculture production (tonnes)	
Country	2005	2018
Australia	42,669	96,799
Canada	154,587	191,323
Chile	723,875	1,266,054
China	28,120,690	47,559,074
Denmark	39,012	36,453
Greece	106,208	107,700
Indonesia	1,197,109	5,426,943
Ireland	60,050	36,896
Japan	746,372	642,854
Norway	661,877	1,354,941
Philippines	557,251	826,060
Republic of Korea	436,571	568,350
Turkey	119,567	311,681
UK	172,813	197,618
Vietnam	1,437,300	4,134,000

Table 1.2. Total aquaculture production (2005 and 2018) of selected marine species generally produced using cage culture. (From FAO, 2021.)

	Total global production (tonnes)	
Species	2005	2018
Atlantic salmon	1,267,297	2,435,948
Asian sea bass	31,388	50,186
Sea bream	193,216	294,587
Yellow tail	–	122
Cobia	20,457	43,356
Groupers	60,560	208,874
Snappers	3,904	16,881

Table 1.3. Cage-culture salmon production (2005, 2015, 2018) in major salmon-producing countries. (From FAO, 2021.)

	Salmon production (tonnes, live weight)		
Country	2005	2015	2018
Norway	582,043	1,303,346	1,282,003
Chile	374,387	608,546	661,138
Scotland	129,823	172,137	156,025
Canada	83,653	121,926	123,184
Faroe Islands	18,962	80,600	78,900
Australia	16,033	48,331	61,227
Russian Federation	204	10,834	20,566
USA	9,401	18,719	16,107
Iceland	6,488	3,260	13,448
Ireland	13,764	13,116	11,984
Denmark	18	428	1,030
Switzerland	–	–	100
Korea, Democratic People's Republic	–	20	60
France	1,190	–	–
Greece	6	–	–
Total	1,237,977	2,381,279	2,425,773

continue to be the primary method of utilizing suitable open-water resources in several Asian countries, contributing to the food, nutritional and economic security of millions of rural farming communities as access to technological advancements to increase productivity and mitigate negative impacts continues to be extended to them. Besides the environmental degradation caused by open-water farming practices, cage culture is also negatively impacted by climate change.

1.4 Impacts of Climate Change on Aquaculture

Climate refers to long-term weather patterns that characterize a place, whereas weather refers to the state of the atmosphere at a certain time at a specific location (Rummukainen, 2013; Adedeji *et al.*, 2014; IPCC, 2021). Whether it is a temperature spike, a hydrological event or an anomaly in air circulation, the more the climatic variable goes from its regular fluctuation range, the more it resembles an extreme incidence (Seneviratne *et al.*, 2012; Camuffo *et al.*, 2020). Natural internal processes (e.g. major fluctuations in ocean currents, volcanic eruptions, etc.) inside the climate system, as well as anthropogenic external forcing, such as the release of greenhouse gases (GHGs), all have the potential to produce climatic variability (Stern and Kaufmann, 2014).

The impacts of climate change on aquaculture have been extensively studied. Maulu *et al.* (2021) evaluated the effects of climate change on aquaculture productivity and its implications for sustainability, mitigation and adaptation. Climate change has both direct and indirect effects on aquaculture (Handisyde *et al.*, 2006; De Silva and Soto, 2009). Direct effects affect the stock's physiology, while indirect effects may occur as a result of changes in primary and secondary productivity, ecosystem structure, input supplies or product prices, fishmeal and fish oil costs, and costs of other goods and services required by fisherfolk and aquaculture producers (Handisyde *et al.*, 2006; De Silva and Soto, 2009; Freeman, 2017; Adhikari *et al.*, 2018). Aquaculture production does not occur in isolation as it is inextricably linked to other modes of food production (De Silva and Soto, 2009; Troell *et al.*, 2014).

Studies have documented the harmful impact of abrupt climate changes on agriculture and aquaculture, as the food production system is primarily affected by environmental changes. For example, Raza *et al.* (2019) analysed stresses produced by climate change, and they included effects on crop/agriculture production, current breeding strategies and biotechnology tactics to cope with climate change in developing climate-resilient crops. The study also highlighted the relevance of climate-smart agriculture in reducing the impact of climate change on crop production. Galappaththi *et al.* (2020) analysed published case studies from the previous two decades to assess climate change adaptation in aquaculture. The study explored the effects of abrupt climate change on aquaculture and identified three areas of climate change adaptation: (i) local coping mechanisms (e.g. water quality management); (ii) multilevel adaptive strategies (e.g. shifting cultural practices); and (iii) managerial approaches (e.g. community-based adaptation).

The global climate risk index (CRI) (Eckstein *et al.*, 2021) uses several extreme weather event parameters (i.e. storms, floods and heatwaves, but not the rise in sea level, polar ice melt or ocean acidification) together with the number of fatalities (total and per 100,000 inhabitants) and losses (total in US$ million and per unit gross domestic product (in %) resulting from weather events to rank countries according to their exposure and vulnerability and their capacity to cope with them. Using the average CRI scores for the period 2000–2019 (Eckstein *et al.*, 2021), the top ten most affected countries ranked by their scores are Myanmar (10.00 – most affected), Haiti (13.67), the Philippines (18.17), Mozambique (25.83), Bangladesh (28.33), Pakistan (29.00), Thailand (29.83), Nepal (31.33), Madagascar (34.67) and Vietnam (35.67). A low CRI score (e.g. 10) suggests that a country is extremely vulnerable and likely to be impacted by climatic events compared with a country with a higher CRI score (e.g. 50), which is less vulnerable or likely to be impacted by extreme weather events. The CRI scores, however, are not a comprehensive score of climate vulnerability and should not be used in isolation as a predictor of future climate impacts.

CRI scores can serve as a warning to countries to ensure that the potential risks of an extreme weather event are effectively communicated to the communities and industries most at risk as part of a national preparedness strategy. Also, it motivates them to act to put in place the necessary measures to mitigate the event's potential impacts. When these CRI ratings are compared with the tonnage rank of aquaculture production, the countries with the most vulnerable aquaculture sectors are Myanmar (greatest risk), Bangladesh, Philippines, Vietnam, India, Thailand, Cambodia, Pakistan, China and Nepal.

1.5 Aquaculture Impacts on Climate Change

Since food production systems are dependent on non-renewable energy sources and increasingly compromised ecosystem services (Rasmussen *et al.*, 2018), they also contribute significantly to climate change through both direct and indirect emissions of GHGs (Springmann *et al.*, 2018), estimated to account for 20 to 37% of anthropogenic GHG emissions annually (Poore and Nemecek, 2018; Rosenzweig *et al.*, 2020). Large variability exists in the GHGs emitted per portion of protein produced, both within and between food production sectors (Hilborn *et al.*, 2018). The GHG emissions per unit of protein produced by aquaculture generally compare favourably with most livestock production and some wild-caught fisheries (Tilman and Clark, 2014; MacLeod *et al.*, 2020). It has been estimated that 245 million tonnes of CO_2 equivalent (i.e. approximately 0.49% of the total global anthropogenic GHG emissions in 2017) originated from aquaculture (MacLeod *et al.*, 2020). This is substantially lower than the GHG emissions footprint of terrestrial farming. According to Jones *et al.* (2022), although mariculture is typically reported as having a smaller GHG footprint than other food production industries (Hilborn *et al.*, 2018; MacLeod *et al.*, 2020), there are large differences in the median GHG emissions footprint of the fed finfish, bivalve and seaweed sectors and considerable variability within sectors. This variation, and the critical need to rapidly reduce GHG emissions at a global level, substantiate the need to identify and implement strategies that advance the climate-friendly capacity of mariculture, regardless of its current GHG emissions profile. As the importation of seafood products becomes more viable in emerging economies, there is the potential for GHG emissions associated with transport to increase substantially. Shortening supply chains and building regional markets could reduce GHG emissions at the same time, potentially contributing to greater food security (Belton *et al.*, 2018) and industry resilience in times of crisis (Froehlich *et al.*, 2021).

1.6 Changes in Climatic Conditions

Aquaculture cannot be sustainable without addressing the implications of climate change. The major elements of climate change threatening aquaculture production and sustainability include rising temperatures, ocean acidification, diseases, harmful algal blooms, changes in rainfall/precipitation patterns, sea-level rise, uncertainty about external input supplies, changes in sea-surface salinity and severe climatic events (Handisyde *et al.*, 2006; Brander, 2007; Ficke *et al.*, 2007; Barange *et al.*, 2018). Not all these factors need to affect aquaculture production in the same way as they may act in concert or independently, making it challenging to predict their impact, dependent on several factors such as season, location and size of the specific segment in the sector. Their impact may also be visible throughout key stages of a fish's life. The effect of these variables on aquaculture production is also not consistent. Some of the important features of climate change impacts on aquaculture are explored in this section.

1.6.1 Temperature

Surface land temperatures for the last four decades based on HadCRUT global temperature data from the Climatic Research Unit (University of East Anglia) and the Hadley Centre (UK Meteorological Office) show a global rate of increase. Using the HadCRUT5 data set generated by Morice *et al.* (2021), which uses a combination of sea-surface temperatures and near-surface air temperatures over land, shows that global temperatures (2.5–97.5% confidence interval) have increased by an average of 0.354°C (0.320–0.388°C) over the period of 1970 to 2021. There does, however, appear to be doubling each decade in the rate of increase (i.e. 1970–1979 = –0.063°C (–0.102 to –0.025°C); 1980–1989 = 0.160°C (0.126–0.194°C); 1990–1999 = 0.320°C (0.286–0.354°C); 2000–2009 = 0.521°C (0.489–0.553°C); 2010–2019 = 0.735°C (0.720–0.786°C); 2020–2021 = 0.824°C (0.807–0.878°C)) when compared with the 30-year reference period of 1961–1990. However, when the rate of decadal increase is compared with the preceding decade, then the rate of increase is variable (i.e. 1980s = 0.224°C; 1990s = 0.160°C; 2000s = 0.201°C; 2010s = 0.214°C).

When sea-surface temperatures are considered in isolation, then these are lower with an average increase of 0.286 ± 0.254°C across 1970–2021, with decadal increments when compared with the preceding decade of 0.188°C (1980s), 0.160°C (1990s), 0.111°C (2000s) and 0.171°C (2010s). While the deceleration in the increase of temperature throughout 1970–2009 looks promising, it is unknown if this is due to variations in the volume of polar meltwater being added or due to other factors.

1.6.2 Sea-level rise and saltwater intrusion

Since *c*.2000, global sea levels have risen by 61.74 mm when compared with the 1993–2008 average sea level (Church and White, 2006; Our World in Data, 2022; Sweet *et al.*, 2022). The current rate of global average sea-level rise between 1993 and 2010 was 3.2 (range 2.8–3.6) mm per year (IPCC, 2014). The changes are linked to thermal expansion and waters from land-based ice sheets and glaciers (Mengel *et al.*, 2016). Between 1993 and 2018, the thermal expansion of the oceans contributed 42% to sea-level rise, while the melting of temperate glaciers accounted for 21%, 15% due to the reduction in the size of the ice sheets in Greenland and 8% due to changes to the ice sheets of Antarctica. The decrease in Arctic sea-ice extent (1979–2012) is estimated to be 3.5–4.1% per decade, with the summer sea-ice minimum decreasing by an estimated 9.4–13.6% per decade (i.e. 0.73–1.07 million km^2 per decade). At the same time, the annual mean Antarctic sea-ice extent increased by an estimated 1.2–1.8% per decade (i.e. 0.13–0.20 million km^2 per decade) (IPCC, 2014).

1.6.3 Ocean acidification

The World Meteorological Organization recently indicated that between 1970 and 2019 there were 11,000 extreme weather events, with climate change driving a fivefold increase in the frequencies. The number of severe tropical cyclones is expected to increase with every 0.1°C rise in global temperature (Eckstein *et al.*, 2021).

1.6.4 Changes in sea-surface salinity

Changes in sea salinity occur due to increased evaporation resulting from rising temperature and ocean circulation changes or induced directly by climate change (Cooper, 1988; Robinson *et al.*, 2005; Cochrane *et al.*, 2009). All aquatic organisms have a range of salinity tolerance within which they can survive; any alterations may lead to mortalities and production losses. Salinity levels above optimal tolerance can reduce survival, impact growth and reduce immunity (Jahan *et al.*, 2019), making fish more susceptible to pathogens and diseases. In general, variation in water salinity will lead to increased mortalities for several species, which may affect the economic and social sustainability of the sector through increased species losses and higher management costs. However, most of the current knowledge on the effect of climate-related changes on salinity in aquaculture has been biased towards reporting the effect of higher salinity. There is a need for research into the effect of salinity levels lower than the optimal requirement for certain finfish and shellfish species. Furthermore, the response of several species of commercial importance to climate-induced salinity changes is poorly understood. This information is especially useful for aquaculture adaptation, as salinity changes may favour the cultivation of tolerant species (Jahan *et al.*, 2019).

1.6.5 Other factors

Changes in rainfall patterns can cause flooding or drought; both can be detrimental to aquaculture production. Numerous reports on losses due to flooding in many South and Southeast Asian countries are available. Risks of flooding include losing fish from ponds during floods, invasion of ponds by unwanted species, and pond damage resulting from infilling and washing away of walls (Rutkayová *et al.*, 2017). The mixing of pond water and fish with those in the wild could negatively affect the environmental sustainability of aquaculture production, mainly through the introduction of invasive fish species and water quality deterioration. Furthermore, fish losses from ponds threaten the social and economic dimensions of aquaculture sustainability by lowering the economic gains of the producers and inducing poverty in communities. It is also reported that macroalgal (e.g. kelp) productivity may be affected by heavier rainfall bringing various nutrient loadings to nearshore environments (Collins *et al.*, 2020).

Drought may lead to water stress, such as shortages and quality deterioration, negatively affecting aquaculture production (Hambal *et al.*, 1994). The predicted water shortages and competition for resources from common pools driven by climate change will lead to increased conflicts for water among the different user groups, such as aquaculture, agriculture, domestic and industries (Handisyde *et al.*, 2006; Barange *et al.*, 2018). This will affect all the dimensions of aquaculture sustainability. However, there is a need to investigate further how different species and life stages of fish, especially those of economic importance, will respond to changes under a reduced or altered precipitation pattern.

Cyclones, waves and storms can also impact aquaculture, mainly in marine and coastal environments. Rough sea conditions can affect fish and shellfish in cages, rafts and ropes, and cause coastal erosion (Hamdan *et al.*, 2012). The increased number of storm events predicted for certain seasons in certain regions may also increase the risk of aquatic organism escapes due to equipment failure and may require site relocation or changes in production practices that may seriously affect the social and economic sustainability of aquaculture in these areas (Gubbins *et al.*, 2013). According to Barange *et al.* (2018), severe climatic events have increased in several regions in the recent past and are represented by at least 80% of all climate-related disasters. However, these events are predicted to occur more frequently in Africa, particularly in East and Southern Africa (IPCC, 2019). On the contrary, storms may improve mixing of water columns and nutrients and avoid thermal stratification (Seggel *et al.*, 2016), which could improve aquaculture productivity.

1.7 Cage Culture – Mitigation and Adaptation to Climate Change

Climate change adaptation is a broad term that encompasses efforts to mitigate the negative effects of climate change while capitalizing on potential new possibilities. It entails modifying policies and behaviours in response to observed or projected climate changes. Galappaththi *et al.* (2020) identified several research gaps that must be addressed in the future, including inland fish farming adaptation studies, whether different groups of aquaculture farmers confront and adapt to climate change differently, household studies, and the use of Geographic Information Systems (GIS) and remote sensing for designing adaptive strategies. Adaptation to climate change can be a complicated process that presents numerous problems, more so when competing demands for similar resources frequently result in user conflicts. Bueno and Soto (2017) addressed three categories of hazards in their review entitled *Adaptation Strategies of the Aquaculture Sector to the Impacts of Climate Change*. These were (i) physical, (ii) chemical and (iii) biological. Physical hazards include temperature anomalies such as fluctuating air and water temperatures, sea-surface temperature changes, precipitation anomalies, rising sea levels, floods, droughts and cyclones. They listed lower pH values or acidification, salinity changes and low oxygen levels as important chemical hazards in culture waters. The biological hazards include eutrophication, harmful algal blooms, pathogens and parasites, and pollution. Additionally, present scientific understanding about the effect of specific repercussions varies and is sometimes confined to cumulative effects, making adaptation planning in the aquaculture sector almost impossible (Seggel *et al.*, 2016).

Farming communities, ecosystems and populations, in general, may benefit from climate change mitigation and adaptation strategies that are also effective and efficient in building resilience and in responding to changing climatic conditions. General defence strategies to ensure the security of farming sites might include: tree planting or the use of bunds (i.e. stone circles) to minimize soil loss by wind and rain, siltation, etc.; hedge planting to slow water flow rates; the use of gabions to protect vulnerable river banks; the use of earth dams/reservoirs to hold water for its strategic use during dry periods; the construction of levees/embankments or use of secondary flood channels and/or retention ponds to minimize or to contain the overflow from rivers; restoration of river bends to manage the hydrodynamics of river flow and the transport of sediment; installation of groynes to interrupt river flow or revetments to absorb water energy; and the employment of natural flood management strategies such as the creation of flood plains and coastal saltmarshes to absorb wave energy. For fish cultured in cages or pens, either in ponds, rivers, lakes or in open seawater, periodic assessments of the cage design and its capacity to maintain its position and integrity in preventing escape or loss may be necessary. All mitigation and countermeasures, however, are not without the need for capital investment; yet such investments may be beyond the means of many small to medium-sized producers at the local scale and for low- to low–middle-income countries on a national scale where coordinated strategic plans of area management would be required.

Several studies have investigated the effects of climate change on aquaculture over the last decade, classifying them as direct (e.g. cage destruction due to high water levels) or indirect (e.g. increased disease risk as water temperature rises) and/or as positive and negative consequences (Handisyde *et al.*, 2006; Daw *et al.*, 2009; IFAD, 2014; Ahmed *et al.*, 2019; Galappaththi *et al.*, 2020). Most of these studies have discussed future strategies for adapting aquaculture and fisheries to climate change, focusing on uncertainty or variability of yield, lower profitability and economic concerns. At the same time, many other studies have focused on the overall influence of climatic changes on aquaculture and fisheries (i.e. marine, coastal and inland aquaculture) productivity. Only a few studies have focused on the influence of climatic changes on cage farming and the mitigation and adaptation techniques required (Noble *et al.*, 2007; Lebel *et al.*, 2015a,b,c; Chitmanat *et al.*, 2016; Baleta *et al.*, 2019). This section provides an overview of the impact of climate change specifically on different cage-culture systems and environments and various risk management strategies for climate change mitigation and adaptation. Finally, the climate-smart aquaculture (CSA) paradigm is explained in the context of cage culture, with multiple recommendations for approaches at various levels, including national and regional, local and enterprise, and community and farmer levels.

1.7.1 Abnormal changes in climate and seasons affecting cage culture

There are limited case studies specifically related to the influence of climate change on cage culture and mitigation and adaptation measures published in recent years. The general impacts of climate change on cage culture reported from different countries are summarized in Table 1.4. Most research came from Asia, notably Thailand, where the rapidly expanding freshwater cage culture suffered many losses (Lebel and Chitmanat, 2015; Lebel *et al.*, 2015a,b,c; Lebel *et al.*, 2016, 2018). In addition, unexpected and rapid environmental changes due to shifting weather conditions impair marine cage-cultivation methods in the temperate zone (Zilberg, 2001; Noble *et al.*, 2007). Table 1.4 outlines a series of climatic change effects on cage culture, focusing on cultured species, specific climatic variation or severe conditions, and key implications. Research has also attempted to reveal biochemical or molecular changes in cage-cultured species in response to climate and seasonal variations, which in the past had been recorded as simple fish mortality events, demonstrating the growing importance of understanding the impact of abrupt climatic changes on cage farming. Molecular and biochemical changes in fish, declines or stratification of dissolved oxygen in the culture environment, disease outbreaks in warm water in the tropics or infections following heat stress, and changes in feed efficiency are the leading causes of fish mortalities during seasonal temperature variations. When assessing the influence of climate changes on cage farming, the economic losses and welfare concerns must also be considered. Unexpected fish mortality (cultured and wild), cage damage, disease outbreaks and genetic pollution may result not only in a one-time financial loss but also the farmers having to abandon the farming location until it is ready for another cycle. For a while, the loss of natural stocks may also decrease the supply of seeds for cage farming.

1.7.2 Expected impacts of climatic change on cage-culture systems

Apart from the cases mentioned in Table 1.4, several additional effects specific to the cage-culture systems and environments could be predicted and should be investigated in future research. For instance, fast flows and flooding from upper river discharge with high rainfall may damage river cages, resulting in deformities in cage infrastructure and relocation of cages, damage and stress from flood debris, collisions with cultured species, and inflow of polluted water and sediments, all of which can cause fish stress. Low flows and shallow water, on the other hand, create poor water quality related to lower dissolved oxygen in the water and higher fish densities because of low water levels, causing fish stress and damaging cages. Furthermore, drought or reduced rainfall may necessitate the consolidation of cages in greater densities at specific locations, worsening water quality issues and increasing the risk of infectious diseases. Table 1.5 summarizes the projected cage-culture system-related and operational impacts following severe climatic variabilities.

Salmon are coldwater fish that do not tolerate a warmwater environment. The optimal water temperature for Atlantic salmon is 11–14°C. Salmon eat well and grow rapidly within this thermal range but are stressed when water temperatures exceed 16°C as they eat less and have a lower growth rate. Salmon die if the temperature is above 23°C. Mortality events in farmed stocks have been recorded at 20°C, but other mitigating factors may also have been involved (Falconer *et al.*, 2020). Warming of the seas is predicted to be the most serious threat to salmon cage farming from climate change. Other climatic factors that may affect marine aquaculture include sea-level rise, stronger and more frequent storms, ocean acidification, and changes in marine biological processes such as the increased frequency of harmful plankton bloom events and increases in parasite populations (e.g. salmon lice are more abundant in the summer when sea temperatures are higher due to an acceleration of their life cycle).

1.7.3 Risk management approaches and their effectiveness

Risk management techniques or practices must be used at different spatial and temporal levels to limit the negative effects of climate change on

Table 1.4. Impacts of climate change on cage culture (2005 to 2020).

Country/region	Species cultured in cages	Climatic variations	Climatic impacts/observations	References
Montalegre, Portugal	*Oncorhynchus mykiss* (3N)	Seasonal temperature fluctuations	Peroxidase activity increased in the warmer months, lysozyme and antiprotease activity increased in the cold season, antioxidant defence mechanisms were substantially changed seasonally, increasing oxidative stress at higher temperatures, and warmer waters caused a drop in energy production assessed in the liver and variations in feed efficiency	Rodrigues *et al.* (2020)
Oyan Lake, Nigeria	*Oreochromis niloticus*	Increased rainfall fluctuations and declining rainfall	Cage damage, fish escape and genetic pollution	Oyebola and Fada (2020)
Magat Reservoir, Philippines	*O. niloticus*	Seasonal temperature fluctuations	Fish mortality with increasing temperature	Baleta *et al.* (2019)
Northern Thailand	*Oreochromis* sp. (red hybrid) *O. niloticus* *Ictalurus punctatus* *Clarias batrachus*	Drought or low water levels, heatwaves, cold spells and cloud cover	Fish mortality caused by fluctuations in temperature and dissolved oxygen level	Lebel *et al.* (2016)
Northern Thailand	*Oreochromis* spp. *O. niloticus*	Drought, flood, rapid variations in temperature, heavy rainfall, persistent cloud cover, late or early wet seasons	Disease problems triggered by changes in weather, seasons or extreme climate events, which stress fish, making them more susceptible to diseases	Lebel *et al.* (2015b)
Northern Thailand	*Oreochromis* sp. (red hybrid) *O. niloticus* Bagrid catfish	Seasonal flow-related constraints	Damage to cages, fish deaths, slow growth and disease problems	Lebel *et al.* (2015a)
Penghu Archipelago, Southern Taiwan	*Rachycentron canadum*	Seasonal current circulation patterns/low water temperature event (2008/2011)	Fish mortality with a sudden decrease in temperature	Chang *et al.* (2013)
Scotland	*Salmo salar*	Storms	Damage to cages, hybridization with wild species and loss of genetic diversity	Callaway *et al.* (2012)
Scotland	*S. salar*	Change in day length, water temperature, water clarity, daily rainfall and mean daily wind speed	Daily feeding rhythms changed with season, from a morning peak in summer to a midday peak in winter	Noble *et al.* (2007)

Table 1.5. Expected impacts of climate change on different cage-culture systems. (Modified from Handisyde *et al.*, 2006.)

Environment	Drivers of change	Culture system-related effects	Impacts on operations
Marine cage culture	Variation in sea-surface temperature References: Sumaila *et al.* (2011); Tout *et al.* (2015); Geraldi *et al.* (2019); Trombetta *et al.* (2019); Maulu *et al.* (2021)	• An increase in toxic algal blooms, leaving toxins in the water and killing fish • Decrease in dissolved oxygen in water • Increased disease and parasite incidences • Growing seasons that are longer than usual • Natural winter mortality becoming lower • Changes in the location and extent of a species' appropriate range • Enhanced primary productivity to aid filter-feeder production • Increased metabolic rate and better feed conversions may be positive • Changing competitors and predators in local ecosystems • Exotic species causing competition, parasitism and predation	• Infrastructure and operational expenses will change • Fouling organisms, pests, unwanted species and predators will become more prevalent • Distribution, abundance and range of aquatic animals for cultivation will expand • Production levels could fluctuate and become unpredictable
	Other oceanographic phenomena such as wind velocity, currents and wave actions References: Price *et al.* (2015); Maulu *et al.* (2021); Karathanasi *et al.* (2022)	• Reduced flushing rate, which may limit the availability of food for shellfish • Water exchange and waste dispersion are both changing • Changes in the number and range of capture fisheries species used in fishmeal and fish oil production	• Waste accumulating beneath the pens or cages will increase • Costs of operating will increase
	Increasing sea levels References: Kibria (2016); Maulu *et al.* (2021); Stoltz *et al.* (2021)	• Aquaculture-friendly environments will disappear • Loss of habitats such as mangroves, which may offer wave protection and serve as nursery sites for aquaculture seed • Flooding might become more severe as sea levels rise in tandem with storm surges	• Infrastructure damage will occur • Aquaculture zoning will change • Ecosystems that provide coastal defence functions (such as mangroves) compete for space • Insurance rates will increase • Fresh water availability will decrease
	Storms becoming more frequent and intense References: Maulu *et al.* (2021); Muhala *et al.* (2021)	• Large waves • Surges caused by storms • Flooding as a result of heavy rain • Risk of damage to the cage structure and escape of stock • Fluctuating water salinity • Faster transmission of diseases or predators during floods	• Stock depreciation[a] • Facilities will be damaged • Increased capital expenses, as well as the necessity to build cages, moorings and jetties that can withstand storms • Insurance premiums will increase

River, reservoir and brackish-water cage culture	Changes in air temperature, solar radiation intensity and wind speed result in warmer inland water temperatures References: Van Vliet *et al*. (2011); Lebel and Chitmanat (2015); Mugwanya *et al*. (2022)	• Water quality deterioration, particularly in terms of dissolved oxygen • Increased incidences of diseases and parasites • Production may also benefit from increased primary productivity • Change in the location and size of a species' appropriate range • If quality of water and dissolved oxygen levels are acceptable, enhanced metabolic rate leads to increased feeding rate, better food conversion ratio and growth Otherwise, feeding and growth performance may lower	• Variations in production levels • Variations in operating expenses will increase capital expenses, e.g. aeration and improved cage structures
	Changes in rainfall intensity, frequency, seasonality and variability cause floods References: Lebel *et al*. (2015a); Lebel *et al*. (2018); Maulu *et al*. (2021)	• The salinity of the water may fluctuate • Diseases or predators will be introduced • Damage to the structure and stock can escape	• Stock depreciation[a] • Damage to the facilities • Engineering flood resistance will have higher capital expenses • Increased insurance rates
	Drought as an extreme event rather than a steady decrease in water supply References: Lebel *et al*. (2015b); Lebel *et al*. (2018)	• The salinity of the water may fluctuate • Water quality deterioration • Water supply will be limited	• Stock depreciation[a] • Limited productivity and loss of opportunity
	Water stress or progressive decrease in water availability because of higher evaporation rates and lower rainfall References: Patrick (2016); Maulu *et al*. (2021)	• Reduced water quality leads raised incidence and severity of diseases • Reduce water levels in the pond • Changed and decreased freshwater supplies – if operating near the water supply limit, there will be a larger impact of drought	• Costs of artificially maintaining pond levels • Increased conflicts with other users of water resources • Stock depreciation[a] • Production capacity will be reduced • Costs of production per unit will increase • Species change in culture may become necessary

[a]Stock depreciation: if a fish stock is overexploited relative to maximum economic yield (MEY) or maximum sustainable yield (MSY), the depreciation of the stock may be estimated as: Depreciation = (stock at MEY – current stock) × resource rent/stock unit (Frost *et al*., 2018).

cage farming. Adopting a combination of technical, economic and social risk management measures is also crucial for a better and more sustainable outcome. Specific risks are routinely handled in cage farming using a range of strategies and methodologies; nevertheless, a single management strategy may have an enormous impact on numerous risks (Lebel *et al.*, 2016).

The first step in developing a risk management strategy is to identify the risk. Farmers, for example, may observe and determine that, as the summer progresses, the greatest concern may be a lack of water, which may be exacerbated if the ambient temperature is unusually high. Additionally, having a clear understanding of how concerned different groups of individuals are about specific scenarios is critical for developing risk communications and other risk management interventions. Using in-depth interviews with 662 fish farmers rearing tilapia in cages in rivers in northern Thailand, Lebel *et al.* (2015c) investigated how fish farmers perceive major climate risks and acknowledge climate change. They concluded that a good understanding of the critical perception is likely to play a role in enhancing climate risk management and, thereby, climate change adaptation. In another study, Lebel *et al.* (2016) found that risk perceptions of survey participants in northern Thailand with small to large farms (cages) were comparable, with two obvious exceptions: a sudden rise in temperature was regarded as a relevant factor in disease risk by almost all fish farmers interviewed (97%). Most cage farmers recognized that weather and seasonal fluctuations have a significant role in the risk of infections such as streptococcosis in cage-cultured fish.

It is necessary to determine the mechanisms, probability of occurrence and the potential magnitude of negative impacts on cage culture caused by climate change events in each aquaculture region. With this knowledge, strategies can be implemented to mitigate the potential impacts of climate-related events. For instance, the impact of heavy rainfall and runoff (climate-related risk) on river cage culture can be elaborated as an influx of polluted water or sediments that stress fish, consequentially increasing their susceptibility to disease, leading to higher mortality rates. In addition, key climatic drivers include intense rainfall and first rains after the dry period; the climate change concern for this scenario is an increase in frequency, while the interacting factors are watershed land uses and riparian-bank conditions (Lebel and Chitmanat, 2015). As another example, the impact mechanism of rapid changes in temperature on cage culture in reservoirs can be explained as thermal destratification, exposing fish to low dissolved oxygen stress after sharp decreases in temperature, leading to fish mortalities. In this scenario, the key climate drivers are the seasonal transition into the cool or dry season and windy conditions mixing upper and lower water layers; here, the climate change concern is an increased likelihood of sharp temperature changes, while the interacting factors include reservoir management operations that influence mixing.

Farmers can decide on optimal management strategies to be implemented after understanding the climate-related risk and its mechanism. Also, the effectiveness of available options should be evaluated while considering their benefits, cost of operations, resource availability (e.g. electricity for aerators) and long-term impact on the culture cycle (e.g. whether early harvest affects the product quality and demand). Aeration, avoiding some seasons, early harvesting, relocating cages closer to banks, reducing stocking density and moving fish from cages to ponds are some of the common risk management practices against climate change (Lebel and Chitmanat, 2015; Lebel *et al.*, 2016). After evaluation of the effectiveness of available risk management options, best and economically, socially and environmentally sustainable management techniques can be selected following the willingness of all farmers, and an action plan should be developed giving priority for large-scale cage farmers or stronger farming communities as it would be beneficial in managing many risks. After selection, the action can be implemented in a well-organized manner.

1.7.4 Adaptation strategies

Climate change adaptation is required to produce more fish with minimal environmental impact (Shelton, 2014; Ahmed *et al.*, 2019). Cage aquaculture productivity must be expanded in a sustainable manner while its environmental impacts must be significantly reduced (Lebel *et al.*, 2015b). Cage aquaculture adaptation must be flexible enough to allow methods to be discontinued, switched, merged, or new ones incorporated when conditions and knowledge shift. Furthermore, adaptive strategies must meet a number of criteria, including being effective enough to significantly reduce risk, robust enough to effectively function under a variety of conditions, flexible enough to accommodate future modifications, free of unwanted side effects, cost-effective and supportive of all stakeholders, particularly small farmers and vulnerable households (Lebel and Chitmanat, 2015). Table 1.6 shows several long-term adaptation strategies for cage farming to abrupt climate changes.

Table 1.6. Adaptive measures to mitigate the climate-related risks of cage farming. (Modified from Shelton, 2014; Lebel and Chitmanat, 2015; Lebel *et al.*, 2016.)

Strategy for adaptation	Remarks
Recognizing prospects for diversification of revenue sources	New options, such as aquaculture-based livelihoods where agriculture is not an option, should be identified to reduce fish farms as a source of income. Alternative feed sources such as agricultural wastes and insect meals to replace fishmeal should be considered to save costs
With new financial structures, savings and loans may be used to cope and recover	When a fish crop collapses, savings and loans should be used to protect quality of life. Insurance and other innovative tools at all levels should be considered effective incentives
Farmer cooperatives, clubs and capacity development should be strengthened	Credit may increase ability to negotiate with enterprises and government agencies to get lower-cost inputs. For effective climate-resilient planning, partnerships among enterprises, the public, civil society and NGOs are also essential
Encourage the sharing of best practices and the integration of local, national and regional policies and programmes	Government and corporate extension services regularly integrate and distribute best practice information and policies related to poverty reduction, food security and infrastructure improvement, considering both farming and business management qualities
Ecosystem services must be restored and maintained	Wetlands reduce the severity of flooding, increase above- and below-ground water storage during the dry season, reduce water quality issues during drought by digesting extra nutrients, reduce pollutant and sediment runoff into aquaculture water, and provide some shade and relief from intense heatwaves
Production zoning in accordance with carrying capacity	The maximum number of cages in a river reach, reservoir region or pond in an irrigation or catchment area is regulated to provide acceptable water quality – particularly critical during dry seasons with low flow
External stresses on natural systems should be reduced	Reduce pollution from land-based sources (e.g. agricultural and urban runoff)
Initiatives for integrated water resource management and mainstreaming	Aquaculture is considered a net water consumer during droughts when water allocation is planned. This means that cage aquaculture should be a part of all national programmes for climate change mitigation and adaptation, as well as food security programmes, including trade-offs, concessions and planning with other businesses that impact cage aquaculture at the national level
Improve climate information systems, early warning systems and education	Early warning systems should be established for floods, cold spells, heatwaves and drought in particular areas. Using a climate information management system, farmers may learn about the 'normal' degree of risk. Seasonal forecasting is based on the start of the monsoon or the ENSO phase. This shows whether the total amount of rainy season precipitation will be high or low

Continued

Table 1.6. Continued.

Strategy for adaptation	Remarks
Develop international trade and alternative export markets	General and aquaculture-specific trade and economic regulations boost the export potential of farmed fish, resulting in a higher price. There is no one-size-fits-all climate-related threat or cultural system, and it might happen simultaneously as aquaculture-specific regulations are implemented
Research and development to advance technologies	To cope with limited water supplies, functional aerators, cheaper water quality monitoring devices and measures to enhance water productivity are required. Flood prevention measures, such as enhancing the durability of cages and shading techniques against heatwaves, must also be included
Establish voluntary sustainability criteria	Farmers who apply best practices should be rewarded with increased access to markets, investment policies and other types of assistance under an incentive structure
Breeding and species selection	Testing and improving various strains for high-temperature tolerance or disease resistance might benefit a range of climatic change challenges

NGO, non-governmental organization; ENSO, El Niño–Southern Oscillation.

However, several of the adaptation strategies in Table 1.6 may not meet all the requirements. Ecosystem restoration and infrastructure, for example, are less flexible than other options. In addition, the robustness of trade development and implementation standards varies across nations. Techniques such as species or strain selection have an intermediate level of flexibility, equitability, cost and side effects while still being very effective and robust. Most of the other solutions meet practically all the requirements and are well-matched for use in cage farming.

Surface-based cages are the most common way to grow marine finfish. Some problems with surface-based aquaculture, such as extreme weather, poor environmental conditions, parasites and conflicts with people who use the coast, make it more difficult to do this kind of farming. Many of these problems can be solved using submerged cages, and commercial interest in them has grown. Modern cage technologies include the introduction of submersible 'Atlantis' cages in Norway, stocked with 2 kg size salmon that can be sunk to a depth of 30 m. Some of the other scientific and technological developments we can expect in salmonid aquaculture include selective breeding for more temperature-tolerant salmon, salmonid vaccines to protect farmed fish from diseases, and improvements in salmonid feed nutrition and husbandry. Other technologies include submersible offshore salmon cages, advances in RAS for onshore salmon farming, and expansion and improvement of environmental monitoring to increase salmon cage-farming sustainability (Sievers *et al.*, 2021). Fish with open swim bladders, like salmonids, make submergence more difficult because they need surface air to refill their swim bladders and keep them buoyant. The growth and well-being of open swim bladder fish can be impacted by long-term submergence, but addition of underwater air domes to submerged cages can help. Even though this is a step forward, we still need to learn more about how to combine submerged culture with favourable environmental conditions to improve fish growth and welfare over the course of a commercial production cycle (Sievers *et al.*, 2021).

1.7.5 Climate-smart cage-farming protocols

CSA (climate-smart aquaculture) aims to improve food security while considering the need to adapt and ways to reduce the effects of climate change (FAO, 2013). CSA tackles the difficulties of achieving synergies among the associated goals of mitigating climate change, adaptation, and increased production and income while reducing possible negative consequences. CSA addresses three major goals, according to the FAO. The first aim is linked to the overall goal of developing sustainable food production, and it includes environmental, social and economic elements of commercial and artisanal fishing, as well as aquaculture. The second goal focuses on the need to reduce the sector's vulnerability to climate change impacts and increase its resilience so that it would cope with the effects of climate variability on resource availability, as well as with natural disasters caused by an enhanced incidence of severe climate incidents. The third goal is to allow the sector to contribute to GHG emissions reduction at the harvesting and production phases, as well as across the whole value chain, which is critical given the high degree of processing, transportation and marketing activities engaged in the sector. Figure 1.9 depicts the components of CSA.

Formulation, implementation and management of adaptation strategies at key organizational levels that can mitigate the impacts of climate change within the context of CSA are outlined below.

National- and regional-level approaches

- Develop strategic options for cage culture mitigation and adapt them at national and international levels. Connect this effort to other development programmes and financing systems, both sectoral and indeed including climate change and GHG options. Define the most important methods, mobilization timelines and result indicators.
- Develop better insights on the trade-offs among energy prices and policy and the cage culture sector's profitability, production, competitiveness and trade. Investigate and identify the connections between food security and the industry.

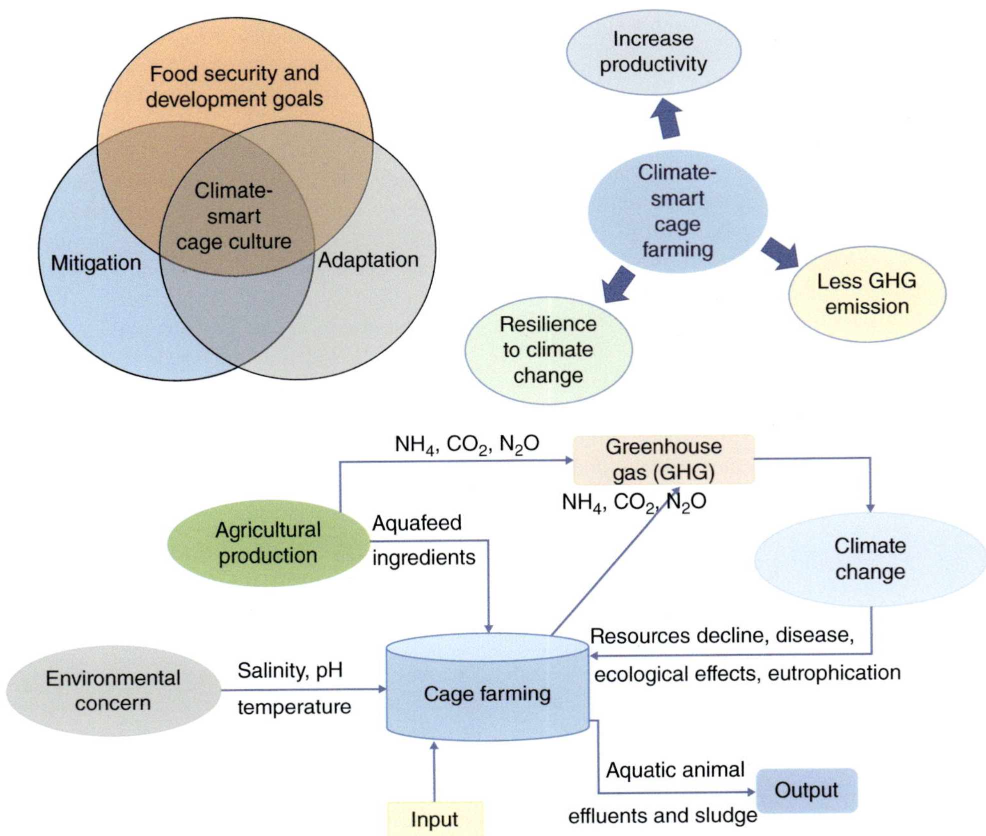

Fig. 1.9. Components of climate-smart cage culture.

- Realistic insights on national and regional cage culture-based aquatic food supply opportunities under climate change scenarios should be provided to promote economic and trade strategies. The following questions should be addressed. (i) Are adaptation costs too high to ensure productivity and profits? (ii) Is it more strategic to boost imports or invest in manufacturing? (iii) Funds are needed to take advantage of more supply, but what trade and value-added options are available?
- Conducting strategic evaluations of institutional and human capital development, determining potential data and information sources, improving the awareness of decision making at different levels, increasing awareness, readiness and skills, and increasing the effectiveness of political relations may be helpful. Financial derivatives (insurance) with product and market views are needed to improve supply and demand flexibility and make the supply chain more resilient in the event of risk.

Strategic industry and subsector levels

- Assess and specify key investments, such as infrastructure, to safeguard and improve production and supply chain aspects. Define trade-offs between 'hard' and 'soft' engineering techniques (cages) and macro-level cage-culture protection, water supply and drainage systems versus low-cost adaptive initiatives.
- To mitigate risk, there is a need to launch subsector activities to discover and share best practices, resilient supply methods, and direct targeted research and development.

Local and enterprise levels

- Define the local socio-economic and policy framework, including economic dependency, livelihood possibilities, human, social and financial assets, institutional framework and capacity-building strategies.
- Investigate the effects of various scenarios and use capital and operational cost models (e.g. temperature, salinity, disease transmission and aquatic system features). Develop location-based (e.g. GIS) risk mapping for important features and create viability profiles under various scenarios.
- Physical alterations (e.g. physical structures of cages and storm and wave protection), modified operational methods, environmental management, varied and segmented supply chains, cross-investments, diverse species mixtures, input possibilities and marketplaces are all practical components of risk reduction. Connect the old knowledge to the new possibilities and expand the capabilities.
- Define and identify relocation options, policies and investment options, job options and ways to diversify income as needed.

Individual and community levels

- Identify key stakeholders, their roles and responsibilities, important risks and choices, and their links to ongoing and prospective development goals.
- Define and improve local learning procedures and information exchanges that can be supported by defined performance metrics.
- Develop and implement information and communication technologies for specialized ideas, information and choice exchanges. Establish contacts with other communities whenever feasible to allow for comparisons and the creation of best practices.
- Encourage active engagement in cross-sectoral planning and discussions.
- Form external collaborations to offer specialized technical and social assistance as well as resilience improvement.

1.7.6 Recommendations (for farmers, researchers and policy developers)

Farmers, researchers, academics and policy makers are important players in developing approaches and policies to manage climate risks in cage culture. Consequently, Table 1.7 includes several recommendations that may be relevant for future studies on the successful implementation of adaptation techniques in cage farming in the presence of climatic risks.

Table 1.7. Recommendations for farmers, researchers and policy makers.

Farmers	Researchers	Policy makers
As part of the business planning, identify and analyse climate-related risks, and keep track of important indices of fish farm productivity so the activities can be quickly shifted	Establish climate risk management techniques for fish producers based on reliable research results	Improve the availability of weather and climatic information to fish farmers and alter current aquaculture development policies, strategies and plans to accommodate climate change
Assist in developing and adopting standards that will make aquaculture more climate resilient	Look into ecosystem-based adaptability alternatives	Enhance the sharing of excellent risk management techniques among fish farmers and raise awareness of aquaculture's role in aquatic resource management
As needed, invest in farm-level risk reduction initiatives	Investigate long-term aquaculture industry adaptation solutions	Collaborate with commercial companies and fish farmers to develop new risk-sharing alternatives and develop and execute an aquaculture zoning strategy
Stay updated on new fish-farming methods to see if there is any potential to increase resilience	Assess aquaculture's contribution to household resilience	Promote and encourage enhanced pollution control and management in rivers utilized for cage aquaculture and investment in climate-resilient aquaculture technology and rearing methods
To effectively represent fish farming interests in water management decisions, establish and enhance aquacultural organizations and networks	Extend aquaculture evaluation operations to additional species and locations	Take advantage of the knowledge of risk perception in communication to create a climate and water strategy for cage farming

1.8 Prospects and Conclusions

Due to the stagnation of capture fisheries, aquaculture productivity must be enhanced to fulfil the demand for affordable protein from an increasing population. Cage culture is an effective method to enhance aquaculture productivity. Climate change poses a severe threat to aquaculture, especially cage culture. Several studies have looked at potential techniques for aquaculture and fisheries adaptation to climate change, with an emphasis on production unpredictability, lower profitability and economic issues. However, most research has focused on the overall impact of climatic changes on aquaculture and fisheries production. Only a few studies have focused explicitly on the impact of climate change on cage farming and the required mitigation and adaptation measures. Using recently published case studies and literature reviews, possible adaptation techniques for cage culture to mitigate the impact of climatic change have been reviewed. Additionally, the potential difficulties that cage culture may encounter imminently as a result of abrupt climate change were examined.

There is a paucity of information on specific adaptation and mitigation strategies for tackling climate change impacts on cage culture, with limited case studies related to climate-associated risks in cage farming. Diversification of revenue sources, connections to new financial structures, farmer cooperation and capacity development, sharing of best practices, ecosystem restoration, production zoning, reduction of external stressors, early warning systems, and research and development were identified as general adaptive measures that can be successfully applied to mitigate negative climatic impacts on cage culture and secure the farmer's livelihood.

In addition, the adaptation of novel approaches to cage farming such as CSA, as well as various recommendations for new policy development, research and sustainable cage culture, were briefly reviewed.

From the current work, several research gaps have been identified. It emphasizes the urgency for ongoing studies to fill current knowledge gaps necessary to formulate robust biosecurity measures for cage culture and strategies to mitigate the impacts of climate change on production. Ongoing research is needed to better understand the tolerance of intensively farmed species to environmental stressors such as extremes of temperature, salinity, dissolved oxygen, etc.; the tolerance of new genetic lines; and the evaluation of new species or strains for aquaculture with better resilience to these changes. Studies on the effects of climate change on molecular-level alterations in fish, for example, are scarce, and such research might promote the genetic selection of resistant strains suitable for cage culture and more easily adapted to climate change impacts. Additionally, research on the impact of ocean acidification on marine cage-cultured species is limited, and data on other climatic impacts such as temperature fluctuations, storms and rainfall fluctuations must be expanded across multiple regions and ecosystems to develop mitigation and adaptive strategies that can be implemented with minor modifications in any region.

References

Adedeji, O., Reuben, O. and Olatoye, O. (2014) Global climate change. *Journal of Geoscience and Environment Protection* 2, 114–122. https://doi.org/10.4236/gep.2014.22016

Adhikari, S., Keshav, C.A., Barlaya, G., Rathod, R., Mandal, R.N., *et al.* (2018) Adaptation and mitigation strategies of climate change impact in freshwater aquaculture in some states of India. *Journal of Fisheries Science* 12(1). https://doi.org/10.21767/1307-234x.1000142

Ahmed, N., Thompson, S. and Glaser, M. (2019) Global aquaculture productivity, environmental sustainability, and climate change adaptability. *Environmental Management* 63(2), 159–172. https://doi.org/10.1007/s00267-018-1117-3

Apostle, R. (2012) Closed–containment aquaculture in Atlantic Canada. *Maritime Studies* 11(1), 13. https://doi.org/10.1186/2212-9790-11-13

Asmah, R., Falconer, L., Telfer, T.C., Karikari, A.Y., Al Wahaibi, M., *et al.* (2021) Waterbody scale assessment using spatial models to identify suitable locations for cage aquaculture in large lake systems: a case study in Volta Lake, Ghana. *Aquaculture Research* 52(8), 3854–3870. https://doi.org/10.1111/are.15230

Baleta, F.N., Bolaños, J.M. and Medrano, W.C. (2019) Assessment of tilapia cage farming practices in relation to the occurrence of fish mortality along the fish cage belt at Magat Reservoir, Philippines. *Journal of Fisheries and Environment* 43(2), 1–13.

Barange, M., Bahri, T., Beveridge, M.C.M., Cochrane, K.L., Funge-Smith, S. and Poulain, F. (2018) *Impacts of Climate Change on Fisheries and Aquaculture: Synthesis of Current Knowledge, Adaptation and Mitigation Options*. FAO Fisheries and Aquaculture Technical Paper No. 627. Food and Agriculture Organization of the United Nations, Rome. Available at: http://www.fao.org/3/i9705en/i9705en.pdf (accessed 11 September 2022).

Belton, B., Bush, S.R. and Little, D.C. (2018) Not just for the wealthy: Rethinking farmed fish consumption in the Global South. *Global Food Security* 16, 85–92.

Beveridge, M.C.M. (2004) *Cage Aquaculture*, 3rd edn. Blackwell Publishing, Oxford, UK.

Blow, P. and Leonards, S. (2007) A review of cage culture: sub-Saharan Africa. In: Halwart, H., Soto, D. and Arthur, J.R. (eds) *Cage Culture: Regional Reviews and Global Overview*. FAO Fisheries and Aquaculture Technical Paper No. 498. Food and Agriculture Organization of the United Nations, Rome, pp. 191–210.

Brander, K.M. (2007) Global fish production and climate change. *Proceedings of the National Academy of Sciences USA* 104(50), 19709–19714. https://doi.org/10.1073/pnas.0702059104

Brugère, C. and De Young, C. (eds) (2015) *Assessing Climate Change Vulnerability in Fisheries and Aquaculture: Available Methodologies and Their Relevance for the Sector*. FAO Fisheries and Aquaculture Technical Paper No. 597. Food and Agriculture Organization of the United Nations, Rome.

Bueno, P.B. and Soto, D. (2017) *Adaptation Strategies of the Aquaculture Sector to the Impacts of Climate Change*. FAO Fisheries and Aquaculture Circular No. 1142. Food and Agriculture Organization of the United Nations, Rome. Available at: https://www.fao.org/3/i6943e/i6943e.pdf (accessed 11 September 2022).

Callaway, R., Shinn, A.P., Grenfell, S.E., Bron, J.E., Burnell, G., *et al.* (2012) Review of climate change impacts on marine aquaculture in the UK and Ireland. *Aquatic Conservation: Marine and Freshwater Ecosystems* 22(3), 389–421.

Camuffo, D., Becherini, F. and della Valle, A. (2020) Relationship between selected percentiles and return periods of extreme events. *Acta Geophysica* 68(4), 1201–1211. https://doi.org/10.1007/s11600-020-00452-x

Chang, Y., Lee, M.A., Lee, K.T. and Shao, K.T. (2013) Adaptation of fisheries and mariculture management to extreme oceanic environmental changes and climate variability in Taiwan. *Marine Policy* 38, 476–482. https://doi.org/10.1016/j.marpol.2012.08.002

Chitmanat, C., Lebel, P., Whangchai, N., Promya, J. and Lebel, L. (2016) Tilapia diseases and management in river-based cage aquaculture in northern Thailand. *Journal of Applied Aquaculture* 28(1), 9–16. https://doi.org/10.1080/10454438.2015.1104950

Chu, Y.I., Wang, C.M., Park, J.C. and Lader, P.F. (2020) Review of cage and containment tank designs for offshore fish farming. *Aquaculture* 529, 734928. https://doi.org/10.1016/j.aquaculture.2020.734928

Church, J.A. and White, N.J. (2006) A 20th century acceleration in global sea-level rise. *Geophysical Research Letters* 33(1), 1–4. https://doi.org/10.1029/2005GL024826

Cochrane, K., De Young, C., Soto, D. and Bahri, T. (eds) (2009) *Climate Change Implications for Fisheries and Aquaculture: Overview of Current Scientific Knowledge*. FAO Fisheries and Aquaculture Technical Paper No. 530. Food and Agriculture Organization of the United Nations, Rome.

Collins, C., Bresnan, E., Brown, L., Falconer, L., Guilder, J., *et al.* (2020) Impacts of climate change on aquaculture. *MCCIP Science Review* 2020, 482–520. Available at: https://www.mccip.org.uk/sites/default/files/2021-08/21_aquaculture_2020.pdf (accessed 11 September 2022).

Cooper, N.S. (1988) The effect of salinity in tropical ocean models. *Journal of Physical Oceanography* 18, 1333–1347. https://doi.org/10.1175/1520-0485(1988)018<0697:TEOSOT>2.0.CO;2

CSAS (Canadian Science Advisory Secretariat) (2008) *Assessing Potential Technologies for Closed-Containment Saltwater Salmon Aquaculture*. Science Advisory Report 2008/001. CSAS, National Capital Region, Fisheries and Oceans Canada, Ottawa. Available at: https://waves-vagues.dfo-mpo.gc.ca/Library/332156.pdf (accessed 11 September 2022).

Das, A.K., Vass, K.K., Shrivastava, N.P. and Katiha, P.K. (2009) *Cage Culture in Reservoirs in India (A Handbook)*. WorldFish Center Technical Manual No. 1948. The WorldFish Center, Penang, Malaysia. Available at: https://www.worldfishcenter.org/publication/cage-culture-reservoirs-india-handbook (accessed 11 September 2022).

Daw, T., Adger, W., Brown, K. and Badjeck, M. (2009) Climate change and capture fisheries: potential impacts, adaptation and mitigation. In: Cochrane, K., De Young, C., Soto, D. and Bahri, T. (eds) *Climate Change Implications for Fisheries and Aquaculture: Overview of Current Scientific Knowledge*. FAO Fisheries and Aquaculture Technical Paper No. 530. Food and Agriculture Organization of the United Nations, Rome, pp. 107–150.

De Silva, S.S. and Phillips, M.J. (2007) A review of cage aquaculture: Asia (excluding China). In: Halwart, M., Soto, D. and Arthur, J.R. (eds) *Cage Aquaculture: Regional Reviews and Global Overview*. FAO Fisheries and Aquaculture Technical Paper No. 498. Food and Agriculture Organization of the United Nations, Rome, pp. 21–48.

De Silva, S.S. and Soto, D. (2009) Climate change and aquaculture: potential impacts, adaptation and mitigation. In: *Climate Change Implications for Fisheries and Aquaculture: Overview of Current Scientific Knowledge*. FAO Fisheries and Aquaculture Technical Paper No. 530. Food and Agriculture Organization of the United Nations, Rome, pp. 151–213.

Dempster, T. and Sanchez-Jerez, P. (2008) Aquaculture and coastal space management in Europe: an ecological perspective. In: Holmert, M., Black, K., Duarte, C.M., Marbà, N. and Karakassis, I. (eds) *Aquaculture in the Ecosystem*. Springer, Dordrecht, The Netherlands, pp. 87–116.

Eckstein, D., Künzel, V. and Schäfer, L. (2021) *Global Climate Risk Index 2021. Who Suffers Most from Extreme Weather Events? Weather-Related Loss Events in 2019 and 2000–2019*. Germanwatch eV, Bonn, Germany. Available at: https://germanwatch.org/sites/default/files/Global%20Climate%20Risk%20Index%202021_1.pdf (accessed 11 September 2022).

Ellis, T., Turnbull, J.F., Knowles, T.G., Lines, J.A. and Auchterlonie, N.A. (2016) Trends during development of Scottish salmon farming: an example of sustainable intensification? *Aquaculture* 458, 82–99. https://doi.org/10.1016/j.aquaculture.2016.02.012

Eng, C.T. and Tech, E. (2002) Introduction and history of cage culture. In: Woo, P.T.K., Bruno, D.W. and Lim, L.H.S. (eds) *Diseases and Disorders of Finfish in Cage Culture*, 1st edn. CAB International, Wallingford, UK. pp. 1–39.

Falconer, L., Middelboe, A.L., Kaas, H., Ross, L.G. and Telfer, T.C. (2020) Use of geographic information systems for aquaculture and recommendations for development of spatial tools. *Reviews in Aquaculture* 12(2), 664–677. https://doi.org/10.1111/raq.12345

FAO (Food and Agriculture Organization of the United Nations) (2013) *Climate-Smart Agriculture Sourcebook*. FAO, Rome. Available at: https://www.fao.org/3/i3325e/i3325e.pdf (accessed 11 September 2022).

FAO (Food and Agriculture Organization of the United Nations) (2020) *The State of World Fisheries and Aquaculture 2020: Sustainability in Action*. FAO, Rome. https://doi.org/10.4060/ca9229en

FAO (Food and Agriculture Organization of the United Nations) (2021) *FAO Yearbook of Fisheries and Aquaculture Statistics 2019*. FAO, Rome.

FAO (Food and Agriculture Organization of the United Nations) (2022) Fishery and Aquaculture Statistics. Global aquaculture production 1950–2019 (FishstatJ). FAO, Rome. Available at: https://www.fao.org/fishery/statistics/software/fishstatj/en (accessed 28 March 2022).

Ficke, A.D., Myrick, C.A. and Hansen, L.J. (2007) Potential impacts of global climate change on freshwater fisheries. *Reviews in Fish Biology and Fisheries* 17, 581–613. https://doi.org/10.1007/s11160-007-9059-5

Freeman, O.E. (2017) Impact of climate change on aquaculture and fisheries in Nigeria; a review. *International Journal of Multidisciplinary Research and Development* 4(1), 53–59.

Froehlich, H.E., Gentry, R.R., Lester, S.E., Cottrell, R.S., Fay, G., *et al.* (2021) Securing a sustainable future for US seafood in the wake of a global crisis. *Marine Policy* 124, 104328. https://doi.org/10.1016/j.marpol.2020.104328

Frost, H., Ravensbeck, L., Hoff, A. and Andersen, P. (2018) The economics of ecosystem-based fisheries management. In: Islam, M.N. and Jørgensen, S.E. (eds) *Environmental Management of Marine Ecosystems*. CRC Press, Boca Raton, Florida, pp. 157–207. https://doi.org/10.1201/9781315153933-7

Galappaththi, E.K., Ichien, S.T., Hyman, A.A., Aubrac, C.J. and Ford, J.D. (2020) Climate change adaptation in aquaculture. *Reviews in Aquaculture* 12(4), 2160–2176. https://doi.org/10.1111/raq.12427

Gentry, R.R., Froehlich, H.E., Grimm, D., Kareiva, P., Parke, M., *et al.* (2017) Mapping the global potential for marine aquaculture. *Nature Ecology & Evolution* 1(9), 8. https://doi.org/10.1038/s41559-017-0257-9

Geraldi, N.R., Kellison, G.T. and Bacheler, N.M. (2019) Climate indices, water temperature, and fishing predict broad scale variation in fishes on temperate reefs. *Frontiers in Marine Science* 6, 30. https://doi.org/10.3389/fmars.2019.00030

Gubbins, M., Bricknell, I. and Service, M. (2013) Impacts of climate change on aquaculture. *MCCIP Science Review* 2013, 318–327. Available at: https://www.mccip.org.uk/sites/default/files/2021-08/2013arc_sciencereview_33_aqua_final.pdf (accessed 11 September 2022).

Halwart, M. and Moehl, J.F. (eds) (2006) *Regional Technical Expert Workshop on Cage Culture in Africa. Entebbe, Uganda, 20–23 October 2004*. FAO Fisheries Proceedings No. 6. Food and Agriculture Organization of the United Nations, Rome.

Halwart, M., Soto, D. and Arthur, J.R. (eds) (2007) *Cage Aquaculture: Regional Reviews and Global Overview*. FAO Fisheries and Aquaculture Technical Paper No. 498. Food and Agriculture Organization of the United Nations, Rome.

Hambal, H., Mohd. Akhir, A. and Saniah, Y. (1994) *Environmental Issues on Aquaculture Development in Malaysia*. Working Papers of the Fisheries Research Institute. Department of Fisheries, Kuala Lumpur.

Hamdan, O., Khali Aziz, H., Shamsudin, I. and Raja Barizan, R. (2012) *Status of Mangroves in Peninsular Malaysia*. Forest Research Institute Malaysia, Kuala Lumpur.

Handisyde, N., Ross, L.G. and Allison, E.H. (2006) *The Effects of Climate Change on World Aquaculture: A Global Perspective*. Report to the Department for International Development, London. Available at: https://www.researchgate.net/publication/265497647 (accessed 11 September 2022).

Hannesson, R. (2015) World fisheries in crisis? *Marine Resource Economics* 30(3), 251–260. https://doi.org/10.1086/680443

Hasimuna, O.J., Maulu, S., Monde, C. and Mweemba, M. (2019) Cage aquaculture production in Zambia: assessment of opportunities and challenges on Lake Kariba, Siavonga district. *Egyptian Journal of Aquatic Research* 45(3), 281–285. https://doi.org/10.1016/j.ejar.2019.06.007

Hilborn, R., Banobi, J., Hall, S.J., Pucylowski, T. and Walsworth, T.E. (2018) The environmental cost of animal source foods. *Frontiers in Ecology and the Environment* 16(6), 329–335. https://doi.org/10.1002/fee.1822

IFAD (International Fund for Agricultural Development) (2014) *Guidelines for Integrating Climate Change Adaptation into Fisheries and Aquaculture Projects*. IFAD, Rome.

IPCC (Intergovernmental Panel on Climate Change) (2014) *Climate Change 2014, Synthesis Report, Summary for Policymakers*. IPCC, Geneva, Switzerland. Available at: https://www.ipcc.ch/site/assets/uploads/2018/02/AR5_SYR_FINAL_SPM.pdf (accessed 11 September 2022).

IPCC (Intergovernmental Panel on Climate Change) (2019) *The IPCC's Special Report on Climate Change and Land: What' s In It for Africa?* IPCC, Geneva, Switzerland. Available at: https://reliefweb.int/report/world/ipcc-s-special-report-climate-change-and-land-what-s-it-africa (accessed 11 September 2022).

IPCC (Intergovernmental Panel on Climate Change) (2021) *Climate Change 2021: The Physical Science Basis. Contribution of Working Group I to the Sixth Assessment Report of the Intergovernmental Panel on Climate Change* (Masson-Delmotte, V., Zhai, P., Pirani, A., Connors, S.L., Péan, C., *et al.*, eds]. Cambridge University Press, Cambridge, UK.

Iversen, A., Asche, F., Hermansen, Ø. and Nystøyl, R. (2020) Production cost and competitiveness in major salmon farming countries 2003–2018. *Aquaculture* 522, 735089. https://doi.org/10.1016/j.aquaculture.2020.735089

Jackson, D., Drumm, A., McEvoy, S., Jensen, Ø., Mendiola, D., *et al.* (2015) A pan-European valuation of the extent, causes and cost of escape events from sea cage fish farming. *Aquaculture* 436, 21–26. https://doi.org/10.1016/j.aquaculture.2014.10.040

Jahan, I., Hossain, M.M. and Karim, M.R. (2019) Effect of salinity stress on plant growth and root yield of carrot. *Progressive Agriculture* 3 (January), 263–274. https://doi.org/10.3329/pa.v30i3.45151

Jensen, Ø., Dempster, T., Thorstad, E.B., Uglem, I. and Fredheim, A. (2010) Escapes of fishes from Norwegian sea-cage aquaculture: causes, consequences and prevention. *Aquaculture Environment Interactions* 1, 71–83. https://doi.org/10.3354/aei00008

Joffre, O., So, N. and Chheng, P. (2016) *Aquaculture Production in Cambodia: Trends and Patterns in Recent Years*. Inland Fisheries Research and Development Institute (Fisheries Administration) and WorldFish, Phnom Penh. Available at: https://www.researchgate.net/publication/308063867_Aquaculture_production_in_Cambodia_trends_and_patterns_in_recent_years (accessed 11 September 2022).

Jones, A.R., Alleway, H.K., McAfee, D., Reis-Santos, P., Theuerkauf, S.J. and Jones, R.C. (2022) Climate-friendly seafood: the potential for emissions reduction and carbon capture in marine aquaculture. *BioScience* 72(2), 123–143. https://doi.org/10.1093/biosci/biab126

Karathanasi, F.E., Soukissian, T.H. and Hayes, D.R. (2022) Wave analysis for offshore aquaculture projects: a case study for the eastern Mediterranean Sea. *Climate* 10(1), 2. https://doi.org/10.3390/cli10010002

Kibria, G. (2016) Sea-level rise and its impact on wetlands, water, agriculture, fisheries, aquaculture, public health, displacement, infrastructure and adaptation. Available at: https://www.researchgate.net/publication/266794121_Sea-level_rise_and_its_impact_on_wetlands_water_agriculture_fisheries_aquaculture_public_health_displacement_infrastructure_and_adaptation (accessed 11 September 2022).

Lebel, L. and Chitmanat, C. (eds) (2015) Climate change and aquaculture in northern Thailand: an assessment of risks and adaptation options. AQUADAPT Project. Unit for Social and Environmental Research, Faculty of Social Sciences, Chiang Mai University. Available at: https://idl-bnc-idrc.dspacedirect.org/bitstream/handle/10625/55974/IDL-55974.pdf (accessed 11 September 2022).

Lebel, L., Lebel, P. and Lebel, B. (2016) Impacts, perceptions and management of climate-related risks to cage aquaculture in the reservoirs of northern Thailand. *Environmental Management* 58(6), 931–945. https://doi.org/10.1007/s00267-016-0764-5

Lebel, L., Lebel, P., Chitmanat, C., Uppanunchai, A. and Apirumanekul, C. (2018) Managing the risks from the water-related impacts of extreme weather and uncertain climate change on inland aquaculture in northern Thailand. *Water International* 43(2), 257–280. https://doi.org/10.1080/02508060.2017.1416446

Lebel, P., Whangchai, N., Chitmanat, C. and Lebel, L. (2015a) Risk of impacts from extreme weather and climate in river-based tilapia cage culture in northern Thailand. *International Journal of Global Warming* 8(4), 534–554. https://doi.org/10.1504/IJGW.2015.073054

Lebel, P., Whangchai, N., Chitmanat, C. and Lebel, L. (2015b) Climate risk management in river-based tilapia cage culture in northern Thailand. *International Journal of Climate Change Strategies and Management* 7(4), 476–498. https://doi.org/10.1108/IJCCSM-01-2014-0018

Lebel, P., Whangchai, N., Chitmanat, C., Promya, J. and Lebel, L. (2015c) Perceptions of climate-related risks and awareness of climate change of fish cage farmers in northern Thailand. *Risk Management* 17(1), 1–22. https://doi.org/10.1057/rm.2015.4

MacLeod, M.J., Hasan, M.R., Robb, D.H.F. and Mamun-Ur-Rashid, M. (2020) Quantifying greenhouse gas emissions from global aquaculture. *Scientific Reports* 10(1), 11679. https://doi.org/10.1038/s41598-020-68231-8

Maulu, S., Hasimuna, O.J., Haambiya, L.H., Monde, C., Musuka, C.G., *et al.* (2021) Climate change effects on aquaculture production: sustainability implications, mitigation, and adaptations. *Frontiers in Sustainable Food Systems* 5, 609097. https://doi.org/10.3389/fsufs.2021.609097

Mengel, M., Levermann, A., Frieler, K., Robinson, A., Marzeion, B. and Winkelmann, R. (2016) Future sea level rise constrained by observations and long-term commitment. *Proceedings of the National Academy of Sciences USA* 113(10), 2597–2602. https://doi.org/10.1073/pnas.1500515113

Morice, C.P., Kennedy, J.J., Rayner, N.A., Winn, J.P., Hogan, E., *et al.* (2021) An updated assessment of near-surface temperature change from 1850: the HadCRUT5 data set. *Journal of Geophysical Research: Atmospheres* 126(3), e2019JD032361. https://doi.org/10.1029/2019JD032361

Mugwanya, M., Dawood, M.A.O., Kimera, F. and Sewilam, H. (2022) Anthropogenic temperature fluctuations and their effect on aquaculture: a comprehensive review. *Aquaculture and Fisheries* 7(3), 223–243. https://doi.org/10.1016/j.aaf.2021.12.005

Muhala, V., Chicombo, T.F., Macate, I.E., Guimarães-Costa, A., Gundana, H., *et al.* (2021) Climate change in fisheries and aquaculture: analysis of the impact caused by Idai and Kenneth cyclones in Mozambique. *Frontiers in Sustainable Food Systems* 5, 714187. https://doi.org/10.3389/fsufs.2021.714187

Musinguzi, L., Lugya, J., Rwezawula, P., Kamya, A., Nuwahereza, C., *et al.* (2019) The extent of cage aquaculture, adherence to best practices and reflections for sustainable aquaculture on African inland waters. *Journal of Great Lakes Research* 45(6), 1340–1347. https://doi.org/10.1016/j.jglr.2019.09.011

Nilsen, A., Nielsen, K.V. and Bergheim, A. (2020) A closer look at closed cages: growth and mortality rates during production of post-smolt Atlantic salmon in marine closed confinement systems. *Aquacultural Engineering* 91, 102124. https://doi.org/10.1016/j.aquaeng.2020.102124

Noble, C., Kadri, S., Mitchell, D.F. and Huntingford, F.A. (2007) The impact of environmental variables on the feeding rhythms and daily feed intake of cage-held 1+ Atlantic salmon parr (*Salmo salar* L.). *Aquaculture* 269(1–4), 525–531. https://doi.org/10.1016/j.aquaculture.2007.04.079

Our World in Data (2022) Sea level rise. Available at: https://ourworldindata.org/explorers/climate-change?facet=none&pickerSort=desc&pickerMetric=sea_level_rise_cw2011&Metric=Sea+level+rise&Long-run+series%3F=false&country=OWID_WRL~ATA~Gulkana+Glacier~Lemon+Creek+Glacier~North+America~South+Cascade+Glacier~Wolverine+Glacier (accessed 9 March 2022).

Oyebola, O.O. and Fada, P.O. (2020) Climate change: a potential risk to cage culture and genetic structure of native fisheries in fish cage-installed inland lakes. *Lakes & Reservoirs* 25(3), 283–291. https://doi.org/10.1111/lre.12331

Patrick, A.E.S. (2016) Influence of rainfall and water level on inland fisheries production: a review. *Archives of Applied Science Research* 8(6), 44–51.

Poore, J. and Nemecek, T. (2018) Reducing food's environmental impacts through producers and consumers. *Science* 360(6392), 987–992. https://doi.org/10.1126/science.aaq0216

Price, C., Black, K.D., Hargrave, B.T. and Morris, J.A. (2015) Marine cage culture and the environment: effects on water quality and primary production. *Aquaculture Environment Interactions* 6, 151–174. https://doi.org/10.3354/aei00122

Rasmussen, L.V., Coolsaet, B., Martin, A., Mertz, O., Pascual, U., *et al.* (2018) Social–ecological outcomes of agricultural intensification. *Nature Sustainability* 1, 275–282.

Raza, A., Razzaq, A., Mehmood, S., Zou, X., Zhang, X., *et al.* (2019) Impact of climate change on crops adaptation and strategies to tackle its outcome: a review. *Plants* 8(2), 34. https://doi.org/10.3390/plants8020034

Robinson, R.A., Learmonth, J.A., Hutson, A.M., Macleod, C.D., Sparks, T.H., *et al.* (2005) *Climate Change and Migratory Species: A Report for Defra Research Contract CR0302*. BTO Research Report 414. British Trust for Ornithology, Thetford, UK. Available at: https://www.bto.org/sites/default/files/shared_documents/publications/research-reports/2005/rr414.pdf (accessed 11 September 2022).

Rodrigues, A.C.M., Gravato, C., Silva, C.J.M., Pires, S.F.S., Costa, A.P.L., *et al.* (2020) Seasonal temperature fluctuations differently affect the immune and biochemical parameters of diploid and triploid *Oncorhynchus mykiss* cage-cultured in temperate latitudes. *Sustainability* 12(21), 8785. https://doi.org/10.3390/su12218785

Rosenzweig, C., Mbow, C., Barioni, L.G., Benton, T.G., Herrero, M., *et al.* (2020) Climate change responses benefit from a global food system approach. *Nature Food* 1(2), 94–97. https://doi.org/10.1038/s43016-020-0031-z

Rummukainen, M. (2013) Climate change: changing means and changing extremes. *Climatic Change* 121, 3–13. https://doi.org/10.1007/s10584-013-0888-z

Rutkayová, J., Vácha, F., Maršálek, M., Beneš, K., Civišová, H., *et al.* (2017) Fish stock losses due to extreme floods – findings from pond-based aquaculture in the Czech Republic. *Journal of Flood Risk Management* 11(3), 351–359. https://doi.org/10.1111/jfr3.12332

Salin, K.R. and Ataguba, G.A. (2018) Aquaculture and the environment: towards sustainability. In: Hai, F., Visvanathan, C. and Boopathy, R. (eds) *Sustainable Aquaculture.* Springer, Cham, Switzerland, pp. 1–62. https://doi.org/10.1007/978-3-319-73257-2_1

Seggel, A., De Young, C. and Soto, D. (2016) *Climate Change Implications for Fisheries and Aquaculture. Summary of the Findings of the IPCC Fifth Assessment Report.* FAO Fisheries and Aquaculture Circular No. C1122. Food and Agriculture Organization of the United Nations, Rome. Available at: http://www.fao.org/3/a-i5707e.pdf (accessed 11 September 2022).

Seneviratne, S.I., Neville, N., David, E., Goodess, C., Kanae, S., *et al.* (2012) Changes in climate extremes and their impacts on the natural physical environment. In: Field, C.B., Barros, V., Stocker, T.F., Qin, D., Dokken, D.J., *et al.* (eds) *Managing the Risks of Extreme Events and Disasters to Advance Climate Change Adaptation: A Special Report of Working Groups I and II of the Intergovernmental Panel on Climate Change (IPCC).* Cambridge University Press, Cambridge, UK, pp. 109–230.

Shelton, C. (2014) *Climate Change Adaptation in Fisheries and Aquaculture: Compilation of Initial Examples.* FAO Fisheries and Aquaculture Circular No. 1088. Food and Agriculture Organization of the United Nations, Rome. Available at: https://www.fao.org/3/i3569e/i3569e.pdf (accessed 11 September 2022).

Sievers, M., Korsøen, Ø., Warren-Myers, F., Oppedal, F., Macaulay, G., *et al.* (2021) Submerged cage aquaculture of marine fish: a review of the biological challenges and opportunities. *Reviews in Aquaculture* 14(1), 106–119. https://doi.org/10.1111/raq.12587

Springmann, M., Clark, M., Mason-D'Croz, D., Wiebe, K., Bodirsky, B.L., *et al.* (2018) Options for keeping the food system within environmental limits. *Nature* 562(7728), 519–525. https://doi.org/10.1038/s41586-018-0594-0

Stern, D.I. and Kaufmann, R.K. (2014) Anthropogenic and natural causes of climate change. *Climatic Change* 122, 257–269. https://doi.org/10.1007/s10584-013-1007-x

Stoltz, A.D., Shivlani, M. and Glazer, R. (2021) Fishing industry perspectives on sea-level rise risk and adaptation. *Water* 13(8), 1124. https://doi.org/10.3390/w13081124

Subasinghe, R. (2017) *World Aquaculture 2015: A Brief Overview.* FAO Fisheries and Aquaculture Circular No. 1140. Food and Agriculture Organization of the United Nations, Rome.

Sumaila, U.R., Cheung, W.W.L., Lam, V.W.Y., Pauly, D. and Herrick, S. (2011) Climate change impacts on the biophysics and economics of world fisheries. *Nature Climate Change* 1, 449–456. https://doi.org/10.1038/nclimate1301

Sweet, W.V., Hamlington, B.D., Kopp, R.E., Weaver, C.P., Barnard, P.L., *et al.* (2022) *Global and Regional Sea Level Rise Scenarios for the United States: Updated Mean Projections and Extreme Water Level Probabilities along US Coastlines.* NOAA Technical Report NOS 01. National Oceanic and Atmospheric Administration, National Ocean Service, Silver Spring, Maryland. Available at: https://oceanservice.noaa.gov/hazards/sealevelrise/noaa-nos-techrpt01-global-regional-SLR-scenarios-US.pdf (accessed 11 September 2022).

Thong, N.T., Ankamah-Yeboah, I., Bronnmann, J., Nielsen, M., Roth, E. and Schulze-Ehlers, B. (2020) Price transmission in the pangasius value chain from Vietnam to Germany. *Aquaculture Reports* 16, 100266.

Tilman, D. and Clark, M. (2014) Global diets link environmental sustainability and human health. *Nature* 515(7528), 518–522. https://doi.org/10.1038/nature13959

Tout, J., Siboni, N., Messer, L.F., Garren, M., Stocker, R., *et al.* (2015) Increased seawater temperature increases the abundance and alters the structure of natural *Vibrio* populations associated with the coral *Pocillopora damicornis*. *Frontiers in Microbiology* 6, 432. https://doi.org/10.3389/fmicb.2015.00432

Troell, M., Naylor, R.L., Metian, M., Beveridge, M., Tyedmers, P.H., *et al.* (2014) Does aquaculture add resilience to the global food system? *Proceedings of the National Academy of Sciences USA* 111(37), 13257–13263. https://doi.org/10.1073/pnas.1404067111

Trombetta, T., Vidussi, F., Mas, S., Parin, D., Simier, M. and Mostajir, B. (2019) Water temperature drives phytoplankton blooms in coastal waters. *PLoS ONE* 14(4), e0214933. https://doi.org/10.1371/journal.pone.0214933

Van Vliet, M.T.H., Ludwig, F., Zwolsman, J.J.G., Weedon, G.P. and Kabat, P. (2011) Global river temperatures and sensitivity to atmospheric warming and changes in river flow. *Water Resources Research* 47(2), W02544. https://doi.org/10.1029/2010WR009198

World Bank (2013) *Fish to 2030: Prospects for Fisheries and Aquaculture.* World Bank Report No. 83177-GLB. World Bank, Washington, DC. Available at: https://documents1.worldbank.org/curated/en/458631468152376668/pdf/831770WP0P11260ES003000Fish0to02030.pdf (accessed 11 September 2022).

2 Infectious Diseases of Coldwater Fish in Marine and Brackish Waters

Jo-Ann C. Leong[1]*, Oystein Evensen[2] and Patrick T.K. Woo[3]
[1]*University of Hawaii at Manoa, Kaneohe, Hawaii, USA;* [2]*Norwegian University of Life Sciences, Ås, Norway;* [3]*University of Guelph, Guelph, Ontario, Canada*

2.1 Introduction

Aquaculture provides 52% of the fish destined for human consumption (82 million tonnes, valued at US$250 billion) (FAO, 2020). With the population predicted to increase to about 9.7 billion by 2050, the demand for animal protein including fish will only increase. Aquaculture must increase production especially since the annual catch fisheries is declining (UN, 2019) and it must do this sustainably in the ongoing climate change. In this chapter, we focus on infectious diseases because disease outbreaks have major impacts on production by causing mortalities, decreasing feed conversion, and altering stocking densities and harvest time, requiring testing, sampling and decontamination strategies. Research has provided some insights on the impact of increased temperatures on diseases. However, other climate-induced changes in aquatic acidification, oxygen levels, salinity and migration patterns of possible carriers have not been well characterized for their effects on disease virulence and prevalence in marine aquaculture. This chapter examines available information on viral, bacterial, parasitic and fungal infectious diseases of cage-cultured fish in marine and estuarine waters.

VIRAL DISEASES

2.2 Infectious Haematopoietic Necrosis (IHN)

2.2.1 Name of disease, distribution and fish

Infectious haematopoietic necrosis (IHN) is caused by the novirhabdovirus, infectious haematopoietic necrosis virus (IHNV), family *Rhabdoviridae*, and outbreaks in cage-cultured salmon can result in 50–90% mortality (Bootland and Leong, 1999; Dixon *et al.*, 2016; OIE, 2021).

IHNV is a bullet-shaped, enveloped, single-stranded, minus-sense RNA virus with a genome size ranging from 11,098 to 11,144 bp (https://www.ncbi.nlm.nih.gov/nuccore/?term=Infectious+hematopoietic+necrosis+virus+genome, accessed 21 September 2022). The genome encodes six viral genes in the order: 3′-nucleocapsid (N)–phosphoprotein (P)–matrix (M)–glycoprotein (G)–non-virion protein (Nv)–RNA polymerase (L)-5′. These genes have been characterized to construct vaccines (Evensen and Leong, 2013; Ma *et al.*, 2019), examine virus evolution (Kurath *et al.*, 2016) and identify virulence determinants

*joannleo@hawaii.edu

 Climate Change on Diseases and Disorders of Finfish in Cage Culture, 3rd Edition (eds P.T.K. Woo and R.P. Subasinghe)
DOI: 10.1079/9781800621640.0002

(Abbadi *et al.*, 2021). The phylogenetic analysis of the IHNV glycoprotein gene has identified five genogroups named U (upper), M (middle), L (lower), E (European) and J (Japanese) (Kurath *et al.*, 2016).

Originally identified in western North America, the disease has spread to Europe and Asia. Countries reporting detection or suspicion of IHN to the World Organisation for Animal Health (OIE) include the USA, Canada, Japan, Republic of Korea, China and most European countries. It has not been reported in Africa or the southern hemisphere (OIE, 2021).

Susceptible species include:

- *Oncorhynchus* spp. *(Oncorhynchus mykiss, Oncorhynchus tshawytscha, Oncorhynchus kisutch, Oncorhynchus nerka, Oncorhynchus keta, Oncorhynchus rhodurus, Oncorhynchus masou, Oncorhynchus clarkii)*;
- *Salmo* spp. *(Salmo salar, Salmo trutta, Salmo marmoratus, Salmo namaycush, Salmo labrax)*;
- *Salvelinus* spp. *(Salvelinus alpinus, Salvelinus fontinalis, Salvelinus leucomaenis)*;
- *Esox lucius*;
- *Acipenser transmontanus*;
- *Cymatogaster aggregata*;
- *Clupea pallasii*;
- *Gadus morhua*;
- *Dicentrarchus labrax;* and
- *Scophthalmus maximus*.

2.2.2 Impacts and the environment

IHN outbreaks are costly and estimates from epizootics in Canada from 1992–1996 and 2001–2003 resulted in combined economic losses of CAN\$40 million in inventory and CAN\$200 million in lost sales (Dixon *et al.*, 2016).

Outbreaks usually involve juvenile rainbow trout at water temperatures from 8 to 15°C, with cumulative mortality reaching 20–90%. The virus can cause significant losses among older Atlantic salmon in seawater.

2.2.3 Diagnoses

Clinical signs include lethargy interspersed with bouts of frenzied activity. The infected fish can display spiral swimming, show trailing faecal casts, darkening of the skin, exophthalmia, distended abdomen and external haemorrhages (OIE, 2021). Internally, petechial haemorrhages are evident (Fig. 2.1).

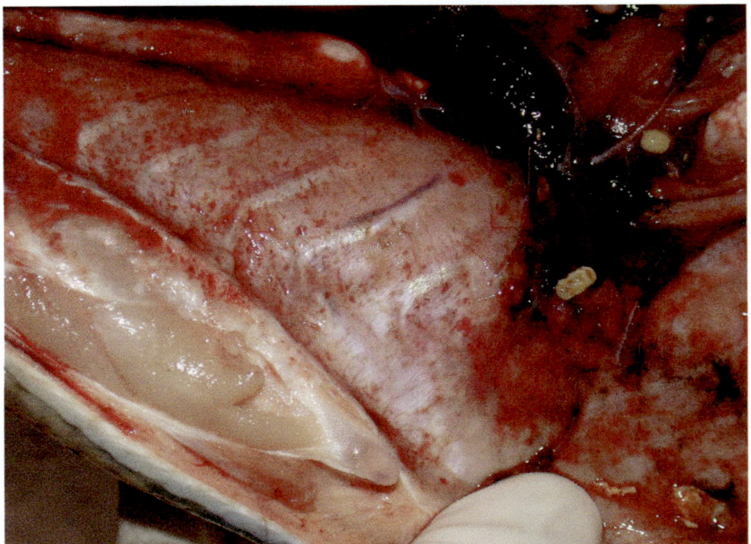

Fig. 2.1. IHN lesions in Atlantic salmon. Farmed Atlantic salmon with clinical IHN disease from a marine net-pen in British Columbia, Canada. Petechial haemorrhages on the surfaces of internal organs are characteristic lesions. (Photograph courtesy of Kyle Garver, Pacific Biological Station, DFO, Canada.)

The viral agent's presence can be confirmed using immunocytopathology, serum neutralization and reverse transcription-polymerase chain reaction (RT-PCR) assay (OIE, 2021).

2.2.4 Transmission, spread and pathogen dynamics

IHNV is transmitted primarily by fish-to-fish contact from infected fish that shed virus in faeces, urine, sexual fluids and mucus. The virus can survive for at least 14 days in seawater at 15°C (Toranzo and Hetrick, 1982). The virus was probably introduced to Asia and Europe via virus-contaminated eggs. Salmon lice (*Lepeophtheirus salmonis*) can transmit the virus from infected Atlantic salmon to susceptible animals. IHNV has also been isolated from mayflies (*Callibaetis* spp.), leeches (*Piscicola salmositica*) and a copepod (*Salminicola* sp.) collected from spawning sockeye salmon (Bootland and Leong, 1999).

IHN outbreaks in Europe, Asia and North America have occurred in fresh water except for Atlantic salmon in saltwater pens in British Columbia, Canada. Fish (60 g to 6.8 kg) were infected with IHNV U genogroup, common in sockeye salmon (Kurath *et al.*, 2016).

2.2.5 Effects of temperature

IHNV is a coldwater disease and rising ocean temperatures may diminish the number of IHN epizootics. However, studies have shown that even transient exposure to warmer temperatures results in stress-induced immunosuppression in salmon (Home and Tveten, 2020). Also, higher temperatures (10–15°C) result in faster viral replication, more neutralizing antibodies and lower fish mortality. Viral RNA persists longer at the higher temperatures (Paez *et al.*, 2021).

2.2.6 Prevention/control strategies

The primary cause of IHN epizootics is importation of infected eggs or fry. In areas where the disease is endemic, good biosecurity is important and involves the disinfection of eggs with iodophor, virus-free water for egg incubation and larval rearing. Feed formulated from fish carcasses should be sterilized and sanitation of tanks after use is important. A DNA vaccine for IHNV has been licensed for use in Canada for net-pen-reared Atlantic salmon (Alonso and Leong, 2013).

2.3 Viral Haemorrhagic Septicaemia (VHS)

2.3.1 Name of disease, distribution and fish

Viral haemorrhagic septicaemia (VHS) is caused by a novirhabdovirus, viral haemorrhagic septicaemia virus (VHSV), in the family *Rhabdoviridae*. The disease affects many economically important species with 85% mortality in marine rainbow trout.

The virus, like IHNV, has six viral genes arranged in the order 3′-N–P–M–G–Nv–L-5′ with a genome length of 10,845 to 11,144 bp (155 complete genomes annotated in GenBank; https://www.ncbi.nlm.nih.gov/nuccore/?term=Viral+Hemorrhagic+Septicemia+virus+genome, accessed 1 February 2022). The virus can be genotyped into four main groups by analysing the viral glycoprotein gene: I, II, III and IV, with nine subtypes (Panzarin *et al.*, 2020; Stepien *et al.*, 2020). Virulence markers in the P, N and Nv genes have been identified (Panzarin *et al.*, 2020).

Genotype Ia is predominantly found in Salmonidae in Europe and Asia; genotypes Ib, II and III in *Clupea harengus*, *Sprattus sprattus* and *Trisopterus esmarkii* in Europe and Asia. *C. pallasii* is the natural host for genotype IVa. These shoaling fish can transmit the virus to cage-reared fish.

The OIE (2021) lists 71 species of fish susceptible to VHSV and 48 additional species with incomplete evidence of susceptibility along with the infectious virus genotype. Economically important species include:

- *O. mykiss*;
- *S. maximus*;
- *S. salar*; and
- *Paralichthys olivaceus*.

2.3.2 Impacts and the environment

Since its identification in rainbow trout in Europe in the 1930s, VHSV has spread to many species in both marine and freshwater environments. An outbreak can be devastating. For example, VHS outbreaks in marine-cultured *P. olivaceus* can lead to 40–60% mortality in market-sized fish in Korea. The disease is responsible for 12.5% of all *P. olivaceus* mortalities in Korea where the industry was valued at US$443 million in 2018 (Chun *et al.*, 2021).

The virus produces its highest mortalities at temperatures from 9 to 12°C and stops replicating above 15°C.

2.3.3 Diagnoses

VHS is characterized by a rapid onset of mortality with lethargy, darkening of the skin, exophthalmia, pale gills, and haemorrhages at the base of the fins or in the gills, skin and eyes. There is marked necrosis of haematopoietic tissue in the anterior kidney. In some fish, there is abnormal swimming with spiralling and flashing, and peritoneal oedema.

Virus isolation on cell culture and specific detection with antibody neutralization or immunostaining are used for diagnosis. Molecular tests with RT-PCR are rapid, sensitive and specific for diagnostic confirmation (OIE, 2021).

2.3.4 Transmission, spread and pathogen dynamics

Transmission of the virus occurs horizontally through water contaminated with excreted virus in the urine of infected fish and virus released from skin lesions. Oral transmission can also occur when fish prey upon infected fish. Intra-ovum vertical transmission in *O. mykiss*, *Sander vitreus* and *E. lucius* has been ruled out in several studies. Potential vectors of transmission have been reported and include *Diporeia* spp. (amphipod), leeches and freshwater turtles (OIE, 2021).

VHSV genotypes Ib, Id, Ie, II, III, IVa, IVc and IVd have been isolated from outbreaks in finfish in marine and estuarine environments in Europe, North America and Asia (Korea and Japan). The virus appears to be less stable in seawater than fresh water (Hawley and Garver, 2008). Survivors of epizootics can become long-term carriers of the virus, for example *C. pallasii* and other wild species (OIE, 2021). VHSV, with its observed ecological plasticity, has been identified as a global risk factor for fish species within the Perciformes, Salmoniformes and Gadiformes orders (Escobar *et al.*, 2018). These authors identified 19 bioclimatic variables to develop ecological niche models for risk of VHSV occurrence. These models predict that VHSV may have favourable physical (temperature, runoff), chemical (salinity, pH, phosphate) and biotic (chlorophyll) conditions for spread to freshwater and marine ecosystems of Africa, Latin America, Australia and China (Fig. 2.2). This ecological niche modelling can also incorporate projected climate changes, but that work has not been carried out (L. Escobar, Virginia, 2022, personal communication).

2.3.5 Prevention/control strategies

There is no commercial vaccine for VHSV despite decades of research (Evensen and Leong, 2013). A broad-spectrum antiviral, JL122, has been shown to block infection and inhibit transmission of aquatic rhabdoviruses like IHNV, VHSV and spring viraemia of carp virus (SVCV) (Balmer *et al.*, 2018). Disinfection of newly fertilized or eyed eggs is still the most efficient and cost-effective preventive measure for stopping the spread of these diseases in salmonid fish. Efforts to selectively breed disease-resistant strains of rainbow trout and Japanese flounder are underway (Verrier *et al.*, 2013; Liyanage *et al.*, 2022).

2.4 Infectious Pancreatic Necrosis (IPN)

2.4.1 Name of disease, distribution and fish

Infectious pancreatic necrosis (IPN), a disease of young salmonid fry, is caused by *Infectious pancreatic necrosis virus* (IPNV) (family *Birnaviridae*, genus *Aquabirnavirus*) (Dhar *et al.*, 2017; Dopazo, 2020a,b).

This small, non-enveloped, icosahedral virus has a genome consisting of two double-stranded RNA segments. The larger A segment encodes a polyprotein, NH_2–Vp2–Vp4–Vp3–COOH, that is

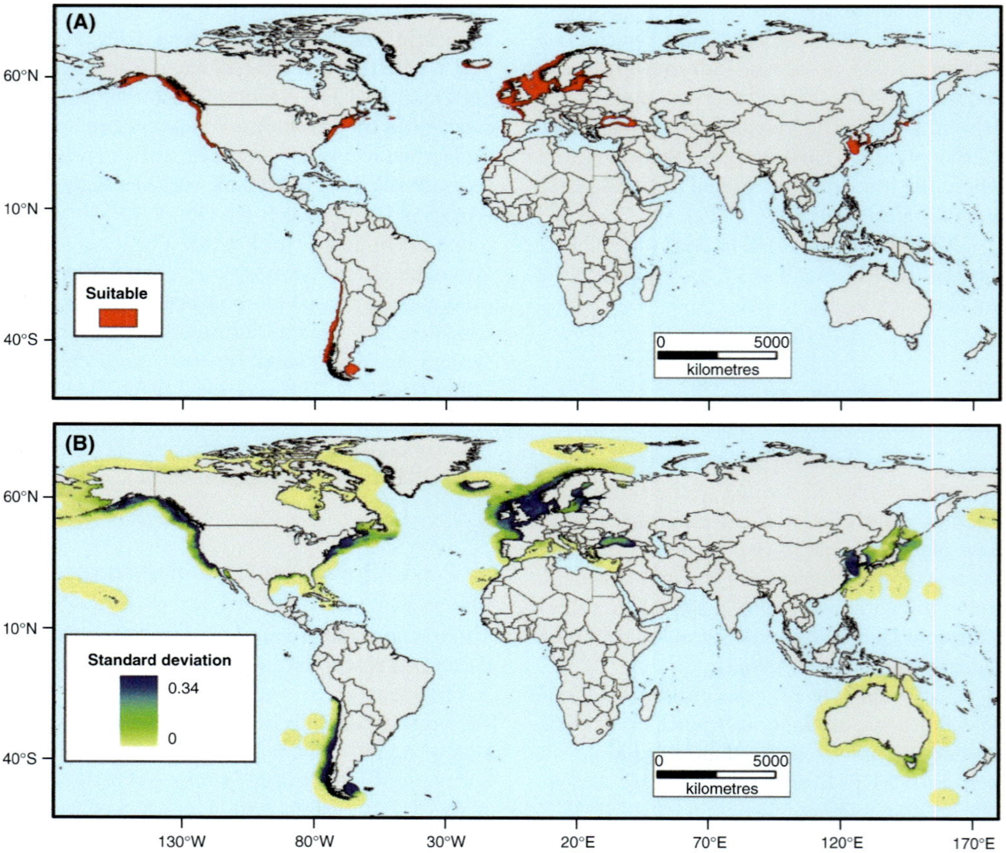

Fig. 2.2. Potential global sites for VHSV dispersal. Global ecological niche model for marine coastal regions, showing where VHSV dispersal is potentially possible in red (Escobar *et al.*, 2018). This model is based on the ecological plasticity of VHSV and favourable physical (temperature, runoff), chemical (salinity, pH, phosphate) and biotic (chlorophyll) conditions for establishing VHSV in marine aquaculture. The model includes regions that are currently disease free. (Courtesy of *Reviews in Fish Biology and Fisheries* and the author, Luis Escobar, Virginia Tech, USA.)

subsequently cleaved by the internal protease, VP4, to produce VP2, external capsid protein, and VP3, internal capsid protein. A second open reading frame (ORF) on segment A encodes a small, non-structural protein, VP5, found only in infected cells. The virulence determinants for IPNV are located on the VP2 gene. RNA segment B encodes the viral RNA-dependent RNA polymerase (VP1). Early serotype analysis identified two major serogroups, A and B; in A, there were nine subtypes. Since then, seven genotypes have been identified (Dopazo, 2020a,b).

IPN outbreaks have been reported in salmon farms in Scotland, Norway, Chile and Asia where typical mortalities reach approximately 25%, and severe outbreaks kill as many as 80–90% of farmed fish.

IPNV infects over 60 species of fish and 15 species of freshwater and marine invertebrates (Reno, 1999; Munro and Midtlyng, 2011; Dopazo, 2020a). Economically important marine species include:

- *S. salar*;
- *Anguilla anguilla*;
- *Anguilla japonica*;
- *Seriola quinqueradiata*;
- *S. maximus*;
- *Limanda limanda*;
- *Hippoglossus hippoglossus*; and
- *G. morhua*.

2.4.2 Impacts and the environment

IPNV was once a major pathogen for the marine aquaculture industry and historically was responsible for high losses in Atlantic salmon smolts after transfer to seawater. For countries such as Italy and Scotland, even vaccinated IPN-QTL (quantitative trait locus) salmon have outbreaks with new IPNV isolates (Panzarin *et al.*, 2018; Benkaroun *et al.*, 2021).

2.4.3 Diagnoses

IPN is a disease of young fry and is characterized by skin darkening, erratic corkscrew swimming, anorexia and trailing faecal casts. On necropsy, viral necrosis of the pancreas and thick mucus surrounding the small intestine are often observed.

Microscopic signs include focal necrosis of the pancreatic acinar and islet cells and the haematopoietic cells of the kidney. Confirmatory diagnosis is usually by cell culture methods and subsequent serological testing. RT-PCR diagnostic methods are employed as well.

2.4.4 Transmission, spread and pathogen dynamics

IPNV is transmitted both horizontally through virus contaminated faeces and mucus, and vertically in ovarian and seminal fluids. The virus is highly infectious; just 1 $TCID_{50}$/ml is enough to infect salmon (Munro and Midtlyng, 2011). Persistently infected asymptomatic survivors of IPNV infection serve as reservoirs of the virus, spreading it through faeces and reproductive fluids. The virus enters through the gills, skin and intestinal epithelium.

Wild fish are potential reservoirs: *G. morhua*, *Pollachius pollachius* and *Merluccius capensis*, as are shellfish and birds (Reno, 1999).

2.4.5 Effects of temperature

The virus is highly stable under various environmental conditions. It has been isolated from cold (20°C and below), cool (20–28°C) and warm waters (28°C and above). Although the virus has an optimal replication temperature of 10–15°C, it has been isolated from outbreaks in waters between 5 and 25°C. These characteristics ensure that this virus will persist and might adapt under the warmer-water scenarios of climate change (Panzarin *et al.*, 2018).

2.4.6 Prevention/control strategies

Control of the spread of IPNV relies on testing and restrictions of movement of fish and fish eggs from infection sites. Vaccines based on the IPNV VP2 protein induce virus neutralizing antibodies (Dhar *et al.*, 2010). DNA vaccines have been reported using the VP2 gene as the basis for vaccination and oral formulations for DNA vaccines, killed virus and biotechnologically produced VP2 protein have shown some effectiveness in vaccine trials (Bogwald and Dalmo, 2021; Hua *et al.*, 2021; Tamer *et al.*, 2021).

IPN-resistant fish have also been developed through selective breeding based on the gene for epithelia cadherin that provides resistance to IPNV (Moen *et al.*, 2015). Use of IPN-resistant salmon strains has reduced outbreaks of IPN in Norway from 223 to 19 from 2009 to 2019 (Sommerset *et al.*, 2020). This success, however, has been tempered by reports of new IPNV variants that can cause mortality in genetically resistant fish (Hillestad *et al.*, 2021).

2.5 Infectious Salmon Anaemia (ISA)

2.5.1 Name of disease, distribution and fish

Infectious salmon anaemia (ISA) is a severe disease in seawater-cultured *S. salar*, *S. trutta* and *O. mykiss*, causing severe anaemia and haemorrhages and necrosis in the gills, heart, liver, kidney and spleen (OIE, 2021).

The disease is caused by the orthomyxovirus, infectious salmon anaemia virus (ISAV) (family *Orthomyxoviridae*, genus *Isavirus*). Its genome contains eight segments of minus-sense, single-stranded RNA encoding ten proteins. The structural proteins – nucleoprotein, matrix protein, haemagglutinin-esterase (HE) protein responsible for receptor binding and a surface glycoprotein with fusion (F) activity – are encoded by genome segments 3, 8, 6 and 5 respectively. Segments 1, 2 and 4 encode the viral polymerases PB2, PB1 and PA (Rimstad and Markussen, 2019). There are two groups of ISAV: (i) the pathogenic HPR-deleted (highly polymorphic region) ISAV which is missing a 35-amino-acid region of the HE gene; and (ii) the non-pathogenic HPR0 (non-deleted HPR) ISAV which retains this region. A virulence marker has been identified on the F gene where a single amino acid change near the protein's cleavage site is the suggested virulence requirement (Kibenge *et al.*, 2007). The descriptions that follow refer to the HPR-deleted strains of ISAV.

Outbreaks have occurred in Norway, Chile, Scotland, Ireland, the Faroe Islands, North Atlantic Canada and the USA. The virus is a reportable disease (OIE).

2.5.2 Impacts and the environment

ISA epizootics in pen-reared Atlantic salmon can be economically devastating since the disease usually occurs post seawater transfer when the fish are 8–16 months old. Mortalities vary from 5 to 90% (Ritchie *et al.*, 2009).

Although outbreaks can occur year-round, ISA is most common in the spring/early summer and late autumn.

2.5.3 Diagnoses

Clinical signs include pale gills, exophthalmia, distended abdomen, blood in the anterior eye chamber and skin haemorrhages. The disease is found in all life stages. Pathological signs are yellowish or blood-tinged fluid in the peritoneal and pericardial cavities, oedema of the swim bladder, small haemorrhages of the peritoneum, pinpoint haemorrhages of the skeletal muscle, dark red kidney, dark red spleen and dark red liver. The diagnosis is confirmed by cytopathogenic effect in tissue culture, serological confirmation with anti-ISA antibody and RT-PCR techniques (OIE, 2021).

2.5.4 Transmission, spread and pathogen dynamics

ISAV is spread by contaminated equipment and fish movement. Horizontal transmission of the virus does occur, but waterborne spread over distances is limited because ISAV does not survive more than 8 days in seawater and is no longer infectious after 3 h in seawater exposed to natural ultraviolet (UV) radiation (Vike *et al.*, 2014). Wild salmonids are potentially carriers of the virus and sea lice (*Caligus rogercresseyi*) can act as mechanical vectors. Vertical transmission has been suggested but evidence for transmission is inconclusive (Christiansen *et al.*, 2021).

Susceptible species are *C. harengus* and *O. masou*. The virus has been found using PCR in *O. kisutch*. When fish species were investigated for the ISAV receptor, all salmonids expressed the receptor; only some cod-like and perch-like fish did; and all flat fish were negative (Aamelfot *et al.*, 2014). These studies suggest other host reservoirs (*G. morhua*, *Mallotus villosus*, *T. esmarkii*, *Merlangius merlangus*, etc.) (Aamelfot *et al.*, 2014). HPR-deleted ISAV only causes disease in Atlantic salmon (Rimstad and Markussen, 2019).

2.5.6 Effects of temperature

ISAV has been characterized for its growth at different temperatures in Atlantic salmon head kidney cells, SHK-1, and was found to grow well at 10°C, optimally at 15°C, less well at 20°C and not at all at 25°C (Falk *et al.*, 1997). Disease outbreaks in farmed salmon in Norway occurred at water temperatures that ranged from 4 to 13°C (Lyngstad *et al.*, 2018).

2.5.7 Prevention/control strategies

Good biosecurity and mandatory health control, transport and slaughterhouse regulations have reduced the incidence of ISAV outbreaks. Vaccination with an inactivated ISA vaccine has been in use since 1999 mainly in Chile, the Faroe Islands and Canada (OIE, 2021). Centrovet produces a yeast-based subunit vaccine for ISA that is available in Chile (Ma *et al.*, 2019). Parmaq and Elanco sell whole virus-inactivated vaccines. Selective breeding programmes in Norway have produced ISA-resistant Atlantic salmon since 2001 (Chase-Topping *et al.*, 2021) and these fish have been found to be less likely to transmit infection after exposure to ISAV.

2.6 Salmonid Alphavirus (SAV) Infections: Pancreas Disease (PD) or Sleeping Disease (SD)

2.6.1 Name of disease, distribution and fish

Pancreas disease (PD) or sleeping disease (SD) in Atlantic salmon is caused by salmonid alphavirus (SAV) (family *Togaviridae*, genus *Alphavirus*). Outbreaks in marine salmon farms begin with a sudden drop in appetite, fish swimming slowly at the surface and weak 'sleeping' fish at the bottom. Mortalities can reach 50%, but the mean mortality is 6.9% (Karlsen and Johansen, 2017; OIE, 2021).

SAV is an enveloped, single-stranded, positive-sense RNA virus. Its genome consists of a single RNA segment of 11,900+ nucleotides, encoding eight proteins in two ORFs: four structural capsid proteins (E1, E2, E3 and 6K) at the 3′ end and four non-structural proteins (nsP1–4) at the 5′ end. There are possibly seven genotypes of SAV (SAV-1 to SAV-7) based on the nucleic acid sequences of E2 and nsP3 proteins. The possible seventh genotype is a new SAV isolated from wild-caught marine *Labrus bergylta* (Tighe *et al.*, 2020). The genotypes SAV-1 and SAV-2 cause disease in fish both in fresh water and seawater, while the four genotypes SAV-3 to SAV-6 are found only in seawater.

Susceptible host species for SAV include *S. salar*, *L. limanda*, *O. mykiss* and *S. alpinus*. The virus has also been isolated from *C. harengus*, *Myoxocephalus octodecemspinosus*, *Melanogrammus aeglefinus*, *T. esmarkii*, *Pollachius virens*, *M. merlangus*, *G. morhua*, *Merluccius hubbsi*, *Platichthys flesus* and *S. trutta*. These fish in the wild may also be reservoirs for the virus.

2.6.2 Impacts and the environment

The direct costs of a PD outbreak in 2010 was estimated at US$1.74 million per 500,000 smolts annually for Norwegian farms (Aunsmo *et al.*, 2010). This estimate included biological losses due to reduced growth, mortality, increased feed conversion ratio and carcass downgrading. Included as well were the extra expenditures associated with disease prevention and treatment. In 2015, the estimated costs were US$3.09 million/500,000 smolts (Pettersen *et al.*, 2015).

2.6.3 Diagnoses

Clinical signs include lethargy, anorexia, and mortality that begins after 1–2 weeks and continues for 32 weeks. Preliminary diagnosis based on clinical signs and histopathology/cytopathology can be confirmed using cell culture isolation and serum neutralization and/or nucleic acid amplification techniques (OIE, 2021).

2.6.4 Transmission, spread and pathogen dynamics

The spread of SAV by horizontal transmission is well established. Experimental studies have shown that both faeces and mucus from infected salmon contain the virus which spreads through water currents over long distances. Vertical transmission is negligible (OIE, 2021).

All life stages of *S. salar* and *O. mykiss* are susceptible to experimental infection. There is evidence that survivors of an SAV outbreak become carriers of the virus, so farmed Atlantic salmon and rainbow trout are considered the main reservoir of SAV (Taksdal and Sindre, 2016).

2.6.5 Effects of temperature

SAV-3 from infected fish survives in seawater at 6, 12 and 16°C. The virus is not infectious after 1 week at 12 and 16°C, with a dramatic reduction by week 2 at 6°C (Jarungsriapisit *et al.*, 2020). Yet, PD outbreaks occur with greater frequency when there is increasing seawater temperature while decreasing temperatures lengthen the incubation period (Stene *et al.*, 2014).

2.6.6 Prevention/control strategies

Good husbandry practices include standard disinfection of eggs, use of UV inactivation for rearing of the young fish, reduced stocking densities, and regular cleaning of tanks and pens, and are all recommended to reduce the incidence of PD. Commercial inactivated PD vaccines are available, and in 2017 the European Commission authorized a DNA vaccine against PD for *S. salar* (Thorarinsson *et al.*, 2021).

A QTL for resistance to SAV-3 has been identified (Gonen *et al.*, 2015; Aslam *et al.*, 2020). Breeding programmes in Ireland and Norway have produced fish that are genetically less susceptible to PD and these fish are commercially available.

2.7 Viral Encephalopathy and Retinopathy (VER)

2.7.1 Name of disease, distribution and fish

Viral encephalopathy and retinopathy (VER) or viral nervous necrosis (VNN) is a disease of more than 50 species of marine fish around the world (Olveira *et al.*, 2013). The disease is characterized by vacuolating lesions of the central nervous system and the retina (OIE, 2019; Bandin and Souto, 2020; Toffan and Panzarin, 2020).

The causative agent is a betanodavirus in the family *Nodaviridae*, a group of non-enveloped viruses containing two segments of single-stranded, positive-sense RNA: RNA1 encodes the non-structural RNA-dependent RNA polymerase and RNA2 encodes the capsid protein. A subgenomic RNA3 is produced from RNA1 during replication intracellularly and it encodes proteins B1 and B2. B1 has an anti-necrotic death function and B2 has an RNA silencing-suppression function. Betanodaviruses are classified into four genotypes based on the hypervariable T4 region of RNA2: *Tiger puffer nervous necrosis virus* (TPNNV); *Redspotted grouper nervous necrosis virus* (RGNNV), *Barfin flounder nervous necrosis virus* (BFNNV) and *Striped jack nervous necrosis virus* (SJNNV) (Huang *et al.*, 2022). RGNNV exhibits the widest host range of warmwater fish species; TPNNV infects only tiger puffer fish; BFNNV infects a range of coldwater marine fish species; and SJNNV was initially restricted to Japanese waters, but it has been found to infect *Solea senegalensis*, *Sparus aurata* and *D. labrax* in Southern Europe. The four genotypes also have different optimum growth temperatures (Table 2.1).

2.7.2 Impacts and the environment

VER is a serious disease of economically important farmed marine fish because it affects many species over a wide range of temperatures with a worldwide distribution (OIE, 2019; Bandin and Souto, 2020). Most fish are infected at early development with 100% mortality. In *Lates calcarifer* for example, 100% mortalities have

Table 2.1. Genotype and serotype characterization of the different nervous necrosis viruses of fish. (Modified from OIE, 2019.)

Genotype	Serotype	Target host fish	Optimum growth temperature (°C)	Reassortants
SJNNV	A	Striped jack	20–25	
TPNNV	B	Tiger puffer	20	
BFNNV	C	Coldwater fish: Atlantic halibut, Atlantic cod, flounders, etc.	15–20	
RGNNV	C	Warmwater fish: Asian sea bass, European sea bass, groupers, etc.	25–30	
RGNNV/SJNNV	A	Gilthead sea bream, European sea bass	25–30	RNA1 from RGNNV
SJNNV/RGNNV	C	Gilthead sea bream, European sea bass	20–25	RNA1 from SJNNV

accounted for huge economic losses for the industry and models for assessing the economic impact of the disease in European sea bass highlight the more severe impact on smaller farms (Fernandez Sanchez *et al.*, 2022). The ability of these betanodaviruses to undergo reassortment increases the potential for these viruses to adapt to new hosts and environments.

2.7.3 Diagnoses

Affected fish swim with a swirling and spinning movement with periods of apathy, swim bladder hyperinflation and blindness. Histopathological findings include vacuolation and necrosis of nervous cells of the brain, spinal cord and retina. Electron microscopic examination will reveal subspherical viral particles, 25 nm in diameter, in nervous tissue. RT-PCR is the most rapid and convenient method for diagnosing the virus infection. Immunohistochemistry and enzyme-linked immunosorbent assay (ELISA)-based methods are also used (OIE, 2019).

2.7.4 Transmission, spread and pathogen dynamics

Although the disease is primarily found in the larval and juvenile stages, market-sized and adult fish are killed by the virus, especially *H. hippoglossus*, *Epinephelus septemfasciatus* and *D. labrax*. The virus is spread through contaminated water and equipment, and interspecies transmission has been documented. Since VER is found in the gills, skin, fins and intestine of infected fish, virus release from these organs is probable. Vertical transmission is also implicated since the virus is found in the gonads and sexual fluids. Whether the virus is transmitted *in ovo* has not been determined (Bandin and Souto, 2020).

VER episodes have been reported in marine fish reared in Asian, Australian and European waters since the 1990s. Routine surveys have revealed the existence of RGNNV carriers and high prevalence of the RGNNV genotype in wild fish in Asia and the Mediterranean Basin. Nervous necrosis virus (NNV) has also been detected in 21 species of diverse marine invertebrates (OIE, 2019).

2.7.5 Effects of temperature

For RGNNV, studies have shown that when water temperature was increased from 16 to 22°C, the viral load and mortalities increased in infected sole (Souto *et al.*, 2015). BFNNV causes disease at much lower temperatures; consequently, global warming may reduce the incidence of outbreaks.

The virus is sensitive to warmer temperatures in seawater with its highest reported viability at 15°C for all genotypes (Vazquez-Salgado *et al.*, 2021). Viral persistence drops in seawater as temperatures increase. Thus, in the absence of susceptible or shedder fish, NNV should not survive long in seawater.

2.7.6 Prevention/control strategies

Control of VER outbreaks relies mainly on biosecurity and good sanitation practices. Proadifen hydrochloride (a cytochrome P450 inhibitor), ribavirin (an antiviral) and amantadine have been shown to strongly inhibit NNV replication both *in vivo* and *in vitro* (Zhu *et al.*, 2022). Inactivated virus, subunit vaccines, DNA and live, attenuated vaccines have been developed for NNV and injectable killed virus vaccines are commercially available for RGNNV in the Mediterranean Basin.

Genetic selection for resistance to NNV has been initiated for Atlantic cod, Asian sea bass and European sea bass, but there has been no report of specific genes for resistance (Yang *et al.*, 2022).

2.8 Cardiomyopathy Syndrome (CMS)

2.8.1 Name of disease, distribution and fish

Cardiomyopathy syndrome (CMS) in Atlantic salmon is a severe cardiac disease caused by piscine myocarditis virus (PMCV), a member of the *Totiviridae* family of double-stranded RNA viruses. The disease primarily affects adult Atlantic salmon at 14 to 18 months after transfer to seawater.

PMCV is a non-enveloped virus with a genome of 6688 nucleotides encoding three proteins. ORF1 encodes the virion capsid protein, ORF2 encodes the RNA polymerase and ORF3 encodes a putative non-structural protein that is only seen in totiviruses infecting vertebrate hosts (Tighe *et al.*, 2019; Sandlund *et al.*, 2021a).

2.8.2 Impacts and the environment

CMS was first described in Norway in 1985 and since then has spread to the Faroe Islands, Scotland, Ireland and Canada. In the last decade, CMS was diagnosed in 191 farms in Norway (Fritsvold and Bang Jensen, 2021). Since the disease affects adult Atlantic salmon close to harvest, the losses are economically costly. The problem is that fish appear normal without clinical signs until mortalities begin 4–8 weeks after infection.

2.8.3 Diagnoses

Infected fish exhibit exophthalmia and ventral petechiae. Internally, the fish have ascites, an enlarged or ruptured atrium, blood clots in the cardiac cavity, and a dark or uneven coloured liver.

Necropsy, histological findings and RT-PCR of heart samples are used to confirm the diagnosis. There is no efficient cell culture method to isolate and amplify the virus (Haugland *et al.*, 2011; Garseth *et al.*, 2018).

2.8.4 Transmission, spread and pathogen dynamics

CMS is spread horizontally in seawater presumably through infected mucus and other excretions. When infected fish cohabit with non-infected fish, PCMV is transmitted to the naïve fish, but the route of entry is unknown (Haugland *et al.*, 2011). Effective disinfection of fertilized eggs has ruled out vertical transmission by *in ovo* virus transfer (Mikalsen *et al.*, 2020).

Prevalence of PCMV in wild Atlantic salmon is low and suggests that wild salmon are not major reservoirs (Garseth *et al.*, 2018). Instead, escaped infected farmed salmon may be the source of infections between farms. PCMV is also found in *Argentina silus*, *Symphodus melops* and *L. bergylta*. The wrasses are susceptible to PCMV infection (Scholz *et al.*, 2017). The virus was not found in sediments, plankton, biofilm and bottom-dwelling organisms around fish cages in a farm with a CMS outbreak, but was found in the mucus, faeces and salmon lice from infected fish (Hellebø *et al.*, 2014).

Risk factors for CMS mortalities include length of time in the sea, increasing cohort size, previous outbreaks at the site and lice removal treatments (Bang Jensen *et al.*, 2020).

2.8.5 Effects of temperature

Atlantic salmon have a normal thermal preference of ~14°C, but production sites in the northern hemisphere experience peak summer temperatures around 18–20°C. Long-term exposure (56 days) to 19°C had a negative impact on fish growth performance and a 50% reduction in feed intake, with cardiac gene and protein expression changes (Hevroy *et al.*, 2012). The cardiac transcriptional changes suggested vascularization changes and altered innate immune responses. These findings suggest salmon would be more susceptible to CMS with warmer temperatures, yet CMS outbreaks increase during the cold periods.

2.8.6 Prevention/control strategies

Control and prevention measures include biosecurity to prevent virus introduction to aquaculture facilities, appropriate husbandry practices to reduce CMS-associated mortalities, selective breeding for CMS resistance, and diet alterations (Garseth *et al.*, 2018). There is no vaccine available for CMS, but a candidate plant-produced subunit vaccine has been reported (Su *et al.*, 2021). A QTL for resistance to CMS has been identified on Atlantic salmon chromosome 27 (Ssa27), a genome segment that also contains several immune-related genes (MHC II) (Hillestad *et al.*, 2020). QTL-selected eggs have been available commercially since 2013 (Garseth *et al.*, 2018).

2.9 Heart and Skeletal Muscle Inflammation (HSMI) Syndrome

2.9.1 Name of disease, distribution and fish

Heart and skeletal muscle inflammation (HSMI) affects farmed Atlantic salmon 5 to 9 months after transfer to seawater. There are no obvious external signs and mortalities vary from insignificant to 20%. Internally, affected fish have a pale heart, yellowish liver, ascites, swollen spleen and petechiae in the perivisceral fat. In heart and skeletal muscle, there is diffuse infiltration of lymphocytic cells and myocardial degeneration.

The disease agent is *Piscine orthoreovirus* (PRV) (family *Reoviridae*, genus *Orthoreovirus*) with a genome of ten segments that sort into three size groups: large (L1–L3), medium (M1–M3) and small (S1–S4), encoding ten to 13 ORFs. The S1- and M2-encoded proteins $\sigma 3$ and $\mu 1$ form a heterohexamer, $(\mu 1)_3(\sigma 3)_3$, in the outer capsid and mutations in these genes form the basis for virulence in PRV strains. Three subtypes of PRV, called PRV-1, -2 and -3, have been identified: PRV-1 causes HSMI in *S. salar* and jaundice syndrome in *O. tshawytscha*; PRV-2 causes erythrocytic inclusion body syndrome (EIBS) in *O. kisutch*; and PRV-3 causes pathological heart lesions in *O. mykiss* (Dahle and Olsen, 2021).

2.9.2 Impacts and the environment

Since its discovery in 1999, HSMI syndrome has become an increasing problem for the Norwegian farming industry, with some farms experiencing yearly outbreaks and subsequent economic losses. HMSI syndrome is now present in the UK, Ireland, Chile, the USA and Canada where accidental releases of infected farmed *S. salar* have spread the disease to wild populations (Polinski *et al.*, 2020).

Phylogeographic analyses of PRV-1 suggest that *S. salar* aquaculture facilitated the spread of HMSI syndrome from Europe to the critically endangered wild Pacific salmon (Mordecai *et al.*, 2021).

2.9.3 Diagnoses

Diagnostic criteria for HSMI syndrome are based on histopathological findings in the heart and skeletal muscles that distinguish this disease from other known diseases in salmon. Moderate to severe myocarditis is found in the compact and spongy layer of the heart. This myocarditis is characterized by infiltration of lymphocytes and macrophages. The skeletal muscle also exhibits inflammatory infiltration leading to severe myodegeneration and necrosis (Dahle and Olsen, 2020). Confirmation is determined using RT-PCR and immunohistochemistry of the heart and muscle tissue.

2.9.4 Transmission, spread and pathogen dynamics

HSMI syndrome is transmitted via cohabitation with fish infected with the virus and since the virus is resistant to high temperatures and UV treatment, it can easily spread by waterborne transmission. There are no reports verifying vertical transmission of PRV.

Survey work has expanded the known host range of PRV to include: *Oncorhynchus* spp. (*O. clarkii*, *O. tshawytscha*, *O. nerka*, *O. mykiss*, *O. kisutch*, *O. keta*, *Oncorhynchus gorbuscha*) and *S. salar* (Polinski and Garver, 2019). In North Atlantic waters PRV has been detected in both salmonid and non-salmonid fish (*S. salar*, *S. trutta*, *A. silus*, *Trachurus trachurus*, *C. harengus* and *M. villosus*) (Wiik-Nielsen *et al.*, 2012).

2.9.5 Effects of temperature

Salmon infected with PRV-1 have reduced cardiac performance and thermal tolerance (Lund *et al.*, 2017). Projected climate effects include episodes of higher temperatures which might impact the mortality rates for PRV-infected fish.

2.9.6 Prevention/control strategies

Since PRV are non-enveloped viruses, they are resistant to detergent-based cleaning routines, and strong acids or bases are recommended for decontamination. There are no vaccines available on the market, although moderate levels of protection have been demonstrated with killed virus. Treatment with anti-inflammatory agents has some ameliorating effects and HSMI-QTL strains of salmon are available (Dahle and Olsen, 2020).

BACTERIAL DISEASES

2.10 Vibriosis

Vibriosis is caused by bacteria belonging the family *Vibrionaceae* and includes *Vibrio anguillarum*, *Vibrio ordalii*, *Aliivibrio salmonicida* (formerly *Vibrio salmonicida*) and *Photobacterium damselae* (formerly *Vibrio damsela*) in coldwater marine aquaculture (Toranzo *et al.*, 2017; Mohamad *et al.*, 2019). *A. salmonicida* infections are responsible for coldwater vibriosis 'hitra' at temperatures below 10°C. This section focuses on *V. anguillarum* as it is considered one of the main pathogens for salmonid aquaculture during their first year in saltwater although outbreaks have been controlled by vaccination and antimicrobial therapy.

Vibrio anguillarum

2.10.1 Name of disease, distribution and fish

Classical vibriosis caused by *V. anguillarum* is characterized by haemorrhages at the ventral and lateral areas, at the base of fins, mouth, operculum and eyes. Exophthalmia is common. Infected fish have pale gills because of anaemia, and they stop feeding. Internally, there is splenomegaly, distended intestines, haemorrhages in the spleen, liver and kidneys, and large necrotic lesions in the musculature (Toranzo *et al.*, 2017).

V. anguillarum is a Gram-negative, rod-shaped bacterium with a polar flagellum. The genus *Vibrio* is characterized by the presence of arginine hydrolase and the capacity to produce acid from sucrose, maltose, mannose, trehalose and mannitol. It grows best between 30 and 34°C with a maximum growth temperature of 38.5°C, and its growth is inhibited at pH values above 9 and below 6. There are 23 O serotypes of the bacterium, with O1, O2 and O3 causing mortalities in fish. The genome is 4.2–4.3 Mbp and consists of two circular chromosomes (Holm *et al.*, 2018).

Outbreaks have occurred in more than 50 species of fish including *A. anguilla*, *Mugil cephalus*, *G. morhua*, *S. salar*, *Oncorhynchus* spp., *P. virens*, *D. labrax*, *S. aurata*, *Cynoglossus semilaevis*, *H. hippoglossus*, *S. maximus* and *Morone saxatilis* (Hickey and Lee, 2017). This pathogen has a global distribution in temperate waters.

2.10.2 Impacts and the environment

This pathogen is in marine and estuarine environments and without vaccines or antimicrobial therapy, outbreaks at aquaculture facilities can lead to the complete loss of production for salmonids, cod, lumpfish, sea bream and turbot within 5 days post-exposure (Hickey and Lee, 2017). Temperatures above 41°C and salinity over 7% will kill the bacteria. The bacteria can remain viable in seawater for over 50 months.

2.10.3 Diagnoses

V. anguillarum is halophilic and grows in blood agar or tryptic soy agar containing 1% NaCl. Biochemical tests that distinguish the bacteria from other *Vibrio* spp. are extensive (see Toranzo *et al.*, 2017) and can be verified with PCR tests that detect the *rpoN* gene (sigma factor 54), the *rpoS* gene (sigma factor 38), the *empA* gene (EmpA extracellular zinc metalloprotease), the *amiB* gene (peptidoglycan hydrolase) or the *groEL* gene (bacterial chaperonins).

2.10.4 Transmission, spread and pathogen dynamics

Transmission can be from fish to fish through open lesions and faeces, as well as contaminated sediment. The bacteria can survive for long periods in the sediment and studies with other *Vibrio* spp. indicate that survivors of the disease can harbour the pathogen. Contaminated tools, feed, water, rotifers and trash fish are all potential sources of the pathogen (see Mohamad *et al.*, 2019).

2.10.5 Effects of temperature

V. anguillarum is eurythermic (5 to >40°C) and outbreaks occur during summer at water temperatures above 15°C (Lillehaug and Colquhoun, 2020). The increase in temperature is associated with increased bacterial virulence (Lages *et al.*, 2019) and reduced immune capacity of the host fish.

2.10.6 Prevention/control strategies

Vibriosis vaccines are effective and provide protection by both immersion and injection vaccination. Florfenicol and oxolinic acid are mainly used to treat vibriosis in cod fry production in Norway (Frans *et al.*, 2011). Antibiotic resistance reported in *V. anguillarum* isolates from marine fishes in South China (Deng *et al.*, 2020) indicates widespread antibiotic use in aquaculture still occurs in some countries. Biosecurity measures are also important control/prevention methods for vibriosis.

2.11 Piscirickettsiosis (*Piscirickettsia salmonis*)

2.11.1 Name of disease, distribution and fish

Piscirickettsia salmonis causes piscirickettsiosis or salmon rickettsia syndrome (SRS) and is an important disease in salmonid species in Chile. *P. salmonis* is a Gram-negative, facultative intracellular γ-proteobacterium (Ortiz-Severín *et al.*, 2019; Pérez-Stuardo *et al.*, 2020).

2.11.2 Impacts and the environment

P. salmonis infection causes high mortalities in Atlantic salmon (*S. salar*), rainbow trout (*O. mykiss*) and coho salmon (*O. kisutch*) (Otterlei *et al.*, 2016). The disease is the most important infection among salmonid species in Chile and causes heavy losses to the industry annually (Rozas and Enríquez, 2014). SRS does not cause significant disease problems in other salmon production areas, although infections have been detected in Norway, Scotland and Ireland.

2.11.3 Diagnoses

Presumptive diagnosis is based on clinical and pathological observations. As shown in Fig. 2.3, small grey foci in the liver and gills with greyish surfaces are common findings in infected fish. *P. salmonis* infection is confirmed using histopathological examination, isolation of the pathogen using culture which is combined with identification by either immunofluorescence or immunoperoxidase staining, as well as dot blot DNA hybridization (Smith and Mardones, 2020). In addition, PCR techniques are available for the rapid identification of *P. salmonis* in clinically affected animals.

2.11.4 Transmission, spread and pathogen dynamics

Transmissions occur via skin and the gills but less frequently through the intestine (Smith *et al.*, 2004). The bacterium can penetrate intact skin or gills (Smith *et al.*, 1999). SRS has been reported primarily in marine fish farms but has also been found in freshwater facilities (Bravo, 1994; Gaggero *et al.*, 1995). Transmission is horizontal both in sea and fresh waters (Smith *et al.*, 1999; Rozas-Serri *et al.*, 2017). No vector or intermediate host has been identified (Mauel and Miller, 2002).

The most important determinants for occurrence of SRS are the number of infected farms in the area, seawater salinity and temperature (Bravo *et al.*, 2020), and increasing seawater temperatures increase the prevalence of the disease. Prevalence of SRS was 70% at water temperatures above 9°C while below this temperature it drops to 20% (Bravo *et al.*, 2020). Climate change and increasing seawater temperatures can pose a risk for increased occurrence of SRS but generally it is influenced by a complex interaction

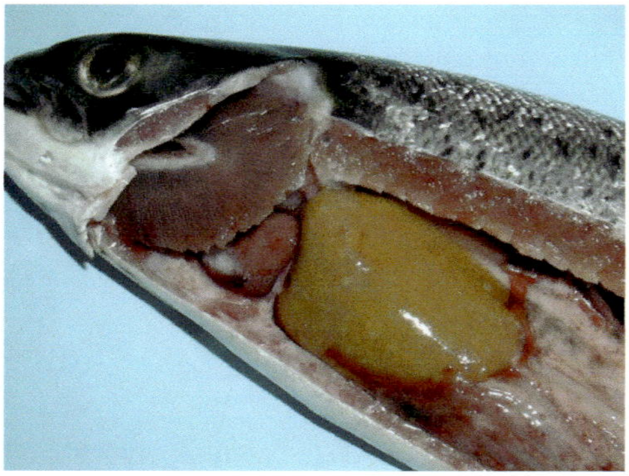

Fig. 2.3. Piscirickettsiosis in Atlantic salmon. Macroscopic changes in Atlantic salmon infected with *Piscirickettsia salmonis* showing small, grey foci in the liver parenchyma of varying sizes. Gills have a greyish surface, and the heart is pale. (Photograph courtesy of Professor Sandra Bravo, Universidad de Austral, Chile.)

of different factors where the number of infected farms in an area plays a significant role (Bravo *et al.*, 2020).

2.11.5 Prevention/control strategies

Disease prevention strategies include reduced stress on fish, better and improved husbandry practices with focus on fish density, avoiding transport/contact between farms, regulating or restricting movements by use of well-boats, separation of year classes, screening of brood stock to prevent vertical transmission, and vaccination (Smith and Mardones, 2020). Use of antibiotics has also been part of the control strategy particularly since *P. salmonis* is susceptible to antimicrobials, although resistance has been recorded (Yuksel *et al.*, 2006).

Vaccines are commercially available on the Chilean market, currently more than 25 different vaccines are available (Bravo and Midtlyng, 2007). Vaccines are mainly inactivated, whole-cell bacteria in oil-adjuvanted (water-in-oil emulsions) or non-adjuvanted, live attenuated vaccines. Current vaccines confer relatively good short-term protection against disease and mortality, but they provide inefficient long-term protection. Oral vaccines against SRS used for boosting a primary injection vaccination are also available.

2.12 Pasteurellosis (*Pasteurella skyensis*)

2.12.1 Name of disease, distribution and fish

Pasteurellosis is a disease of Atlantic salmon (*S. salar*) and lumpsuckers (*Cyclopterus lumpus*) that has been increasing in occurrence in Scotland and Norway (Nilsen *et al.*, 2021). The disease was first identified in Norway in 1989. The bacterial agents are *Pasteurella skyensis* in Scotland, *Pasteurella atlantica* genomovar *salmonicida* (Gulla *et al.*, 2020) in Norway and most recently, *P. skyensis* in Norway (Strom and Nilsen, 2021). Please note that the term pasteurellosis was once used to characterize the disease caused by *Photobacterium damselae* subsp. *piscicida* (formerly *Pasteurella piscicida*). Genomic analyses of *P. skyensis* and *P. atlantica* place these pathogens in the family *Pasteurellaceae*, genus *Pasteurella* (Ellul *et al.*, 2021) (https://www.ncbi.nlm.nih.gov/data-hub/genome/?taxon=97481, https://www.ncbi.nlm.nih.gov/nuccore/?term=Pasteurella+atlantica, accessed 11 June 22). The *Pasteurella* are Gram-negative, non-motile, ovoid to rod-shaped bacteria.

The disease in salmon is associated with 'blood eye', caused by infection of the eye and eye cavity, along with bloody boils in the skeleton and heart musculature. There is bleeding in the internal organs and pus accumulation from the inflammatory response. In lumpfish, white spots around the eyes and petechial haemorrhages on the skin are common (Strom and Nilsen, 2021).

2.12.2 Impacts and the environment

Although the pathogen was isolated in salmon in Norway in the late 1980s, the first appearance in lumpsuckers occurred in 2012, and since then outbreaks among these cleaner fish have increased. In the spring of 2018, the incidents of pasteurellosis in salmon increased as well, from none in 2017 to 60 farms in 2020 (Nilsen *et al.*, 2021). Disease outbreaks in Scotland with 2.5–13% mortalities were reported in 2017 (Scottish Government, 2018).

Approximately 42 million lumpsuckers were used in 2019 to control sea lice in farmed salmon in Norway. This biocontrol programme is important since the frequent use of chemotherapeutants in the past decade have resulted in resistant lice (Sandlund *et al.*, 2021b). Recent studies showed that lumpsuckers were more susceptible to pasteurellosis than Atlantic salmon in bath challenges with isolates from salmon and from lumpsuckers (Sandlund *et al.*, 2021b).

2.12.3 Diagnoses

Preliminary diagnoses using clinical and microbiological examinations are confirmed by bacterial growth on blood agar supplemented with 1.5% NaCl (final concentration of 2% NaCl) and incubation at 18°C for 24–48 h. Matrix-assisted laser desorption/ionization–time-of-flight mass spectrometry (MALDI-TOF-MS) and/or PCR analysis provides the final confirmation of the species.

2.12.4 Transmission, spread and pathogen dynamics

The origin of the pathogen is unclear since *P. skyensis* in Scotland is genetically distinct from *P. atlantica* in Norway. For lumpsuckers, the bacterial pathogen is found at all life stages including in the milt and eggs of infected fish; thus, vertical transmission is suspected for this species. Moreover, lumpsuckers recovering from pasteurellosis can be carriers of the bacteria. Thus, rigorous screening of lumpsuckers before their use in the biocontrol of sea lice is important (Sandlund *et al.*, 2021b).

2.12.5 Effects of temperature

P. skyensis can grow at temperatures from 14 to 32°C in the laboratory (Birkbeck *et al.*, 2002). The 'warmer' temperature range is in keeping with the observations that mortalities in infected fish spike during the warmest seawater temperatures (14–19°C) and decrease when water temperatures fall below 12°C (Strom and Nilsen, 2021).

2.12.6 Prevention/control strategies

Currently, there are no approved commercial vaccines for pasteurellosis in lumpsucker and salmon. Two inactivated whole-cell vaccines against *Pasteurella* spp. isolated from lumpsucker were developed, tested and found to have induced antibodies, but little protection (Ellul *et al.*, 2019).

2.13 Winter Ulcer of Salmon (*Moritella viscosa*)

2.13.1 Name of disease, distribution and fish

Moritella viscosa (formerly *Vibrio viscosus*) is a Gram-negative, facultative anaerobic, marine bacterium. It belongs to the genus *Moritella*, a group of bacteria mainly associated with cold seawater. There is a significant degree of genetic variability within the species *M. viscosa* (Grove *et al.*, 2010). Winter ulcer is caused by a group of closely related strains, a clonal complex, in salmon in Norway, Scotland and the Faroe Islands. More heterogenic strains have been isolated from salmon in Iceland, from rainbow trout and from marine fish species with wide geographic distribution.

Winter ulcer or winter sore occurs at low temperatures in seawater and the name of the disease is based on clinical manifestation and time of year when it occurs.

Experimental challenge studies of *M. viscosa* infection have shown that marine species like Atlantic cod (*G. morhua*) and Atlantic halibut (*H. hippoglossus*) are susceptible to infection (Gudmundsdóttir *et al.*, 2006).

2.13.2 Impacts and the environment

M. viscosa infections occur throughout the sea phase (on-growth phase), mainly in the cold season, winter and spring. Wounds of varying sizes are typical findings, mainly on the sides of the fish, extending below the subcutaneous tissues and exposing underlying muscle (Bruno *et al.*, 1998). Mortalities are linked to osmotic stress. Surviving fish are downgraded on the slaughter line. Winter sores pose a significant welfare problem.

Mortality is usually moderate to low, but individual cases of high mortality have been reported. The disease raises welfare concerns and reduces quality of the final product due to wounds, where healed wounds can cause scars and downgrading of the final product as a result.

2.13.3 Diagnoses

The diagnosis is made by observation of clinical signs, gross pathology and bacteriological examinations. Wounds can occur for many reasons: physical injuries related to handling, transport, severe weather conditions, storms, etc. Winter ulcer caused by *M. viscosa* infection is diagnosed by culturing from the kidney or from the wounds (periphery) on salt-containing agar (2%). Bacteria grow as yellow to grey, haemolytic and filamentous colonies, of 2–3 mm in diameter by 2–3 days of culture at 15°C (Benediktsdóttir and Heidarsdóttir, 2007). *M. viscosa* is susceptible to vibriostatic agents.

2.13.4 Transmission, spread and pathogen dynamics

The bacterium is part of the marine microflora and infections are waterborne. Winter ulcer is defined as an infectious disease, but predisposing factors play a role for disease outbreaks. It is likely there is a combination of mechanically imposed skin lesions that pave the way for opportunistic infections. Physical handling (mechanical delousing), cage density and also feeding regimes can play a role.

M. viscosa produces proteolytic enzymes which are likely to be important in the development of disease (Bjornsdottir *et al.*, 2011). Finally, other bacterial species like *Tenacibaculum* spp. and *Aliivibrio wodanis* (formerly *Vibrio wodanis*) are frequently detected together with *M. viscosa* infections.

2.13.5 Effects of temperature

M. viscosa infection typically occurs at temperatures below 8–10°C. Two different variants of *M. viscosa* have been detected. The new variant causes infection at higher water temperatures than what has been typical for this species and there are indications of differences in antigenic determinants (Grove *et al.*, 2010). Work is ongoing to develop new vaccine formulations that cover both variants.

2.13.6 Prevention/control strategies

Formalin-inactivated cultures of *M. viscosa* are included in multicomponent, oil-adjuvanted vaccines. Vaccination is done prior to sea transfer. Vaccines are available for use against winter ulcer in Atlantic salmon in the UK, the Faroe Islands and Norway. These vaccines contain components of the classical salmon variant. Outbreaks of disease occur however despite use of vaccines and these ambiguous results in the field have raised some concerns as to the efficacy of the vaccine. New variants of the bacterium have been identified and work is ongoing to develop new and, it is hoped, improved vaccine formulations. Despite clinical disease occurring in vaccinated fish, there is a general perception that vaccination contributes to a lower incidence of disease.

2.14 Mycobacteriosis

2.14.1 Name of disease, distribution and fish

Mycobacteriosis is caused by bacteria belonging to the family *Mycobacteriaceae*, genus *Mycobacterium*, which includes 188 recognized or proposed species (Gupta *et al.*, 2018). These are aerobic, non-spore forming, Gram positive, acid-fast bacilli. They are non-motile, straight or slightly curved rods of 0.2–0.6 μm in diameter and 1–10 μm in length. Fish pathogens include *Mycobacterium marinum*, *Mycobacterium fortuitum*, *Mycobacterium chelonae*, *Mycobacterium shottsii* and *Mycobacterium pseudoshottsii* (Whipps *et al.*, 2020). A proposal to change the *Mycobacterium* nomenclature is under consideration and transfers *M. chelonae* and *Mycobacterium salmoniphilum* to the genus *Mycobacteroides* and

M. fortuitum to the genus *Mycolicibacterium*. *M. marinum* and *M. shottsii* will remain in the genus *Mycobacterium* (Janda, 2020).

Mycobacteriosis is found worldwide in over 150 species of wild and cultured fish including salmonids and other marine fish (Decostere *et al.*, 2004). In colder, marine waters, mycobacteriosis occurs in cultured salmonids, *D. labrax*, *S. aurata*, *S. maximus* and *Sciaenops ocellatus* (Delghandi *et al.*, 2020; Whipps *et al.*, 2020).

The disease's first clinical signs are chronic wasting and granulomatous skin lesions. Internal examination shows grey or white nodules on organs, and enlarged spleen, kidney and/or liver. Histological staining reveals acid-fast mycobacteria in granulomas.

2.14.2 Impacts and the environment

Mycobacteriosis is non-notifiable in Norway, so the actual number of outbreaks there is unknown. The Norwegian Veterinary Institute diagnosed the disease at 11 sites in 2006/2007 and outbreaks were registered in 2008 and 2009. In 2020, fish in five farms were diagnosed with the disease and *M. salmoniphilum* was confirmed in three. In Chesapeake Bay, USA, *M. marinum* was isolated in 3% of wild striped bass (*M. saxatilis*) (Rhodes *et al.*, 2004). Outbreaks of *M. marinum* have been reported in marine cultured fish in Italy, China and Australia (Hashish *et al.*, 2018).

Mycobacteria readily form surface biofilms and can persist without a host in fresh, brackish and seawater environments in a wide range of temperatures. *M. salmoniphilum* grows well at 20–30°C but not at 37°C; *M. shottsii* grows at 23°C, weakly at 30°C and not at 37°C. *M. marinum* will grow at 30°C and not at 37°C in media culture but can grow in mice and in macrophage cell culture at 37°C (Whipps *et al.*, 2020).

2.14.3 Diagnoses

Mycobacteria are fastidious, slow-growing organisms and require selective media for cultivation (Delghandi *et al.*, 2020). Immunohistochemistry tests are available as well as molecular diagnostic tests using RT-PCR.

2.14.4 Transmission, spread and pathogen dynamics

Horizontal transmission by direct contact and ingestion of infected feed or through contaminated water are the most probable causes for the spread of the disease. Vertical transmission has been reported for some viviparous fish species (Whipps *et al.*, 2020).

The disease has a long incubation period and infected, subclinical fish may be spreaders of the disease at a farm site weeks before detection. Infected fish may have no clinical signs for several years following infection (Erkinharju *et al.*, 2021).

2.14.5 Effects of temperature

Most *Mycobacterium* spp. grow better as temperatures increase with optimal growth at 30–32°C. For many of the cage-cultured coldwater marine fish, these temperatures are stressful and since many fish have subclinical chronic infections, temperature increases may lead to more severe cases of mycobacteriosis as a result of immunosuppression (Whipps *et al.*, 2020). However, studies examining the effects of temperature on *M. shottsii* and *M. pseudoshottsii* disease in wild striped bass in Chesapeake Bay show that increasing temperatures from 20, 25 to 30°C attenuated the densities of bacteria in the spleen and kidneys of infected fish (Gauthier *et al.*, 2021).

2.14.6 Prevention/control strategies

Currently, there are no effective treatments for mycobacteriosis. Antibiotics are not useful because it is difficult for the antibiotic to breach the granuloma and the bacterial cell wall. Destruction of infected fish stocks is one way to limit the spread of infection. A DNA vaccine encoding the AG85A gene of *M. marinum* has some effectiveness in trials with hybrid striped bass (Whipps *et al.*, 2020).

2.15 Yersiniosis (*Yersinia ruckeri*)

2.15.1 Name of disease, distribution and fish

Yersinia ruckeri is a member of the family *Enterobacteriaceae*. It is a Gram-negative, rod-shaped bacterium that is the aetiological agent of systemic infection predominantly in salmonids. Diseased fish show external signs of bleeding in the skin and internal signs include bleeding on serosal surfaces, often with ascites. Yersiniosis in rainbow trout is referred to as enteric redmouth disease (ERM), because of the clinical signs. Disease is primarily seen in young fish but yersiniosis is reported in larger fish and over the last years also in sea-transferred Atlantic salmon in Norway (Gulla *et al.*, 2018). The disease and the pathogen have global distribution.

2.15.2 Impacts and the environment

Yersiniosis is most commonly in farmed rainbow trout, but over the last years an increasing number of clinical cases have been reported in Norway after sea transfer. Diseases in salmon are also found in Australia, Scotland and Chile (Kumar *et al.*, 2015).

2.15.3 Diagnoses

The bacterium grows well on different agars with an optimum temperature between 20 and 28°C and can be detected using immunological (ELISA, agglutination tests or immunofluorescence tests) or molecular techniques based on PCR methods (Tobback *et al.*, 2007; Kumar *et al.*, 2015). *Y. ruckeri* has been subdivided into two biotypes, where biotype 1 strains are motile and secrete lipase, and biotype 2 is negative for both traits (Kumar *et al.*, 2015).

2.15.4 Transmission, spread and pathogen dynamics

The infection occurs primarily in fresh water for both rainbow trout and Atlantic salmon. Transmission occurs via direct contact and carrier states are known where bacteria are shed in faeces for months with increased shedding during stress periods (Kumar *et al.*, 2015). Biofilm represents a further source of the bacterium where it survives for extended periods on surfaces and in sediments in aquatic environments. There has been a change in manifestation of disease over the last years where disease outbreaks have occurred in Atlantic salmon after sea transfer and often in large, farmed salmon in Norway. These disease outbreaks have been linked to mechanical sea lice treatments which can represent significant stress by itself, and handling can also result in stress-induced activation of a subclinical infection. The bacterium produces a number of virulence factors of which iron-acquiring components (such as the catecholate siderophore ruckerbactin) (Fernández *et al.*, 2004) play an important role.

2.15.5 Effects of temperature

Infection typically occurs during warmer periods of the year. Carriers of the bacterium start to shed the bacteria from the gut during periods of higher temperatures while no shedding was observed in carriers not subject to temperature stress (Hunter *et al.*, 1980).

2.15.6 Prevention/control strategies

Y. ruckeri is separated into serotypes, biotypes and outer-membrane protein types. In Norway, serotype O1b (also referred to as serovar III) has been detected in yersiniosis in Atlantic salmon and thus serotype O1b is used in vaccines for Atlantic salmon. Vaccines are prepared from formalin-inactivated cultures of *Y. ruckeri*, produced by several vaccine manufacturers, are water-based without adjuvant, and can be administered by injection or immersion. Immersion is used during the freshwater phase while injection vaccination is given prior to sea transfer simultaneously with the oil-based vaccines. Immersion vaccines are used when disease has been detected in hatcheries (Kumar *et al.*, 2015).

2.16 Furunculosis (*Aeromonas salmonicida*)

2.16.1 Name of disease, distribution and fish

Furunculosis is characterized by haemorrhaging internally and externally with furuncules (boils on the surface), exophthalmia, distended abdomen and petechiae at the base of fins. The disease is classified into four categories based on severity: acute, subacute, chronic or latent. The disease is also known as tail rot. It is predominantly found in fresh water but is also in brackish water.

The causative agent, *Aeromonas salmonicida*, is a Gram-negative, non-spore forming bacillus that can grow as single or paired rods. *A. salmonicida* includes five subspecies: *Aeromonas salmonicida* subsp. *salmonicida* is considered the typical species and the subsp. *masoucida*, *smithia*, *achromogenes* and *pectinolytica* are atypical (Vasquez *et al.*, 2022). Comparative genomics reveals high genome identity among the subspecies; plasmid content and insertion sequences formed the basis for the observed differences in virulence and host range (Vasquez *et al.*, 2022). It has a surface A-layer composed of a 50 kD protein and lipopolysaccharide which are virulence factors. The bacterium grows optimally between 22 and 25°C. A few strains can grow at 37°C (Vincent and Charette, 2022).

Outbreaks have occurred in salmonids (*Oncorhynchus* spp., *Salmo* spp., *Salvelinus* spp.), cyprinids, *E. lucius*, *Perca flavescens*, Ictaluridae, *P. olivaceus*, *Verasper variegatus*, *L. bergylta*, *G. morhua*, *H. hippoglossus*, *S. maximus*, *Petromyzon marinus*, *C. lumpus*, *Anguilla rostrata*, *A. japonica*, *Anarhichas lupus*, *Anarhichas minor* and *S. aurata* (Menanteau-Ledouble *et al.*, 2016).

2.16.2 Impacts and the environment

A. salmonicida causes diseases over a broad range of temperatures in freshwater and marine environments worldwide. The impact of the disease is exacerbated by low oxygenation and increased pollution. Losses at fish farms have been cited as having a major economic impact on the aquaculture industry (Austin and Austin, 2012). For example, in Quebec, Canada, this pathogen causes 25–60% of all infections reported in farmed fish.

A. salmonicida can maintain its pathogenicity in fresh water for 6–9 months and in saltwater for 10 days without a host.

2.16.3 Diagnoses

Isolation of the pathogen on trypticase soy agar or blood agar at 20–25°C and the formation of characteristic brown colonies is the first step in laboratory diagnosis of the disease. Confirmatory analyses with immunoassays, RT-PCR and MALDI-TOF-MS have also been developed (Menanteau-Ledouble *et al.*, 2016).

2.16.4 Transmission, spread and pathogen dynamics

Transmission occurs mainly through fish-to-fish contact by skin or by ingestion of infected material/water. Transmission by airborne droplets and cohabitation with infected molluscs has also been demonstrated. Rainbow trout survivors of the disease can carry the pathogen for up to 2 years and can serve as a source of infection (Austin and Austin, 2012). DNA probes used to examine wild Atlantic salmon in three Irish river systems found 87% of the fish positive for *A. salmonicida* (Mooney *et al.*, 1995).

2.16.5 Effects of temperature

A. salmonicida subsp. *salmonicida* and the subspecies *smithia*, *achromogenes* and *masoucida* are strictly psychrophilic, unable to grow at 37°C with few exceptions. Then, in 2000 the subspecies *pectinolytica* was isolated from a polluted river in Argentina with no known host and it grew efficiently at 37°C. Genomic analysis confirmed it as *A. salmonicida*. Soon thereafter, *pectinolytica* was isolated from humans, birds, diseased fish and a leech. These mesophilic *pectinolytica* strains also grew well at 18 and 7°C. Genomic analyses revealed greater diversity among the *pectinolytica* than among the psychrophilic strains (Vincent and Charette, 2022).

These studies highlight the capacity of pathogens to adapt to different environmental conditions that will occur with climate change.

2.16.6 Prevention/control strategies

Control of this highly contagious pathogen is difficult with the rise of antibiotic resistance in recent isolates. A bacterin vaccine administered with an oil adjuvant intraperitoneally was adopted early, but cumbersome intraperitoneal injections of individual fish slowed growth in the vaccinees, and granulomatous lesions and scarification at the injection site led aquaculture farms to seek other control methods. These include recombinant vaccines, low-frequency ultrasound as an adjuvant, probiotics, immunostimulants, and selection of *S. salar* and *O. mykiss* breeds resistant to *A. salmonicida* (Menanteau-Ledouble *et al.*, 2016; Marana *et al.*, 2021).

2.17 Francisellosis (*Francisella noatunensis* subsp. *noatunensis*)

2.17.1 Name of disease, distribution and fish

Francisellosis is a serious disease with worldwide distribution in farmed and captured fish species in marine and fresh waters. Previously, two *Francisella noatunensis* subspecies were recognized, *F. noatunensis* subsp. *noatunensis* (affecting colder-water species) and *F. noatunensis* subsp. *orientalis* (affecting warmer-water species). In 2020, comparative genomics analyses led investigators to propose *F. orientalis* sp. nov. as a separate species and *F. noatunensis* now includes a new subspecies *F. noatunensis* subsp. *chilensis* (Ramirez-Paredes *et al.*, 2020). The new subspecies was isolated from diseased farmed Atlantic salmon in Chile and was distinct from those from Atlantic cod in Northern Europe.

Francisellosis in coldwater fish is characterized as a systemic, chronic, granulomatous infection with white, protruding lesions on the spleen, liver and kidney. These are commonly found in infected *G. morhua*, *S. salar* and *O. mykiss*.

Members of the genus *Francisella* are Gram-negative, non-motile, aerobic facultative intracellular organisms. *F. noatunensis* subsp. *noatunensis* requires cysteine in the culture media and is unable to growth at temperatures above 35°C (Colquhoun and Duodu, 2011).

2.17.2 Impacts and the environment

F. noatunensis is identified as the most important disease problem in Norwegian cod farming and is considered the reason for the collapse of the Norwegian cod industry in 2010 (Puvanendran *et al.*, 2021). The disease affects large cod primarily and mortality is associated with high seawater temperatures (above 14°C) that result in temperature-related stress and immunosuppression (Colquhoun and Duodu, 2011).

2.17.3 Diagnoses

Confirmatory diagnosis of francisellosis is made by bacterial isolation on selective growth media and specific PCR analyses combined with DNA sequencing (Colquhoun and Duodu, 2011).

2.17.4 Transmission, spread and pathogen dynamics

Horizontal transmission by fish-to-fish contact or through infected water is well documented. Vertical transmission is also possible since diseased fish still spawn and the bacterium is in egg batches and fry (Puvanendran *et al.*, 2021). Farmed and wild cod, a wide variety of marine fish species, mussels and crabs carry the bacterium and have the potential to transmit to susceptible hosts (Colquhoun and Duodu, 2011).

2.17.5 Prevention/control strategies

There is no efficient commercial vaccine against francisellosis. The challenge has been to develop a cell-mediated response against an intracellular pathogen (Bakkemo *et al.*, 2016). Vaccine trials with or without different oil adjuvants have had only limited effectiveness. Attenuated live vaccines are difficult to license because of possible reversion to virulence. Antibiotic therapy with oxytetracycline has worked (Colquhoun and Duodu, 2011).

2.18 Tenacibaculosis (*Tenacibaculum maritimum*)

2.18.1 Name of disease, distribution and fish

Tenacibaculosis is characterized by ulcerated skin lesions, frayed or eroded fins and tail, with moderate to severe erosions of the mouth. Infected skin is white to pale yellow from bacterial mats. Infected gills produce excessive mucus, are pale and exhibit lamellar hyperplasia.

Tenacibaculosis is caused by *Tenacibaculum maritimum*; however, other species of the genus, *Tenacibaculum soleae*, *Tenacibaculum discolor*, *Tenacibaculum gallicum*, *Tenacibaculum dicentrarchi* and *Tenacibaculum finnmarkense*, have also been isolated from diseased fish with similar clinical signs (Fernández-Álvarez and Santos, 2018). The pathogen is a Gram-negative, aerobic, gliding, filamentous bacterium of the family *Flavobacteriaceae*. *T. maritimum* requires at least 30% seawater in the medium for growth.

The bacterium has worldwide distribution and infects a wide variety of marine fish including *Solea solea*, *S. senegalensis*, *Dicologlossa cuneata*, *S. maximus*, *P. olivaeus*, *Rhombosolea tapirina*, *Aldrichetta forsteri*, *S. quinqueradiata*, *Pagrus major*, *S. aurata*, *D. labrax*, *S. salar*, *O. mykiss*, *O. tshawytscha*, *Sardinops sagax*, *Engraulis mordax*, *C. lumpus*, *Carcharias taurus*, etc. (ICES, 2019).

2.18.2 Impacts and the environment

Estimates of the costs of tenacibaculosis in Atlantic salmon in Canada were CAN$1.6 million in 2014 and in subsequent years about CAN$500,000 per outbreak (Nowlan *et al.*, 2020). In Spain, marine tenacibaculosis has caused economic losses in turbot, Atlantic salmon, Senegalese sole and gilthead sea bream (Fernández-Álvarez and Santos, 2018). Outbreaks of this disease in Norwegian Atlantic salmon have been primarily associated with *T. finnmarkense* (Spilsberg *et al.*, 2022).

Increased prevalence and disease severity are associated with temperatures above 15°C and salinities above 30% (ICES, 2019).

2.18.3 Diagnoses

The bacterium can be isolated on marine agar (30% seawater or 1.5% NaCl) between 10 and 34°C with optimal growth at 30°C. Colonies are often pale yellow, catalase- and oxidase-positive. However, the slow growth of the organisms makes this a difficult process when a quick diagnosis is required. DNA and MALDI-TOF-MS diagnostic methods have been developed (ICES, 2019).

2.18.4 Transmission, spread and pathogen dynamics

The pathogen is transmitted through the water from fish to fish, generally requiring minor abrasions or other environmental stressors. Vertical transmission has not been observed. Since the pathogen is in many species of wild marine fish, these animals may serve as sources of infection.

2.18.5 Prevention/control strategies

Antibiotic therapy and surface-acting disinfectants administered by immersion have been employed as prophylactic measures. No vaccine is available.

PARASITIC DISEASES

2.19 Amoebiosis (*Neoparamoeba perurans*)

2.19.1 Name of disease, distribution and fish

Amoebic gill disease (AGD) in Atlantic salmon is caused by a free-living marine amoeba, *Neoparamoeba perurans* (Young *et al.*, 2007; Crosbie *et al.*,

2012). Most references on 'amoebiosis' are gleaned from Sokołowska and Nowak (2020). In a few countries AGD has developed into a complex gill disease (CGD) when it is associated with another pathogen (Herrero *et al.*, 2018).

The parasite becomes rounded and produces pseudopodia under *in vitro* conditions. Its nucleus is close to the parasome in stained smears and in gill sections (Nowak, 2012).

AGD has caused high fish mortalities (10–60%) in many countries (see Table 2.2).

2.19.2 Impacts and the environment

AGD outbreaks can be costly; for example, a one-year outbreak in Scotland was estimated to cost about US$81 million (Kube *et al.*, 2012; Shinn *et al.*, 2015).

Outbreaks are associated with elevated temperatures (e.g. Rodger, 2014; Oldham *et al.*, 2016) and increased water salinity (Oldham *et al.*, 2016). The ongoing climate change will increase the intensity and frequently of outbreaks.

2.19.3 Diagnoses

i. Morphology and clinical signs are used for tentative identification of AGD. Clinical signs of AGD initially appear as tiny white lesions (Taylor *et al.*, 2009) which will spread to cover the gill (Zilberg and Munday, 2000; Adams and Nowak, 2003). Other signs include lethargy and increased opercular movements. Increases in opercular movement may be absent in some infected fish (Powell, 2006).

ii. Molecular techniques are used to confirm tentative identification. Young *et al.* (2008) first used PCR and it is still used (Mouton *et al.*, 2014; Kim *et al.*, 2016). There are at least four other molecular techniques including RT-qPCR (real-time quantitative RT-PCR) (Steigen *et al.*, 2018).

2.19.4 Transmission, spread and pathogen dynamics

Naïve fish get infected directly from the water column or from infected fish. The parasite multiplies rapidly in warm waters and spreads via migrating fish and/or water currents to new areas as climate change continues to increase temperature in 'cooler' regions. Also, current amoeba-infested areas may become too warm and with decreased salinity may see a drop in numbers of amoebae.

2.19.5 Prevention/control strategies

i. Good husbandry includes reducing stocking density, fallowing, removal of dead and dying fish, and preventing net fouling with regular cleaning (see Nowak, 2012). Cages should be located in areas with steady water currents and be well separated.
ii. There is no vaccine so the use of immunostimulants needs further investigation as innate immune response against the pathogen is important (see Nowak, 2012).
iii. Outbreaks are reduced with regular gill inspection and freshwater or hydrogen peroxide baths. These procedures need to be repeated, are

Table 2.2. Hosts and countries with outbreaks of amoebic gill disease. (Adapted from Sokołowska and Nowak, 2020.)

Fish/organism	Country/countries
Salmo salar	Australia, Chile, Norway, Scotland
Oncorhynchus kisutch	USA, Korea
Oncorhynchus mykiss	Australia
Scophthalmus maximus	Spain
Plecoglossus altivelis	Japan
Acanthopagrus schlegelii	Korea
Oplegnathus fasciatus	Korea

time-consuming and add to production costs (see Nowak, 2012).
iv. Selective breeding of AGD-resistant salmon based on genomic data is reliable (Kube *et al.*, 2012) and is better than fish pedigree (Robledo *et al.*, 2018; Boison *et al.*, 2019). AGD-resistant fish also need to have tolerance for higher temperatures and lower dissolved oxygen.
v. Offshore cage culture needs to be considered as bluefin tuna, *Thunnus maccoyii*, in offshore cages do not have parasites (Kirchhoff *et al.*, 2011). Submerged cages lowered AGD gill scores (Wright *et al.*, 2017, 2018); however, cages need to include air domes (see Section 2.24.6.i).

2.20 Salmonid Cryptobiosis (*Cryptobia salmositica*)

2.20.1 Name of disease, distribution and fish

Cryptobiosis is caused by the haemoflagellate *Cryptobia salmositica*. The pathogen infects all Pacific *Oncorhynchus* spp., *S. trutta* and seven species of sculpins (*Cottus* spp.) in freshwater streams and spawning beds in western North America (see Woo, 2006).

Cryptobia is an elongated flagellate with an anterior free flagellum, an undulating membrane that ends as a posterior free flagellum. Its nucleus and kinetoplast are at the anterior end (Woo, 1979). It is in the blood and on the body surface of infected fish (Woo and Wehnert, 1983).

2.20.2 Diagnoses

i. Parasitological. In acute infections *Cryptobia* can be detected in blood or ascites using wet mounts or stained smears. The haematocrit centrifuge technique (Woo, 1969) was modified (Woo and Wehnert, 1983) for use to detect early and chronic infections.
ii. Clinical signs include bilateral exophthalmia (Fig. 2.4), general oedema, abdominal distension with ascites, anorexia and microcytic hypochromic anaemia (Woo, 1979).

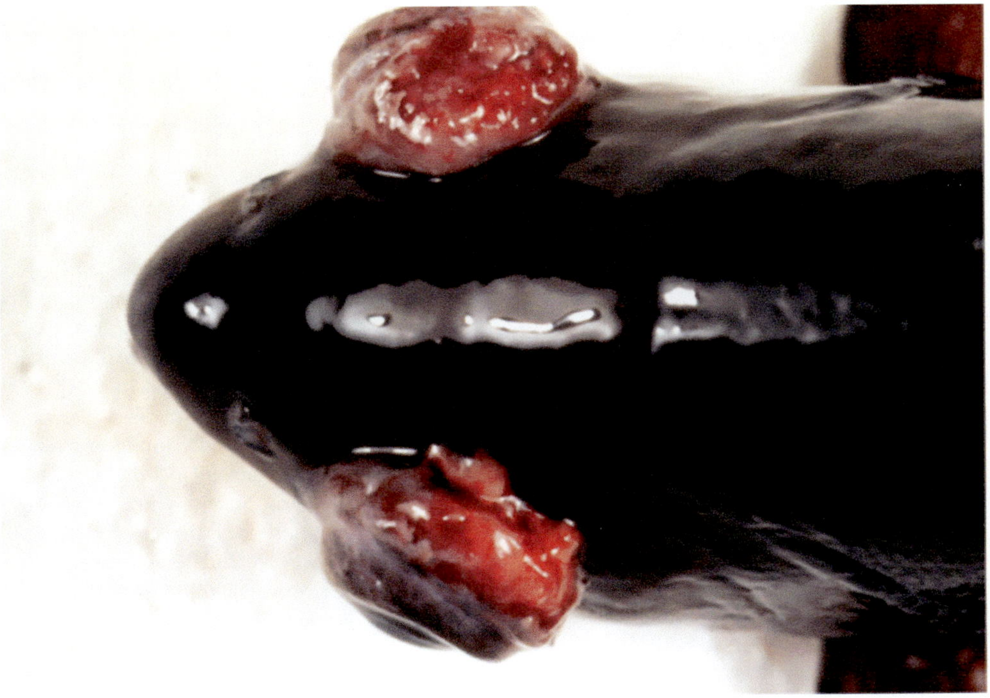

Fig. 2.4. Cryptobiosis in rainbow trout. Dorsal view of bilateral exophthalmia in rainbow trout experimentally infected with *Cryptobia salmositica*. (From Woo, 2006; courtesy of CABI, Wallingford, UK.)

iii. Antibody-capture ELISA is sensitive and is used routinely in experimental studies. Blood or sera adsorbed on filter papers are stored at −20°C (see Woo, 2006).

2.20.3 Transmission, spread and pathogen dynamics

i. Indirect. Transmission in freshwater streams and hatcheries is by the leech, *P. salmositica*. The parasite ingested in a blood meal multiplies in the leech and is transmitted when it feeds again (see Woo, 2006).
ii. Direct. Woo and Wehnert (1983) found *Cryptobia* on the body surface of infected rainbow trout. In two experiments, 67 and 80% of naïve fish became infected 20 weeks after cohabitation in fresh water with infected fish.

Paterson and Woo (1983) found a functional contractile vacuole in *Cryptobia* which allowed it to survive on the body surface. Bower and Margolis (1983) also showed direct transmission. Naïve fish became infected when infected and uninfected fish were held briefly in dip nets. Mortalities were 65–89% when fish were maintained in fresh water and 94% in seawater.

2.20.4 Effects of temperature

i. *P. salmositica* breeds at between 5 and 12°C (Becker and Katz, 1965). Temperatures higher than those preferred temperatures may reduce reproduction; consequently, climate change may reduce leech reproduction and limit 'indirect' transmission of *Cryptobia* (see Section 2.20.3.i).
ii. *C. salmositica* under *in vitro* conditions becomes sluggish at 15°C and is not active at 22°C. Elevated temperatures will prevent 'direct' transmissions (see Section 2.20.3.ii). Juvenile *Oncorhynchus* spp. adapted to 20°C enhanced their survival on infection (Bower and Margolis, 1985).
iii. *Oncorhynchus*. Selective breeding of rainbow trout produced fish with higher temperature tolerance (Ineno *et al.*, 2005, 2018). Climate change will reduce the geographical range of salmon unless they have tolerance for higher temperatures and lower dissolved oxygen; these attributes may be achievable via selective breeding.

2.20.5 Prevention/control strategies

i. Innate immunity (see Woo and Ardelli, 2014):
a. *Cryptobia*-resistant brook trout (*S. fontinalis*) are available and can be bred. Protection is via the alternative pathway of complement activation and is controlled by a dominant Mendelian locus (Forward *et al.*, 1995; Forward and Woo, 1996).
b. *Cryptobia*-tolerant brook trout can maintain high levels of α2-macroglobulins which neutralize metalloprotease (Zuo and Woo 1997a), the disease-causing factor secreted by *Cryptobia* (Zuo and Woo, 1997b). Development of transgenic *Cryptobia*-tolerant salmonids can be done with further research.
ii. Adaptive immunity (see Woo and Ardelli, 2014). Recovered *Oncorhynchus* spp. are resistant to reinfection and there are two experimental vaccines to protect fish from the pathogen (a) or disease (b):
a. An attenuated *Cryptobia* vaccine is protective (Woo and Li, 1990). It has no bioenergetic cost to trout and one injection protects them from infection (in fresh or seawater) for at least 24 months (Li and Woo, 1995, 1997).
b. The metalloprotease-DNA vaccine does not prevent infection but antibodies in vaccinated fish neutralize the metalloprotease secreted by *Cryptobia* (Zuo and Woo, 1997b). Infected vaccinated fish do not have cryptobiosis and recover faster (Tan *et al.*, 2008).
iii. Chemotherapy (Ardelli and Woo, 2001). Isometamidium chloride (2.5 mg/kg) inoculated into infected Atlantic salmon eliminated *Cryptobia* infection in 30% of fish and significantly reduced parasitaemias in remaining fish. All infected juvenile chinook salmon inoculated with the drug (1.0 mg/kg) survived while 100% of non-treated fish died with massive infections. Also, the drug is prophylactic, and it does not affect growth nor other biological functions.
iv. Selective breeding. Species and strains of salmonids have differences in susceptibility to cryptobiosis (Bower *et al.*, 1995); consequently, breeding more resistant salmonids in hatcheries for release is a good option.
v. Outbreaks in sea cages (see Woo, 2006). In one outbreak mortalities of smolts and preharvest chinook salmon varied between cages (3.3–24.0%) after fish from a hatchery were transferred to sea cages. It was suggested that some transferred

fish had chronic infections that relapsed and *Cryptobia* was transmitted 'directly' to resident fish (see Section 2.20.3.ii).

Fish for transfer should be quarantined for 4–5 weeks, monitored for clinical signs (see Section 2.20.2.ii) and random fish examined for parasites and antibodies (see Sections 2.20.2.i and 2.20.2.iii) before transferring them to cages that have resident fish.

2.21 Spironucleosis (*Spironucleus salmonicida*)

2.21.1 Name of disease, distribution and fish

Spironucleus salmonicida causes spironucleosis. Earlier reports of outbreaks in Atlantic salmon, grayling (*Thymallus thymallus*) and Arctic charr (*S. alpinus*) in Norway (Mo *et al.*, 1990; Poppe *et al.*, 1992; Sterud *et al.*, 1997, 1998, 2003), chinook salmon in Canada (Kent *et al.*, 1992) and experimental studies (Guo and Woo, 2004 a,b) were under different species names.

Spironucleus is pear-shaped or rounded. It has two teardrop-shaped nuclei, six anterolateral flagella and two recurrent flagella that emerge from two posterior cytostomes (Sterud *et al.*, 1998).

2.21.2 Impacts and the environment

Outbreaks in sea cages were found in Europe and North America. Guo and Woo (2004b) showed experimentally that the pathogen has a blood and a tissue phase in Atlantic salmon. Parasites are present in the blood for 1–8 weeks post-infection after which they are found in organs, eye sockets and/or muscles. Fish mortality (38 of 40) occurred during the tissue phase with no mortality in control fish.

There was no difference in susceptibility between three genetic families of hatchery-raised Atlantic salmon. All 30 experimentally infected salmon (ten per family) developed the disease, and had similar blood parasitaemias, clinical signs and gross pathology. Mortality was 29 of 30 infected fish with no mortality in the 30 control fish.

2.21.3 Diagnoses

i. Parasitological. The haematocrit centrifuge technique (Woo and Wehnert, 1983) is more sensitive than wet mount examinations of blood especially during the tissue phase. It is routinely used to detect infections (Guo and Woo, 2004a).
ii. Serological. ELISA was modified to detect antibodies against *S. salmonicida*. It is sensitive and is useful for epidemiological studies or to confirm low infections (Guo and Woo, 2004a).
iii. Clinical signs and gross pathology. Anorexia, lethargy, anaemia, skin blisters, muscle ulcerations (Fig. 2.5) and unilateral exophthalmia are the clinical signs. Gross pathologies include haemorrhaging of internal organs and splenomegaly with lesions in the spleen and liver.

2.21.4 Transmission, spread and pathogen dynamics

Flagellates are present on the body surface of infected fish. Naïve fish cohabiting with infected fish became infected after 4 weeks and the parasite was found on the body surface (Guo and Woo, 2004b).

The parasite is also found in internal organs (micro-aerobic environments). Its genome also indicates it can adapt to fluctuating environments (Xu *et al.*, 2014). Hence, the pathogen may be transmitted if environmental conditions (e.g. elevated temperatures, lower dissolved oxygen) are within its physiological range.

Future studies to include host immune response, vaccine development (see Section 2.20.5.ii), use of immunostimulants and pathobiology of spironucleosis are suggested.

2.22 Microsporidial Gill Disease of Salmon (*Loma salmonae*)

2.22.1 Name of disease, distribution and fish

Microsporidial gill disease of salmon (MGDS) in Pacific *Oncorhynchus* spp. is caused by *Loma salmonae*. The pathogen was first reported in juvenile coho salmon transferred from sea to fresh water (Becker and Speare, 2007). Most references in this review are gleaned from Speare (2020).

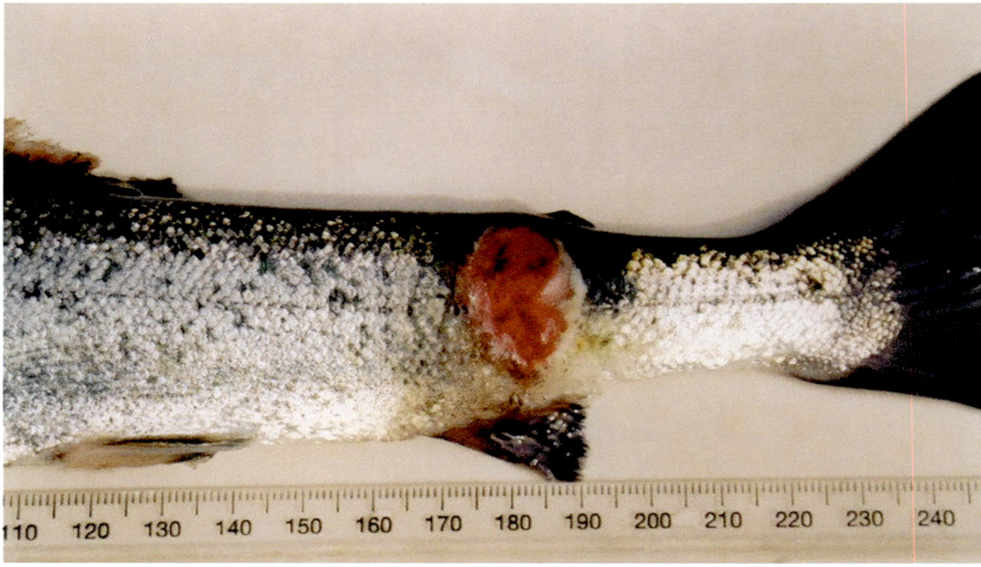

Fig. 2.5. Spironucleosis in Atlantic salmon. A larger ulcer on the body surface of an Atlantic salmon experimentally infected with *Spironucleus salmonicida*. (From Guo and Woo, 2004a; courtesy of *Diseases of Aquatic Organisms*.)

It is an important pathogen of salmon in British Columbia, Canada. The parasite has not been reported in Pacific salmon in most other regions (e.g. Russia, Japan, Patagonia); however, Bruno *et al.* (1995) found the parasite in farmed rainbow trout in Scotland.

There are degrees of susceptibility and persistence of the pathogen in salmon (Ramsay *et al.*, 2002). *L. salmonae* may also have two variants (Sanchez *et al.*, 2001b); gill materials from salmon caused high infections in rainbow trout and low infections in brook trout. However, spores from brook trout were not infective to rainbow trout but caused high infections in brook trout and low infections in chinook salmon.

2.22.2 Impacts and the environment

Clinical disease (see Section 2.22.3.iii) appears in July and August during the second summer fish are in sea cages (Becker and Speare, 2007). Morbidity and mortality persist until late autumn. Outbreaks are costly as fish are near market weight and mortality can be up to 80% (Constantine, 1999).

Rainbow trout and chinook salmon will be protected after an infection (Speare and Daley, 2003). Consequently, further research on a vaccine is warranted (see Section 2.22.6.i).

Salmon at sea are lightly infected but pre-spawners in rivers/spawning grounds have large numbers of xenomas resulting in high fish mortality (Kent *et al.*, 1998). Future studies may include elucidation of disease activation in pre-spawning migrants.

2.22.3 Diagnoses

i. An *L. salmonae*-specific single-stranded DNA labelled with digoxigenin (for *in situ* hybridization) can detect pre-xenomas in gills at 2 weeks after infection (Sanchez *et al.*, 1999).
ii. Gill xenomas appear as white dots. Under light microscopy unstained xenomas appear as granular masses and slight pressure on the cover slip will release spores.

There is no host response to intact xenomas in gills (Fig. 2.6), but there is a significant inflammatory response around xenomas with released spores (Kent *et al.*, 1995). Also, xenomas are found in organs in chinook salmon (Kent *et al.*, 1995, 1998; Ramsay *et al.*, 2002).

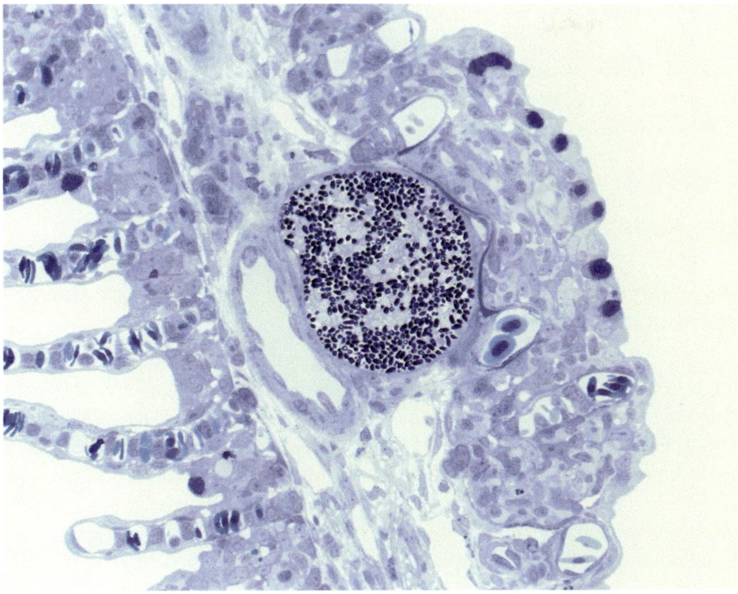

Fig. 2.6. Microsporidial gill disease in chinook salmon. Histological section of a *Loma salmonae* xenoma in the gill of an infected chinook salmon. Note the absence of host response to the intact xenoma. (Image courtesy of Dr Jan Lovy, New Jersey Division of Fish and Wildlife, USA.)

iii. Clinical signs and gill examinations are used to diagnose MGDS. Fish with high infections have rapid opercular movements and are found near the surface or along the edge of sea cages.

2.22.4 Transmission, spread and pathogen dynamics

The pathogen is transmitted directly to naïve fish via ingestion of spores from xenomas (Ramsay *et al.*, 2002; Becker *et al.*, 2005). Presumably, infections in sea-cage fish come from nearby wild salmon. A 1 h exposure of naive fish to infected fish was sufficient for successful transmission.

2.22.5 Effects of temperature

L. salmonae survives and develops in fish at 13–17°C (Beaman and Speare, 1999; Becker *et al.*, 2003; Becker and Speare, 2004). If water temperature is reduced to 11°C or increased to 19°C, numbers of xenomas are significantly lower. Xenoma development will resume at preferred temperatures (Beaman and Speare, 1999).

Spores are viable for up to 100 days at 4°C (Shaw *et al.*, 2000) and those in fish tissues remain infective for extended periods (Ramsay *et al.*, 2002, 2003). Blue mussels (*Mytilus edulis*) in the proximity of cages may serve as reservoirs of infection (McConnachie *et al.*, 2013). Investigations on the long-term survival of spores under natural conditions are warranted.

2.22.6 Prevention/control strategies

i. Cell-mediated immunity in recovered rainbow trout protected them from reinfection (Speare and Daley, 2003):

a. Rainbow trout inoculated with heat-killed spores had 85% fewer xenomas (Speare *et al.*, 2007).

b. Fish vaccinated with a low virulence *L. salmonae* developed 14 times fewer xenomas per gill filament when challenged with the pathogen (Sanchez *et al.*, 2001a).

ii. Chemotherapy and prophylactic treatments:

a. Fumagillin reduced xenomas in infected fish (Kent and Dawe, 1994).

b. β-Glucans, an immunostimulant, would be effective if treatment was initiated weeks

before fish were exposed to spores (Guselle *et al.*, 2006, 2010).
iii. Offshore and/or submerged cages (see Section 2.24.6.i) to prevent or reduce infections are suggested for future studies.

2.23 *Kudoa thyrsites*

2.23.1 Name of disease, distribution and fish

Kudoa thyrsites is a metazoan parasite. It has been reported from numerous fish species (see Moran and Kent, 1999; Moran *et al.*,1999a) in North and South America, Europe, Africa, Asia and Australia.

Spores from infected fish are stellate-shaped with four unequal polar capsules and four valves.

2.23.2 Impacts and the environment

The parasite has significant effects on both aquaculture (e.g. in sea cages) and commercial fisheries and these are discussed in Moran *et al.* (1999a). The parasite is more prevalent in sexually mature Atlantic salmon and reconditioned grilses than in immature market-sized fish (St-Hilaire *et al.*, 1998).

The parasite does not cause clinical disease in live fish, but it secretes a proteolytic enzyme in dead fish or when fillets are stored for 3–6 days on ice or smoked (see Kent and Poppe, 1998). With large numbers of parasites the enzyme causes softening of muscles and fillets, which are not marketable, and myoliquefaction correlates well with severity of infections (St-Hilaire *et al.*, 1997; Dawson-Coates *et al.*, 2003). Lost revenue due to the parasite in British Columbia in 2002 was estimated to be CAN$50 million (Funk *et al.*, 2007).

2.23.3 Diagnoses

i. Spores are visible in wet mounts from freshly cut fillets or crushed pieces of muscle, or in Giemsa-stained histological sections of muscles (see Kent and Poppe, 1998).
ii. The standard PCR technique is sensitive (Hervio *et al.*, 1997) and is used routinely to detect the 18S subunit ribosomal DNA in *K. thyrsites*. However, DNA-based qPCR is also recommended as it is 'readily transferred among laboratories, is less labour intensive' (Funk *et al.*, 2007).

2.23.4 Transmission, spread and pathogen dynamics

Spores were detected 5–6 months after fish were housed in sea cages (Moran *et al.*, 1999b,c). Briefly, in fish the sarcoplasm from the actinosporean stage migrates to muscle fibres where it forms a pseudocyst to produce spores. Its complete life cycle has not been elucidated.

2.23.5 Prevention/control strategies

i. Preventing biofouling and good husbandry help to reduce transmission of the pathogen.
ii. Removing sexually mature and reconditioned grilses from sea cages prior to harvesting the immature market-sized fish may reduce pathogen transmission.
iii. Another option is to consider using submerged cages (see Section 2.24.6.i).

2.24 Lepeophtheirosis (*Lepeophtheirus salmonis*)

2.24.1 Name of disease, distribution and fish

L. salmonis is an ectoparasitic copepod that infects wild and cultured salmonids in Canada, the USA, Scotland, Ireland and Norway. The parasite is relatively specific to salmonids. Its eggs are in elongated strings attached to the genital segment. Eggs become darker prior to hatching because of nauplii in them (Kabata, 1973).

Most references are gleaned from Fast and Dalvin (2020), including information from national and international documents which are not readily available. *L. salmonis* occurs in ocean-migrating salmonids; but there are biological differences between the Pacific and Atlantic

L. salmonis (Fast *et al.*, 2003; Bricknell *et al.*, 2006; Saksida *et al.*, 2011). It consists of two subspecies: *Lepeophtheirus salmonis oncorhynchi* in the Pacific (infects both *Oncorhynchus* spp. and Atlantic salmon) and *Lepeophtheirus salmonis salmonis* in the Atlantic (infects Atlantic salmon and sea trout, *S. trutta*).

2.24.2 Impacts and the environment

L. salmonis salmonis is a major disease in *S. salar* in Atlantic Canada. There are about 15 million sea cages in New Brunswick and infections can be 20–30 lice/fish (see Chang *et al.*, 2011; ACF-FA, 2016). Recreational fishing is well established with 50,000–60,000 fish caught per year in Newfoundland and Labrador (DFO, 2016).

According to Beamish *et al.* (2005), nearly all salmon in central British Columbia waters were infected with *L. salmonis oncorhynchi* and the prevalence was 100% in pink (*O. gorbuscha*) and chum salmon (*O. keta*) in Queen Charlotte Strait and Smith and Rivers inlets. Pink, chum and sockeye salmon had on average 41.5–52.0 lice/fish while chinook and coho salmon averaged 16.1–18.5 lice/fish. However, the current impact may have shifted to sea-cage culture of Atlantic salmon.

2.24.3 Diagnosis

The parasite can be seen with the naked eye on the body surface of infected fish. It may cause small grey patches with melanization of attached areas. Lesions are more pronounced as inflammation and fibrosis occur. Severe erosion may extend deep into the tissue. Fish with skin damage likely will have osmotic problems and are also exposed to other pathogens (Boxaspen, 2006).

2.24.4 Transmission, spread and pathogen dynamics

Briefly, nauplii hatch from an egg and undergo two moults to the infective copepodid stage. Copepodids need to locate a fish before their energy reserves are depleted. Attachment to the host is low and unattached copepodids die (Brooker *et al.*, 2018). Attached copepodids utilize chemosensory receptors to determine the suitability of the host. Infected salmonids show increased jumping and rolling and changed swimming patterns (Bui *et al.*, 2016).

2.24.5 Effects of temperature

Development of attached copepodid and egg production will be more rapid at high temperatures while low water temperatures like 5°C will result in low infections and poor egg production (Brooker *et al.*, 2018; Hamre *et al.*, 2019). The parasite does not develop to adult stage at 24°C.

In some aquaculture areas climate change with decreased rainfall and runoff will increase salinity, which will be beneficial for *L. salmonis*. However, low salinity in other regions due to heavy rainfall or massive melting of snow will be problematic for larval stages (Wright *et al.*, 2016). Variations in the thermal and saline environments will affect *L. salmonis oncorhynchi* (Ljungfeldt *et al.*, 2017) as eggs will not hatch at 10 ppt (10°C), while 15 and 20 ppt salinity will increase hatch rate to 70 and 78% (Johnson and Albright, 1991). Hatching of *L. salmonis salmonis* eggs also increases (3–80%) with salinity (Bricknell *et al.*, 2006).

2.24.6 Prevention/control strategies

i. Non-chemical:

a. Good husbandry includes removal of diseased and dead fish, as well as keeping age classes in separate cages.

b. Moving cages offshore is beneficial as lower water temperatures will decrease parasite reproduction and result in exposure of fish to fewer lice. Cages can be relocated northwards as coastal waters become warmer, but this may introduce sea lice to new areas (Abolofia *et al.*, 2017; Klinger *et al.*, 2017).

c. Submerged cages significantly reduce sea lice infections. However, cages need modifications to include air domes as salmonids have open swim bladders and have to re-fill them (Sievers *et al.*, 2022).

d. Adding cleaner fish (lumpfish or wrasse) to cages works as they will feed on lice attached to salmonids. However, cages need modifications and establishment of programmes to breed cleaner fish (Imsland *et al.*, 2016; Hvas *et al.*, 2018). Lice infestations are also reduced if infected fish are given a freshwater and/or thermal bath(s). This approach will also gradually adapt fish to ongoing climate change.

ii. Chemical:

a. Numerous chemicals have been used against *L. salmonis* and these include organophosphates, pyrethroids, emamectin benzoate and hydrogen peroxide. However, when a chemical is used constantly and extensively sea lice will develop resistance (Denholm *et al.*, 2002; Boxaspen, 2006). Parasites and drug-resistant parasites spread to nearby farms via water currents. Fish farmers and officials monitor sea lice infestations on farms and design the most effective treatment regime. Perhaps, outbreaks and sea lice infestations can be better regulated using an integrated strategy (i and ii).

References

Aamelfot, M., Dale, O.B., Weli, S.C., Koppang, E.O. and Falk, K. (2014) The *in situ* distribution of glycoprotein-bound 4-*O*-acetylated sialic acids in vertebrates. *Glycoconjugates Journal* 31, 327–335.

Abbadi, M., Gastaldelli, M., Pascoli, F., Zamperin, G., Buratin, A., *et al.* (2021) Increased virulence of Italian infectious hematopoietic necrosis virus (IHNV) associated with the emergence of new strains. *Virus Evolution* 7(2), veab056.

Abolofia, J., Wilen, J.E., Asche, F. (2017) The cost of lice: quantifying the impacts of parasitic sea lice on farmed salmon. *Marine Resource Economics* 32, 329–349.

ACFFA (Atlantic Canada Fish Farmers Association) (2017) 2016 New Brunswick Annual Sea Lice Management Report. Available at: https://static1.squarespace.com/static/56e827cb22482efe36420c65/t/58e3b26b893fc0130751d5ce/1491317360506/2016+Sea+Lice+Mgt+Report.pdf (accessed 23 September 2022).

Adams, M.B. and Nowak, B.F. (2003) Amoebic gill disease: sequential pathology in cultured Atlantic salmon, *Salmo salar* L. *Journal of Fish Diseases* 26, 601–614.

Alonso, M. and Leong, J.C. (2013) Licensed DNA vaccines against infectious hematopoietic necrosis virus (IHNV). *Recent Patents on DNA and Gene Sequences* 7(1), 62–65.

Ardelli, B.F. and Woo, P.T.K. (2001) Therapeutic and prophylactic effects of isometamidium chloride (Samorin) against the haemoflagellate *Cryptobia salmositica* in chinook salmon (*Oncorhynchus tshawytscha*). *Parasitology Research* 87, 18–26.

Aslam, M.L., Robledo, D., Krasnov, A., Moghadam, H.K., Hillestad, B., *et al.* (2020) Quantitative trait loci and genes associated with salmonid alphavirus load in Atlantic salmon: implications for pancreas disease resistance and tolerance. *Scientific Reports* 10, 10393.

Aunsmo, A., Valle, P.S., Sandberg, M., Midtlyng, P.J. and Bruheim, T. (2010) Stochastic modelling of direct costs of pancreas disease (PD) in Norwegian farmed Atlantic salmon (*Salmo salar* L.). *Preventive Veterinary Medicine* 93, 233–241.

Austin, B. and Austin, D.A. (2012) *Bacterial Fish Pathogens: Disease of Farmed and Wild Fish*, 5th edn. Springer, New York.

Bakkemo, K.R., Mikkelsen, H., Johansen, A., Robertsen, B. and Seppola, M. (2016) *Francisella noatunensis* subsp. *noatunensis* invades, survives and replicates in Atlantic cod cells. *Diseases of Aquatic Organisms* 121, 149–159.

Balmer, B.F., Getchell, R.G., Powers, R.L., Lee, J., Zhang, T., *et al.* (2018) Broad-spectrum antiviral JL122 blocks infection and inhibits transmission of aquatic rhabdoviruses. *Virology* 525, 143–149.

Bandin, I. and Souto, S. (2020) Betanodavirus and VER disease: a 30-year research review. *Pathogens* 9, 106.

Bang Jensen, B., Martensson, A. and Kristoffersen, A.B. (2020) Estimating risk factors for the daily risk of developing clinical cardiomyopathy syndrome (CMS) on a fish group level. *Preventive Veterinary Medicine* 175, 104852.

Beaman, H.J. and Speare, D.J. (1999) Regulatory effects of water temperature on *Loma salmonae* (Microspora) development in rainbow trout. *Journal of Aquatic Animal Health* 11, 237–245.

Beamish, R.J., Neville, C.M., Sweeting, R.M. and Ambers, N. (2005) Sea lice on adult Pacific salmon in the coastal waters of Central British Columbia, Canada. *Fisheries Research* 76, 198–208.

Becker, C.D. and Katz, M. (1965) Distribution, ecology and biology of the salmonid leech, *Piscicola salmositica* (Rhynchobdella: Piscicolidae). *Journal of the Fisheries Research Board of Canada* 22, 1175–1195.

Becker, J.A. and Speare, D.J. (2004) Impact of water temperature shift on xenoma clearance and recovery time during a *Loma salmonae* (Microsporidia) infection in rainbow trout *Oncorhynchus mykiss*. *Diseases of Aquatic Organisms* 58, 185–191.

Becker, J.A. and Speare, D.J. (2007) Transmission of the microsporidian gill parasite, *Loma salmonae*. *Animal Health Research Reviews* 8, 59–68.

Becker, J.A., Speare, D.J. and Dohoo, I.R. (2003) Effect of water temperature and flow rate on the transmission of microsporidial gill disease caused by *Loma salmonae* in rainbow trout *Oncorhynchus mykiss*. *Fish Pathology* 38, 105–112.

Becker, J.A., Speare, D.J. and Dohoo, I.R. (2005) Effects of the number of infected fish and acute exposure period on the horizontal transmission of *Loma salmonae* (Microsporidia) in rainbow trout, *Oncorhynchus mykiss*. *Aquaculture* 244, 1–9.

Benediktsdóttir, E. and Heidarsdóttir, K.J. (2007) Growth and lysis of the fish pathogen *Moritella viscosa*. *Letters in Applied Microbiology* 45, 115–120.

Benkaroun, J., Muir, K.F., Allshire, R., Tamer, C. and Weidmann, M. (2021) Isolation of a new infectious pancreatic necrosis virus (IPNV) variant from a fish farm in Scotland. *Viruses* 13, 385.

Birkbeck, T.H., Laidler, L.A., Grant, A.N. and Cox, D.L. (2002) *Pasteurella skyensis* sp. nov., isolated from Atlantic Salmon (*Salmo salar* L.). *International Journal of Systematic and Evolutionary Microbiology* 52(Pt 3), 699–704.

Bjornsdottir, B., Gudmundsdottir, T. and Gudmundsdottir, B.K. (2011) Virulence properties of *Moritella viscosa* extracellular products. *Journal of Fish Diseases* 34, 333–343.

Bogwald, J. and Dalmo, R.A. (2021) Protection of teleost fish against infectious diseases through oral administration of vaccines: update 2021. *International Journal of Molecular Sciences* 22, 10932.

Boison, S.A., Gjerde, B., Hillestad, B., Makvandi-Nejad, S. and Moghadam, H.K. (2019) Genomic and transcriptomic analysis of amoebic gill disease resistance in Atlantic salmon (*Salmo salar* L.) *Frontiers in Genetics* 10, 68.

Bootland, L.M. and Leong, J.C. (2011) Infectious haematopoietic necrosis virus. In: Woo, P.T.K. and Bruno, D.W. (eds) *Fish Diseases and Disorders.* Volume 3: *Viral, Bacterial and Fungal Infections*, 2nd edn. CAB International, Wallingford, UK, pp. 66–109.

Bower, S.M. and Margolis, L. (1983) Direct transmission of the haemoflagellate *Cryptobia salmositica* among Pacific salmon (*Oncorhynchus* spp.). *Canadian Journal of Zoology* 61, 1242–1250.

Bower, S.M. and Margolis, L. (1985) Effects of temperature and salinity on the course of infection with the haemoflagellate *Cryptobia salmositica* in juvenile Pacific salmon, *Oncorhynchus* spp. *Journal of Fish Diseases* 8, 25–33.

Bower, S.M., Withler, R.E. and Riddell, B.E. (1995) Genetic variation in resistance to the hemoflagellate *Cryptobia salmositica* in coho and sockeye salmon. *Journal of Aquatic Animal Health* 7, 185–194.

Boxaspen, K. (2006) A review of the biology and genetics of sea lice. *ICES Journal of Marine Science* 63, 1304–1316.

Bravo, F., Sidhu, J., Bernal, P., Bustamante, R., Condie, S., *et al.* (2020) Hydrodynamic connectivity, water temperature, and salinity are major drivers of piscirickettsiosis prevalence and transmission among salmonid farms in Chile. *Aquaculture Environment Interactions* 12, 263–279.

Bravo, S. (1994) Piscirickettsiosis in freshwater. *Bulletin of the European Association of Fish Pathologists* 14, 137–138.

Bravo, S. and Midtlyng, P.J. (2007) The use of fish vaccines in the Chilean salmon industry 1999–2003. *Aquaculture* 270, 36–42.

Bricknell, I.R., Dalesman, S.J., O'Shea, B., Pert, C.C. and Mordue Luntz, A.J. (2006) Effect of environmental salinity on sea lice *Lepeophtheirus salmonis* settlement success. *Diseases of Aquatic Organisms* 71, 201–212.

Brooker, A.J., Skem-Mauritzen, R. and Bron, J.E. (2018) Production, mortality and infectivity of planktonic larval sea lice, *Lepeophthelirus salmonis* (Krøyer, 1837): current knowledge and implications for epidemiological modeling. *ICES Journal of Marine Science* 75, 1214–1234.

Bruno, D.W., Collins, R.O. and Morrison, C.M. (1995) The occurrence of *Loma salmonae* (Protozoa: Microspora) in farmed rainbow trout, *Oncorhynchus mykiss* Walbaum, in Scotland. *Aquaculture* 133, 341–344.

Bruno, D.W., Griffiths, J., Petrie, J. and Hastings, T.S. (1998) *Vibrio viscosus* in farmed Atlantic salmon *Salmo salar* in Scotland: field and experimental observations. *Diseases of Aquatic Organisms* 34, 161–166.

Bui, S., Oppedal, F., Stien, L. and Dempster, T. (2016) Sea lice infestation level alters salmon swimming depth in sea-cages. *Aquaculture Environment Interactions* 8, 429–435.
Chang, B.D., Page, F.H., Beattie, M.J. and Hill, B.W.H. (2011) Sea louse abundance on farmed salmon in Southwestern New Brunswick area of the Bay of Fundy. In: Jones, S.R.M. and Beamish, R. (eds) *Salmon Lice: An Integrated Approach to Understanding Parasite Abundance and Distribution*. Wiley, Chichester, UK, pp. 83–115.
Chase-Topping, M.E., Pooley, C., Moghadam, H.K., Hillestad, B., Lillehammer, M., *et al.* (2021) Impact of vaccination and selective breeding on the transmission of infectious salmon anemia virus. *Aquaculture* 535, 736365.
Christiansen, D.H., Petresen, P.E., Dahl, M.M., Vest, N., Aamelfot, M., *et al.* (2021) No evidence of the vertical transmission of non-virulent infectious salmon anaemia virus (ISAV-HPRO) in farmed Atlantic salmon. *Viruses* 13, 2428.
Chun, J.H., Jung, J.W., Kim, Y.R., Lazarte, J.M.S., Kim, S.W., *et al.* (2021) Poly (I:C)-potentiated vaccination enhances T cell response in olive flounder (*Paralichthys olivaceus*) providing protection against viral hemorrhagic septicemia virus (VHSV). *Vaccines* 9, 482.
Colquhoun, D.J. and Duodu, S. (2011) *Francisella* infections in farmed and wild aquatic organisms. *Veterinary Research* 42(1), 47.
Constantine, J. (1999) *Estimating the Cost of* Loma salmonae *to BC Aquaculture*. Animal Health Branch, Ministry of Agriculture and Food, Abbotsford, British Columbia, Canada.
Crosbie, P.B.B., Bridle, A.R., Cadoret, K. and Nowak, B.F. (2012) *In vitro* cultured *Neoparamoeba perurans* causes amoebic gill disease in Atlantic salmon and fulfils Koch's postulates. *International Journal for Parasitology* 42, 511–515.
Dahle, M.K. and Olsen, A.B. (2021) Heart and skeletal muscle inflammation (HSMI) in Atlantic salmon and HSMI-like disease in rainbow trout. In: Sommerset, I., Bang Jensen, B., Bornø, B., Haukaas, A. and Brun, E. (eds) *The Health Situation in Norwegian Aquaculture 2020*. Norwegian Veterinary Institute, Oslo, pp. 68–73.
Dawson-Coates, J.A., Chase, J.C., Funk, V., Booy, M.H., Haines L.R., *et al.* (2003) The relationship between flesh quality and numbers of *Kudoa thyrsites* plasmodia and spores in farmed Atlantic salmon, *Salmo salar* L. *Journal of Fish Diseases* 26, 451–459.
Decostere, A., Hermans, K. and Haesebrouch, F. (2004) Piscine mycobacteriosis: a literature review covering the agent and the disease it causes in fish and humans. *Veterinary Microbiology* 99, 159–166.
Delghandi, M.R., El-Matbouli, M. and Menanteau-Ledouble, S. (2020) Mycobacteriosis and infections with non-tuberculous mycobacteria in aquatic organisms: a review. *Microorganisms* 8(9), 1368.
Deng, Y., Xu, L., Chen, H., Liu, S., Guo, Z., *et al.* (2020) Prevalence, virulence genes, and antimicrobial resistance of *Vibrio* species isolated from diseased marine fish in South China. *Scientific Reports* 10(1), 14329.
Denholm, L., Devine, G., Horsberg, T.E., Sevaldal, S., Fallang, A., *et al.* (2002) Analysis and management of resistance to chemotherapeutics in salmon lice, *Lepeophtheirus salmonis* (Copepoda: Caligidae). *Pest Management Science* 56, 528–536.
DFO (Fisheries and Oceans Canada) (2016) *DFO sea laudits of BC marine finfish aquaculture sites*. DFO, Ottawa. Available at: http://open.canada.ca/data/en/dataset/5cfd93bd-b3ee-4b0b-8816-33d388f6811d (accessed 11 April 2020).
Dhar, A.K., Bowers, R.M., Rowe, C.G. and Allnutt, F.T. (2010) Expression of a foreign epitope on infectious pancreatic necrosis virus VP2 capsid protein subviral particle (SVP) and immunogenicity in rainbow trout. *Antiviral Research* 85, 525–531.
Dhar, A.K., LaPatra, S., Orry, A., and Allnutt, F.C.T. (2017) Infectious pancreatic necrosis virus. In: Woo, P.T.K. and Cipriano, R.C. (eds) *Fish Viruses and Bacteria: Pathobiology and Protection*, 1st edn. CAB International, Wallingford, UK, pp. 1–12.
Dixon, P., Paley, R., Alegria-Moran, R. and Oldtmann, B. (2016) Epidemiological characteristics of infectious hematopoietic necrosis virus (IHNV): a review. *Veterinary Research* 47, 63.
Dopazo, C.P. (2020a) The infectious pancreatic necrosis virus (IPNV) and its virulence determinants: what is known and what should be known. *Pathogens* 9(2), 94.
Dopazo, C.P. (2020b) Aquatic birnavirosis (infectious pancreatic necrosis virus). In: Woo, P.T.K., Leong, J. and Buchmann, K. (eds) *Climate Change and Infectious Fish Diseases*. CAB International, Wallingford, UK, pp. 102–123.
Ellul, R.M., Bulla, J., Brudal, E., Colquhoun, D., Wergeland, H., *et al.* (2019) Protection and antibody reactivity in lumpsucker (*Cyclopterus lumpus* L.) following vaccination against *Pasteurella* sp. *Fish & Shellfish Immunology* 95, 650–658.

Erkinharju, T., Hilsen, H. and Funesvik, L. (2021) Mycobacterioisis in salmonids. In: Sommerset, I., Bang Jensen, B., Bornø, B., Haukaas, A. and Brun, E. (eds) *The Health Situation in Norwegian Aquaculture 2020*. Norwegian Veterinary Institute, Oslo, pp. 96–97.

Escobar, L.E., Escobar-Dodero, J. and Phelps, N.B.D. (2018) Infectious disease in fish: global risk of viral hemorrhagic septicemia virus. *Reviews in Fish Biology and Fisheries* 28, 637–655.

Evensen, O. and Leong, J.C. (2013) DNA vaccines against viral diseases of farmed fish. *Fish & Shellfish Immunology* 35(6), 1751–1758.

Falk, K., Namork, E., Rimstad, E., Mjaaland, S. and Dannevig, B.H. (1997) Characterization of infectious salmon anemia virus, an orthomyxo-like virus isolated from Atlantic salmon (*Salmo salar* L.). *Journal of Virology* 71(12), 9016–9023.

FAO (Food and Agriculture Organization of the United Nations) (2020) *The State of World Fisheries and Aquaculture 2020*. FAO, Rome.

Fast, M.D. and Dalvin, S. (2020) Lepeophtheirosis (*Lepeophtheirus salmonis*). In: Woo, P.T.K., Leong, J. and Buchmann, K. (eds) *Climate Change and Infectious Fish Diseases*. CAB International, Wallingford, UK, pp. 471–488.

Fast, M.D., Burka, J.F., Johnson, S.C. and Ross, N.W. (2003) Enzymes released from *Lepeophtheirus salmonis* in response to mucus from different salmonids. *Journal of Parasitology* 89, 7–13.

Fernández, L., Marquez, I. and Guijarro, J. (2004) Identification of specific *in vivo*-induced (ivi) genes in *Yersinia ruckeri* and analysis of ruckerbactin, a catecholate siderophore iron acquisition system. *Applied and Environmental Microbiology* 70, 5199–5207.

Fernández-Álvarez, C. and Santos, Y. (2018) Identification and typing of fish pathogenic species of the genus *Tenacibaculum*. *Applied Microbiology and Biotechnology* 102(23), 9973–9989.

Fernandez Sanchez, J.L., Breton, A.L., Brun, E., Vendramin, N., Spilipoulos, G., *et al.* (2022) Assessing the economic impact of diseases in Mediterranean grow-out farms culturing European sea bass. *Aquaculture* 547, 73730.

Forward, G.M. and Woo, P.T.K. (1996) An *in vitro* study on the mechanism of innate immunity on *Cryptobia*-resistant brook charr (*Salvelinus fontinalis*) against *Cryptobia salmositica*. *Parasitology Research* 82, 238–241.

Forward, G.M., Ferguson, M.M. and Woo, P.T.K. (1995) Susceptibility of brook charr, *Salvelinus fontinalis* to the pathogenic haemoflagellate, *Cryptobia salmositica*, and the inheritance of innate resistance by progenies of resistant fish. *Parasitology* 111, 337–345.

Frans, I., Michiels, C.W., Bossier, P., Willems, K.A., Lievens, B. and Rediers, H. (2011) *Vibrio anguillarum* as a fish pathogen: virulence factors, diagnosis and prevention. *Journal of Fish Diseases* 34, 643–661.

Fritsvold, C. and Bang Jensen, B. (2021) Cardiomyopathy syndrome (CMS). In: Sommerset, I., Bang Jensen, B., Bornø, B., Haukaas, A. and Brun, E. (eds) *The Health Situation in Norwegian Aquaculture 2020*. Norwegian Veterinary Institute, Oslo, pp. 74–79.

Funk, V.A., Raap, M., Sojonky, K., Jones, S., Robinson, J., *et al.* (2007) Development and validation of an RNA- and DNA-based quantitative PCR assay for determination of *Kudoa thyrsites* infection levels in Atlantic salmon, *Salmo salar*. *Diseases of Aquatic Organisms* 75, 239–249.

Gaggero, A., Castro, H., and Sandino, A.M. (1995) First isolation of *Piscirickettsia salmonis* from coho salmon, *Oncorhynchus kisutch* (Walbaum), and rainbow trout, *Oncorhynchus mykiss* (Walbaum), during the freshwater stage of their life cycle. *Journal of Fish Diseases* 18, 277–279.

Garseth, A.H., Fritsvold, C., Svendsen, J.C., Jensen, B.B. and Mikalsen, A.G. (2018) Cardiomyopathy syndrome in Atlantic salmon *Salmo salar* L.: a review of the current state of knowledge. *Journal of Fish Diseases* 41, 11–26.

Gauthier, D.T., Haines, A.N. and Vogelbein, W.K. (2021) Elevated temperatures inhibits *Mycobacterium shottsii* infection and *Mycobacterium pseduoshottsii* disease in striped bass *Morone saxatillis*. *Diseases of Aquatic Organisms* 144, 159–174.

Gonen, S., Baranski, M., Thorland, I., Norris, A., Grove, H., *et al.* (2015) Mapping and validation of a major QTL affecting resistance to pancreas disease (salmonid alphavirus) in Atlantic salmon (*Salmo salar*). *Heredity* 115, 405–414.

Grove, S., Wiik-Nielsen, C.R., Lunder, T., Tunsjo, H.S., Tandstad, N.M., *et al.* (2010) Previously unrecognised division within *Moritella viscosa* isolated from fish farmed in the North Atlantic. *Diseases of Aquatic Organisms* 93, 51–61.

Gudmundsdóttir, B.K., Björnsdóttir, B., Gudmundsdóttir, S. and Bambir, S.H. (2006) A comparative study of susceptibility and induced pathology of cod, *Gadus morhua* (L.), and halibut, *Hippoglossus hippoglossus* (L.), following experimental infection with *Moritella viscosa*. *Journal of Fish Diseases* 29, 481–487.

Gulla, S., Barnes, A.C., Welch, T.J., Romalde, J.L., Ryder, D., *et al.* (2018) Multilocus variable-number tandem-repeat analysis of *Yersinia ruckeri* confirms the existence of host specificity, geographic endemism, and anthropogenic dissemination of virulent clones. *Applied and Environmental Microbiology* 84, e00730-18.

Gulla, S., Nilsen, H., Olsen, A.B. and Collquhoun, D. (2020) Fiskepatogene Pasteurella I Norge. *Norsk Fiskeoppdrett* 11, 46–47.

Guo, F.C. and Woo, P.T.K. (2004a) Detection and quantification of *Spironucleus barkhanus* in experimentally infected Atlantic salmon *Salmo salar*. *Diseases of Aquatic Organisms* 61, 175–178.

Guo, F.C. and Woo, P.T.K. (2004b) Experimental infections of Atlantic salmon *Salmo salar* with *Spironucleus barkhanus*. *Diseases of Aquatic Organisms* 61, 59–66.

Gupta, R.S., Lo, B. and Son, J. (2018) Phylogenomics and comparative genomic studies robustly support division of the genus *Mycobacterium* into an emended genus *Mycobacterium* and four novel genera. *Frontiers in Microbiology* 9, 67.

Guselle, N.J., Markham, R.J.F. and Speare, D.J. (2006) Intraperitoneal administration of β-1,3/1,6-glucan to rainbow trout, *Oncorhynchus mykiss* (Walbaum), protects against *Loma salmonae*. *Journal of Fish Diseases* 29, 375–381.

Guselle, N.J., Speare, D.J. and Markham, R.J.F. (2010) Efficacy of intraperitoneally and orally administered ProVale, a yeast β-(1,3)/(1,6)-d-glucan product, in inhibiting xenoma formation by the microsporidian *Loma salmonae* on rainbow trout gills. *North American Journal of Aquaculture* 72, 65–72.

Hamre, L.A., Bui, S., Oppedal, F., Skern-Mauritzen, R. and Dalvin, S. (2019) Development of the salmon louse, *Lepeophtheirus salmonis* parasitic stages in temperatures ranging from 3 to 24°C. *Aquaculture Environment Interactions* 11, 429–433.

Hashish, E., Merwad, A., Elgami, S., Amer, A., Karmal, H., *et al.* (2018) *Mycobacterium marinum* infection in fish and man: epidemiology, pathophysioloogy and management; a review. *Veterinary Quarterly* 38(1), 35–46.

Haugland, O., Mikalsen, A.B., Nilsen, P., Lindmo, K., Thu, B.J., *et al.* (2011) Cardiomyopathy syndrome of Atlantic salmon (*Salmo salar* L.) is caused by a double-stranded RNA virus of the *Totiviridae* family. *Journal of Virology* 85(11), 5275–5286.

Hawley, L.M. and Garver, K.A. (2008) Stability of viral hemorrhagic septicemia virus (VHSV) in freshwater and seawater at various temperatures. *Diseases of Aquatic Organisms* 82, 171–178.

Hellebø, A., Stene, A. and Aspehaug, V. (2014) *Potensielle reservoarer for SAV og PMCV på marine akvakulturanlegg*. Møreforskning MARIN MA, Ålesund, Norway, pp. 1–23.

Herrero, A., Thompson, K.D., Ashby A., Rodger, H.D. and Dagieish, M.F. (2018) Complex gill disease: an emerging syndrome in farmed Atlantic salmon (*Salmo salar* L.). *Journal of Comparative Pathology* 163, 23–28.

Hervio, D.M.L, Kent, M.L., Khattra, J., Sakanari, J., Yokoyama, H., *et al.* (1997) Taxonomy of *Kudoa* species (Myxosporea), using a small subunit DNA sequence. *Canadian Journal of Zoology* 75, 2112–2119.

Hevroy, E.M., Waagbo, R., Torstensen, B.E., Takle, H., Stubhaug, I., *et al.* (2012) Ghrelin is involved in voluntary anorexia in Atlantic salmon raised at elevated sea temperatures. *General and Comparative Endocrinology* 175(1), 118–134.

Hickey, M.E., and Lee, J.L. (2018) A comprehensive review of *Vibrio* (*Listonella*) *anguillarum* ecology, pathology and prevention. *Reviews in Aquaculture* 10(3), 585–610.

Hillestad, B., Kristjansson, O.H., Makvandi-Nejad, S. and Moghadam, H.K. (2020) Genome-wide association study confirms previous findings of major loci affecting resistance to *Piscine myocarditis virus* in Atlantic salmon (*Salmo salar* L.). *Genes* 11, 608.

Hillestad, B., Johannessen, S., Melingen, G.O. and Moghadam, H.K. (2021) Identification of a new infectious pancreatic necrosis virus (IPNV) variant in Atlantic salmon (*Salmo salar* L.) that can cause high mortality even in genetically resistant fish. *Frontiers in Genetics* 12, 635185.

Holm, K.O., Bækkedal, C., Söderberg, J.J. and Haugen, P. (2018) Complete genome sequences of seven *Vibrio anguillarum* strains as derived from PacBio sequencing. *Genome Biology and Evolution* 10(4), 1127–1131.

Home, K.S. and Tveten, A.-K. (2020) Current approaches in decoding the molecular mechanisms of long-term stress in adult farmed Atlantic salmon (*Salmo salar*). *Reviews in Aquaculture* 12, 1708–1720.

Hua, X., Zhou, Y., Feng, Y., Duan, K., Ren, X., *et al.* (2021) Oral vaccine against IPNV based on antibiotic-free resistance recombinant *Lactobacillus casei* expressing CK6-VP2 fusion protein. *Aquaculture* 535, 736425.

Huang, S., Wu, Y., Su, L., Su, T., Zhou, Q., *et al.* (2022) A single-chain variable fragment antibody exerts anti-nervous necrosis virus activity by irreversible binding. *Aquaculture* 552, 738001.

Hunter, V.A., Knittel, M. and Fryer, J. (1980) Stress-induced transmission of *Yersinia ruckeri* infection from carriers to recipient steelhead trout *Salmo gairdneri* Richardson. *Journal of Fish Diseases* 3, 467–472.

Hvas, M., Folkedal, O., Imsland, A., Oppedal, F. (2018) Metabolic rates, swimming capabilities, thermal niche and stress response of the lumpfish, *Cyclopterus lumpus*. *Biology Open* 7, bio036079.

ICES (International Council for the Exploration of the Sea) (2019) *Tenacibaculum maritimum*, causal agent of tenacibaculosis in marine fish. Original by Y. Santos, F. Paxoz and J.L. Barja (No. 55), Revised by Simon S.R. Jones and Lone Madsen. ICES Identification Leaflets for Diseases and Parasites of Fish and Shellfish No. 70. ICES, Copenhagen. Available at: http://doi.org/10.17895/ices.pub.4681 (accessed 14 September 2022).

Imsland, A.K., Reynolds, P., Nytro, A.V., Eliassen, G., Hangstad, T.A., *et al.* (2016) Effects of lumpfish size on foraging behaviour and co-existence with sea lice infected Atlantic salmon in sea cages. *Aquaculture* 465, 19–27.

Ineno, T., Tsuchida, S., Kanada, M. and Watabe, S. (2005) Thermal tolerance of a rainbow trout, *Oncorhynchus mykiss* strain selected by high temperature breeding. *Fisheries Science* 71, 767–775.

Ineno, T., Tamaki, K., Yamada, K., Kodama, K., Ischida, S., *et al.* (2018) Thermal tolerance of a thermally selected strain of rainbow trout, *Oncorhynchus mykiss* and the pedigrees of its F_1 and F_2 generations indicated by their critical thermal maxima. *Fisheries Science* 84, 671–679.

Janda, J.M. (2020) Proposed nomenclature or classification changes for bacteria of medical importance: taxonomic update 5. *Diagnostic Microbiology and Infectious Disease* 97(3), 115047.

Jarungsriapisit, J., Nunez-Ortiz, N., Nordbo, J., Moore, L.J., Maehle, S., *et al.* (2020) The effect of temperature on the survival of salmonid alphavirus analysed using *in vitro* and *in vivo* methods. *Aquaculture* 516, 734647.

Johnson, S.C. and Albright, L.J. (1991) The developmental stages of *Lepeophtheirus salmonis* (Kreyer 1837) (Copepoda: Caligidae). *Canadian Journal of Zoology* 69, 929–950.

Jorgensen, A. and Sterud, E. (2006) The marine pathogenic genotype of *Spironucleus barkhanus* from farmed salmonids is redescribed as *Spironucleus salmonicida* n. sp. *Journal of Eukaryotic Microbiology* 53, 531–541.

Kabata, Z. (1973) The species of *Lepeophtheirus* (Copepoda: Caligidae) from fishes of British Columbia. *Journal of the Fisheries Research Board of Canada* 30, 729–759.

Karlsen, M. and Johansen, R. (2017) Alphaviruses in salmonids. In: Woo, P.T.K. and Cipriano, R.C. (eds) *Fish Viruses and Bacteria: Pathobiology and Protection*, 1st edn. CAB International, Wallingford, UK, pp. 160–172.

Kent, M.L. and Dawe, S.C. (1994) Efficacy of Fumagillin DCH against experimentally induced *Loma salmonae* (Microsporea) infections in chinook salmon *Oncorhynchus tshawytscha*. *Diseases of Aquatic Organisms* 20, 231–233.

Kent, M.L. and Poppe, T.T. (1998) *Kudoa thyrsites*. In: *Diseases of Seawater Netpen-Reared Salmonid Fishes*. Quadra Printers Ltd, Nanaimo, British Columba, Canada, pp. 59–60.

Kent, M.L., Ellis, J., Fournie, J.W., Dawe, S.C., Bagshaw, J.W., *et al.* (1992) Systemic hexamitid (Protozoa: Diplomonadida) infection in seawater pen-reared chinook salmon, *Oncorhynchus tshawytscha*. *Diseases of Aquatic Organisms* 14, 81–89.

Kent, M.L., Dawe, S.C. and Speare, D.J. (1995) Transmission of *Loma salmonae* (Microsporea) to chinook salmon in sea water. *Canadian Veterinary Journal* 36, 98–101.

Kent, M.L., Traxler, G.S., Kieser, D., Richard, J., Dawe, S.C., *et al.* (1998) Survey of salmonid pathogens in ocean-caught fishes in British Columbia, Canada. *Journal of Aquatic Animal Health* 10, 211–219.

Kibenge, F.S.B., Kibenge, M.J.T., Wang, Y., Qian, B., Hariharan, S., *et al.* (2007) Mapping of putative virulence motifs on infectious salmon anemia virus surface glycoprotein genes. *Journal of General Virology* 88, 3100–3111.

Kim, W.S., Kong, K.H., Kim, J.O. and Oh, M.J. (2016) Amoebic gill infection in coho salmon *Oncorhynchus kisutch* farmed in Korea. *Diseases of Aquatic Organisms* 121, 75–78.

Klinger, D.H., Levin, S.A. and Watson, J.R. (2017) The growth of finfish in global open-ocean aquaculture under climate change. *Proceedings of the Royal Society B* 284, 20170834.

Kirchhoff, N.T., Rough, K. and Nowak, B.F. (2011) Moving cages further offshore: effects on southern bluefin tuna, *T. maccoyii*, parasites, health and performance. *PLoS ONE* 6(8), e23705.

Kube, P.D., Taylor, R.S. and Elliott, N.G. (2012) Genetic variation in parasite resistance of Atlantic salmon to amoebic gill disease over multiple infections. *Aquaculture* 364–365, 165–172.

Kumar, G., Menanteau-Ledouble, S., Saleh, M. and El-Matbouli, M. (2015) *Yersinia ruckeri*, the causative agent of enteric redmouth disease in fish. *Veterinary Research* 46, 103.

Kurath, G., Winton, J.R., Dale, O.B., Purcell, M.K., Falk, K., *et al.* (2016) Atlantic salmon, *Salmo salar* L. are broadly susceptible to isolates representing the North American genogroups of infectious hematopoietic necrosis virus. *Journal of Fish Diseases* 39(1), 55–67.

Lages, M.A., Balado, M. and Lemos, M.L. (2019) The expression of virulence factors in *Vibrio anguillarum* is dually regulated by iron levels and temperature. *Frontiers in Microbiology* 10, 2335.

Li, S. and Woo, P.T.K. (1995) Efficacy of a live *Cryptobia salmositica* vaccine, and its mechanism of protection in vaccinated *Oncorhynchus mykiss* (Walbaum) against cryptobiosis. *Veterinary Immunology and Immunopathology* 48, 343–353.

Li, S. and Woo, P.T.K. (1997) Vaccination of rainbow trout, *Oncorhynchus mykiss* (Walbaum) against cryptobiosis: efficacy of the vaccine in fresh and sea water. *Journal of Fish Diseases* 20, 369–374.

Lillehaug, A. and Colquhoun, D. (2020) *Vibriosis caused by Vibrio anguillarum and V. ordalii*. ICES Identification Leaflets for Diseases and Parasites in Fish and Shellfish No. 71. International Council for the Exploration of the Sea, Copenhagen.

Liyanage, D.S., Lee, S., Yang, H., Lim, C., Omeka, W.K.M., *et al.* (2022) Genome-wide association study of VHSV-resistance trait in *Paralichthys olivaceus*. *Fish & Shellfish Immunology* 124, 391–400.

Ljungfeldt, L.E.R., Quintela, M., Besnier, F., Nilsen, F. and Glover, K.A. (2017) A pedigree-based experiment reveals variation in salinity and thermal tolerance in the salmon louse, *Lepeophtheirus salmonis*. *Evolutionary Applications* 10, 1007–1019.

Lund, M., Dahle, M.R., Timmerhaus, G., Alarcon, M., Powell, M., *et al.* (2017) Hypoxia tolerance and responses to hypoxic stress during heart and skeletal muscle inflammation in Atlantic salmon (*Salmo salar*). *PLoS ONE* 12(7), e0181109.

Lyngstad, T.M., Quviller, L., Sindre, H., Brun, E. and Kristoffrsen, A.B. (2018) Risk factors associated with outbreaks of infectious salmon anemia (ISA) with unknown source of infection in Norway. *Frontiers in Veterinary Sciences* 5, 308.

Ma, J., Bruce, T.J., Jones, E.M. and Cain, K.D. (2019) A review of fish vaccine development strategies: conventional methods and modern biotechnological approaches. *Microorganisms* 7, 569.

Marana, M.H., Karami, A.M., Ødegård, J., Zuo, S., Jaafar, R.M., *et al.* (2021) Whole-genome association study searching for QTL for *Aeromonas salmonicida* resistance in rainbow trout. *Scientific Reports* 11(1), 17857.

Mauel, M.J. and Miller, D.L. (2002) Piscirickettsiosis and piscirickettsiosis-like infections in fish: a review. *Veterinary Microbiology* 87, 279–289.

McConnachie, S.H., Guselle, N.J. and Speare, D.J. (2013) Retention of viable microsporidial (*Loma salmonae*) spores within the blue mussel (*Mytilus edulis*): use of an experimental laboratory model probing pathogen transfer within a multi-trophic aquaculture setting. *Aquaculture* 376–379, 1–5.

Menanteau-Ledouble, S., Kumar, G., Saleh, M. and El-Matbouli, M. (2016) *Aeromonas salmonicida*: updates on an old acquaintance. *Diseases of Aquatic Organisms* 120, 49–68.

Mikalsen, A.B., Lund, M., Maji, F., Kjonstad, M.V., Bergtun, P.H., *et al.* (2020) Lack of evidence of vertical transmission of piscine myocarditis virus in Atlantic salmon (*Salmo salar* L.). *Journal of Fish Diseases* 43, 715–718.

Mo, T.A., Poppe, T.T. and Iversen, L. (1990) Systemic hexamitosis in salt-water reared Atlantic salmon (*Salmo salar* L.). *Bulletin of the European Association of Fish Pathologists* 10, 69–70.

Moen, T., Torgersen, J., Santi, N., Davidson, W.S., Baranski, M., *et al.* (2015) Epithelial cadherin determines resistance to infectious pancreatic necrosis virus in Atlantic salmon. *Genetics* 200, 1313–1326.

Mohamad, N., Amal, M.N.A., Yasin, I.S.M., Saad, M.Z., Nasruddin, N.S., *et al.* (2019) Vibriosis in cultured marine fishes: a review. *Aquaculture* 512, 734289.

Mooney, J., Powell, E., Clabby, C. and Powell, R. (1995) Detection of *Aeromonas salmonicida* in wild Atlantic salmon using a specific DNA probe test. *Diseases of Aquatic Organisms* 21, 131–135.

Moran, J.D.W. and Kent, M.L. (1999) *Kudoa thyrsites* (Myzozoa: Myxosporea) infections in pen-reared Atlantic salmon in the northeast Pacific Ocean with a survey of potential non-salmonid reservoir hosts. *Journal of Aquatic Animal Health* 11, 101–109.

Moran, J.D.W., Whitaker, K.J. and Kent, M.L. (1999a) A review of the myxosporean genus *Kudoa* Meglitsch, 1947, and its impact on the international aquaculture industry and commercial fisheries. *Aquaculture* 172, 163–196.

Moran, J.D.W., Margolis, L., Webster, J.M. and Kent, M.L. (1999b) Sequential development of *Kudoa thyrsites* (Myxozoa: Myxosporea) in netpen-reared Atlantic salmon determined by light microscopy and a polymerase chain reaction text. *Diseases of Aquatic Organisms* 37, 185–193.

Moran, J.D.W. Whitaker, D.J. and Kent, M.L. (1999c) Natural and laboratory transmission of the marine myxosporean parasite *Kudoa thyrsites* to Atlantic salmon. *Journal of Aquatic Animal Health* 11, 110–115.

Mordecai, G., Miller, K.M., Bass, A.L., Bateman, A.W., Teffer, A.K., *et al.* (2021) Aquaculture mediates global transmission of a viral pathogen to wild salmon. *Science Advances* 7, eabe2592.

Mouton, A., Crosbie, P., Cadoret, K. and Nowak, B. (2014) First record of amoebic gill disease caused by *Neoparamoeba perurans* in South Africa. *Journal of Fish Diseases* 37, 407–409.

Munro, E.S. and Midtlyng, P.J. (2011) Infectious pancreatic necrosis and associated aquatic birnaviruses. In: Woo, P.T.K. and Bruno, D.W. (eds) *Fish Diseases and Disorders*. Volume 3: *Viral, Bacterial and Fungal Infections*, 2nd edn. CAB International, Wallingford, UK, pp. 1–65.

Nilsen, H.K., Colquhoun, D. and Gulla, S. (2021) Pasteurellosis in salmon. In: Sommerset, I., Bang Jensen, B., Bornø, B., Haukaas, A. and Brun, E. (eds) *The Health Situation in Norwegian Aquaculture 2020*. Norwegian Veterinary Institute, Oslo, pp. 98–99.

Nowak, B.F. (2012) *Neoparamoeba perurans*. In: Woo, P.T.K. and Buchman, K. (eds) *Fish Parasites: Pathobiology and Protection*. CAB International, Wallingford, UK, pp. 1–18.

Nowlan, J.P., Lumsden, J.S. and Russell, S. (2020) Advancements in characterizing *Tenacibaculum* infections in Canada. *Pathogens* 9(12), 1029.

OIE (World Organisation for Animal Health) (2019) Chapter 2.3.12: Viral Encephalopathy and Retinopathy. Manual of Diagnostic Tests for Aquatic Animals – 14/11/2019. OIE, Paris. Available at: https://www.woah.org/fileadmin/Home/eng/Health_standards/aahm/current/chapitre_viral_encephalopathy_retinopathy.pdf (accessed 14 September 2022).

OIE (World Organisation for Animal Health) (2021) Manual of Diagnostic Tests for Aquatic Animals, 8th edn. OIE, Paris. Available at: https://www.oie.int/en/what-we-do/standards/codes-and-manuals/aquatic-manual-online-access/ (accessed 18 January 2022).

Oldham, T., Rodger, H. and Nowak, B.F. (2016) Incidence and distribution of amoebic gill disease (AGD) – an epidemiological review. *Aquaculture* 457, 35–42.

Olveira, J.G., Souto, S., Dopazo, C.P. and Bandín, I. (2013) Isolation of betanodavirus from farmed turbot *Psetta maxima* showing no signs of viral encephalopathy and retinopathy. *Aquaculture* 406, 125–130.

Ortiz-Severín, J., Travisany, D., Maass, A.P., Chavez, F. and Cambiazo, V. (2019) *Piscirickettsia salmonis* cryptic plasmids: source of mobile DNA and virulence factors. *Pathogens* 8, 269.

Otterlei, A., Brevik, O.J., Jensen, D., Duesund, H., Sommerset, I., *et al.* (2016) Phenotypic and genetic characterization of *Piscirickettsia salmonis* from Chilean and Canadian salmonids. *BMC Veterinary Research* 12, 55.

Paez, D.J., Powers, R.L., Jia, P., Ballesteros, N., Kurath, G., *et al.* (2021) Temperature variation and host immunity regulate viral persistence in a salmonid host. *Pathogens* 10, 855.

Panzarin, V., Holmes, E.C., Abbadi, M., Zamperin, G., Quartesan, R., *et al.* (2018) Low evolutionary rate of infectious pancreatic necrosis virus (IPNV) in Italy is associated with reduced virulence in trout. *Virus Evolution* 4(2), vey019.

Panzarin, V., Cuenca, A., Gastaldelli, M., Alencar, A.L.F., Pascoli, F., *et al.* (2020) VHSV single amino acid polymorphisms (SAPs) associated with virulence in rainbow trout. *Frontiers in Microbiology* 11, 1984.

Paterson, W.B and Woo, P.T.K. (1983) Electron microscope observations of the bloodstream form of *Cryptobia salmositica* Katz 1951 (*Kinetoplastida*, *Bodonina*). *Journal of Protozoology* 39, 431–437.

Pérez-Stuardo, D., Espinoza, A., Tapia, S., Morales-Reyes, J., Barrientos, C., *et al.* (2020) Non-specific antibodies induce lysosomal activation in Atlantic salmon macrophages infected by *Piscirickettsia salmonis*. *Frontiers in Immunology* 11, 544718.

Pettersen, J.M., Rich, K.M., Jensen, B.B. and Aunsmo, A. (2015) The economic benefits of disease triggered early harvest: a case study of pancreas disease in farmed Atlantic salmon from Norway. *Preventive Veterinary Medicine* 121, 314–324.

Polinski, M. and Garver, K. (2019) *Characterization of Piscine Orthoreovirus (PRV) and Associated Diseases to Inform Pathogen Transfer Risk Assessments in British Columbia*. Science Advisory Report 2019/035. Canadian Science Advisory Secretariat, National Capital Region, Fisheries and Oceans Canada, Ottawa.

Polinski, M., Vendramin, N., Cuenca, A. and Garver, K.A. (2020) *Piscine orthoreovirus*: biology and distribution in farmed and wild fish. *Journal of Fish Diseases* 43, 1331–1352.

Poppe, T.T., Mo, T.A. and Iversen, L. (1992) Disseminated hexamitosis in sea-caged Atlantic salmon *Salmo salar*. *Diseases of Aquatic Organisms* 14, 91–97.

Powell, M.D. (2006) Physiological and clinical pathology. In: Ferguson, H.W. (ed.) *Systemic Pathology of Fish*, 2nd edn. Scotian Press, London, pp. 336–355.

Puvanendran, V., Mortensen, A., Johansen, L.H., Kettunen, A., Hansen, Ø.J., *et al.* (2022) Development of cod farming in Norway: past and current biological and market status and future prospects and directions. *Reviews in Aquaculture* 14(1), 308–342.

Ramirez-Paredes, J.G., Larsson, P., Thompson, K.D., Penman, D.J., Busse, H.J., *et al.* (2020) Reclassification of *Francisella noatunensis* subsp. *orientalis* Ottem *et al.* 2009 as *Francisella orientalis* sp. nov., *Francisella noatunensis* subsp. *chilensis* subsp. nov. and emended description of *Francisella noatunensis*. *International Journal of Systematic and Evolutionary Microbiology* 70(3), 2034–2048.

Ramsay, J.M., Speare, D.J., Dawe, S.C. and Kent, M.L. (2002) Xenoma formation during microsporidial gill disease of salmonids caused by *Loma salmonae* is affected by host species (*Oncorhynchus tshawytscha*, *O. kisutch*, *O. mykiss*) but not by salinity. *Diseases of Aquatic Organisms* 48, 125–131.

Ramsay, J.M., Speare, D.J., Becker, J.A. and Daley, J. (2003) *Loma salmonae*-associated xenoma onset and clearance in rainbow trout, *Oncorhynchus mykiss* (Walbaum): comparisons of *per os* and cohabitation exposure using survival analysis. *Aquaculture Research* 34, 1329–1335.

Reno, P.W. (1999) Infectious pancreatic necrosis virus and associated aquatic birnavirus. In: Woo, P.T.K. and Bruno, D.W. (eds) *Fish Diseases and Disorders*. Volume 3: *Viral, Bacterial and Fungal Infections*, 1st edn. CAB International, Wallingford, UK, pp. 1–55.

Rhodes, M.W., Kator, H., Kaattari, I., Gauthier, D., Vogelbein, W., *et al.* (2004) Isolation and characterization of mycobacteria from striped bass *Morone saxatilis* from the Chesapeake Bay. *Diseases of Aquatic Organisms* 61(1–2), 41–51.

Rimstad, E. and Markussen, T. (2019) Infectious salmon anaemia virus – molecular biology and pathogenesis of the infection. *Journal of Applied Microbiology* 129, 85–97.

Ritchie, R.J., McDonald, J.T., Glebe, B., Young-Lai, W., Johnsen, E., *et al.* (2009) Comparative virulence of infectious salmon anaemia virus isolates in Atlantic salmon, *Salmo salar* L. *Journal of Fish Diseases* 32, 157–171.

Robledo, D., Matika, O. and Hamilton, A. (2018) Genome-wide association and genomic selection for resistance to amoebic gill disease in Atlantic salmon. *G3: Genes, Genomes, Genetics* 8, 1195–1203.

Rodger, H.D. (2014) Amoebic gill disease (AGD) in farmed salmon (*Salmo salar*) in Europe. *Fish Veterinary Journal* 14, 16–27.

Rozas, M. and Enríquez, R. (2014) Piscirickettsiosis and *Piscirickettsia salmonis* in fish: a review. *Journal of Fish Diseases* 37(3), 163–188.

Rozas-Serri, M., Ildefonso, R., Pena, A., Enriquez, R., Barrientos, S., *et al.* (2017) Comparative pathogenesis of piscirickettsiosis in Atlantic salmon (*Salmo salar* L.) post-smolt experimentally challenged with LF-89-like and EM-90-like *Piscirickettsia salmonis* isolates. *Journal of Fish Diseases* 40, 1451–1472.

Saksida, S.M., Morrison, D., Sheppard, M. and Keith, I. (2011) Sea lice management on salmon farms in British Columbia, Canada. In: Jones, S.R.M. and Beamish, R. (eds) *Salmon Lice: An Integrated Approach to Understanding Parasite Abundance and Distribution*. Wiley, Chichester, UK, pp. 235–278.

Sanchez, J.G., Speare, D.J. and Markham, R.J.F. (1999) Nonisotopic detection of *Loma salmonae* (Microspora) in rainbow trout (*Oncorhynchus mykiss*) gills by *in-situ* hybridization. *Veterinary Pathology* 36, 610–612.

Sanchez, J.G., Speare, D.J., Markham, R.J.F. and Jones, S.R.M. (2001a) Experimental vaccination of rainbow trout against *Loma salmonae* using a live low-virulence variant of *L. salmonae*. *Journal of Fish Biology* 59, 427–441.

Sanchez, J.G., Speare, D.J., Markham, R.J.F. and Jones S.R.M. (2001b) Isolation of a *Loma salmonae* variant: biological characteristics and host range. *Journal of Fish Biology* 59, 427–441.

Sandlund, L., Mor, S.K., Singh, V.K., Padhi, S.K., Phelps, N.B.D., *et al.* (2021a) Comparative molecular characterization of novel and known piscine toti-like viruses. *Viruses* 13, 1063.

Sandlund, N., Ronneseth, A., Ellul, R.M., Nylund, S. and Sandlund, L. (2021b) *Pasteurella* spp. infections in Atlantic salmon and lumpsucker. *Journal of Fish Diseases* 44, 1201–1214.

Scholz, F., Ruane, N.M., Morrissey, T., Marcos-Lopez, M., Mitchell, S., *et al.* (2017) Piscine myocarditis virus detected in corkwing wrasse (*Symphoduc melops*) and ballan wrasse (*Labrus bergylta*). *Journal of Fish Diseases* 21, 1–6.

Scottish Government (2018) Salmon farming industry, fish infected by *Pasturella skyensis: EIR release*. FOI-18-00401 – Mortality relating to *Pasteurella skyensis*.pdf. Available at: https://www.gov.scot/publications/foi-18-00315/ (accessed 14 September 2022).

Shaw, R.W., Kent, M.L. and Adamson, M.L. (2000) Viability of *Loma salmonae* (Microsporidia) under laboratory conditions. *Parasitology Research* 86, 978–981.

Shinn, A.P., Pratoomyot, J., Bron, J.E., Paladini, G., Brooker, E.E., *et al.* (2015) Economic costs of protistan and metazoan parasites to global mariculture. *Parasitology* 142, 196–270.
Sievers, M., Korsoen, O., Warren-Myers, F., Oppedal, F., Macaulay, G., *et al.* (2022) Submerged cage aquaculture of marine fish: a review of the biological challenges and opportunities. *Reviews in Aquaculture* 14, 106–119.
Smith, P.A. and Mardones, F.O. (2020) Piscirickettsiosis (*Piscirickettsia salmonis*). In: Woo, P.T.K., Leong, J. and Buchmann, K. (eds) *Climate Change and Infectious Fish Diseases*. CAB International, Wallingford, UK, pp. 280–286.
Smith, P.A., Pizarro, P., Ojeda, P., Contreras, J., Oyanedel, S., *et al.* (1999) Routes of entry of *Piscirickettsia salmonis* in rainbow trout *Oncorhynchus mykiss*. *Diseases of Aquatic Organisms* 37, 165–172.
Smith, P.A., Rojas, M.E., Guajardo, A., Contreras, J., Morales, M.A., *et al.* (2004) Experimental infection of coho salmon *Oncorhynchus kisutch* by exposure of skin, gills and intestine with *Piscirickettsia salmonis*. *Diseases of Aquatic Organisms* 61, 53–57.
Sokołowska, J. and Nowak, B.F. (2020) Amoebiosis (*Neoparamoeba perurans*). In: Woo, P.T.K., Leong, J. and Buchmann, K. (eds) *Climate Change and Infectious Fish Diseases*. CAB International, Wallingford, UK, pp. 303–318.
Sommerset, I., Walde, C.S., Bang Jensen, B., Bornø, B., Haukaas, A., *et al.* (2020) *Fiskehelserapporten 2019*. Norwegian Veterinary Institute, Oslo.
Souto, S., Olveira, J.G. and Bandín, I. (2015) Influence of temperature on Betanodavirus infection in Senegalese sole (*Solea senegalensis*). *Veterinary Microbiology* 179(3–4), 162–167.
Speare, D.J. (2020) Microsporidiosis (*Loma salmonae*). In: Woo, P.T.K., Leong, J. and Buchmann, K. (eds) *Climate Change and Infectious Fish Diseases*. CAB International, Wallingford, UK, pp. 363–380.
Speare, D.J. and Daley, J. (2003) Failure of vaccination in brook trout *Salvelinus fontinalis* against *Loma salmonae* (Microspora). *Fish Pathology* 38, 27–28.
Speare, D.J., Markham, R.J.F. and Guselle, N.J. (2007) Development of an effective whole-spore vaccine using a low-virulence strain of *Loma salmonae* to protect against microsporidial gill disease in rainbow trout (*Oncorhynchus mykiss*). *Clinical and Vaccine Immunology* 14, 12–18.
Spilsberg, B., Nilsen, H.K., Tavornpanich, S., Gulla, S., Jansen, M.D., *et al.* (2022) Tenacibaculosis in Norwegian Atlantic salmon (*Salmo salar*) cage-farmed in cold sea water is primarily associated with *Tenacibaculum finnmarkense* genomovar *finnmarkense*. *Journal of Fish Diseases* 45(4), 523–534.
St-Hilaire, S., Hill, M., Kent, M.L., Whitaker, D.J. and Ribble, C. (1997) A comparative study of muscle texture and intensity of *Kudoa thyrsites* infection in farmed-reared Atlantic salmon *Salmo salar* on the Pacific coast of Canada. *Diseases of Aquatic Organisms* 31, 221–225.
St-Hilaire, S., Ribble, C. and Kent, M.L. (1998) Prevalence of *Kudoa thyrsites* in sexually mature and immature Atlantic salmon (*Salmo salar*) in British Columbia, Canada. *Aquaculture* 162, 69–77.
Steigen, A., Nylund, A., Plarre, H., Watanabe, K., Karisbakk, B., *et al.* (2018) Presence of selected pathogens on the gills of five wrasse species in western Norway. *Diseases of Aquatic Organisms* 128, 21–35.
Stene, A., Viljugrein, H., Yndestad, H., Tavornpanich, S. and Skjerve, E. (2014) Transmission dynamics of pancreas disease (PD) in a Norwegian fjord: aspects of water transport, contact networks and infection pressure among salmon farms. *Journal of Fish Diseases* 37(2), 123–134.
Stepien, C., Leamon, D. and Niner, M.D. (2020) Rhabdovirosis (viral haemorrhagic septicaemia virus). In: Woo, P.T.K., Leong, J. and Buchmann, K. (eds) *Climate Change and Infectious Fish Diseases*. CAB International, Wallingford, UK, pp. 58–84.
Sterud, E., Mo, T.A. and Poppe, T.T. (1997) Ultrastructure of *Spironucleus barkhanus* n.sp. (Diplomodadida: Hexamitidae) from grayling *Thymallus thymallus* (L.) (Salmonidae) and Atlantic salmon *Salmo salar* L. (Salmonidae). *Journal of Eukaryotic Microbiology* 44, 399–407.
Sterud, E., Mo, T.A. and Poppe, T.T. (1998) Systemic spironucleosis in sea-farmed Atlantic salmon *Salmo salar*, caused by *Spironucleus barkhanus* transmitted from feral Arctic char *Salvelinus alpinus*? *Diseases of Aquatic Organisms* 33, 63–66.
Sterud, E., Poppe, T.T. and Bornø, G. (2003) Intracellular infection with *Spironucleus barkhanus* (Diplomonadida: Hexamitidae) in farmed Arctic char *Salvelinus alpinus*. *Diseases of Aquatic Organisms* 56, 155–161.
Strom, S.B. and Nilsen, H. (2021) *Pasteurella skyensis* in Atlantic salmon (*Salmo salar* L.) in Western Norway. *Bulletin of the European Association of Fish Pathologists* 41(4), 160–168.
Su, H., Eerde, A.V., Steeen, H.S., Heldal, I., Haugslien, S., *et al.* (2021) Establishment of a piscine myocarditis virus (PMCV) challenge model and testing of a plant-produced subunit vaccine candidate against cardiomyopathy syndrome (CMS) in Atlantic salmon *Salmo salar*. *Aquaculture* 541, 736706.

Taksdal, T. and Sindre, H. (2016) Togaviruses of fish. In: Kibenge, F. and Godoy, M. (eds) *Aquaculture Virology*. Academic Press, San Diego, California, pp. 357–364.
Tamer, C., Cavunt, A., Durmaz, Y., Ozun, E., Kai, H., *et al.* (2021) Inactivated infectious pancreatic necrosis virus (IPNV) vaccine and *E. coli*-expressed recombinant IPNV-VP2 subunit vaccine afford protection against IPNV challenge in rainbow trout. *Fish & Shellfish Immunology* 115, 2205–2011.
Tan, C.W., Jesudhasan, P.R.R. and Woo, P.T.K. (2008) Towards a metalloprotease-DNA vaccine against piscine cryptobiosis caused by *Cryptobia salmositica*. *Parasitology Research* 102, 265–275.
Taylor, R.S., Muller, W.J., Cook, M.T., Kube, P.D. and Elliott, N.G. (2009) Gill observations in Atlantic salmon (*Salmo salar*) during repeated amoebic gill disease (AGD) field exposure and survival challenge. *Aquaculture* 200(1–2), 1–8.
Tighe, A.J., Carlsson, J., Morrissey, T., Swords, F. and Ruane, N.M. (2019) Genetic diversity of piscine myocarditis virus in Atlantic salmon *Salmo salar* L. in Ireland. *Journal of Fish Diseases* 42(8), 1161–1168.
Tighe, A.J., Gallagher, M.D., Carlsson, J., Matejusova, I., Swords, F., *et al.* (2020) Nanopore whole genome sequencing and partitioned phylogenetic analysis supports a new salmonid alphavirus genotype (SAV7). *Diseases of Aquatic Organisms* 142, 203–211.
Thorarinsson, R., Wolf, J.C., Inami, M., Phillips, L., Jones, G., *et al.* (2021) Effect of a novel DNA vaccine against pancreas disease caused by salmonid alphavirus subtype 3 in Atlantic salmon (*Salmo salar*). *Fish & Shellfish Immunology* 108, 116–126.
Tobback, E., Decostere, A., Hermans, K., Haesebrouck, F. and Chiers, K. (2007) *Yersinia ruckeri* infections in salmonid fish. *Journal of Fish Diseases* 30(5), 257–268.
Toffan, A. and Panzarin, V. (2020) Viral encephalopathy and retinopathy/viral nervous necrosis (VER/VNN). In: Zrncic, S. (ed.) *Diagnostic Manual for the Main Pathogens in European Seabass and Gilthead Seabream Aquaculture (Options Méditerranéennes B*, no. 75). CHHEAM, Zaragoza, Spain, pp. 45–60.
Toranzo, A.E. and Hetrick, F.M. (1982) Comparative stability of two salmonid viruses and poliovirus in fresh, estuarine and marine waters. *Journal of Fish Diseases*, 5(3), 223–231.
Toranzo, A.E., Magarinos, B. and Avendano-Herrera, R. (2017) Vibriosis: *Vibrio anguillarum*, *V. ordalii* and *Allivibrio salmonicida*. In: Woo, P.T.K. and Cipriano, R.C. (eds) *Fish Viruses and Bacteria: Pathobiology and Protection*, 1st edn. CAB International, Wallingford, UK, pp. 314–333.
UN (United Nations) (2019) *The World Population Prospects 2019 Highlights*. UN Department of Economic and Social Affairs, New York.
Vasquez, I., Hossain, A., Gnanagobal, H., Valderrama, K., Campbell, B., *et al.* (2022) Comparative genomics of typical and atypical *Aeromonas salmonicida* complete genomes revealed new insights into pathogenesis evolution. *Microorganisms* 10(1), 189.
Vazquez-Salgado, L., Olveira, J.G. and Bandin, I. (2022) Nervous necrosis virus viability modulation by water salinity and temperature. *Journal of Fish Diseases* 45, 561–568.
Verrier, E.R., Dorson, M., Mauger, S., Torhy, C., Ciobotaru, C., *et al.* (2013) Resistance to a rhabdovirus (VHSV) in rainbow trout: identification of a major QTL related to innate mechanisms. *PLoS ONE* 8(2), e55302.
Vike, S., Oelckers, K., Duesund, H., Erga, S.R., Gonzalez, J., *et al.* (2014) Infectious salmon anemia (ISA) virus: infectivity in seawater under different physical conditions. *Journal of Aquatic Animal Health* 26 (1), 33–42.
Vincent, A.T. and Charette, S.J. (2022) To be or not to be mesophilic, that is the question for *Aeromonas salmonicida*. *Microorganisms* 10(2), 240.
Whipps, C.M., Gauthier, D.T. and Kent, M.L. (2020) Fish mycobacteriosis. In: Woo, P.T.K., Leong, J. and Buchmann, K. (eds) *Climate Change and Infectious Fish Diseases*. CAB International, Wallingford, UK, pp. 265–279.
Wiik-Nielsen, C.R., Løvol, L.M., Sandlund, N., Faller, R., Wiik-Nielen, J., *et al.* (2012) First detection of piscine reovirus (PRV) in marine fish species. *Diseases of Aquatic Organisms* 97, 255–258.
Woo, P.T.K. (1969) The haematocrit centrifuge technique for the detection of trypanosomes in blood. *Canadian Journal of Zoology* 47, 921–923.
Woo, P.T.K. (1979) *Trypanoplasma salmositica*: experimental infections in rainbow trout, *Salmo gairdneri*. *Experimental Parasitology* 47, 36–48.
Woo, P.T.K. (2006) Diplomonadida (phylum Parabasalia) and Kinetoplastea (phylum Euglenozoa). In: Woo, P.T.K. (ed.) *Fish Diseases and Disorders*. Volume 1: *Protozoan and Metazoan Infections*, 2nd edn. CAB International, Wallingford, UK, pp. 46–115.
Woo, P.T.K. and Ardelli, B.F. (2014) Immunity against selected piscine flagellates. *Developmental and Comparative Immunology* 43, 268–279.

Woo, P.T.K. and Li, S. (1990) *In vitro* attenuation of *Cryptobia salmositica* and its use as a live vaccine against cryptobiosis in *Oncorhynchus mykiss*. *Journal of Parasitology* 76, 752–755.

Woo, P.T.K. and Wehnert, S.D. (1983) Direct transmission of a haemoflagellate, *Cryptobia salmositica* Katz,1951 (Kinetoplastida, Bodonina) between rainbow trout under laboratory conditions. *Journal of Protozoology* 39, 334–337.

Wright, D.W., Oppedal, F. and Dempster, T. (2016) Early-stage sea lice recruits on Atlantic salmon are freshwater sensitive. *Journal of Fish Diseases* 39, 1179–1186.

Wright, D., Nowak, B., Oppedal, F., Bridle, A. and Dempster, I. (2017) Free-living *Neoparamoeba perurans* depth distribution is mostly uniform in salmon cages but reshaped by stratification and potentially extreme fish crowding. *Aquaculture Environment Interactions* 9, 269–279.

Wright, D., Geitung, L., Karsbakk, E., Stien, L.H., Dempster, I., *et al.* (2018) Surface environment modification in Atlantic salmon sea-cages: effects on amoebic gill disease, salmon lice growth and welfare. *Aquaculture Environment Interactions* 10, 255–265.

Xu, F.F., Hultqvist, J.J., Einarsson, E., Astvaldsson, A., Svard, S.G., *et al.* (2014) The genome of *Spironucleus salmonicida* highlights a fish pathogen adapted to fluctuating environments, *PLoS Genetics* 10(2), e1004053.

Yang, Z., Yue, G.H. and Wong, S.-M. (2022) VNN disease and status of breeding for resistance to NNV in aquaculture. *Aquaculture and Fisheries* 7, 147–157.

Young, N.D., Crosbie, P.B.B., Adams, M.B., Nowak, B.F. and Morrison, R.N. (2007) *Neoparamoeba perurans* n. sp., an agent of amoebic gill disease of Atlantic salmon (*Salmo salar*). *International Journal for Parasitology* 37, 1469–1481.

Young, N.D., Dyková, I., Nowak, B.F. and Morrison, R.H. (2008) Development of a diagnostic PCR to detect *Neoparamoeba perurans*, agent of amoebic gill disease. *Journal of Fish Diseases* 31, 285–295.

Yuksel, S.A., Thompson, K.D. and Adams, A. (2006) Rickettsial infections of fish. *Turkish Journal of Fisheries and Aquatic Sciences* 6(1), 63–78.

Zhu, S., Miao, B., Zhang, Y.-Z., Wang, D.-S. and Wang, G.-X. (2021) Amantadine, a promising therapeutic agent against viral encephalopathy and retinopathy. *Journal of Fish Diseases* 45, 451–459.

Zilberg, D. and Munday, B.L. (2000) Pathology of experimental amoebic gill disease in Atlantic salmon, *Salmo salar* L., and the effect of pre-maintenance of fish in sea water on the infection. *Journal of Fish Diseases* 23, 401–407.

Zuo, X. and Woo, P.T.K. (1997a) Natural antiproteases in rainbow trout, *Oncorhynchus mykiss*, and brook charr, *Salvelinus fontinalis*, and the *in vitro* neutralization of fish alpha2-macroglobulin by the metalloprotease from the pathogenic haemoflagellate, *Cryptobia salmositica*. *Parasitology* 114, 375–382.

Zuo, X. and Woo, P.T.K. (1997b) The *in vitro* neutralization of proteases from *Cryptobia salmositica* by the alpha2-macroglobulin in the blood of rainbow trout, *Oncorhynchus mykiss* and brook charr, *Salvelinus fontinalis*. *Diseases of Aquatic Organisms* 29, 67–72.

3 Infectious Diseases of Coldwater Fish in Fresh Water

Kenneth D. Cain[1]* **and Mark P. Polinski**[2]
[1]*University of Idaho, Moscow, Idaho, USA;* [2]*US Department of Agriculture–Agricultural Research Service, Franklin, Maine, USA*

3.1 Introduction

The culture of coldwater fish in freshwater environments is practised globally and primarily involves salmonids. Cage culture of coldwater species in fresh water occurs in many areas but is limited in terms of capacity and revenue when compared with marine net-pen farming. Freshwater cage culture generally occurs in lakes, reservoirs, large rivers or man-made canals (Fig. 3.1A and B). Disease concerns are related to the life stages but would mimic those experienced at other freshwater aquaculture sites that rear fish in earthen ponds, tanks or raceways.

In this chapter, the primary disease problems that affect different life stages of freshwater fish reared in a variety of water sources (ranging from groundwater and springs to rivers, lakes and reservoirs) are highlighted. This includes diseases that are current problems in freshwater cage culture of coldwater species or that could be potential problems in the future. Disease impacts and risks can be assumed to translate directly to cage and pen culture in a freshwater environment.

Diseases discussed in this chapter occur regularly in coldwater species at approximately 15°C or less and manifest primarily in a freshwater environment. In some cases, the pathogen may be contracted in fresh water but becomes problematic and disease occurs once fish are transferred or migrate to seawater. Examples of disease agents that may occur or be transmitted in fresh water but continue to cause problems in seawater are *Aeromonas salmonicida* (causative agent of furunculosis) and *Renibacterium salmoninarum* (causing bacterial kidney disease). In some cases a disease may have few reports in fresh water (e.g. salmonid rickettsial septicaemia) or impact coldwater fish to a lesser extent than warmwater species (e.g. columnaris disease), and these are only briefly discussed in this chapter.

DISEASES CAUSED BY VIRAL PATHOGENS

3.2 Viral haemorrhagic septicaemia

Viral haemorrhagic septicaemia (VHS) is a serious disease in both freshwater and marine fish culture. It is caused by viral haemorrhagic septicaemia virus (VHSV) and is known to affect more than 80 fish species (Smail and Snow,

*kcain@uidaho.edu

 Climate Change on Diseases and Disorders of Finfish in Cage Culture, 3rd Edition (eds P.T.K. Woo and R.P. Subasinghe)

DOI: 10.1079/9781800621640.0003

Fig. 3.1. (A) Freshwater rainbow trout cage-culture operation located on the Columbia River in Washington State, USA. (B) Freshwater cage culture of chinook salmon on man-made glacial-fed canal, New Zealand. (Columbia River image courtesy of J. Bielka.)

2011; Kurath, 2012). The high virulence, wide range of susceptible host species and ability for rapid spread of this disease are of significant concern for many global regulatory agencies and disease prevalence is closely monitored by the World Organisation for Animal Health (Office International des Epizooties) (OIE). VHS manifests in fresh water with more severe pathology and associated mortality than in marine waters, thus making the disease of primary concern in freshwater culture systems.

3.2.1 Impact

A wide range of species are susceptible to VHSV (Skall *et al.*, 2005), with the most serious effects occurring in cultured rainbow trout in Europe. Although accurate estimates of economic impact are not available, VHS has been regarded as a major concern in endemic areas since the 1950s and has caused significant economic losses in many European countries (Olesen and Korsholm, 1997). The most prevalent strain in European aquaculture has been genotyped Ia, which has been the most virulent and pathogenic form of this virus. However, a severe epidemic in the Great Lakes region of the USA caused by the IVb genotype resulted in massive losses throughout the region. Although the majority of North American fish mortality occurred in wild populations, there is a well-founded concern for its potential impact in aquaculture, particularly with regard to cage culture in areas where the virus now appears to be endemic or is likely to spread. In Asia, the disease has primarily been restricted to marine-culture species (Skall *et al.*, 2005), but given the likely marine origins for the current virulent freshwater European and American strains (Einer-Jensen *et al.*, 2004) coupled with growing aquaculture in the region, the possibility for novel VHS identification in fresh water in Asian countries must be considered.

3.2.2 Disease characterization and diagnosis

VHSV is an enveloped, negative-sense, single-stranded RNA virus in the *Rhabdoviridae* family, similar to infectious haematopoietic necrosis virus (IHNV) discussed later in this chapter. Viral replication occurs in the cytoplasm, ultimately producing bullet-shaped virions of approximately 180 nm × 60 nm in size. As with most rhabdoviruses of vertebrates, the viral genome codes for five proteins which include a large matrix protein (M), nucleocapsid protein (N), polymerase-associated phosphoprotein (P), surface glycoprotein (G) and virus polymerase (L). Additionally, the VHSV genome codes for a sixth non-virion protein (Nv) which is unique to a specific set of the aquatic rhabdoviruses known as novirhabdoviruses (Betts and Stone, 2000). Phylogenetic analyses based on *N*, *P*, *G* and *Nv* genes has grouped VHSV into four major genogoups that roughly correlate to their original geographic location: European freshwater and north European marine isolates (genogroup I); marine isolates from the Baltic Sea (genogroup II); isolates of the North Sea (genogroup III); and all North American isolates (genogroup IV) (Skall *et al.*, 2005; He *et al.* 2014). Genogroup IV has further been subdivided by sequence analysis to delineate marine (IVa) from freshwater isolates (IVb) (Elsayed *et al.*, 2006). It is currently believed that all freshwater isolates have originated from marine ancestors, most of which have arisen in the past 50–300 years in multiple adaptive events (Einer-Jensen *et al.*, 2004; He *et al.* 2014). This may help to explain at least partially the juxtaposition between the rather stable relationship and relative low virulence of marine viral isolates with their hosts and the highly pathogenic and lethal effects caused by the recently evolved freshwater strains, as the host–pathogen relationship has had more time to equilibrate in the marine environment. These factors may at least partially be explained in host entry by the virus, as Brudeseth *et al.* (2008) demonstrated gill epithelium of rainbow trout (*Oncorhynchus mykiss*) to be far more susceptible to a freshwater genogroup I virus relative to its marine counterpart, which also helps in explaining the refractivity of rainbow trout to the marine strain of this virus by immersion that would otherwise cause mortality if administered by intraperitoneal injection (Skall *et al.*, 2004).

VHS can manifest as both an acute or a chronic disease in freshwater fish, and both juvenile and adult fish may become infected and present clinical signs. General non-specific symptoms in chronic infections include lethargy, dark coloration and moderate exophthalmia. Fish may also appear anaemic and internal organs may show some haemorrhaging and oedema. Chronic infections apparent in stressful environments may develop into latent carrier infections with no abnormal presentation apart from potential hyperactivity. Acute infections with heavy mortality can approach 100% in susceptible species such as rainbow trout. The characteristic signs of severe infections include haemorrhaging in the ocular tissue, skin and fin bases. In some cases, intermittent spiral swimming may be observed due to infection of brain and nervous tissue (Olesen and Skall, 2009). Internal signs will often include swollen, anaemic kidneys and spleen, as well as pinpoint haemorrhaging throughout the viscera and occasionally in the skeletal muscle.

The 'gold standard' in VHS diagnostics is laboratory culture on a susceptible cell line followed by specific nucleotide or protein identification. Bluegill fry (BF-2) and rainbow trout gonad (RTG-2) cell lines have been shown to be the most sensitive for detection of freshwater isolates (Lorenzen *et al.*, 1999). However, other cell lines such as epithelioma papulosum cyprini (EPC) and chinook salmon embryo (CHSE-214) are also susceptible and have frequently been used for routine diagnostics (Wolf, 1988) as a result of their availability in diagnostic facilities. Following cell culture isolation, confirmation of VHSV protein can be accomplished using enzyme-linked immunosorbent assay (ELISA) (Way and Dixon, 1988) or nucleic acid identity can be confirmed using viral-specific reverse transcription polymerase chain reaction (RT-PCR) (Miller *et al.*, 1998). A quantitative RT-PCR method has also been developed that has shown higher sensitivity for detection of viral RNA in tissue than conventional cell culture techniques and shortens the time required to identify pathogen presence (Hope *et al.*, 2010).

3.2.3 Transmission

VHSV primarily infects new hosts via lateral waterborne transmission. Specifically, virus shed into water via infected urine disseminates to gill epithelium and/or skin of a nearby host (Kim and Faisal, 2011). Although ingestion of infected fish or parasitic transfer via infected leeches are also possible alternative horizontal transfer methods, they have not been confirmed and are likely a secondary means of transmission at best. There is currently no evidence for vertical intra-ovum transfer from mother to offspring of this virus.

3.2.4 Host population dynamics

Host age is a primary factor in defining VHS severity. In general, VHS and associated mortality are greatest in young fish (fry–juvenile) (Kim and Faisal, 2011). If a fish survives an initial acute infection with VHSV it is typically immune to reinfection.

3.2.5 Climate change impacts

The optimal temperature range for VHSV-associated disease is relatively narrow, between 9 and 12°C. Thus, changes in climate conditions resulting in environments above, below or moving into this temperature range may have dramatic consequences on the potential impacts associated with VHSV infection of cultured freshwater fish in endemic areas. Regional temperature changes will also likely geographically shift impacts associated with this disease to reflect maximization of water temperatures within this range.

3.2.6 Control

Prevention or eradication continues to be the most appropriate method in combating this disease in farmed fish. Eradication of infected stocks has been used to effectively eliminate this pathogen from several European countries (Stone *et al.*, 2008; Dale *et al.*, 2009; Olesen and Skall, 2009). The disease rarely manifests above 18°C, and the enveloped nature of the virus results in VHSV having relatively poor environmental stability. Virions have been shown to lose infectivity after approximately 2 weeks at 15°C in fresh water (Hawley and Garver, 2008), although it should also be noted that virus held in purified water at 4°C remained infective for more than a year. Ultraviolet (UV) radiation, hypochlorite and iodophor-based disinfection are all effective at eliminating pathogenic virus from water and equipment (Enzmann, 1983; Yoshimizu *et al.*, 1986). There is currently no commercial vaccine or therapy for the treatment of VHS. Experimental DNA vaccination against VHSV has been demonstrated to be extremely effective (Lorenzen *et al.*, 1998, 2001; Byon *et al.*, 2006); however, their injection-based delivery has not been deemed economically feasible for large-scale fish production at the early juvenile life stage where resistance is most necessary. A potentially more cost-effective oral delivery method for administering a DNA vaccine has been developed (Adelmann *et al.*, 2008); whether this becomes a common vaccination strategy is yet to be seen. Additionally, increasing water temperature to 21°C during a VHSV immersion challenge of Japanese flounder *Paralichthys olivaceus* precluded mortality and provided strong protection in subsequent re-challenge of fish at 15°C (Nishizawa *et al.*, 2011). More studies will be needed; however these data may rekindle the possibility for live vaccination to combat this disease.

3.3 Infectious Haematopoietic Necrosis

Infectious haematopoietic necrosis (IHN) caused by IHNV is the other serious rhabdoviral disease discussed in this chapter. Like VHS, IHN can cause severe mortality in naïve freshwater salmonid populations and has been a major concern for Pacific salmon aquaculture in North America for more than 60 years. Most salmonid species are known to be highly susceptible, and endemic populations are widespread in North America, Europe and Asia (Bootland and Leong, 1999). Large losses of revenue and fish have occurred in salmonid aquaculture as a result of this virus (Congleton, 1988), and IHNV is of specific concern to the OIE due to its high pathogenicity and possibility for transmission within global salmon fisheries.

3.3.1 Impact

IHN has been a problem to the freshwater culture of salmonid fish in North America since the 1950s. By the 1980s the disease had spread to Europe and Asia through the transportation of fish and eggs, and it continues to impact salmon aquaculture in both regions (Bootland and Leong, 2011). The disease has been known to cause high mortality in naïve fish, and spinal deformities can occur among surviving fish potentially limiting growth and subsequent value.

3.3.2 Disease characterization and diagnosis

Natural epizootics have historically been observed only in salmonids. Nevertheless, non-salmonids have been infected in experimental studies and some wild species are known to harbour viable pathogen with no clinical disease (Castric and Jeffroy, 1991; LaPatra *et al.*, 1995). Like VHSV, the virus genome consists of a single-stranded, negative-sense RNA with L, G, N, P, M and Nv coding regions as previously described for VHSV. It is believed that IHNV originated in North America, and phylogenetic analysis based on the G-coding nucleic acid sequence has led to the description of three distinct genogroups which correspond roughly to original endemic geographic distribution: the upper (U) genogroup found in the Columbia River basin of Washington state extending north through Canada to Alaska; the middle (M) genogroup found in the north-western US Columbia River basin inland to the Snake River of Idaho; and the lower (L) genogroup found in northern California and southern Oregon (Kurath *et al.*, 2003). Subsequent spread of the virus to Europe in the 1980s has led to the evolution of a distinct European genotype, shown to be originally derived from the North American M genogroup (Enzmann *et al.*, 2005). Similarly in Japan, a JRt genotype has been identified with ancestral ties to the North American U genogroup introduced in the 1970s (Nishizawa *et al.*, 2006).

IHNV targets the endothelial cells of blood capillaries and haematopoietic tissues for replication. As a result, clinical signs typically include oedema and haemorrhaging in the visceral organs, particularly the kidney, spleen and liver. General anaemia may be observable by pale gills and a low haematocrit. Non-specific signs of acute infection are similar to VHSV and include lethargy interspersed with increased activity, darkening of the skin, pale gills, ascites, distended abdomen, exophthalmia, and in some cases external petechial haemorrhaging. Fish will go off feed, and a trailing faecal cast may become evident.

Traditional detection of IHNV is based on virus isolation in cell culture. Both EPC and CHSE cell lines are known to be susceptible and are typically used for screening of this virus. Confirmatory identification may be achieved by use of immunological-based methods such as plaque neutralization (Jorgensen *et al.*, 1991), an indirect fluorescent antibody test (FAT) (LaPatra *et al.*, 1989a; Arnzen *et al.*, 1991) and ELISA (Dixon and Hill, 1984). Molecular methods such as PCR (Arakawa *et al.*, 1990; Purcell *et al.*, 2006) or use of a DNA probe (Deering *et al.*, 1991) can also be used for confirmatory testing following cell culture, and alternatively applied to directly analysing tissue. Further, a reverse transcription loop-mediated isothermal amplification (RT-LAMP) protocol has been developed for detection of IHNV that has been shown to be more sensitive for identifying viral RNA than RT or nested PCR (Gunimaladevi *et al.*, 2005). This detection method may also provide a useful 'on-site' screening and monitoring tool for cage-culture facilities, as reaction conditions require only a single-temperature heat block and UV light source to amplify and identify targeted nucleic acid sequences (Notomi *et al.*, 2000).

3.3.3 Transmission

IHNV appears to enter susceptible fish through the fin bases, gill or oral/gastrointestinal tract (Drolet *et al.*, 1994; Harmache *et al.*, 2006). It then disseminates through the circulatory system to virtually all bodily organs with the highest loads developing in the anterior kidney and spleen (Drolet *et al.*, 1994). It is then shed into the environment in high quantities during primary amplification, most likely via external mucus (LaPatra *et al.*, 1989b; Wargo *et al.*, 2017).

3.3.4 Host population dynamics

All salmonids are believed to be susceptible to the virus; however, the ability of IHNV to cause disease among different species is often genotype specific (LaPatra *et al.*, 1990a,b, 1993; Garver *et al.*, 2006). For example, LaPatra *et al.* (1990b) showed an isolate from the M genogroup to incur 64% cumulative mortality in rainbow trout following immersion challenge; yet isolates from either the U or L genogroup only produced 4 and 6% mortality, respectively, under similar conditions. In general, isolates of the U genogroup have the highest virulence in sockeye salmon (*Oncorhynchus nerka*), M isolates in rainbow trout and L isolates in chinook salmon (*Oncorhynchus tshawytscha*) (LaPatra, 1998; Garver *et al.*, 2006). However, IHNV susceptibility is also known to vary between stocks of fish within a species (Garver *et al.*, 2006).

Adult fish are susceptible and can transmit IHNV; however, clinical IHN rarely manifests in adult populations. Juvenile fish (less than 20 g) often less than 2 months of age appear to be the most susceptible and experience the highest rates of clinical disease and mortality (Dixon *et al.*, 2016). Fish density is also a significant factor in assessing IHNV effects in a population, with high density as seen in many culture systems allowing for higher viral densities to accrue and incur disease compared with systems with low fish density (Ogut and Reno, 2004). Following acute infection with the disease, survivors are generally immune to reinfection (Lorenzen and LaPatra, 1999).

3.3.5 Climate change impacts

IHN manifestations usually occur at water temperatures of 8–14°C with optimal viral conditions being between 10 and 12°C (Dixon *et al.*, 2016). Thus, changes in climate conditions resulting in environmental temperatures above, below or moving into this temperature range may have dramatic consequences on the potential impacts associated with IHN. Regional temperature changes will also likely geographically shift impacts associated with this disease.

3.3.6 Control

Avoidance is currently the best control method for this disease. Iodophor disinfection of eggs is highly effective in stopping egg-associated transmission, and other typical disinfectants including hypochlorite and UV radiation are effective for sanitizing equipment (Winton, 1991). Both whole-inactivated cell and DNA vaccines for IHN have been shown to be efficacious in preventing disease (Lorenzen and LaPatra, 2005) and are currently licensed for commercial use in Atlantic salmon net-pen aquaculture on the west coast of North America (Kurath, 2008). In a freshwater setting however, the injection-based vaccines can be costly and time-intensive to administer to small juvenile fish at typical commercial production numbers. Selective breeding has shown promise in producing resistance to disease in rainbow trout (Purcell *et al.*, 2010) and is a potential method for control in areas where IHNV is known to occur.

3.4 Infectious Pancreatic Necrosis

Infectious pancreatic necrosis virus (IPNV) is a member of the ubiquitous birnavirus family which causes pancreatic necrosis in salmonid fish worldwide (Reno, 1999). The virus infects a wide range of host species; however, the major epizootics in aquaculture revolve around salmonids. Once considered a freshwater disease, in recent decades IPN has also become a significant disease in the saltwater culture of salmonids (see Leong *et al.*, Chapter 2, this volume, 2023). The focus in the present chapter is on the freshwater aspect of the disease.

3.4.1 Impact

Rainbow trout and brook trout are most susceptible, although all salmonid species can be infected. Typically acute infection becomes apparent within a week following exposure and the disease outbreak is concluded within another week (Reno *et al.*, 1978), during which time severe mortality occurs. The most severe outbreaks occur at temperatures between 10 and 14°C. Due to its environmental stability and ability to chronically persist in adult fish, eradication is extremely difficult once an outbreak has occurred.

3.4.2 Disease characterization and diagnosis

IPNV has a single-shelled, non-enveloped, icosahedral virion of approximately 60 nm diameter. The genome consists of a double-stranded RNA with two segments – one coding for the RNA-directed RNA polymerase (VP1), the other coding for structural associated proteins (VP2–VP4) as well as an anti-apoptosis protein (VP5). The structural protein VP2 has been shown to be highly immunogenic (Heppell *et al.*,1995), the diversity and replication efficiency of which have been directly correlated to virulence and the ability to cause disease (Coulibaly *et al.*, 2010; Skjesol *et al.*, 2011). Historically, isolates causing IPN have been classified into two major serogroups, A and B, containing ten distinct serotypes (nine from A and one from B) (Hill and Way, 1995). Phylogenetic analysis has alternatively categorized isolates into seven genogroups (Blake *et al.*, 2001; Nishizawa *et al.*, 2005), which roughly correspond to the previously described serotypes and original geographic distributions with some modification. Isolates from both A and B serogroups (all seven genogroups) infect fish that live in or migrate to freshwater environments.

Acute infection of IPN in young fish is rapid. Typically within a week of becoming infected fish go off feed and become darker in colour. Often fish lose equilibrium and swim with abnormal body rotation. Trailing faecal casts are common, and petechial haemorrhaging on the ventral surface and exophthalmia are also common non-specific signs. A pale and anaemic liver is a hallmark of this disease, although general anaemia throughout the viscera is also common.

Primary isolation from tissues has typically relied on cell culture, and many teleost cell lines have been shown to be susceptible (Reno, 1999), including BF-2, EPC and CHSE-214. Confirmation or direct testing of tissue can be conducted by a FAT (Swanson and Gillespie, 1981), real-time RT-PCR (Orpetveit *et al.*, 2010) or RT-LAMP (Soliman *et al.*, 2009). Molecular assays can also be used directly on fish tissue to screen for viral presence.

3.4.3 Transmission

IPNV is extremely environmentally stable and is known to remain infective for months in aqueous solutions (Smail *et al.*, 1993), thus making waterborne horizontal transmission extremely effective. Infected fish shed IPNV through faeces into the environment (Bootland *et al.*, 1986) which then enters naïve hosts through epithelial surfaces (skin, gill or intestine). The virus has also been demonstrated to be transmitted vertically from infected adult brook trout to their progeny (Bootland *et al.*, 1991), although this method of transmission appears to be of limited efficacy. Survivors of acute infections, although asymptomatic in presentation, are often chronic carriers and shedders of the virus.

3.4.4 Host population dynamics

Outbreaks of IPN in fresh water are typically observed in young fish less than 6 months of age (LaPatra *et al.*, 2000), although persistent chronic low levels of infection have been observed in fish during all life stages. Susceptibility appears to be at least partially contingent on host genotype (Guy *et al.*, 2009).

3.4.5 Climate change impacts

IPNV can tolerate a wide range of temperatures, salinities and pH manipulations. Although disease manifestations typically occur within a rather narrow temperature range (10–14°C), this may be more to do with the limited temperature tolerance of host juvenile salmonids than with the virus. As such, although climate change may have direct implications on salmonid cage culture, its effects on IPNV are likely to be secondary.

3.4.6 Control

Strict biosecurity measures will aid in avoidance of the pathogen. This should include measures to prevent both horizontal and vertical transmission such as egg disinfection, screening and separation of new stocks, and disinfection of all potentially contaminated equipment. The virus is environmentally stable and is also considered one of the most resistant viruses to disinfection (Munro and Midtlyng, 2011). For example, the UV-C dose required to inactivate IPNV (>100 mW s/cm^2) is 50 times higher than what is required to inactivate aquatic rhabdoviruses such as VHSV, IHNV and spring viraemia of carp virus (SVCV) (Skall and Olesen, 2011). The development of at least some resistance to IPNV in rainbow trout has been shown by selective breeding (Guy *et al.*, 2009), although the process requires multiple generations. Injection-based commercial vaccines are available against IPNV and have successfully been applied in various aspects of aquaculture for both adult and juvenile fish (Ramstad *et al.*, 2007). Unfortunately, their use as a preventive treatment for salmonid fry typically infected in fresh water is virtually impossible due to the fry's small size and fragility. de las Heras *et al.* (2010) showed promising results from oral administration of a DNA vaccine during early feeding of young (1 g) rainbow trout by encapsulating viral DNA within an alginate particle to allow better uptake through the intestine without exposing the antigen to digestive degradation. These findings, coupled with the identification and immunogenic role of the crystalline capsid structure of IPNV (Coulibaly *et al.*, 2010), may lead to more effective vaccination strategies against this virus.

3.5 Sleeping Disease

Salmonid alphaviruses (SAVs) are a serious danger to Atlantic salmon and rainbow trout culture in Europe and cause significant economic loss to the European aquaculture industry. Two diseases are attributed to SAVs: pancreas disease (PD) in Atlantic salmon and sleeping disease (SD) in rainbow trout. The occurrence of PD in salmon is associated with the marine environment and is covered in Chapter 2 (this volume). SD is an infection of farmed rainbow trout in fresh water and is the focus of the present chapter. Nevertheless, both PD and SD are caused by closely related alphavirus subtypes and there are many overlaps in aetiology, detection and control for these two diseases.

3.5.1 Impact

SD was first described in 1994 in cultured rainbow trout in France (Boucher and Baudin Laurencin, 1994), but has since spread to many areas in Europe (McLoughlin and Graham, 2007). Atlantic salmon, rainbow trout and brown trout (*Salmo trutta*) are susceptible (Boucher *et al.*, 1995), although rainbow trout are most affected by this disease in fresh water. The disease can cause variable mortality, but losses of up to 43% have been recorded in cage culture of rainbow trout in the UK (Graham *et al.*, 2007).

3.5.2 Disease characterization and diagnosis

SD is caused by salmon alphavirus-2 (SAV-2), a single-stranded, positive-sense RNA virus with a spherical enveloped capsid of approximately 65 nm diameter. There are two other subgroups of SAV with relevant disease association – SAV-1 and SAV-3 – of which both cause disease in marine Atlantic salmon (see Section 2.6 and Chapter 2, this volume) (McLoughlin and Graham, 2007). Amino acid identity for both structural and non-structural proteins between these subgroups is over 93% (Weston *et al.*, 2002), suggesting high similarity between both freshwater and marine isolates regarding replication and host interactions. For terrestrial alphaviruses, an arthropod vector (such as a mosquito) is involved in the transmission of disease. Currently no such vector has been definitively identified for SAV. Sea lice have been suggested as a possible vector for the marine SAVs (Petterson *et al.*, 2009); however, horizontal fish-to-fish transmission has been demonstrated (Boucher *et al.*, 1995), thus precluding the necessity of a vector.

Both SD and PD cause sequential necrosis in the pancreatic tissue, cardiac muscle and skeletal muscle. Characteristic ‘sleeping’ behaviour in rainbow trout experiencing a severe disease outbreak is lying motionless on their sides along the bottom of a holding area as a consequence of damage to skeletal red muscle fibres (Castric *et al.*, 1997). Other non-specific signs may include exophthalmia, bloated abdomen, faecal casting and lack of appetite. Internally, lesions in the cardiac and skeletal muscles may be visible. In cage culture, clinical signs and mortalities have been observed 8 to 15 weeks following infection which resolved after an additional 10–20 days (Graham *et al.*, 2007).

Following preliminary diagnosis based on clinical signs and histopathology, a variety of confirmatory techniques are available including virus isolation, serology and RT-PCR, which are reviewed elsewhere (McLoughlin and Graham, 2007). Specifically, real-time PCR protocols using SYBR Green chemistry (Graham *et al.*, 2006) or TaqMan probes (Christie *et al.*, 2007) have been shown to be sensitive and specific for identifying SAV by subtype.

3.5.3 Transmission

The main mode of SAV transmission is considered to be horizontal though water as demonstrated by cohabitation experiments and epizootic modelling (Jansen *et al.*, 2017). Infectious SAV particles have been identified in mucus and faeces of PD-affected Atlantic salmon (Graham *et al.*, 2012), indicating that SAV is shed through faeces and possibly skin mucus. Similarly the virus may also enter/re-enter hosts via skin mucus. SAV does not appear to be transmitted vertically intra-ovum from mother to offspring (Kongtorp *et al.*, 2010).

3.5.4 Host population dynamics

At least some heritable genetic resistance to SAV infections has been identified in Atlantic salmon (Gonen *et al.*, 2015) in both fresh and salt water and it is likely that a similar case is true for freshwater rainbow trout. Moreover, survivors of SAV infection generally show immunity to further disease and vaccines have shown variable rates of successful protection, indicating that both innate and adaptive immune responses can be effectively employed to combat this disease.

3.5.5 Climate change impacts

Temperature is known to play a significant role in the infectivity of this virus (Metz *et al.*, 2011) and warm (>15°C) water temperatures may aid in minimizing the severity of SD and fish mortality.

3.5.6 Control

There are commercial inactivated virus vaccines administered via injection that are efficacious in minimizing effects of SAV and new vaccine development is ongoing (Karlsen *et al.* 2012; Veenstra *et al.*, 2020). Clinical disease is often associated with stress conditions, and therefore in areas where the virus is endemic, good management and hygienic practices can further aid in minimizing the impacts of disease.

3.6 Salmonid Herpes Virus Infections

Four viruses from the *Alloherpesviridae* family have been identified to cause disease in cultured salmonid fish: herpesvirus salmonis (HPV or SalHV-1); *Oncorhynchus masou* virus (OMV or SalHV-2); epizootic epitheliotropic disease (EEDV or SalHV-3); and Atlantic salmon papillomatosis virus (ASPV or SalHV-4). A fifth alloherpesvirus has also been identified in lake trout (*Salvelinus namaycush*) from the Great Lakes region of North America known as Namaycush herpesvirus (NamHV or SalHV-5) but with unknown disease association (Glenney *et al.*, 2016a). These viruses appear to have limited host and geographic distributions and generally are latent in nature without overt disease manifestations. Nevertheless, significant disease outbreaks can and do occur in salmonid fish culture, specifically in association with SalHV-2 and -3 which are endemic in Japan and the Great Lakes region, respectively.

3.6.1 Impact

SalHV-1 was originally isolated from rainbow trout from Washington state in 1978 (Wolf *et al.*, 1978) and again in 1986 in northern California from anadromous populations of the same species (Hedrick *et al.*, 1986). This genotype has been shown to cause cytopathic effects in cell culture (Wolf *et al.*, 1978); however, mortality events in susceptible cultured species have never been directly attributed to this virus. SalHV-2 has been documented in Japanese aquaculture since the late 1970s (Kimura *et al.*, 1981), but somewhat surprisingly has not spread to any other global region. The disease particularly affects young fish in freshwater environments, and although recent outbreaks have been limited, severe mortality has previously been attributed to this disease where mortality events have reached in excess of 80% of cultured stocks (Furihata *et al.*, 2003). Multiple species including rainbow trout, and masou, sockeye, coho and chum salmon are known to be susceptible to SalHV-2, which has been observed in both wild and cage-cultured fish in Japan (Furihata *et al.*, 2003). The endemic nature of this virus in Japan and the known infectivity of ocean-run populations suggest that this virus is also present in other coastal areas of Asia where natural runs of Pacific salmon occur, but it is relatively unconfirmed. SalHV-3 caused severe mortality in stocks of lake trout in the Great Lakes region in the late 1980s resulting in over 15 million mortalities of cultured fish in a five-year span (Bradley *et al.*, 1989). Disease outbreaks were limited through the early 2000s; however, SalHV-1 was confirmed to be persisting in spawning adult lake trout in 2009 (Kurobe *et al.*, 2009) and a disease resurgence was observed in lake trout from Lake Michigan from 2012 to at least 2017 (Faisal *et al.*, 2019). SalHV-4 is associated with a relatively benign freshwater skin condition in Atlantic salmon of Russia's Kola Peninsula, Scandinavia and Scotland (Doszpoly *et al.*, 2013).

3.6.2 Disease characterization and diagnosis

Salmon herpes viruses are enveloped, double-stranded DNA viruses of approximately 200 nm diameter. For all five genotypes, infectivity is optimal around 10°C. Taxonomic and genetic classification of herpes-like viruses has been somewhat unsettled in the past decade. The discovery of herpes viruses in fish and molluscs that are distantly related to those in birds and mammals required a reordering of herpes viruses and the formation of two new families – the *Alloherpesviridae* family which incorporates bony fish and amphibian viruses, and the *Malacoherpesviridae* family which contains those of molluscs (Davison, 2010). The *Alloherpesviridae* has been further phylogenetically subdivided into three clades, with the five salmon herpes viruses branching together (approximately 80% amino acid similarity of viral DNA polymerase between genotypes) (Waltzek *et al.*, 2009; Boutier *et al.*, 2021). Although genetically similar, each SalHV genotype has rather stringent host specificity. SalHV-1 has only been known to infect Pacific-run rainbow trout (Wolf *et al.*, 1978; Hedrick *et al.*, 1986) and SalHV-3 has exclusively caused disease in lake trout (*S. namaycush*) of the Great Lakes region, USA (Bradley *et al.*, 1989). SalHV-2 has somewhat wider prevalence and distribution but has historically caused disease only in Pacific (*Oncorhynchus* spp.) salmon in Japan (Yoshimizu *et al.*, 1995).

Infection with SalHV-1 has only been identified in adults with no clinical signs. Both SalHV-2 and SalHV-3 are known to cause acute disease and mortality in young (<1 year old) fish. Onset of acute disease is rapid, with mortality occurring within 1 to 2 weeks following infection and is associated with systemic haemorrhaging. Pathology may be observed in haematopoietic tissues and non-specific signs may include lethargy, dark body colour, and abnormal corkscrew or hyperactive swimming. The most definitive macroscopic signs include oncogenic and ulcerative conditions of epithelial tissues of the mouth and jaw as the infection becomes chronic. Lesions may also develop on the skin and fin bases, although to a lesser extent than on buccal-associated tissue. These ulcerative conditions have been known to persist for up to a year following acute infection. Gross internal signs during latent infection are minimal.

Historically, SalHV-1 was identified by cell culture isolation using RTG-2 or CHSE-214 cell lines followed by microscopy or immunological identification (Wolf *et al.*, 1978; Kimura *et al.*, 1981; Hedrick *et al.*, 1986). Unfortunately, inoculation of infected fish tissues onto established cell lines has not always resulted in the isolation of virus (McAllister, 1993), making false negatives a concern. Currently, viral DNA can be specifically detected using PCR-based techniques for both SalHV-2 (Aso *et al.*, 2001) and SalHV-3 (Kurobe *et al.*, 2009). Multiple quantitative PCR (qPCR) assays have also been developed to differentially identify SalHV-3, -4, and -5 (Glenney *et al.*, 2016b).

3.6.3 Transmission

Salmonid herpes viruses are vertically transmitted as evidenced by viral DNA detection in ovarian fluid (Kurobe *et al.*, 2009). Virus can also be transmitted horizontally, where other more well-studied alloherpesviruses have demonstrated skin and to a lesser extent gill or digestive track as portals of entry and suggest skin-to-skin or waterborne dissemination as likely mechanisms for transmission (Boutier *et al.*, 2015).

3.6.4 Host population dynamics

As discussed above, SalHVs appear to have rather stringent host species requirements as well as geographical and temperature limitations. In endemic areas, latent infections can nevertheless be common (Glenney *et al.*, 2016b) and developed immunity is unlikely. Genotypic or phenotypic resistance may also be attainable (Yoshimizu, 2009).

3.6.5 Climate change impacts

Disease manifestations associated with SalHVs are highly constricted to temperatures from 9 to 12°C. Thus, environmental conditions trending above, below or into this range will have a significant impact on disease presentation. As with many herpesviruses, host stress appears to be associated with disease presentation and may be another factor to consider as a result of changing climates.

3.6.6 Control

SalHV-2 is known to be inactivated by UV and iodophor treatment (Hisae *et al.*, 2002). A formalin-killed SalHV-2 vaccine has been used successfully to reduce viral titres in ovarian fluid of infected fish (Yoshimizu, 2009), and selective breeding has also been successful in eliminating severe disease outbreaks in rainbow trout following four to five generations. An antiviral agent, acyclovir, was also shown to be effective in inhibiting replication of SalHV-2 *in vitro* as well as *in vivo* during experimental infections of chum salmon fry (Kimura *et al.*, 1983a,b).

DISEASES CAUSED BY BACTERIAL PATHOGENS

3.7 Furunculosis

Furunculosis is one of the oldest known fish diseases and is caused by the Gram-negative bacterium, *A. salmonicida*. It has been studied extensively and its name is derived from its clinical sign – large boils (furuncles) under the skin of infected fish. It has a wide distribution and has been detected in fish from many different countries worldwide. Early reports were from cultured and wild fish in Europe (Emmerich and Weibel, 1894). The host range of *A. salmonicida* is extensive and it occurs in both freshwater and marine environments, affecting fish (salmonid and non-salmonid) of all ages (Bernoth, 1997; Wiklund and Dalsgaard, 1998). It is infective to almost all fish species and many may serve as carriers, making eradication of the disease difficult (Herman, 1968). Furunculosis causes significant problems in salmonids. Brook trout (*Salvelinus fontinalis*), Atlantic salmon and brown trout are particularly susceptible. Rainbow trout are less affected and somewhat resistant (McCarthy, 1977). For additional background on furunculosis, there are a number of reviews available (Bernoth, 1977; Wiklund and Dalsgaard, 1998; Hiney and Olivier, 1999).

3.7.1 Impact

A. salmonicida subsp. *salmonicida* is widespread and furunculosis can occur in all life stages of salmonids. It is thought that most disease outbreaks are a result of movement of infected stocks or introduction from wild carrier fish. *A. salmonicida* can affect non-salmonid stocks; however, the focus here is primarily on the 'typical' strains and their impacts on salmonid species. Serious losses due to furunculosis have been reported in farmed and wild fish stocks (Roberts, 2012), and this disease causes major impacts to both commercial salmonid aquaculture and public resource hatcheries aimed at stocking of public waters. Originally, furunculosis occurred almost exclusively in fresh water, but intensive culture of species such as Atlantic salmon resulted in severe outbreaks at the smolt stage when these fish were moved to seawater. Roberts (2012) reported a more acute and highly contagious form of the disease which came about due to increasing intensive salmon culture. This was a particular problem in Atlantic salmon smolts in their first year at sea. Although many life stages are susceptible, Munro and Hastings (1993) suggested that young fry are less commonly infected.

3.7.2 Disease characterization and diagnosis

A. salmonicida is readily isolated from internal organs and most isolates produce a distinct brown, diffusing, melanin-like, water-soluble pigment when cultured on agar plates containing appropriate media. Early literature referred to the bacterium causing furunculosis as *Bacterium* or *Bacillus salmonicida* (McCraw, 1952), but later this was named *Aeromonas salmonicida* (Griffin *et al.*, 1953). It is generally accepted that there are four subspecies of *A. salmonicida* in fish. The subspecies *salmonicida*, which is considered the 'typical' strain that causes furunculosis, is focused upon in this section. It is characterized as a non-motile, non-sporulating, fermentative, Gram-negative aerobic bacillus which reduces nitrate, liquefies gelatin, hydrolyses starch and produces cytochrome oxidase. Other subspecies are *masoucida*, *achromogenes* and *smithia*, and these are often referred to as 'atypical' stains (Plumb and Hanson, 2011). The taxonomy of *A. salmonicida* is not always clear and discrepancies exist in the literature regarding subspecies classification. For example, another subspecies, *nova*, is included in a classification scheme summarized by Munro and Hastings (1993). This classification placed subspecies into three groups:

- Group 1 – includes *A. salmonicida* subsp. *salmonicida* as the 'typical' strain derived from salmonid fishes;
- Group 2 – includes *A. salmonicida* subsp. *achromogenes* as an 'atypical' strain from salmonids that shows variation in some biochemical properties and includes former subspecies *masoucida*; and
- Group 3 – includes *A. salmonicida* subsp. *nova* as an 'atypical' strain that is associated with disease in non-salmonid fishes.

This classification has not been widely accepted and there is a lack of reliable traits for subspecies discrimination. According to Wiklund and Dalsgaard (1998), further studies have to be based on larger numbers of strains and use techniques such as polynucleotide sequencing and DNA–DNA or RNA–DNA hybridization.

The classic boil-like furuncle is observed in some fish, but often fish die from an acute infection without any obvious clinical features. Furuncles may be present on the sides or dorsum of fish, appear as raised or haemorrhagic areas (Fig. 3.2), and can ulcerate and release necrotic cells and tissue along with bacteria. This is thought to increase the risk of horizontal transmission by increasing bacteria in the water column. Depending on the nature of the infection, fish can become dark, lethargic and have petechial haemorrhaging at the base of fins (Munro and Hastings, 1993). If the disease occurs in very young fish, high mortality may be observed, but otherwise limited signs may be apparent beyond some dark fish, anorexia and congregation near outlets.

Internal signs associated with furunculosis can include ascites and an empty intestine filled with mucus and cellular debris. Haematocrit values are often very low and blood vessels around the pyloric caeca and intestine can become inflamed. Toxins are often released by *A. salmonicida* and will cause liquefaction of tissues and severe inflammation.

There are a number of effective ways to diagnose furunculosis through culture of *A. salmonicida* or identification by histological sections. Typically a definitive diagnosis would be based on a combination of clinical signs and associated isolation of *A. salmonicida* from affected fish organs/tissues. The bacterium is most often cultured on tryptic soy agar (TSA) but will grow on brain heart infusion agar (BHIA). Within 48 h of culture at 22–25°C, typical small, raised circular colonies appear that are non-motile, oxidase-positive and fermentative (Roberts, 2012). The brown pigment associated with the typical strain will usually develop within 10 days, but atypical strains generally lack this pigment. Colonies can easily be collected on a loop and bacteria examined on a slide following a variety of staining procedures. The appearance of short Gram-negative rods measuring 0.8–1.3-2.0 µm that often occur in pairs, chains or clumps is characteristic (Roberts, 2012). Confirmatory diagnosis of *A. salmonicida* can be accomplished rapidly by serological assays such as FAT or a range of other immunological tests (Thoesen, 1994). Molecular methods based on PCR have been developed and could be used to confirm

Fig. 3.2. Rainbow trout showing furuncle-type lesions under the skin attributed to infection with *Aeromonas salmonicida*.

isolates or for identification of carrier fish in a population (Hiney *et al.*, 1992; Oakey *et al.*, 1998).

3.7.3 Transmission

The primary mode of transmission is horizontal, and evidence suggests that ingestion of bacteria and subsequent transfer across the intestinal wall is the most common route of infection (Jutfelt *et al.*, 2006). The organism can be shed in reproductive fluids, but typical egg disinfection procedures should eliminate vertical transmission risks due to egg surface-associated bacteria.

3.7.4 Host population dynamics

Based on the ability of fish to be carriers and the ease of horizontal transmission, spread between wild and cage-cultured species in the natural environment can occur. The typical strain described here caused large impacts on salmonids, with 'atypical' strains occurring in a wide range of non-salmonid species.

3.7.5 Climate change impacts

As temperatures increase in regions where fish species are considered susceptible to *A. salmonicida*, it is likely that furunculosis will become more widespread. Outbreaks occur often at temperatures above 10°C and this bacterium grows well at temperatures above 20°C. The stress of warmer water temperatures on the host combined with pathogen growth characteristics could result in further widespread establishment of *A. salmonicida* subsp. *salmonicida*.

3.7.6 Control

Furunculosis is often controlled by either treatment of infected fish using feed-delivered antibiotics or prevention through vaccination. Proper fish culture methods and strict attention to biosecurity measures are also essential. Fish infected with *A. salmonicida* may pose a risk to other cultured or wild fish stocks and movement of such fish may be restricted through various regulatory authorities. The main method of controlling furunculosis is to prevent or eliminate *A. salmonicida* from water sources. This is much more feasible in facilities where spring or well water sources are available, but it becomes more difficult if wild fish inhabit water supplies. Most cage-culture operations would be at risk. If exposure of fish is anticipated, the best approach to controlling disease is through the implementation of a vaccination programme.

Some of the very first reports on the development of fish vaccines were against furunculosis (Duff, 1942). Today, there are a range of vaccines commercially available for this disease and the use of oil-adjuvanted vaccines in the Atlantic salmon industry is now almost universal. This, combined with improved husbandry and a variety of other precautions, has dramatically reduced problems and limited the need for antibiotic treatments. If fish become infected, losses can be minimized by removing fish showing clinical signs such as furuncles, by improvement of environmental conditions to reduce stress and/or by treatment with antibiotics. Treatment should be considered a last resort as many *A. salmonicida* isolates have developed resistance to certain antibiotics including oxytetracycline, oxolinic acid, trimethoprimsulfadiazine and amoxicillin (Richards *et al.*, 1992). Antibiotic sensitivity testing should be completed and is essential prior to treatment, especially due to the potential of resistant bacterial strains. Oxytetracycline has commonly been applied and is one of the most used antibiotics in aquaculture (Austin and Austin, 1993). However, in many regions, oxytetracycline is no longer routinely used for treatment of furunculosis due to the development of resistant strains of *A. salmonicida* (Smith *et al.*, 1994). Another antibiotic known as Romet™, which contains Ormeto-primsulfadimethoxine, was the antibiotic of choice for furunculosis in the past. However, florfenicol (Aquaflor®) is most often prescribed for food fish in the USA under a veterinary feed directive (VFD).

Other potential control options include the use of various immunostimulants which usually contain some form of β-1,3-glucan from yeast or bacteria added to the feed. Such immunostimulants have also been tested as adjuvants within vaccines and in some cases shown to enhance antibody formation and protection (Midtlyng *et al.*, 1996). Another prevention strategy includes immunization of broodstock in a way to passively transfer immunity to salmonid fry (Kawahara *et al.*, 1993), but the practical application and benefit of such methods are not known. Recent efforts have increased to find alternative approaches that would aid in disease control and promote overall health benefits in fish. This has included the use of naturally occurring bacteria from the gut of fish (probiotics) that exhibit antagonistic activity to selected fish pathogens. Such probiotics have shown promise for control of furunculosis (Irianto and Austin, 2002).

3.8 Motile Aeromonad Septicaemia

Motile aeromonad septicaemia (MAS) is often associated with the *Aeromonas hydrophila* but can be caused by other ubiquitous species including *Aeromonas caviae* and *Aeromonas sobria*. *A. hydrophila* is a common environmental bacterium that is found in clean and polluted freshwater systems as well as some marine systems (Roberts, 2012). It is considered the cause of many severe disease outbreaks in wild, freshwater and pond-cultured fish. Infections and associated disease result from the opportunistic nature of this bacterium and are exacerbated by stress and water temperature. Both coldwater and warmwater fish species are susceptible and if conditions are favourable a bacterial haemorrhagic septicaemia can occur.

3.8.1 Impact

Most fish in fresh water are susceptible to *A. hydrophila* including tilapia (*Oreochromis* spp.), brown trout, striped bass (*Morone saxatilis*) and carp (Fijan, 1972; Thune *et al.*, 1982). When fish are under stress, *A. hydrophila* and other motile aeromonads can cause MAS which may also be referred to as red sore or red pest disease. Tail and fin rot may be common and, if stressful conditions persist, be followed by a rapid rise in mortality in susceptible fish (Fijan, 1972). Since *A. hydrophila* is a common ubiquitous bacterium found naturally in soil and freshwater environments, it is considered opportunistic and outbreaks can occur seasonally when temperatures increase. It is widespread and found in Europe, the Americas and Asia where acute losses and greater than 80% mortality have been reported between 20 and 22°C (Brown and Bruno, 2002). Handling, transfer, crowding, low dissolved oxygen and nutritional status are all stressors that can affect susceptibility of fish species to *A. hydrophila*. Internally, it appears that *A. hydrophila* can multiply in the epithelium of the intestine resulting in heavy shedding of the virus in the faeces (Brown and Bruno, 2002).

3.8.2 Disease characterization and diagnosis

A. hydrophila is a short, motile, Gram-negative rod (0.4–1.0 μm in length). It produces circular, convex, pale white to cream-coloured colonies on TSA at 15°C (Brown and Bruno, 2002) within 24 h at 22–28°C (Roberts, 2012). It may be cultured at warmer temperatures and Roberts (2012) suggested isolation on selective Rimler–Shotts (RS) agar media containing novobiocin, which has been found useful for putative identification from potentially contaminated material. It can be identified based on biochemical profiles (Newman, 1982; Popoff, 1984), is isolated from organs such as the kidney, and may be found in the blood of infected fish (Roberts, 2012). It is aerobic and oxidase-, catalase- and aesculin-positive.

The most apparent clinical sign of MAS is the development of a widespread haemorrhaging that is irregular in shape and may be present from the gills to the vent. Fish will often be dark and lethargic and exhibit tail rot, haemorrhaging, or ulcerated shallow necrotic lesions on the skin. Since toxins and other extracellular products are linked to virulence, internal organs can be congested, and haemorrhage is often observed on the viscera. Internally, clear to blood-stained ascites fluid can be found along with anaemic conditions and a swollen kidney or spleen. This bacterium was previously referred to as *Aeromonas liquefaciens* most likely due to the leakage of fluid from affected internal organs such as the kidney or spleen following incision (Roberts, 2012).

Diagnosis of MAS due to *A. hydrophila* will often be presumptive based on clinical signs and previous occurrence. However, definitive diagnosis typically requires culture and confirmation through biochemical and other tests. Media that will support growth of *A. hydrophila* include TSA, BHIA and RS agar. Serological and molecular methods such as PCR are available for identification of *A. hydrophila* (Cascón *et al.*, 1996), but these should be used only to confirm culture results. It should be emphasized that due to the ubiquitous nature of *A. hydrophila* and the other aeromonads, they can often be found as contaminants in culture due to their natural presence on skin or other internal organs. Therefore, pure cultures from affected fish and their organs are typically needed to confirm the diagnosis.

3.8.3 Transmission

Transmission appears to be almost entirely horizontal in the water. Due to the ubiquitous nature of the bacterium, it is often found on the skin and gills of fish and can be part of the normal intestinal flora of healthy fish (Newman, 1982; Holmes *et al.*, 1996).

3.8.4 Host population dynamics

Fish may be at risk of disease due to *A. hydrophila* at any time of the year due to its ubiquitous nature; however, epizootics are more frequent in the spring in species such as carp due to increasing temperatures and poor condition as fish recover from winter (Bullock *et al.*, 1971). Salmonids are susceptible and acute outbreaks can occur when handling or crowding stress is combined with elevated water temperatures. Interestingly, *A. hydrophila* can infect other species besides fish and has been isolated from frogs, alligators, turtles, shrimp and humans (Newman, 1982).

3.8.5 Climate change impacts

Occurrence of MAS is affected by temperature and in general, warmer temperatures result in greater disease manifestation. Therefore, climate change and shifts in temperatures will likely impact host and geographic ranges and affect MAS occurrence.

3.8.6 Control

The best approach to control MAS will often include improvements in husbandry or changes in rearing practices to eliminate specific stressors. However, antibiotics can be effective if sensitivity is properly determined prior to treatment. The development of antibiotic-resistant strains is widespread and, in many cases, has limited the usefulness of some treatments in the past (Mitchell and Plumb, 1980). Vaccination is a possible preventive strategy that could limit MAS in specific populations. Experimental vaccines have been developed and early formulations met with varying levels of success due to antigenic diversity of *A. hydrophila* (Ramadan *et al.*, 1994; Yin *et al.*, 1996). However, recent development of attenuated live strains (Pridgeon and Klesius, 2011) may have greater promise for aquaculture than earlier vaccines.

3.9 Enteric Redmouth Disease/ Yersiniosis

Enteric redmouth (ERM) is an important disease that has resulted in large economic impacts in aquaculture worldwide. ERM or yersinosis is caused by the bacterium, *Yersinia ruckeri*. This bacterium was originally isolated in the Hagerman Valley of Idaho, USA from diseased rainbow trout (Bullock *et al.*, 1971; Tobback *et al.*, 2007). The disease primarily affects rainbow trout but other salmonids can be impacted and outbreaks can lead to high losses. There are a variety of *Y. ruckeri* strains or biotypes and the severity of disease may depend on this and the salmonid host (Brown and Bruno, 2002). Although this disease can be devastating to an operation and fish can become carriers of the pathogen, much success in the control of ERM has been realized through the use of commercial vaccines.

3.9.1 Impact

Y. ruckeri was considered a major disease and severely impacted the commercial trout industry in Idaho, USA (Ross *et al.*, 1966). In the past, ERM resulted in losses of up to 35% for the US trout industry with an estimated annual impact of US$2.5 million in the Hagerman Valley of Idaho (Busch, 1978). Following initial isolation, the bacterium was reported from other areas in the north-west USA and Canada. It was reported in Europe in the mid-1980s and can now be found in Norway, Denmark, the UK, France, Germany, Italy, South Africa and Australia (Llewellyn, 1980; Bragg and Henton, 1986). It continues to cause problems in many areas and preventive strategies to limit disease outbreaks are often implemented. *Y. ruckeri* is considered an obligate pathogen and rainbow trout are the species that is most often impacted by ERM/yersiniosis, but *Y. ruckeri* appears to be able to infect all salmonids reared in fresh water.

3.9.2 Disease characterization and diagnosis

Y. ruckeri is a Gram-negative bacterium and is characterized as a short, motile rod of dimensions 0.5–0.8 µm × 1.0–3.0 µm (Roberts, 2012). Optimum temperatures for growth range from 22 to 25°C and white- to cream-coloured colonies 1–2 mm in diameter should form within 24–48 h (Plumb and Hanson, 2011). There are five most commonly recognized serovars; however, Stevenson *et al.* (1993) described six serotypes based on whole-cell analysis. Serovar Type I (Hagerman strain) is the most commonly isolated and is often considered the most virulent. Vaccine efficacy has been linked to biotype and much work has been done on typing various strains of *Y. ruckeri*. At least two clones may have emerged separately in Europe and North America (Wheeler *et al.*, 2009). Furthermore, atypical biotypes have been associated with mass mortality and vaccine failure in Atlantic salmon in Australia (Bridle *et al.*, 2012).

In small fry, acute infections can result in heavy losses. *Y. ruckeri* causes a septicaemic infection and the disease can manifest in a range of forms from subclinical to acute infections. Chronic forms of the disease may linger in larger fish where clinical signs include dark coloration, lethargy and blindness. Affected fish will appear weak and may congregate near the surface or edges of pond surfaces or at the downstream end of raceways. As the disease name suggests, a common clinical sign of ERM that is often associated with subacute infections involves ulceration and haemorrhage in the oral cavity leading to reddening in the mouth, jaw and on the head. In many cases, however, such signs never develop, and fish may just appear dark and die without other apparent external clinical pathology. *Y. ruckeri* infections in Atlantic salmon, as described by Frerichs *et al.* (1985), did not exhibit classic reddening of the mouth or opercula. Bloody ascites, splenomegaly, inflammation in the vent area and yellowish fluid in the intestine may be common. Venous and capillary congestion of brain and blood vessels is often observed along with intestinal haemorrhaging and petechial haemorrhage in the musculature (Brown and Bruno, 2002). Bacteria may spread from the gills to other tissues such as muscle and liver, leading to tissue oedema and focal areas of necrosis. Exophthalmia and haemorrhage of the ocular cavity may lead to rupture of the eye.

Diagnosis of ERM is usually made based on clinical signs and confirmed through histology and/or culture of *Y. ruckeri* from infected tissues of moribund fish. Typical isolation is on general-purpose media such as TSA or BHIA and colonies appear circular and non-pigmented. *Y. ruckeri* is fermentative, citrate- and catalase-positive, oxidase- and cytochrome oxidase-negative, and does not produce indole in tryptone broth (Brown and Bruno, 2002). It should be noted that variation in some biochemical tests has been reported for some isolates leading to false positives (Hastings and Bruno, 1985). Therefore, immunodiagnostic procedures utilizing specific antibodies or molecular techniques such as PCR can be incorporated for confirmatory diagnosis. It is possible to detect *Y. ruckeri*-infected fish using PCR and sampling kidney tissue directly (Argenton *et al.*, 1996). This can be useful for detecting carrier fish and can even be applied non-lethally using blood samples (Altinok *et al.*, 2001).

3.9.3 Transmission

Sources of infection have been linked to carrier fish in water sources but, interestingly, the bacterium has been isolated from non-fish hosts such as crawfish, mammals (muskrat), humans, seagulls, and even from sewage and river water

(Mitchel *et al.*, 1986; Stevenson *et al.*, 1993). It has been found that infected fish with no clinical signs may not necessarily transmit the infection; however, Hunter *et al.* (1980) found that if carrier steelhead with no clinical signs were stressed at 25°C, *Y. ruckeri* was transmitted.

3.9.4 Host population dynamics

Horizontal transmission of *Y. ruckeri* is most common and fish can readily become carriers of this pathogen. It has been isolated from wild salmonids and non-salmonid fish including walleye (*Stizostedion vitreum vitreum*), sturgeon (*Acipenser baerii*), carp and goldfish (*Carassius auratus*) (McArdle and Dooley-Martin, 1985; Mitchel *et al.*, 1986; Enriquez and Zamora, 1987; Vuillaume *et al.*, 1987).

3.9.5 Climate change impacts

The temperature range where this pathogen can survive and cause disease is wide and optimal growth has been stated to be at 28°C (Fernández *et al.*, 2007). Interestingly, it has been observed that some virulence-associated genes show higher expression at 18°C and disease typically occurs at such lower temperatures. This may be related to the temperature ranges of the primary salmonid host rather than the broad growth capabilities of the pathogen under *in vitro* culture conditions.

3.9.6 Control

Control of ERM is primarily achieved through prevention methods utilizing commercial vaccines. Vaccination is viewed as highly effective and the single most important tool for limiting impact to the industry. In areas where salmonids are impacted by ERM or yersiniosis, correct implementation of a vaccination programme should limit major disease problems. Generally, healthy fish are immersion vaccinated in a suspension of killed bacteria at a size of 4 g or larger, and if necessary, a booster immunization may follow (Larsen and Pedersen, 1997). The efficacy of such vaccines is affected by bacteria serotype, host species, weight and temperature. In some cases, immunity is longer lasting and more effective if fish can be vaccinated by injection. As with any vaccine, proper administration and fish culture practices that eliminate environmental or other infections will greatly affect efficacy.

When ERM is diagnosed and treatment considered necessary, antibiotics such as oxytetracycline, ciprofloxacin (oxolinic acid) and amoxicillin (Busch, 1983) have been shown to be effective against *Y. ruckeri*. In general, antibiotic use for treatment of ERM is not widespread (presumably due to the success of commercial vaccines) and antibiotic-resistant strains of *Y. ruckeri* have decreased over the years (Brown and Bruno, 2002).

3.10 Bacterial Coldwater Disease

Rainbow trout fry syndrome (RTFS) or bacterial coldwater disease (CWD) is caused by the Gram-negative bacterium, *Flavobacterium psychrophilum*. The disease was first referred to as 'peduncle disease' and described in rainbow trout displaying lesions on the caudal peduncle (Davis, 1946). The disease-causing bacterium, however, was not isolated and identified until 1948 from coho salmon displaying similar lesions (Borg, 1948). The disease affects fish ranging in size from early sac fry to fingerling and production size salmonids. *F. psychrophilum* is capable of infecting many species of fish, but generally salmonids are considered the most susceptible. RTFS is often used to describe this disease when early hatched alevins are affected by *F. psychrophilum*. When fish are impacted at these early stages, mortality may be more than 50% (Holt *et al.*, 1993).

There has been considerable debate about the sources of infection due to *F. psychrophilum*, especially when early life stages are affected. Although it is a common bacterium considered ubiquitous in the environment, there is growing evidence that *F. psychrophilum* can be transmitted vertically and is capable of surviving within the egg, which may influence outbreaks in early life stages. *F. psychrophilum* has been isolated from salmonid milt (Kumagai and Nawata, 2011), ovarian fluid (Madsen *et al.*, 2005; Chen *et al.*, 2008), egg surfaces (Vatsos *et al.*, 2001, 2006), and the contents of both unfertilized and eyed eggs (Brown *et al.*, 1997). Although true (intra-ovum) vertical transmission of the bacterium in salmonids has not been conclusively demonstrated, many factors that affect *F. psychrophilum* survival within eggs have been investigated. In fact, Ekman *et al.* (2003) demonstrated through nanoinjection that *F. psychrophilum* can survive within the yolk of fertilized rainbow trout eggs. Furthermore, it is known that the bacterium can survive when exposed to high levels of lysozyme and there is evidence that iodophor concentrations routinely used for egg surface disinfection do not effectively kill *F. psychrophilum* (Brown *et al.*, 1997).

3.10.1 Impact

Most salmonid-producing areas are affected by RTFS. *F. psychrophilum* has been isolated from fish in the USA, Canada, Chile, Japan, Korea and Australia, and it is widespread throughout Europe from a wide range of salmonid and non-salmonid species. However, disease outbreaks most notably affect salmonids. RTFS occurs in freshwater cage culture of salmonids, but the greatest impacts have been noted in commercial trout culture utilizing flow through raceway systems. Impacts result not only from mortalities during an outbreak, but also from poor performance and increased deformities in survivors that reduce product market value of food fish. Another aquaculture sector that is heavily impacted is the public steelhead and salmon mitigation hatcheries in the north-west USA where RTFS causes greater overall losses than any other fish disease (J. Varney, Washington, 2016, personal communication).

3.10.2 Disease characterization and diagnosis

The pathogen causing RTFS has undergone several taxonomic changes. It was initially referred to as *Flexibacter psychrophilus* and *Cytophaga psychrophila*, but DNA–rRNA hybridization studies have resulted in reclassification as *Flavobacterium psychrophilum* (Bernardet *et al.*, 1996; Bader and Shotts, 1998). *F. psychrophilum* is a rod-shaped Gram-negative bacterium that does not produce fruiting bodies. Cell morphology is typical of other *Flavobacterium* spp. and cell size is in the range of 0.2–0.75 μm × 2–7 μm depending on the age and growth environment of the culture (Pacha, 1968; Lorenzen *et al.*, 1997; Kondo *et al.*, 2001; Vatsos *et al.*, 2003). *F. psychrophilum* lacks pili and flagella but moves by gliding motility, which has been implicated as a potential virulence factor. LaFrentz *et al.* (2011) demonstrated that at least one *F. psychrophilum* gliding motility protein, GldN, can be highly immunogenic and hypothesized that gliding motility could aid the bacterium in gaining entry to cells or fin tissue.

The outer membrane of *F. psychrophilum* consists of proteins, lipopolysaccharide and a glycocalyx layer of varying thickness that appears to be loosely associated with the cell, is expressed differentially between strains and is immunogenic (LaFrentz *et al.*, 2007). *F. psychrophilum* forms yellow-pigmented colonies and colony morphology is often described as 'fried egg', meaning that it is a round, convex colony with a thin, spreading margin. However, a number of strains do not have such colony morphology and form a convex colony with smooth edges. *F. psychrophilum* strains commonly secrete enzymes that can degrade collagen, chondroitin sulfate, gelatin and casein (Bertolini *et al.*, 1994; Ostland *et al.*, 2000). *F. psychrophilum* is unable to hydrolyse starch or break down simple and complex carbohydrates, and all strains appear to be catalase-positive but variation in cytochrome oxidase utilization has been noted (Nematollahi *et al.*, 2003). Biochemically, *F. psychrophilum* is relatively homogeneous, but many different strains or serotypes have been characterized. In general, serological characteristics have suggested linkages to different geographic regions or different host species, and strains range from highly virulent to non-virulent. Genetic characterization of isolates, however, suggests greater heterogeneity and less correlation between *F. psychrophilum* and host specificity, making it difficult to develop a standardized classification system.

Clinical signs of *F. psychrophilum* infections can vary. In general, if RTFS manifests as an acute septicaemic infection, mortality may be high with limited external clinical signs. If coho salmon are infected during the fry stage, erosion of the epithelial layer covering the yolk sac may be evident due to septicaemia (Holt, 1988). As fish increase to fingerling size and older, erosion of the caudal peduncle region may be observed, but this clinical sign may not be apparent. Infected fish may exhibit spiral swimming behaviour, go off feed, exhibit frayed fins, pale gills and exophthalmia, haemorrhage, and display dark pigmentation in the caudal region. Internally, splenomegaly is common along with ascites and general septicaemia as bacteria infiltrate into internal organs. Fish that survive outbreaks of RTFS often show poor performance and may have a range of deformities including lordosis, scoliosis and posterior spinal compression (Fig. 3.3) affecting the caudal region (Conrad and DeCew, 1967; Madsen *et al.*, 2001). In Japan, ayu (*Plecoglossus altivelis*) are affected with clinical signs including ulcerative lesions in the caudal region and lower jaw, anaemia and haemorrhaging (Miwa and Nakayasu, 2005).

Microscopic examination can reveal long, thin, Gram-negative bacteria from externally affected areas. Following sampling of moribund fish, imprints and/or histological sections from the spleen and other organs including the liver and kidney often show presence of long filamentous rods (Fig. 3.4). The liver can show vascular degeneration and necrotic hepatocytes, but *F. psychrophilum* can be easily observed in spleen or kidney imprints from heavily infected fish.

Diagnosis is often based on clinical signs followed by isolation of yellow-pigmented bacteria characteristic of *F. psychrophilum*. However, yellow-pigmented bacteria are common, and it is important to implement confirmatory tests to identify *F. psychrophilum* as the causative agent of RTFS. This is usually accomplished through culture combined with other biochemical, molecular or serological methods to specifically identify *F. psychrophilum*. Spleen, kidney or other organs such as the brain are sampled and plated on the appropriate agar media. *F. psychrophilum* is a slow-growing aerobic bacterium that is generally incubated at 15–18°C. Yellow-pigmented colonies usually appear within 4–7 days. Similar to other *Flavobacterium* species, *F. psychrophilum* requires low-nutrient media for successful culture. One of the most common is tryptone-yeast extract-salts (TYES), but others such as Cytophaga media, Shjieh agar, Anacker & Ordal, and Hsu–Shotts have been used (Cain and LaFrentz, 2007).

Biochemical confirmation of *F. psychrophilum* is not always reliable. Therefore, definitive diagnosis should utilize immuno- or PCR-based assays. *F. psychrophilum* can be confirmed from culture or from infected fish tissues such as the kidney using an ELISA or a FAT. Lindstrom *et al.* (2009) developed a highly specific monoclonal antibody (FL43) against the outer membrane fraction of *F. psychrophilum* for routine detection or broodstock screening using an ELISA and FAT. Assay protocols are available and FL43 has been produced for commercial availability. Molecular assays based on PCR are also available, and a nested PCR that can be used on culture, tissues or reproductive fluids is highly sensitive (Taylor, 2004).

Fig. 3.3. Posterior spinal compression present in rainbow trout surviving an outbreak of RTFS. (Photograph courtesy of S.L. LaPatra.)

Fig. 3.4. Long filamentous rods of *Flavobacterium psychrophilum*.

3.10.3 Transmission

F. psychrophilum can be transmitted horizontally and vertically through survival on the surface and potentially within the egg. Studies have shown it capable of surviving within the egg when microinjected (Ekman *et al.*, 2003) and it has been found within eggs following immersion in a bacterial solution just prior to water hardening (Kumagai, 2005). The ubiquitous nature, high rate of shedding from infected mortalities and risk of vertical transmission from broodstock lend to the persistence and ease of transmission of *F. psychrophilum*.

3.10.4 Host population dynamics

Important species such as coho salmon along with rainbow and steelhead trout are considered most susceptible, but many other important species including Atlantic salmon (Valdebenito and Avendaño-Herrera, 2009) are impacted. Mortality in susceptible species can range from 20 to 90% (Bruno, 1992) and reports of RTFS impacts have increased over the past two decades. In the USA this may reflect adaptations to higher temperatures (>10°C) for some *F. psychrophilum* strains and increased resistance to commonly applied antibiotics.

3.10.5 Climate change impacts

As indicated by the name and psychrophilic nature of this pathogen, it causes impacts in colder temperatures, but disease now regularly occurs at temperatures up to and above 15°C. Clearly, strains have evolved and adapted over time suggesting that climate change impacts will potentially alter the geographic range of this pathogen and its hosts.

3.10.6 Control

Control of RTFS is achieved through proper culture and management techniques aimed at reducing stress, promoting strict biosecurity and maintaining high water quality. If an outbreak occurs, several options are available. Removing all dead fish is important to limit the spread of *F. psychrophilum* as it is shed into the water column. Immersion bath treatments using salt, antibiotics (water-soluble oxytetracycline) or potassium permanganate have been shown to be effective in some cases, but they must be administered before fin or caudal erosion is evident (Cipriano and Holt, 2005). Once an outbreak is confirmed, the most effective treatment for RTFS involves antibiotic administration through the feed. In the USA, there are currently two antibiotics licensed for use against *F. psychrophilum* in freshwater-reared salmonids: oxytetracycline dihydrate and florfenicol. Florfenicol (Aquaflor®) was recently approved, and all antibiotic treatments now require veterinary approval prior to feed incorporation and sale. Although antibiotic treatment is effective, there is always concern that resistant strains of *F. psychrophilum* may develop. The occurrence of strains resistant to oxytetracycline is well documented in a number of countries. Cases of *F. psychrophilum* strains acquiring resistance to florfenicol have not been reported; however, the minimum inhibitory concentration of florfenicol has been documented to be higher than average for isolates at some farms (del Cerro *et al.*, 2010; Hesami *et al.*, 2010; Henríquez-Núñez *et al.*, 2012).

Ideally, preventive methods such as vaccination would be preferred for RTFS and would limit the risk of antibiotic-resistant strains. Early work focused on killed whole-cell vaccine formulations, but limited success was achieved in the absence of an adjuvant (LaFrentz *et al.*, 2002; Rahman *et al.*, 2003). Recent work has led to the development of live attenuated *F. psychrophilum* strains that, when used as vaccines, are capable of conferring protection in fish. LaFrentz *et al.* (2008) developed such a strain using a rifampicin resistance strategy and showed that fish immunized by injection or immersion were significantly protected from *F. psychrophilum* infection. Field trials on rainbow trout (1 g initial weight) immersion immunized have shown the vaccine to be safe and provide protection from natural outbreaks of RTFS (K.D. Cain, 2015, unpublished results).

Promising alternative methods of control have been suggested and one option may be to incorporate naturally occurring gut bacteria as probiotics in the feed. Burbank *et al.* (2011) isolated two *Enterobacter* strains from the intestinal tract of healthy fish that could inhibit *F. psychrophilum* growth *in vitro*. When mixed into a commercial feed and fed to rainbow trout, they were capable of significantly reducing mortality following disease challenge. Other management-based approaches to control RTFS may need to focus on reducing overall bacterial prevalence at facilities and reducing the risk of vertical transmission by screening broodstock and culling eggs originating from heavily infected broodstock.

3.11 Bacterial Kidney Disease

Bacterial kidney disease (BKD) is a systemic disease that originates in fresh water but can affect salmonids during both freshwater and seawater phases of their lifecycle. It is caused by the Gram-positive bacterium *R. salmoninarum* and was first described in Scotland in 1930, where it was documented in Atlantic salmon from the Dee and Spey rivers (Mackie *et al.*, 1933). With the exception of a few salmonid-producing countries such as Australia and New Zealand, BKD is widespread.

3.11.1 Impact

Juvenile salmonids are severely impacted by BKD, but in many cases significant mortality can occur in adults. BKD is a concern in freshwater and marine cage culture. The disease has been documented as causing significant losses in wild Pacific salmonids (Kent *et al.*, 1998), linked to large epizootics of chinook salmon in the Great Lakes, USA (Holey *et al.*, 1998) and has been identified in wild Atlantic salmon in the north-eastern USA (Smith, 1964). In North America, losses in Pacific salmon stocks have reached as high as 80% historically (Evenden *et al.*, 1993). Prevalence of BKD in such stocks has been attributed to the mode of transmission and the persistence of *R. salmoninarum* in the egg following iodophor disinfection. This emphasizes the critical need to establish *R. salmoninarum*-free broodstock whenever possible.

Disease severity may also be influenced by water quality, and Warren (1963) suggested that hatcheries with soft water had greater incidences of BKD compared with hatcheries with hard water. Smolts transferred from fresh to seawater are particularly susceptible to BKD; for example, *R. salmoninarum*-infected coho salmon smolts held in seawater experienced 17% mortality over 150 days compared with 4% for siblings held in fresh water (Fryer and Sanders, 1981).

3.11.2 Disease characterization and diagnosis

R. salmoninarum was first characterized as a coryneform bacterium that was a strongly Gram-positive, non-motile rod measuring 0.3–1.0 μm × 1.0–1.5 μm that might occur in pairs or V formations (Sanders and Fryer, 1980). It is extremely fastidious with a strict requirement for L-cysteine in the growth media (Evelyn *et al.*, 1990). It is acid-fast, non-sporulating, and periodic acid Schiffs (PAS)-positive. Growth in culture is slow and primary isolation may take from 8 to 12 weeks to produce white to yellowish circular colonies at 15°C (Brown and Bruno, 2002). Improved methods of culturing *R. salmoninarum* have been developed (Evelyn *et al.*, 1990; McIntosh *et al.*, 1997) and Faisal *et al.* (2010b) developed an improved culture method that results in colony growth within 5–7 days. In general, serological (ELISA and FAT) and PCR-based molecular assays have become important for detection and quantification of infection levels in fish and are widely utilized for diagnostic and management purposes (Roberts, 2012).

A range of clinical signs have been reported for BKD which include darkening of fish, exophthalmia and lethargy. Spawning salmon may exhibit haemorrhaging at the base of fins, and in farmed trout, petechial haemorrhaging or raised vesicles may be found on the side or lateral line of fish. Upon necropsy, classic signs that include white-grey granulomatous lesions may be observed in the kidney of affected fish. Other internal signs include bloody ascites and enlargement of the kidney. Granulomatous lesions are common for BKD and may be found in organs such as the heart, liver, spleen, kidney, and even the gill and muscle of affected fish (Bruno, 1986).

Histological characterization of BKD often focuses on lesion development in the kidney and other organs. Necrosis of tissues occurs and extends to areas between kidney tubules where granulomas containing *R. salmoninarum* may be observed along with leucocytes and other cellular debris. Bacteria can often be observed in organs and tissues and macrophage proliferation may be evident. *R. salmoninarum* is able to survive in macrophages, which represents a potential mechanism to avoid the immune response (Bruno, 1986; Grayson *et al.*, 2002). In the kidney, the bacterium may also be observed within endothelial cells lining the glomerular blood vessels as well as the lumen of collecting ducts, but it is typically not observed within the proximal tubules (Brown and Bruno, 2002). Bruno (1986) noted that the nuclei of endothelial cells are diffusely stained, slightly cloudy and may contain bacteria.

For most bacterial pathogens culture is the primary method for identification and diagnostics. However, this is not typically the case for *R. salmoninarum*. Although culture is sensitive, it has not been a primary diagnostic tool due to the slow growth and impractical application when treatment or management decisions need to be implemented. This may change with improvements in culture methods; however, histology and immunoassays such as FATs and ELISAs will continue to be important tools for diagnosing *R. salmoninarum*. In North America, broodstock populations of farmed and wild fish (returning for stock enhancement or mitigation programmes) are screened using a commercially available ELISA and procedures have been standardized among many diagnostic laboratories. Until recently, the ELISA was the most sensitive assay available to screen tissue samples for *R. salmoninarum*; however, qPCR assays have been developed (Powell *et al.*, 2005) and their use may become routine in the future. Each assay has limitations when confirmatory diagnosis is required. For example, the ELISA and FAT are most often carried out on kidney and ovarian fluid, respectfully. These assays rely on polyclonal (and sometimes monoclonal) antibodies that are most often directed against a soluble antigen (p57) of the bacterium. Such antibodies have been shown to cross-react with other bacterial species (Brown *et al.*, 1995) and, due to the soluble nature of p57, may react to antigen in the tissue even in the absence of an active infection. PCR methods developed to the p57-encoding gene or other genes also have potential problems in that they are detecting bacterial DNA and again may not effectively report the presence of viable bacteria.

3.11.3 Transmission

Although *R. salmoninarum* can be transmitted horizontally from fish to fish, intra-ovum vertical transmission plays a major role and is important to consider for disease management of exposed stock (Bruno and Munro, 1986; Evelyn *et al.*, 1990).

3.11.4 Host population dynamics

R. salmoninarum is considered an obligate pathogen of fish in the family Salmonidae. It is generally found in trout and salmon (subfamily Salmoninae); however, Faisal *et al.* (2010a) isolated *R. salmoninarum* from wild populations of whitefish, *Coregonus* spp. (subfamily Coregoninae), in the Great Lakes, USA. It often occurs as a slow chronic infection that results in serious losses at times of physiological stress, such as smoltification.

3.11.5 Climate change impacts

Impacts on BKD due to shifts in climate are likely, primarily due to host-level impacts and the stress of changing temperatures. It is known that if *R. salmoninarum* is endemic in a region, epizootics can occur seasonally when stressors such as increasing or declining water temperatures occur or if fish are undergoing smoltification.

3.11.6 Control

The ideal method of controlling BKD would be to limit exposure of fish to *R. salmoninarum*. Care should be taken if utilizing water sources containing wild fish stocks, or whenever fish are introduced into a farm or hatchery. In areas where the pathogen is endemic outbreaks may occur. Antibiotic therapy can, in some cases, limit the severity of an outbreak but is typically not satisfactory and may require long-term treatment.

Interestingly, prophylactic feeding of erythromycin to juvenile fish along with antibiotic injection in broodstock has been reported in some hatchery programmes as a management strategy for BKD in Pacific salmon. However, this strategy does not effectively limit outbreaks of BKD, and only when management strategies incorporated a culling and segregation programme based on screening broodstock for high levels of *R. salmoninarum* antigen did such programmes achieve success (Munson *et al.*, 2010). Such screening programmes utilize a polyclonal ELISA against the heat-stable, soluble p57 antigen of *R. salmoninarum* (Pascho and Mulcahy, 1987) to evaluate antigen levels in kidney tissue samples. The great success of such programmes is due to the ability to quantify infection levels in broodstock based on ELISA optical density (OD) values. By standardizing infection levels based on 'high', 'medium' or 'low' OD values, managers can cull eggs from heavily infected fish or segregate progeny from infected adults if stocks are highly valuable. Such a strategy could be applied to any population where broodstock are regularly detected with *R. salmoninarum*. Indeed, all fish or eggs entering a facility should be tested for the presence of *R. salmoninarum* and if detected those stocks should not be used unless absolutely necessary.

Development of a BKD vaccine has met with only modest success. There is, however, at least one commercial vaccine. This vaccine 'Renogen' is a live formulation that consists of a closely related soil bacterium, *Arthrobacter davidanieli*, which elicits the production of cross-reactive antibodies to *R. salmoninarum* following immunization (Griffiths *et al.*, 1998). In field trials, this vaccine elicited significant protection against BKD in Atlantic salmon and is most effective when administered by injection (Salonius *et al.*, 2005). However, Alcorn *et al.* (2005) reported a lack of protective immunity in chinook salmon following administration of this vaccine along with five experimental vaccines. Such results suggest that further work is required in the area of BKD vaccines and species-specific responses must be considered.

3.12 Flavobacteriosis Caused by Emerging *Flavobacteriaceae* Species

Disease risks and challenges from an emerging group of bacteria in the *Flavobacteriaceae* family have also been increasing in recent years (Loch and Faisal, 2014, 2015). Molecular methods to better identify bacterial groups have confirmed that a range of novel flavobacterial pathogens in the genera *Flavobacterium* and *Chryseobacterium* are being isolated from disease outbreaks in freshwater aquaculture facilities. Several novel flavobacterial species such as *Flavobacterium chilense* and *Flavobacterium araucananum* have been isolated along with *F. psychrophilum* from salmonid fish suffering from mixed infections (Kämpfer *et al.*, 2012). Similarly, new species such as *Flavobacterium frigidarium*, *Flavobacterium plurextorum*, *Flavobacterium tructae*, *Flavobacterium piscis* (Bowman and Nowak, 2004; Zamora *et al.*, 2012, 2014) and an increasing number of partially characterized *Flavobacterium*-like organisms have been isolated from diseased rainbow trout (Holliman *et al.*, 1991; Shotts and Starliper, 1999; Austin and Austin, 2007). In addition, several other new and emerging members of *Flavobacteriaceae* including *Chryseobacterium piscicola*, *Chryseobacterium oncorhynchi*, *Chryseobacterium joostei*, *Chryseobacterium viscerum*, *Chryseobacterium aahli*, *Chryseobacterium shigense* and *Chryseobacterium indologenes* are being discovered and implicated in infections and serious disease outbreaks of salmonid fish (Ilardi *et al.*, 2009, 2010; Zamora *et al.*, 2012; Kämpfer *et al.*, 2013; Loch and Faisal, 2014, 2015). The isolation of these emerging bacterial species is concerning in that treatment and prevention strategies have not been developed and they appear to commonly be linked to misdiagnosis of diseases such as CWD or columnaris.

3.13 Bacterial Diseases with Secondary Impact on Coldwater Fish Reared in Fresh Water

Please refer to Cain and Polinski (2014) for information on bacterial diseases with secondary impact on coldwater fish reared in fresh water, including salmonid rickettsial septicaemia (*Piscirickettsia salmonis*) and columnaris (*Flavobacterium columnare*).

DISEASES CAUSED BY FUNGAL/ PSEUDOFUNGAL PATHOGENS

3.14 Saprolegniosis

Saprolegnia is a genus of freshwater moulds that is a ubiquitous part of most aquatic freshwater environments. Although *Saprolegnia* can tolerate a wide temperature range (0 to 35°C), cold (3–15°C) conditions are preferred and are where disease manifestations become most apparent in aquaculture (van West, 2006). Generally, *Saprolegnia* are saprotrophic organisms, feeding on dead and decaying organic matter. However, they can often become opportunistic pathogens when organic debris or necrotic tissue provides a foundation from which growing filaments or zoospores can easily spread to living tissue. In some instances, specific strains or species are also believed to directly target living hosts during sporulation (Willoughby and Pickering, 1977). Given the right environmental conditions, it is likely that any fish species would be susceptible to infection by this pathogen at any life stage; however, the most severe infections in a culture environment often have occurred during egg incubation, larval rearing, and in post-spawning adult salmon broodstock. Saprolegniosis is ultimately fatal for both eggs and fish if left untreated and is a serious and widespread concern in freshwater fish culture.

3.14.1 Impact

A practical assumption is that any freshwater fish may be opportunistically parasitized by *Saprolegnia* within its tolerated temperature rage (approximately 2–35°C). At least three species of *Saprolegnia* are known to infect either fish or fish eggs (*Saprolegnia parasitica*, *Saprolegnia diclina* and *Saprolegnia ferax*) and infectivity by additional species is probable. *S. parasitica* is often viewed to be the species of most concern and estimated costs during the aquaculture of salmon and catfish are in the tens of millions of dollars annually (van West, 2006). However, this appears to be a matter of context as both *S. diclina* and *S. ferax* have been shown to be more pathogenic during egg incubation of Atlantic salmon eggs than *S. parasitica* (Thoen *et al.*, 2011), which more likely targets adult organisms rather than eggs. A financial value is harder to define regarding loss during eggs and larvae production, but in developing culture of burbot (*Lota lota*), mortality of both eggs and larvae has approached 100% without administration of chemical prophylactics (Polinski *et al.*, 2010).

3.14.2 Disease characterization and diagnosis

The genus *Saprolegnia* belongs to a class (Oomycota) of 'fungal-like' organisms which exhibit many characteristics, such as filamentous hyphae, sporulation and utilization of saprotrophic nutrition by extracellular digestion, similar to those of many fungi. Indeed, the 'cotton mould' appearance of growing *Saprolegnia* filaments is visually quite similar to that of many terrestrial fungal moulds. However, a diploid life stage, cellulose cell walls and non-septate filaments taxonomically separate *Saprolegnia* from true fungi and classify them more closely with other heterokonts such as diatoms and brown algae. Phylogenetic comparisons also support this classification with protist heterokonts rather than with fungi (Guerriero *et al.*, 2010).

Visualization of cotton-like tufts can be followed by macroscopic observation to identify non-septate filamentous hypha for presumptive diagnosis (Fig. 3.5). Low-nutrient culture media can also be used for continued observation of newly emerging hyphae for sexual differentiation and definitive diagnosis (Stueland *et al.*, 2005). Species confirmation by microscopy is somewhat difficult, as it relies on morphological identification of sexual structures (Wood and Willoughby, 1986). Phylogenetic sequencing is becoming increasing available for species and strains which can be utilized for PCR identification (Thoen *et al.*, 2011), but as yet not all species can be differentiated by this method.

Fig. 3.5. *Saprolegnia* colonizing the eggs of burbot during late-stage embryo development. Hyphae can be seen extending from dead eggs to engulf adjacent live embryos.

Nevertheless, with specific regard to the culture of fish, *Saprolegnia* species identification is fairly inconsequential as both treatment and preventive measures are not species specific.

3.14.3 Transmission

The life cycle of *Saprolegnia* includes both sexual and asexual reproductive phases. In the asexual phase, a spore or sporangium is formed at the end of a hypha which releases motile zoospores (Bruno *et al.*, 2011). These primary zoospores swim (via an apical flagella) for a short time before they encyst (become dormant) and then release a secondary zoospore. Secondary zoospores are motile (via a lateral flagella) for a longer period than primary zoospores and are considered the main dispersion and infective form of *Saprolegnia* (Willoughby, 1994). The secondary spore can also release new zoospores and the repeated cycles of encystment and motile zoospores (called polyplanetism) can allow the organism to survive and persist in the environment for long periods (Beakes, 1982). The secondary zoospores of some *Saprolegnia* species also possess hairs and with many species (or strains within a species) the hairs are hooked, presumably to facilitate attachment to a living host (Beakes, 1982; Burr and Beakes, 1994; Grandes *et al.*, 2000).

3.14.4 Host population dynamics

Infection is characterized by filamentous 'cotton-like' tufts which appear on the external surface of fish and eggs. For fish, infection will often initiate around the site of previous injury and radiate out in a circular, crescent or whorled pattern. Environmental stresses and previous disease infection are predisposing factors that enhance the likelihood of infection with *Saprolegnia*.

3.14.5 Climate change impacts

Temperature can affect the vegetative and asexual life stages of *S. parasitica* as well as its ability to cause host disease. For brown trout, it has been observed that prolonged warmer temperatures alone were not enough to increase mortality associated with saprolegniosis but rather required an acute increased temperature event in concordance with *Saprolegnia* exposure (Matthews, 2019). Therefore, high environmental variability likely poses the most serious risk for *Saprolegnia* infection in coldwater fish aquaculture.

3.14.6 Control

The ubiquitous nature of *Saprolegnia* generally precludes the ability for complete avoidance in coldwater fish culture. As previously stated, minimizing environmental stress, physical injury and external pathogenic diseases can aid in preventing initial infection of fish. During egg and larval rearing it is important to promptly remove detritus such as dead egg casings, faeces and excess feed so as to eliminate the preferred food source of this opportunistic pathogen. If infection occurs, or if pre-emptive measures are desired, chemical therapeutics can be used to effectively eliminate or prevent infection. Malachite green was historically used for treatment of this disease with excellent success but is currently not approved for use in most commercial aquaculture settings due to its carcinogenic and toxicological effects (van West, 2006). Other compounds found to inhibit the growth of *Saprolegnia* include sodium chloride, formalin and hydrogen peroxide, of which hydrogen peroxide has become the chemical of choice for most situations due to its lower environmental and human handling effects relative to formalin, and the reduced quantities needed relative to sodium chloride (>30 g/l) (Marking *et al.*, 1994). Hydrogen peroxide has been applied successfully at 250–500 mg/l as daily treatments of 15 min duration during egg incubation of multiple coldwater species (Barnes and Stephenson, 2003; Barnes and Soupir, 2006; Soupir and Barnes, 2006; Polinski *et al.*, 2010), and shows continued effectiveness when administered during fry and larval development at 50–250 mg/l by immersion for 1 h every other day (Rach *et al.*, 1997; Gaikowski *et al.*, 1998, 1999; Polinski *et al.*, 2012). For adult salmonids, 50–100 mg/l and exposure time of 60 min may be used (FDA, 2007), although treatment below 100 mg/l may not be sufficient to adequately control an established infection (Marking *et al.*, 1994).

DISEASES CAUSED BY PARASITIC PATHOGENS

3.15 Proliferative Kidney Disease

Proliferative kidney disease (PKD) is caused by the extrasporogonic stage of a Myxozoa parasite, *Tetracapsuloides bryosalmonae* (Canning *et al.*, 2000), and affects salmonids in fresh water. The disease was named by Roberts and Shepherd (1974) due to the clinical characteristics of the disease in the kidney and spleen. It is a primary problem in rainbow trout and the disease-causing organism was originally referred to as PKX (Kent and Hedrick, 1985).

3.15.1 Impact

PKD in salmonids was first described and recognized in North America following a disease outbreak at the Hagerman State Fish Hatchery in Idaho in 1981 (Smith *et al.*, 1984). In Europe and the British Isles it has been present for many years and is a major disease that affects rainbow trout production. France and Italy have been impacted heavily (Ferguson and Ball, 1979) and 100% of the fish have been affected on some farms. Both wild and farmed salmonids including grayling (*Thymallus thymallus*) (Wahli *et al.*, 2002) and Arctic char (*Salvelinus alpinus*) (Kent and Hedrick, 1985) can be infected, and if water sources containing infected bryozoans are used for fish culture, then such operations would be at risk. Although PKD is a problem in Europe and North America, evidence based on phlyogeographic studies suggests that spread of the parasite is not typical of other Myxozoa and it appears that fish may be dead-end hosts for this parasite (Henderson and Okamura, 2004).

3.15.2 Disease characterization and diagnosis

PKD causes diffuse and chronic inflammation of the haematopoietic tissue, and the organ is surrounded by inflammatory cells. Later stages of the infection result in formation of granulomatous tissue and sporoblasts of the parasite in the lumen and walls of kidney tubules. PKX cells are often observed in affected tissue and organs, having even been found in the gills of fish.

Clinical signs of fish with PKD vary but typically they have distended abdomens accompanied by longitudinal swelling at the lateral line, dark coloration, exophthalmia, pale gills, and apparent respiratory distress in moribund fish. Fish may be anaemic and show nervous disorders and loss of equilibrium in the water column. Swelling of the kidney is common and may be accompanied by grey bulbous ridges most often near the posterior of this organ. The spleen and liver may be affected and abdominal swelling due to excess ascites fluid in the peritoneum is often apparent (Roberts, 2012).

Clinical signs and gross pathology can aid in diagnosis of PKD if disease history is well documented on farms; however, definitive diagnosis requires recognition of lesions along with examination of kidney and/or other organs to identify *T. bryosalmonae* spores in tissue sections or organ imprints. A variety of stains can enhance identification and monoclonal antibodies specific for *T. bryosalmonae* are available for immunological confirmation (Adams *et al.*, 1992). Additionally, Castagnaro *et al.* (1991) discovered a lectin that could enhance diagnostics by effectively binding to the PKX organism. Histological examination may reveal spores surrounded by macrophages or other phagocytes. However, definitive confirmation of PKD may require molecular techniques such as PCR or serological assays incorporating specific antibody-based reagents.

3.15.3 Transmission

The life cycle of this parasite was described by Canning *et al.* (1999), who confirmed the alternative host and were able to successfully transmit *T. bryosalmonae* from infected bryozoans to rainbow trout. Anderson *et al.* (1999) also confirmed that 18S rDNA sequences from PKX databanks were similar to those identified from freshwater bryozoans. Spores of *T. bryosalmonae* released by freshwater bryozoans penetrate the skin of fish to cause infection. Once in the fish host, these spores are thought to proliferate rapidly and migrate primarily to the kidney and spleen, but they also reach other internal organs

(Roberts, 2012). The extrasporogonic stage is often identified in the kidney, appears as large cells (>20 μm) and may be seen in stained tissue sections or imprints (Brown and Bruno, 2002).

3.15.4 Host population dynamics

With the occurrence of freshwater bryozoans in many systems, the rearing of salmonids in these waters or utilizing such water sources for aquaculture operations could potentially result in PKD outbreaks.

3.15.5 Climate change impacts

PKD is often associated with seasonal temperature changes in hatchery-reared salmonids, particularly rainbow trout, and climate shifts will likely impact hosts and parasite occurrence and geographic ranges.

3.15.6 Control

Effective control methods for PKD are limited and no commercial vaccine is available. It is known that water temperature can change the dynamics of the infection in both the bryozoan and the fish host. Although decreasing water temperatures can limit the effects of PKD, Ferguson (1981) found that prolonged holding of juvenile fish at temperatures higher than 15°C could also minimize the disease. Even after recovery from PKD a portion of the population may still remain chronically infected and show clinical signs of disease. Attempts have been made to control PKD using malachite green, fumagillin and its synthetic analogue TNP-470 (Morris *et al.*, 2003). Some efficacy has been suggested but toxicity and potential environmental risks can be a concern and have limited use of such treatments. Vaccine development may be feasible but limited information or success has been reported in the literature.

Proper fish culture practices that limit stress and maintain good water quality are important to limit effects of PKD. If possible, lowering summer water temperatures or keeping fish in pathogen-free water sources until they are fully immunocompetent is effective and should be considered in areas where *T. bryosalmonae* is present. A preferred option would be to control bryozoans or limit spore release from them; however, this has not been successful due to the number of spores that can be released from only small colonies of bryozoans.

3.16 Whirling Disease

Whirling disease is caused by *Myxobolus cerebralis* and is a widespread parasitic infection originating in Europe, where it was first reported in 1893 (Hofer, 1903). The parasite has a complex life cycle and infects fish primarily during early life stages. In the USA, whirling disease was first diagnosed in 1958 (Hoffman, 1990) and was suspected to be introduced via movement of fish from Europe. Cultured fish can be impacted if the infectious triactinomyxon (TAM) stage is present in the rearing water. Early incidences in cultured fish were associated with earthen pond rearing where the intermediate host (an oligochaete worm, *Tubifex tubifex*) was present and allowed the pathogen to perpetuate. Such observations spurred a move away from earthen pond culture to concrete raceways, which interrupted the life cycle of the parasite. Concern over the disease in cultured fish declined, but in the 1980s it was quickly realized that the parasite could infect and cause disease in wild fish (Nehring and Walker, 1996) and once established could not be eliminated from a water body. Dramatic population declines were documented in wild rainbow trout populations in the Madison River in Montana and the Colorado River in Colorado, USA (Walker and Nehring, 1995). Currently, *M. cerebralis* is well established in many river systems and is widespread across the western USA.

M. cerebralis has an affinity for cartilage in the head and spores develop to cause skeletal lesions. Spores may impact the nerves, resulting in blackening of the caudal region in fish (Fig. 3.6), and can cause improper development of the spine and cranium leading to a range of deformities. The quantity of TAMs during exposure and the life stage of the fish often determine the severity of disease. In alevins, exposure to high doses of the infectious stage can result in 100%

Fig. 3.6. Characteristic black tail often associated with clinical signs of *Myxobolus cerebralis* infection.

mortality, whereas fish exposed at later stages may show limited infection rates and no mortality (Markiw, 1991). This is thought to relate primarily to the level of bone ossification as fish get older, which impacts the parasite's ability to destroy developing cartilage.

3.16.1 Impact

Water source for a cage-culture operation is of great concern and cages must not be located where *M. cerebralis*-infected fish are present. Movement of infected fish is often highly regulated due to the risk of parasite spread. If infected fish are detected in a facility, quarantine and depopulation may be required. If an outbreak of whirling disease occurs, heavy impacts resulting in high mortalities or deformed unmarketable survivors can result. In most cases, whirling disease is considered a chronic infection in fry and fingerlings (El-Matbouli *et al.*, 1995) and hatchery-released or wild fish populations may experience mortality (Markiw and Wolf, 1974; Hedrick, 1998).

3.16.2 Disease characterization and diagnosis

The classic signs of whirling disease include tail chasing (whirling), skeletal deformities (primarily in the head and vertebrae of fish), blackening of the tail and mortality. It is hypothesized that whirling behaviour is linked to damage to cartilage surrounding the organ of equilibrium (Markiw, 1992), while discoloration of the caudal region of the tail may be due to inflammation and compression that impair nerves that influence pigmentation (Rose *et al.*, 2000). Infected fish can often exhibit no clinical signs of disease or, depending on factors described above, mortality can reach 100%. Infected fish may cease feeding and will in almost all cases become lifelong carriers of *M. cerebralis*. Internal examination may reveal normal tissues and organs, and only upon histological examination will tissue damage characteristic of *M. cerebralis* infection be recognized. Staining of cranial sections will often show characteristic areas of tissue damage including inflammation, lysis and digestion of cartilage.

Although disease history and clinical signs may provide a presumptive diagnosis, techniques to diagnose whirling disease and detect *M. cerebralis* most often focus on identification of spores within cartilage of fish. Methods to accomplish this include pepsin–trypsin–dextrose (PTD) digestion of the cartilage of the head (Lorz and Amandi, 1994), histopathology and PCR (Andree *et al.*, 1998).

The PTD digest is aimed at extraction of myxospores from the cartilage of infected fish and involves removal or de-fleshing of soft tissue in the head of fish. The cartilage can then be dissolved and mature myxospores isolated and identified based on shape, size and presence of polar capsules. Although typically considered a 'gold standard', it is possible that other myxospore species similar in size could be identified as *M. cerebralis*. An advantage of PTD is that myxospores can be easily enumerated and provide a relative indication of infection intensity. However, the length of time it takes to perform the assay is a consideration and has led to greater use of molecular PCR-based detection methods for confirmation of *M. cerebralis*.

In younger fish where spores may not have developed, histopathology can identify damage to the cartilage tissue in fish. It can also be useful to demonstrate presence and severity of infection based on degree of inflammation and numbers of granulomatous lesions within the cartilage of the head (Baldwin *et al*., 2000), but is considered less sensitive than PTD.

PCR-based methods are now widely used to detect *M. cerebralis* infections with follow-up methods to confirm the parasite as *M. cerebralis*. *M. cerebralis* detection using PCR has advantages over other methods in that it can be applied to both hosts at all life stages. PCR can also detect the pathogen within environmental samples in some cases and is more sensitive than other diagnostic methods (Andree *et al*., 1998). PCR assays for *M. cerebralis* originally targeted ribosomal sequences (Andree *et al*., 1998), but additional assays utilizing a segment of the heat-shock protein 70 (Hsp70) gene have been developed (Epp *et al*., 2002). PTD and histology have the advantage of providing relative severity of infections; however, a qPCR assay based on Hsp70 has been shown to relate directly to histology score of *M. cerebralis*-infected fish (Cavender *et al*., 2004).

3.16.3 Transmission

M. cerebralis has a two-host life cycle (Fig. 3.7) involving an aquatic oligochaete worm that ingests mature spores following death of the salmonid host. The life cycle is initiated with release of TAMs into the water column from an infected *T. tubifex* and then finding a susceptible host (El-Matbouli *et al*., 1999). TAMs are infective and each TAM has four appendages. Three of these appendages measure between 170 and 200 μm (Markiw, 1992) and the fourth measures approximately 140 μm. The fourth appendage contains the epispore that has a minimum of 64 sporoplasms within it (El-Matbouli *et al*., 1995). Once the TAMs contact fins, skin, gills, oesophagus or the digestive tract in fish, they penetrate and release sporoplasms into the epidermis (Markiw, 1989; El-Matbouli *et al*., 1995). These sporoplasms have a tropism for cartilage and upon release they migrate via peripheral nerves and the central nervous system to host cartilage (El-Matbouli *et al*., 1995, 1999; MacConnell and Vincent, 2002). It takes time for myxospores to develop in the cartilage and, at approximately 860 degree-days (measured as Celsius temperature units (CTUs)), they can be observed in tissue of infected fish (Hedrick and El-Matbouli, 2002). Myxospores are 8–10 μm in diameter and have two polar capsules (Markiw, 1992). Once the fish host dies and decays or is consumed and excreted by pisciverous predators, myxospores settle out to the sediment and are ingested by *T. tubifex* where they then multiply within the intestine. Prior to ingestion by *T. tubifex*, spores are extremely tough and can remain viable in the environment for many years. Following ingestion, TAMs are released by death of the worm or intermittent egestion, which can occur for greater than 12 months (El-Matbouli *et al*., 1995; El-Matbouli and Hoffmann, 1998).

3.16.4 Host population dynamics

Although all salmonids are susceptible, infection onset and intensity depend on many factors including fish species (Hedrick *et al*., 1999, 2001; Baldwin *et al*., 2000; Vincent, 2002), size, age, genetics, water temperature and concentration of TAMs during exposure (Hoffman and Byrne, 1974; Vincent, 2002). Rainbow trout are the most susceptible while coho salmon, brown trout, Arctic grayling and lake trout are considered relatively refractory (Hedrick *et al*., 1999). The intermediate *T. tubifex* host is widespread in a range of

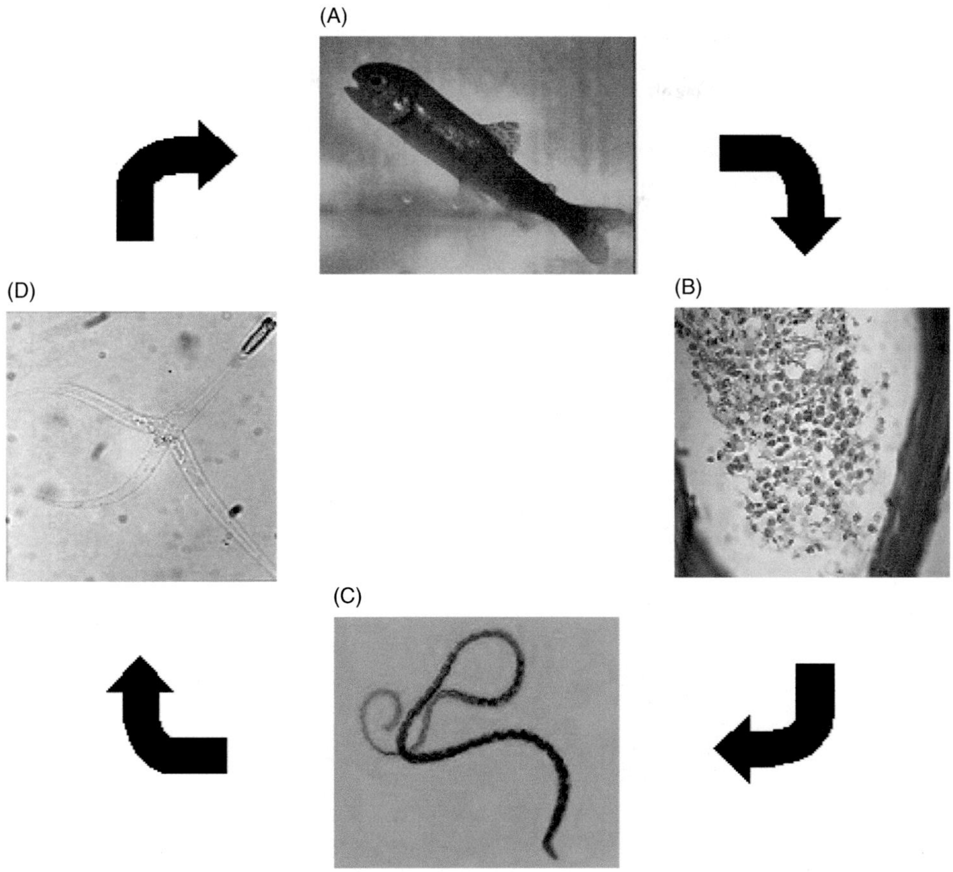

Fig. 3.7. Life cycle of *Myxobolus cerebralis*. (A) Susceptible salmonid host becomes infected by TAM spores. (B) Myxospores form in the head cartilage of infected salmonids. (C) *Tubifex tubifex* ingest myxospores and TAMs develop in the digestive tract. (D) TAMs are released into the environment or consumed by a susceptible host.

environments and is commonly present in habitats where salmonids are found.

3.16.5 Climate change impacts

Due to the relationship and wide range of habitats in which the intermediate host is found, the impacts due to climate change and water temperature fluctuations will likely be related to changes in host geographic ranges. Concern over rearing of fish in freshwater habitats where the parasite occurs is high, but efforts to limit spread of this parasite and the availability of disease-resistant rainbow trout strains should minimize overall risk to aquaculture.

3.16.6 Control

The only way to control whirling disease is to prevent exposure of fish to the infectious stage of *M. cerebralis*. Due to the risk associated with carrier fish, they are often required to be destroyed or their movement is severely restricted if a facility is detected as positive for this parasite. To prevent exposure, an *M. cerebralis*-free water source must be used for rearing of fish, especially during early susceptible life stages. Although various treatments have been investigated to control the effects of whirling disease, none have been satisfactory. A vaccine could theoretically be developed, and antibodies can be produced to TAMs following exposure or immunization of rainbow trout (Adkinson *et al.*, 1997). However, considering the mode of infection and the parasite's use of the nervous system to avoid the immune response, it is doubtful that vaccination would be an option for prevention of this disease. Facilities that have been diagnosed as positive for *M. cerebralis* have had some success by depopulating and completely disinfecting if they have a closed water source.

3.17 Gyrodactylosis

Gyrodactylus is a genus of ectoparasite known to infect more than 150 species of fish ranging from salt to fresh water under many different temperatures and environmental conditions (Bakke *et al.*, 2002, 2007). With regard to coldwater aquaculture and cage culture, a single species, *Gyrodactylus salaris*, has undoubtedly had the most significant impact by causing high mortality in both wild and cultured freshwater salmonids in Europe, specifically Atlantic salmon in Norway. Although most *Gyrodactylus* infect only a single host species (Bakke *et al.*, 2002), *G. salaris* is known to infect multiple salmonid hosts, of which Atlantic salmon appear to be highly susceptible.

3.17.1 Impact

G. salaris has been reported from at least 13 European countries and will likely spread further via the trade of infected salmonids, particularly rainbow trout (Bakke *et al.*, 2007). Norway has been the hardest hit by this pathogen, and it has been estimated that the direct cost associated with *G. salaris* in Norway has been over US$600 million during over past 40+ years (Bakke *et al.*, 2007). Natural salmonid populations have been decimated by nearly 90% in many Norwegian rivers (Johnsen *et al.*, 1999), and the annual loss in production during the culture of salmon associated with this pathogen is thought be as high as 15–20% in some years (Bakke *et al.*, 2007).

3.17.2 Disease characterization and diagnosis

G. salaris is a small monogenean ectoparasite (flatworm) about 0.5 mm in length. Although *Gyrodactylus* species are known to inhabit both fresh and salt water, most species are not considered eurohaline and *G. salaris*, for example, survives only in fresh water. Although most species of *Gyrodactylus* currently identified (71%) are specific to a single host species (Bakke *et al.*, 2002), *G. salaris* is known to infect multiple hosts including Atlantic salmon, rainbow trout, Arctic char, brook trout, grayling, lake trout and brown trout (listed in order of presumed susceptibility). The biology, reproduction and host specificity of *Gyrodactylus* have been discussed in detail in previous reviews (Bakke *et al.*, 2002, 2007; Cable and Harris, 2002).

Usually there are no clinical signs of disease in fish with a low (<50–100 parasites per fish) level of infection. As infection increases, flashing is typically observed, and fish may also become greyish due to increased mucus production. Heavily parasitized fish become lethargic and are usually found in slower-moving water (Mo, 2009). During heavy infection, the dorsal and pectoral fins may become whitish as a result of increased thickening of the epidermis. Secondary fungal infections (*Saprolegnia* spp.) are also commonly observed due to tissue destruction and necrosis caused by *G. salaris* (Johnsen, 1978; Bakke *et al.*, 2007).

The parasite is extremely difficult to see with the naked eye but can be seen with a handheld lens *in situ* or from fin clippings using a dissection microscope. Scrapings (wet mounts) from skin or fins are used to detect *Gyrodactylus* specimens on fish by compound microscopy; however, preparations of wet mounts are usually not suitable for identification to the species level and low levels of infection are often missed (Mo, 2009). For surveillance monitoring or a suspected low-level infection, fin examination by dissection microscopy is the most straightforward solution as even low-level infections of *Gyrodactylus* can usually be observed on fins if at all present on the fish (Mo, 2009). If species confirmation is required, it has historically been obtained based on morphology and morphometry of hamuli (anchoring hooks) and bars in the opisthaptor attachment organ (Harris *et al.*, 1999). Additionally, several strains of *G. salaris* have been identified on the basis of genotyping with the mitochondrial cytochrome oxidase 1 (COI) marker (Meinila *et al.*, 2002, 2004; Hansen *et al.*, 2003, 2007) and PCR sequencing for species confirmation is becoming a more prevalent definitive diagnostic tool.

3.17.3 Transmission

The parasite attaches to the fish by a large specialized posterior attachment organ, the haptor, which has 16 hooks around its margin. Initial infections are usually found around the fins and head of the host fish, which can spread to the entire body surface. All *Gyrodactylus* have an unusual mode of reproduction in that adult worms contain several generations of embryos boxed one inside another akin to a 'Russian doll'. Each parasite gives birth to a fully grown worm which attaches to the host alongside its parent or moves to nearby hosts via waterborne or contact transmission.

3.17.4 Host population dynamics

As with virtually all pathogens, high host densities typically associated with cage-culture environments provide opportunities for rapid transmission and development of high parasite densities, which can lead to substantial mortality if mitigation is not undertaken. The rather long infectious period of these parasites (typically 90 days) and sometimes low-level infections can also make movement of infected fish between sites a concern.

3.17.5 Climate change impacts

Warmer water temperatures are thought to lead to faster maturation times of *G. salaris* (Denholm *et al.*, 2013) and field observations of *G. salaris* in a Norwegian river system appeared positively correlated with water temperature (Jansen and Bakke, 1993). However, modelling *G. salaris* infectivity in association with temperature has also identified biological trade-offs associated with number of offspring produced versus timing of first birth, which may limit temperature-associated effects on parasite abundance (Denholm *et al.*, 2013).

3.17.6 Control

G. salaris is sensitive to most disinfection chemicals (e.g. compounds containing hypochlorite or iodine), which should be used to disinfect equipment associated with infected or potentially infected stock. Acidified aluminium sulfate has been used to effectively eliminate *G. salaris* from fish in laboratory trials (Poleo *et al.*, 2004); attempts have been made to use this chemical for eradication of the parasite in river systems in Norway, but its overall effectiveness on such a large scale is uncertain at best (Soleng *et al.*, 2005; Bakke *et al.*, 2007). Rotenone has also been used to eradicate infected stocks; however, as this chemical can indiscriminately kill all gilled aquatic animals and invertebrates including fish at the concentrations needed to kill *G. salaris*, its use in natural environments has caused controversy (Bakke *et al.*, 2007). Avoidance or separation of infected stocks, if possible, remains the best control practice with regard to minimizing the impact of this aquatic pathogen.

3.18 Climate Change Impacts

The main climate change impact on cold freshwater fish pathogens derives from temperature fluctuation. There is evidence that increasing environmental temperature will make hosts more susceptible and pathogens more virulent, causing more disease outbreaks. Climate change has and will undoubtably continue to incur instability and force adaptation of both host and pathogens of cage-cultured fish. Unfortunately, anticipating the degree, direction or variance associated with future climate change involves extreme uncertainty. Add to this further uncertainty of how fish and pathogens will adapt to such changes, and the future seems near unpredictable. It is therefore recommendable to promote general resilience characteristics in coldwater cultured fish, both in the direction of disease resistance and climate tolerances, to give freshwater cage culture the best chances for sustainably moving forward into the coming decades.

3.19 Conclusions

The major pathogenic diseases of coldwater fish in fresh water have remained relatively consistent for over 20 years. Although new emerging flavobacterial pathogens and various viral pathogens are concerning, conversely minimal progress has been made in eradicating established diseases affecting production. It is not surprising, therefore, that prudent disease mitigation approaches have remained consistent for decades: implementation of strict biosecurity, vaccination use when possible, possession of up-to-date knowledge of location-specific pathogen risks, and establishment of a mitigation plan for dealing with endemic pathogens.

References

Adams, A., Richards, R.H. and Marin de Mateo, M. (1992) Development of monoclonal antibodies to PKX, the causative agent of proliferative kidney disease. *Journal of Fish Diseases* 15, 515–521.

Adelmann, M., Kollner, B., Bergmann, S.M., Fishcher, U., Lange, B., *et al.* (2008) Development of an oral vaccine for immunization of rainbow trout (*Oncorhynchus mykiss*) against viral haemorrhagic septicaemia. *Vaccine* 26, 837–844.

Adkinson, M.A., Levine, R.P. and Hedrick, R.P. (1997) Investigations into the immune response of rainbow trout to *Myxobolus cerebralis*, the causative agent of whirling disease. In: *Proceedings of the National Whirling Disease Symposium, Logan, Utah*. Montana Water Center, Bozeman, Montana, pp. 141–150.

Alcorn, S., Murray, A.L., Pascho, R.J. and Varney, J. (2005) A cohabitation challenge to compare the efficacies of vaccines for bacterial kidney disease (BKD) in chinook salmon *Oncorhynchus tshawytscha*. *Diseases of Aquatic Organisms* 63, 151–160.

Altinok, I., Grizzle, J.M. and Liu, Z. (2001) Detection of *Yersinia ruckeri* in rainbow trout blood by use of the polymerase chain reaction. *Diseases of Aquatic Organisms* 44, 29–34.

Anderson, C.L., Canning, E.U. and Okamura, B. (1999) 18S rDNA sequences indicate that PKX organism parasitizes Bryozoa. *Bulletin of the European Association of Fish Pathologists* 19, 94–97.

Andree, K.B., MacConnell, E. and Hedrick, R.P. (1998) A nested polymerase chain reaction for the detection of genomic DNA of *Myxobolus cerebralis* in rainbow trout *Oncorhynchus mykiss*. *Diseases of Aquatic Organisms* 34, 145–154.

Arakawa, C.K., Deering, R.E., Higman, K.H., Oshima, K.H., O'Hara, P.J. and Winton, J.R. (1990) Polymerase chain reaction (PCR) amplification of a nucleoprotein gene sequence of infectious hematopoietic necrosis virus. *Diseases of Aquatic Organisms* 8, 165–170.

Argenton, F., De Mas, S., Malocco, C., Dalla Valle, L. and Giorgetti, G. (1996) Use of random DNA amplification to generate specific molecular probes for hybridization tests and PCR-based diagnosis of *Yersinia ruckeri*. *Diseases of Aquatic Organisms* 24, 121–127.

Arnzen, J.M., Ristow, S.S., Hesson, C.P. and Lientz, J. (1991) Rapid fluorescent antibody tests for infectious hematopoietic necrosis virus (IHNV) utilizing monoclonal antibodies to the nucleoprotein and glycoprotein. *Journal of Aquatic Animal Health* 3, 109–113.

Aso, Y., Wani, J., Klenner, D.A.S. and Yoshimizu, M. (2001) Detection and identification of *Oncorhynchus masou* virus (OMV) by polymerase chain reaction (PCR). *Bulletin of Fisheries Sciences, Hokkaido University* 52, 111–116.

Austin, B. and Austin, D.A. (1993) *Bacterial Fish Pathogens: Diseases in Farmed and Wild Fish*, Vol. 2. Ellis Horwood Ltd, Chichester, UK.

Austin, B. and Austin, D.A. (2007) *Bacterial Fish Pathogens: Diseases of Farmed and Wild Fish*, 4th edn. Praxis Publishing Ltd, Chichester, UK.

Bader, J.A. and Shotts, E.B. Jr (1998) Determination of phylogenetic relationships of *Flavobacterium psychrophilum* (*Flexibacter psychrophilus*), *Flavobacterium columnare* (*Flexibacter columnaris*), and *Flexibacter maritimus* by sequence analysis of 16S ribosomal RNA genes amplified by polymerase chain reaction. *Journal of Aquatic Animal Health* 10, 320–327.

Baldwin, T.J., Vincent, E.R., Silflow, R.M. and Stanek, D. (2000) *Myxobolus cerebralis* infection in rainbow trout (*Oncorhynchus mykiss*) and brown trout (*Salmo trutta*) exposed under natural stream conditions. *Journal of Veterinary Diagnostic Investigation* 12, 312–321.

Bakke, T.A., Harris, P.D. and Cable, J. (2002) Host specificity dynamics: observations on gyrodactylid monogeneans. *International Journal for Parasitology* 32, 281–308.

Bakke, T.A., Cable, J. and Harris, P.D. (2007) The biology of gyrodactylid monogeneans: the 'Russian-doll killers'. *Advances in Parasitology* 64, 161–376.

Barnes, M.E. and Soupir, C.A. (2006) Evaluation of formalin and hydrogen peroxide treatment regimes on rainbow trout eyed eggs. *North American Journal of Aquaculture* 69, 5–10.

Barnes, M.E. and Stephenson, H. (2003) Use of hydrogen peroxide and formalin treatments during incubation of landlocked fall chinook salmon eyed eggs. *North American Journal of Aquaculture* 65, 151–154.

Beakes, G. (1982) A comparative account of cyst coat ontogeny in saprohytic and fish-lesion (pathogenic) isolates of the *Saprolegnia diclina–parasitica* complex. *Canadian Journal of Botany* 61, 603–625.

Bernardet, J.F., Segers, P., Vancanneyt, M., Berthe, F., Kersters, K. and Vandamme, P. (1996) Cutting a Gordian knot: emended classification and description of the genus *Flavobacterium*, emended description of the family Flavobacteriaceae, and proposal of *Flavobacterium hydatis* nom. nov. (basonym, *Cytophaga aquatilis* Strohl and Tait 1978). *International Journal of Systematic Bacteriology* 46, 128–148.

Bernoth, E.-M. (1997) Furnuculosis: the history of the diseases and of disease research. In: Bernoth, E.-M., Ellis, A.E., Midtyng, P.J., Olivier, G. and Smith, P. (eds) *Furunculosis – Multidisciplinary Fish Disease Research*. Academic Press, London, pp. 1–20.

Bertolini, J.M., Wakabayashi, H., Watral, V.G., Whipple, M.J. and Rohovec, J.S. (1994) Electrophoretic detection of proteases from selected strains of *Flexibacter psychrophilus* and assessment of their variability. *Journal of Aquatic Animal Health* 6, 224–233.

Betts, A.M. and Stone, D.M. (2000) Nucleotide sequence analysis of the entire coding regions of virulent and avirulent strains of viral haemorrhagic septicaemia virus. *Virus Genes* 20, 259–262.

Blake, S., Ma, J.Y., Caporale, D.A., Jairath, S. and Nicholson, B.L. (2001) Phylogenetic relationships of aquatic birnaviruses based on deduced amino acid sequences of genome segment A cDNA. *Diseases of Aquatic Organisms* 45, 89–102.

Bootland, L.M. and Leong, J.C. (1999) Infectious haematopoietic necrosis virus. In: Woo, P.T.K. and Bruno, D.W. (eds) *Fish Diseases and Disorders*. Vol. 3. *Viral, Bacterial and Fungal Infections*, 1st edn. CAB International, Wallingford, UK, pp. 57–121.

Bootland, L.M. and Leong, J.C. (2011) Infectious haematopoietic necrosis virus. In: Woo, P.T.K. and Bruno, D.W. (eds) *Fish Diseases and Disorders*, Vol. 3. *Viral, Bacterial and Fungal Infections*, 2nd edn. CAB International, Wallingford, UK, pp. 66–109.

Bootland, L.M., Dobos, P. and Stevenson, R.M.W. (1986) Experimental induction of the carrier state in yearling brook trout: a model challenge protocol for IPNV immunization. *Veterinary Immunology and Immunopathology* 12(1–4), 365–372.

Bootland, L.M., Dobos, P. and Stevenson, R.M. (1991) The IPNV carrier state and demonstration of vertical transmission in experimentally infected brook trout. *Diseases of Aquatic Organisms* 10, 13–21.

Borg, A.F. (1948) Studies on myxobacteria associated with diseases in salmonid fish. PhD dissertation. University of Washington, Seattle, Washington.

Boucher, P. and Baudin Laurencin, F. (1994) Sleeping disease (SD) of salmonids. *Bulletin of the European Association of Fish Pathologists* 14, 179–180.

Boucher, P., Raynard, R.S., Houghton, G. and Baudin Laurencin, F. (1995) Comparative experimental transmission of pancreas disease in Atlantic salmon, rainbow trout and brown trout. *Diseases of Aquatic Organisms* 22, 19–24.

Boutier, M., Ronsmans, M., Rakus, K., Jazowiecka-Rakus, J., Vancsok, C., *et al.* (2015) Cyprinid herpesvirus 3: an archetype of fish alloherpesviruses. *Advances in Virus Research* 93, 161–256.

Boutier, M., Morvan, L., Delrez, N., Origgi, F., Doszpoly, A. and Vanderplasschen, A. (2021) Fish and amphibian alloherpesviruses (Herpesviridae). In: Bamford, D. and Zuckerman, M. (eds) *Encyclopedia of Virology*, Vol. 2. *Viruses as Infectious Agents: Human and Animal Viruses*, 4th edn. Academic Press, London, pp. 306–315.

Bowman, J. and Nowak, B. (2004) Salmonid gill bacteria and their relationship to amoebic gill disease. *Journal of Fish Diseases* 27, 483–492.

Bradley, T.M., Medina, D.J., Chang, P.W. and McClain, J. (1989) Epizootic epitheliotropic disease of lake trout (*Salvelinus namaycush*): history and viral etiology. *Diseases of Aquatic Organisms* 7, 195–201.

Bragg, R.R., and Henton, M.M. (1986) Isolation of *Yersinia ruckeri* from rainbow trout in South Africa. *Bulletin of the European Association of Fish Pathologists* 6, 5–6.
Bridle, A.R., Koop, B.F. and Nowak, B.F. (2012) Identification of surrogates of protection against yersiniosis in immersion vaccinated Atlantic salmon. *PLoS ONE* 7(7), e40841.
Brown, L.L. and Bruno, D.W. (2002) Infectious diseases of coldwater fish in fresh water. In: Woo, P.T.K., Bruno, D.W. and Lim, L.H.S. (eds) *Diseases and Disorders of Finfish in Cage Culture*. CAB International, Wallingford, UK, pp. 107–169.
Brown, L.L., Evelyn, T.P.T., Iwama, G.K., Nelson, W.S. and Levine, R.P. (1995) Bacterial species other than *Renibacterium salmoninarum* cross-react with antisera against *R. salmoninarum* but are negative for the p57 gene of *R. salmoninarum* as detected by the polymerase chain reaction (PCR). *Diseases of Aquatic Organisms* 21, 227–231.
Brown, L.L., Cox, W. and Levinel, R.P. (1997) Evidence that the causal agent of bacterial cold-water disease *Flavobacterium psychrophilum* is transmitted within salmonid eggs. *Diseases of Aquatic Organisms* 29, 213–218.
Brudeseth, B.E., Skall, H.F. and Evensen, O. (2008) Differences in virulence of marine and fresh water isolates of viral haemorrhagic septicemia virus *in vivo* correlate with *in vitro* ability to infect gill epithelial cells and macrophages of rainbow trout (*Oncorhynchus mykiss*). *Journal of Virology* 82, 10359–10365.
Bruno, D.W. (1986) Histopathology of bacterial kidney disease in laboratory infected rainbow trout, *Salmo gairdneri*, Richardson, and Atlantic salmon, *Salmo salar* L. with reference to naturally infected fish. *Journal of Fish Diseases* 9, 523–537.
Bruno, D.W. (1992) *Cytophaga psychrophila* (*Flexibacter psychrophilus*) histopathology associated with mortalities among farmed rainbow trout, *Oncorhynchus mykiss* Walbaum, in the UK. *Bulletin of the European Association of Fish Pathologists* 12, 215–216.
Bruno, D.W. and Munro, A.L.S. (1986) Observations on *Renibacterium salmoninarum* and the salmonid egg. *Diseases of Aquatic Organisms* 1, 83–87.
Bruno, D.W., van West, P. and Beakes, G.W. (2011) *Saprolegnia* and other Oomycetes. In: *Fish Diseases and Disorders*. Vol. 3. *Viral, Bacterial and Fungal Infections*, 2nd ed. CAB International, Wallingford, UK, pp. 669–720.
Bullock, G.L., Conroy, D.A. and Snieszko, S.F. (1971) Bacterial diseases of fishes. In: Snieszko, S.F. and Axelrod, H.R. (eds) *Book 2A of Diseases of Fishes*. T.F.H. Publications, Jersey City, New Jersey, pp. 151–192.
Burbank, D.R., Shah, D.H., LaPatra, S.E., Fornshell, G. and Cain, K.D. (2011) Enhanced resistance to coldwater disease following feeding of probiotic bacterial strains to rainbow trout (*Oncorhynchus mykiss*). *Aquaculture* 321, 185–190.
Burr, A.W. and Beakes, G.W. (1994) Characterization of zoospore and cyst surface structure in saprophytic and fish pathogenic *Saprolegnia* species (oomycete fungal protists). *Protoplasma* 181, 142–163.
Busch, R.A. (1978) Protective vaccines for mass immunisation of trout. *Salmonid* 1, 10–22.
Busch, R.A. (1983) Enteric redmouth disease (*Yersinia ruckeri*). In: Anderson, D.P., Dorson, M. and Dubourget, Ph. (eds) *Antigens of Fish Pathogens*. Collection Fondation Marcel Merieux, Lyon, France, pp. 201–222.
Byon, J.Y., Ohira, T., Hirono, I. and Aoki, T. (2006) Comparative immune responses in Japanese flounder *Paralichthys olivaceus* after vaccination with viral haemorrhagic septicaemia virus (VHSV) recombinant glycoprotein and DNA vaccine using a microarray analysis. *Vaccine* 24, 921–930.
Cain, K.D. and LaFrentz, B.R. (2007) Laboratory maintenance of *Flavobacterium psychrophilum* and *Flavobacterium columnare*. *Current Protocols in Microbiology* 6, 13B.1.1–13B.1.12. doi: 10.1002/9780471729259.mc13b01s6.
Cain, K.D. and Polinski, M.P. (2014) Infectious diseases of coldwater fish in fresh water. In: Woo, P.T.K. and Bruno, D.W. (eds) *Diseases and Disorders of Finfish in Cage Culture*, 2nd edn. CAB International, Wallingford, UK, pp. 60–113.
Cable, J. and Harris, P.D. (2002) Gyrodactylid developmental biology: historical review, current status and future trends. *International Journal for Parasitology* 32, 255–280.
Canning, E.U., Curry, A., Feist, S.W., Longshaw, M. and Okamura, B. (1999) *Tetracapsula bryosalmonae* n.sp. for PKX organism, the cause of PKD in salmonid fish. *Bulletin of the European Association of Fish Pathologists* 19, 203–206.
Canning, E.U., Curry, A., Feist, S.W., Longshaw, M. and Okamura, B. (2000) A new class and order of myxozoans to accommodate parasites of bryozoans with ultrastructural observations on *Tetracapsula bryosalmonae* (PKX organism). *Journal of Eukaryotic Microbiology* 47, 456–468.

Cascón, A., Anguita, J., Hernanz, C., Sánchez, M., Fernandez, M. and Naharro, G. (1996) Identification of *Aeromonas hydrophila* hybridization group 1 by PCR assays. *Applied and Environmental Microbiology* 62, 1167–1170.
Castagnaro, M., Marin de Mateo, M., Ghittino, C. and Hedrick, R.P. (1991) Lectin histochemistry and ultrastructure of the rainbow trout *Oncorhynchus mykiss* affected by proliferative kidney disease. *Diseases of Aquatic Organisms* 10, 173–183.
Castric, J. and Jeffroy, J. (1991) Experimentally induced diseases in marine fish with IHNV and a rhabdovirus of eel. *European Aquaculture Society Special Publication* 14, 54–55.
Castric, J., Baudin Laurencin, F., Bremont, M., Jeffroy, J., Le Ven, A. and Bearzotti, M. (1997) Isolation of the virus responsible for sleeping disease in experimentally infected rainbow trout (*Oncorhynchus mykiss*). *Bulletin of the European Association of Fish Pathologists* 17, 27–30.
Cavender, W.P., Wood, J.S., Powell, M.S., Overturf, K. and Cain, K.D. (2004) Real-time quantitative PCR (QPCR) to identify *Myxobolus cerebralis* in rainbow trout (*Oncorhynchus mykiss*). *Diseases of Aquatic Organisms* 60, 205–213.
Chen, Y.C., Davis, M.A., LaPatra, S.E., Cain, K.D., Snekvik, K.R. and Call, D.R. (2008) Genetic diversity of *Flavobacterium psychrophilum* recovered from commercially raised rainbow trout, *Oncorhynchus mykiss* (Walbaum), and spawning coho salmon, *O. kisutch* (Walbaum). *Journal of Fish Diseases* 31, 765–773.
Christie, K.E., Graham, D.A., McLoughlin, M.F., Villoing, S., Todd, D. and Knappskog, D. (2007) Experimental infection of *Salmo salar* pre-smolts by i.p. injection with new Irish and Norwegian pancreas disease virus (PDV) isolates – a comparative study. *Diseases of Aquatic Organisms* 75, 13–22.
Cipriano, R.C. and Holt, R.A. (2005) Flavobacterium psychrophilum*, Cause of Bacterial Cold-water Disease and Rainbow Trout Fry Syndrome*. Fish Disease Leaflet No. 86. US Department of the Interior, US Geological Survey, National Fish Health Research Laboratory, Kearneysville, West Virginia.
Congleton, J.L. (1988) Getting in control of deadly IHN. *Focus on Renewable Resources* 14, 18–19.
Conrad, J.F. and DeCew, M. (1967) Observations on deformed juvenile coho salmon. *Fisheries Commission of Oregon Briefs* 13, 129.
Coulibaly, F., Chevalier, C., Delmas, B. and Rey, F.A. (2010) Crystal structure of an aquabirnavirus particle: insights into antigenic diversity and virulence determinism. *Journal of Virology* 84, 1792–1799.
Dale, O.B., Orpetveit, I., Lyngstad, T., Kahns, S., Skall, H.F., *et al.* (2009) Outbreak of viral haemorragic septicaemia (VHS) in seawater-farmed rainbow trout in Norway caused by VHS virus genotype III. *Diseases of Aquatic Organisms* 85, 93–103.
Davis, H.S. (1946) *Care and Diseases of Trout*. US Department of the Interior Report No. 12. US Government Printing Office, Washington, DC.
Davison, A.J. (2010) Herpesvirus systematics. *Veterinary Microbiology* 143, 52–69.
Deering, R.E., Arakawa, C.K., Oshima, K.H., O'Hara, P.J., Landolt, M.L. and Winton, J.R. (1991) Development of a biotinylated DNA probe for detection and identification of infectious hematopoietic necrosis virus. *Diseases of Aquatic Organisms* 1, 57–65.
de las Heras, A.I., Rodríguez Saint-Jean, S. and Pérez-Prieto, S.I. (2010) Immunogenic and protective effects of an oral DNA vaccine against infectious pancreatic necrosis virus in fish. *Fish & Shellfish Immunology* 28, 562–570.
del Cerro, A., Márquez, I. and Prieto, J.M. (2010) Genetic diversity and antimicrobial resistance of *Flavobacterium psychrophilum* isolated from cultured rainbow trout, *Oncorhynchus mykiss* (Walbaum), in Spain. *Journal of Fish Diseases* 33, 285–291.
Denholm, S.J., Norman, R.A., Hoyle, A.S., Shinn, A.P. and Taylor, N.G. (2013) Reproductive trade-offs may moderate the impact of *Gyrodactylus salaris* in warmer climates. *PLoS ONE* 8(10), e78909.
Dixon, P.F. and Hill, B.J. (1984) Rapid detection of fish rhabdoviruses by the enzyme-linked immunosorbent assay (ELISA). *Aquaculture* 42, 1–12.
Dixon, P., Paley, R., Alegria-Moran, R. and Oidtmann, B. (2016) Epidemiological characteristics of infectious hematopoietic necrosis virus (IHNV): a review. *Veterinary Research* 47, 63.
Doszpoly, A., Karaseva, T.A., Waltzek, T.B., Kalabekov, I.M. and Shchelkunov, I.S. (2013) Atlantic salmon papillomatosis in Russia and molecular characterization of the associated herpesvirus. *Diseases of Aquatic Organisms* 107, 121–127.
Drolet, B.S., Rohovec, J.S. and Leong, J.C. (1994) The route of entry and progression of infectious haematopoietic necrosis virus in *Oncorhynchus mykiss* (Walbaum): a sequential immunohistochemical study. *Journal of Fish Diseases* 17, 337–344.
Duff, D.C.B. (1942) The oral immunization of trout against *Bacterium salmonicida*. *Journal of Immunology* 44, 87–94.

Einer-Jensen, K., Ahrens, P., Forsberg, R. and Lorenzen, N. (2004) Evolution of the fish rhabdovirus viral haemorrhagic septicaemia virus. *Journal of General Virology* 85, 1167–1179.

Ekman, E., Akerman, G., Balk, L. and Norrgren, L. (2003) Nanoinjection as a tool to mimic vertical transmission of *Flavobacterium psychrophilum* in rainbow trout *Oncorhynchus mykiss*. *Diseases of Aquatic Organisms* 55, 93–99.

El-Matbouli, M. and Hoffmann, R.W. (1998) Light and electron microscopic studies on the chronological development of *Myxobolus cerebralis* to the actinosproean stage in *Tubifex tubifex*. *International Journal for Parasitology* 28, 195–217.

El-Matbouli, M., Hoffmann, R.W. and Mandok, C. (1995) Light and electron microscopic observations on the route of the triactinomyxon-sporoplasm of *Myxobolus cerebralis* for epidermis into rainbow trout cartilage. *Journal of Fish Biology* 46, 919–935.

El-Matbouli, M., Hoffmann, R.W., Shoel, H., McDowell, T. and Hedrick, R.P. (1999) Whirling disease: host specificity and interaction between the actinosporean stage of *Myxobolus cerebralis* and rainbow trout *Oncorhynchus mykiss*. *Diseases of Aquatic Organisms* 35, 1–12.

Elsayed, E., Faisal, M., Thomas, M., Whelan, G., Batts, W. and Winton, J.R. (2006) Isolation of viral haemorrhagic septicaemia virus from muskellunge, *Esox masquinongy* (Mitchill), in Lake St Clair, Michigan, USA reveals a new sublineage of the North American genotype. *Journal of Fish Diseases* 29, 611–619.

Emmerich, R. and Weibel, E. (1894) Ueber eine durch Bakterien erzengte Seuche unter den Forellen. *Archives für Hygiene und Bakteriologie* 21, 1–21.

Enriquez, R. and Zamora, J. (1987) Isolation of *Yersinia ruckeri* in carp (*Cyprinus carpio*) in Valdivia. *Archivos de Medicina Veterinaria* 19, 33–36.

Enzmann, P.J. (1983) Considerations on the effectiveness of VHS vaccination. *Bulletin of the European Association of Fish Pathologists* 5, 54–55.

Enzmann, P.J., Kurath, G., Fichtner, D. and Bergmann, S.M. (2005) Infectious hematopoietic necrosis virus: monophyletic origin of European isolates from North American genogroup M. *Diseases of Aquatic Organisms* 66, 187–195.

Epp, J.K., Wood, J.S. and Milton, J.B. (2002) New PCR test for *Myxobolus cerebralis* based on a heat-shock protein gene. In: *Proceedings of the 8th Annual Whirling Disease Symposium, Denver, Colorado*. Whirling Disease Foundation, Bozeman, Montana, pp. 61–62.

Evelyn, T.P.T., Prosperi-Porta, L. and Ketcheson, J.E. (1990) Two new techniques for obtaining consistent results when growing *Renibacterium salmoninarum* on KDM2 culture medium. *Diseases of Aquatic Organisms* 9, 209–212.

Evenden, A.J., Grayson, T.H., Gilpin, M.L. and Munn, C.B. (1993) *Renibacterium salmoninarum* and bacterial kidney disease – the unfinished jigsaw. *Annual Review of Fish Diseases* 3, 87–104.

Faisal, M., Loch, T.P., Brenden, T.O., Eissa, A.E., Ebener, M.P., *et al.* (2010a) Assessment of *Renibacterium salmoninarum* infections in four lake whitefish (*Coregonus clupeaformis*) stocks from northern Lakes Huron and Michigan. *Journal of Great Lakes Research* 36, 29–37.

Faisal, M., Eissa, A.E. and Starliper, C.E. (2010b) Recovery of *Renibacterium salmoninarum* from naturally infected salmonid stocks in Michigan using a modified culture protocol. *Journal of Advanced Research* 1, 95–102.

Faisal, M., Loch, T.P., Shavalier, M., VanDeuren, M.G., Standish, I., *et al.* (2019) Resurgence of Salmonid herpesvirus-3 infection (epizootic epitheliotropic disease) in hatchery-propagated lake trout in Michigan. *Journal of Aquatic Animal Health* 31, 31–45.

FDA (US Food and Drug Administration) (2007) *FDA/CVM Approved Drugs for Use in Aquaculture*. Center for Veterinary Medicine, US Food and Drug Administration, Blacksburg, Virginia.

Ferguson, H.W. (1981) The effects of water temperature on the development of proliferative kidney disease in rainbow trout, *Salmo gairdneri* Richardson. *Journal of Fish Diseases* 4, 175–177.

Ferguson, H.W. and Ball H.J. (1979) Epidemiological aspects of proliferative kidney disease amongst rainbow trout, *Salmo gairdneri* Richardson, in Northern Ireland. *Journal of Fish Diseases* 2, 219–225.

Fernández, L., Prieto, M. and Guijarro, J.A. (2007) The iron- and temperature-regulated haemolysin YhlA is a virulence factor of *Yersinia ruckeri*. *Microbiology* 153, 483–489.

Fijan, N. (1972) Infectious dropsy in carp – a disease complex. In: Mawdesley-Thomas, L.E. (ed.) *Diseases of Fish. Symposia of the Zoological Society of London*. Academic Press, London, pp. 39–51.

Frerichs, G.N., Stewart, J.A. and Collins, R.O. (1985) Atypical infection of rainbow trout, *Salmo gairdneri* Richardson, with *Yersinia ruckeri*. *Journal of Fish Diseases* 8, 383–387.

Fryer, J.L. and Sanders, J.E. (1981) Bacterial kidney disease of salmonid fish. *Annual Reviews in Microbiology* 35, 273–298.

Furihata, M., Hosoe, A., Takei, K., Kohara, M., Kakamura, J., *et al.* (2003) Outbreak of salmonid herpesviral disease in cultured rainbow trout. *Fish Pathology* 38, 23–25.
Gaikowski, M.P., Rach, J.J., Olson, J.J. and Ramsay, R.T. (1998) Toxicity of hydrogen peroxide treatments to rainbow trout eggs. *Journal of Aquatic Animal Health* 10, 241–251.
Gaikowski, M.P., Rach, J.J. and Ramsay, R.T. (1999) Acute toxicity of hydrogen peroxide treatments to selected lifestages of cold-, cool-, and warmwater fish. *Aquaculture* 178, 191–207.
Garver, K.A., Batts, W.N. and Kurath, G. (2006) Virulence comparisons of infectious hematopoietic necrosis virus U and M genogroups in sockeye salmon and rainbow trout. *Journal of Aquatic Animal Health* 18, 232–243.
Glenney, G.W., Barbash, P.A. and Coll, J.A. (2016a) Initial detection and molecular characterization of namaycush herpesvirus (Salmonid herpesvirus 5) in lake trout. *Journal of Aquatic Animal Health* 28, 46–55.
Glenney, G.W., Barbash, P.A. and Coll, J. A. (2016b) A quantitative polymerase chain reaction assay for the detection and quantification of epizootic epitheliotropic disease virus (EEDV; Salmonid herpesvirus 3). *Journal of Aquatic Animal Health* 28, 56–67.
Gonen, S., Baranski, M., Thorland, I., Norris, A., Grove, H., *et al.* (2015) Mapping and validation of a major QTL affecting resistance to pancreas disease (salmonid alphavirus) in Atlantic salmon (*Salmo salar*). *Heredity* 115, 405–414. doi: org/10.1038/hdy.2015.37.
Graham, D.A., Taylor, C., Rodgers, D., Weston, J., Khalili, M., *et al.* (2006) Development and evaluation of a one-step real-time reverse transcription polymerase chain reaction assay for the detection of salmonid alphaviruses in serum and tissue. *Diseases of Aquatic Organisms* 70, 47–54.
Graham, D.A., Jewhurst, H.L., McLoughlin, M.F., Branson, E.J., McKenzie, K., *et al.* (2007) Serological, virological and histopathological study of an outbreak of sleeping disease in farmed rainbow trout *Oncorhynchus mykiss*. *Diseases of Aquatic Organisms* 74, 191–197.
Graham, D.A., Brown, A., Savage, P. and Frost, P. (2012) Detection of salmon pancreas disease virus in the faeces and mucus of Atlantic salmon, *Salmo salar* L., by real-time RT-PCR and cell culture following experimental challenge. *Journal of Fish Diseases* 35, 949–951.
Grandes, J.M.F., Diez, M.F. and Gancedo, J.M.A. (2000) Ultrastructural analysis of *Saprolegnia* secondary zoospore cyst ornamentation from infected wild brown trout, *Salmo trutta* L, and river water indicates two distinct morphotypes amongst long-spined isolates. *Journal of Fish Diseases* 23, 147–160.
Grayson, T.H., Cooper, F.F., Wrathmell, A.B., Evenden, A.J. and Gilpin, M.L. (2002) Host responses to *Renibacterium salmoninarum* and specific components of the pathogen reveal the mechanisms of immune suppression and activation. *Immunology* 106, 273–283.
Griffin, P.J., Snieszko, S.F. and Friddle, S.B. (1953) A more comprehensive description of *Bacterium salmonicida*. *Transactions of the American Fisheries Society* 82, 129–138.
Griffiths, S.G., Melville, K.J. and Salonis, K. (1998) Reduction of *Renibacterium salmoninarum* culture activity in Atlantic salmon following vaccination with avirulent strains. *Fish & Shellfish Immunology* 8, 607–619.
Guerriero, G., Avino, M., Zhou, Q., Fugelstad, J., Clergeot, P.-H. and Bulone, V. (2010) Chitin synthases from *Saprolegnia* are involved in tip growth and represent a potential target for anti-Oomycete drugs. *PLoS Pathology* 6, e1001070.
Gunimaladevi, I., Kono, T., LaPatra, S.E. and Sakai, M. (2005) A loop mediated isothermal amplification (LAMP) method for detection of infectious hematopoietic necrosis virus (IHNV) in rainbow trout (*Oncorhynchus mykiss*). *Archives of Virology* 150, 899–909.
Guy, D.R., Bishop, S.C., Woolliams, J.A. and Brotherstone, S. (2009) Genetic parameters for resistance to infectious pancreatic necrosis in pedigreed Atlantic salmon (*Salmo salar*) post-smolts using a reduced animal model. *Aquaculture* 290, 229–235.
Hansen, H., Bachmann, L. and Bakke, T.A. (2003) Mitochondrial DNA variation of *Gyrodactylus* spp. (Monogenea, Gyrodactylidae) populations infecting Atlantic salmon, grayling, and rainbow trout in Norway and Sweden. *International Journal for Parasitology* 33, 1471–1478.
Hansen, H., Bakke, T.A. and Bachmann, L. (2007) Mitochondrial haplotype diversity of *Gyrodactylus thymalli* (Platyheminthes; Monogenea): extended geographic sampling in the United Kingdom, Poland, and Norway reveals further lineages. *Parasitology Research* 100, 1389–1394.
Harmache, A., LeBerre, M., Droineau, S., Giovannini, M. and Brémont, M. (2006) Bioluminescence imaging of live infected salmonids reveals that the fin bases are the major portal of entry for *Novirhabdovirus*. *Journal of Virology* 80, 3655–3659.

Harris, P.D., Cable, J., Tinsley, R.C. and Lazarus, C. (1999) Combined ribosomal DNA and morphological analysis of individual gyrodactylid monogeneans. *Journal of Parasitology* 85, 188–191.
Hastings, T.S. and Bruno, D.W. (1985) Enteric redmouth disease: survey in Scotland and evaluation of a new medium, Shotts–Waltman, for differentiating *Yersinia ruckeri*. *Bulletin of the European Association of Fish Pathologists* 54, 2594–2597.
Hawley, L.M. and Garver, K.A. (2008) Stability of viral haemorrhagic septicemia virus (VHSV) in fresh water and seawater at various temperatures. *Diseases of Aquatic Organisms* 82, 171–178.
He, M., Yan, X.C., Liang, Y., Sun, X.W. and Teng, C.B. (2014) Evolution of the viral hemorrhagic septicemia virus: divergence, selection and origin. *Molecular Phylogenetics and Evolution* 77, 34–40.
Hedrick, R.P. (1998) Relationships of the host, pathogen, and environment: implications for disease of cultured and wild fish populations. *Journal of Aquatic Animal Health* 10, 107–111.
Hedrick, R.P. and El-Matbouli, M. (2002) Recent advances with taxonomy, life cycle, and development of *Myxobolus cerebralis* in the fish and oligochaete host. In: Bartholomew, J.L. and Wilson, J.C. (eds) *Whirling Disease: Reviews and Current Topics. American Fisheries Society Symposium 29*. American Fisheries Society, Bethesda, Maryland, pp. 45–53.
Hedrick, R.P., McDowell, T.S., Eaton, W.D., Chan, L. and Wingfield, W. (1986) Herpesvirus salmonis (HPV): first occurrence in anadromous salmonids. *Bulletin of the European Association of Fish Pathologists* 6, 66–67.
Hedrick, R.P., McDowell, T.S., Mukkatira, K. and Georgiadis, M.P. (1999) Susceptibility of selected inland salmonids to experimentally induced infections with *Myxobolus cerebralis*, the causative agent of whirling disease. *Journal of Aquatic Animal Health* 11, 330–339.
Hedrick, R.P., McDowell, T.S., Mukkatira, K. and Georgiadis, M.P. (2001) Susceptibility of three species of anadromous salmonids to experimentally induced infection with *Myxobolus cerebralis*, the causative agent of whirling disease. *Journal of Aquatic Animal Health* 13, 43–50.
Henderson, M. and Okamura, B. (2004) The phylogeography of salmonid proliferative kidney disease in Europe and North America. *Proceedings of the Royal Society of London* 271, 1729–1736.
Henríquez-Núñez, H., Evrard, O., Kronvall, G. and Avendaño-Herrera, R. (2012) Antimicrobial susceptibility and plasmid profiles of *Flavobacterium psychrophilum* strains isolated in Chile. *Aquaculture* 354–355, 38–44.
Heppell, J., Tarrab, E., Lecomte, J., Berthiaume, L. and Arella, M. (1995) Strain variability and localization of important epitopes on the major structural protein (VP2) of infectious pacreatic necrosis virus. *Virology* 214, 40–49.
Herman, R.L. (1968) Fish furunculosis. *Transactions of the American Fisheries Society* 97, 221–230.
Hesami, S., Parkman, J., MacInnes, J.I., Gray, J.T., Gyles, C.L. and Lumsden, J.S. (2010) Antimicrobial susceptibility of *Flavobacterium psychrophilum* isolates from Ontario. *Journal of Aquatic Animal Health* 22, 39–49.
Hill, B.J. and Way, K. (1995) Serological classification of infectious pancreatic necrosis (IPN) virus and other aquatic birnaviruses. *Annual Review of Fish Diseases* 5, 55–77.
Hiney, N. and Olivier, G. (1999) Furunculosis (*Aeromonas salmonidida*). In: Woo, P.T.K. and Bruno, D.W. (eds) *Fish Diseases and Disorders*. Vol. 3. *Viral, Bacterial and Fungal Infections*, 1st edn. CAB International, Wallingford, UK, pp. 341–425.
Hiney, M., Dawson, M.T., Heery, D., Smith, P.R., Gannon, F. and Powell, R. (1992) DNA probe for *Aeromonas salmonicida*. *Applied and Environmental Microbiology* 58, 1039–1042.
Hisae, K., Yoshimizu, M. and Ezura, Y. (2002) Disinfection of water for aquaculture. *Fisheries Science* 68, 821–824.
Hofer, B. (1903) Uber die Drehkrankheit der Regenbogenforelle. *Allgemeine Fischerein Zeitung* 28(1), 1–7.
Hoffman, G.L. (1990) *Myxobolus cerebralis*, a worldwide cause of salmonid whirling disease. *Journal of Aquatic Animal Health* 2, 30–37.
Hoffman, G.L. and Byrne, C.J. (1974) Fish age as related to susceptibility to *Myxosoma cerebralis*, cause of whirling disease. *The Progressive Fish-Culturist* 36, 151.
Holey, M.E., Elliot, R.F., Marcquenski, S.V., Hnath, J.G. and Smith, K.D. (1998) Chinook salmon epizootics in Lake Michigan: possible contributing factors and management implications. *Journal of Aquatic Animal Health* 10, 202–210.
Holliman, A., Austin, B. and Pottinger, T. (1991) A previously unrecognized Cytophage-like bacterium (CLB) causing jaw erosion in farmed rainbow trout (*Oncorhynchus mykiss*). *Bulletin of the European Association of Fish Pathologists* 11, 114–115.

Holmes, P., Niccols, L.M. and Sartory, D.P. (1996) The ecology of mesophilic *Aeromonas* in the aquatic environment. In: Austin, B., Altwegg, M., Gosling, P.J. and Joseph S. (eds) *The Genus Aeromonas*. Wiley, Chichester, UK, pp. 127–150.
Holt, R.A. (1988) *Cytophaga psychrophila*, the causative agent of bacterial cold-water disease in salmonid fish. PhD dissertation. Oregon State University, Corvallis, Oregon.
Hope, K.M., Casey, R.N., Groocock, G.H., Getchell, R.G., Bowser, P.R. and Casey, J.W. (2010) Comparison of quantitative RT-PCR with cell culture to detect viral haemorrhagic septicemia virus (VHSV) IVb infections in the Great Lakes. *Journal of Aquatic Animal Health* 22, 50–61.
Hunter, V.A., Knittel, M.D. and Fryer, J.L. (1980) Stress-induced transmission of *Yersinia ruckeri* infection from carriers to recipient steelhead trout, *Salmo gairdneri* Richardson. *Journal of Fish Diseases* 3, 467–472.
Ilardi, P., Abad, J., Rintamäki, P., Bernardet, J.F. and Avendaño-Herrera, R. (2010) Phenotypic, serological and molecular evidence of *Chryseobacterium piscicola* in farmed Atlantic salmon, *Salmo salar* L., in Finland. *Journal of Fish Diseases* 33, 179–181.
Ilardi, P., Fernández, J. and Avendaño-Herrera, R. (2009) *Chryseobacterium piscicola* sp. nov., isolated from diseased salmonid fish. *International Journal of Systematic and Evolutionary Microbiology* 59, 3001–3005.
Irianto, A. and Austin, B. (2002) Use of probiotics to control furunculosis in rainbow trout, *Oncorhynchus mykiss* (Walbaum). *Journal of Fish Diseases* 25, 333–342.
Jansen, M.D., Bang Jensen, B., McLoughlin, M.F., Rodger, H.D., Taksdal, T., *et al.* (2017) The epidemiology of pancreas disease in salmonid aquaculture: a summary of the current state of knowledge. *Journal of Fish Diseases* 40, 141–155.
Jansen, P.A. and Bakke, T.A. (1993) Regulatory processes in the monogenean *Gyrodactylus salaris* Malmber–Atlantic salmon (*Salmo salar* L.) association. *Fisheries Research* 17, 87–101.
Johnsen, B.O. (1978) The effects of an attach by the parasite *Gyrodactylus salaris* on the population of salmon parr in the river Lakselva, Misvaer in Northern Norway. *Astarte* 11, 7–9.
Johnsen, B.O., Mokkelgjerd, P.I. and Jensen, A.J. (1999) The parasite *Gyrodactylus salaris* on salmon parr in Norwegian rivers, status report at the beginning of year 2000. *NINA Oppdargsmelding* 617, 1–129.
Jorgensen, P.E.V., Olesen, N.J., Lorenzen, N., Winton, J.R. and Ristow, S.S. (1991) Infectious hematopoietic necrosis (IHN) and viral haemorrhagic septicemia (VHS): detection of trout antibodies to the causative viruses by means of plaque neutralization, immunofluorescence, and enzyme-linked immunosorbent assay. *Journal of Aquatic Animal Health* 3, 100–108.
Jutfelt, F., Olsen, R.E., Glette, J., Ringo, E. and Sundell, K. (2006) Translocation of viable *Aeromonas salmonicida* across the intestine of rainbow trout, *Oncorhynchus mykiss* (Walbaum). *Journal of Fish Diseases* 29, 255–262.
Kämpfer, P., Lodders, N., Martin, K. and Avendaño-Herrera, R. (2012) *Flavobacterium chilense* sp. nov. and *Flavobacterium araucananum* sp. nov., isolated from farmed salmonid fish. *International Journal of Systematic and Evolutionary Microbiology* 62, 1402–1408.
Karlsen, M., Tingbo, T., Solbakk, I.T., Evensen, O., Furevik, A. and Aas-Eng, A. (2012) Efficacy and safety of an inactivated vaccine against salmonid alphavirus (family Togaviridae). *Vaccine* 30, 5688–5694.
Kawahara, E., Inarimori, R., Urano, K., Nomura, S. and Takahashi, Y. (1993) Transfer of maternal immunity of white-spotted char *Salvelinus leucomaenis* against furunculosis. *Nippon Suisan Gakkaishi* 59, 567.
Kent, M.L. and Hedrick, R.P. (1985) PKX the causative agent of proliferative kidney disease (PKD) in Pacific salmonid fishes and its affinities with the Myxozoa. *Journal of Protozoology* 32, 254–260.
Kent, M.L., Traxler, G.S., Keiser, D., Richard, J., Dawe, S.C., Shaw, R.W., *et al.* (1998) Survey of salmonid pathogens in ocean-caught fishes in British Columbia, Canada. *Journal of Aquatic Animal Health* 10, 211–219.
Kim, R. and Faisal, M. (2011) Emergence and resurgence of the viral hemorrhagic septicemia virus (Novirhabdovirus, Rhabdoviridae, Mononegavirales). *Journal of Advanced Research* 2, 9–23.
Kimura, T., Yoshimizu, M., Tanaka, M. and Sannohe, H. (1981) Studies on a new virus (OMV) from *Oncorhynchus masou*. I, Characteristics and pathogenicity. *Fish Pathology* 15, 143–147.
Kimura, T., Suzuki, S. and Yoshimizu, M. (1983a) *In vitro* antiviral effect of 9-(2-hydroxyethoymethyl) guanine on the fish herpesvirus *Oncorhynchus masou* virus (OMV). *Antiviral Research* 3, 93–101.
Kimura, T., Suzuki, S. and Yoshimizu, M. (1983b) *In vivo* antiviral effect of 9-(2-hydroxyethoxymethyl) guanine on experimental infection of chum salmon (*Oncorhynchus keta*) fry with *Oncorhynchus masou* virus (OMV). *Antiviral Research* 3, 103–108.

Kondo, M., Kawai, K., Yagyu, K., Nakayama, K., Kurohara, K. and Oshima, S. (2001) Changes in the cell structure of *Flavobacterium psychrophilum* with length of culture. *Microbiology and Immunology* 45, 813–818.

Kongtorp, R.T., Stene, A., Andreassen, P.A., Aspehaug, V., Graham, D.A., *et al.* (2010) Lack of evidence for vertical transmission of SAV 3 using gametes of Atlantic salmon, *Salmo salar* L., exposed by natural and experimental routes. *Journal of Fish Diseases* 33, 879–888.

Kumagai, A. (2005) Bacterial coldwater disease in coho salmon. *Journal of the Japanese Society of Fisheries Science* 71, 645–649.

Kumagai, A. and Nawata, A. (2011) Concentration of *Flavobacterium psychrophilum* in the ovarian fluid and milt of cultured salmonids. *Fish Pathology* 46, 116–119.

Kurath, G. (2008) Biotechnology and DNA vaccines for aquatic animals. *Reviews in Science and Technology* 27, 175–196.

Kurath, G. (2012) Fish novirhabdoviruses. In: Dietzgen, R.G. and Kuzmin, I.V. (eds) *Rhabdoviruses: Molecular Taxonomy, Evolution, Genomics, Ecology, Host–Vector Interactions, Cytopathology and Control*. Caister Academic Press, Norfolk, UK, pp. 89–116.

Kurath, G., Garver, K.A., Troyer, R.M., Emmenegger, E.J., Einer-Jensen, K. and Anderson, E.D. (2003) Phylogeography of infectious haematopoietic necrosis virus in North America. *Journal of General Virology* 84, 803–814.

Kurobe, T., Marcquenski, S. and Hedrick, R.P. (2009) PCR assay for improved diagnostics of epitheliotropic disease virus (EEDV) in lake trout *Salvelinus namaycush*. *Diseases of Aquatic Organisms* 84, 17–24.

LaFrentz, B.R., LaPatra, S.E., Jones, G.R., Congleton, J.L., Sun, B. and Cain, K.D. (2002) Characterization of serum and mucosal antibody responses and relative per cent survival in rainbow trout, *Oncorhynchus mykiss* (Walbaum), following immunization and challenge with *Flavobacterium psychrophilum*. *Journal of Fish Diseases* 25, 703–713.

LaFrentz, B.R., Lindstrom, N.M., LaPatra, S.E., Call, D.R. and Cain, K.D. (2007) Electrophoretic and Western blot analyses of the lipopolysaccharide and glycocalyx of *Flavobacterium psychrophilum*. *Fish & Shellfish Immunology* 23, 770–780.

LaFrentz, B.R., LaPatra, S.E., Call, D.R. and Cain, K.D. (2008) Isolation of rifampicin resistant *Flavobacterium psychrophilum* strains and their potential as live attenuated vaccine candidates. *Vaccine* 26, 5582–5589.

LaFrentz, B.R., LaPatra, S.E., Call, D.R., Wiens, G.D. and Cain, K.D. (2011) Identification of immunogenic proteins within distinct molecular mass fractions of *Flavobacterium psychrophilum*. *Journal of Fish Diseases* 34, 823–830.

LaPatra, S.E. (1998) Factors affecting pathogenicity of infectious hematopoietic necrosis virus (IHNV) for salmonid fish. *Journal of Aquatic Animal Health* 10, 121–131.

LaPatra, S.E., Roberti, K.A., Rohovec, J.S. and Fryer, J.L. (1989a) Fluorescent antibody test for the rapid diagnosis of infectious hematopoietic necrosis. *Journal of Aquatic Animal Health* 1, 29–36.

LaPatra, S.E., Rohovec, J.S. and Fryer, J.L. (1989b) Detection of infectious hematopoietic necrosis virus in fish mucus. *Fish Pathology* 24, 197–202.

LaPatra, S.E., Groberg, W.J., Rohovec, J.S. and Fryer, J.L. (1990a) Size-related susceptibility of salmonids to two strains of infectious hematopoietic necrosis virus. *Transactions of the American Fisheries Society* 119, 25–30.

LaPatra, S.E., Groff, J.M., Fryer, J.L. and Hedrick, R.P. (1990b) Comparative pathogenesis of three strains of infectious hematopoietic necrosis virus in rainbow trout *Oncorhynchus mykiss*. *Diseases of Aquatic Organisms* 8, 105–112.

LaPatra, S.E., Fryer, J.L. and Rohovec, J.S. (1993) Virulence comparison of different electrophero-types of infectious hematopoietic necrosis virus. *Diseases of Aquatic Organisms* 16, 115–120.

LaPatra, S.E., Jones, G.R., Lauda, K.A., McDowell, T.S., Schneider, R. and Hendrick, R.P. (1995) White sturgeon as a potential vector of infectious hematopoietic necrosis virus. *Journal of Aquatic Animal Health* 7, 225–230.

LaPatra, S.E., Barone, L., Jones, G.R. and Zon, L.I. (2000) Effects of infectious hematopoietic necrosis virus and infectious pancreatic necrosis virus infection on hematopoietic precursors of the zebrafish. *Blood Cells, Molecules, and Diseases* 26, 445–452.

Larsen, J.L. and Pedersen, K. (1997) Vaccination strategies in freshwater salmonid aquaculture. *Fish Vaccinology* 90, 391–400.

Leong, J., Evensen, O. and Woo, P.T.K. (2023) Infectious diseases of coldwater fish in marine and brackish waters. In: Woo, P.T.K. and Subasinghe, R.P. (eds) *Climate Change on Diseases and Disorders of Finfish in Cage Culture*, 3rd edn. CAB International, Wallingford, UK, pp. 34–75.

Lindstrom, N.M., Call, D.R., House, M.L., Moffitt, C.M. and Cain, K.D. (2009) A quantitative enzyme-linked immunosorbent assay and filtration-based fluorescent antibody test as potential tools to screen broodstock for infection with *Flavobacterium psychrophilum*. *Journal of Aquatic Animal Health* 21, 43–56.

Llewellyn, L.C. (1980) A bacterium with similarities to the redmouth disease and *Serratia liquifaciens* (Grimes and Hennerty) causing mortalities in hatchery-reared salmonids in Australia. *Journal of Fish Diseases* 3, 29–39.

Loch, T. and Faisal, M. (2014) *Chryseobacterium aahli* sp. nov., isolated from lake trout (*Salvelinus namaycush*) and brown trout (*Salmo trutta*), and emended descriptions of *Chryseobacterium ginsenosidimutans* and *Chryseobacterium gregarium*. *International Journal of Systematic and Evolutionary Microbiology* 64, 1573–1579.

Loch, T. and Faisal, M. (2015) Emerging flavobacterial infections in fish: a review. *Journal of Advanced Research* 6, 283–300.

Lorenzen, E., Dalsgaard, I. and Bernardet, J.F. (1997) Characterization of isolates of *Flavobacterium psychrophilum* associated with coldwater disease or rainbow trout fry syndrome I: phenotypic and genomic studies. *Diseases of Aquatic Organisms* 31, 197–208.

Lorenzen, E., Carstensen, B. and Olesen, N.J. (1999) Inter-laboratory comparison of cell lines for susceptibility to three viruses: VHSV, IHNV and IPNV. *Diseases of Aquatic Organisms* 37, 81–88.

Lorenzen, N. and LaPatra, S.E. (1999) Immunity to rhabdoviruses in rainbow trout: the antibody response. *Fish & Shellfish Immunology* 9, 345–360.

Lorenzen, N. and LaPatra, S.E. (2005) DNA vaccines for aquacultured fish. *Reviews in Science and Technology* 24, 201–213.

Lorenzen, N., Lorenzen, E., Einer-Jensen, K., Heppell, J., Wu, T. and Davis, H. (1998) Protective immunity to VHS in rainbow trout (*Oncorhynchus mykiss*, Walbaum) following DNA vaccination. *Fish & Shellfish Immunology* 8, 261–270.

Lorenzen, N., Lorenzen, E. and Einer-Jensen, K. (2001) Immunity to viral haemorrhagic septicaemia (VHS) following DNA vaccination of rainbow trout at an early life-stage. *Fish & Shellfish Immunology* 11, 585–591.

Lorz, H.V. and Amandi, A. (1994) Parasitic diseases of fishes: whirling disease of salmonids. In: Thoesen, J.C. (ed.) *Blue Book: Suggested Procedures for the Detection and Identification of Certain Finfish and Shellfish Pathogens*, 4th edn. American Fisheries Society, Fish Health Section, Bethesda, Maryland.

MacConnell, E. and Vincent, R.E. (2002) The effects of *Myxobolus cerebralis* on the salmonid host. In: Bartholomew, J.L. and Wilson, J.C. (eds) *Whirling Disease: Reviews and Current Topics. American Fisheries Society Symposium 29*. American Fisheries Society, Bethesda, Maryland, pp. 95–107.

Mackie, T.J., Arkwright, J.A., Pryce-Tannatt, T.E., Mottram, J.C., Johnston, W.D. and Manzies, W.J.M. (1993) *Furunculosis Committee Report, Second Interim*. HMSO, Edinburgh, UK.

Madsen, L., Arnbjerg, J. and Dalsgaard, I. (2001) Radiological examination of the spinal column in farmed rainbow trout *Oncorhynchus mykiss* (Walbaum): experiments with *Flavobacterium psychrophilum* and oxytetracycline. *Aquaculture Research* 32, 235–241.

Madsen, L., Møller, J.D. and Dalsgaard, I. (2005) *Flavobacterium psychrophilum* in rainbow trout, *Oncorhynchus mykiss* (Walbaum), hatcheries: studies on broodstock, eggs, fry and environment. *Journal of Fish Diseases* 28, 39–47.

Marking, L.L., Rach, J.J. and Schreier, T.M. (1994) Evaluation of antifungal agents for fish culture. *The Progressive Fish-Culturist* 56, 225–231.

Markiw, M.E. (1989) Portals of entry for salmonid whirling disease in rainbow trout. *Diseases of Aquatic Organisms* 6, 7–10.

Markiw, M.E. (1991) Whirling disease: earliest susceptible age of rainbow trout to the triactinomyxid of *Myxosoma cerebralis*. *Aquaculture* 92, 1–6.

Markiw, M.E. (1992) *Salmonid Whirling Disease*. US Fish and Wildlife Service Leaflet No. 17. US Fish and Wildlife Service, Washington, DC.

Markiw, M.E. and Wolf, K. (1974) *Myxosoma cerebralis*: isolation and concentration from fish skeletal elements – sequential enzymatic digestions and purification by differential centrifugation. *Journal of the Fisheries Research Board of Canada* 31, 15–20.

Matthews, E. (2019) Environmental factors impacting *Saprolegnia* infections in wild fish stocks. PhD thesis, Cardiff University, Cardiff, UK.

McAllister, P.E. (1993) Salmonid fish viruses. In: Stoskopf, M.K. (ed.) *Fish Medicine*. WB Saunders, Philadelphia, Pennsylvania, pp. 380–408.

McArdle, J.F. and Dooley-Martin, C. (1985) Isolation of *Yersinia ruckeri* type 1 (Hagerman strain) from goldfish *Carassius auratus*. *Bulletin of the European Association of Fish Pathologists* 5, 10–11.
McCarthy, D.H. (1977) Some ecological aspects of the bacterial fish pathogen – *Aeromonas salmonicida*. In: Skinner, F.A. and Shewan, J.M. (eds) *Aquatic Microbiology, Society for Applied Bacteriology Symposium Series 6*. Academic Press, London, pp. 299–324.
McCraw, B.M. (1952) *Furunculosis of Fish*. US Fish and Wildlife Service Special Scientific Report: Fisheries No. 84. US Fish and Wildlife Service, Washington, DC.
McIntosh, D., Flano, E., Grayson, T.H., Gilpin, M.L., Austin, B. and Villena, A.J. (1997) Production of putative virulence factors by *Renibacterium salmoninarum* grown in cell culture. *Microbiology* 143, 3349–3356.
McLoughlin, M.F. and Graham, D.A. (2007) Alphavirus infections in salmonids – a review. *Journal of Fish Diseases* 30, 511–531.
Meinila, M., Kuusela, J., Zietara, M. and Lumme, J. (2002) Primers for amplifying approximately 820 bp of highly polymorphic mitochondrial COI gene of *Gyrodactylus salaris*. *Hereditas* 137, 72–74.
Meinila, M., Kuusela, J., Zietara, M. and Lumme, J. (2004) Initial steps of the speciation by geographic isolation and host switch in salmonid pathogen *Gyrodactylus salaris* (Monogenea: Gyrodactylidae). *International Journal for Parasitology* 34, 515–526.
Metz, S.W., Feenstra, F., Villoing, S., van Hulten, M.C., van Lent, J.W., *et al.* (2011) Low temperature-dependent salmonid alphavirus glycoprotein processing and recombinant virus-like particle formation. *PLoS ONE* 6(10), e25816.
Miller, T.A., Rapp, J., Wastlhuber, U., Hoffmann, R.W. and Enzmann, P.J. (1998) Rapid and sensitive reverse transcriptase-polymerase chain reaction based detection and differential diagnosis of fish pathogenic rhabdoviruses in organ samples and cultured cells. *Diseases of Aquatic Organisms* 34, 13–20.
Midtlyng, P.J., Reitan, L.F., Lillehaug, A. and Ramstad, A. (1996) Protection, immune responses and side effects in Atlantic salmon (*Salmo salar* L.) vaccinated against furunculosis by different procedures. *Fish & Shellfish Immunology* 6, 599–613.
Mitchel, C., Faivre, B. and DeKinkelin, P. (1986) A clinical case of enteric redmouth in minnows (*Pimephales promelas*) imported in Europe as baitfish. *Bulletin of the European Association of Fish Pathologists* 6, 97–99.
Mitchell, A.J. and Plumb, J.A. (1980) Toxicity and efficacy of furanace on channel catfish infected experimentally with *Aeromonas hydrophila*. *Journal of Fish Diseases* 3, 93–100.
Miwa, S. and Nakayasu, C. (2005) Pathogenesis of experimentally induced bacterial coldwater disease in ayu *Plecoglossus altivelis*. *Diseases of Aquatic Organisms* 67, 93–104.
Mo, T.A. (2009) Gyrodactylosis (*Gyrodactylus salaris*). In: *Manual of Diagnostic Tests for Aquatic Animals*. World Organisation for Animal Health, Paris, pp. 200–208.
Morris, D.J., Adams, A., Smith, P. and Richards, R.H. (2003) Effects of oral treatment with TNP-470 on rainbow trout (*Oncorhynchus mykiss*) infected with *Tetracapsuloides bryosalmonae* (Malacosporea), the causative agent of proliferative kidney disease. *Aquaculture* 221, 51–64.
Munro, A.L.S. and Hastings, T.S. (1993) Furunculosis. In: Inglis, V., Roberts, R.J. and Bromage, N.R. (eds) *Bacterial Diseases of Fish*. Blackwell Science Ltd, Oxford, UK, pp. 122–142.
Munro, E.S. and Midtlyng, P.J. (2011) Infectious pancreatic necrosis and associated aquatic birnaviruses. In: Woo, P.T.K. and Bruno, D.W. (eds) *Fish Diseases and Disorders*. Vol. 3. *Viral, Bacterial and Fungal Infections*, 2nd edn. CAB International, Wallingford, UK, pp. 1–65.
Munson, D.A., Elliot, D.G. and Johnson, K. (2010) Management of bacterial kidney disease in chinook salmon hatcheries based on broodstock testing by enzyme-linked immunosorbent assay: a multiyear study. *North American Journal of Fisheries Management* 30, 940–955.
Nehring, R.B. and Walker, P.G. (1996) Whirling disease in the wild: the new reality in the intermountain west. *Fisheries* 21, 28–30.
Nematollahi, A., Decostere, A., Pasmans, F. and Haesebrouck, F. (2003) *Flavobacterium psychrophilum* infections in salmonid fish. *Journal of Fish Diseases* 26, 563–574.
Newman, S.G. (1982) *Aeromonas hydrophila*: a review with emphasis on its role in fish diseases. In: Anderson, D.P., Dorson, M. and Dubouret, P.H. (eds) *Antigens of Fish Pathogens: Development and Production for Vaccines and Serodiagnostics*. Foundation Marcel Merieux, Lyon, France, pp. 87–117.
Nishizawa, T., Kinoshita, S. and Yoshimizu, M. (2005) An approach for genogrouping of Japanese isolates of aquabirnaviruses in a new genogroup, VII, based on the VP2/NS junction region. *Journal of General Virology* 86, 1973–1978.

Nishizawa, T., Kinoshita, S., Kim, W.-S., Higashi, S. and Yoshimizu, M. (2006) Nucleotide diversity of Japanese isolates of infectious hematopietic necrosis virus (IHNV) based on the glycoprotein gene. *Diseases of Aquatic Organisms* 71, 267–272.

Nishizawa, T., Takami, I., Yang, M. and Oh, M.-J. (2011) Live vaccine of viral haemorrhagic septicemia virus (VHSV) for Japanese flounder at fish rearing temperature of 21°C instead of poly(I:C) administration. *Vaccine* 29, 8397–8404.

Notomi, T., Okayama, H., Masubuchi, H., Yonekawa, T., Watanabe, K., *et al.* (2000) Loop-mediated isothermal amplification of DNA. *Nucleic Acids Research* 28, e63.

Oakey, H.J., Ellis, J.T. and Gibson, L.R. (1998) RAPD-PCR used to develop random DNA probes for *Aeromonas salmonicida*. In: Barnes, A.C., Davidson, G.A., Hiney, M.P. and McIntosh, D. (eds) *Methodology in Fish Disease Research*. Fisheries Research Services, Aberdeen, UK, pp. 173–176.

Ogut, H. and Reno, P.W. (2004) Effects of fish density on spread of infectious hematopoietic necrosis virus (IHNV) in rainbow trout, *Oncorhynchus mykiss*. *Israeli Journal of Aquaculture – Bamidgeh* 56, 218–225.

Olesen, N.J. and Korsholm, H. (1997) Control measures for viral diseases in aquaculture: eradication of VHS and IHN. *Bulletin of the European Association of Fish Pathologists* 17, 229–233.

Olesen, N.J. and Skall, H.F. (2009) Viral haemorrhagic septicaemia. In: *Manual of Diagnostic Tests for Aquatic Animals*. World Organisation for Animal Health, Paris, pp. 279–298.

Orpetveit, I., Mikalsen, A.B., Sindre, H., Evensen, O., Dannevig, B.H. and Midtlyng, P.J. (2010) Detection of infectious pancreatic necrosis virus in subclinically infected Atlantic salmon by virus isolation in cell culture or real-time reverse transcription polymerase chain reaction: influence of sample preservation and storage. *Journal of Veterinary Diagnostic Investigation* 22, 886–895.

Ostland, V.E., Byrne, P.J., Hoover, G. and Ferguson, H.W. (2000) Necrotic myositis of rainbow trout, *Oncorhynchus mykiss* (Walbaum): proteolytic characteristics of a crude extracellular preparation from *Flavobacterium psychrophilum*. *Journal of Fish Diseases* 23, 329–336.

Pacha, R.E. (1968) Characteristics of *Cytophaga psychrophila* (Borg) isolated from outbreaks of bacterial cold-water disease. *Applied Microbiology* 16, 97–101.

Pascho, R.J. and Mulcahy, D. (1987) Enzyme-linked immunosorbent assay for a soluble antigen of *Renibacterium salmoninarum*, the causative agent of salmonid bacterial kidney disease. *Canadian Journal of Fisheries and Aquatic Sciences* 44, 183–191.

Petterson, E., Sandberg, M. and Santi, N. (2009) Salmonid alphavirus associated with *Lepeophtheirus salmonis* (Copepoda: Caligidae) from Atlantic salmon, *Salmo salar*. *Journal of Fish Diseases* 32, 477–479.

Plumb, J.A. and Hanson, L.A. (2011) *Health Maintenance and Principal Microbial Diseases of Cultured Fishes*, 3rd edn. Wiley, Hoboken, New Jersey.

Poleo, A.B., Schjolden, J., Hansen, H., Bakke, T.A., Mo, T.A., *et al.* (2004) The effect of various metals on *Gyrodactylus salaris* (Platyhelminthes, Monogenea) infections in Atlantic salmon (*Salmo salar*). *Parasitology* 128, 169–177.

Polinski, M.P., Jensen, N.R., Johnson, K.A., Ireland, S.C. and Cain, K.D. (2010) Assessment of formalin and hydrogen peroxide use during egg incubation of North American burbot. *North American Journal of Aquaculture* 72, 111–117.

Polinski, M.P., Jensen, N.R., Foltz, J., Ireland, S.C. and Cain, K.D. (2012) Hydrogen peroxide treatments administered to hatchery-reared burbot: assessing treatment regimes from embryonic development through juvenile rearing. *North American Journal of Aquaculture* 75, 50–56.

Popoff, M. (1984) Genus III. *Aeromonas* Kluyver and Van Niel 1936, 398[AL]. In: Krieg, N.R. and Holt, J.G. (eds) *Bergey's Manual of Systematic Bacteriology*, Vol. 1. Williams and Wilkins, Baltimore, Maryland, pp. 545–548.

Powell, M., Overturf, K., Hogge, C. and Johnson, K. (2005) Detection of *Renibacterium salmoninarum* in chinook salmon, *Oncorhynchus tshawytscha* (Walbaum), using quantitative PCR. *Journal of Fish Diseases* 28, 615–622.

Pridgeon, J.W. and Klesius, P.H. (2011) Development and efficacy of novobiocin and rifampicin-resistant *Aeromonas hydrophila* as novel vaccines in channel catfish and Nile tilapia. *Vaccine* 29, 7896–7904.

Purcell, M.K., Hart, S.A., Kurath, G. and Winton, J.R. (2006) Strand-specific, real-time RT-PCR assays for quantification of genomic and positive-sense RNAs of the fish rhabdovirus, infectious hematopoietic necrosis virus. *Journal of Virological Methods* 132, 18–24.

Purcell, M.K., LaPatra, S.E., Woodson, J.C., Kurath, G. and Winton, J.R. (2010) Early viral replication and induced or constitutive immunity in rainbow trout families with different resistance to infectious hematopoietic necrosis virus (IHNV). *Fish & Shellfish Immunology* 28, 98–105.

Rach, J.J., Schreier, T.M., Howe, G.E. and Redman, S.D. (1997) Effects of species, life stage, and water temperature on the toxicity of hydrogen peroxide to fish. *The Progressive Fish-Culturist* 59, 41–46.
Rahman, M.H., Ototake, M. and Nakanishi, T. (2003) Water-soluble adjuvants enhance the protective effect of *Flavobacterium psychrophilum* vaccines in ayu *Plecoglossus altivelis*. *Fish Pathology* 38, 171–176.
Ramadan, A., Afifi, N.A., Moustafa, M.M. and Samy, A.M. (1994) The effect of ascogen on the immune response of tilapia fish to *Aeromonas hydrophila* vaccine. *Fish & Shellfish Immunology* 4, 159–165.
Ramstad, A., Romstad, A.B., Knappskog, D.H. and Midtlyng, P.J. (2007) Field validation of experimental challenge models for IPN vaccines. *Journal of Fish Diseases* 30, 723–731.
Reno, P.W. (1999) Infectious pacreatic necrosis and associated aquatic birnaviruses. In: Woo, P.T.K. and Bruno, D.W. (eds) *Fish Diseases and Disorders*. Vol. 3. *Viral, Bacterial and Fungal Infections*, 1st edn. CAB International, Wallingford, UK, pp. 1–55.
Reno, P.W., Darley, S. and Savan, M. (1978) Infectious pancreatic necrosis: experimental induction of a carrier state in trout. *Journal of the Fisheries Research Board of Canada* 35, 1451–1456.
Richards, R.H., Inglis, V., Frerichs, G.N. and Millar, S.D. (1992) Variation in antibiotic resistance patterns of *Aeromonas salmonicida* isolated from Atlantic salmon *Salmo salar* L. in Scotland. In: Michel, C. and Alderman, D.J. (eds) *Chemotherapy in Aquaculture: From Theory to Reality*. Office International des Epizooties (OIE), Paris, pp. 276–284.
Roberts, R.J. and Shepherd, C.J. (1974) *Handbook of Salmon and Trout Diseases*. Fishing News (Books) Ltd, West Byfleet, UK.
Roberts, R.J. (2012) *Fish Pathology*, 4th edn. Wiley, Chichester, UK.
Rose, J.D., Marrs, G.S., Lewis, C. and Schisler, G. (2000) Whirling disease behavior and its relation to pathology of brain stem and spinal cord in rainbow trout. *Journal of Aquatic Animal Health* 12, 107–118.
Ross, A.J., Rucker, R.R. and Ewing, W.H. (1966) Description of a bacterium associated with redmouth disease of rainbow trout (*Salmo gairdneri*). *Canadian Journal of Microbiology* 12, 763–770.
Salonius, K., Siderakis, C., MacKinnon, A.M., and Griffiths, S.G. (2005) Use of *Arthrobacter davidanieli* as a live vaccine against *Renibacterium salmoninarum* and *Piscirickettsia salmonis* in salmonids. In: Midtlyng, P.J. (ed.) *Progress in Fish Vaccinology. Developments in Biological Standardization*, Vol. 121. Karger, Basel, Switzerland, pp. 189–197.
Sanders, J.E. and Fryer, J.L. (1980) *Renibacterium salmoninarum* gen. nov., sp. nov., the causative agent of bacterial kidney disease in salmonid fishes. *International Journal of Systemic Bacteriology* 30, 496–502.
Shotts, E. and Starliper, C. (1999) Flavobacterial diseases: columnaris disease, cold-water disease and bacterial gill disease. In: Woo, P.T.K. and Bruno, D.W. (eds) *Fish Diseases and Disorders*. Vol. 3. *Viral, Bacterial and Fungal Infections*, 1st edn. CAB International, Wallingford, UK, pp. 559–576.
Skall, H.F. and Olesen, N.J. (2011) *Treatment of Wastewater from Fish Slaughterhouses, Evaluation and Recommendations for Hygenisation Methods*. National Veterinary Institute, Copenhagen.
Skall, H.F., Slierendrecht, W.J., King, J.A. and Olesen, N.J. (2004) Experimental infection of rainbow trout *Oncorhynchus mykiss* with viral haemorrhagic septicaemia virus isolates from European marine and farmed fishes. *Diseases of Aquatic Organisms* 58, 99–110.
Skall, H.F., Olesen, N.J. and Mellergaard, S. (2005) Viral haemorrhagic septicaemia virus in marine fish and its implications for fish farming – a review. *Journal of Fish Diseases* 28, 509–529.
Skjesol, A., Skjaeveland, I., Elnaes, M., Timmerhaus, G., Fredriksen, B., *et al.* (2011) IPNV with high and low virulence: host immune responses and viral mutations during infection. *Virology Journal* 8, 396.
Smail, D.A., Huntly, P.J. and Munro, A.L.S. (1993) Fate of four fish pathogens after exposure to fish silage containing fish farm mortalities and conditions for the inactivation of infectious pancreatic necrosis virus. *Aquaculture* 113, 173–181.
Smail, D.A. and Snow, M. (2011) Viral haemorrhagic septicaemia. In: Woo, P.T.K. and Bruno, D.W. (eds) *Fish Diseases and Disorders*. Vol. 3. *Viral, Bacterial and Fungal Infections*, 2nd edn. CAB International, Wallingford, UK, pp. 123–146.
Smith, C.E., Morrison, J.K., Ramsey, H.W. and Ferguson, H.W. (1984) Proliferative kidney disease: first reported outbreak in North America. *Journal of Fish Diseases* 7, 207–216.
Smith, I.W. (1964) The occurrence of pathology of Dee disease. *Department of Agriculture of Fisheries, Scottish Fresh Water Salmon Fisheries Research* 34, 1–12.
Smith, P., Hiney, M.P. and Samuelsen, O.B. (1994) Bacterial resistance to antimicrobial agents using fish farming: a critical evaluation of method and meaning. *Annual Review of Fish Diseases* 4, 273–313.
Soleng, A., Poléo, A. and Bakke, T.A. (2005) Toxicity of aqueous aluminium to the ectoparasitic monogenean *Gyrodactylus salaris*. *Aquaculture* 250, 616–620.

Soliman, H., Midtlyng, P.J. and El-Matbouli, M. (2009) Sensitive and rapid detection of infectious pancreatic necrosis virus by reverse transcription loop mediated isothermal amplification. *Journal of Virological Methods* 158, 77–83.

Soupir, C.A. and Barnes, M.E. (2006) Reduced formalin and hydrogen peroxide treatments during walleye egg incubation. *North American Journal of Aquaculture* 68, 276–280.

Stevenson, R.M.W., Flett, D. and Raymond, B.T. (1993) Enteric redmouth (ERM) and other enterobacterial infections of fish. In: Inglis, V., Roberts, R.J. and Bromage, N.R. (eds) *Bacterial Diseases of Fish*. Blackwell Scientific Publications, Oxford, UK, pp. 80–105.

Stone, D.M., Ferguson, H.W., Tyson, P., Savage, J., Wood, G., *et al.* (2008) The first report of viral haemorrhagic septicaemia in farmed rainbow trout, *Oncorhynchus mykiss* (Walbaum) in the United Kingdom. *Journal of Fish Diseases* 31, 775–784.

Stueland, S., Hatai, K. and Skaar, I. (2005) Morphological and physiological characteristics of *Saprolegnia* spp. strains pathogenic to Atlantic salmon, *Salmo salar* L. *Journal of Fish Diseases* 28, 445–453.

Swanson, R.N. and Gillespie, J.H. (1981) An indirect fluorescent antibody test for the rapid detection of infectious pancreatic necrosis virus in tissues. *Journal of Fish Diseases* 4, 309–315.

Taylor, P.W. (2004) Detection of *Flavobacterium psychrophilum* in eggs and sexual fluids of Pacific salmonids by a polymerase chain reaction assay: implications for vertical transmission of bacterial coldwater disease. *Journal of Aquatic Animal Health* 16, 104–108.

Thoen, E., Evensen, Ø. and Skaar, I. (2011) Pathogenicity of *Saprolegnia* spp. to Atlantic salmon, *Salmo salar* L., eggs. *Journal of Fish Diseases* 34, 601–608.

Thoesen, J.C. (1994) A standard inspection procedure for detection for certain pathogens in salmonid fish. In: Thoesen, J.C. (ed) *Blue Book: Suggested Procedures for the Detection and Identification of Certain Finfish and Shellfish Pathogens*, 4th edn. American Fisheries Society, Fish Health Section, Bethesda, Maryland.

Thune, R.L., Graham, T.E., Riddle, L.M. and Amborski, R.L. (1982) Effects of *Aeromonas hydrophila* extracellular products and endotoxins. *Transactions of the American Fisheries Society* 111, 739–754.

Tobback, E., Decostere, A., Hermans, K., Haesebrouck, F. and Chiers, K. (2007) *Yersinia ruckeri* infections in salmonid fish. *Journal of Fish Diseases* 30, 257–268.

Valdebenito, S. and Avendaño-Herrera, R. (2009) Phenotypic, serological and genetic characterization of *Flavobacterium psychrophilum* strains isolated from salmonids in Chile. *Journal of Fish Diseases* 32, 321–333.

van West, P. (2006) *Saprolegnia parasitica*, an oomycete pathogen with a fishy appetite: new challenges for an old problem. *Mycologist* 20, 99–104.

Vatsos, I.N., Thompson, K.D. and Adams, A. (2001) Adhesion of the fish pathogen *Flavobacterium psychrophilum* to unfertilized eggs of rainbow trout (*Oncorhynchus mykiss*) and n-hexadecane. *Letters in Applied Microbiology* 33, 178–182.

Vatsos, I.N., Thompson, K.D. and Adams, A. (2003) Starvation of *Flavobacterium psychrophilum* in broth, stream water and distilled water. *Diseases of Aquatic Organisms* 56, 115–126.

Vatsos, I.N., Thompson, K.D. and Adams, A. (2006) Colonization of rainbow trout, *Oncorhynchus mykiss* (Walbaum), eggs by *Flavobacterium psychrophilum*, the causative agent of rainbow trout fry syndrome. *Journal of Fish Diseases* 29, 441–444.

Veenstra, K.A., Hodneland, K., Fischer, S., Takehana, K., Belmonte, R. and Fischer, U. (2020) Cellular immune responses in rainbow trout (*Onchorhynchus mykiss*) following vaccination and challenge against salmonid alphavirus (SAV). *Vaccines* 8(4), 725.

Vincent, E.R. (2002) Relative susceptibility of various salmonids to whirling disease with emphasis on rainbow and cutthroat trout. In: Bartholomew, J.L. and Wilson, J.C. (eds) *Whirling Disease: Reviews and Current Topics. American Fisheries Society Symposium 29*. American Fisheries Society, Bethesda, Maryland, pp. 109–115.

Vuillaume, A., Brun, R., Cnene, P. and Lesel, R. (1987) First isolation of *Yersinia ruckeri* from sturgeon, *Acipencer baeri* Brant, in south west of France. *Bulletin of the European Association of Fish Pathologists* 7, 18–19.

Wahli, T., Knuesel, R., Bernet, D., Segner, H., Pugovkin, D., *et al.* (2002) Proliferative kidney disease in Switzerland: current state of knowledge. *Journal of Fish Diseases* 25, 491–500.

Walker, P.G. and Nehring, R.B. (1995) *An Investigation to Determine the Cause(s) of the Disappearance of Young Wild Rainbow Trout in the Upper Colorado River, in Middle Park, Colorado*. Colorado Division of Wildlife Report. Colorado Division of Wildlife, Denver, Colorado.

Waltzek, T.B., Kelley, G.O., Alfaro, M.E., Kurobe, T., Davison, A.J. and Hedrick, R.P. (2009) Phylogenetic relationships in the family *Alloherpesviridae*. *Diseases of Aquatic Organisms* 84, 179–194.

Wargo, A.R., Scott, R.J., Kerr, B. and Kurath, G. (2017) Replication and shedding kinetics of infectious hematopoietic necrosis virus in juvenile rainbow trout. *Virus Research* 227, 200–211.

Warren, J.W. (1963) Kidney diseases of salmonid fishes and the analysis of hatchery waters. *The Progressive Fish-Culturist* 25, 121–131.

Way, K. and Dixon, P.F. (1988) Rapid detection of VHS and IHN viruses by the enzyme-linked immunosorbent assay (ELISA). *Journal of Applied Ichthyology* 4, 182–189.

Weston, J., Villoing, S., Bremont, M., Castric, J., Pfeffer, M., *et al.* (2002) Comparison of two aquatic alphaviruses, salmon pancreas disease virus and sleeping disease virus, by using genome sequence analysis, monoclonal reactivity, and cross-infection. *Journal of Virology* 76, 6155–6163.

Wheeler, R.W., Davies, R.L., Dalsgaard, I., Garcia, J., Welch, T.J., *et al.* (2009) *Yersinia ruckeri* biotype 2 isolates from mainland Europe and the UK likely represent different clonal groups. *Diseases of Aquatic Organisms* 84, 25–33.

Wiklund, T. and Dalsgaard, I. (1998) Occurrence and significance of atypical *Aeromonas salmonicida* in non-salmonid and salmonid fish species: a review. *Diseases of Aquatic Organisms* 32, 49–69.

Willoughby, L.G. (1994) *Fungi and Fish Diseases*. Pisces Press, Stirling, UK,

Willoughby, L.G. and Pickering, A.D. (1977) Viable Saprolegniaceae spores on the epidermis of the salmonid fish *Salmo trutta* and *Salvelinus alpinus*. *Transactions of the British Mycological Society* 68, 91–95.

Winton, J.R. (1991) Recent advances in the detection and control of infectious hematopoietic necrosis virus (IHNV) in aquaculture. *Annual Review of Fish Diseases* 1, 83–93.

Wolf, K. (1988) *Fish Viruses and Fish Viral Diseases*. Cornell University Press, Ithaca, New York.

Wolf, K., Darlington, R.W., Taylor, W.G., Quimby, M.C. and Nagabayashi, T. (1978) Herpes-virus salmonis: characterization of a new pathogen of rainbow trout. *Journal of Virology* 27, 659–666.

Wood, S.E. and Willoughby, L.G. (1986) Ecological observation on the fungal colonization of fish by Saprolegniaceae in Windermere. *Journal of Applied Ecology* 23, 737–749.

Yin, Z., Lam, T.J. and Sin, Y.M. (1996) The role of specific antiserum of catfish, *Clarias gariepinus*, as a defense against *Aeromonas hydrophila*. *Fish & Shellfish Immunology* 6, 57–69.

Yoshimizu, M. (2009) Control strategy for viral diseases of salmonid fish, flounders and shrimp at hatchery and seed production facility in Japan. *Fish Pathology* 44, 9–13.

Yoshimizu, M., Takizawa, H. and Kimura, T. (1986) UV susceptibility of some fish pathogenic viruses. *Fish Pathology* 21, 47–52.

Yoshimizu, M., Fukuda, H., Sano, T. and Kimura, T. (1995) Salmonid herpesvirus 2. Epizootiology and serological relationship. *Veterinary Research* 26, 486–492.

Zamora, L., Fernández-Garayzábal, J.F., Svensson-Stadler, L.A., Palacios, M.A., Domínguez, L., *et al.* (2012) *Flavobacterium oncorhynchi* sp. nov., a new species isolated from rainbow trout (*Oncorhynchus mykiss*). *Systemic and Applied Microbiology* 35, 86–91.

Zamora, L., Vela, A., Sánchez-Porro, C., Palacios, M., Moore, E., *et al.* (2014) *Flavobacterium tructae* sp. nov. and *Flavobacterium piscis* sp. nov., isolated from farmed rainbow trout (*Oncorhynchus mykiss*). *International Journal of Systematic and Evolutionary Microbiology* 64, 392–399.

4 Non-Infectious Disorders of Coldwater Fish

Thomas Wahli*, Heike Schmidt-Posthaus and Helmut Segner
University of Bern, Bern, Switzerland

4.1 Introduction

Coldwater finfish are reared in both marine and freshwater environments. In the marine environment, coldwater finfish aquaculture is dominated by salmonids, mainly Atlantic salmon (*Salmo salar*), coho salmon (*Oncorhynchus kisutch*), rainbow trout (*Oncorhynchus mykiss*) and chinook salmon (*Oncorhynchus tshawytscha*) (FAO, 2020). Other fish species such as turbot (*Scophthalmus maximus*), bastard halibut (*Paralichthys olivaceus*), Atlantic halibut (*Hippoglossus hippoglossus*) and Atlantic cod (*Gadus morhua*) are gaining in importance although their production volume is lower than that of salmonids. These newer aquaculture species are cultured either in onshore tanks or offshore cages. Production of juvenile fish is a major bottleneck for these species due to high mortalities during weaning. Other biological and physiological difficulties include skin malpigmentation, low quality and discontinuous spawning, and high incidence of skeletal deformities (Fernández and Gisbert, 2011). These examples highlight the importance of understanding the environmental needs and nutritional requirements of potential new aquaculture species.

Coldwater freshwater fish are reared in water from rivers, streams or groundwaters. Culture systems are dominated by open or recirculation systems while fish (mainly salmonids) are kept in tanks or cages. The focus here is on diseases and disorders of rainbow trout (*O. mykiss*), brown trout (*Salmo trutta*), brook trout (*Salvelinus fontinalis*) and Arctic char (*Salvelinus alpinus*). Non-infectious disorders are often regarded as economically less devastating. However, infectious diseases are often promoted either directly or indirectly by a weakened immune response due to adverse environmental conditions (Barton, 1997). Therefore, knowledge on adverse physical, chemical or biological factors, and imbalances in nutrition, is crucial for good fish husbandry.

Of particular importance are climate change-related effects on farmed fish. Climate-related environmental change can influence diverse production-relevant factors including fish growth, feed utilization, physiological performance or fish health, be it through an impact on the immune system and pathogen susceptibility of fish or through an impact on the distribution and abundance of pathogens. Evaluating the role of climate change in such effects is generally difficult as single factors can impose different effects on different life stages of fish and there is a complex interplay between different factors (Falconer *et al.*, 2022). Information on

*thomas.wahli@unibe.ch

 Climate Change on Diseases and Disorders of Finfish in Cage Culture, 3rd Edition (eds P.T.K. Woo and R.P. Subasinghe)
DOI: 10.1079/9781800621640.0004

direct effects of climate change on non-infectious diseases is limited. Where available, respective data are integrated in this chapter. Detailed discussions of climate change impacts on non-infectious disorders and infectious diseases of fish are available (see respectively Woo and Iwama, 2020; Woo *et al.*, 2020).

4.2 Production Problems

4.2.1 Smolt failure in salmonids

Salmonid species with fresh and seawater life cycle phases undergo a physiological, morphological and behavioural transformation known as smoltification, which allows them to adapt from freshwater to marine conditions. Atlantic salmon are prepared physiologically for saltwater transfer only temporarily (Hoar, 1988; McCormick, 2012). Within this time window, fish must prepare for a significant osmoregulatory challenge and thus are at high risk. Under natural conditions, the main triggers for smoltification are parr size, internal nervous and endocrine signals, and external photoperiod and water temperature. Intensification of salmon aquaculture and the increasing market demand for a continuous fish supply have led to production changes in order to decrease economic costs and increase throughput all year round. These changes have mainly focused on manipulating the duration of the freshwater stage. In captivity, where manipulations of daylength, temperature and food supply are possible, these mechanisms are used to produce different age classes and season-related smolts allowing transfers to sea out of the natural season (Duncan *et al.*, 2002). This desynchronization can be associated with pathologies (Speare, 2002).

CLINICAL SIGNS, GROSS AND HISTOPATHOLOGICAL LESIONS. Inappropriate time point of sea transfer of salmon may have detrimental consequences such as poor growth, an affected immune system (Johansson *et al.*, 2016) and higher mortalities (Berge *et al.*, 1995). Even under good transfer conditions, the acclimatization period and the stress caused by the transfer are likely to influence fishes' feeding response and immune system. Failed smolts are smaller and retain the characteristic parr marks (vertical dark stripes) instead of adopting the silver coloration shown by smolts due to deposition of guanine. In addition, failed smolts have fewer chloride cells in the gills.

CONTROL AND PREVENTION. Monitoring of smoltification is essential for a successful transfer (Van Rijn *et al.*, 2021). As a routine, levels of Na^+/K^+-ATPase activity are measured to assess osmoregulatory competence and adaptability to seawater (McGowan *et al.*, 2021). Given that, besides photoperiod, temperature is an important trigger for inducing smoltification (Nisembaum *et al.*, 2020), changes in temperature regimes due to climate change have to be considered when defining the correct time point for the sea transfer of salmon.

4.2.2 Stress

Farmed fish are susceptible to a wide range of stressors, and different species display wide variation of physiological and behavioural responses. Changes in environmental conditions (light, temperature, salinity) as well as capture, transport, grading, treatment, crowding, malnutrition, poor water quality, contaminant exposure, inadequate housing, disease, physical trauma, noise and predators can all be a cause of stress (Schreck and Tort, 2016; Martos-Sitcha *et al.*, 2020). Many of these factors are influenced by global climate change, which thus might lead to situations of elevated stress. Temperature stress plays a master role for all stages of farmed fish, with extreme temperatures causing neuroendocrine, oxidative, metabolic, osmotic, molecular, haemato-biochemical and immune responses (Islam *et al.*, 2021) (Fig. 4.1). Culture-related stressors like transportation, grading or density cause stress and can lead to disease outbreaks (Iversen *et al.*, 2005; Zhang *et al.*, 2022). In the culture of Atlantic salmon, seawater transfer is stressful and is associated with disease outbreaks (Iversen *et al.*, 2005). Larvae and young fish appear to be most sensitive to stress (Rehman *et al.*, 2017).

CLINICAL SIGNS, GROSS AND HISTOPATHOLOGICAL LESIONS. The primary stress response involves

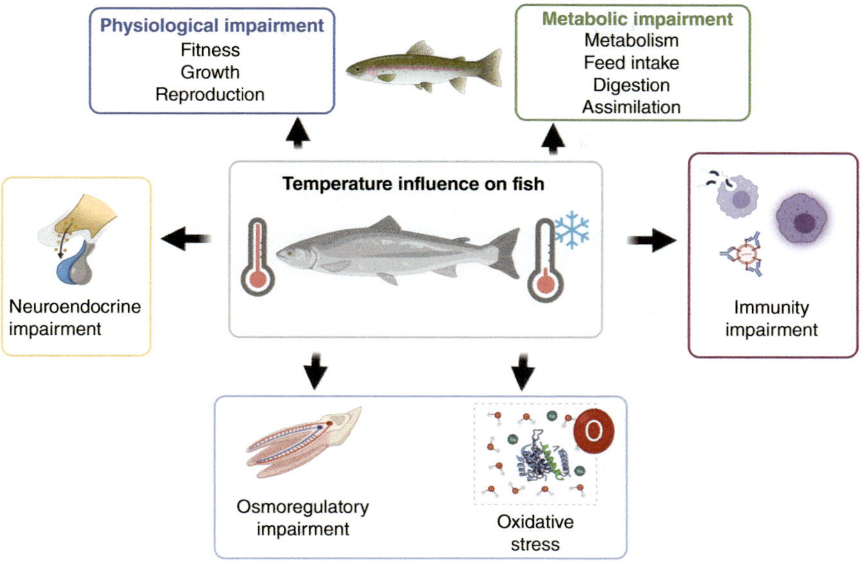

Fig. 4.1. Temperature influence on different body functions in fish. (Adapted from Islam *et al.*, 2021; created with BioRender.com.)

the secretion of adrenergic and glucocorticoid hormones by the chromaffin and inter-renal cells, respectively. This initially triggers activational responses such as increases in alertness, respiration rate, blood pressure and hepatic glycogen catabolism (Schreck and Tort, 2016). If the stress stimuli persist, negative effects develop, such as chronic immune suppression, depletion of energy reserves and osmoregulatory impairment (Schreck and Tort, 2016). Gross and histological changes due to stress are difficult to diagnose in fish, as they are usually unspecific. Externally, fish may show increased mucus production, darker coloration and skin erosion (mainly on the fins). Histological changes may be observed in organs such as the skin, gills, liver and kidney (Harper and Wolf, 2009). Chronic stress-induced changes include skeletal muscle atrophy, hepatocyte atrophy due to glycogen catabolism (Wolf and Wolfe, 2005), gill lamellar epithelium hypertrophy, hyperplasia and/or oedema, and increased numbers of pigmented macrophage centres (Agius and Roberts, 1981; Wolke, 1992).

CONTROL AND PREVENTION. When handling fish, care must be taken to avoid abrasions, removal of scales and skin mucus, and excessive air exposure. A 24 h recovery period between handling is recommended (Gatica *et al.*, 2010). Using well-boats for transfers or treatments is extremely important to avoid overcrowding, keep oxygen levels high and minimize the build-up of metabolic wastes. Temporary starvation (24 h) is recommended before certain management procedures to reduce stress. This reduces metabolism, oxygen demand and waste production (Ashley, 2007).

4.2.3 Behaviour-related problems

Aggression and cannibalistic behaviour occur in some species and can cause losses or deformities in cultured fish (Baras and Jobling, 2002; Ashley, 2007; Martos-Sitcha *et al.*, 2020). Stocking densities and feeding methods have a strong influence on the levels of social interactions and dominance hierarchies (Ashley, 2007). In species showing strong social hierarchies (e.g. salmonids), dominance can lead to aggression, chronic stress, reduced feeding and growth as well as histopathological changes in submissive individuals (Ejike and Schreck, 1980; Alanärä and Brännäs, 1996). Aggression can result in fin, skin and eye damage (Speare, 2002), facilitating the entrance of secondary infectious organisms. For example, these problems occur more frequently in chinook than in Atlantic salmon (Speare, 2002). Also ploidy of salmonid species was found to play a role in brown trout, for example, where diploid individuals showed a higher aggressive behaviour than triploid animals (Preston *et al.*, 2014). Fin rot, a common problem in farmed salmon and rainbow trout, can result from abrasions with containment structures (e.g. nets, cages) or due to aggressive interactions (Ottesen *et al.*, 2007). High stocking densities can increase competition for resources and therefore aggression levels. However, in salmonids like trout and salmon, too low stocking densities can also induce aggressive behaviour (Ellis *et al.*, 2002; Speare, 2002). Furthermore, the method of feeding, such as hand feeding or automatic feeders, as well as the feeding level influence aggression among the cultured fishes (Greaves and Tuene, 2001). Together, food availability and feeding methods, stocking densities and appropriate housing conditions are key factors to prevent behavioural problems (Cooke, 2016; Macaulay *et al.*, 2021).

4.2.4 Predators

Aquatic mammals and birds are the main predators of cage-reared fish. Direct or indirect losses due to predators can be significant if preventive measures are not taken. Control methods include cage nettings and visual and acoustic scaring devices. Predators can kill the fish (Fig. 4.2) or cause wounds facilitating entry for pathogens. Besides health and welfare considerations, damaged fish will have a lower market value. Predators such as seals can also cause net damages allowing escapes, which will have economic, environmental and legal consequences. The presence of predators in the surroundings of the cage can also induce a fear and stress response in the fish.

4.3 Environmental Problems

4.3.1 Algal blooms

The impact of algal blooms on coldwater finfish is covered extensively in Shahmohamadloo *et al.* (Chapter 10, this volume, 2023).

Fig. 4.2. Farmed Atlantic salmon (*Salmo salar*) with severe lesion due to seal attack. (Photograph courtesy of Mar Marcos-Lopez.)

4.3.2 Harmful zooplankton – jellyfish

Jellyfish cause significant fish losses worldwide. Different modes of action have been described and ascribed to different taxa. While in fisheries the majority of jellyfish species causing problems belong to the scyphozoans, in aquaculture nearly 50% of reported cases are attributed to hydrozoans (Bosch-Belmar *et al.*, 2020). In the case of free-swimming stages, small jellyfish (e.g. class Hydrozoa) can pass through nets and reach the fish inside cages, while bigger species (e.g. class Scyphozoa) tend to break in contact with the nets and the freed tentacles sting fish. Representing an important part of biological fouling on submerged structures, hydrozoans can impact aquaculture in different ways, particularly in the case of blooms: occlusion of nets resulting in reduced water flow and subsequently limited oxygen availability, and direct contact of fish with free-living stages from seasonal budding with large polyp colonies attached to inner net-pen structures. Direct contact can result in gill damage and/or toxicity via nematocysts discharge and release of haemolytic, cytotoxic and/or neurotoxic chemicals (Bosch-Belmar *et al.*, 2020). Jellyfish can sting skin, eyes or gills, but if ingested they can also induce damage in the gastrointestinal tract (Bruno and Ellis, 1985). In some areas jellyfish have been involved in complex gill disorders resulting in major losses of farmed salmon (Baxter *et al.*, 2011a). In sublethal cases, stinging may result in gill and skin damage or ulceration which can lead to secondary bacterial infections. Jellyfish have been suggested as vectors for certain bacterial diseases, like *Tenacibaculum maritimum* affecting farmed Atlantic salmon (Tørud and Håstein, 2008; Ferguson *et al.*, 2010).

Low-level mortalities in salmon farms were until recently not considered in connection with jellyfish effects but addressed as 'waterborne irritant damage' (Marcos-López *et al.*, 2016). This may be due to the difficulty in observing small jellyfish, particularly when water samples or evidence of jellyfish in fresh or histology preparations are lacking. Only severe events with coinciding jellyfish observations were reported. Thus, huge blooms of *Pelagia noctiluca* and *Aurelia* spp. caused massive losses to the Atlantic salmon industry in northern Europe (Baxter *et al.*, 2011b; Mitchell and Rodger, 2011), for example in Ireland, killing all the stock (~250,000 salmon) from the only Northern Irish salmon farm (Doyle *et al.*, 2008). During that time, *P. noctiluca* swarms were also reported from the Scottish coast (Doyle *et al.*, 2008; Hay and Murray, 2008). *P. noctiluca* occurs worldwide in both warm and temperate waters, but global warming may allow its expansion to northern waters. *Muggiaea atlantica* caused high losses (>100,000 salmon) in Norway (Fossa *et al.*, 2003) and was also suggested as the cause of a catastrophic event (>1,000,000 salmon) in Ireland in 2003 (Cronin *et al.*, 2004). Other jellyfish species associated with fish kills in cold waters are *Phialella quadrata* in Scotland (Bruno and Ellis, 1985; Ferguson *et al.*, 2010), *Apolemia uvaria* in Norway (Båmstedt *et al.*, 1998) and *Aurelia aurita* in Scotland and Ireland (Mitchell and Rodger, 2011; Bruno *et al.*, 2013). An increasing problem in northern European aquaculture is the fouling hydroid *Ectopleura larynx* (Guenther *et al.*, 2010; Baxter *et al.*, 2012; Bosch-Belmar *et al.*, 2019).

CLINICAL SIGNS, GROSS AND HISTOPATHOLOGICAL LESIONS. Affected fish show abnormal behaviour including decreased feeding, lethargy, jumping, gasping, head shaking and/or collision with cage walls. After a bloom, residual effects (i.e. decreased growth and performance) may also occur. At inspection, gill haemorrhage, necrosis and excess mucus production, eye damage, darker or burn-like marks in the skin and local inflammatory response can be observed (Schmidt-Posthaus and Marcos-López, 2014; Marcos-López *et al.*, 2016). Histopathology can reveal lamellar fusion, epithelial hyperplasia, ballooning degeneration, spongiosis, hydropic degeneration and necrosis of the affected epithelia. Haemorrhage, focal dermal necrosis and exocytosis of eosinophilic granular cells can also be noted in the skin (Schmidt-Posthaus and Marcos-López, 2014; Powell *et al.*, 2018). Remains of, or whole, jellyfish are sometimes observed between filaments in the gills (Fig. 4.3). As a result of prolonged nematocyst discharge, secondary bacterial infections may occur (Schmidt-Posthaus and Marcos-López, 2014).

RISK FACTORS. Jellyfish swarms may occur year-round, but blooms are most common from spring to autumn. Increased frequency of jellyfish blooms has been attributed to climate change

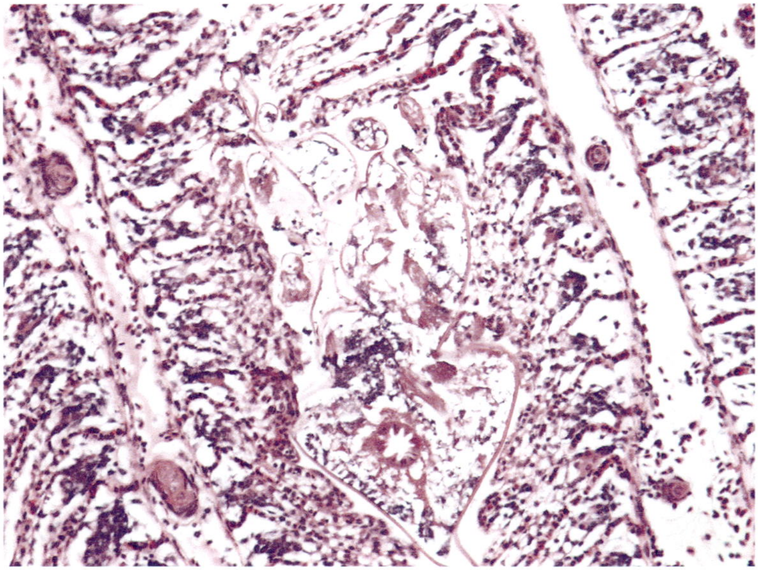

Fig. 4.3. Farmed Atlantic salmon (*Salmo salar*) with small jellyfish (species not identified) between gill filaments. Note localized lamellar epithelium necrosis, sloughing and focal thrombosis. Haematoxylin and eosin stain. (Image courtesy of Hamish Rodger.)

(Attrill *et al.*, 2007; Lynam *et al.*, 2011). However, this is controversial. According to a review (Condon *et al.*, 2013), jellyfish populations undergo worldwide oscillatory waves with an approximate 20-years periodicity. Reliable global data to elucidate the impact of climate change on jellyfish blooms are still lacking. Recommendations to fill this gap and to standardize methods have been proposed (Gibbons and Richardson, 2013). Increased eutrophication due to anthropogenic activities including overfishing may favour jellyfish multiplication (Goldstein and Steiner, 2020). Floating aquaculture structures also provide a suitable surface for polyp settlement (Purcell *et al.*, 2007).

CONTROL AND PREVENTION. Blooms are difficult to predict and therefore control measures are difficult to put in place. As with algal blooms, site location, routine monitoring and reduction of eutrophication causes, when possible, are important factors to consider. In any case, key to understanding the adverse results of jellyfish exposure is knowledge on harmful species and the effect of population exposure densities (Clinton *et al.*, 2021). Remotely operated aerial surveillance methods have been shown to be helpful to identify blooms (Schaub *et al.*, 2018) although these techniques do not allow to discriminate species. Predictive models may be beneficial to establish preventive measures (Elzeir *et al.*, 2005). Mitigation measures such as oxygenation and stopping feeding can be helpful. Other protective devices, such as bubble curtains, have been proposed but they are still experimental (Rodger *et al.*, 2011; Haberlin *et al.*, 2021). Coated nets to prevent initial settlement and build-up of biofouling organisms have been used (Guenther *et al.*, 2011). Short-term tarpaulin wrapping can be applied around net cages (Clinton *et al.*, 2021). Regular net-washing operations should be performed, however considering that these actions might be stressful to fish (Bloecher *et al.*, 2015), an optimized cleaning frequency to keep nets clean and minimize harm to fish is important (Floerl *et al.*, 2016).

4.3.3 Physicochemical parameters

Gas bubble disease

Gas bubble disease is seen in a wide range of fish species and under various circumstances

(Rucker, 1975; Machova *et al.*, 2017). Influencing factors include age, species, metabolic rate, water temperature, management and water source. Salmonids, especially larval stages, are very sensitive (Schmidt-Posthaus and Marcos-López, 2014). The disease occurs when the total pressure of dissolved gases (oxygen or nitrogen) in the water is higher than their atmospheric pressure. Under this circumstance, the excess of aqueous gas tends to leave the solution to equilibrate the gas concentration in both phases. If this occurs in the fish blood vessels or tissues, gas bubble disease results (Noga, 2010). Most gas emboli are produced by excess nitrogen since oxygen is assimilated metabolically and is less likely to form persistent bubbles. The acute form occurs at gas saturations of 110 to 115%, with the chronic form occurring at 103 to 105% (Hoffmann, 2005). Even small changes in temperature and pressure can induce the diffusion of liquid gas into gas nuclei, as solubility is decreased with increasing water temperature. Despite temperature having been recognized as an important factor for the occurrence of gas bubble disease, there are no clear indications for a climate change-related increase of cases of the condition in farmed coldwater fish.

CLINICAL SIGNS, GROSS AND HISTOPATHOLOGICAL LESIONS. The diagnosis is carried out on the farm because gas bubbles can disappear during transport. The presence of gas emboli is pathognomonic for gas bubble disease. Determination of gas supersaturation is based on the measurement of the total concentration of dissolved gas in the water.

The acute form is often characterized by increased mortality without any overt clinical signs. In cases with clinical signs, agitation, darkening of the skin and increased breathing rates can be observed. In larval fish, gas bubbles are most common in the subcutis and the yolk sac, but in larval flatfish the edges of the body fins seem to be particularly predisposed (Noga, 2010). In older fish, pinhead-sized gas bubbles can be visible under the skin, in the fins, mouth epithelium, jaw, gills, swim bladder, peritoneum, and in all chambers of the eye, where damage can lead to blindness and phthisis (Espmark *et al.*, 2010; Noga, 2010) (Fig. 4.4A). Intravascular gas emboli and occlusion of large branchial vessels are one of the major causes of death (Edsall and Smith, 1991) (Fig. 4.4B) with up to 100% mortality (Colt, 1986). Histologically, various sizes of sub-epithelially located gas bubbles can be seen. On the gill arches the main location are gill rakers, while in the skin, bubbles are seen between scales and epithelium (Espmark *et al.*, 2010). Oedema of the gill lamellae with degeneration of overlaying epithelium, oedema and bullous degeneration of buccal and intestinal mucosa, and vacuolar degeneration of the renal tubular epithelium have been recorded (Roberts, 2012a; Schmidt-Posthaus and Marcos-López, 2014). Gas bubbles can also be found in other parenchyma and in the central nervous system (Hoffmann, 2005). During recovery, secondary infectious diseases of the gills are common (Speare, 2010).

Clinical signs of the chronic form are mainly seen in salmonid larvae (Hoffmann, 2005). They are also unspecific, including chronic low

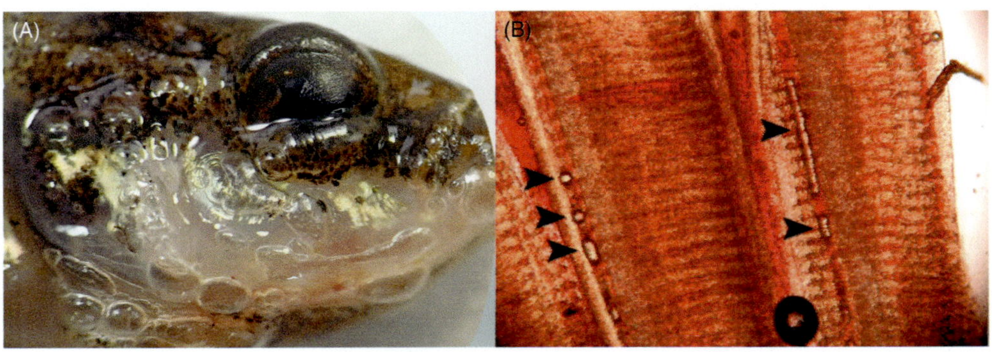

Fig. 4.4. Gas bubble disease. (A) Gas-filled dermal vesicles on the head in a stickleback (*Apeltes quadracus*). (B) Gas emboli in filament vessels of a rainbow trout (*Oncorhynchus mykiss*) (arrowheads).

mortalities ($<5\%$), hyperinflation of the swim bladder, and extravascular emboli in the gastrointestinal tract and mouth (Noga, 2010). Secondary effects, like increased sensitivity to infectious diseases and decreased growth, are also reported (Baur *et al.*, 2010).

PREVENTION AND CONTROL. Usually the condition is brought about by problems in filtration and circulation. Also rapid changes in temperature may lead to gas supersaturation. Monitoring of dissolved gas pressure and saturation can help to prevent the problem. Piping and pumping components should be regularly checked for leaks and other problems to prevent the disease. Another measure is to check the temperature of new water that is added to the culture system.

Oxygen deficiency

Oxygen demand depends on fish species, age and metabolic rate. Salmonids have higher oxygen requirements; young stages are more sensitive than adults and oxygen demand increases with metabolic rate. The latter is influenced by feeding and water temperature. Warmer temperatures increase metabolism and reduce dissolved oxygen levels. Increase of water temperature in rivers and brooks that serve as water source for landlocked coldwater fish farms might become a problem. For coldwater fish species living in the wild, range contraction due to climate change-related increase in water temperature has been demonstrated or predicted (Poff *et al.*, 2002; Van Zuiden *et al.*, 2016). In cage culture, anoxic conditions can result from eutrophication, algal and zooplankton blooms, high stocking densities, or when fish are shipped (for transfer or treatment) in insufficiently aerated containers. The decrease in oxygen availability to tissues can lead to necrotic or apoptotic lesions, although adaptive responses have been reported (van der Meer *et al.*, 2005).

CLINICAL SIGNS AND GROSS LESIONS. Clinical signs of oxygen deficiency are increased breathing rate and accumulation of the affected fish near the surface or the water inflow. At necropsy, fish show abducted opercula and open mouth. In chronic cases, low oxygen levels can lead to stress with subsequent decreased performance and increased susceptibility to diseases (Hoffmann, 2005; Noga, 2010).

Waterborne irritants

i. Ammonia

Ammonia is the primary metabolic waste product of fish. Further, it originates from the bacterial decay of nitrogenous compounds from faeces or uneaten food. High ammonia levels can also originate from agricultural and mining operation runoffs or other human activities, excessive biological waste accumulation due to high stocking densities and/or excessive feeding, or insufficient water aeration (Schmidt-Posthaus and Marcos-López, 2014). Climate change-related droughts may reduce the dilution effects of anthropogenic nitrogen compounds in water (Hosseini *et al.*, 2017) and thus have detrimental effects on fish farms depending on affected water sources. In marine cage culture at appropriate stocking densities and site locations, ammonia toxicity should not occur. Ammonia is present in two forms; the unionized form (NH_3) is much more toxic to fish than the ionized form (NH_4^+). The concentration of each form depends on temperature, pH and salinity. High pH, high water temperature and low salinity favour the formation of NH_3 and H_2O, while low pH, low water temperature and high salinity accelerate the formation of NH_4^+ and OH^-. Sensitivity is different among fish species. In salmonids, unionized ammonia levels should not increase above 0.002 mg/l H_2O, whereas marine species could tolerate up to 0.05 mg/l H_2O (Hoffmann, 2005). Ammonia toxicity may be exacerbated by increased pH or temperature, excessive exercise, starvation and stress (Schmidt-Posthaus and Marcos-López, 2014). The precise pathogenesis of ammonia poisoning is still unresolved; however, high ammonia levels in water will increase ammonia concentration in blood and tissue, causing elevated blood pH, osmoregulatory disturbance, increased tissue oxygen consumption and decreased blood oxygen levels (Ip and Chew, 2010; Schmidt-Posthaus and Marcos-López, 2014). An elevated susceptibility to pathogens in salmon exposed to subacute levels of ammonia has been described (Ackerman *et al.*, 2006).

CLINICAL SIGNS, GROSS AND HISTOPATHOLOGICAL LESIONS. Acute ammonia toxicity can cause behavioural abnormalities, such as hyperexcitability and appetite suppression. In the most severe cases, convulsions, coma and death are reported. Chronic toxicity has been associated with gill epithelial hyperplasia and hypertrophy (Schmidt-Posthaus and Marcos-López, 2014). However, it is unclear if these changes are directly related to ammonia or induced by other poor water-quality parameters that usually accompany high ammonia levels. In a study carried out by Daoust and Ferguson (1983), chronic ammonia toxicity in rainbow trout did not produce histological gill changes despite obvious neurological signs. Ammonia-related lesions have also been reported in kidney, liver, intestine and ovary of fish (Schmidt-Posthaus and Marcos-López, 2014). Sublethal levels can be associated with poor swimming performance (McKenzie *et al.*, 2003).

PREVENTION AND CONTROL. Ammonia concentration can be determined using commercial kits. Ammonia levels can be reduced over the short term by addition of fresh water. However, in cultured fish, prevention of ammonia toxicity is preferable to therapy. Good-quality feed high in digestible protein will reduce ammonia production. Also reducing stocking densities or feeding rates and increasing water flow are effective. Treating water with zeolite is another possibility (Noga, 2010). Supplementation of vitamin C has been reported to reduce the stress induced by chronic high ammonia levels (Liu *et al.*, 2008).

ii. Nitrite

In the presence of oxygen, ammonia is converted into nitrite by *Nitrosomonas* bacteria and into nitrate by *Nitrobacter* bacteria. Nitrite is toxic to fish while nitrate, which can be used by plants or converted into nitrogen, is mostly unproblematic. Nitrite intoxication usually follows an increase in ammonia, especially in the autumn when low water temperature inhibits the activity of *Nitrobacter* bacteria. Nitrite poisoning does not occur in flow-through systems and cage cultures with high water interchange as there is no significant conversion from ammonia to nitrite during the short time the water is present in the system (Noga, 2010). More at risk for nitrite intoxication are recirculation systems due to insufficient removal of nitrogenic waste (Ciji and Akhtar, 2020). As for ammonia, toxicity of nitrite varies highly between fish species; salmonids appear to be most sensitive (Kocour Kroupová *et al.*, 2018). In salmonids, values of 96 h LC_{50} range from 0.2 to 12 mg/l (Kocour Kroupová *et al.*, 2018). In theory, saltwater fish are also sensitive, but extremely high levels are required which will not occur under culture conditions (Daniels *et al.*, 1987). Experiments have shown that ammonia oxidation and nitrite oxidation may become uncoupled at water temperatures between 20 and 30°C, resulting in increased nitrite levels (Schaefer and Hollibaugh, 2017).

CLINICAL SIGNS, GROSS AND HISTOPATHOLOGICAL LESIONS. Nitrite is transported through the gills into the bloodstream where it causes oxidation of haemoglobin (Hb) to methaemoglobin (MetHb) (Ciji and Akhtar, 2020). MetHb cannot bind oxygen and therefore oxygen uptake is restricted even when water oxygen levels are normal. Affected animals show respiratory distress including hyperventilation and swollen, flared gills, lethargy and jumping to increase oxygen content in the water (Ciji and Akhtar, 2020). MetHb is brown and at concentrations around 40%, gills can show a pale-brownish discoloration.

PREVENTION AND CONTROL. Nitrite concentrations in the culture system should be regularly monitored. A complete diagnosis includes measurement of nitrite in the water and detection of methaemoglobinaemia (Saunders *et al.*, 2012), although the practicalities of testing should be assessed for each particular case. MetHb concentrations in the blood above 25% are considered abnormal, but susceptibility to toxicity varies among species.

Toxic compounds

Fishes are exposed, either intentionally or unintentionally, to a vast array of chemical and particulate contaminants, both of natural and man-made origin. Examples include pharmaceuticals, agricultural chemicals, manufacturing by-products, animal or human waste materials, and mining effluents. Often fish are exposed to mixtures of contaminants, which makes it difficult or impossible to identify specific responses. Water-soluble contaminants are taken up primarily by gills and skin, whereas intestinal uptake becomes important with increasing lipophilicity of the contaminants. Temperature can elevate the toxicity of toxic substances (Patra *et al.*, 2015). Because of the large variety of potential toxic components in the water, here only the general reactions of fish to acute and chronic toxicity are discussed; for a more detailed treatment of the subject, the reader is referred to textbooks (Schlenk and Benson, 2001; Di Giulio and Hinton, 2008).

In cases of suspected intoxication, recording of case history is crucial for diagnosis. Water samples and affected fish have to be analysed toxicologically. Toxic substances can usually be found in gills and skin or in the alimentary tract (Schlenk and Benson, 2001; Noga, 2010).

i. Acute intoxications

CLINICAL SIGNS AND GROSS LESIONS. Acute intoxications are characterized by sudden high mortality, usually in all age classes and different species. In addition to direct chemical intoxications, indirect effects can also occur, for instance oxygen deprivation caused by slurry rainwash in the water supply.

ii. Chronic intoxications

CLINICAL SIGNS AND GROSS LESIONS. Chronic intoxications are usually difficult to diagnose as mortality is low and in line with the diversity of toxic compounds and their modes of action, different target organs can be impacted and fish may display a diversity of pathological responses. A number of substances can lead to immune depression predisposing fish to secondary parasitic or bacterial infections, which in turn mask the original/primary problem. Histologically, alterations in blood cell counts, degeneration and atrophy of hematopoietic and lymphopoietic tissue, or an increase of pigmented macrophage centres in spleen, kidney and liver can be visible (Carlson and Zelikoff, 2008).

4.4 Nutritional and Feed-Related Problems

Adequate nutrition is essential for animal maintenance and health, including stress and disease resistance. While in extensive or semi-intensive farming systems fish obtain at least part of their nutrient needs from naturally available food organisms, fish maintained in intensive culture systems rely totally on the provision of a nutritionally complete and balanced diet. A balanced nutrition directly influences the intestinal microbiome and is of critical relevance for fish immunity and disease resistance (e.g. Blazer, 1992; Dawood *et al.*, 2020). Nutritional diseases can develop as a result of deficiency (undernutrition), excess (overnutrition) or imbalance (malnutrition) of nutrients present in the artificial diets used in aquaculture (Trichet, 2010; Oliva-Teles, 2012). Also toxic feed ingredients, such as pesticides contained in plant materials, or antinutritional factors, such as protease inhibitors, can lead to disease. In addition, adequate nutrition is not only a matter of a balanced diet, but also depends on the nutrient bioavailability from the diet.

For fish species that have long been cultured like rainbow trout, nowadays in-depth information on their nutritional needs is available. The situation is more difficult for fish species that have been taken into culture more recently (Gatlin, 2008). However, also for well-studied fish species the nutritional needs may vary depending on rearing and environmental conditions, so that a theoretically appropriate diet still can be associated with malnutrition symptoms.

Nutrition-related diseases are often difficult to diagnose. This is partly due to the fact the clinical symptoms of nutritional diseases like reduced food intake, skin darkening, lethargy or decreased growth are rather unspecific (Hardy, 2012; Rašković and Berillis, 2022). Also, the disease may develop gradually because animals have body reserves that compensate to a certain extent for nutritional deficiencies or imbalances.

4.4.1 Absolute nutritional deficiency: starvation

Starvation may be due to a complete deprivation of food, to inadequate feeding levels, or to behavioural, physiological or mechanical factors that prevent fish from feeding (Hardy, 2012; Roberts, 2022). Behavioural starvation is an issue in the weaning of fish larvae from a natural plankton diet to an artificial dry diet (Segner and Roesch, 1992; Laczynska *et al.*, 2020).

CLINICAL SIGNS GROSS AND HISTOPATHOLOGICAL LESIONS. Starved fish, as to be expected, become thinner and lose weight. The skin darkens, and gills may show anaemia. At necropsy no abdominal fat tissue is visible and the gall bladder is enlarged due to bile retention. Also changes in morphological body ratios take place; for instance, fish larvae experience a decrease in body width, resulting in 'pin heads' with very slender body and normal size head (Ehrlich *et al.*, 1976). Histologically, atrophy of muscle fibres can be observed, including reduction of sarcoplasmic content and myofibre vacuolation. Prominent alterations also include a decrease in height of the intestinal epithelia and in liver size, together with a shrinkage of liver cell volume and a loss of glycogen (Ehrlich *et al.*, 1976; Segner and Möller, 1984; Segner *et al.*, 1987; Gwak *et al.*, 1999). Due to the catabolic breakdown and accumulation of metabolic end products, like ceroid and lipofuscin, there is proliferation of the pigmented macrophage centres (Agius and Roberts, 1981; Hur *et al.*, 2006).

4.4.2 Deficiencies and imbalances of macronutrients

Proteins

Dietary protein requirements of fish vary with their natural feeding habits, with omnivorous and herbivorous species generally having lower protein requirements than carnivorous species. In a literature meta-analysis, Teles *et al.* (2020) found that dietary protein requirements range between 24 and 70%, depending on species, life stage and trophic level. If dietary protein supply is too low, this will result in reduced growth and health of the fish. If the dietary protein levels are too high, the fish will utilize part of the protein not for growth but for energy production (Oliva-Teles, 2012). This is economically unsound, since protein is an expensive feed ingredient, and it is environmentally unsound since it results in elevated nitrogenous waste (Gatlin, 2008). Protein sources vary in their nutritional value, depending on their amino acid (AA) composition and AA availability.

Quantitative requirements of AA have been determined for several fish species (Cowey, 1994; Waagbø, 2008; Mai *et al.*, 2022a). The essential AAs (arginine, histidine, isoleucine, leucine, lysine, methionine, phenylalanine, threonine, tryptophan and valine) are the same for all fish species. Cysteine and tyrosine can be synthesized from their precursors, methionine and phenylalanine, and are therefore regarded as semi-essential. Taurine is not classified as essential but seems to be necessary to larval and juvenile marine fish. Fishmeal is especially rich in taurine, whereas plant meal contains only low or no detectable levels (Gaylord *et al.*, 2006). When fishmeal is substituted by alternative protein sources, it is essential that their AA composition and protein digestibility are still adequate (Gatlin *et al.*, 2007).

CLINICAL SIGNS, GROSS AND HISTOPATHOLOGICAL LESIONS. Impaired growth and poor feed conversion are the most obvious indications of a protein or AA deficiency. Furthermore, protein malnutrition compromises immune defence mechanisms (Oliva-Teles, 2012). Malnutrition signs have also been described for individual AAs. Deficiency in arginine, threonine, tryptophan, valine, leucine, isoleucine, lysine, arginine or histidine leads to vertebral deformities in salmonids (Tacon, 1992; Schmidt-Posthaus and Marcos-López, 2014). Methionine and histidine deficiencies may lead to cataract formation (Waagbø *et al.*, 2010), while for tryptophan and valine deficiencies, fin erosion and kidney calcinosis have been reported (Tacon, 1992; Gatlin, 2008).

Carbohydrates

Complex carbohydrates are poorly digested by many fish species, but digestion and absorption can be improved by technologies such as heat treatment of the feeds (Krogdahl *et al.*, 2005; Kaushik *et al.*, 2022). Carbohydrates can be a source of energy in fish nutrition, although the ability to utilize carbohydrates varies strongly between fish species (Krogdahl *et al.*, 2005; Gatlin *et al.*, 2007; Kaushik *et al.*, 2022). Carnivorous species like salmonids have a limited capacity for carbohydrate utilization, whereas omnivorous or herbivorous species like many cyprinids or catfish species can utilize carbohydrates for energy generation and thereby spare dietary protein. Again in a species-dependent manner, fishes have the enzyme equipment for conversion of excess carbohydrates into lipids (Segner and Böhm, 1994; Sargent *et al.*, 2002).

CLINICAL SIGNS, GROSS AND HISTOPATHOLOGICAL LESIONS. Excessive dietary carbohydrate may affect disease resistance and stress tolerance (Oliva-Teles, 2012). In fish species with low carbohydrate needs, increased dietary carbohydrate levels result in enhanced hepatic glycogen and lipid deposition and eventually lead to hepatocyte degeneration (Hemre *et al.*, 2002; Prisingkorn *et al.*, 2017; Magalhães *et al.*, 2022).

Lipids

Fish diets must supply sufficient amounts of neutral lipids as a concentrated source of energy as well as sufficient amounts of essential fatty acids (EFAs). The latter have a number of important physiological functions in larval development, in the formation of cell membranes, or as precursors of eicosanoids, prostaglandins and leucotrienes, which have functions in inflammatory processes, among others (Sargent *et al.*, 2002; Tocher, 2003; Wall *et al.*, 2010; Oliva-Teles, 2012; Turchini *et al.*, 2022). EFAs cannot be synthesized by the organism but must be supplied with the diet. They include polyunsaturated fatty acids (PUFAs) of the 18:3*n*-3 and 18:2*n*-6 series and long-chain highly unsaturated fatty acids (HUFAs) such as 20:5*n*-3 and 22:6*n*-3 fatty acids. Fish species differ in their total lipid as well as EFA requirements. Many freshwater species possess the ability to elongate and desaturate PUFAs to HUFAs; in contrast, most marine fish species require HUFAs from the diet (Sargent *et al.*, 2002; Turchini *et al.*, 2022). Besides EFAs, phospholipids have to be provided by fish diets, particularly in diets for early life stages (Oliva-Teles, 2012).

There is a tendency in fish feed development to increase dietary energy density by including higher lipid levels. This helps also to spare dietary protein. This practice may result in extensive lipid storage in the adipose tissue and liver (Caballero *et al.*, 1999; Gatlin, 2008) which, however, does not necessarily represent a pathological condition (Hinton *et al.*, 2008). A number of fish species like cod (*G. morhua*) have naturally high lipid storage in the liver and this does not represent a pathological condition (Hinton *et al.*, 2008). On the other hand, high-fat diets can induce chronic inflammatory responses (Dai *et al.*, 2019). Thus, it needs profound knowledge of the nutritional and metabolic physiology of a fish species to interpret the presence of elevated lipid storage in the liver.

Fatty acids are particularly susceptible to (auto)oxidation, giving rise to rancid diets. Exposure to sunlight and high temperatures increase the risk for food rancidity. Supplementation of diets with compounds such as vitamin E, an antioxidant, can reduce the risk of lipid autoxidation.

CLINICAL SIGNS, GROSS AND HISTOPATHOLOGICAL LESIONS. EFA deficiency can result in increased mortality and retarded growth, particularly in early life stages of fish (Tacon, 1992; Sargent *et al.*, 2002). In addition, it can lead to swollen, pale livers with fatty infiltration, alterations of adipocyte structure, anaemia, reduced reproductive success, decreased swimming activity and impaired immune function (Tacon, 1992, 1996; Oliva-Teles, 2012; Roberts, 2022).

An unbalanced dietary EFA composition can cause pathologies. In Atlantic salmon, this condition was found to cause cardiovascular disorders (Bell *et al.*, 1993), and in flatfish it led to malpigmentation and impaired eye migration during larval development (Hamre *et al.*, 2007).

The most serious problem associated with rancid diets, at least in salmonid culture, is lipoid liver disease (Tacon, 1992, 1996; Bell *et al.*, 2003; Roberts, 2022). Clinical signs of the disease include anaemia, as well as swelling and paleness of the liver. Histologically, there is extreme lipid infiltration of the hepatocytes, together with a distortion of the typical muralia-like arrangement of the parenchyma. The hepatocytes eventually degenerate, followed by focal necrosis, pigmented macrophage immigration and deposition of ceroid, a breakdown product of phospholipid metabolism. The haematopoietic tissue in the spleen and kidney may degenerate, which leads to anaemia (Castell *et al.*, 1972). If the disease is still mild, the animals can recover completely but this is unlikely in the case of advanced disease manifestation. Other pathologies associated with lipoid liver diseases include skeletal myopathy as well as increased susceptibility to fin erosions in association with infection by *Flavobacterium* spp. (Tacon, 1992; Schmidt-Posthaus and Marcos-López, 2014).

4.4.3 Deficiencies and imbalances of micronutrients

Vitamins

Vitamins are required in trace amounts for normal growth, reproduction and disease resistance (Mai *et al.*, 2022b). In times of high metabolic activity, like growth or spawning, requirements can increase. Supplementation of certain vitamins, particularly vitamins A, E and C, is supposed to have a positive effect on the immune system (e.g. Bendich, 1990; Wahli *et al.*, 1998, 2003; Hernandez *et al.*, 2007; Cerezuela *et al.*, 2009), although this is still debatable (e.g. Lim *et al.*, 2010; Oliva-Teles, 2012). Water-soluble vitamins include thiamine, riboflavin, pyridoxine, pantothenic acid, niacin, biotin, folic acid, vitamin B_{12}, ascorbic acid/vitamin C, inositol, choline, *p*-aminobenzoic acid and lipoic acid. Fat-soluble vitamins include vitamin A, vitamin D, vitamin E and vitamin K. The fat-soluble vitamins are stored in the body and are metabolized only slowly, while water-soluble vitamins are rapidly excreted. Diagnosis of a vitamin deficiency can be difficult, as clinical signs and histopathological features are often unspecific and diets are usually not deficient in only one micronutrient (Mai *et al.*, 2022b; Roberts, 2022). Hypervitaminoses are mainly reported in association with fat-soluble vitamins, although they are unlikely under practical conditions (Tacon, 1992; Halver, 2002). In the following, vitamin-associated diseases are reported for a few selected vitamins only; for a comprehensive discussion, the reader is referred to Mai *et al.* (2022b).

i. Vitamin A (retinol)

Vitamin A (fat-soluble) is crucial for epithelial cells, body growth including bone development, eye vision, corticosterone synthesis and immune function/disease resistance (Thompson *et al.*, 1994; Cuesta *et al.*, 2002; Hardy, 2012). Carotenoids are precursors of vitamin A and also enhance immune response (Amar *et al.*, 2004). As a fat-soluble vitamin, vitamin A can be stored in the body, for instance in the perisinusoidal cells of Ito or fat-storing cells of the fish liver (Hinton *et al.*, 2008).

CLINICAL SIGNS, GROSS AND HISTOPATHOLOGICAL LESIONS. Vitamin A deficiency leads to reduced growth, depigmentation, and eye changes including exophthalmia, clouding and thickening of the corneal epithelium, keratomalacia and retinal degeneration (Tacon, 1992; Moren *et al.*, 2004). In salmon, vitamin A and astaxanthin, a carotenoid, are essential during the first-feeding period. Undersupply induces poor growth and low survival rates (Christiansen *et al.*, 1994). In contrast, hypervitaminosis results in epithelial squamous metaplasia, hepatomegaly, splenomegaly, osteopathy, and choroidal and corneal inflammation (Poston, 1971; Ørnsrud *et al.*, 2002). Additionally, in flounder, vitamin A hypervitaminosis is characterized by increased mortality, retarded growth, abnormal vertebral growth and changes in pigmentation (Dedi *et al.*, 1995; Martinez *et al.*, 2007). Unless the condition is extreme, it is readily reversible. Exceptions are early life stages of Atlantic salmon, where vitamin A stress or toxicity occurs at relatively low levels (Ørnsrud *et al.*, 2002).

ii. Vitamin E (tocopherol)

Vitamin E acts as an antioxidant; for instance, it can protect against oxidation of PUFAs and HUFAs. Therefore, to prevent oxidative stress, vitamin E is usually supplemented in high amounts (Martínez-Álvarez *et al.*, 2005). The vitamin acts as a free-radical trap to stabilize the unsaturated carbon bonds. The metabolism of vitamin E is closely related to selenium, which catalyses the regeneration of oxidized tocopherol (Lin and Shiau, 2009). Fish feeds usually contain a protected vitamin E form (α-tocopherol acetate) to prevent oxidation during storage (Hardy, 2012). Feeding diets containing oxidized fatty acids can affect the liver antioxidant status and dietary vitamin E can attenuate this process (Mourente *et al.*, 2002; Tocher *et al.*, 2002). Other roles of vitamin E are modulation of immune functions, like enhancement of antibody production and macrophage phagocytic activity (Bendich, 1990), and, together with vitamin C and selenium, vitamin E is important for normal reproductive function (Halver, 2002).

CLINICAL SIGNS, GROSS AND HISTOPATHOLOGICAL LESIONS. Vitamin E deficiency can be associated with increased mortality, reduced growth, reduced egg hatching rate, muscular degeneration, exophthalmos, ascites, pale gills with clubbed epithelial proliferation, epicarditis, erythrocyte fragility, reduced antibody response, impaired immunity and increased susceptibility to infections, and/or ceroid deposition in liver and spleen (Obach and Laurencin, 1992; Tacon, 1992; Wahli *et al.*, 1998; Halver, 2002). Associated with the enhanced erythrocyte fragility, anaemia and splenic haemosiderosis can occur.

iii. Vitamin C (ascorbic acid)

Vitamin C is a cofactor for prolyl-4-hydroxylase, which is responsible for hydroxylation of proline to hydroxyproline. Furthermore, vitamin C is involved in hydroxylation of l-lysine to hydroxylysine. Both components are essential parts of collagen, which is present in connective tissue, bone matrix and scar tissue in wound repair. The vitamin is also involved in reproduction, immune response and stress responses (Dabrowski *et al.*, 2004; Falcon *et al.*, 2007; Liu *et al.*, 2008; Mai *et al.*, 2022b). Another important function of vitamin C is its antioxidant activity, especially in combination with vitamin E. The synergistic effects of vitamin C and vitamin E may be due to the ability of vitamin C to regenerate vitamin E and the ability of vitamin E to spare the requirements of vitamin C (Sealey and Gatlin, 2002; Lim *et al.*, 2010). Vitamin C deficiencies are age dependent, with young stages being usually more sensitive than adults (Dabrowski *et al.*, 1988, 1989; Frischknecht *et al.*, 1994).

CLINICAL SIGNS, GROSS AND HISTOPATHOLOGICAL LESIONS. Clinical findings associated with vitamin C deficiency include reduced growth, darkening and decreased reproduction success (Tacon, 1992). In salmonids, skeletal deformities including lordosis and scoliosis, stress fractures, opercular deformation and gill filament distortion (Fig. 4.5A and B) are common (Tacon, 1992; Halver, 2002; Hardy, 2012). These changes are due to a deforming diathesis of the

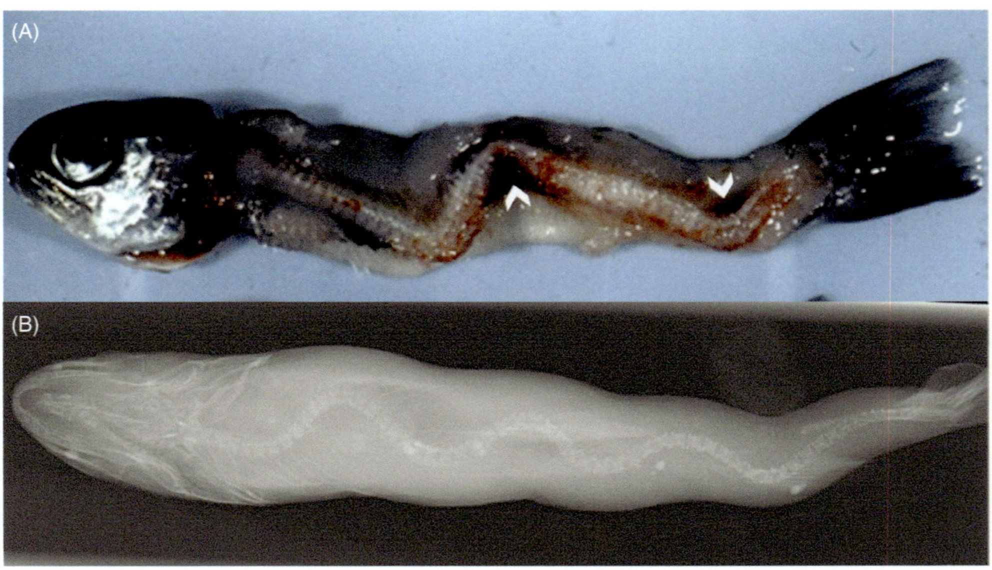

Fig. 4.5. Vitamin C deficiency in rainbow trout (*Oncorhynchus mykiss*). (A) Skeletal deformities, including lordosis and scoliosis, stress fractures and accompanied haemorrhages (arrowheads). (B) Osteoid changes in the vertebrae are associated with a decrease in radio-opacity.

cartilage, with replacement of bony tissue by osteoid. Additionally, fish fed vitamin C-deficient diets lose the capacity to initiate fibroplasia after wound induction, including repair of dermal fibres, revascularization and re-establishment of normal dermal and muscle structure (Halver *et al.*, 1969; Wahli *et al.*, 2003). The epidermis is not affected, and epidermal cells immediately migrate over the damaged area, independent of vitamin C levels. Particularly in combination with vitamin E deficiency, vitamin C-deficient fish may show reduced immune responses and increased susceptibility to infectious diseases (Obach and Laurencin, 1992; Wahli *et al.*, 1998; Montero *et al.*, 1999). In marine species like turbot, vitamin C-deficient diets lead to blockage of tyrosine catabolism and to hypertyrosinaemia. Tyrosine crystals accumulate in different parenchymal organs, especially kidney, liver and spleen. The crystals induce a granulomatous reaction of foreign body type with epithelioid macrophages (Baudin Laurencin *et al.*, 1989).

Minerals

Minerals are important for diverse physiological and immunological functions of fish, including bone and tooth formation, energy production, enzyme activity, muscle and brain functioning, antioxidant responses and antibacterial defence (Herrera *et al.*, 2019; Lall, 2022). Fish obtain minerals from both the diet and the aqueous environment; therefore deficiencies are less common than in terrestrial animals (Hardy, 2012; Lall, 2022). Moreover, mineral supplementation is not expensive, consequently deficiency can be avoided. However, pathologies and disease may arise from imbalances of minerals in fish, as has been suggested, for instance, for the hole-in-the-head disease (Amesberger-Freitag *et al.*, 2019). Also deficiencies of minerals may occur due to reduced bioavailability; that is, because of complex formation with other dietary elements like vitamins, fibre or phytic acid, or because of an antagonism with other minerals (Hilton, 1989; Tacon, 1992; Satoh, 2007; Lall and Kaushik, 2021). For instance, excess dietary calcium may inhibit absorption of other trace elements like zinc, iron, manganese and magnesium (Watanabe *et al.*, 1980; Tacon, 1992; Sugiura *et al.*, 2000). The dietary availability of minerals to fish might be improved by supplying chelate- or nanoparticle-bound minerals, or by the addition of probiotics (e.g. Terova *et al.*, 2018; Ahire *et al.*, 2019).

One of the few minerals for which fish cannot obtain their nutritional needs from the water is phosphorus, since aquatic concentrations are usually low, both in marine and fresh water. Therefore, elevated quantities are needed in the diet. However, too high amounts of phosphorus in the diet can have negative effects on growth and zinc availability (Tacon, 1992). The availability of phosphorus is dependent on the diet source with inorganic and animal phosphorus being more available than phosphorus from plant sources (Pimentel-Rodrigues and Oliva-Teles, 2007) because plant phosphorus is stored as phytate and animals lack the enzyme phytase (Oliva-Teles *et al.*, 1998). With the increasing need to replace fishmeal by plant meal as a protein source, the probability of phosphorus deficiency will increase.

CLINICAL SIGNS, GROSS AND HISTOPATHOLOGICAL LESIONS. With the exception of iron and iodine, where deficiencies are characterized by specific syndromes such as goitre formation, typical clinical signs associated with mineral deficiencies are reduced growth and feed efficiency, anorexia and increased mortality (Tacon, 1992). In the presence of low phosphorus levels, bone demineralization with deformed head, ribs and vertebrae is reported in salmonids (Ogino and Takeda, 1976; Skonberg *et al.*, 1997). Combined deficiencies of phosphorus and selenium in fast-growing strains of Atlantic salmon, particularly in post-smolts, can cause failure to ossify and resorption of phosphorus from the bones, which become deformed, especially at stress sites like the jaw, giving the animals a characteristic appearance (Bruno, 1990; Roberts, 2022). They also show spinal deformities, like kyphosis or scoliosis (Sullivan *et al.*, 2007a). Inadequate phosphorus levels in fry or fingerlings result in twisted or curled dorsal vertebral spines that are often encrusted with mineralized plaques (Sullivan *et al.*, 2007b).

Cataracts are reported with magnesium, zinc, manganese or copper deficiency (Ketola, 1978; Tacon, 1992). Magnesium deficiency additionally can lead to nephrocalcinosis, vertebral curvature, and degeneration of muscle fibres, epithelial cells of pyloric caeca and gill filaments. Zinc deficiency is also associated with short-body dwarfism, fin erosion and poor egg hatchability (Tacon, 1992; Hardy, 2012). Additionally, zinc and manganese deficiencies lead to depressed activity of natural killer cells and therefore to reduced immune function (Inoue *et al.*, 1998). Low selenium levels can provoke muscular dystrophy as well as depressed glutathione peroxidase activity leading to cell membrane instability due to oxidative damage and reduced immune response (Tacon, 1992). Iron deficiency is characterized by hypochromic microcytic anaemia. Iodine-deficient fish can exhibit – partly proliferative – thyroid follicular hyperplasia (Murray *et al.*, 2018). Deficiency of iodine, in combination with energy limitation and unbalanced fatty acid composition, has been suggested as a contributing factor for malpigmentation and impaired eye migration in flatfish larvae (Hamre *et al.*, 2007). It also has been suspected in the increased prevalence of thyroid goitres in salmonids from the Great Lakes region, USA (Moccia *et al.*, 1977). Supplementation of fish diets with iodine was shown to have a positive effect on the immune system in Atlantic salmon (Lall, 2022).

4.4.4 Toxic dietary components

Minerals

Excessive levels of certain minerals, in particular heavy metals such as zinc or copper, can cause toxicity and disease (Tacon, 1992; Lall and Kaushik, 2021). The origin of such excess metal levels can stem from contaminated feed ingredients such as zinc in feather meal or copper in fermentation residues, but also from leachates of storage vessels (Handy, 1996). High calcium levels in the diet can reduce the toxicity of metals such as cadmium, zinc or lead, partly because calcium impairs the intestinal/branchial absorption and cellular uptake of toxic metals (Hogstand, 1996).

CLINICAL SIGNS, GROSS AND HISTOPATHOLOGICAL LESIONS. Clinical signs associated with mineral toxicity are summarized by Tacon (1992). In rainbow trout, reduced growth and feed efficiency have been associated with selenium, copper and chromium toxicities. Selenium toxicity is characterized by nephrocalcinosis, increased susceptibility to infections or high mortality in rainbow trout (Hilton *et al.*, 1980; Hilton and Hodson, 1983; Hicks *et al.*, 1984). High copper exposure can lead to increased intestinal cell proliferation and apoptosis (Berntssen *et al.*, 1999) and higher susceptibility to bacterial infections (Hilton and Hodson, 1983). Cadmium toxicity is associated with scoliosis, hyperactivity and decreased bone calcium content. High levels of lead induce scoliosis, lordosis, black tail, anaemia and degeneration of the caudal fin in rainbow trout (Tacon, 1992). For minerals, the level of toxicity can vary depending on the food uptake together with water temperature and metabolic activity as well as the calcium content of the diet.

Calcium, because of its interaction with other minerals or because of excess concentrations in the diet, can lead to calcifications. As in mammals, metastatic or dystrophic calcification also occurs in fish. Metastatic calcification is normally caused by hypercalcinosis, while dystrophic calcification occurs most frequently in degenerated areas, like chronic granuloma, due to changes in tissue pH. In salmonids, a specific syndrome, nephrocalcinosis, is associated with dystrophic calcification (Harrison and Richards, 1979). The cause of this chronic condition often remains undetermined, but high CO_2 levels as well as magnesium deficiency and selenium toxicity are implicated (Hilton and Hodson, 1983; Hicks *et al.*, 1984). Nephrocalcinosis is characterized by widespread calcium deposits in renal tubules, collecting ducts and ureters, in the lumen and along the epithelial cells. This condition is especially common in intensively reared rainbow trout and brook trout. Macroscopically, chalky white material can be seen on the surface of the kidney. In the early stages, lesions are usually restricted to thick, white ureters (Fig. 4.6A). If the condition exacerbates, the kidney becomes swollen and greyish, and the ureters become more sinuous and thickened. In severe cases, the dorsal musculature may also be affected. Histologically, mineral deposits evoke a granulomatous inflammation affecting both excretory and interstitial parts of the kidney, followed by interstitial fibrosis (Fig. 4.6B). Obstruction of urine flow leads to distended tubuli and collecting ducts as well as tubular and glomerular degeneration (Harrison and Richards, 1979) (Fig. 4.6B).

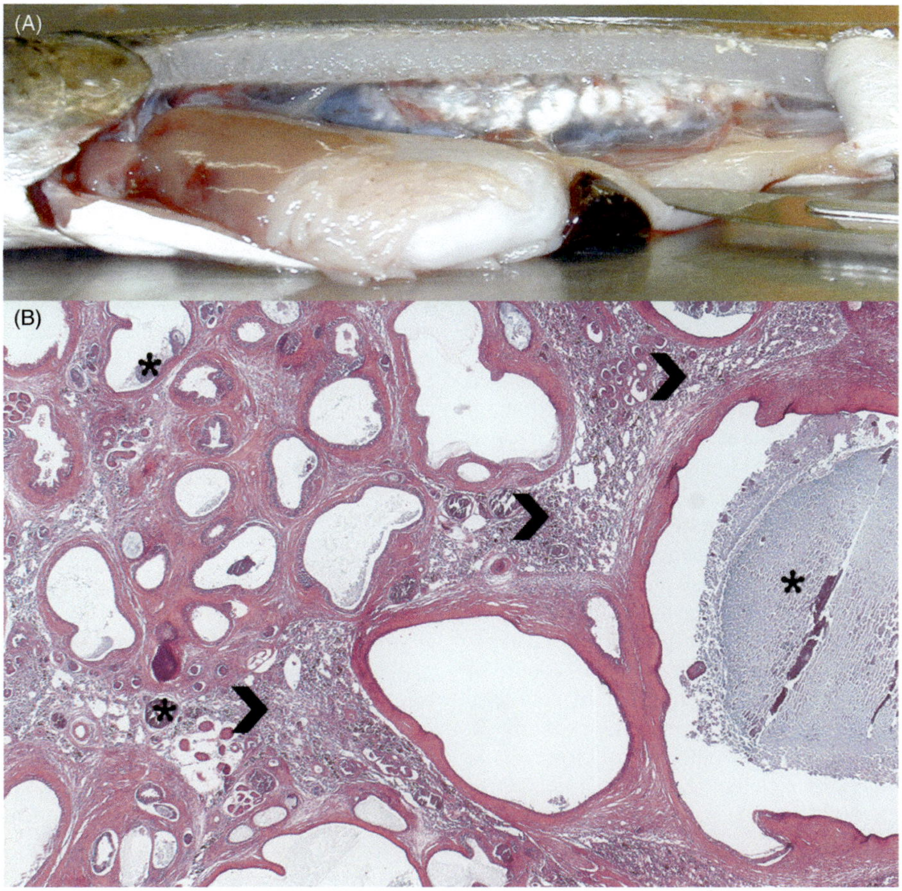

Fig. 4.6. Nephrocalcinosis in rainbow trout (*Oncorhynchus mykiss*). (A) In the early stages, lesions are usually restricted to thick, white ureters on the surface of the kidney. (B) Obstruction of urine flow leads to extended tubules and collecting ducts and tubular and glomerular degeneration, calcium deposits are found in renal tubules and collecting ducts (asterisks), deposits are located in the lumen and along the epithelial cells, mineral deposits evoke a severe granulomatous inflammation affecting both tubules and interstitial parenchyma, followed by interstitial fibrosis (arrowheads). Haematoxylin and eosin stain.

Often the lesions are associated with gastric submucosal granuloma.

Anthropogenic organic contaminants

Fish feeds can contain a variety of anthropogenic organic contaminants. One critical group of contaminants present in fish diets are the persistent organic pollutants (POPs) such as dioxins or polychlorinated biphenyls (PCBs) (Antunes and Gil, 2004; Jacobs *et al.*, 2004; Bell *et al.*, 2005). They originate mainly from fishmeal and fish oil prepared from wild fish catch. POPs are ubiquitously present in the aquatic environment. Therefore, wild fishes are exposed to POPs and due to the lipophilic nature of these chemicals, they are readily bioaccumulated. Another critical group of contaminants in fish feeds are pesticides which originate, in the case of organochlorines such as DDT, from fish oil (Jacobs *et al.*, 2004) or from plant materials used for diet preparation (Schlechtriem *et al.*, 2016; Berntssen *et al.*, 2021). Usually, the dietary levels of such contaminants are too low to cause overt toxicity, but there is a risk for sublethal effects such as alterations of lipid metabolism (Berntssen *et al.*, 2021).

Mycotoxins

The increased use of plant-based ingredients in fish feeds results in an increased risk of contamination by fungi and mycotoxins and a higher incidence of mycotoxicosis in fish (Oliveira and Vasconcelos, 2020; Pietsch, 2020). Toxicogenic fungi can be categorized into two groups: field fungi (e.g. *Fusarium* spp.), which grow on the plant before harvest; and storage fungi (e.g. *Aspergillus* spp., *Penicillium* spp.), which mostly contaminate the crop postharvest. In particular, aflatoxins are a long-recognized problem in aquaculture. Aflatoxin is a mycotoxin produced by *Aspergillus flavus*, a blue-green mould that is a common contaminant on oilseeds under warm humid conditions (Motalebi *et al.*, 2008). Four major aflatoxins (AFB1, AFB2, AFG1 and AFG2) have been identified. Improper storage is one of the most important factors favouring the growth of aflatoxin-producing moulds and it is the main element that can be controlled by the fish producer.

In rainbow trout, an extremely sensitive species, AFB1 is a potent carcinogen. This condition is rare in temperate regions but in countries where storage of grains involves exposure to humid and warm conditions, aflatoxicosis is still a serious problem (Motalebi *et al.*, 2008; Pietsch, 2020).

CLINICAL SIGNS, GROSS AND HISTOPATHOLOGICAL LESIONS. Clinical signs associated with aflatoxicosis in fish include pale gills, impaired blood clotting, poor growth rates and lack of weight gain (Oliveira and Vasconcelos, 2020). Visible signs of severe intoxication might be reduced survival rate, darkening/yellowing of the body and abnormal behaviour. Aflatoxin induces neoplastic changes after a few months in rainbow trout fed a contaminated diet with levels as low as 0.01 ppb (Fig. 4.7). Histologically, the lesion is characterized by a malignant invasive hepatocellular or bile duct carcinoma, often with a high amount of fibrovascular stroma. This is one of the few examples of a neoplastic lesion in fish which regularly metastasizes, often to the kidney, spleen or gills (Majeed *et al.*, 1984). At higher concentrations, the toxin induces acute toxicity with hepatic necrosis, branchial oedema and generalized haemorrhage (Hardy, 2012).

Antinutritional dietary compounds

A number of dietary compounds cause nutritional problems in fish (Gatlin, 2008; Krogdahl *et al.*, 2022; Roberts, 2022). They include particularly the plant-derived substances that are synthesized by plants as protection against predation. Examples are the lectin ricin in castor beans (*Ricinus communis*), gossypol in cotton seeds and mimosine in the protein-rich leguminous plants *Leucaena* spp. (Vogt *et al.*, 1994; Krogdahl *et al.*, 2022). Legume seeds and also soybean often contain protease inhibitors which are able to inhibit the endogenous proteases in the fish gut, but heat treatment during

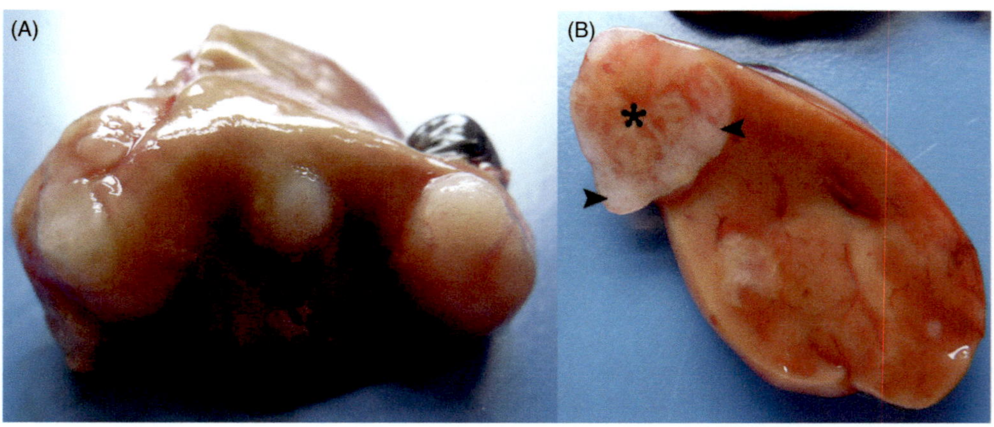

Fig. 4.7. Aflatoxin intoxication in a rainbow trout (*Oncorhynchus mykiss*), inducing mixed hepatocellular carcinoma and bile duct carcinoma. (A) Multiple neoplastic nodules in the liver. (B) Cut section showing whitish hepatocellular carcinoma (arrowheads) and central, more yellowish bile duct carcinoma (asterisk).

diet preparation mostly destroys this peptidergic antinutritional factor.

Soybean-induced enteritis

Fishmeal is traditionally used as the main protein source in fish feed for carnivorous species such as salmonids. However, the costs and availability of fishmeal can be a limiting factor, which has led to increased research on alternative protein sources. Soybean protein is of good quality and is highly digestive, but inclusion of soybean into the diet can induce an inflammatory response of the intestine that is known as soybean-induced enteritis (Urán *et al.*, 2008; Krogdahl *et al.*, 2015; Gu *et al.*, 2016). The exact cause and mechanisms behind the inflammation are not fully understood, but lectins and soy saponins have been discussed as potential causative factors (Hardy, 2012; Krogdahl *et al.*, 2015). Salmonids are highly susceptible, while Atlantic cod is unaffected (Hardy, 2012). Nowadays the disease rarely occurs, but it can still be a problem in situations where access to fishmeal is restricted or in new aquaculture species when the status of sensitivity is unknown. The severity of the inflammation depends on the fish species, soybean meal origin and processing, soybean meal levels and temperature (Urán *et al.*, 2008, 2009). Also dietary ingredients other than soybean, for instance corn gluten meal, are able to induce an enteritis, at least at high supplemental levels (Bai *et al.*, 2019).

CLINICAL SIGNS, GROSS AND HISTOPATHOLOGICAL LESIONS. The induced enteritis affects mainly the distal intestine and can lead to a disturbed digestive process and reduced health status. The intestinal folds are widened and shortened, apical epithelial vacuolation is lost, and the lamina propria is infiltrated with a mixed population of inflammatory cells (Ferguson, 2006; Urán *et al.*, 2008; Krogdahl *et al.*, 2015). Urán *et al.* (2008) also observed blockage of endocytosis and a strong decrease of the microvilli length. The enteritis can be induced within a few days (2 to 7 days, depending on the dose) and can resolve completely within 3 weeks after removal of soybean meal. Recovery changes can start as early as 2 days after feed change (Ferguson, 2006; Urán *et al.*, 2009).

4.5 Vaccination- and Treatment-Related Problems

4.5.1 Vaccines

Vaccines against bacterial and viral diseases can be administrated orally (via feed), by immersion or by intraperitoneal (i.p.) injection. The type of vaccine used depends on the species and the size of the fish; for example, i.p. vaccines are used in fish over 30 g. The vast majority of vaccination-related problems are described for fish treated by i.p. injection.

CLINICAL SIGNS, GROSS AND HISTOPATHOLOGICAL LESIONS. The i.p. administration of oil-adjuvanted vaccines has been reported to cause side effects including retarded growth (Midtlyng and Lillehaug, 1998), granulomatous peritonitis, intra-abdominal adhesions and melanization (Midtlyng, 1997; Ferguson, 2006; Berg *et al.*, 2007). Retardation in growth can be transient, vaccinated fish might have the same weight as non-vaccinated specimens at harvest (Mutoloki *et al.*, 2004). Numerous pigmented macrophages and remnants of vaccine seen as clear vacuoles can be observed in the granulomatous response (Fig. 4.8). The extensive granulomatous inflammation is likely to impact the exocrine pancreas and subsequent production of digestive enzymes (Ferguson, 2006). In severe cases, melanized foci can penetrate into the abdominal wall and occasionally into the skin (Koppang *et al.*, 2005). Muscle necrosis, fibrosis and granulomatous inflammation can be seen in these lesions. As a possible mechanism for part of these reactions, development of a systemic autoimmunity has been proposed (Koppang *et al.*, 2008; Haugarvoll *et al.*, 2010). Bjørge *et al.* (2011) observed in Atlantic salmon that behavioural changes, including reduced feeding response and decreased activity, correlated with the severity of the peritonitis. The retarded growth observed in post-vaccinated fish is likely related to both the poor feeding response and the chronic enzymatic insufficiency induced (Ferguson *et al.*, 2010). Higher water temperatures and changing environment could result in a shift of known or emergence of new pathogens in formerly colder areas. Thus the need for vaccines will rise in the future (Thomas *et al.*, 2022).

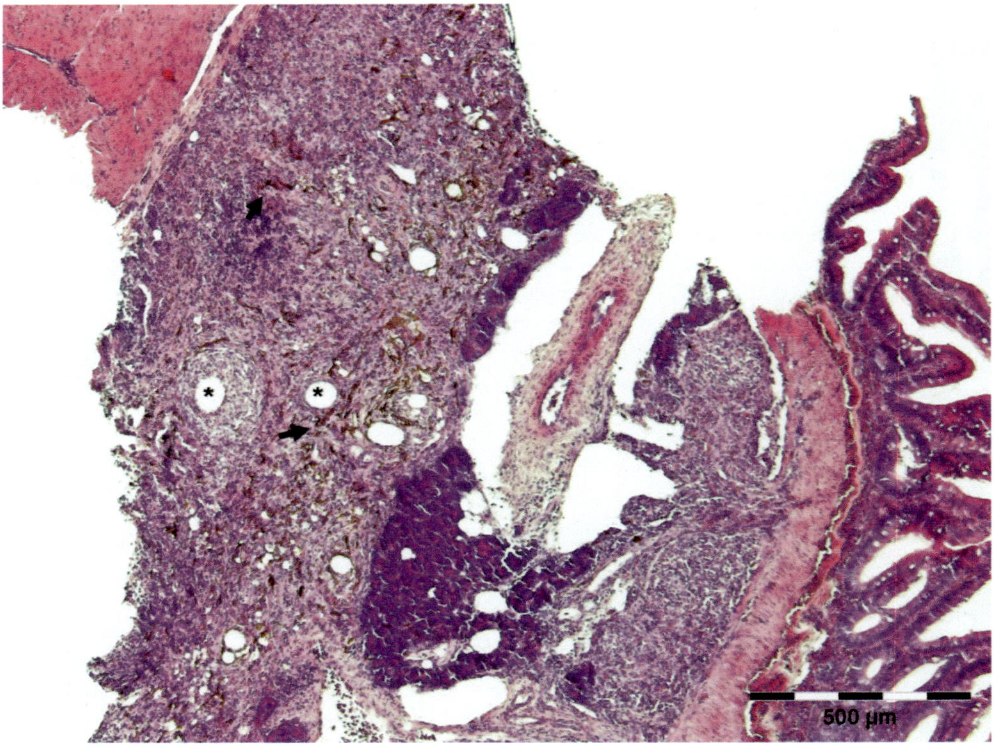

Fig. 4.8. Farmed Atlantic salmon (*Salmo salar*) showing severe granulomatous peritonitis around pancreas and pyloric caeca after i.p. vaccination. Note the presence of melanomacrophages (arrow) and the clear vacuoles (asterisks), remnants of vaccine. Haematoxylin and eosin stain, scale bar = 500 µm. (Image courtesy of Mar Marcos-Lopez.)

Nowadays, many of the abovementioned side effects of vaccine administration have been cleared through research, and the use of proper vaccination and vaccine delivery procedures do not really harm fish. For example, micro-dose vaccines have played an important role in minimizing side effects from fish vaccination, including local inflammation and pigmentation in the abdominal cavity, prolonged period before fish return to normal feeding and potential downgrading of the fish fillets. Any adverse reactions are typically in a more confined region for fish vaccinated with the micro-dose compared with a normal 0.1 ml injection dose.

4.5.2 Chemical treatments

In cages, chemical bath treatments (such as hydrogen peroxide, formalin, chloramine T and commercial sea lice treatments) applied to combat fungal infections and external parasites can be administrated within well-boats or inside the cages by fitting tarpaulins around them. To decrease the volume of chemical needed, the nets of the cages can be raised in the tarpaulin system. This, however, increases the fish density, leading to an increase in oxygen consumption and demand. Oxygen levels should be monitored, and extra oxygenation supplied when required. Other considerations are tides and currents, which can affect the final concentration and distribution of the chemical in the water. To avoid heterogeneous distribution of the chemical, several inlet pumps at different cage locations and water agitation systems should be installed. In landlocked raceways, chemicals can be added for a limited time in a given concentration by stopping the water flow or by continuous addition in a flow-through system. In both situations chemical concentrations and oxygen levels have to be surveyed. When using chemicals in recirculation systems, potential harmful effects on biofilter systems have to be considered (Pedersen and Pedersen, 2012). The margin between effective and toxic concentrations is narrow and will depend on the fish species, health status, stocking density, exposure time and water temperature. For example, temperatures higher than 13°C increase hydrogen peroxide toxicity, thus Bruno and Raynard (1994) observed 35% mortalities at 13.5°C but none at 10°C in Atlantic salmon. Increased temperatures due to climate change have to be considered when treating fish.

CLINICAL SIGNS, GROSS AND HISTOPATHOLOGICAL LESIONS. Effects of chemical bath treatments in treated fish can result in a reduction of the amount of oxygen bound to Hb and gill damage, including lamellar epithelial swelling, lifting, hyperplasia, necrosis and intercellular oedema (Schmidt-Posthaus and Marcos-López, 2014). Several adverse effects have been associated with the excessive use of antibiotics in fish, such as body malformation and damage, compromised immune response, oxidative stress, hepatotoxicity, impaired haematological parameters, neurotoxicity and genotoxicity (Yang *et al.*, 2020; Limbu *et al.*, 2021). Other effects are development of resistance, environmental problems (i.e. residues in fishfarm sediments) and drug residues in fish products (Yonar, 2012).

4.6 Neoplasia

Neoplasia is uncontrolled proliferation and dispersal of autologous cells throughout the host body. It is based on accumulated DNA mutations. Mutations can occur spontaneously or be triggered by dietary or environmental pollution (chemicals), radiation or infectious agents (da Rocha *et al.*, 2018). In contrast, temperature as a cause of tumour induction is not reported. There is an increasing chance of mutations and therefore of spontaneous neoplasms with increasing age. As fish in culture usually do not survive to old age, incidental tumours are rare. Classification of fish tumours is based on criteria developed for human medicine, which are based on phenomena that characterize the identity and behaviour of neoplastic cells (WHO, 2013; Meuten, 2020).

There is only very limited information available on non-pathogen-related neoplasms in cultured coldwater fish. In addition, it is often difficult to distinguish between infectious and non-infectious causes of neoplasms. Even metastases are far less common in fish neoplasms than in mammals (Martineau and Ferguson, 2006), thus judging the stage of progression (from preneoplastic foci to anaplastic, invasive growing neoplasms) might be useful in respect of estimating exposure duration to carcinogens for example. In environmentally induced tumours, the long time between exposure to environmental chemical carcinogens and cancer development often makes it difficult to prove a causative relationship. Many man-made carcinogens, like some polycyclic aromatic hydrocarbons (PAHs), may be present in fishmeal and fish oil. They are hydrophobic and therefore efficiently absorbed by fish.

4.6.1 Common neoplasms in coldwater finfish

Skin

PAPILLOMAS. The most common skin tumours are papillomas. Many are flat and show a discrete expansive growth of altered epidermal cells. Specialized cells, such as goblet cells, are usually missing or rare. The proliferated epithelium is often spongiotic and infiltrated with mainly lymphocytes. 'Papillomatosis' of Atlantic salmon is most common in young fish in freshwater stocks during mid-summer. Clinically, these neoplasms appear as blue-grey raised plaques (Shchelkunov *et al.*, 1992) and they usually regress spontaneously; but sometimes they become secondarily infected following ulceration (Carlisle and Roberts, 1977). According to Harshbarger *et al.* (2021), epidermal plaque-forming neoplasms might be of viral aetiology, namely herpesviruses (Doszpoly *et al.*, 2013), while lobular-type neoplasms are suspected to have a chemical aetiology.

SQUAMOUS CELL CARCINOMAS. Squamous cell carcinomas are less common, but in Atlantic salmon a progression from the papillomatous lesion has been recorded in individual animals (Grizzle and Goodwin, 2010). The epidermal papillomas of Pacific flatfish are characterized by the presence of X-cells in addition to the Malpighian cell proliferation. These lesions were classified as pseudo-neoplasm. The X-cells are of parasitic, namely protist, origin (Miwa *et al.*, 2004). Mainly young fish are affected at the upper pigmented side of the body. Atlantic species of flatfish occasionally show epizootic epithelial hyperplasia. These are usually homogeneous, flat plaques of proliferating Malpighian cells up to 15 mm in diameter.

PIGMENT CELL TUMOURS. Pigment cell tumours include melanomas, erythrophoromas, xanthophoromas, guanophoromas and iridophoromas. Melanomas are most common; however, they are mainly seen in aquaria fish. In haematoxylin and eosin stain, pigments are visible in melanomas and iridophoromas. As in mammals, some melanomas can be amelanotic and show variation in their cellular morphology from round to fusiform or spindle shaped (e.g. Gimenez-Conti *et al.*, 2001). Iridophoromas are composed of spindle-shaped cells with a distinct olive-green pigment (Schmidt-Posthaus *et al.*, 2005) (Fig. 4.9A and B).

FIBROMAS AND FIBROSARCOMAS. The most common mesenchymal tumours affecting a wide variety of different species are fibromas and fibrosarcomas. Fibromas can be loosely organized and soft – called myxoma – or hard, as found in the dermal sarcoma of walleye caused by a retrovirus (Martineau *et al.*, 1990, 1992). Generally, they are enclosed by a covering layer of epithelium. The cut surface is usually firm and shiny, pale pink or whitish, and occasionally with central necrosis. Fibrosarcomas are characterized by their invasiveness rather than metastases. Environmental and infectious agents are suggested as possible causes for fibroma or fibrosarcoma. A fibrosarcoma has been described in a farmed sockeye salmon (*Oncorhynchus nerka*) (Meyers and Hendricks, 1983).

PERIPHERAL NERVE SHEATH TUMOURS. Peripheral nerve sheath tumours are also common neoplasms, especially in goldfish and damselfish, but have also been described in farmed European eel (*Anguilla anguilla*) (Gjurčević *et al.*, 2014) and coho salmon (Masahito *et al.*, 1985). Generally these tumours are benign.

LIPOMAS. Lipomas are occasionally seen in the dermis as soft, greasy masses. Typically, they appear as well-differentiated, rounded and encapsulated masses. Liposarcomas are not described in fish. Haemangiomas, benign proliferations of small blood vessels, are also sometimes seen in the dermis (Meyers and Hendricks, 1983).

Gills and pseudobranches

Tumours of the gills are rare. In Atlantic and Pacific cod, but also in dab from the North Sea, tumorous lesions in the pseudobranch are

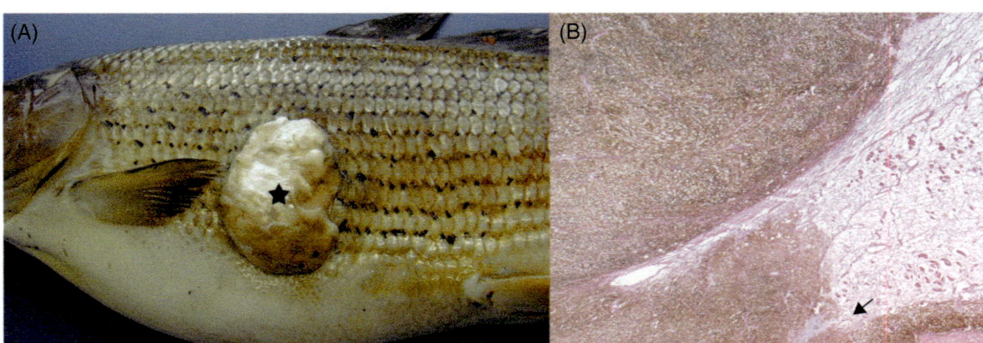

Fig. 4.9. Iridophoroma in a grayling (*Thymallus thymallus*). (A) Macroscopically, a firm whitish mass (star) is visible on the right flank. (B) Histologically, the neoplastic cells are arranged in bundles, the cytoplasm is filled with an olive-green pigment, the arrow demonstrates the transition between the hyperplastic pigment cell layer and the neoplastic lesion. Haematoxylin and eosin stain.

described mostly as bilateral masses at the dorsolateral part of the gill cavity, containing large pale cells similar to the X-cells seen in Pacific flatfish papilloma (Schmidt-Posthaus and Marcos-López, 2014). As structures resembling X-cells have been found, the lesions are considered to be pathogen induced. But branchioblastoma and gill epithelial cell tumours can be chemically induced (e.g. medaka) (Brittelli *et al.*, 1985) or occur spontaneously in brown and rainbow trout and Atlantic salmon (Roberts, 2012b; Schmidt-Posthaus and Marcos-López, 2014).

Kidney

In nephroblastomas, reported in salmonid fish (Lumsden and Marshall, 2003), pluripotent embryonic cells proliferate to form most of the excretory elements usually present in the trunk kidney, like tubuli and glomeruli. Fibrous tissue and cartilage may be present in variable amounts. These tumours can be seen protruding from the posterior kidney in adult salmon, with none or minimal clinical consequences.

Haematopoietic tissue

The majority of tumours in haematopoietic tissues originate from lymphoid cells (Harada *et al.*, 1990). Spontaneous lymphosarcomas have been diagnosed in Atlantic and chinook salmon (Speare, 2002) infiltrating different inner organs, particularly the kidney interstitium. Lymphosarcomas of thymic origin (thymomas) are reported in salmonids (Warr *et al.*, 1984; Bowser *et al.*, 1987; Bruno and Smail, 1998), also with a leukaemic component involving the kidney. Plasmacytoid leukaemia principally affects cage-cultured chinook salmon (Kent *et al.*, 1990; Kent and Dawe, 1993). For some of these tumours a retroviral aetiology was identified (Kent and Dawe, 1990, 1993; Eaton and Kent, 1992). Occasionally, leukaemia with involvement of other organs, such as the liver, can be seen in rainbow trout (Schmidt-Posthaus and Marcos-López, 2014).

Liver

Hepatocellular adenomas and carcinomas, as well as bile duct adenomas and carcinomas, are relatively common in different fish species. Both tumours, originating from hepatocytes and bile ducts, can occur in the same individual. As mentioned in Section 4.4.4, rainbow trout are especially sensitive to aflatoxin intoxication inducing hepatocellular and bile duct carcinoma (Fig. 4.7). This is one of the few examples where neoplasms in fish metastasize, often to the spleen and kidney (Schmidt-Posthaus and Marcos-López, 2014). Another example for a metastatic process was demonstrated by Dale *et al.* (2009), who found tumours in the liver of Atlantic salmon deriving from primary tumours in the intestinal tract. The primary tumour development was associated with the diet.

Liver tumours can be used to assess environmental pollution (Harshbarger and Clark, 1990) and rainbow trout can serve as a model for liver tumour induction by environmental toxicants (Bailey *et al.*, 1996; Williams, 2012). In wild fish, the presence of hepatocellular neoplasms, from benign to highly malignant forms, could be correlated to levels of PAHs and other environmental contaminants (Schmidt-Posthaus and Marcos-López, 2014; Meier *et al.*, 2020).

Gastrointestinal tract and swim bladder

Neoplasms arising from the gastrointestinal tract are rare in fish. Teeth tumours, like ameloblastomas, are reported in salmonids (Gorlin, 1972). An epizootic of fibrosarcomas arising from the swim bladder and associated with a retrovirus infection was reported in caged Atlantic salmon in Scotland (Duncan, 1978). A similar condition, also caused by a retrovirus, was described in Atlantic salmon on the western seaboard of the Atlantic Ocean (Black and Baumann, 1991). Inappropriate diet causing intestinal inflammation can be followed by intestinal adenocarcinoma (Dale *et al.*, 2009). Adenocarcinomas in conjunction with intussusception were found in rainbow trout. In advanced stages metastasis in liver and gill was seen (Hoitsy *et al.*, 2021).

Reproductive system

Tumours arising from the different cell components of the ovary and testis are known in fish. However, these tumours are far more common in warmwater species than in coldwater finfish.

Endocrine system

Distinguishing thyroid tumours, like adenomas or carcinomas, from thyroid follicular hyperplasia (goitre) can be challenging, especially as transition is often blurred, thyroid follicles in fish are non-encapsulated and ectopic follicles can exist in various organs (Baker *et al.*, 1955; Harada *et al.*, 1996).

4.7 Deformations

4.7.1 Pigmentation abnormalities

Pigmentation abnormalities occurring at larval stages are a common and important problem in flatfish hatcheries. The market value and consumer acceptability of the affected individuals are decreased. During metamorphosis, flatfish undergo a substantial transformation involving shift of the body axis, eye migration and pigmentation changes. Flatfish larvae have pigmented cells on both sides of the body, but during metamorphosis larval melanophores disappear and adult melanophores develop on the ocular (dorsal) but not on the blind (ventral) side (Bolker and Hill, 2000). Pigmentation abnormalities can occur on both sides (Kang *et al.*, 2012). Non-pigmented areas on the ocular side are known as hypomelanosis or pseudo-albinism, while hypermelanosis or ambicoloration refers to the presence of pigmented areas on the blind surface (Venizelos and Benetti, 1999). Both types of malpigmentation may be related to inappropriate hatchery conditions such as diet and lighting (Bolker and Hill, 2000).

4.7.2 Skeletal and eye deformities

The most common external physical deformities in fish are skeletal (vertebral, opercular, jaw and fins) and eye deformities (Fig. 4.10A). Consequences depend on the degree and type of deformity, ranging from asymptomatic to poor growth and performance, poor welfare, decreased market value and mortality. Shortened opercula, for example, compromises the fish's capacity for pumping water and increases gill exposure to environmental and infectious stressors. Factors causing physical deformities may be pathogens, genetic, environmental (Fig. 4.10B), nutritional, or a combination of these (Silverstone and Hammell, 2002). Stages with unfinished ossification are more susceptible to malformations. Traumatic injuries and infections (e.g. rainbow trout fry syndrome caused by *Flavobacterium psychrophilum* or infection with *Myxobolus cerebralis*) can also lead to deformities. Congenital physical deformities have low incidence but are more common in captive than in wild fish due to the absence of natural selective pressures (Sadler *et al.*, 2001). In salmon aquaculture, the induction of triploids is used to avoid problems associated with sexual maturation, such as lower growth rates, increased susceptibility to diseases and deterioration of organoleptic properties (Piferrer *et al.*, 2009). However, triploidy in Atlantic salmon has been associated with higher prevalence of skeletal deformities and cataracts after sea transfer (Taylor *et al.*, 2011, 2013). O'Flynn *et al.* (1997) reported that Atlantic salmon triploids and diploids showed similar freshwater growth and survival; however, in seawater, triploids grew better but had higher deformities and lower survival. Similar findings were reported by Taylor *et al.* (2013). The faster growth showed by triploids in seawater may explain some of the observed deformities. High growth rate is known to induce inferior bone quality in other animal species (e.g. broilers) (Julian, 1998). Also in Atlantic cod, the high growth rate during juvenile stages has been proposed as a potential factor associated with the high number of skeletal deformities under culture conditions (Fjelldal *et al.*, 2009).

Temperature conditions during egg development and rearing of rainbow trout (Crichigno *et al.*, 2021) and the freshwater stage of salmon have been identified as an important factor for causing skeletal deformities (Ørnsrud *et al.*, 2002; Takle *et al.*, 2005; Ytteborg *et al.*, 2010; Grini *et al.*, 2011; Fraser *et al.*, 2015). A range of malformations was induced in Atlantic salmon exposed to temperatures above 8°C during egg incubation (Takle *et al.*, 2005), thus 8°C is recognized as the upper limit for normal embryonic development for this species. Triploids have proven to be more susceptible to higher temperatures and low oxygen conditions (Ojolick *et al.*, 1995; Sadler *et al.*, 2001; Fraser *et al.*, 2015; Clarkson *et al.*, 2021).

Fig. 4.10. (A) Farmed Atlantic salmon (*Salmo salar*) showing skeletal deformity due to fused vertebrae. The origin in this case is unknown. (Photograph courtesy of David Bruno.) (B) Farmed rainbow trout (*Oncorhynchus mykiss*) showing skeletal deformities due to injury in barriers.

Sánchez *et al.* (2011a,b) showed that exposure to hypoxia (60% O_2 saturation) during the early post-hatching period induced a higher frequency of vertebral column deformities in Atlantic salmon. Hypoxia altered the type and distribution of collagen in the notochord sheath and retarded the mineralization process of the vertebral body. The structural properties of the notochord allow movement and provide anchoring points for the myotomal segments (Sánchez *et al.*, 2011a,b). Avoiding hypoxic conditions by controlling oxygenation, water temperature and stock density is recommended to prevent potential skeletal deformities. Deficiencies or toxicities with minerals (calcium, phosphorus, zinc, selenium and manganese) and vitamins (A, D, C, E and K), as well as their interactions and lipid peroxidation, can also cause skeletal deformities (see Sections 4.4.3 and 4.4.4) (Lall and Lewis-McCrea, 2007; Fernández and Gisbert, 2011; Baeverfjord *et al.*, 2019).

4.7.3 Internal malformations

Malformation of the swim bladder in Atlantic salmon has been recognized as the underlying cause for aberrant swimming behaviour. In affected fish a shortened swim bladder in the anterior part of the body cavity was seen, while the pneumatic duct typical for this physoclistous species was dramatically elongated (Poppe *et al.*, 1997). Swim bladder inflation is a key step during the early larval stage of most fish and occurs when air gulped from the surface is transferred to the swim bladder by the pneumatic duct. Water with excessive surface oil will impair the ability to fill the swim bladder (Ferguson, 2006). The use of surface skimmers to prevent build-up of surface oils can help to avoid this.

Internal malformations of the soft tissues are also reported from cultured fish, such as aplasia of the septum transversum (Poppe *et al.*, 1998) and abnormal-shaped hearts (Poppe *et al.*,

2003). Takle *et al.* (2006) suggested an effect of hyperthermia during embryo development as a possible causative factor, while no effect of temperature on heart morphology was found in Atlantic salmon in the grow-out phase (Foddai *et al.*, 2022). Heart deformities, such as ventricular hypoplasia or rounder hearts, can cause poor cardiac function, reduced tolerance to stress and subsequent mortalities (Poppe *et al.*, 2003). Also, *situs inversus* was demonstrated (Poppe *et al.*, 1997).

With inappropriate temperatures being an important trigger for deformities, increase in water temperatures might cause problems in farming of coldwater species in future. This might be a particular problem in farms which rely on water of ambient temperature.

References

Ackerman, P.A., Wicks, B.J., Iwama, G.K. and Randall, D.J. (2006) Low levels of environmental ammonia increase susceptibility to disease in chinook salmon smolts. *Physiological and Biochemical Zoology* 79(4), 695–707.

Agius, C. and Roberts, R.J. (1981) Effects of starvation on the melano-macrophage centres of fish. *Journal of Fish Biology* 19(2), 161–169.

Ahire, J.J., Mokashe, N.U. and Chaudhari, B.L. (2019) Effect of dietary probiotic *Lactobacillus helveticus* on growth performance, antioxidant levels, and absorption of essential trace elements in goldfish (*Carassius auratus*). *Probiotics and Antimicrobial Proteins* 11(2), 559–568.

Alanärä, A. and Brännäs, E. (1996) Dominance in demand-feeding behaviour in Arctic charr and rainbow trout: the effect of stocking density. *Journal of Fish Biology* 48(2), 242–254.

Amar, E.C., Kiron, V., Satoh, S. and Watanabe, T. (2004) Enhancement of innate immunity in rainbow trout (*Oncorhynchus mykiss* Walbaum) associated with dietary intake of carotenoids from natural products. *Fish & Shellfish Immunology* 16(4), 527–537.

Amesberger-Freitag, A., Tichy, A., El-Matbouli, M. and Lewisch, E. (2019) Hole-in-the-head disease in discus fish, *Symphysodon* (Heckel, 1840): is it a consequence of a dietary Ca/P imbalance? *Journal of Fish Diseases* 42(8), 1133–1142.

Antunes, P. and Gil, O. (2004) PCB and DDT contamination in cultivated and wild sea bass from Ria de Aveiro, Portugal. *Chemosphere* 54(10), 1503–1507.

Ashley, P.J. (2007) Fish welfare: current issues in aquaculture. *Applied Animal Behaviour Science* 104(3–4), 199–235.

Attrill, M.J., Wright, J. and Edwards, M. (2007) Climate-related increases in jellyfish frequency suggest a more gelatinous future for the North Sea. *Limnology and Oceanography* 52(1), 480–485.

Baeverfjord, G., Antony Jesu Prabhu, P., Fjelldal, P.G., Albrektsen, S., Hatlen, B., *et al.* (2019) Mineral nutrition and bone health in salmonids. *Reviews in Aquaculture* 11(3), 740–765.

Bai, N., Gu, M., Liu, M., Jia, Q., Pan, S. and Zhang, Z. (2019) Corn gluten meal induces enteritis and decreases intestinal immunity and antioxidant capacity in turbot (*Scophthalmus maximus*) at high supplementation levels. *PLoS ONE* 14(3), e0213867.

Bailey, G.S., Williams, D.E. and Hendricks, J.D. (1996) Fish models for environmental carcinogenesis: the rainbow trout. *Environmental Health Perspectives* 104(Suppl. 1), 5–21.

Baker, K.F., Berg, O., Gorbman, A., Nigrelli, R.F. and Gordon, M. (1955) Functional thyroid tumors in the kidneys of platyfish. *Cancer Research* 15(2), 118–123.

Båmstedt, U., Fosså, J.H., Martinussen, M.B. and Fosshagen, A. (1998) Mass occurrence of the physonect siphonophore *Apolemia uvaria* (Lesueur) in Norwegian waters. *Sarsia* 83(1), 79–85.

Baras, E. and Jobling, M. (2002) Dynamics of intracohort cannibalism in cultured fish. *Aquaculture Research* 33(7), 461–479.

Barton, B.A. (1997) Stress in finfish: past, present and future – a historical perspective. In: Iwama, G.K., Pickering, A.D., Sumpter, J.P. and Schreck, C.B. (eds) *Fish Stress and Health in Aquaculture*. Cambridge University Press, Cambridge, UK, pp. 1–33.

Baudin Laurencin, F., Messager, J. and Stephan, G. (1989) Two examples of nutritional pathology related to vitamins E and C deficiencies. *Advances in Tropical Aquaculture, Workshop at Tahiti, French Polynesia, 20 February–4 March 1989. Actes de Colloque* 9, 171–181.

Baur, W.H., Bräuer, G. and Rapp, J. (2010) *Nutzfische und Krebse: Lebensraum, Erkrankungen und Therapie*. Georg Thieme Verlag, Leipzig, Germany.

Baxter, E.J., Rodger, H.D., McAllen, R. and Doyle, T.K. (2011a) Gill disorders in marine-farmed salmon: investigating the role of hydrozoan jellyfish. *Aquaculture Environment Interactions* 1(3), 245–257.

Baxter, E.J., Sturt, M.M., Ruane, N.M., Doyle, T.K., McAllen, R., *et al.* (2011b) Gill damage to Atlantic salmon (*Salmo salar*) caused by the common jellyfish (*Aurelia aurita*) under experimental challenge. *PLoS ONE* 6(4), e18529.

Baxter, E.J., Sturt, M.M., Ruane, N.M., Doyle, K., McAllen, R. and Rodger, H.D. (2012) Biofouling of the hydroid *Ectopleura larynx* on aquaculture nets in Ireland: implications for finfish health. *Fish Veterinary Journal* 13, 17–29.

Bell, J.G., Dick, J.R., McVicar, A.H., Sargent, J.R. and Thompson, K.D. (1993) Dietary sunflower, linseed and fish oils affect phospholipid fatty acid composition, development of cardiac lesions, phospholipase activity and eicosanoid production in Atlantic salmon (*Salmo salar*). *Prostaglandins, Leukotrienes and Essential Fatty Acids* 49(3), 665–673.

Bell, J.G., McEvoy, L.A., Estevez, A., Shields, R.J. and Sargent, J.R. (2003) Optimising lipid nutrition in first-feeding flatfish larvae. *Aquaculture* 227(1–4), 211–220.

Bell, J.G., McGhee, F., Dick, J.R. and Tocher, D.R. (2005) Dioxin and dioxin-like polychlorinated biphenyls (PCBs) in Scottish farmed salmon (*Salmo salar*): effects of replacement of dietary marine fish oil with vegetable oils. *Aquaculture* 243(1–4), 305–314.

Bendich, A. (1990) Antioxidant vitamins and their functions in immune responses. In: Bendich, A., Phillips, M. and Tengerdy, R.P. (eds) *Antioxidant Nutrients and Immune Functions*. Advances in Experimental Medicine and Biology, Vol. 262. Springer, Boston, Massachusetts, pp. 35–55.

Berg, A., Rødseth, O.M. and Hansen, T. (2007) Fish size at vaccination influence the development of side-effects in Atlantic salmon (*Salmo salar* L.). *Aquaculture* 265(1), 9–15. https://doi.org/10.1016/j.aquaculture.2007.02.014

Berge, Å.I., Berg, A., Barnung, T., Hansen, T., Fyhn, H.J. and Stefansson, S.O. (1995) Development of salinity tolerance in underyearling smolts of Atlantic salmon (*Salmo salar*) reared under different photoperiods. *Canadian Journal of Fisheries and Aquatic Sciences* 52(2), 243–251.

Berntssen, M.H.G., Hylland, K., Bonga, S.E.W. and Maage, A. (1999) Toxic levels of dietary copper in Atlantic salmon (*Salmo salar* L.) parr. *Aquatic Toxicology* 46(2), 87–99.

Berntssen, M.H.G., Rosenlund, G., Garlito, B., Amlund, H., Sissener, N.H., *et al.* (2021) Sensitivity of Atlantic salmon to the pesticide pirimiphos-methyl, present in plant-based feeds. *Aquaculture* 531, 735825.

Bjørge, M.H., Nordgreen, J., Janczak, A.M., Poppe, T., Ranheim, B. and Horsberg, T.E. (2011) Behavioural changes following intraperitoneal vaccination in Atlantic salmon (*Salmo salar*). *Applied Animal Behaviour Science* 133(1–2), 127–135.

Black, J.J. and Baumann, P.C. (1991) Carcinogens and cancers in freshwater fishes. *Environmental Health Perspectives* 90, 27–33.

Blazer, V.S. (1992) Nutrition and disease resistance in fish. *Annual Review of Fish Diseases* 2, 309–323.

Bloecher, N., Floerl, O. and Sunde, L.M. (2015) Amplified recruitment pressure of biofouling organisms in commercial salmon farms: potential causes and implications for farm management. *Biofouling* 31(2), 163–172.

Bolker, J.A. and Hill, C.R. (2000) Pigmentation development in hatchery-reared flatfishes. *Journal of Fish Biology* 56(5), 1029–1052.

Bosch-Belmar, M., Escurriola, A., Milisenda, G., Fuentes, V.L. and Piraino, S. (2019) Harmful fouling communities on fish farms in the SW Mediterranean Sea: composition, growth and reproductive periods. *Journal of Marine Science and Engineering* 7(9), 288.

Bosch-Belmar, M., Milisenda, G., Basso, L., Doyle, T.K., Leone, A. and Piraino, S. (2020) Jellyfish impacts on marine aquaculture and fisheries. *Reviews in Fisheries Science and Aquaculture* 29(2), 242–259.

Bowser, P.R., Wolfe, M.J. and Wallbridge, T. (1987) A lymphosarcoma in an Atlantic salmon (*Salmo salar*). *Journal of Wildlife Diseases* 23(4), 698–701.

Brittelli, M.R., Chen, H.H.C. and Muska, C.F. (1985) Induction of branchial (gill) neoplasms in the medaka fish (*Oryzias latipes*) by *N*-methyl-*N′*-nitro-*N*-nitrosoguanidine. *Cancer Research* 45(7), 3209–3214.

Bruno, D.W. (1990) Jaw deformity associated with farmed Atlantic salmon (*Salmo salar*). *Veterinary Record* 126(16), 402–403.

Bruno, D.W. and Ellis, A.E. (1985) Mortalities in farmed Atlantic salmon associated with the jellyfish *Phialella quadrata*. *Bulletin of the European Association of Fish Pathologists* 5(3), 64–65.

Bruno, D.W. and Raynard, R.S. (1994) Studies on the use of hydrogen peroxide as a method for the control of sea lice on Atlantic salmon. *Aquaculture International* 2(1), 10–18.

Bruno, D.W. and Smail, D.A. (1998) Lymphoma in farmed rainbow trout, *Oncorhynchus mykiss* (Walbaum): a case study. *Journal of Fish Diseases* 21(5), 381–386.
Bruno, D.W., Noguera, P.A. and Poppe, T.T. (2013) *A Colour Atlas of Salmonid Diseases*. Springer, Dordrecht, The Netherlands.
Caballero, M.J., López-Calero, G., Socorro, J., Roo, F.J., Izquierdo, M.S. and Férnandez, A.J. (1999) Combined effect of lipid level and fish meal quality on liver histology of gilthead seabream (*Sparus aurata*). *Aquaculture* 179(1–4), 277–290.
Carlisle, J.C. and Roberts, R.J. (1977) An epidermal papilloma of the Atlantic salmon I: epizootiology, pathology and immunology. *Journal of Wildlife Diseases* 13(3), 230–234.
Carlson, E. and Zelikoff, J.T. (2008) The immune system of fish: a target organ of toxicity. In: Di Giulio, R.T. and Hinton, D.E. (eds) *The Toxicology of Fishes*, 1st edn. CRC Press, Boca Raton, Florida, pp. 489–529.
Castell, J.D., Sinnhuber, R.O., Wales, J.H. and Lee, D.J. (1972) Essential fatty acids in the diet of rainbow trout (*Salmo gairdneri*): growth, feed conversion and some gross deficiency symptoms. *The Journal of Nutrition* 102(1), 77–85.
Cerezuela, R., Cuesta, A., Meseguer, J. and Esteban, M.A. (2009) Effects of dietary vitamin D3 administration on innate immune parameters of seabream (*Sparus aurata* L.). *Fish & Shellfish Immunology* 26(2), 243–248.
Christiansen, R., Lie, Ø. and Torrissen, O.J. (1994) Effect of astaxanthin and vitamin A on growth and survival during first feeding of Atlantic salmon, *Salmo salar* L. *Aquaculture Research* 25(9), 903–914.
Ciji, A. and Akhtar, M.S. (2020) Nitrite implications and its management strategies in aquaculture: a review. *Reviews in Aquaculture* 12(2), 878–908.
Clarkson, M., Taylor, J.F., McStay, E., Palmer, M.J., Clokie, B.G.J. and Migaud, H. (2021) A temperature shift during embryogenesis impacts prevalence of deformity in diploid and triploid Atlantic salmon (*Salmo salar* L.). *Aquaculture Research* 52(3), 906–923.
Clinton, M., Ferrier, D.E.K., Martin, S.A.M. and Brierley, A.S. (2021) Impacts of jellyfish on marine cage aquaculture: an overview of existing knowledge and the challenges to finfish health. *ICES Journal of Marine Science* 78(5), 1557–1573.
Colt, J. (1986) Gas supersaturation – impact on the design and operation of aquatic systems. *Aquacultural Engineering* 5(1), 49–85.
Condon, R.H., Duarte, C.M., Pitt, K.A., Robinson, K.L., Lucas, C.H., *et al.* (2013) Recurrent jellyfish blooms are a consequence of global oscillations. *Proceedings of the National Academy of Sciences USA* 110(3), 1000–1005.
Cooke, M. (2016) Animal Welfare in Farmed Fish. Business Benchmark on Farm Animal Welfare Investor Briefing No. 23. Available at: https://www.bbfaw.com/media/1432/investor-briefing-no-23-animal-welfare-in-farmed-fish.pdf (accessed 20 September 2022).
Cowey, C.B. (1994) Amino acid requirements of fish: a critical appraisal of present values. *Aquaculture* 124(1–4), 1–11.
Crichigno, S.A., Orellana, M., Larraza, R., Mirenna, G. and Cussac, V.E. (2021) Thermal effects in rainbow trout (*Oncorhynchus mykiss*) F_1 embryos (farmed female × wild thermal-resistant male). *Journal of Fish Biology* 99(1), 197–205. https://doi.org/10.1111/jfb.14711
Cronin, M., Cusack, C., Geoghegan, F., Jackson, D., McGovern, E., *et al.* (2004) Salmon Mortalities at Inver Bay and McSwyne's Bay Finfish Farms, County Donegal, Ireland, during 2003. *Marine Environment and Health Series No. 15*. Marine Institute, Galway, Ireland.
Cuesta, A., Ortuno, J., Rodriguez, A., Esteban, M.A. and Meseguer, J. (2002) Changes in some innate defence parameters of seabream (*Sparus aurata* L.) induced by retinol acetate. *Fish & Shellfish Immunology* 13(4), 279–291.
Dabrowski, K., Hinterleitner, S., Sturmbauer, C., El-Fiky, N. and Wieser, W. (1988) Do carp larvae require vitamin C? *Aquaculture* 72(3–4), 295–306.
Dabrowski, K., Segner, H., Dallinger, R., Hinterleitner, S., Sturmbauer, C. and Wieser, W. (1989) Rearing of cyprinid fish larvae: the vitamin C–minerals interrelationship and nutrition-related histology of the liver and intestine of roach (*Rutilus rutilus* L.). *Journal of Animal Physiology and Animal Nutrition* 62(1–5), 188–202.
Dabrowski, K., Lee, K.-J., Guz, L., Verlhac, V. and Gabaudan, J. (2004) Effects of dietary ascorbic acid on oxygen stress (hypoxia or hyperoxia), growth and tissue vitamin concentrations in juvenile rainbow trout (*Oncorhynchus mykiss*). *Aquaculture* 233(1–4), 383–392.
Dai, Y.-J., Cao, X.-F., Zhang, D.-D., Li, X.-F., Liu, W.-B. and Jiang, G.-Z. (2019) Chronic inflammation is a key to inducing liver injury in blunt snout bream (*Megalobrama amblycephala*) fed with high-fat diet. *Developmental and Comparative Immunology* 97, 28–37.

Dale, O.B., Tørud, B., Kvellestad, A., Koppang, H.S. and Koppang, E.O. (2009) From chronic feed-induced intestinal inflammation to adenocarcinoma with metastases in salmonid fish. *Cancer Research* 69(10), 4355–4362.

Daniels, H.V., Boyd, C.E. and Minton, R.V. (1987) Acute toxicity of ammonia and nitrite to spotted seatrout. *The Progressive Fish-Culturist* 49(4), 260–263.

Daoust, P.Y. and Ferguson, H.W. (1983) Gill diseases of cultured salmonids in Ontario. *Canadian Journal of Comparative Medicine* 47(3), 358–362.

da Rocha, C.A.M., Moreira-Nunes, C.A., da Rocha, S.M., da Silva, M.A.S. and Burbano, R.R. (2018) A review on occurrence of neoplasia in fish. *Acta of Fisheries and Aquatic Resources* 5(2), 19–24.

Dawood, M.A.O., Metwally, A.E.-S., El-Sharawy, M.E., Atta, A.M., Elbialy, Z.I., *et al.* (2020) The role of β-glucan in the growth, intestinal morphometry, and immune-related gene and heat shock protein expressions of Nile tilapia (*Oreochromis niloticus*) under different stocking densities. *Aquaculture* 523, 735205.

Dedi, J., Takeuchi, T., Seikai, T. and Watanabe, T. (1995) Hypervitaminosis and safe levels of vitamin A for larval flounder *Paralichthys olivaceus* fed *Artemia* nauplii. *Aquaculture* 133(2), 135–146.

Di Giulio, R.T. and Hinton, D.E. (eds) (2008) *The Toxicology of Fishes*. 1st edn. CRC Press, Boca Raton, Florida.

Doszpoly, A., Karaseva, T.A., Waltzek, T.B., Kalabekov, I.M. and Shchelkunov, I.S. (2013) Atlantic salmon papillomatosis in Russia and molecular characterization of the associated herpesvirus. *Diseases of Aquatic Organisms* 107(2), 121–127.

Doyle, T.K., De Haas, H., Cotton, D., Dorschel, B., Cummins, V., Houghton, J.D.R., *et al.* (2008) Widespread occurrence of the jellyfish *Pelagia noctiluca* in Irish coastal and shelf waters. *Journal of Plankton Research* 30(8), 963–968.

Duncan, I.B. (1978) Evidence for an oncovirus in swimbladder fibrosarcoma of Atlantic salmon *Salmo salar* L. *Journal of Fish Diseases* 1(1), 127–131.

Duncan, N.J., Thrush, M.A., Elliott, J.A.K. and Bromage, N.R. (2002) Seawater growth and maturation of Atlantic salmon (*Salmo salar*) transferred to sea at different times during the year. *Aquaculture* 213(1–4), 293–309.

Eaton, W.D. and Kent, M.L. (1992) A retrovirus in chinook salmon (*Oncorhynchus tshawytscha*) with plasmacytoid leukemia and evidence for the etiology of the disease. *Cancer Research* 52(23), 6496–6500.

Edsall, D.A. and Smith, C.E. (1991) Oxygen-induced gas bubble disease in rainbow trout, *Oncorhynchus mykiss* (Walbaum). *Aquaculture Research* 22(2), 135–140.

Ehrlich, K.F., Blaxter, J.H.S. and Pemberton, R. (1976) Morphological and histological changes during the growth and starvation of herring and plaice larvae. *Marine Biology* 35(2), 105–118.

Ejike, C. and Schreck, C.B. (1980) Stress and social hierarchy rank in coho salmon. *Transactions of the American Fisheries Society* 109(4), 423–426.

Ellis, T., North, B., Scott, A.P., Bromage, N.R., Porter, M. and Gadd, D. (2002) The relationships between stocking density and welfare in farmed rainbow trout. *Journal of Fish Biology* 61(3), 493–531.

Elzeir, M., Hansen, I.S. and Hay, S. (2005) Predicting jellyfish outbreaks around Shetland using MIKE 3. *WIT Transactions on the Built Environment*, 78. Available at: https://www.witpress.com/elibrary/wit-transactions-on-the-built-environment/78/14673 (accessed 20 September 2022).

Espmark, Å.M., Hjelde, K. and Baeverfjord, G. (2010) Development of gas bubble disease in juvenile Atlantic salmon exposed to water supersaturated with oxygen. *Aquaculture* 306(1–4), 198–204.

Falcon, D.R., Barros, M.M., Pezzato, L.E., Sampaio, F.G. and Hisano, H. (2007) Physiological responses of Nile tilapia, *Oreochromis niloticus*, fed vitamin C- and lipid-supplemented diets and submitted to low-temperature stress. *Journal of the World Aquaculture Society* 38(2), 287–295.

Falconer, L., Telfer, T.C., Garrett, A., Hermansen, Ø., Mikkelsen, E., *et al.* (2022) Insight into real-world complexities is required to enable effective response from the aquaculture sector to climate change. *PLoS Climate* 1(3), e0000017.

FAO (Food and Agriculture Organization of the United Nations) (2020) The State of Fisheries and Aquaculture 2020. FAO, Rome. Available at: https://www.fao.org/documents/card/en/c/ca9229en (accessed 20 September 2022).

Ferguson, H.W. (2006) *Systemic Pathology of Fish: A Text and Atlas for Normal Tissues in Teleosts and Their Responses in Disease*, 2nd edn. Scotian Press, London.

Ferguson, H.W., Christian, M.J.D., Hay, S., Nicolson, J., Sutherland, D. and Crumlish, M. (2010) Jellyfish as vectors of bacterial disease for farmed salmon (*Salmo salar*). *Journal of Veterinary Diagnostic Investigation* 22(3), 376–382.

Fernández, I. and Gisbert, E. (2011) The effect of vitamin A on flatfish development and skeletogenesis: a review. *Aquaculture* 315(1–2), 34–48.
Fjelldal, P.G., van der Meeren, T., Jørstad, K.E. and Hansen, T.J. (2009) A radiological study on vertebral deformities in cultured and wild Atlantic cod (*Gadus morhua*, L.). *Aquaculture* 289(1–2), 6–12.
Floerl, O., Sunde, L.M. and Bloecher, N. (2016) Potential environmental risks associated with biofouling management in salmon aquaculture. *Aquaculture Environment Interactions* 8, 407–417.
Foddai, M., Carter, C.G., Hilder, P.E., Gurr, H. and Ruff, N. (2022) Combined effects of elevated rearing temperature and dietary energy level on heart morphology and growth performance of Tasmanian Atlantic salmon (*Salmo salar* L.). *Journal of Fish Diseases* 45(2), 301–313.
Fossa, J.H., Flood, P.R., Olsen, A.B. and Jensen, F. (2003) Sma og usynlige, men plagsomme maneter av arten *Muggiaea atlantica*. *Fisken Og Havet* 2, 99–103.
Fraser, T.W.K., Hansen, T., Fleming, M.S. and Fjelldal, P.G. (2015) The prevalence of vertebral deformities is increased with higher egg incubation temperatures and triploidy in Atlantic salmon *Salmo salar* L. *Journal of Fish Diseases* 38(1), 75–89.
Frischknecht, R., Wahli, T. and Meier, W. (1994) Comparison of pathological changes due to deficiency of vitamin C, vitamin E and combinations of vitamins C and E in rainbow trout, *Oncorhynchus mykiss* (Walbaum). *Journal of Fish Diseases* 17(1), 31–45. https://doi.org/10.1111/j.1365-2761.1994.tb00343.x
Gatica, M.C., Monti, G.E., Knowles, T.G., Warriss, P.D. and Gallo, C.B. (2010) Effects of commercial live transportation and preslaughter handling of Atlantic salmon on blood constituents. *Archivos de Medicina Veterinaria* 42(1), 73–78.
Gatlin, D.M. III (2008) Non-infectious diseases: nutritional factors. In: Eiras, J.C., Segner, H., Wahli, T. and Kapoor, B.G. (eds) *Fish Diseases (2 Vols)*. CRC Press, Boca Raton, Florida, pp. 1215–1250.
Gatlin, D.M. III, Barrows, F.T., Brown, P., Dabrowski, K., Gaylord, T.G., *et al.* (2007) Expanding the utilization of sustainable plant products in aquafeeds: a review. *Aquaculture Research* 38(6), 551–579.
Gaylord, T.G., Teague, A.M. and Barrows, F.T. (2006) Taurine supplementation of all-plant protein diets for rainbow trout (*Oncorhynchus mykiss*). *Journal of the World Aquaculture Society* 37(4), 509–517.
Gibbons, M.J. and Richardson, A.J. (2013) Beyond the jellyfish joyride and global oscillations: advancing jellyfish research. *Journal of Plankton Research* 35(5), 929–938. https://doi.org/10.1093/plankt/fbt063
Gimenez-Conti, I., Woodhead, A.D., Harshbarger, J.C., Kazianis, S., Setlow, R.B., *et al.* (2001) A proposed classification scheme for *Xiphophorus* melanomas based on histopathologic analyses. *Marine Biotechnology* 3(1), S100–S106.
Gjurčević, E., Sfacteria, A. and Marino, F. (2014) Spontaneous multicentric myxoma of the dermal nerve sheaths in farmed European eels *Anguilla anguilla*. *Diseases of Aquatic Organisms* 111(2), 173–176.
Goldstein, J. and Steiner, U.K. (2020) Ecological drivers of jellyfish blooms – the complex life history of a 'well-known' medusa (*Aurelia aurita*). *Journal of Animal Ecology* 89(3), 910–920.
Gorlin, R.J. (1972) Odontogenic tumors in mammals and fish. *Oral Surgery, Oral Medicine, Oral Pathology* 33(1), 86–90.
Greaves, K. and Tuene, S. (2001) The form and context of aggressive behaviour in farmed Atlantic halibut (*Hippoglossus hippoglossus* L.). *Aquaculture* 193(1–2), 139–147.
Grini, A., Hansen, T., Berg, A., Wargelius, A. and Fjelldal, P.G. (2011) The effect of water temperature on vertebral deformities and vaccine-induced abdominal lesions in Atlantic salmon, *Salmo salar* L. *Journal of Fish Diseases* 34(7), 531–546.
Grizzle, J.M. and Goodwin, A.E. (2010) Neoplasms and related disorders. In: Leatherland, J.F. and Woo, P.T.K. (eds) *Fish Diseases and Disorders*. Vol. 2. *Non-infectious Disorders*, 2nd edn. CAB International, Wallingford, UK, pp. 19–84.
Gu, M., Bai, N., Zhang, Y. and Krogdahl, Å. (2016) Soybean meal induces enteritis in turbot *Scophthalmus maximus* at high supplementation levels. *Aquaculture* 464, 286–295.
Guenther, J., Misimi, E. and Sunde, L.M. (2010) The development of biofouling, particularly the hydroid *Ectopleura larynx*, on commercial salmon cage nets in mid-Norway. *Aquaculture* 300(1–4), 120–127.
Guenther, J., Fitridge, I. and Misimi, E. (2011) Potential antifouling strategies for marine finfish aquaculture: the effects of physical and chemical treatments on the settlement and survival of the hydroid *Ectopleura larynx*. *Biofouling* 27(9), 1033–1042.
Gwak, W.-S., Seikai, T. and Tanaka, M. (1999) Evaluation of starvation status of laboratory-reared Japanese flounder *Paralichthys olivaceus* larvae and juveniles based on morphological and histological characteristics. *Fisheries Science* 65(3), 339–346.
Haberlin, D., McAllen, R. and Doyle, T.K. (2021) Field and flume tank experiments investigating the efficacy of a bubble curtain to keep harmful jellyfish out of finfish pens. *Aquaculture* 531, 735915.

Halver, J.E. (2002) The vitamins. In: Halver, J.E. and Hardy, R.W. (eds) *Fish Nutrition*, 3rd edn. Academic Press, New York, pp. 61–141.
Halver, J.E., Ashley, L.M. and Smith, R.R. (1969) Ascorbic acid requirements of coho salmon and rainbow trout. *Transactions of the American Fisheries Society* 98(4), 762–771.
Hamre, K., Holen, E. and Moren, M. (2007) Pigmentation and eye migration in Atlantic halibut (*Hippoglossus hippoglossus* L.) larvae: new findings and hypotheses. *Aquaculture Nutrition* 13(1), 65–80.
Handy, R.D. (1996) Dietary exposure to toxic metals in fish. In: Taylor, E.W. (ed.) *Toxicology of Aquatic Pollution: Physiological, Molecular and Cellular Approaches*. Cambridge University Press, Cambridge, UK, pp. 29–60.
Harada, T., Hatanaka, J., Kubota, S.S. and Enomoto, M. (1990) Lymphoblastic lymphoma in medaka, *Oryzias latipes* (Temminck et Schlegel). *Journal of Fish Diseases* 13(2), 169–173.
Harada, T., Itoh, H., Hatanaka, J., Kamiya, S. and Enomoto, M. (1996) A morphological study of a thyroid carcinoma in a medaka, *Oryzias latipes* (Temminck and Schlegel). *Journal of Fish Diseases* 19(4), 271–277.
Hardy, R.W. (2012) The nutritional pathology of teleosts. In: Roberts, R.J. (ed.) *Fish Pathology*, 4th edn. Wiley, Chichester, UK, pp. 402–424.
Harper, C. and Wolf, J.C. (2009) Morphologic effects of the stress response in fish. *ILAR Journal* 50(4), 387–396.
Harrison, J.G. and Richards, R.H. (1979) The pathology and histopathology of nephrocalcinosis in rainbow trout *Salmo gairdneri* Richardson in fresh water. *Journal of Fish Diseases* 2(1), 1–12.
Harshbarger, J.C. and Clark, J.B. (1990) Epizootiology of neoplasms in bony fish of North America. *Science of the Total Environment* 94(1–2), 1–32.
Harshbarger, J.C., Spero, P.M. and Wolcott, N.M. (2021) Neoplasms in wild fish from the marine ecosystem emphasizing environmental interactions. In: Couch, J.A. and Fournie, J.W. (eds) *Pathobiology of Marine and Estuarine Organisms*. CRC Press, Boca Raton, Florida, pp. 157–176.
Haugarvoll, E., Bjerkås, I., Szabo, N.J., Satoh, M. and Koppang, E.O. (2010) Manifestations of systemic autoimmunity in vaccinated salmon. *Vaccine* 28(31), 4961–4969. https://doi.org/10.1016/j.vaccine.2010.05.032
Hay, S. and Murray, A. (2008) Jellyfish problems faced by the aquaculture industry. *Fish Farmer* 5, 40–41.
Hemre, G., Mommsen, T.P. and Krogdahl, Å. (2002) Carbohydrates in fish nutrition: effects on growth, glucose metabolism and hepatic enzymes. *Aquaculture Nutrition* 8(3), 175–194.
Hernandez, L.H.H., Teshima, S., Koshio, S., Ishikawa, M., Tanaka, Y. and Alam, M.S. (2007) Effects of vitamin A on growth, serum anti-bacterial activity and transaminase activities in the juvenile Japanese flounder, *Paralichthys olivaceus*. *Aquaculture* 262(2–4), 444–450.
Herrera, M., Mancera, J.M. and Costas, B. (2019) The use of dietary additives in fish stress mitigation: comparative endocrine and physiological responses. *Frontiers in Endocrinology* 10, 447.
Hicks, B.D., Hilton, J.W. and Ferguson, H.W. (1984) Influence of dietary selenium on the occurrence of nephrocalcinosis in the rainbow trout, *Salmo gairdneri* Richardson. *Journal of Fish Diseases* 7(5), 379–389.
Hilton, J.W. (1989) The interaction of vitamins, minerals and diet composition in the diet of fish. *Aquaculture* 79(1–4), 223–244.
Hilton, J.W. and Hodson, P.V. (1983) Effect of increased dietary carbohydrate on selenium metabolism and toxicity in rainbow trout (*Salmo gairdneri*). *The Journal of Nutrition* 113(6), 1241–1248.
Hilton, J.W., Hodson, P.V. and Slinger, S.J. (1980) The requirement and toxicity of selenium in rainbow trout (*Salmo gairdneri*). *The Journal of Nutrition* 110(12), 2527–2535.
Hinton, D.E., Segner, H., Au, D.W.T., Kullman, S.W. and Hardmann, R.C. (2008) Liver toxicity. In: Di Giulio, R.T. and Hinton, D.E. (eds) *The Toxicology of Fishes*, 1st edn. CRC Press, Boca Raton, Florida, pp. 327–400.
Hoar, W.S. (1988) The physiology of smolting salmonids. In: Hoar, W.S. and Randall, D.J. (eds) *Fish Physiology*. Vol. 11, Part B. *The Physiology of Developing Fish: Viviparity and Posthatching Juveniles*. Academic Press, San Diego, California, pp. 275–343.
Hoffmann, R.W. (2005) *Fischkrankheiten*. Ulmer, Stuttgart, Germany.
Hogstand, C. (1996) The physiology and toxicology of zinc in fish. In: Taylor, E.W. (ed.) *Toxicology of Aquatic Pollution: Physiological, Molecular and Cellular Approaches*. Cambridge University Press, Cambridge, UK, pp. 61–84.
Hoitsy, M., Hoitsy, G., Jakab, C., Molnár, T., Gál, J. and Baska, F. (2021) Intussusception caused by intestinal neoplasia in mature rainbow trout (*Oncorhynchus mykiss*, Walbaum 1792). *Journal of Fish Diseases* 44(7), 893–898.

Hosseini, N., Johnston, J. and Lindenschmidt, K.-E. (2017) Impacts of climate change on the water quality of a regulated prairie river. *Water* 9(3), 199. https://doi.org/10.3390/w9030199
Hur, J.W., Woo, S.R., Jo, J.H. and Park, I. (2006) Effects of starvation on kidney melano-macrophage centre in olive flounder, *Paralichthys olivaceus* (Temminck and Schlegel). *Aquaculture Research* 37(8), 821–825.
Inoue, M., Satoh, S., Maita, M., Kiron, V. and Okamoto, N. (1998) Recovery from derangement of natural killer-like activity of leucocytes due to Zn or Mn deficiency in rainbow trout, *Oncorhynchus mykiss* (Walbaum), by the oral administration of these elements. *Journal of Fish Diseases* 21(3), 233–236. https://doi.org/10.1046/j.1365-2761.1998.00080.x
Ip, A.Y.K. and Chew, S.F. (2010) Ammonia production, excretion, toxicity, and defense in fish: a review. *Frontiers in Physiology* 1, 134.
Islam, M.J., Kunzmann, A. and Slater, M.J. (2021) Responses of aquaculture fish to climate change-induced extreme temperatures: a review. *Journal of the World Aquaculture Society* 53(2), 314–366. https://doi.org/10.1111/jwas.12853
Iversen, M., Finstad, B., McKinley, R.S., Eliassen, R.A., Carlsen, K.T. and Evjen, T. (2005) Stress responses in Atlantic salmon (*Salmo salar* L.) smolts during commercial well boat transports, and effects on survival after transfer to sea. *Aquaculture* 243(1–4), 373–382.
Jacobs, M.N., Covaci, A., Gheorghe, A. and Schepens, P. (2004) Time trend investigation of PCBs, PBDEs, and organochlorine pesticides in selected *n*-3 polyunsaturated fatty acid rich dietary fish oil and vegetable oil supplements; nutritional relevance for human essential *n*-3 fatty acid requirements. *Journal of Agricultural and Food Chemistry* 52(6), 1780–1788.
Johansson, L.-H., Timmerhaus, G., Afanasyev, S., Jørgensen, S.M. and Krasnov, A. (2016) Smoltification and seawater transfer of Atlantic salmon (*Salmo salar* L.) is associated with systemic repression of the immune transcriptome. *Fish & Shellfish Immunology* 58, 33–41.
Julian, R.J. (1998) Rapid growth problems: ascites and skeletal deformities in broilers. *Poultry Science* 77(12), 1773–1780.
Kang, D.-Y., Lee, J.-H., Kim, W.-J. and Kim, H.-C. (2012) Morphological specificity in cultured starry flounder *Platichthys stellatus* reared in artificial facility. *Fisheries and Aquatic Sciences* 15(2), 117–123.
Kaushik, S.J., Panserat, S. and Schrama, J.W. (2022) Carbohydrates. In: Hardy, R.W. and Kaushik, S.J. (eds) *Fish Nutrition*, 4th edn. Academic Press, London, pp. 555–591. https://doi.org/10.1016/B978-0-12-819587-1.00008-2
Kent, M.L. and Dawe, S.C. (1990) Experimental transmission of a plasmacytoid leukemia of chinook salmon, *Oncorhynchus tshawytscha*. *Cancer Research* 50(17 Suppl.), 5679S–5681S.
Kent, M.L. and Dawe, S.C. (1993) Further evidence of viral etiology in plasmocytoid leukemia of chinook salmon *Oncorhynchus tshawytscha*. *Diseases of Aquatic Organisms* 15(2), 115–121.
Kent, M.L., Groff, J.M., Traxler, G.S., Zinkl, J.G. and Bagshaw, J.W. (1990) Plasmacytoid leukemia in seawater reared chinook salmon *Oncorhynchus tshawytscha*. *Diseases of Aquatic Organisms* 8(3), 199–209.
Ketola, H.G. (1978) Dietary zinc prevents cataract in trout. *Federation Proceedings* 37(3), 584.
Kocour Kroupová, H., Valentová, O., Svobodová, Z., Šauer, P. and Máchová, J. (2018) Toxic effects of nitrite on freshwater organisms: a review. *Reviews in Aquaculture* 10(3), 525–542.
Koppang, E.O., Haugarvoll, E., Hordvik, I., Aune, L. and Poppe, T.T. (2005) Vaccine-associated granulomatous inflammation and melanin accumulation in Atlantic salmon, *Salmo salar* L., white muscle. *Journal of Fish Diseases* 28(1), 13–22.
Koppang, E.O., Bjerkås, I., Haugarvoll, E., Chan, E.K.L., Szabo, N.J., *et al.* (2008) Vaccination-induced systemic autoimmunity in farmed Atlantic salmon. *The Journal of Immunology* 181(7), 4807–4814.
Krogdahl, Å., Hemre, G. and Mommsen, T.P. (2005) Carbohydrates in fish nutrition: digestion and absorption in postlarval stages. *Aquaculture Nutrition* 11(2), 103–122.
Krogdahl, Å., Gajardo, K., Kortner, T.M., Penn, M., Gu, M., *et al.* (2015) Soya saponins induce enteritis in Atlantic salmon (*Salmo salar* L.). *Journal of Agricultural and Food Chemistry* 63(15), 3887–3902.
Krogdahl, Å., Kortner, T.M. and Hardy, R.W. (2022) Antinutrients and adventitious toxins. In: Hardy, R.W. and Kaushik, S.J. (eds) *Fish Nutrition*, 4th edn. Academic Press, London, pp. 775–821. https://doi.org/10.1016/B978-0-12-819587-1.00001-X
Laczynska, B., Siddique, M.A.M., Ziomek, E., Shelton, W.L. and Fopp-Bayat, D. (2020) Early weaning effects on survival, growth, and histopathology of larval sterlet *Acipenser ruthenus*. *North American Journal of Aquaculture* 82(2), 181–189.
Lall, S.P. (2022) The minerals. In: Hardy, R.W. and Kaushik, S.J. (eds) *Fish Nutrition*, 4th edn. Academic Press, London, pp. 469–554. https://doi.org/10.1016/B978-0-12-819587-1.00005-7

Lall, S.P. and Kaushik, S.J. (2021) Nutrition and metabolism of minerals in fish. *Animals* 11(9), 2711.
Lall, S.P. and Lewis-McCrea, L.M. (2007) Role of nutrients in skeletal metabolism and pathology in fish – an overview. *Aquaculture* 267(1–4), 3–19.
Lim, C., Yildirim-Aksoy, M., Welker, T., Klesius, P.H. and Li, M.H. (2010) Growth performance, immune response, and resistance to *Streptococcus iniae* of Nile tilapia, *Oreochromis niloticus*, fed diets containing various levels of vitamins C and E. *Journal of the World Aquaculture Society* 41(1), 35–48. https://doi.org/10.1111/j.1749-7345.2009.00311.x
Limbu, S.M., Chen, L., Zhang, M. and Du, Z. (2021) A global analysis on the systemic effects of antibiotics in cultured fish and their potential human health risk: a review. *Reviews in Aquaculture* 13(2), 1015–1059.
Lin, Y.-H. and Shiau, S.-Y. (2009) Mutual sparing of dietary requirements for alpha-tocopherol and selenium in grouper, *Epinephelus malabaricus*. *Aquaculture* 294(3–4), 242–245.
Liu, H., Xie, S., Zhu, X., Lei, W., Han, D. and Yang, Y. (2008) Effects of dietary ascorbic acid supplementation on the growth performance, immune and stress response in juvenile *Leiocassis longirostris* Günther exposed to ammonia. *Aquaculture Research* 39(15), 1628–1638.
Lumsden, J.S. and Marshall, S. (2003) Sporadic neoplasms of farmed chinook salmon, *Oncorhynchus tshawytscha* (Walbaum), from New Zealand. *Journal of Fish Diseases* 26(7), 393–399.
Lynam, C.P., Lilley, M.K.S., Bastian, T., Doyle, T.K., Beggs, S.E. and Hays, G.C. (2011) Have jellyfish in the Irish Sea benefited from climate change and overfishing? *Global Change Biology* 17(2), 767–782.
Macaulay, G., Bui, S., Oppedal, F. and Dempster, T. (2021) Challenges and benefits of applying fish behaviour to improve production and welfare in industrial aquaculture. *Reviews in Aquaculture* 13(2), 934–948.
Machova, J., Faina, R., Randak, T., Valentova, O., Steinbach, C., *et al.* (2017) Fish death caused by gas bubble disease: a case report. *Veterinární Medicína* 62(4), 231–237.
Magalhães, R., Martins, N., Fontinha, F., Couto, A., Serra, C.R., *et al.* (2022) Effects of dietary ARA, DHA, and carbohydrates levels on gilthead sea bream liver and intestine oxidative stress, tissue histomorphology, and gut microbiota. *Aquaculture* 552, 738014.
Mai, K., Xue, M., He, G., Xie, S.Q. and Kaushik, S.J. (2022a) Protein and amino acids. In: Hardy, R.W. and Kaushik, S.J. (eds) *Fish Nutrition*, 4th edn. Academic Press, London, pp. 181–302. https://doi.org/10.1016/B978-0-12-819587-1.00012-4
Mai, K., Waagbø, R., Zhou, X.Q., Ai, Q.H. and Feng, L. (2022b) Vitamins. In: Hardy, R.W. and Kaushik, S.J. (eds) *Fish Nutrition*, 4th edn. Academic Press, London, pp. 57–179. https://doi.org/10.1016/B978-0-12-819587-1.00014-8
Majeed, S.K., Jolly, D.W. and Gopinath, C. (1984) An outbreak of liver cell carcinoma in rainbow trout, *Salmo gairdneri* Richardson, in the UK. *Journal of Fish Diseases* 7(2), 165–168.
Marcos-López, M., Mitchell, S.O. and Rodger, H.D. (2016) Pathology and mortality associated with the mauve stinger jellyfish *Pelagia noctiluca* in farmed Atlantic salmon *Salmo salar* L. *Journal of Fish Diseases* 39(1), 111–115.
Martineau, D. and Ferguson, H.W. (2006) Neoplasia. In: Ferguson, H.W. (ed.) *Systemic Pathology of Fish: A Text and Atlas for Normal Tissues in Teleosts and Their Responses in Disease*, 2nd edn. Scotian Press, London, pp. 313–335.
Martineau, D., Bowser, P.R., Wooster, G.A. and Armstrong, L.D. (1990) Experimental transmission of a dermal sarcoma in fingerling walleye (*Stizostedion vitreum vitreum*). *Veterinary Pathology* 27(4), 230–234.
Martineau, D., Bowser, P.R., Renshaw, R.R. and Casey, J.W. (1992) Molecular characterization of a unique retrovirus associated with a fish tumor. *Journal of Virology* 66(1), 596–599.
Martínez-Álvarez, R.M., Morales, A.E. and Sanz, A. (2005) Antioxidant defenses in fish: biotic and abiotic factors. *Reviews in Fish Biology and Fisheries* 15(1), 75–88.
Martinez, G.M., Baron, M.P. and Bolker, J.A. (2007) Skeletal and pigmentation defects following retinoic acid exposure in larval summer flounder, *Paralichthys dentatus*. *Journal of the World Aquaculture Society* 38(3), 353–366.
Martos-Sitcha, J.A., Mancera, J.M., Prunet, P. and Magnoni, L.J. (2020) Welfare and stressors in fish: challenges facing aquaculture. *Frontiers in Physiology* 11, 162.
Masahito, P., Ishikawa, T., Yanagisawa, A., Sugano, H. and Ikeda, K. (1985) Neurogenic tumors in coho salmon (*Oncorhynchus kisutch*) reared in well water in Japan. *Journal of the National Cancer Institute* 75(4), 779–790.
McCormick, S.D. (2012) Smolt physiology and endocrinology. In: McCormick, S.D., Farrell, A.P. and Brauner, C.J. (eds) *Fish Physiology*. Vol. 32. *Euryhaline Fishes*. Elsevier, Amsterdam, pp. 199–251.

McGowan, M., MacKenzie, S., Steiropoulos, N. and Weidmann, M. (2021) Testing of NKA expression by mobile real time PCR is an efficient indicator of smoltification status of farmed Atlantic salmon. *Aquaculture* 544, 737085. https://doi.org/10.1016/j.aquaculture.2021.737085
McKenzie, D.J., Shingles, A. and Taylor, E.W. (2003) Sub-lethal plasma ammonia accumulation and the exercise performance of salmonids. *Comparative Biochemistry and Physiology Part A: Molecular & Integrative Physiology* 135(4), 515–526. https://doi.org/10.1016/S1095-6433(03)00116-8
Meier, S., Karlsen, Ø., Le Goff, J., Sørensen, L., Sørhus, E., *et al.* (2020) DNA damage and health effects in juvenile haddock (*Melanogrammus aeglefinus*) exposed to PAHs associated with oil-polluted sediment or produced water. *PLoS ONE* 15(10), e0240307.
Meuten, D.J. (ed.) (2020) *Tumors in Domestic Animals*, 5th edn. Wiley, Hoboken, New Jersey.
Meyers, T.R. and Hendricks, J.D. (1983) Histopathology of four spontaneous neoplasms in three species of salmonid fishes. *Journal of Fish Diseases* 6(6), 481–499.
Midtlyng, P.J. (1997) Vaccinated fish welfare: protection versus side-effects. *Developments in Biological Standardization* 90, 371–379.
Midtlyng, P.J. and Lillehaug, A. (1998) Growth of Atlantic salmon *Salmo salar* after intraperitoneal administration of vaccines containing adjuvants. *Diseases of Aquatic Organisms* 32(2), 91–97.
Mitchell, S.O. and Rodger, H.D. (2011) A review of infectious gill disease in marine salmonid fish. *Journal of Fish Diseases* 34(6), 411–432.
Miwa, S., Nakayasu, C., Kamaishi, T. and Yoshiura, Y. (2004) X-cells in fish pseudotumors are parasitic protozoans. *Diseases of Aquatic Organisms* 58(2–3), 165–170.
Moccia, R.D., Leatherland, J.F. and Sonstegard, R.A. (1977) Increasing frequency of thyroid goiters in coho salmon (*Oncorhynchus kisutch*) in the Great Lakes. *Science* 198(4315), 425–426.
Montero, D., Marrero, M., Izquierdo, M.S., Robaina, L., Vergara, J.M. and Tort, L. (1999) Effect of vitamin E and C dietary supplementation on some immune parameters of gilthead seabream (*Sparus aurata*) juveniles subjected to crowding stress. *Aquaculture* 171(3–4), 269–278.
Moren, M., Opstad, I., Berntssen, M.H.G., Infante, J.-L.Z. and Hamre, K. (2004) An optimum level of vitamin A supplements for Atlantic halibut (*Hippoglossus hippoglossus* L.) juveniles. *Aquaculture* 235(1–4), 587–599.
Motalebi, A.A., Ardalani, K. and Jamili, S. (2008) Effect of temperature on the produced aflatoxins in the rainbow trout feed in West Azerbaijan Province. *Journal of Fisheries and Aquatic Science* 3(6), 392–397.
Mourente, G., Díaz-Salvago, E., Bell, J.G. and Tocher, D.R. (2002) Increased activities of hepatic antioxidant defence enzymes in juvenile gilthead sea bream (*Sparus aurata* L.) fed dietary oxidised oil: attenuation by dietary vitamin E. *Aquaculture* 214(1–4), 343–361.
Murray, K.N., Wolf, J.C., Spagnoli, S.T., Lains, D., Budrow, N. and Kent, M.L. (2018) Reversibility of proliferative thyroid lesions induced by iodine deficiency in a laboratory zebrafish colony. *Zebrafish* 15(6), 558–565.
Mutoloki, S., Alexandersen, S. and Evensen, Ø. (2004) Sequential study of antigen persistence and concomitant inflammatory reactions relative to side-effects and growth of Atlantic salmon (*Salmo salar* L.) following intraperitoneal injection with oil-adjuvanted vaccines. *Fish & Shellfish Immunology* 16(5), 633–644.
Nisembaum, L.G., Martin, P., Fuentes, M., Besseau, L., Magnanou, E., *et al.* (2020) Effects of a temperature rise on melatonin and thyroid hormones during smoltification of Atlantic salmon, *Salmo salar*. *Journal of Comparative Physiology B* 190(6), 731–748.
Noga, E.J. (2010) *Fish Disease: Diagnosis and Treatment*, 2nd edn. Wiley, Hoboken, New Jersey.
O'Flynn, F.M., McGeachy, S.A., Friars, G.W., Benfey, T.J. and Bailey, J.K. (1997) Comparisons of cultured triploid and diploid Atlantic salmon (*Salmo salar* L.). *ICES Journal of Marine Science* 54(6), 1160–1165.
Obach, A. and Laurencin, F.B. (1992) Effects of dietary oxidized fish oil and deficiency of anti-oxidants on the immune response of turbot, *Scophthalmus maximus*. *Aquaculture* 107(2–3), 221–228.
Ogino, C. and Takeda, H. (1976) Mineral requirements in fish. III. Calcium and phosphorus requirements in carp. *Bulletin of the Japanese Society for the Science of Fish* 42(7), 793–799.
Ojolick, E.J., Cusack, R., Benfey, T.J. and Kerr, S.R. (1995) Survival and growth of all-female diploid and triploid rainbow trout (*Oncorhynchus mykiss*) reared at chronic high temperature. *Aquaculture* 131(3–4), 177–187.
Oliva-Teles, A. (2012) Nutrition and health of aquaculture fish. *Journal of Fish Diseases* 35(2), 83–108.
Oliva-Teles, A., Pereira, J.P., Gouveia, A. and Gomes, E. (1998) Utilisation of diets supplemented with microbial phytase by seabass (*Dicentrarchus labrax*) juveniles. *Aquatic Living Resources* 11(4), 255–259.

Oliveira, M. and Vasconcelos, V. (2020) Occurrence of mycotoxins in fish feed and its effects: a review. *Toxins* 12(3), 160.
Ørnsrud, R., Graff, I.E., Høie, S., Totland, G.K. and Hemre, G. (2002) Hypervitaminosis A in first-feeding fry of the Atlantic salmon (*Salmo salar* L.). *Aquaculture Nutrition* 8(1), 7–13.
Ottesen, O.H., Noga, E.J. and Sandaa, W. (2007) Effect of substrate on progression and healing of skin erosions and epidermal papillomas of Atlantic halibut, *Hippoglossus hippoglossus* (L.). *Journal of Fish Diseases* 30(1), 43–53.
Patra, R.W., Chapman, J.C., Lim, R.P., Gehrke, P.C. and Sunderam, R.M. (2015) Interactions between water temperature and contaminant toxicity to freshwater fish. *Environmental Toxicology and Chemistry* 34(8), 1809–1817. https://doi.org/10.1002/etc.2990
Pedersen, L.-F. and Pedersen, P.B. (2012) Hydrogen peroxide application to a commercial recirculating aquaculture system. *Aquacultural Engineering* 46, 40–46.
Pietsch, C. (2020) Risk assessment for mycotoxin contamination in fish feeds in Europe. *Mycotoxin Research* 36(1), 41–62.
Piferrer, F., Beaumont, A., Falguière, J.-C., Flajšhans, M., Haffray, P. and Colombo, L. (2009) Polyploid fish and shellfish: production, biology and applications to aquaculture for performance improvement and genetic containment. *Aquaculture* 293(3–4), 125–156.
Pimentel-Rodrigues, A. and Oliva-Teles, A. (2007) Phosphorus availability of inorganic phosphates and fish meals in European sea bass (*Dicentrarchus labrax* L.) juveniles. *Aquaculture* 267(1–4), 300–307.
Poff, N.L., Brinson, M.M. and Day, J.W. (2002) *Aquatic Ecosystems and Global Climate Change*. Pew Center on Global Climate Change, Arlington, Virginia.
Poppe, T.T., Hellberg, H., Griffiths, D. and Meldal, H. (1997) Swimbladder abnormality in farmed Atlantic salmon *Salmo salar*. *Diseases of Aquatic Organisms* 30(1), 73–76.
Poppe, T.T., Midtlyng, P.J. and Sande, R.D. (1998) Examination of abdominal organs and diagnosis of deficient septum transversum in Atlantic salmon, *Salmo salar* L., using diagnostic ultrasound imaging. *Journal of Fish Diseases* 21(1), 67–72. https://doi.org/10.1046/j.1365-2761.1998.00074.x
Poppe, T.T., Johansen, R., Gunnes, G. and Tørud, B. (2003) Heart morphology in wild and farmed Atlantic salmon *Salmo salar* and rainbow trout *Oncorhynchus mykiss*. *Diseases of Aquatic Organisms* 57(1–2), 103–108.
Poston, H.A. (1971) Effect of feeding excess vitamin A on the carbohydrate and lipid metabolism and growth of brook trout. *Fisheries Research Bulletin* 34, 22–26.
Powell, M.D., Åtland, Å. and Dale, T. (2018) Acute lion's mane jellyfish, *Cyanea capillata* (Cnideria: Scyphozoa), exposure to Atlantic salmon (*Salmo salar* L.). *Journal of Fish Diseases* 41(5), 751–759.
Preston, A.C., Taylor, J.F., Adams, C.E. and Migaud, H. (2014) Surface feeding and aggressive behaviour of diploid and triploid brown trout *Salmo trutta* during allopatric pair-wise matchings. *Journal of Fish Biology* 85(3), 882–900.
Prisingkorn, W., Prathomya, P., Jakovlić, I., Liu, H., Zhao, Y.-H. and Wang, W.-M. (2017) Transcriptomics, metabolomics and histology indicate that high-carbohydrate diet negatively affects the liver health of blunt snout bream (*Megalobrama amblycephala*). *BMC Genomics* 18(1), 856.
Purcell, J.E., Uye, S. and Lo, W.-T. (2007) Anthropogenic causes of jellyfish blooms and their direct consequences for humans: a review. *Marine Ecology Progress Series* 350, 153–174.
Rašković, B. and Berillis, P. (2022) Special Issue on the Histopathology of Aquatic Animals. *Applied Sciences* 12(3), 971.
Rehman, S., Gora, A.H., Ahmad, I. and Rasool, S.I. (2017) Stress in aquaculture hatcheries: source, impact and mitigation. *International Journal of Current Microbiology and Applied Sciences* 6(10), 3030–3045.
Roberts, R.J. (ed.) (2012a) Miscellaneous non-infectious diseases. In: *Fish Pathology*, 4th edn. Wiley, Chichester, UK, pp. 425–438.
Roberts, R.J. (ed.) (2012b) Neoplasia in teleosts. In: *Fish Pathology*, 4th edn. Wiley, Chichester, UK, pp. 167–185.
Roberts, R.J. (2022) Nutritional pathology. In: Hardy, R.W. and Kaushik, S.J. (eds) *Fish Nutrition*, 4th edn. Academic Press, London, pp. 823–855. https://doi.org/10.1016/B978-0-12-819587-1.00009-4
Rodger, H.D., Henry, L. and Mitchell, S.O. (2011) Non-infectious gill disorders of marine salmonid fish. *Reviews in Fish Biology and Fisheries* 21(3), 423–440.
Rucker, R.R. (1975) Gas bubble disease: mortalities of coho salmon, *Oncorhynchus kisutch*, in water with constant total gas pressure and different oxygen–nitrogen ratios. *Fishery Bulletin* 73(4), 915–918.
Sadler, J., Pankhurst, P.M. and King, H.R. (2001) High prevalence of skeletal deformity and reduced gill surface area in triploid Atlantic salmon (*Salmo salar* L.). *Aquaculture* 198(3–4), 369–386.

Sánchez, R.C., Obregón, E.B. and Rauco, M.R. (2011a) Hypoxia is like an ethiological factor in vertebral column deformity of salmon (*Salmo salar*). *Aquaculture* 316(1–4), 13–19.
Sánchez, R.C., Obregón, E.B., Rauco, M.R., Castro, S.R., Bustos Obregon, E. and Rojas, R.M. (2011b) Vertebral column deformity and hypoxia in *Salmo salar*. *International Journal of Morphology* 29(4), 1291–1295.
Sargent, J.R., Tocher, D.R. and Bell, J.G. (2002) The lipids. In: Halver, J.E. and Hardy, R.W. (eds) *Fish Nutrition*, 3rd edn. Academic Press, New York, pp. 181–257.
Satoh, S. (2007) Minerals. In: Nakagawa, H., Sato, M. and Gatlin, D.M. III (eds) *Dietary Supplements for the Health and Quality of Cultured Fish*. CAB International, Wallingford, UK, pp. 74–85.
Saunders, J., Speare, D.J. and McConkey, S. (2012) Evaluation of co-oximetry for the measurement of methemoglobin in rainbow trout (*Oncorhynchus mykiss*) and values in three salmonid species. *Veterinary Clinical Pathology* 41(4), 471–477.
Schaefer, S.C. and Hollibaugh, J.T. (2017) Temperature decouples ammonium and nitrite oxidation in coastal waters. *Environmental Science and Technology* 51(6), 3157–3164. https://doi.org/10.1021/acs.est.6b03483
Schaub, J., Hunt, B.P.V., Pakhomov, E.A., Holmes, K., Lu, Y. and Quayle, L. (2018) Using unmanned aerial vehicles (UAVs) to measure jellyfish aggregations. *Marine Ecology Progress Series* 591, 29–36.
Schlechtriem, C., Bischof, I., Atorf, C., Bergendahl, E., Seymour, P. and Whalley, P. (2016) Development of a regulatory testing procedure to study the metabolism of pesticides in farmed fish. *Pest Management Science* 72(2), 362–370.
Schlenk, D. and Benson, W.H. (eds) (2001) *Target Organ Toxicity in Marine and Freshwater Teleosts: Systems*. CRC Press, Boca Raton, Florida.
Schmidt-Posthaus, H. and Marcos-López, M. (2014) Non-infectious disorders of coldwater fish. In: Woo, P.T.K. and Bruno, D.W. (eds) *Diseases and Disorders of Finfish in Cage Culture*, 2nd edn. CAB International, Wallingford, UK, pp. 114–154.
Schmidt-Posthaus, H., Pugovkin, D. and Wahli, T. (2005) Iridophoroma in a European grayling (*Thymallus thymallus*) in Switzerland. *Bulletin of the European Association of Fish Pathologists* 25(1), 36–39.
Schreck, C.B. and Tort, L. (2016) The concept of stress in fish. In: Schreck, C.B., Tort, L., Farrell, A.P. and Brauner, C.J. (eds) *Fish Physiology*. Vol. 35. *Biology of Stress in Fish*. Elsevier, Amsterdam, pp. 1–34.
Sealey, W.M. and Gatlin, D.M. III (2002) Dietary vitamin C and vitamin E interact to influence growth and tissue composition of juvenile hybrid striped bass (*Morone chrysops*♀ × *M. saxatilis*♂) but have limited effects on immune responses. *The Journal of Nutrition* 132(4), 748–755.
Segner, H. and Böhm, R. (1994) Enzymes of lipogenesis. In: Hochachka, P. and Mommsen, T.P. (eds) *Biochemistry and Molecular Biology of Fishes*. Vol. 3. *Analytical Techniques*. Elsevier Science, Amsterdam, pp. 313–326.
Segner, H. and Möller, H. (1984) Electron microscopical investigations on starvation-induced liver pathology in flounders *Platichthys flesus*. *Marine Ecology Progress Series* 19, 193–196.
Segner, H. and Roesch, R. (1992) Larvae of *Coregonus lavaretus* fed with dry diet show reversible starvation symptoms during early rearing. *Polskie Archiwum Hydrobiologii* 39(3–4), 623–632.
Segner, H., Burkhardt, P., Avila, E.M., Juario, J.V. and Storch, V. (1987) Nutrition-related histopathology of the intestine of milkfish *Chanos chanos* fry. *Diseases of Aquatic Organisms* 2(2), 99–107.
Shahmohamadloo, R.S., Frenken, T., Rudman, S.M., van West, P., Ibelings, B.W. and Trainer, V.L. (2023) Diseases and disorders in fish due to harmful algal blooms. In: Woo, P.T.K. and Subasinghe, R.P. (eds) *Climate Change on Diseases and Disorders of Finfish in Cage Culture*, 3rd edn. CAB International, Wallingford, UK, pp. 387–429.
Shchelkunov, I.S., Karseva, T.A. and Kadoshnikov, Y.U.P. (1992) Atlantic salmon papillomatosis: visualization of herpesvirus-like particles in skin growths of affected fish. *Bulletin of the European Association of Fish Pathologists* 12(1), 28–31.
Silverstone, A.M. and Hammell, L. (2002) Spinal deformities in farmed Atlantic salmon. *The Canadian Veterinary Journal* 43(10), 782–784.
Skonberg, D.I., Yogev, L., Hardy, R.W. and Dong, F.M. (1997) Metabolic response to dietary phosphorus intake in rainbow trout (*Oncorhynchus mykiss*). *Aquaculture* 157(1–2), 11–24.
Speare, D.J. (2002) Non-infectious disorders of coldwater fish. In: Woo, P.T.K., Bruno, D.W. and Lim, L.H.S. (eds) *Diseases and Disorders of Finfish in Cage Culture*, 1st edn. CAB International, Wallingford, UK, pp. 171–193.
Speare, D.J. (2010) Disorders associated with exposure to excess dissolved gases. In: Leatherland, J.F. and Woo, P.T.K. (eds) *Fish Diseases and Disorders*. Vol. 2. *Non-infectious Disorders*, 2nd edn. CAB International, Wallingford, UK, pp. 342–356.

Sugiura, S.H., Babbitt, J.K., Dong, F.M. and Hardy, R.W. (2000) Utilization of fish and animal by-product meals in low-pollution feeds for rainbow trout *Oncorhynchus mykiss* (Walbaum). *Aquaculture Research* 31(7), 585–593.

Sullivan, M., Hammond, G., Roberts, R.J. and Manchester, N.J. (2007a) Spinal deformation in commercially cultured Atlantic salmon, *Salmo salar* L.: a clinical and radiological study. *Journal of Fish Diseases* 30(12), 745–752.

Sullivan, M., Reid, S.W.J., Ternent, H., Manchester, N.J., Roberts, R.J., *et al.* (2007b) The aetiology of spinal deformity in Atlantic salmon, *Salmo salar* L.: influence of different commercial diets on the incidence and severity of the preclinical condition in salmon parr under two contrasting husbandry regimes. *Journal of Fish Diseases* 30(12), 759–767.

Tacon, A.G.J. (1992) *Nutritional Fish Pathology: Morphological Signs of Nutrient Deficiency and Toxicity in Farmed Fish*. FAO Fish Technical Paper No. 330. Food and Agriculture Organization of the United Nations, Rome.

Tacon, A.G.J. (1996) Lipid nutritional pathology in farmed fish. *Archives of Animal Nutrition* 49(1), 33–39.

Takle, H., Baeverfjord, G., Lunde, M., Kolstad, K. and Andersen, Ø. (2005) The effect of heat and cold exposure on HSP70 expression and development of deformities during embryogenesis of Atlantic salmon (*Salmo salar*). *Aquaculture* 249(1–4), 515–524.

Takle, H., Baeverfjord, G., Helland, S., Kjorsvik, E. and Andersen, O. (2006) Hyperthermia induced atrial natriuretic peptide expression and deviant heart development in Atlantic salmon *Salmo salar* embryos. *General and Comparative Endocrinology* 147(2), 118–125.

Taylor, J.F., Preston, A.C., Guy, D. and Migaud, H. (2011) Ploidy effects on hatchery survival, deformities, and performance in Atlantic salmon (*Salmo salar*). *Aquaculture* 315(1–2), 61–68.

Taylor, J.F., Sambraus, F., Mota-Velasco, J., Guy, D.R., Hamilton, A., *et al.* (2013) Ploidy and family effects on Atlantic salmon (*Salmo salar*) growth, deformity and harvest quality during a full commercial production cycle. *Aquaculture* 410–411, 41–50.

Teles, A.O., Couto, A., Enes, P. and Peres, H. (2020) Dietary protein requirements of fish – a meta-analysis. *Reviews in Aquaculture* 12(3), 1445–1477.

Terova, G., Rimoldi, S., Izquierdo, M., Pirrone, C., Ghrab, W. and Bernardini, G. (2018) Nano-delivery of trace minerals for marine fish larvae: influence on skeletal ossification, and the expression of genes involved in intestinal transport of minerals, osteoblast differentiation, and oxidative stress response. *Fish Physiology and Biochemistry* 44(5), 1375–1391.

Thomas, S., Abraham, A., Rodríguez-Mallon, A., Unajak, S. and Bannantine, J.P. (2022) Challenges in veterinary vaccine development. In: Thomas, S. (ed.) *Vaccine Design. Methods in Molecular Biology*, Vol. 2411. Humana Press, New York, pp. 3–34.

Thompson, I., Fletcher, T.C., Houlihan, D.F. and Secombes, C.J. (1994) The effect of dietary vitamin A on the immunocompetence of Atlantic salmon (*Salmo salar* L.). *Fish Physiology and Biochemistry* 12(6), 513–523.

Tocher, D.R. (2003) Metabolism and functions of lipids and fatty acids in teleost fish. *Reviews in Fisheries Science* 11(2), 107–184.

Tocher, D.R., Mourente, G., Van der Eecken, A., Evjemo, J.O., Diaz, E., *et al.* (2002) Effects of dietary vitamin E on antioxidant defence mechanisms of juvenile turbot (*Scophthalmus maximus* L.), halibut (*Hippoglossus hippoglossus* L.) and sea bream (*Sparus aurata* L.). *Aquaculture Nutrition* 8(3), 195–207.

Tørud, B. and Håstein, T. (2008) Skin lesions in fish: causes and solutions. *Acta Veterinaria Scandinavica* 50(Suppl. 1), S7.

Trichet, V.V. (2010) Nutrition and immunity: an update. *Aquaculture Research* 41(3), 356–372.

Turchini, G.M., Francis, D.S., Du, Z.-Y., Olsen, R.E., Ringø, E. and Tocher, D.R. (2022) The lipids. In: Hardy, R.W. and Kaushik, S.J. (eds) *Fish Nutrition*, 4th edn. Academic Press, London, pp. 303–467. https://doi.org/10.1016/B978-0-12-819587-1.00003-3

Urán, P.A., Schrama, J.W., Rombout, J., Obach, A., Jensen, L., *et al.* (2008) Soybean meal-induced enteritis in Atlantic salmon (*Salmo salar* L.) at different temperatures. *Aquaculture Nutrition* 14(4), 324–330.

Urán, P.A., Schrama, J.W., Rombout, J., Taverne-Thiele, J.J., Obach, A., *et al.* (2009) Time-related changes of the intestinal morphology of Atlantic salmon, *Salmo salar* L., at two different soybean meal inclusion levels. *Journal of Fish Diseases* 32(9), 733–744.

van der Meer, D.L.M., van den Thillart, G.E., Witte, F., de Bakker, M.A.G., Besser, J., *et al.* (2005) Gene expression profiling of the long-term adaptive response to hypoxia in the gills of adult zebrafish. *American Journal of Physiology: Regulatory, Integrative and Comparative Physiology* 289(5), R1512–R1519.

Van Rijn, C.A., Jones, P.L., Evans, B.S. and Afonso, L.O.B. (2021) Physiological and growth responses of juvenile Atlantic salmon (*Salmo salar*) transferred to seawater during different stages of smolt development. *Aquaculture* 538, 736527.

Van Zuiden, T.M., Chen, M.M., Stefanoff, S., Lopez, L. and Sharma, S. (2016) Projected impacts of climate change on three freshwater fishes and potential novel competitive interactions. *Diversity and Distributions* 22(5), 603–614.

Venizelos, A. and Benetti, D.D. (1999) Pigment abnormalities in flatfish. *Aquaculture* 176(1–2), 181–188.

Vogt, G., Böhm, R. and Segner, H. (1994) Mimosine-induced cell death and related chromatin changes. *Journal of Submicroscopic Cytology and Pathology* 26(3), 319–330.

Waagbø, R. (2008) Reducing production-related diseases in farmed fish. In: Lie, Ø. (ed.) *Improving Farmed Fish Quality and Safety*. Woodhead Publishing Ltd, Cambridge, pp. 363–398.

Waagbø, R., Tröße, C., Koppe, W., Fontanillas, R. and Breck, O. (2010) Dietary histidine supplementation prevents cataract development in adult Atlantic salmon, *Salmo salar* L., in seawater. *British Journal of Nutrition* 104(10), 1460–1470.

Wahli, T., Verlhac, V., Gabaudan, J., Schüep, W. and Meier, W. (1998) Influence of combined vitamins C and E on non-specific immunity and disease resistance of rainbow trout, *Oncorhynchus mykiss* (Walbaum). *Journal of Fish Diseases* 21(2), 127–137. https://doi.org/10.1046/j.1365-2761.1998.00088.x

Wahli, T., Verlhac, V., Girling, P., Gabaudan, J. and Aebischer, C. (2003) Influence of dietary vitamin C on the wound healing process in rainbow trout (*Oncorhynchus mykiss*). *Aquaculture* 225(1–4), 371–386. https://doi.org/10.1016/S0044-8486(03)00302-8

Wall, R., Ross, R.P., Fitzgerald, G.F. and Stanton, C. (2010) Fatty acids from fish: the anti-inflammatory potential of long-chain omega-3 fatty acids. *Nutrition Reviews* 68(5), 280–289.

Warr, G.W., Griffin, B.R., Anderson, D.P., McAllister, P.E., Lidgerding, B. and Smith, C.E. (1984) A lymphosarcoma of thymic origin in the rainbow trout, *Salmo gairdneri* Richardson. *Journal of Fish Diseases* 7(1), 73–82.

Watanabe, T., Takeuchi, T. and Ogino, C. (1980) Effect on rainbow trout and chum salmon of deletion of trace elements from fish meal diet. *Bulletin of the Japanese Society of Scientific Fisheries* 46(12), 1521–1525.

WHO (World Health Organization) (2013) *International Classification of Diseases for Oncology (ICD-O)*, 3rd edn, 1st revision. WHO, Geneva, Switzerland.

Williams, D.E. (2012) The rainbow trout liver cancer model: response to environmental chemicals and studies on promotion and chemoprevention. *Comparative Biochemistry and Physiology Part C: Toxicology & Pharmacology* 155(1), 121–127.

Wolf, J.C. and Wolfe, M.J. (2005) A brief overview of nonneoplastic hepatic toxicity in fish. *Toxicologic Pathology* 33(1), 75–85.

Wolke, R.E. (1992) Piscine macrophage aggregates: a review. *Annual Review of Fish Diseases* 2, 91–108.

Woo, P.T.K. and Iwama, G.K. (eds) (2020) *Climate Change and Non-infectious Fish Disorders*. CAB International, Wallingford, UK.

Woo, P.T.K., Leong, J. and Buchmann, K. (eds) (2020) *Climate Change and Infectious Fish Diseases*. CAB International, Wallingford, UK.

Yang, C., Song, G. and Lim, W. (2020) A review of the toxicity in fish exposed to antibiotics. *Comparative Biochemistry and Physiology Part C: Toxicology & Pharmacology* 237, 108840. https://doi.org/10.1016/j.cbpc.2020.108840

Yonar, M.E. (2012) The effect of lycopene on oxytetracycline-induced oxidative stress and immunosuppression in rainbow trout (*Oncorhynchus mykiss*, W.). *Fish & Shellfish Immunology* 32(6), 994–1001.

Ytteborg, E., Baeverfjord, G., Torgersen, J., Hjelde, K. and Takle, H. (2010) Molecular pathology of vertebral deformities in hyperthermic Atlantic salmon (*Salmo salar*). *BMC Physiology* 10(1), 12.

Zhang, B., Shao, C., Chen, S. and Bao, B. (2022) Understanding asymmetrical malpigmentation in flatfishes and improvement measures for aquaculture. *Reviews in Aquaculture* 14(3), 1333–1344. https://doi.org/10.1111/raq.12652

5 Infectious Diseases of Warmwater Fish in Marine and Brackish Waters

Christine A. Parker-Graham[1], Taylor I. Heckman[2], Matt J. Griffin[3] and Esteban Soto[2]*

[1]*US Fish and Wildlife Service, Lacey, Washington, USA;* [2]*University of California, Davis, California, USA;* [3]*Mississippi State University, Stoneville, Mississippi, USA*

5.1 Introduction

Warmwater cage culture is the rearing of finfish in enclosed containers in natural bodies of water, which can range from open-top net pens to submerged cages. They are typically in brackish, estuarine or marine habitats where the water rarely (if ever) drops below 20°C (Colorni and Diamant, 2014). The close relationship between culture systems and the environment is a key feature of cage culture, as water, feed, chemical treatments and waste products can move freely through systems. Because of their intimate relationship with the open ocean, cage-culture systems face unique challenges in the face of climate change. Cage-culture systems are exposed to oceanic conditions, including sea-level rise, salinity changes, storm surges, acidification, rising temperatures, and increased incidents of extreme weather (Hossain *et al.*, 2021). Not only do these challenges pose considerable risks to fish health, but also the effects of climate change deleteriously impact fish welfare by altering their rearing environment in such a way as to inhibit their ability to engage in normal behaviours and to increase allostatic stress loads (Hvas *et al.*, 2020; Martos-Sitcha *et al.*, 2020).

Fish are kept at high densities in intensive rearing systems, which can quickly concentrate pathogens. Disease can be amplified by the physiological stress inherent in intensive production systems (e.g. increased conspecific aggression, increased stocking density, increased waste products, etc.). The open nature of these systems allows for cultured fish to regularly contact wild fish of different ages and life stages, thus of different disease susceptibilities. Prior to the advent of strict biosecurity regulations and surveillance of finfish pathogens, many culture systems were started with unscreened stock that inadvertently introduced diseases into naïve wild fish populations. In some regions wild fish have become reservoirs for diseases (Colorni and Diamant, 2014). This underlines the importance of pre-stocking disease screening and prompt, thorough management of disease issues that can arise within cage-cultured fish populations.

Cage-based mariculture is an important component of global aquaculture which produces over 72.5 million tonnes of edible protein annually (Ahmed *et al.*, 2019b). Responsible mariculture presents an opportunity to conserve wild fish populations by alleviating fisheries pressure on imperiled species. Building resilient

*sotomartinez@ucdavis.edu

DOI: 10.1079/9781800621640.0005

culture systems that take current and future climate conditions, biosecurity, disease and fish welfare into account is critical to supporting biodiversity, food security and continually productive rearing systems.

5.2 Viral Diseases

There are over 125 viruses that infect finfish (Noga, 2010). With the advent of molecular diagnostics novel fish viruses are detected and characterized each year. The pathogenicity of many teleost viruses is dependent on host immune proficiency and environmental temperature, making viral disease of particular concern in a changing climate. Two viral families of importance to marine warmwater aquaculture discussed here are *Iridoviridae* and *Nodaviridae*.

5.2.1 *Iridoviridae*

Iridoviridae is an important family of double-stranded DNA viruses. There are two subfamilies of *Iridoviridae*: *Alphairidovirinae* and *Betairidovirinae*, the former being of primary concern to fish and herpetofauna and the latter being hosted by insects. Two genera of *Alphairidovirinae* manifest clinical disease in multiple fish species: *Megalocytivirus* and *Lymphocystivirus*.

Megalocytiviruses

Megalocytiviruses comprise an important genus in the *Iridoviridae* family. Megalocytiviruses were first detected in disease outbreaks in Japan in 1990, since then they have been found globally and have severe impacts on many cultured fish species (Johan and Zainathan, 2020). *Infectious spleen and kidney necrosis virus* (ISKNV) is the type species in the genus and includes genotypes such as the red sea bream iridovirus (RSIV) and the turbot reddish body iridovirus (TRBIV). Recently, a novel megalocytivirus, *Scale drop disease virus*, was identified in barramundi (*Lates calcarifer*) and yellowfin sea bream (*Acanthopagrus latus*) in South-East Asia (Domingos *et al.*, 2021; Fu *et al.*, 2021). A novel megalocytivirus has been documented in freshwater-reared tilapia (*Oreochromis* spp.) in North America; however, the salinity tolerance and virulence to brackish and marine fish are unknown (Shahin *et al.*, 2021a). Many warmwater finfish species are susceptible to megalocytiviruses, including members of the family Osphronemidae and red drum (*Sciaenops ocellatus*) (López-Porras *et al.*, 2018).

Megalocytiviruses can cause widespread mortality (70–100%) in infected populations, depending on the species of fish and strain of virus (Yanong and Waltzek, 2019). Detections can have regulatory and export implications as RSIV is notifiable to the World Organisation for Animal Health (OIE) and voluntarily reportable to the US Department of Agriculture under the National Animal Health Reporting System. Increasing water temperature is a significant risk for megalocytiviral disease because higher water temperatures enhance viral replication and suppress fish immune responses. Megalocytiviruses can survive in seawater up to 32°C. Yanong and Terrell (2003) demonstrated temperature spikes associated with warmer environmental conditions led to spikes in megalocytiviral disease in oscars (*Astronotus ocellatus*) in Florida, USA.

Not all fish infected with megalocytiviruses show clinical disease or mortality and importers should be aware of carrier fish. Clinical signs are typically non-specific like lethargy, abnormal buoyancy, anorexia, skin darkening, coelomic distension, skin ulceration, gill pallor, subcutaneous haemorrhages and abnormal swimming patterns. Gross pathological findings are similarly non-specific and include hepatomegaly, splenomegaly, pallor and ascites. Megalocytiviral infection produces characteristic cytopathological lesions called inclusion body-bearing cells (IBCs) (Fig. 5.1). IBCs are large, spherical, basophilic inclusions surrounded by a limiting membrane found within hypertrophied cells. IBCs are commonly found in the spleen, gills, alimentary tract and haematopoietic tissues of infected fish (Noga, 2010). IBCs and increased mortality are strongly indicative of megalocytiviral infection, but diagnosis should be confirmed via cell culture, polymerase chain reaction (PCR), immunofluorescent antibody techniques (IFATs) or loop-mediated isothermal amplification (LAMP) assays (Johan and Zainathan, 2020).

Transmission is typically horizontal via contaminated water and equipment or ingestion of infected waste or fish. Vertical transmission of megalocytiviruses has not been reported (Johan

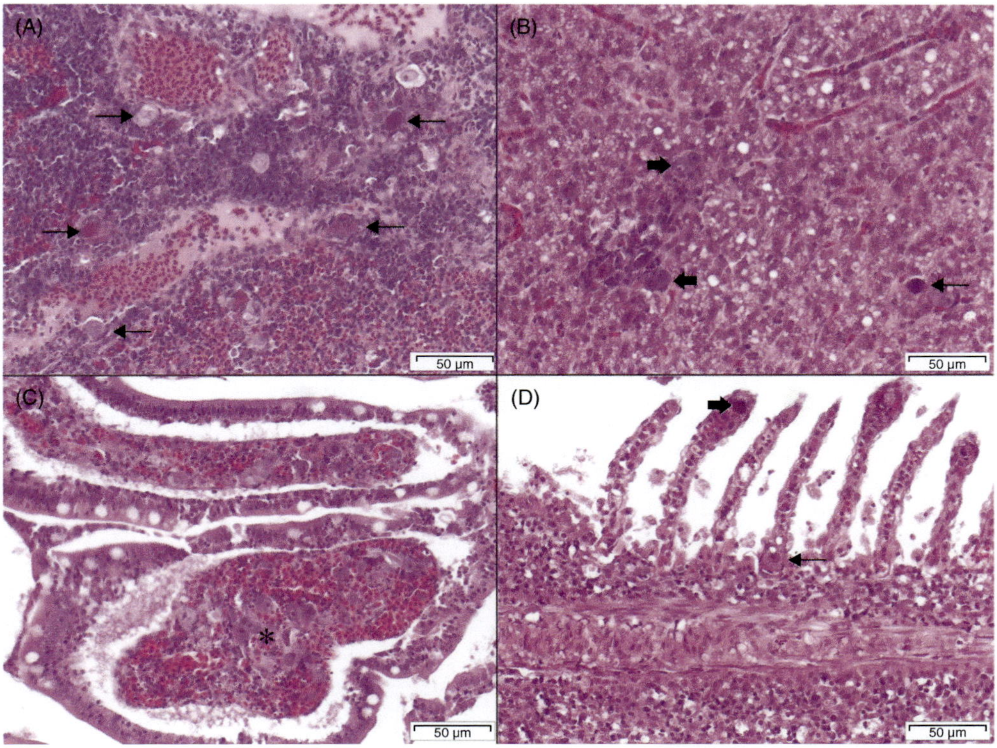

Fig. 5.1. Red sea bream iridovirus (family *Iridoviridae*) in cultured pompano (*Trachinotus* spp.), haematoxylin and eosin stain. (A) Splenic section with inclusion-bearing cells (IBCs) (arrows). (B) Liver section with IBCs (arrows). (C) Intestine section. Note congestion and haemorrhage of lamina propria, with multiple IBCs (asterisk). (D) Gill section with mononuclear cellular infiltrates and an immature IBC at the lamellar base (arrow). (From López-Porras *et al*., 2018, doi: 10.3354/dao03267.)

and Zainathan, 2020). Risk factors for disease transmission include any event that induces host stress and subsequent immune suppression, such as handling, transport, spawning and poor water quality. Movement and translocation pose a risk of introducing megalocytiviral disease to naïve populations because of the presence of subclinical carriers.

There are commercial oral, immersion and injectable vaccines for some strains of megalocytiviruses but these products are not labelled for use nor available to producers in many countries (Fu *et al.*, 2021). In the absence of vaccination, control of megalocytiviruses centres on strict biosecurity and optimal husbandry. Screening incoming lots of fish with molecular methods is recommended because of the potential for subclinical carrier fish which may inadvertently introduce viral infections into naïve populations. The OIE currently recommends virus culture and identification by IFAT or PCR, PCR of kidney/spleen, IFAT of a stamp smear (spleen, heart or kidney) or sequencing as presumptive and confirmatory tests. There are currently no recommendations for antemortem diagnosis of megalocytiviruses, the assays listed above require kidney and/or spleen tissue (Nakajima, 2021). Because of their outer envelope megalocytiviruses are susceptible to many physical and chemical inactivation methods, including potassium permanganate, 5% sodium hypochlorite and ultraviolet radiation (Yanong and Waltzek, 2019).

Lymphocystiviruses

Lymphocystis disease is a chronic disease of freshwater and marine fish caused by genus *Lymphocystivirus* (lymphocystis disease viruses), which are epitheliotropic enveloped viruses in the

Iridoviridae family. Lymphocystiviruses are globally distributed and affect over 125 species of teleosts; primitive teleost orders such as Siluriformes, Cypriniformes and Salmoniformes do not appear to be susceptible to clinical infection (Noga, 2010).

Unlike other *Iridoviridae* genera, *Lymphocystivirus* is mildly pathogenic and typically does not cause mortality in affected fish. Viral infection causes unsightly self-limiting nodules on fins, skin and gills, which can affect the cosmetic value of fish raised for food or ornamental trade. *Lymphocystivirus* incubation is temperature dependent and warmer water significantly reduces incubation time. In bluegill (*Lepomis macrochirus*) *Lymphocystivirus* replicates at temperatures as high as 23–25°C with an incubation period of 12–15 days at 25°C compared with 37 days at 12.5°C (Cascarano *et al.*, 2021).

Lymphocystis disease typically presents as macroscopic nodular lesions on fins, skin and/or gills but internal lesions can develop. Lesions range from 0.3 to 2 mm in diameter and appear white to grey or pink, depending on the affected epithelial tissue. Gross lesions are individual or clustered fibroblasts that enlarge 50,000 to 100,000 times their normal size and produce infective viral particles (Noga, 2010). Diagnosis of lymphocystis disease is usually made by visualization of hypertrophic fibroblasts on wet mount or histological sections of lesions but can be confirmed by PCR or LAMP assays (Fig. 5.2). *Lymphocystivirus* is difficult to grow in cell culture and requires homologous cell lines for successful isolation. Several serological techniques, such as indirect immunofluorescence, flow cytometry and immunoblot, have been successful in identifying carrier fish without clinical signs (Borrego *et al.*, 2017). Important differential diagnoses include *Ichthyophthirius* spp., *Cryptocaryon* spp., *Epistylis* spp., *Saprolegnia* spp., epitheliocystis, infestation by digenean trematodes and neoplasia.

Transmission occurs horizontally via direct contact with infective particles. Ingestion of infected feed items (e.g. *Artemia* metanauplii,

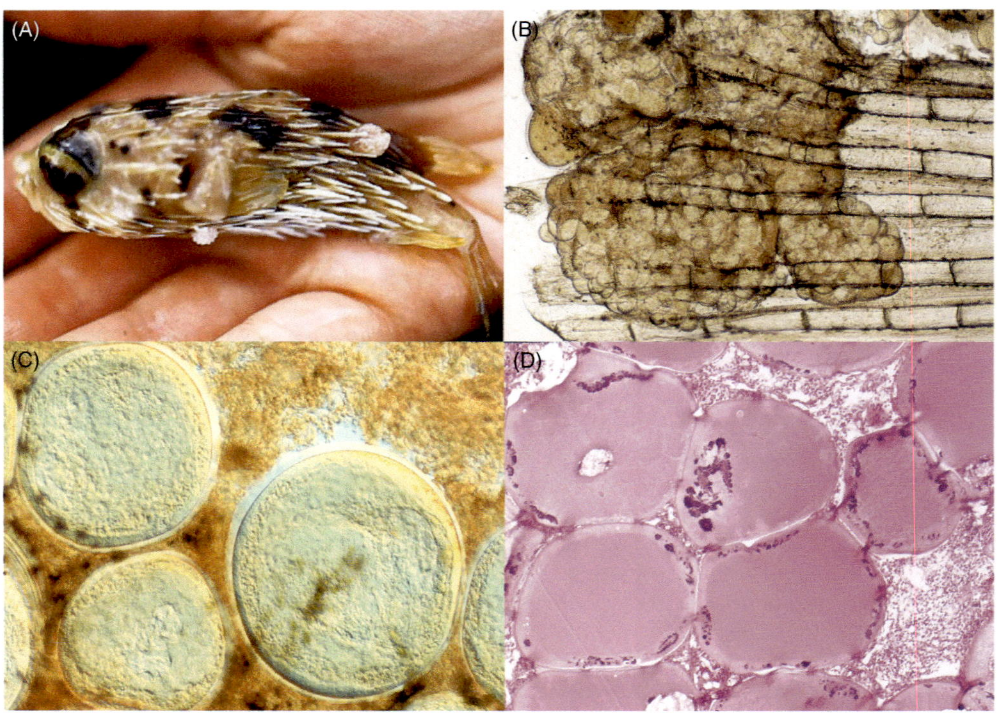

Fig. 5.2. Lymphocystivirus symptoms in fish. (A) White nodular lesion in fin of porcupine fish. (B) Wet mount of a fin clip with notable fibroblast hypertrophy. (C) Wet mount of a skin scrape with hypertrophied fibroblasts. (D) Hypertrophied fibroblasts revealed with haematoxylin and eosin stain. (Images courtesy of Dr Juan A. Morales.)

fishmeal) or infective particles from the water column can lead to infection as well (Carbello *et al.*, 2019; Valverde *et al.*, 2019). Infection *per os* typically leads to development of lesions in internal organs, while infection from the water column typically leads to external epithelial infection. External trauma, parasitic infestations and improper handling can create microlesions in the protective mucus cuticle and underlying epithelium to predispose fish to infection. Environmental stressors, including decreased dissolved oxygen, temperatures outside the preferred optimal temperature zone and poor water quality, can result in population-wide immunosuppression and lead to lymphocystis disease outbreaks. In intensive aquaculture systems lymphocystis disease outbreaks can affect up to 70% of the population (Borrego *et al.*, 2017). Lymphocystis virus does not appear to be vertically transmissible.

There are no chemotherapeutics or vaccines commercially available for lymphocystis virus. Lesions are self-limiting and clear more quickly in warmer water. Lymphocystis disease prevention requires strict biosecurity and hygiene at all rearing ages. This is difficult for operations using raw seawater for rearing as lymphocystis disease virus remains viable in seawater for extended periods; the virus remains viable in seawater for 242 days at 22°C (Leiva-Rebello *et al.*, 2020). As with many infectious diseases, reducing fish stress, optimizing water quality and environment, and reducing trauma and handling are important to reducing the risk of lymphocystis disease in populations. Culling severely affected fish may reduce the risk of transmission by reducing the volume of infectious units discharged into the environment (Colorni and Diamant, 2014). The virus is stable through multiple freeze–thaw cycles and under a variety of storage conditions; it is inactivated by diethyl ether and chloroform, intense heat (56–60°C) or a pH of 3.0 or lower (Stoskopf, 2015).

5.2.2 *Nodaviridae*

Nodaviruses are small, neurotropic, non-enveloped, positive-sense single-stranded RNA viruses within the family *Nodaviridae* and cause viral nervous necrosis (VNN), also known as viral encephalopathy and retinopathy (VER), in susceptible fish species (Noga, 2010). There are two genera: *Alphanodavirus*, which is of concern for insects, and *Betanodavirus*, which primarily infects fish. Betanodaviruses are distributed globally and affect at least 40 teleost species including striped jack (*Pseudocaranx dentex*), grouper (*Epinephelus* spp.), red drum (*S. ocellatus*), red sea bream (*Pagrus major*), amberjack (*Seriola* spp.) and barramundi (*L. calcarifer*). In susceptible species and life stages, mortality can be up to 100% (Bandín and Souto, 2020).

Clinical signs of VNN/VER tend to manifest in the life stages most susceptible to infection (larvae, juveniles) and not all infected fish exhibit clinical disease (Kai *et al.*, 2010). Non-specific signs like anorexia, emaciation, skin darkening, abnormal swimming patterns, lethargy, exophthalmia and abnormal buoyancy are common in infection. Gross lesions are uncommon. Histological lesions are primarily seen in neural tissue, but may be apparent in liver and spleen, and include gliosis, vacuolation and degeneration, particularly in the retina (Noga, 2010). Diagnosis should be confirmed by cell culture, PCR or enzyme-linked immunosorbent assay (ELISA) (Bandín and Souto, 2020).

Transmission of betanodaviruses occurs both horizontally and vertically. Like many viral diseases, transmission and induction of clinical disease are favoured by environmental conditions causing immune suppression in host populations. Replication of betanodavirus is favoured in warmer water (20–30°C), which is higher than many species' preferred optimal thermal zone, allowing the virus to take advantage of accelerated replication and host immune suppression simultaneously (Doan *et al.*, 2017). Surveys of wild fish in Asia and the Mediterranean have revealed a number of wild infected fish without clinical signs. Souto *et al.* (2015) demonstrated a rapid rise in mortalities among Senegalese sole (*Solea senegalensis*) due to VNN when temperature was increased from 16 to 22°C, suggesting warming water temperatures could incite clinical disease in asymptomatic fish. Warming water temperatures could drive carrier animals into higher latitudes and risk exposing naïve wild populations to VNN (Bandín and Souto, 2020).

Betanodavirus particles are very stable in aquatic environments and difficult to eliminate once introduced into an aquaculture system (Bandín and Souto, 2020). There are no chemotherapeutics available against betanodaviruses. Producers can minimize the risk of VNN infection by selecting VNN-free broodstock and by utilizing broodstock vaccinated against VNN (Kai *et al.*, 2010). Species-specific formalin-killed vaccines are available commercially, but licensing and availability vary by country (Doan *et al.*, 2017). A variety of widely available disinfectants can effectively inactivate the virus on equipment and on environmental surfaces, including sodium hypochlorite, benzalkonium chloride, chloroquine and iodine (Bandín and Souto, 2020).

5.3 Bacterial Diseases

Bacterial diseases in finfish are often secondary and take advantage of immune suppression, physiological stress, physical trauma or other stressors. Climate change has profound impacts on the virulence and geographic ranges of primary and secondary bacterial pathogens, but also increases environmental stress and immune suppression, making fish more susceptible to infection and clinical disease.

5.3.1 Vibriosis

Vibriosis describes multiple diseases caused by pathogenic members of the family *Vibrionaceae*, primarily in the *Vibrio* genus. *Vibrio* species are ubiquitous in aquatic environments, particularly temperate, tropical and subtropical areas. Virtually all species of fish are susceptible to at least one *Vibrio* species (Noga, 2010). The wide range of habitats, hosts and regions impacted make vibriosis one of the most common fish diseases in euryhaline and marine environments. *Vibrio anguillarum* ('classical' vibriosis), *Vibrio ordalii* ('atypical' vibriosis), *Vibrio vulnificus* ('warmwater' vibriosis) and *Vibrio harveyi* ('marine' vibriosis) are the best described species affecting aquaculture, but a wide range of other species have been reported in association with diseased fish as primary pathogens or contributing opportunists in mixed infections. More species are anticipated to be described from novel isolations or further taxonomic rearrangements. Species may contain virulent and avirulent strains, depending on highly mobile virulence factor genes. Several members also have zoonotic potential, for example *V. vulnificus*. Clinical vibriosis depends on host and bacterial type, but all acute vibrioses result in a septicaemia that often leads to death (Amaro *et al.*, 2020).

Vibriosis has severe adverse effects for mariculture. Economic losses include direct animal loss in addition to reduced production efficiency and output, treatment expenditures and associated labour. Contemporary economic estimates for fish culture are scarce but a recent study estimated the cost of vibriosis in cage-cultured sea bass on the east coast of Malaysia at €0.19/tail without diagnosis and treatment costs, resulting in thousands of euros lost per cycle in this region alone (Mohd Yazid *et al.*, 2021). Global costs are expected to increase with climate change because the life cycle of *Vibrio* depends on temperature and salinity (Toranzo *et al.*, 2005; Amaro *et al.*, 2015).

General clinical signs of vibriosis are reflective of septicaemia, including lethargy, anorexia, haemorrhage at the bases of fins, exophthalmia, pale gills and abnormal swimming (Fig. 5.3). Coelomic distension, corneal opacity, and elevated opercular rate, piping and respiratory distress from gill damage may also be observed (Ina-Salwany *et al.*, 2019). Clinical vibriosis may progress differently depending on host type, bacterial species/strain and environmental conditions. For example, *V. anguillarum* may cause darkened skin with internal liquefaction of the spleen and kidney in its acute form, or granulomatous cutaneous lesions with internal haemorrhaging and visceral fibrinous adhesions in chronic infections. Because *Vibrio* spp. are found in seawater and in the microbiota of healthy fish, isolation from kidney or other internal organs is preferred for diagnosis of vibriosis. Isolated *Vibrio* spp. grow well on tryptic soy agar (TSA) with 1% NaCl (TSA-1) or the selective thiosulfate–citrate–bile salts–sucrose (TCBS) and *V. anguillarum* media (VAM). Optimal temperature for growth is species dependent but ranges between 15 and 30°C (Ina-Salwany *et al.*, 2019). *Vibrio* spp. are facultative anaerobes, and generally motile. They are catalase- and oxidase-positive, ferment glucose without producing gas, and stain as Gram-negative curved or straight rods.

Phenotypic methods have traditionally been applied to characterize *Vibrio* spp., but similarities between species and lack of inter-laboratory

Fig. 5.3. Vibriosis in tilapia hybrids (*Oreochromis* spp.) raised in brackish water due to *Vibrio vulnificus*. (A) Hyperaemic skin. (B) Severe exophthalmia.

reproducibility limit their reliability (Ina-Salwany *et al.*, 2019). A number of molecular methods have been developed to identify different species of fish-pathogenic *Vibrio*, with varied resolving power and practicality. Several protocols for conventional, multiplex or real-time PCR assays have been developed targeting different genes (Hickey *et al.*, 2015; Toranzo *et al.*, 2017). Sequencing of the 16S rRNA gene has limited discriminatory power, but incorporating multiple housekeeping genes by multilocus sequence analysis (MLSA) can improve resolution to the strain level (Gabriel *et al.*, 2014). The *toxR* gene is also a useful marker for species-specific identification (Pang *et al.*, 2006). Restriction and amplified fragment length polymorphism (R/AFLP), LAMP and fluorescence *in situ* hybridization (FISH) assays have also been employed (Amaro *et al.*, 2020).

Vibrio spp. are prevalent and persist in the water column as planktonic free-living forms or in a sessile, biofilm-associated state (Gomez-Gil *et al.*, 2014; Amaro *et al.*, 2015). Free-living bacteria can survive in water microcosms for over 50 weeks, dependent on water temperature and salinity (Hickey and Lee, 2018). Temperature and environmental factors influence transition between these forms and the viable but non-culturable (VBNC) state. Transmission is primarily horizontal through the water, gills, skin or intestine, although vertical transmission via egg contamination has been reported; the infectious route may be species or strain specific (Amaro *et al.*, 2015). Oral transmission may occur by consumption of contaminated zooplankton: *Vibrio* biofilms form readily on chitin and chitinous zooplankton are suggested to act as a motile reservoir for *Vibrio* spp. (Gomez-Gil *et al.*, 2014). Copepods and other chitinous zooplankton are expected to increase with warming water, potentially increasing disease dissemination (Vezzulli *et al.*, 2015). Spread of *Vibrio* by bird predation on diseased fish has also been described and may be of concern as avian population dynamics change (Senderovich *et al.*, 2010).

Seasonality of vibriosis in temperate regions is well established, with outbreaks increasing in warmer months. This is, in part, due to the life strategy of *Vibrio* spp. Lower temperatures induce a VBNC state facilitating bacterial persistence through the colder months and higher temperatures induce resuscitation and rapid bacterial growth and spread (Amaro *et al.*, 2015). Increasing water temperatures may, therefore, increase *Vibrio* spp. populations and occurrence of disease. Temperature and salinity optimums vary between species, but many of the major pathogens favour warmer (>15°C) temperatures and moderate salinity (1–2%). Polar ice cap melting may decrease salinity in certain regions, which could favour such *Vibrio* spp. (Roux *et al.*, 2015). High organic loads also favour growth and persistence of *Vibrio* spp. (Yilmaz *et al.*, 2022).

When applicable, standard good management practices (e.g. optimal water quality and temperature, reasonable stocking densities, prompt removal of dead fish and excess food) can limit vibriosis in culture systems. Various means of treatment and prevention have been developed for many of the fish-pathogenic *Vibrio* spp. Antimicrobials are still widely used to general effect, although there is substantiated concern for antimicrobial resistance and environmental contamination (Yilmaz *et al.*, 2022). Antimicrobial use regulations vary between jurisdictions and practitioners should be aware of laws and regulations governing antimicrobial use in food-producing animals; for example, there are no antimicrobials currently licensed for the treatment of vibriosis in food-producing fish in the USA. Bacteriophages have been used in experimental treatment of *V. anguillarum* infections with promising results (Castillo *et al.*, 2019). A number of prebiotics, probiotics and combined synbiotics have been investigated, with some evidence of increased survival in experimental challenges (Yilmaz *et al.*, 2022). Commercial inactivated vaccines are available protecting against *V. anguillarum*, *V. ordalii* and *Vibrio salmonicida* in certain fish and countries (Ma *et al.*, 2019). Experimental vaccines in a variety of formulations have been developed against vibriosis, including traditional inactivated vaccines, live attenuated vaccines, subunit vaccines, DNA vaccines and live vector vaccines (Amaro *et al.*, 2015). Ample room remains for improvement in comprehensive protection against different *Vibrio* species, strains and serotypes.

5.3.2 Photobacteriosis

Photobacteriosis, also known as 'fish pasteurellosis' or 'pseudotuberculosis', is a septicaemic disease caused by the Gram-negative, halophilic bacterium *Photobacterium damselae* subsp. *piscicida* (PDSP), which is in the family *Vibrionaceae*. PDSP is a highly virulent, fish-specialized pathogen. *Photobacterium damselae* subsp. *damselae* (PDSD) is a generalist and emerging pathogen of marine animals that is considered opportunistic (Andreoni and Magnani, 2014; Osorio *et al.*, 2018). Disease caused by PDSD is considered a type of vibriosis and is very different in clinical presentation from PDSP (Osorio *et al.*, 2018). PDSP has a wide piscine host range and global distribution. Larvae and juvenile fish are considered more susceptible to photobacteriosis, with mortality of rates up to 100% in affected populations. Fish over 50 g are more resistant, which is attributed to more efficient phagocytotic killing (Andreoni and Magnani, 2014).

Photobacteriosis is responsible for major economic losses in cultured fish worldwide, particularly in the Mediterranean and Japan (Andreoni and Magnani, 2014; Pečur Kazazić *et al.*, 2019). Strains of PDSP have caused epizootics in Asia, Oceania, the Americas and Europe with remarkably high mortality rates, even compared with other *Vibrionaceae* (Vallecillos *et al.*, 2021). Many valuable commercial species are impacted by PDSP, including yellowtail (*Seriola quinqueradiata*), gilthead sea bream (*Sparus aurata*), sea bass (*Dicentrarchus labrax*), hybrid striped bass (*Morone saxatilis* × *Morone chrysops*) and cobia (*Rachycentron canadum*) (Andreoni and Magnani, 2014). Warmer water temperatures (18–25°C) favour outbreaks, with proliferation peaks during summer seasons (Cascarano *et al.*, 2021). Outbreaks of PDSD in fish have also been correlated with unusual increases in water temperature (Osorio *et al.*, 2018).

Photobacteriosis caused by PDSP may cause an acute or chronic form of disease in younger and older fish, respectively (Carraro *et al.*, 2018). General clinical signs of photobacteriosis include lethargy and abnormal swimming behaviour (Andreoni and Magnani, 2014). In the acute form multifocal necrosis may be present in kidney, liver and spleen but fish may show few to no clinical signs or gross pathology. Histologically, large quantities of bacteria can be observed in phagocytes, capillaries and interstitial spaces. In the chronic form white nodules or 'tubercles' form in multiple internal organs. PDSD, in contrast, is associated with ulcerative disease (Osorio *et al.*, 2018). Swabs from kidney or spleen can be inoculated on to marine agar 2216E (MA) or TSA and nutrient blood agar supplemented with 1–2% NaCl, then incubated at 22–25°C for 48–72 h. Shiny, yellow-grey, convex colonies develop on conventional media (Pečur Kazazić *et al.*, 2019). The rod-shaped bacterium is a facultative anaerobe, non-motile, and oxidase- and catalase-positive. It usually exhibits bipolar staining with Gram and Giemsa stains. Most strains are phenotypically homogeneous and may be identified by commercial biochemical panels such as the API 20 E and API 20 NE systems (Pham *et al.*, 2020). Identification techniques that distinguish PDSP and PDSD are necessary, especially as the two subspecies have overlapping host ranges and have been found in concurrent infections (Essam *et al.*, 2016). The *toxR* gene has improved discrimination compared with the 16S rRNA gene for this purpose and has been used alongside the *bamB* gene to develop a discriminatory reverse transcription (RT)-PCR protocol (Carraro *et al.*, 2018; Pham *et al.*, 2020). A multiplex PCR based on the 16S and PDSD-specific *ureC* gene is also used, as well as AFLP and PCR-RFLP assays to differentiate at the species level (Pečur Kazazić *et al.*, 2019). Matrix-assisted laser desorption/ionization–time-of-flight mass spectrometry (MALDI-TOF-MS) has been proposed, but reliable identification is dependent on culture incubation time and plate preparation with inter-replicate variability (Pečur Kazazić *et al.*, 2019).

While widely distributed, PDSP may have limited extra-host persistence in water (Baseggio *et al.*, 2021). VBNC forms may exist or PDSP may rely on carrier fish, invertebrate vectors or biofilms (Remuzgo-Martínez *et al.*, 2014). Transmission has been proposed through the gills, intestine or skin and vertically through gonadal fluid (Vallecillos *et al.*, 2021). PDSP is weakly adherent to various fish cell lines but has high binding capacity to fish intestines (Andreoni and Magnani, 2014). PDSD is capable of surviving for long periods of time in seawater while retaining virulence (Osorio *et al.*, 2018). An arthropod disease vector (e.g. sea lice) has been suggested and is supported by the absence of the urease (*ureC*) gene and presence of a pesticide gene in PDSP genomes (Baseggio *et al.*, 2021). Temperature governs sea lice development and populations are expected to increase with warmer conditions, thus vector-borne transmission holds additional potential to increase with climate change (Godwin *et al.*, 2021).

Warmer water temperatures favour growth of photobacteria. While outbreaks may occur between 14 and 29°C, the optimum range for acute disease is 18–25°C, which overlaps the optimal growth range of 22.5–30°C (Hawke, 2017). Increasing water temperatures from 15 to 18–20°C during experimental challenge exacerbated mortality levels, while decreasing water temperature from 20 to 15°C decreased mortality. Manipulation of temperature may act as a disease control mechanism (Cascarano *et al.*, 2021). While fish survival may increase at temperatures <21°C, survivors can become carriers (Toranzo *et al.*, 2005). Optimal salinity ranges for growth and disease outbreaks are in the range of 1–2.5% and 5–15 ppt, respectively. The pH range for optimal growth is near neutral: 6.47–7.24 (Hawke, 2017).

Traditional inactivated whole-cell or cellular product vaccines have limited effect, most likely due to the intracellular nature of PDSP infection. A successful vaccine will need to robustly stimulate both cellular and humoral immunity to protect against this intracellular bacterium. An extracellular product-enriched bacterin is commercially available in Europe but has had mixed results and requires multiple doses (Andreoni and Magnani, 2014). Experimental vaccines have been developed using subunit formulations of lipopolysaccharide or recombinant proteins and live attenuated strains (Hawke, 2017). PDSP evades phagocytic killing and chemotherapeutics. Antimicrobials may still be employed early in infection, but resistance may develop rapidly because of transferable genetic elements carrying genes for resistance against florfenicol, tetracycline and other antibiotics (Hawke, 2017). Alternative methods of prevention such as probiotics and selective breeding have potential. Resistance to PDSP showed moderate heritability in gilthead sea bream (*S. aurata*) (Vallecillos *et al.*, 2021). Garlic has antibacterial effects on PDSP and has potential to decrease mortalities, but high doses (3%) over 28 days are necessary for significant protection (Guo *et al.*, 2015). In addition to being aware of antimicrobial use restrictions, practitioners should also be aware that some jurisdictions regulate the use of non-drug compounds like garlic in food-producing fish.

5.3.3 Edwardsiellosis

While the original description of the *Edwardsiella* genus was based on isolates collected from humans, it is now primarily associated with fish disease. The genus was composed of three species (*Edwardsiella tarda*, *Edwardsiella hoshinae* and *Edwardsiella ictaluri*) until 2013 when it was reorganized based upon evidence from advanced molecular methods. Two novel species were described: *Edwardsiella piscicida* (Abayneh *et al.*, 2013) and *Edwardsiella anguillarum* (Shao *et al.*, 2015), both containing isolates previously classified as fish-associated *E. tarda* (Reichley *et al.*, 2017). As such, literature regarding *E. tarda* in fish previous to, and even since, this reorganization must be recognized in the frame of the new taxonomy. Of the five currently described species, *E. hoshinae* is the only species not known to cause disease in fish; *E. hoshinae* is associated with avian and reptilian hosts (Griffin *et al.*, 2017). *E. ictaluri* is an important pathogen of catfish and *E. anguillarium* is pathogenic to eels, sea bream, tilapia and grouper (Griffin *et al.*, 2017). The major agent of edwardsiellosis in fish, *E. piscicida*, has been isolated from an increasing number of freshwater, brackish and marine species (Loch *et al.*, 2017; Griffin *et al.* 2020); over 25 host species have been reported and that number is likely an underestimate due to its previous classification under *E. tarda* (Buján *et al.*, 2018). Classical *E. tarda* is still a recognized fish pathogen but causes disease in a wider range of hosts including birds, reptiles, amphibians, humans and other mammals. Humans have been infected with *E. tarda* while handling or ingesting fish or shellfish, making it a zoonotic agent of concern. Geographically, edwardsiellosis has been reported on all seven continents (Griffin *et al.*, 2020).

Economic losses due to *Edwardsiella* spp. are substantial. The recent reclassification suggests that *E. piscicida* is the most problematic agent of edwardsiellosis in global finfish aquaculture, but the wide host and geographic range of *E. piscicida* and *E. tarda* make them both pathogens of concern. Epizootics caused by *E. piscicida* have been reported in Japan, China, the USA, and Northern and Southern Europe, and the disease is still expanding in reported range (Griffin *et al.*, 2020). Edwardsiellosis is considered a warmwater disease. High temperatures and high concentration of organic material are associated with increased disease risk (Griffin *et al.*, 2020). There is potential for shifting climate patterns to impact edwardsiellosis, but this impact has not been clearly defined.

Common clinical signs of edwardsiellosis include erratic swimming, bottom dwelling, anorexia, haemorrhage of skin and internal organs, dermal ulcerations, discoloration of skin and coelomic distension. *E. tarda* and *E. piscicida* had similar clinical signs in an outbreak of multiple edwardsiellosis in farmed barramundi (*L. calcarifer*) (Loch *et al.*, 2017). External signs included ventral haemorrhage, gill pallor, and fin and oral cavity congestion. Internally, *E. tarda*-infected fish had hepatic pallor with mottling, gastrointestinal erythema and swim bladder haemorrhage. In contrast, *E. piscicida* presented with splenomegaly, renomegaly, and small whiteish miliary nodules in liver, spleen and kidney, supported histologically by granulomatous splenitis, hepatitis and nephritis with intrahistiocytic intracytoplasmic bacteria (Loch *et al.*, 2017). Both species can cause acute and chronic mortality with characteristic septicaemic disease signs, and both have been linked to granulomatous responses in affected fish (Loch *et al.*, 2017).

Isolated *Edwardsiella* spp. can be cultured on a variety of media, including TSA with or without 5% bovine or sheep blood, marine agar, brain heart infusion agar (BHIA) and Mueller–Hinton medium. Differential media for *Enterobacteriaceae* enrichment such as *E. tarda* agar, MacConkey or Salmonella Shigella agar can also be used (Buján *et al.*, 2018). The optimal temperature for growth ranges from 27 to 37°C (Abayneh *et al.*, 2013; Griffin *et al.*, 2020). Small, whitish pinpoint colonies will form within 24 h of incubation on conventional media. Phenotypically, *Edwardsiella* spp. are short, variably motile, Gram-negative rods that are catalase-positive and oxidase-negative. Accurate identification of isolates to at least the species level is crucial to understanding the epidemiology of edwardsiellosis. This has been complicated by the high phenotypic and genetic similarities between some species, such as *E. tarda* and *E. piscicida*. Whole-genome sequencing is the most reliable taxonomic tool, but MLSA offers fine resolution, and even the *gyrB*, *sodB* and *dnaJ* genes alone have higher discriminatory power than the 16S rRNA gene (Buján *et al.*, 2018; Griffin *et al.*, 2020). Primers specific to *E. piscicida* were adapted into a multiplex PCR assay that can discriminate the fish-associated *Edwardsiella* spp. Repetitive sequence-based PCR and AFLP were able to differentiate strains at the intraspecies level, while Enterobacterial Repetitive Intergenic Consensus-PCR lacked resolution (Buján *et al.*, 2018).

Edwardsiella spp. are abundant in aquatic environments worldwide, including freshwater, brackish and marine systems. The bacteria may exist in either planktonic or biofilm forms. Intestine and abraded skin are the most likely sites for bacterial colonization. Carrier fish and other animals have been implicated in transmission and dissemination particularly for *E. tarda*. For example, an outbreak in Brazil affected both fish and resident aquatic birds (Miniero Davies *et al.*, 2019). High temperatures, poor water quality and high levels of organic matter favour edwardsiellosis and outbreaks occur during warmer months in most regions. Fluctuating temperatures may favour disease (Griffin *et al.*, 2020). A range of genetic factors involved in environmental adaptation and survival have been proposed, including the molecular switches *RpoS* and *EsrB* which regulate planktonic to biofilm-forming shifts (Leung *et al.*, 2022).

Edwardsiella spp. can survive and multiply in *Edwardsiella*-containing vacuoles (ECVs) within host phagocytic or epithelial cells, complicating treatment and prevention (Leung *et al.*, 2022). A number of antibiotics are effective against most strains of *Edwardsiella* spp., but plasmid-mediated antimicrobial resistance is a concern. *E. tarda* has potential for carrying and transferring antimicrobial resistance genes within the water and soil microbiomes (Leung *et al.*, 2022). There is no commercial vaccine available for the species of *Edwardsiella* causing disease in brackish and marine fish, but an inactivated *E. ictaluri* vaccine has been licensed in some countries for catfish. Vaccine development is an active area of research and experimental formulations of bacterins, live attenuated, recombinant protein and DNA vaccines are under investigation (Buján *et al.*, 2018). Studies on probiotics or selective breeding are limited.

5.3.4 Tenacibaculosis

Tenacibaculosis is a marine-restricted disease referred to by a variety of different names in different host species, such as 'saltwater columnaris disease', 'gliding bacterial disease', 'yellow mouth' or 'black patch necrosis'. The primary aetiological agent of tenacibaculosis, *Tenacibaculum maritimum*, has been assigned several different names historically, previously being described as *Flexibacter marinus*, *Flexibacter maritimus* and *Cytophaga marina* (Boerlage *et al.*, 2020). While *T. maritimum* is the well-characterized agent of tenacibaculosis, several additional members of the genus have been associated with disease in fish, including *Tenacibaculum discolor*, *Tenacibaculum dicentrarchi*, *Tenacibaculum finnmarkense*, *Tenacibaculum gallacium*, *Tenacibaculum piscium* and *Tenacibaculum solae*. The true pathogenic potential and range of most of these agents remain to be fully explored. It is likely more species will be proposed with the advent of molecular genomics.

All marine fish are considered susceptible to tenacibaculosis. *T. maritimum* has been reported to cause disease in over 30 different species worldwide, including salmonids, sole, sea bass, turbot, bream, halibut and sardines (Fernández-Álvarez and Santos, 2018; Nowlan *et al.*, 2020). The other agents of tenacibaculosis have less expansive known ranges. *T. discolor* and *T. solae* were first isolated from diseased sole (*S. senegalensis*) and *T. gallacium* from holding tanks for turbot (*Psetta maxima*). All three species are virulent in sole and turbot (Piñeiro-Vidal *et al.*, 2012). Outbreaks of *T. solae* in wedge sole (*Dicologoglossa cuneata*), brill (*Scophthalmus rhombus*), sea bass (*D. labrax*) and wrasse have also been reported (Fernández-Álvarez and Santos, 2018). *T. dicentrarchi* was isolated from diseased European sea bass (*D. labrax*) in Spain and later from outbreaks in Atlantic salmon (*Salmo salar*) and cod (*Gadus morhua*) in Chile and Norway, as well as red conger eel (*Genypterus chilensis*), sea bass (*D. labrax*), lumpfish (*Cyclopterus lumpus*) and wrasse species (Piñeiro-Vidal *et al.*, 2012; Habib *et al.*, 2014; Avendaño-Herrera *et al.*, 2016). Virulence of *T. dicentrarchi* in Atlantic salmon and rainbow trout was confirmed by experimental challenge (Avendaño-Herrera *et al.*, 2016). *T. finnmarkense* appears the primary agent of tenacibaculosis in Norwegian Atlantic salmon, from which it was first described along with *T. piscium* (Småge *et al.*, 2018; Olsen *et al.*, 2020). The host range of *T. finnmarkense* includes lumpsuckers (*C. lumpus*), coho salmon (*Oncorhynchus kisutch*), cleaner fish (*Symphodus melops*), cod (*G. morhua*) and halibut (*Hippoglossus hippoglossus*) (Nowlan *et al.*, 2020). *T. piscium* has been isolated from skin ulcers of several host species but pathogenic potential has not been confirmed (Olsen *et al.*, 2020).

While there is no global report on the cost of tenacibaculosis, damages from the disease are estimated to amount to millions of dollars annually. In addition to direct mortalities, skin lesions caused by tenacibaculosis reduce commercial value and may lead to culling of stocks for welfare concerns (Småge *et al.*, 2017). Skin lesions may be colonized by other opportunistic pathogens such as *Vibrio* spp. (Nowlan *et al.*, 2020). In Canada, the annual cost associated with outbreaks is estimated to be CAN\$1.8 million (Nowlan *et al.*, 2020). The economic importance of tenacibaculosis appears to be increasing. Warming waters may increase risk of outbreaks, as *T. maritimum* increases in prevalence during warmer seasons (Downes *et al.*, 2018). Other biotic factors that can influence risk, such as algal or sea jelly blooms, may also play a more significant role as plankton distributions shift (Ferguson *et al.*, 2010; Apablaza *et al.*, 2017).

Tenacibaculum spp. are Gram-negative, aerobic, rod-shaped bacteria that generally exhibit gliding motility. They form yellow colonies that may be adherent to agar and are catalase- and oxidase-positive. *Tenacibaculum* spp. may be recovered on marine agar or *Flexibacter maritimus* medium in addition to other non-selective, low-nutrient media with some modifications, such as *Cytophaga* or tryptone–yeast extract–salts (TYES) agar with seawater. *Tenacibaculum* spp. grow slowly and may be outcompeted by coexisting bacteria, selection can be enhanced by including an aminoglycoside into agar preparations (Nowlan *et al.*, 2020). The different species of *Tenacibaculum* share similar morphologies and many growth requirements, but they may vary in haemolytic profile and temperature and salinity range (Olsen *et al.*, 2020). All characterized species of concern can grow between 15 and 20°C, but some may grow at as low as 4°C or as high as 40°C. Salinity tolerance ranges from 0 to 10% NaCl, with fish disease-associated strains tolerating within 1.8–3.5% (Nowlan *et al.*, 2020).

Depending on strain and host, clinical signs of tenacibaculosis may include ulcerative skin lesions, frayed fins, tail erosions, gill necrosis, oral erosions or scale loss with general lethargy, anorexia or increased respiratory rate. Histopathological assessment of dermal lesions shows necrosis and low-grade inflammation with infiltration of the bacteria and possible epidermal loss (Fig. 5.4). Yellowish bacterial mats or margins may be visible grossly with external lesions (Boerlage *et al.*, 2020). Organs may be pale (Fernández-Álvarez and Santos, 2018). Long, thin rods can be visualized in wet mounts of affected tissue. Current identification methods range in their ability to distinguish between the species of *Tenacibaculum* and are largely culture dependent. Biochemical tests such as the API ZYM and API 50 CH strips may not differentiate to the species level (Fernández-Álvarez and

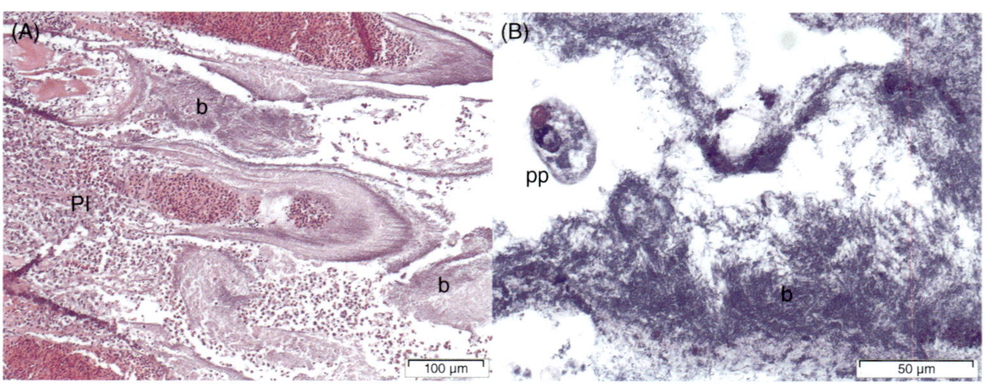

Fig. 5.4. Histopathological changes in gill of cultured spotted rose snapper (*Lutjanus guttatus*) infected with *Tenacibaculum* spp.-suspect. (A) Large areas of necrosis admixed with pyogranulomatous inflammatory infiltrate (PI). The infiltrate is dominated by macrophages and heterophils and the lesions are associated with filamentous bacteria (*Tenacibaculum* spp.) (b). Haematoxylin and eosin stain. (B) Section of gills highlighting the large aggregates of filamentous bacteria (b) and a ciliated protozoan parasite morphologically consistent with *Brooklynella* spp. (pp). Giemsa stain. (Images courtesy of Dr Juan A. Morales.)

Santos, 2018). Multiple PCR methods have been developed to detect *Tenacibaculum* spp. from pure or mixed cultures and tissues, including several based on the 16S rRNA gene. A 16S RT-PCR method coupled with melting curve analysis was able to differentiate *T. maritimum* in fish and seawater samples (Fernández-Álvarez and Santos, 2018; Nowlan *et al.*, 2020). Multiplex PCRs have also been developed for various combinations of the subspecies but need further validation. More species and isolates need to be tested to confirm if MALDI-TOF-MS can distinguish between closely related and newly described species. MLSA appears to be the most reliable method for species-level differentiation (Habib *et al.*, 2014). Both MLSA and MALDI-TOF-MS typing have suggested some degree of geographic clustering, without significant long-distance dissemination linked to international fish movements (Habib *et al.*, 2014; Bridel *et al.*, 2020).

Clarification is needed regarding *Tenacibaculum* spp. transmission and pathogenesis. The genus is widespread in marine environments and may exist in the water column as free-living bacteria or as sessile biofilms associated with biotic and abiotic surfaces (Levipan *et al.*, 2019). Fish-pathogenic *Tenacibaculum* spp. have been isolated from sediment, tank surfaces and water samples and from macroalgae and healthy fish epithelia (Levipan *et al.*, 2019; Boerlage *et al.*, 2020). Co-infections with other agents of skin and gill disease are common (Downes *et al.*, 2018; Fernández-Álvarez and Santos, 2018). They are, therefore, considered mostly opportunistic pathogens although virulence is likely strain specific. Transmission has been suggested through seawater or directly between hosts, although little to no fish-to-fish transfer was observed in a *T. finnmarkense* outbreak and free-living *T. maritimus* has been shown to be outcompeted by microbiota in seawater (Småge *et al.*, 2018; Levipan *et al.*, 2019). The body surface is the primary site of infection and *T. maritimum* has been shown to have strong attachment to, and survival on, external mucosal surfaces (Mabrok *et al.*, 2016). The oral cavity, gills and skin are all potential sites of infection. *Tenacibaculum* spp. have been isolated from several proposed vector species, including sea jellies, sea lice (*Lepeophtheirus salmonis*) and bioagents of sea lice control such as lumpsuckers (Ferguson *et al.*, 2010; Apablaza *et al.*, 2017; Småge *et al.*, 2017; Downes *et al.*, 2018). Small jellies may pass into sea cages and damage fish gills and skin to provide easy points of access for *Tenacibaculum* spp. infection. Whether the jellies also carry and introduce pathogenic *Tenacibaculum* spp. to fish populations remains to be determined (Ferguson *et al.*, 2010; Småge *et al.*, 2017). Ocean warming has been associated with increasing populations of sea jellies globally and further work on their role in tenacibaculosis transmission is warranted.

As *Tenacibaculum* spp. are generally considered opportunistic pathogens, host resilience and environmental stressors play an important role in disease. High salinity, elevated ammonia, and physical or toxic insults increase outbreak risk. High water temperatures (>15°C) are considered a risk factor for tenacibaculosis but changes either increasing or decreasing the typical water temperature may encourage infection (Boerlage *et al.*, 2020; Nowlan *et al.*, 2020). Detections of *T. maritimum* in Atlantic salmon are significantly correlated with temperature and show direct seasonality, with increases in summer and autumn (Downes *et al.*, 2018). *Tenacibaculum* spp. levels are positively correlated with elevated levels of organic material and phytoplankton blooms (Apablaza *et al.*, 2017; Småge *et al.*, 2017).

Control of tenacibaculosis relies on antimicrobials. Florfenicol is most widely applied because it is effective in low doses and demonstrates rapid pharmacokinetics in studied species (Nowlan *et al.*, 2020). Vaccination is a desirable method for disease control but the diverse nature of the *Tenacibaculum* genus and coexistence of multiple pathogenic species in aquatic environments complicate vaccine design. An inactivated *T. maritimum* vaccine has been licensed for use in turbot in Spain and experimental vaccines have been developed to limited effect (Ma *et al.*, 2019). A whole-cell killed *T. finnmarkense* formulation failed to protect Atlantic salmon (*S. salar*) despite stimulating a systemic antibody response (Småge *et al.*, 2018). Stimulation of mucosal immunity will likely be important for this surface-associated pathogen, necessitating different formulations and mucosal administration methods.

5.3.5 Mycobacteriosis

Mycobacteriosis in fish is caused by members of the non-tuberculous *Mycobacterium* genus in the family *Mycobacteriaceae*. The species *Mycobacterium marinum*, *Mycobacterium chelonae* and *Mycobacterium fortuitum* are the best characterized piscine pathogens within the genus, with *M. marinum* as the most significant agent. However, the number of described species associated with fish disease is expanding. The full diversity of fish-pathogenic *Mycobacterium* spp. is likely underappreciated. Disease has been reported in an extensive range of marine, brackish and freshwater fish worldwide in cold-, temperate- and warmwater habitats. Most teleosts are considered susceptible to mycobacteriosis and several agents of the disease in fish also pose a zoonotic risk (Gauthier and Rhodes, 2017). *M. marinum* is the cause of 'fish tank granuloma' skin infections in humans and may cause invasive disease in immunocompromised individuals (Mugetti *et al.*, 2021).

Mycobacteriosis is associated with significant losses in aquaculture. Acute forms of disease have led to outbreaks of high mortality in pen-reared turbot (*Scophthalmus maximus*), sea bream (*S. aurata*), sea bass (*D. labrax*) and Atlantic salmon (*S. salar*) (Whipps *et al.*, 2020). The chronic form may cause chronic low-level mortality, decreased survival rates and reduced growth rates. The management of mycobacteriosis is expensive and no effective treatments or preventive strategies exist. Infected systems must be emptied and disinfected when feasible, but the widespread environmental nature of *Mycobacterium* spp. complicates eradication from open systems. It is likely *Mycobacterium* spp. will expand in range and prevalence with increases in water temperature that favour bacterial growth and host immunosuppression.

Non-tuberculous *Mycobacterium* species are divided into fast-growing and slow-growing groups. Cultures of fast-growing *Mycobacterium* spp. like *M. fortuitum* may grow within 5–7 days while slow-growing strains like *Mycobacterium shottsii* and *Mycobacterium pseudoshottsii* can take >6 weeks (Gauthier and Rhodes, 2017). Culture requires specialized media like Middlebrook 710 or Löwenstein–Jensen agar to support *Mycobacterium* spp. and limit overgrowth from other environmental bacteria. Sample collection from internal sites using aseptic techniques is preferred. The temperature optimum differs between species and incubation around 20°C or the environmental temperature of the fish host is recommended for suspected mycobacteriosis (Gauthier and Rhodes, 2017). *Mycobacterium* spp. are Gram-positive, non-motile bacilli to rod-shaped bacteria containing 3-hydroxy long-chain mycolic fatty acids, which makes mycobacteria acid-fast (Martínez-Lara *et al.*, 2021). Some species, like *M. marinum*, are photo- or scotochromogenic and produce yellow-orange pigments under exposure to light or constitutively. Others, like *M. shottsii*, are non-pigmented. The high diversity of the genus and close genetic and phenotypic relationships between species have made identification to species level difficult and no biochemical test or single gene assay is known to differentiate between all known *Mycobacterium* species. The *hsp65* gene is more discriminatory than the 16S rRNA gene, but MLSA most reliably discriminates strains and species (Gauthier and Rhodes, 2017; Mugetti *et al.*, 2021). A combination of histology, culture and such molecular methods is useful for detecting and identifying *Mycobacterium* spp. in fish tissues.

Mycobacteriosis commonly causes chronic mortality in a population with episodes of acute disease and higher mortality. The chronic presentation is characterized by granulomatous inflammation, while the acute form typically causes significant bacteraemia (Gauthier and Rhodes, 2017). External clinical signs may not be present but can include non-specific changes like lethargy, anorexia, loss of equilibrium or buoyancy control, and exophthalmia. A lack of granulomas but high infiltration of intracellular and extracellular acid-fast bacteria with widespread necrosis are typical of the acute form of mycobacteriosis. In the chronic form inflammation appears as greyish-white nodules in muscles and visceral organs, such as kidney, liver and spleen but all tissues can be affected (Gauthier and Rhodes, 2017; Martínez-Lara *et al.*, 2021) (Fig. 5.5). Granulomas are composed of epithelioid macrophages surrounded by inflammatory leucocytes and may be cellular or necrotic with caseation (Gauthier and Rhodes, 2017). Mycobacterial infections may remain dormant for years. Reactivation can lead to rapid growth, overflow and dissemination of granuloma-encapsulated bacteria (Martínez-Lara *et al.*, 2021).

Transmission of *Mycobacterium* spp. has been suggested through the oral route. Oral transmission has been demonstrated experimentally

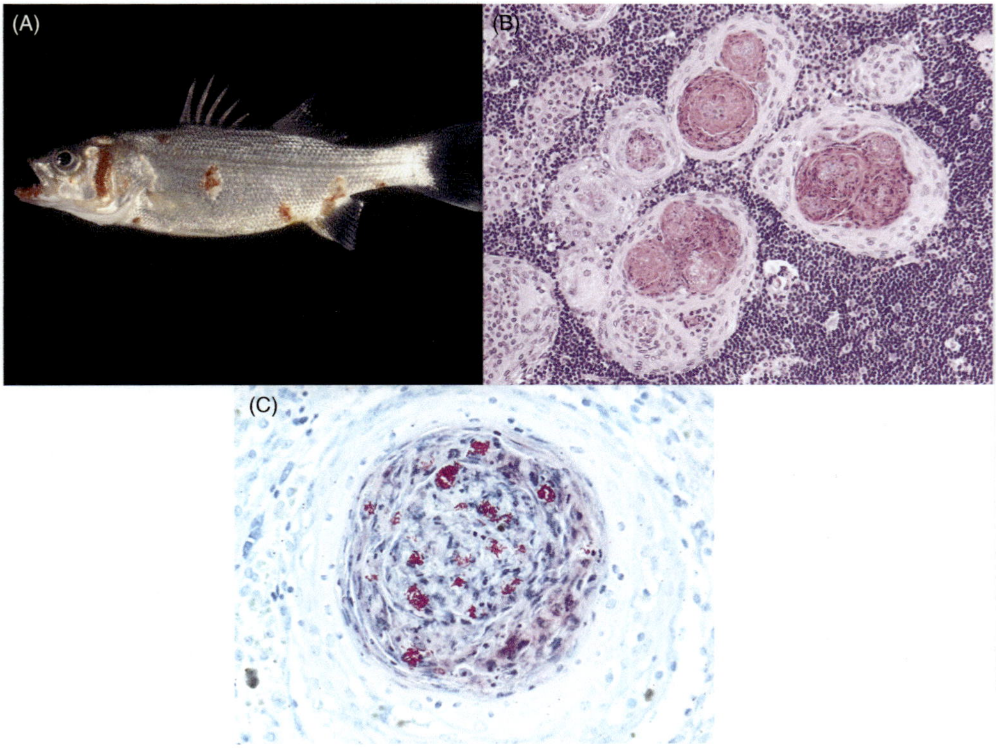

Fig. 5.5. Piscine mycobacteriosis. (A) Hybrid striped bass with skin ulcerations associated with systemic *Mycobacterium marinum* infection. (B) Haematoxylin and eosin-stained histological section of head kidney from a largemouth bass with multifocal mycobacterial granulomas formed by concentric layers of epithelioid macrophages surrounding central necrotic cores. (C) Histological section of spleen from a hybrid striped bass. Ziehl–Neelsen stain reveals numerous acid-fast positive bacteria within a well-organized granuloma. (Images courtesy of Dr Alvin C. Camus.)

through feeding contaminated live feeds and infected carcasses to zebrafish and salmon, respectively (Chang *et al.*, 2019a). Piscine-pathogenic *Mycobacterium* spp. are present in water, sediments and periphyton, and infected fish can shed a large number of bacteria (Mugetti *et al.*, 2021). Extra-host persistence in aquatic systems is facilitated by robust biofilm formation (Whipps *et al.*, 2020). Zebrafish were found to be infected by *M. chelonae* within 2 weeks of introduction to tanks where biofilms had been allowed to form (Chang *et al.*, 2019b). Uptake and persistence by free-living phagocytic protozoa have also been suggested to enhance bacterial dispersal (Whipps *et al.*, 2020). *Mycobacterium* spp. have also been found in commercial fish feed samples and ingredients, although the feed-associated species was determined not to be the same species causing an outbreak at a relevant farm (Mataragka *et al.*, 2022).

Enhanced mortality from mycobacteriosis is associated with host stress and poor water quality. Concentrations of *Mycobacterium* spp. in the water column are positively correlated with temperature, total nitrogen levels and total phosphorus, but negatively correlated with dissolved oxygen levels (Groner *et al.*, 2018). Increased temperatures increase mortality in wild fish but a direct relationship between high mycobacterial concentrations and fish health under warming conditions has not been drawn. It is likely that high temperatures, hypoxia and mycobacterial infection interact to contribute to mortality. Acute high-level mortalities are associated with warmwater, high-density systems (Whipps *et al.*, 2020).

There is no efficacious treatment nor sufficiently protective vaccine available for piscine mycobacteriosis. Rifampin and ethambutol have been used with limited success but are not approved for use in cultured fish in many countries (Gauthier and Rhodes, 2017; Whipps *et al.*, 2020). *Mycobacterium* spp. easily generate resistance to antibiotics and may additionally avoid their effects by residence in host phagocytic cells. The intracellular aspect of mycobacterial infection also complicates vaccine development because strong stimulation of both cellular and humoral immunity is necessary for protection. Experimental live attenuated and DNA vaccines have been developed against *M. marinum* in zebrafish and hybrid striped bass, respectively, but were unable to provide complete protection (Whipps *et al.*, 2020). With such limited options, control of disease relies mainly on quarantine and screening, maintaining good water quality and stocking conditions, and rapid and thorough response to disease outbreaks. Destruction of infected stocks and decontamination of the system are necessary to eradicate the pathogen, but *Mycobacterium* spp. are resistant to some commonly used disinfectants. Ethyl alcohol (50–70%), benzyl-4-chlorophenol/phenylphenol and sodium chlorite formulations are most effective against *M. marinum*. Sodium hypochlorite required extended contact times to reduce bacterial concentrations (Mainous and Smith, 2005).

5.3.6 Streptococcosis

Streptococcosis is used to describe diseases caused by Gram-positive lactic acid bacteria (order Lactobacillales), encompassing the genera *Streptococcus*, *Lactococcus*, *Vagococcus* and *Enterococcus*. *Streptococcus iniae* is the major agent of piscine streptococcosis and is pathogenic to nearly 100 different fish species (Heckman, 2021). *Streptococcus agalactiae* is most prevalent in tilapia, but both species can infect farmed brackish and marine species from a range of habitats worldwide (Shoemaker *et al.*, 2017). *Streptococcus parauberis* has caused outbreaks of high mortality in olive flounder and turbot in Asia and *Vagococcus salmoninarum* has caused large losses of Atlantic salmon in Europe (Mishra *et al.*, 2018). *Streptococcus dysgalactiae*, *Lactococcus garvieae* and *Lactococcus petauri* are emerging diseases in aquaculture impacting an increasing number of cultured marine and freshwater fish over an expanding geographic range (Abdelsalam *et al.*, 2013; Meyburgh *et al.*, 2017; Shahin *et al.* 2021b). All agents of piscine streptococcosis are considered to have zoonotic potential and transmission from infected fish has been clearly implicated in clinical cases of *S. iniae*, *S. agalactiae*, *S. dysgalactiae*, *L. garvieae* and *L. petauri* infections of humans (van Samkar *et al.*, 2016; Brouwer *et al.*, 2017; Meyburgh *et al.*, 2017; Mishra *et al.*, 2018; Porcellato *et al.*, 2021).

Outbreaks of streptococcosis can be devastating and are associated with high mortality rates. The annual cost of the disease to global aquaculture is accordingly high and estimated to amount to billions of dollars (Shoemaker *et al.*, 2017). Economic costs stem from direct losses due to mortality in market-size and larval seed fish as well as management costs from increased labour and treatment expenditures. While outbreaks can occur in cold and temperate systems, streptococcosis is primarily a warmwater disease. Rising water temperatures and rapid temporal changes in water temperatures can increase prevalence and severity of outbreaks. It is reasonable to expect streptococcosis to become a disease of increasing significance with climate change.

The aetiological agents of piscine streptococcosis are Gram-positive, non-motile, facultatively anaerobic, and catalase- and oxidase-negative bacteria. The ovoid or spherical cocci form pairs or chains in liquid media and white to off-white, slightly translucent colonies on solid media. These species grow well on nutrient-rich media, such as BHIA or TSA. Supplementation with 5% sheep's blood is suggested for observation of haemolytic profile, which may differ between species and strains. Although *S. iniae* is considered a β-haemolytic streptococcus, strains from certain genetic groups may display α-haemolysis under aerobic conditions (Heckman *et al.*, 2020). Depending on strain type and culture conditions, fish isolates of *S. agalactiae* range from γ- to β-haemolytic and of *S. dysgalactiae* from α- to β-haemolytic (Abdelsalam *et al.*, 2013; Soto *et al.*, 2015). *Lactococcus* spp. are generally less haemolytic and display generally α- to γ-haemolysis, with occasional reports of β-haemolysis (Meyburgh *et al.*, 2017).

Bacterial strain and species impact clinical presentations of disease, as do host type and environment. Streptococcosis typically causes mortality rates of 30–50% but can reach upwards of 70% in acute outbreaks (Bromage and Owens, 2002). Chronic forms of streptococcosis can cause persistent low-level mortality (Chideroli *et al.*, 2017). The common clinical signs of streptococcosis include external lesions and behavioural changes indicative of central nervous system infection. Fish may exhibit lethargy, anorexia, erratic swimming, spiralling or corkscrewing. Uni- or bilateral exophthalmia with corneal opacity and congestion is often present. Other external clinical signs include ulcers on the caudal peduncle, body curvature, coelomic distension, darkened skin and haemorrhage at the base of fins (Shoemaker *et al.*, 2017) (Fig. 5.6).

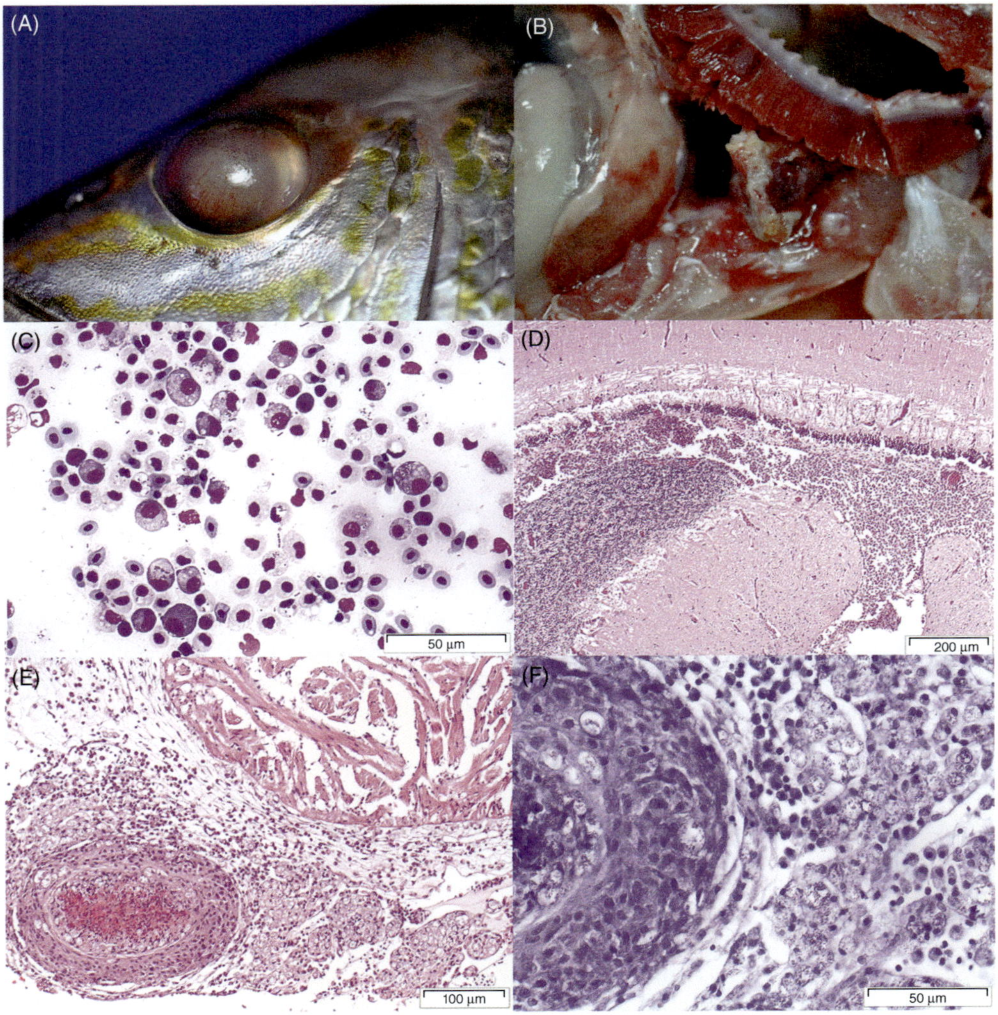

Fig. 5.6. (A) Gross lesions observed in moribund spotted rose snapper (*Lutjanus guttatus*) infected with *Streptococcus iniae* include opaqueness of the eye with severe exophthalmia. (B) Severely infected spotted rose snapper (*L. guttatus*) presenting fibrinous pericarditis, with focal areas of fibrous adhesions covered with white-opaque material. (C) Cytology of heart imprint showing intracellular and extracellular cocci sometimes forming short chains. (D–F) Severe encephalitis and meningitis (D) and pericarditis (E, F), predominated by granulomatous inflammatory infiltrates dominated by macrophages with cytoplasmic vacuoles containing bacteria. Haematoxylin and eosin stain (D, E). Giemsa stain (F). (Images courtesy of Dr Juan A. Morales.)

Mortality may occur without external clinical signs. Internal organs may be congested, and splenomegaly and hepatomegaly are common. The brain may be soft, congested and oedematous (Vendrell *et al.*, 2006; Shoemaker *et al.*, 2017; Abu-Elala *et al.*, 2020). Haemorrhage in external and internal organs is especially common in *L. garvieae* and *L. petauri* infections. Histopathology is consistent with systemic disease. Infected fish display meningitis, encephalitis, myocarditis, pericarditis, epicarditis, endocarditis, and endo- or panophthalmitis (Shoemaker *et al.*, 2017; Abu-Elala *et al.*, 2020). Gram-positive cocci are observable in affected tissues, either in extracellular spaces or within macrophages (Baums *et al.*, 2013; Soto *et al.*, 2015; Abu-Elala *et al.*, 2020).

Recovery or observation of bacteria morphologically consistent with streptococci in tissues of diseased fish necessitates further biochemical and molecular analysis for identification. The combination of high intraspecific diversity and common intraspecific genetic and phenotypic characteristics may lead to misidentification from biochemical profiles or single gene analyses. Sequencing of the 16S rRNA gene, 16S–23S intergenic regions or regions targeted in specific species generally allows characterization to the species level but is culture dependent (Meyburgh *et al.*, 2017; Mishra *et al.*, 2018; Maekawa *et al.*, 2020). Quantitative PCR (qPCR) methods have been developed for better sensitivity and detection from tissue and environmental samples (Shoemaker *et al.*, 2017). Resolution to the strain level may have important implications for disease management and requires multigene analysis. Pulsed-field gel electrophoresis (PFGE) has high discriminatory power and has been used to type *S. iniae*, *S. agalactiae*, *S. dysgalactiae* and *L. garvieae* fish isolates (Weinstein *et al.*, 1997; Tsai *et al.*, 2012; Costa *et al.*, 2014; Sun *et al.*, 2016). MLSA and the closely related multilocus sequence typing (MLST) are preferable as more replicable methods that can be combined with whole-genome analysis (Jones *et al.*, 2003; Ferrario *et al.*, 2013; Heckman *et al.*, 2020). This is especially important for *S. dysgalactiae* and *Lactococcus* spp. which are undergoing taxonomic redefinition based on comprehensive molecular analysis of fish-associated strains.

The agents of streptococcosis are transferred horizontally through the water by the faecal–oral route. *Streptococcus* spp. have been reported in water and sediments adjacent to fish farms (Shoemaker *et al.*, 2017) and both *Lactococcus* and *Streptococcus* spp. have been demonstrated to form biofilms which may enhance persistence in aquatic environments (Chideroli *et al.*, 2017; Akoğlu, 2020; Heckman and Soto, 2021; Shahin *et al.* 2021b). *S. iniae* and *S. agalactiae* have been isolated from skin and tissues of inapparently infected farmed and wild fish, implicating carrier fish in the spread of streptococcosis (Zlotkin *et al.*, 1998; Bromage and Owens, 2002; Sun *et al.*, 2016). Ingestion of contaminated feed and cannibalism of infected conspecifics are additional routes that may introduce or perpetuate disease (Kim *et al.*, 2007; Anshary *et al.*, 2014). Shedding of bacteria by orally infected fish may result in sufficient concentrations of the pathogen in water for further infection through skin, gills or nares (Bromage and Owens, 2002; Vendrell *et al.*, 2006; Baums *et al.*, 2013; Shoemaker *et al.*, 2017). Potential vectors for streptococcal species include ichthyophagous birds and external parasites (Vendrell *et al.*, 2006; Chideroli *et al.*, 2017; Shoemaker *et al.*, 2017).

Environmental parameters such as temperature, salinity and water quality influence transmission and pathogenesis of streptococcal species. Temperature and salinity have been shown to have a significant impact on persistence of planktonic and biofilm-associated *S. iniae*. Bacteria persisted longer in marine waters at cooler temperatures (Heckman and Soto, 2021). In contrast, warmer temperatures may increase host susceptibility and bacterial virulence. Both rate and extent of temperature changes affect pathogenesis. Fluctuations in water temperature over a short time can lead to host immunosuppression and decreased resistance to *S. iniae*; mortality events in wild reef fish and farmed fish caused by this pathogen were attributed to abrupt rises in temperature (Genin *et al.*, 2020; Young *et al.*, 2020). Such mortality events may intensify with global climate change. Low dissolved oxygen, high pH (>8) and high ammonia also predispose fish to streptococcosis (Shoemaker *et al.*, 2017).

Good management protocols are the first line of defence against streptococcosis. Maintaining optimal water quality and stocking densities limits host stress and proper biosecurity and screening can prevent introduction of some sources of streptococcal species. Dietary supplementation with probiotics and immunostimulants, such as garlic and β-glucans, can enhance resistance to streptococcal infection (Guo *et al.*, 2012; Pilarski *et al.*, 2017). Selective breeding shows potential for propagating host varieties with reduced susceptibility to *S. iniae* (LaFrentz *et al.*, 2016). Rapid response to an outbreak is crucial. Promptly removing moribund and dead fish can reduce bacterial proliferation and pathogen load. Medicated feeds should be employed before infected populations become anorectic. Several antimicrobials are used for streptococcosis in cultured fish, including florfenicol, oxytetracycline and erythromycin. Aquaflor® (florfenicol; Merck Animal Health) is the only approved antibiotic for this disease in the USA and can only be used under a veterinary feed directive (VFD). Disease-induced anorexia and the facultatively intracellular nature of streptococci can limit effectiveness of these antimicrobials, necessitating multiple treatments and favouring development of carrier fish. Development of multidrug resistance in streptococci, lactococci and enterococci is a serious and growing issue for animal and human health, and there are numerous reports of increasing antimicrobial resistance in fish isolates (Chideroli *et al.*, 2017; Meyburgh *et al.*, 2017; Abu-Elala *et al.*, 2020).

Vaccination can reduce reliance on antimicrobials and a variety of vaccines have been formulated against different aetiological agents of streptococcosis. Several commercial vaccines are available, but there is still a demand for vaccines that can protect against diverse streptococcal strains and species in the long term (Mishra *et al.*, 2018; Ma *et al.*, 2019). In countries where no commercial vaccines for streptococcosis are available, such as the USA, autogenous bacterins can be used for limited protection against homologous strains under veterinary guidance. Experimental vaccines seeking to improve upon traditional, whole-cell killed preparations have been developed for many aetiological agents of streptococcosis, including modified bacterins, live attenuated vaccines, recombinant subunit vaccines, DNA vaccines and bacterial ghosts (Ma *et al.*, 2020; Heckman *et al.*, 2022). A successful vaccine will stimulate both arms of immunity and offer protection against diverse strains and potentially species of *Streptococcus* or *Lactococcus*. Such a cross-protective vaccine may be better informed by whole-genome and bioinformatic analysis.

5.4 Parasitic Diseases

Parasites can be exceptionally problematic in intensive culture systems, as high densities of confined fish increase the probability for host–parasite interactions (Nowak, 2007). Increased average global temperatures are expected to exacerbate parasitic disease in warmwater marine and brackish-water fish. Climate change will significantly impact these environments, most notably coastal regions where cage mariculture is prevalent. In addition to potential for increased geographic ranges of warm- and temperate-water parasitic fish pathogens, there is potential for increased biofouling of net pens, as fouling organisms (macroalgae, barnacles, etc.) also increase their geographic ranges and growing seasons (Poloczanska and Butler, 2010). Biofouling in marine aquaculture is a significant impediment to cage culture, although true economic impacts of biofouling are complicated and the costs to global mariculture are uncertain (Bannister *et al.*, 2019). Regardless, without adequate de-fouling programmes, build-up on net pens may provide substrates in sea cages that can be exploited by parasites which previously required close proximity to the sea floor to complete their life cycles. Still, the impacts of a dynamic climate on temperate- and warmwater mariculture is speculative, as there is presently a knowledge gap regarding downstream effects of climate change on ecology of parasitic diseases and what impacts, if any, environmental change will have on host–parasite dynamics in warmwater marine and brackish-water fish (Godwin *et al.*, 2020). Multiple parasitic organisms have been attributed to significant losses in temperate- and warmwater cage culture. If contemporary models indicating significant increases in global temperatures hold true and current trends continue uninterrupted, it is conceivable these parasitic agents could increase in prevalence and intensity. Several of these are discussed here.

5.4.1 Apicomplexa

Apicomplexans are one of the most important groups of animal pathogens worldwide, characterized by a unique organelle, the apical complex, which facilitates host cell invasion. They are largely known for diseases they cause in humans and domestic animals. Given their medical, veterinary, economic and agricultural importance, apicomplexans receive a great deal of attention in terms of diagnostic testing and development of prophylactic and treatment regimens.

In fish, high infection rates can lead to anorexia, emaciation, muscle atrophy and poor body condition, sometimes accompanied with mortality (Alvarez-Pellitero and Sitjà-Bobadilla, 2002). However, despite these rare cases, most infections in fish produce limited pathological or clinical signs and go unnoticed on most operations (Alvarez-Pellitero and Sitjà-Bobadilla, 2002). Therefore, many reports of apicomplexan infections in fish are the result of parasitological surveys and not necessarily associated with disease (Alvarez-Pellitero and Sitjà-Bobadilla, 2002; Bahri, 2012; Molnár *et al.*, 2012; Certad *et al.*, 2015; Yang *et al.*, 2015). Due to their relatively low pathogenicity and limited mortality in affected populations, fish apicomplexans receive little attention compared with their terrestrial counterparts and their impacts on marine fish farms are largely unknown (Molnár, 2006; Colorni and Diamant, 2014).

5.4.2 Dinoflagellates/ciliates

Two important parasitic diseases of warmwater marine and brackish-water fish are 'marine velvet' caused by the dinoflagellate *Amyloodinium ocellatum* and 'white spot disease' caused by the ciliate *Cryptocaryon irritans* (Fig. 5.7). Amyloodiniosis has been reported in a wide range of marine and estuarine fish and can spread quickly in crowded systems. Springtime spikes in *A. ocellatum* have been reported in cage-reared sea bream and sharp-snout sea bream cultured in the Adriatic Sea (Mladineo, 2005). The parasite has a broad host and geographic range and can tolerate a range of salinities (10–60 ppt) and temperatures (15–30°C) (Francis-Floyd and Floyd, 2011). *C. irritans* is widely distributed and has been reported in more than 45 species of marine fish. Initially classified as *Ichthyophthirius marinus* due to its morphological congruence with the freshwater fish parasite *Ichthyophthirius multifiliis*, later ultrastructural and molecular studies revealed *C. irritans* to form its own genus (Li *et al.*, 2022). Like *A. ocellatum*, *C. irritans* can tolerate a wide range of temperatures (15–30°C) and salinity (20–40 ppt) (Chu *et al.*, 2001). While there are treatments available (e.g. copper sulfate pentahydrate, formalin, hydrogen peroxide, hyposalinity, etc.), treatment applications in open-ocean pens are logistically complicated. Fortunately, *A. ocellatum* and *C. irritans* are not generally a concern for cage-cultured fish, especially in deep water where there is a significant distance between the net pen and the seabed which precludes efficient completion of the parasite life cycle. The density of infectious stages is significantly reduced by the sheer volume of seawater exchanged through ocean currents in open-water systems (Li *et al.*, 2022).

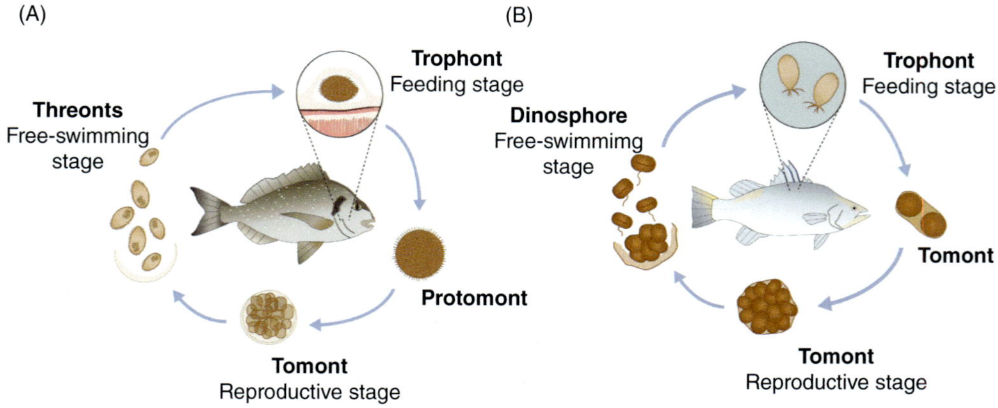

Fig. 5.7. Life cycle diagram of *Cryptocaryon irritans* (A) and *Amyloodinium ocelatum* (B).

However, biofouling on cages can provide substrates for dormant life stages and restrict water flow, which can increase concentrations of infective stages within the pen. It is thought increased global temperatures will expand geographic ranges of many fouling species, which may increase rates of fouling on submerged structures, including aquaculture cages (Poloczanska and Butler, 2010). As such, oceanic eutrophication and climate-driven biofouling may result in greater prevalence of *A. ocellatum* and *C. irritans* in open-water systems if cage biofouling is not managed.

In addition to *A. ocellatum* and *C. irritans*, several species of ciliates have been associated with disease, but these species are often secondary to a greater problem and generally considered opportunistic. *Trichodina* spp., *Brooklynella* spp. and the scuticociliates (*Anophryoides*, *Mesanophrys*, *Miamiensis*, *Philasterides*, *Pseudocohnilembus*, *Tetrahymena* and *Uronema* spp.) are widely distributed protistan ectoparasites affecting a wide range of fish hosts. They are generally indicators of poor water quality, crowding or substandard host conditions. Intensive culture conditions subject fish to various stressors, including infectious disease, which weakens the host and provides windows of opportunity for opportunistic protists.

Trichodinids are common protozoan parasites of marine, brackish-water and freshwater fishes, with more than 300 species reported globally. Generally considered ectocommensal scavengers, they are found most often on gills and skin. Heavy infestations and resultant irritation can lead to epidermal hypertrophy, hyperplasia and destruction of gill architecture, severely inhibiting gill function (Mitchell and Rodger, 2011; Blazhekovikj-Dimovska and Stojanovski, 2020). *Brooklynella* spp. are ciliated protozoan ectoparasites that also target the gills and skin of marine teleosts, resulting in lethargy and reduced feeding activity (Fig. 5.8). The ciliate outwardly resembles the freshwater parasite *Chilodonella*, and while most natural infestations are asymptomatic, catastrophic losses can occur from severe infestations (Diamant, 1998). Scuticociliates are free-living histophagous ciliates inhabiting eutrophic marine coastal waters and have been associated with disease outbreaks in commercial and ornamental fisheries. They are globally distributed in marine aquaculture facilities, causing significant losses in flounder, turbot, sea bass and tuna. They can be particularly damaging to flatfishes, causing ulcerative lesions on the skin and, in severe cases, systemic parasitaemia (Jung and Woo, 2012). While chemotherapeutics are available for treating ciliate infestations, these are mostly applicable to small-scale operations and are often logistically impractical in cage culture. The best approach to controlling ciliates is to maintain optimal fish health through vaccination against primary pathogens, maintaining optimal water quality, minimizing crowding and preventing biofouling.

5.4.3 Microsporidians

Microsporidia are a diverse, cosmopolitan group of obligate intracellular parasites reported from a wide variety of aquatic and terrestrial hosts, ranging across all animal kingdoms. All microsporidia possess a unique polar tube, which facilitates infection of host cells with varying degrees of host specificity (Lom and Dyková, 2005). Microsporidians comprise over 200 genera, the majority of which infect aquatic hosts (Stentiford *et al.*, 2019). Given their global distribution in marine, fresh and brackish waters, coupled with their diverse host range, microsporidians have garnered significant interest as pathogens of wild and cultured fish (Lom, 2002; Rodriguez-Tovar *et al.*, 2011). Fish are exposed to infective spores by consumption of infected fish or spores in the water column, which is facilitated by high stocking densities. Horizontal transmission may be exacerbated by fouling, which can limit water exchange within the cage, leading to increased parasite densities within the pen. Several compounds have been proposed as treatments for microsporidian infections in fish, although most are experimental. Probably the most successful drug for treatment of microsporidiosis in fish is fumagillin, which is best known as an agent to combat the microsporidian *Nosema apis* in honeybee colonies (Kent *et al.*, 2014).

Fish microsporidia embed themselves directly in host cell cytoplasm and can be classified in two general categories: those that form xenomas and those that do not (Lom and Dyková, 2005). Xenomas are tumour-like growths arising from

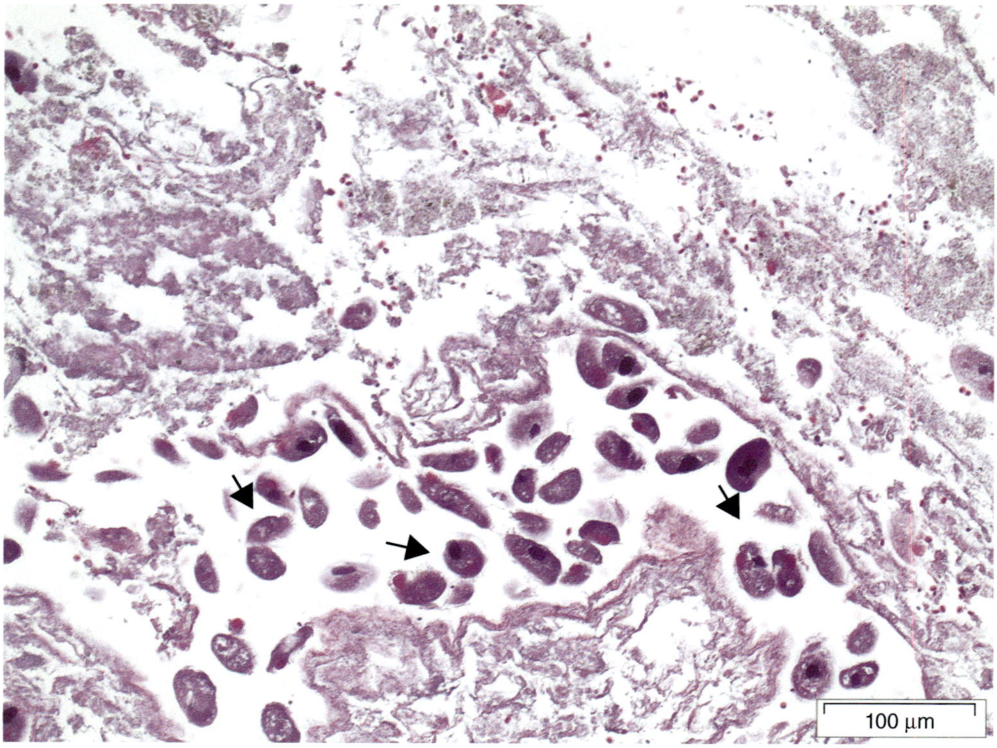

Fig. 5.8. Histopathological changes in protozoal (*Brooklynella* spp.-suspect) and bacterial (*Streptococcus iniae* and *Tenacibaculum* spp.-suspect) co-infected gill of cultured spotted rose snapper (*Lutjanus guttatus*). Pyogranulomatous inflammatory inflammation and necrosis associated with ciliated protozoan parasites and filamentous bacteria morphologically consistent with *Brooklynella* spp. (arrows) and *Tenacibaculum* spp., respectively. (Image courtesy of Dr Juan A. Morales.)

hypertrophic host cells which can cause significant economic losses due to the unsightly appearance of infected fish, particularly if affecting musculature (Lee *et al.*, 2004; Marzouk *et al.*, 2010). Comparably, non-xenoma forming microsporidia do not induce cellular hypertrophy.

Beko disease is a seasonal issue for cage-reared yellowtail (*S. quinqueradiata*) and amberjack (*Seriola dumerli*) in Japanese mariculture (Yokoyama *et al.*, 2011). The causative agent is *Microsporidium seriolae*, which forms cyst-like bodies and myoliquefaction in trunk muscle, resulting in a concave body surface and poor host fitness (Egusa, 1982). Heavy infections result in emaciation and death, but even mild infections can lead to reduced marketability due to consumer rejection of compromised fillets (Yokoyama *et al.*, 2011).

There have been multiple reports of microsporidian-induced losses in cultured gilthead sea bream (*S. aurata*) in the Mediterranean (Colorni and Diamant, 2014). A *Pleistophora* sp. infecting skeletal muscles of cage-cultured sea bream in Greece has been described, resulting in low-level mortality. Administration of fumagillin in feed (15 mg/kg/fish) mitigated mortality, although histological lesions were still present (Athanassopoulou, 1998).

Notably, *Enterospora nucleophila* is a non-xenoma-forming, enteric microsporidian that causes anorexia, cachexia and, in severe cases, mortality in gilthead sea bream from cage and inland aquaculture facilities (Picard-Sánchez *et al.*, 2021). Cited as the cause of emaciative syndrome in farmed gilthead sea bream, initial reports indicated that infected fish failed to thrive, achieving half the size of non-infected hosts, in addition to chronic mortality (0.1–0.3% daily mortality). Along the coast of Spain, prevalence ranged from 5 to 80% and was variable between sea cage and inland systems (Palenzuela *et al.*, 2014). Since these initial reports from Spanish aquaculture facilities, *E. nucleophila* has been reported in Italian and Greek gilthead sea bream farms, both in offshore and inland facilities. Given the effects on production and prevalence across multiple facilities, *E. nucleophila* is considered a significant emergent threat to gilthead sea bream operations in the Mediterranean (Ahmed *et al.*, 2019a; Picard-Sánchez *et al.*, 2020).

5.4.4 Myxozoa

Myxozoa are a globally distributed, specious group of microscopic metazoan parasites reported from a diverse array of terrestrial and aquatic hosts, including reptiles, amphibians, mammals and birds (Prunescu *et al.*, 2007; Fiala *et al.*, 2015). They are mostly known for diseases they cause in commercially important fish, although pathogenic species represent only a fraction of the more than 2200 described taxa (Fiala *et al.*, 2015). By and large, myxozoa do not cause significant disease, although they can be problematic during periods of high stress. Heavy parasite burdens may have immunosuppressive effects, increasing susceptibility of the host to other infectious agents (Alvarez-Pellitero and Sitjà-Bobadilla, 1993).

Generally myxozoa exhibit a two-host life cycle, with an actinospore stage developing in a benthic annelid (definitive host) and a myxospore stage in a fish intermediate host (Fig. 5.9). The pelagic actinospore stage is released into the water column, where it incidentally encounters and infects the fish host. Following a series of intra/intercellular presporogonic stages, a polysporic plasmodium develops within the fish which contains hundreds to thousands of myxospores. When the myxospores are released back into the environment, either by rupture of the plasmodia cyst or death of the host, spores settle to the benthos, where they are ingested by the annelid host and the life cycle is complete (Lom and Dyková, 2006).

Given their economic impact on many globally important aquacultured fishes, an effective anti-myxozoan treatment to prevent infection or

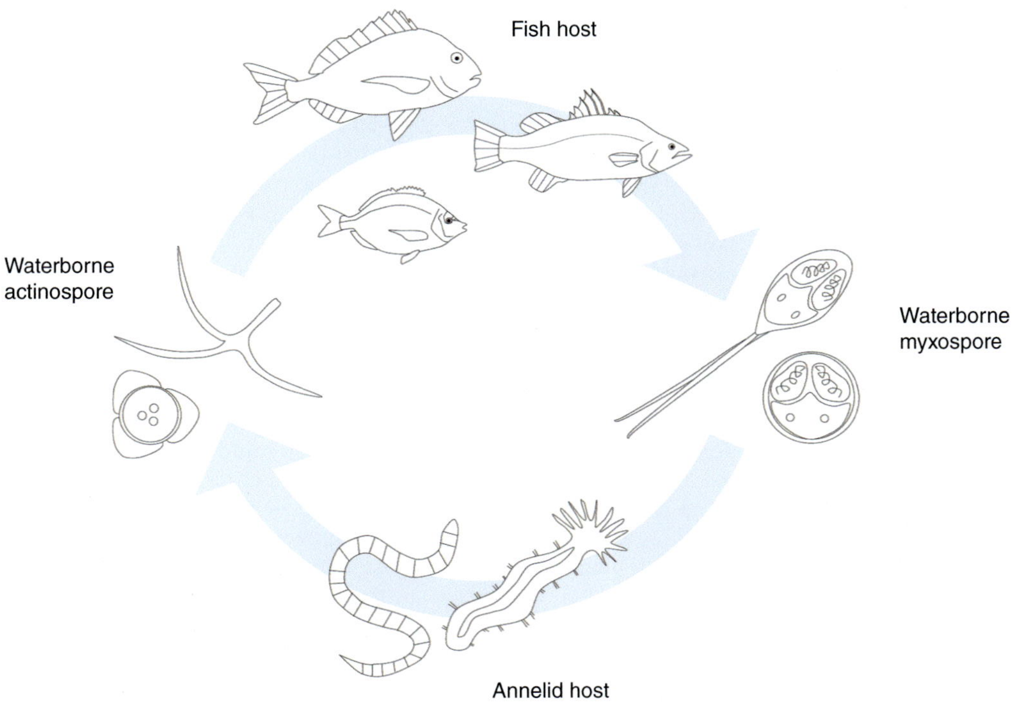

Fig. 5.9. Generalized life cycle of myxozoans.

break the life cycle within the fish host has received considerable research investment by state and federal agencies. Despite these efforts, there are no approved treatments for myxozoan infection. Studies investigating potential antiparasitics against myxozoan parasites of fish are precluded by a lack of viable experimental models for most myxozoans. The majority of experimental treatments have focused on anticoccidial or antimicrosporidian feed additives, to varying degrees of success (Colorni and Diamant, 2014). Most research in this area indicates that timing of administration is critical to successful treatment. A combination of salinomycin and amprolium demonstrated some efficacy against myxozoan infections in sharp-snout sea bream (*Diplodus puntazzo*) (Athanassopoulou *et al.*, 2004; Golomazou *et al.*, 2006). In carp (*Cyprio* spp.), fumagillin was identified as a potential treatment against haemorrhagic thelohanellosis caused by *Thelohanellus hovorkai*, as well as *Sphaerospora renicola* (Molnár *et al.*, 1987; Yokoyama *et al.*, 1999). Fumagillin was shown to reduce presence of *Enteromyxum leei* life stages in *D. puntazzo*, as well as *Myxobolus cerebralis* burden in experimentally infected rainbow trout (*Oncorhynchus mykiss*) (Golomazou *et al.*, 2006). Fumagillin has also shown promise as a treatment for *Myxidium giardi* in European eel (Székely *et al.*, 1988). Moreover, oral administration of quanine or salinomycin appeared to have deleterious effects on developing *Henneguya* spp. in tapir fish (*Gnathonemus petersii*) (Dohle *et al.*, 2002).

Due to conditions required for completion of myxozoan life cycles, most myxozoans are not cause for concern in net-pen mariculture. Unlike inland ponds, where fish and annelid hosts are confined in close proximity within closed systems (Griffin *et al.*, 2009; Rosser *et al.*, 2014), deep-sea net pens preclude completion of most myxozoa life cycles as fish are not impounded adjacent to potential annelid hosts residing on the ocean floor. Further, free exchange of water by ocean currents dilutes or washes away potentially infectious actinospores, preventing build-up and accumulation that can occur with inland pond systems. However, if regular de-fouling programmes are not employed and annelid hosts are established within substrate provided by algal, bryozoan or barnacle build-up on the cage, many of the restrictions against establishment of myxozoan life cycles by sea-cage culture may be lost.

Enteromyxum *spp.*

The genus *Enteromyxum* consists of four species, namely *E. leei* (Diamant *et al.*, 1994), *Enteromyxum fugu* (Tun *et al.*, 2000), *Enteromyxum scophthalmi* (Palenzuela *et al.*, 2002) and *Enteromyxum caesio* (Freeman *et al.*, 2020). Arguably, these are the most significant parasitic groups impacting warmwater mariculture. Myxospore morphology differs from other members within the family *Myxiidae*, with slightly crescent-shaped spores containing large, elongated polar capsules tapering to the distal side which open at the ends of the spore to discharge in opposite directions (Lom and Dyková, 2006). In addition to direct losses due to mortality, most common field observations from infected populations include anorexia, reduced feed conversion and failure to reach target size over the growing season (Sitjà-Bobadilla and Palenzuela, 2012). In contrast to typical myxozoan life cycles, all *Enteromyxum* spp. are believed capable of direct fish-to-fish transmission, which makes them especially problematic in densely stocked aquaculture systems (Diamant, 1997; Yasuda *et al.*, 2002). As such, management strategies focused on removal of dead fish can be effective. In Israel, daily removal of carcasses suppressed *E. leei* infection in an offshore sea bream farm in the Red Sea. Comparably, *E. leei* levels were high in an adjacent facility that did not remove dead fish (Colorni and Diamant, 2014), indicating that carcass removal, while logistically challenging, is an important strategy to mitigate *E. leei* severity in net-pen culture.

Originally described as *Myxidium leei* (Diamant *et al.*, 1994), *E. leei* is perhaps the most important member of the genus, with reports from at least 46 marine fishes, leading to severe emaciation in affected fish (Sitjà-Bobadilla and Palenzuela, 2012). Redondo *et al.* (2002) demonstrated infection could be achieved by feeding fish infected intestines, as well as cohabitation with infected fish, or simply exposure to effluent from a tank harbouring infected fish. Similarly, *E. fugu* has been identified as a significant pathogen causing emaciation disease of farmed tiger puffer (*Takifugu rubripes*) in Japan (Yanagida *et al.*, 2006). Transmission of *Enteromyxum* spp. appears to be temperature dependent, with increased transmission rates observed at increased temperatures (Yanagida *et al.*, 2006;

Picard-Sánchez *et al.*, 2020) and onset of disease is often suppressed at low temperatures (Yanagida *et al.*, 2006; Sitjà-Bobadilla and Palenzuela, 2012). The wide host range and positive correlation between transmission rates and temperatures suggest warmer average global temperatures may exacerbate incidence and severity of *Enteromyxum* spp. infections in global mariculture.

Henneguya spp.

The myxozoan genus *Henneguya* is a globally distributed group parasitizing freshwater and marine fish. Infections are typically benign, with limited pathology or associated mortality. As such, reports of *Henneguya* spp. are often ancillary to other findings (Pote *et al.*, 2012; Colorni and Diamant, 2014; Stilwell *et al.*, 2022). Several species have been attributed to significant losses in warm and temperate mariculture, especially during instances of abnormally heavy parasite burdens. Myxospores can be ellipsoid, spindle-shaped or rounded in valvular view, biconvex in sutural view, with two polar capsules, which are often elongated (Lom and Dyková, 2006). Yokoyama *et al.* (2005) described *Henneguya pagri* as a cause of summertime losses in net-pen-cultured red sea bream (*P. major*) in Japan. Affected fish present with pale gills, enlarged bulbus arteriosus and pericardial haemorrhage (Yokoyama *et al.*, 2005). Similarly, *Henneguya lateolabracis* causes comparable pathology in net-pen-cultured Chinese sea bass (*Lateolabrax* sp.) (Yokoyama *et al.*, 2003). Caffara *et al.* (2003) reported low-level mortality attributed to heart infections by a *Henneguya* sp. in sea bream (*S. aurata*) cultured in Italy, while *Henneguya aegea* is cited as the cause of cardiac henneguyosis in red sea bream (*P. major*) in Greece. In Quangdon province, South China, ovate pompano (*Trachinotus ovatus*) cultured in net pens developed severe enteritis associated with infection by *Henneguya ovata* (Liu *et al.*, 2018).

Kudoa spp.

The genus *Kudoa* consists of approximately 100 multivalvulid species parasitizing a wide range of marine fish (Giulietti *et al.*, 2020). Most are intracellular parasites of muscle cells, although coelozoic species have been reported (Lom and Dyková, 2006). While generally non-pathogenic, *Kudoa* spp. are notorious for eliciting soft, exudative flesh due to postharvest myoliquefaction, resulting in significant losses in marine aquaculture due to reduced product marketability (Henning *et al.*, 2013). Commonly known as 'soft flesh', 'milky flesh' or 'jelly flesh', postmortem myoliquefaction of muscle tissue occurs about 12–48 h after death of the fish host (Giulietti *et al.*, 2022). Perhaps the most well-known 'soft flesh'-inducing species is *Kudoa thyrsites*, which has been reported from at least 37 species of marine teleosts from cold and antitropical mariculture facilities in North America, South America, Europe, Africa, Australia and Japan (Whipps and Kent, 2006). While *K. thyrsites* demonstrates a wide host range and discontinuous distribution, reports appear limited to coldwater species. Alternatively, *Kudoa lateolabracis* has been implicated in postharvest myoliquefaction of Chinese sea bass (*Lateolabrax* sp.) cultured in Japan (Yokoyama *et al.*, 2004).

Several non-myoliquefactive species have been reported. While infections do not result in 'soft flesh', the presence of unsightly macroscopic cysts throughout host muscle also reduces product marketability (Whipps and Kent, 2006; Griffin *et al.*, 2014; Giulietti *et al.*, 2020). In Mediterranean mariculture, *Kudoa iwatai* was reported from gilthead sea bream (*S. aurata*), European sea bass (*D. labrax*) and grey mullet (*Mugil cephalus*). Infections were largely limited to trunk musculature with no observed myoliquefaction (Diamant *et al.*, 2005). It is hypothesized that the presence of cysts in musculature may interfere with swimming behaviour, which could have negative impacts on feeding and growth. Moreover, neurological effects stemming from brain infections may also alter behaviour. The brain-infecting *Kudoa yasunagai* is reported from at least 20 fish species, including red sea bream, olive flounder, tiger puffer and Pacific bluefin tuna, leading to skeletal deformities and abnormal swimming behaviour (Colorni and Diamant, 2014; Ishimaru *et al.*, 2014).

Ceratomyxa/*Sphaerospora* spp. (formerly *Leptotheca*)

Several other myxozoan genera have been implicated in losses in temperate and warmwater mariculture, although to a lesser extent. These reports are complicated by the collapse of the genus *Leptotheca*, of which several taxa were reassigned to *Ceratomyxa* and *Sphaerospora* (Gunter and Adlard, 2010). *Ceratomyxa* spp. generally have elongated, crescent or arcuate-shaped spores with two subspherical polar capsules. Comparably, *Sphaerospora* have spherical or subspherical spores, often with posterior or lateral protuberances, with two subspherical polar capsules (Lom and Dyková, 2006). While generally considered pathogens of nominal concern, there are multiple reports of *Ceratomyxa* spp. and *Sphaerospora* spp. causing emaciative disorders and trickling mortality in maricultured fish. Katharios *et al.* (2007) reported a catastrophic mortality event attributed to *Ceratomyxa diplodae* in sharp-snout sea bream (*D. puntazzo*). Likewise, *C. diplodae* and *Ceratomyxa labrakis* were associated with losses in common dentex (*Dentex dentex*) cultured in Greece (Rigos *et al.*, 1997) and *Ceratomyxa sparusaurati* is present with high prevalence in Spanish sea bream culture systems, sometimes linked with chronic mortality (Palenzuela *et al.*, 1997). Similarly, *C. sparusaurati* was present in high numbers in cage-reared sea bream (*S. aurata*) and red sea bream in the Adriatic Sea, although there were no reports of associated morbidity or mortality (Mladineo, 2005). *Sphaerospora testicularis* reduces reproductive capacity of sea bass (*D. labrax*) (Sitjà-Bobadilla and Álvarez-Pellitero, 1990), while *Sphaerospora sparidarum* (formerly *Leptotheca sparidarum*) (Sitjà-Bobadilla and Álvarez-Pellitero, 2001) infects renal tubules, glomeruli and ureters of multiple maricultured sparids (Rigos and Katharios, 2010). *Sphaerospora fugu* (formerly *Leptotheca fugu*) has been reported from emaciated tiger puffer (*T. rubripes*) in Japan, although presence of *S. fugu* may have been circumstantial as no clear association with host wasting was made (Tun *et al.* 2000).

5.4.5 Monogenea

Monogeneans represent a diverse, cosmopolitan class of platyhelminthic ectoparasites mostly found parasitizing the gills and skin of freshwater and marine fish. All are hermaphroditic with single-host life cycles. The signature identifying character of monogenean flatworms is the opishaptor, which is the primary organ of attachment. The apparatus is localized in the hind part of the worm and possesses specialized hook, clamp or sucker-like structures which secure the worm to the host gills, skin, scales or fins. A second attachment organ, the prohaptor, consisting of adhesive pads and cephalic openings, is present on the fore part of the worm (Buchmann and Bresciani, 2006). As a consequence of their attachment and feeding behaviour, heavy monogenean infestations can result in significant damage to gill architecture and impair gill function. In addition to gill impairment, which alone can lead to morbidity and mortality, mechanical damage at the site of attachment can open hosts up to secondary infections. The high stocking densities employed in mariculture systems facilitate fish-to-fish transmission, leading to rapid build-up within the system. Monogeneans are commonly encountered parasitic agents in sea-pen mariculture and if outbreaks are left untreated, significant losses can occur (Buchmann and Bresciani, 2006; Colorni and Diamant, 2014).

Monogeneans are divided into two subclasses, the Monopisthocotylea (single posterior sucker) and the Polyopisthocotylea (multiple posterior suckers). The Monopisthocotylea are motile scavengers, feeding predominantly on epithelial cells and mucus. The Polyopisthocotylea are blood feeders and largely stationary. These differences in feeding and attachment strategies have important implications in terms of damage inflicted on the host, treatment options, as well as the host's response to parasitism (Buchmann and Bresciani, 2006; Colorni and Diamant, 2014).

Monopisthocotylea

The genus *Gyrodactylus* is well known for the problems it causes in cultured fish and multiple *Gyrodactylus* spp. have been associated with wasting and lethargy in sparid fish cultured in the Mediterranean region (Colorni and Diamant, 2014; Khalil *et al.*, 2018). *Anoplodiscus* spp. are well-known parasites of sparid fish cultured in the Mediterranean, Australia and Japan, particularly red sea bream where heavy infections can cause emaciation and fin erosion (Ogawa, 1994). Of the Monopisthocotylea, members of the Diplectanidae are possibly the most important monogeneans parasitizing sparid fish (Colorni and Diamant, 2014), of which members of the genus *Lamellodiscus* are widely reported from cage-reared fish, where infections result in significant gill pathology (Mladineo and Maršić-Lučić, 2007). During a survey of Adriatic cage-reared fish, *Diplectanum aequans* was present with expected seasonal abundance in sea bass (*D. labrax*) (Mladineo, 2005). Heavy infections have been implicated as a cause of mortality in juvenile sea bass due to gill damage associated with parasite attachment. *Furnestinia echeneis* is highly prevalent among Mediterranean and Red Sea fish farms (Colorni and Diamant, 2014; Mahmoud *et al.*, 2014). *Neobenedenia melleni* and other members of the family Capsalidae (*Benedenia, Capsala, Entobdella, Encotyllabe*) injure the host by feeding and introducing ulcerative lesions that can serve as portals of entry for secondary pathogens. In heavy infestations, they can have adverse effects on cultivated fish (Whittington, 2004; Morsy *et al.*, 2011; Colorni and Diamant, 2014).

Polyopisthocotylea

Sparicotyle chrysophrii is considered an important pathogen of cultured gilthead sea bream in the Mediterranean and has been implicated in catastrophic losses in sea cages (Sitjà-Bobadilla *et al.*, 2006; Rigos and Katharios, 2010). Pathological effects of *S. chrysophrii* have been well documented, with heavy infestations leading to significant gill damage and lethargy attributed to parasite-induced anaemia (Antonelli *et al.*, 2010). In the survey by Mladineo (2005), *S. chrysophrii* (originally *Microcotyle chrysophrii* Vanbeneden and Hesse 1863) was found in high numbers in sea bream (*S. aurata*) and sharp-snout sea bream (*D. puntazzo*) in Adriatic cage culture. Likewise, *Polyabroides multispinosus* is reported to cause restlessness, respiratory distress, increased mucus production and even blindness (Colorni and Diamant, 2014).

5.4.6 Digenea

Digeneans are parasitic platyhelminths requiring one or more intermediate hosts to complete their life cycle (Roberts and Janovy, 2005). Fish can serve as either intermediate or definitive hosts in digenean life cycles. While not typically pathogenic to fish, the presence of intermediate life stages (metacercariae) in the fillet of farm-raised fish can reduce product marketability (Gunn *et al.*, 2021). Alternatively, heavy infestations of some trematode species can reduce feeding activity, resulting in reduced growth and mortality in some cases (Wise *et al.*, 2008; Griffin *et al.*, 2018; Gunn *et al.*, 2021). The logistics of open-sea cage mariculture often preclude digenetic life cycles and as such, digeneans are not typically considered pathogens of concern for sea cage mariculture.

One exception is sanguinicolid infections (blood flukes), which have been implicated in significant mortalities in cage-reared fish in North America, Europe and Asia. Typically found in large blood vessels of susceptible fish, high parasite densities can cause significant harm to the host. Catastrophic mortalities in amberjack (*S. dumerli*) raised in shallow-water cages have been attributed to heavy infestations of *Paradeontacylix* spp. (Ogawa and Egusa, 1986; Ogawa and Fukudome, 1994).

5.4.7 Crustacea

Multiple parasitic crustaceans are found on both freshwater and marine fishes, comprising a diverse and ubiquitous group of metazoan ectoparasites, rivalled only by monogeneans in their diversity. Attachment and feeding can result in significant damage to skin and gills, often increasing host susceptibility to secondary bacterial infections (Colorni and Diamant, 2014). Transmission is direct, which is facilitated by crowding conditions favoured by intensive aquaculture. Chief among crustacean fish parasites are copepods, which have significant economic impacts on both sport fisheries and aquaculture (Boxshall, 2005). While copepod parasitism has rarely been associated with significant mortality in warmwater culture, conditions of cage culture and close proximity of potential fish hosts facilitate proliferation under certain conditions and epizootics still occur. Stressed or compromised fish are particularly vulnerable.

Caligiforms, Ergasiliforms and Lernaeids/Pennellids are the primary groups affecting marine fish, feeding primarily on mucus and epidermis and damaging skin and gills in the process. In Asia, various ergasilids and caligid species have been reported from multiple cultured sparid fish (Yamashita, 1980; Lin *et al.*, 1994; Ogawa and Yokoyama, 1998; Izawa and Choi, 2000). Similarly, Roubal (1995) reported *Ergasilus australiensis*, *Bomolochus stocki*, *Alella macrotrachelus*, *Lernanthropus atrox* and *Lerneanthropus chrysophhrys* in sparid fish cultured in Australia. Treatments for copepods include formalin, organophosphate insecticides, and antiparasitic compounds such as ivermectin, pyrethrum, carbaryl and diflubenzuron. However, different species or different developmental stages within the same species have different degrees of susceptibility to chemical treatments. Further complicating treatment is that there is often a fine line between therapeutic dosage and toxicity to hosts and differing legal statuses of these treatments across jurisdictions (Noga, 2010; Colorni and Diamant, 2014).

Isopods comprise a much smaller group of predominantly marine, warmwater crustacean parasites. Similar to copepods, they can be a significant burden to the host, often resulting in reduced growth. *Ceratothoa parallela* and *Gnathia piscivore*, which parasitize gilthead sea bream, and *Ceratothoa oestroides* from sea bream and European sea bass are common isopods affecting Mediterranean mariculture (Paperna, 1977; Papapanagiotou and Trilles, 2001; Mladineo, 2003).

References

Abayneh, T., Colquhoun, D.J. and Sørum, H. (2013) *Edwardsiella piscicida* sp. nov., a novel species pathogenic to fish. *Journal of Applied Microbiology* 114(3), 644-654. doi: 10.1111/jam.12080.

Abdelsalam, M., Asheg, A. and Eissa, A.E. (2013) *Streptococcus dysgalactiae*: an emerging pathogen of fishes and mammals. *International Journal of Veterinary Science and Medicine* 1(1), 1–6. doi: 10.1016/j.ijvsm.2013.04.002.

Abu-Elala, N.M., Abd-Elsalam, R.M. and Younis, N.A. (2020) Streptococcosis, lactococcosis and enterococcosis are potential threats facing cultured Nile tilapia (*Oreochromis niloticus*) production. *Aquaculture Research* 51(10), 4183–4195. doi: 10.1111/are.14760.

Ahmed, N.H., Caffara, M., Sitjà-Bobadilla, A., Fioravanti, M.L., Mazzone, A., *et al.* (2019a) Detection of the intranuclear microsporidian *Enterospora nucleophila* in gilthead sea bream by in situ hybridization. *Journal of Fish Diseases* 42(6), 809–815.

Ahmed, N., Thompson, S. and Glaser, M. (2019b) Global aquaculture productivity, environmental sustainability, and climate change adaptability. *Environmental Management* 63, 159–172. doi: 10.1007/s00267-018-1117-3.

Akoğlu, A. (2020) The effect of some environmental conditions on planktonic growth and biofilm formation by some lactic acid bacteria isolated from a local cheese in Turkey. *Biotechnology Letters* 42(3), 481–492. doi: 10.1007/s10529-020-02794-4.

Alvarez-Pellitero, P. and Sitjà-Bobadilla, A. (1993) Pathology of Myxosporea in marine fish culture. *Diseases of Aquatic Organisms* 17, 229–229.

Alvarez-Pellitero, P. and Sitjà-Bobadilla, A. (2002) *Cryptosporidium molnari* n. sp. (Apicomplexa: Cryptosporidiidae) infecting two marine fish species, *Sparus aurata* L. and *Dicentrarchus labrax* L. *International Journal for Parasitology* 32(8), 1007–1021.

Amaro, C., Sanjuán, E., Fouz, B., Pajuelo, D., Lee, C.T., *et al.* (2015) The fish pathogen *Vibrio vulnificus* biotype 2: epidemiology, phylogeny, and virulence factors involved in warm-water vibriosis. *Microbiology Spectrum* 3(3). doi: 10.1128/microbiolspec.VE-0005-2014.

Amaro, C., Fouz, B., Sanjuán, E. and Romalde, J.L. (2020) Vibriosis. In: Woo, P.T.K., Leong, J. and Buchmann, K. (eds) *Climate Change and Infectious Fish Diseases*. CAB International, Wallingford, UK, pp. 182–210.

Andreoni, F. and Magnani, M. (2014) Photobacteriosis: prevention and diagnosis. *Journal of Immunology Research* 2014, 793817. doi: 10.1155/2014/793817.

Anshary, H., Kurniawan, R.A., Sriwulan, S., Ramli, R. and Baxa, D.V. (2014) Isolation and molecular identification of the etiological agents of streptococcosis in Nile tilapia (*Oreochromis niloticus*) cultured in net cages in Lake Sentani, Papua, Indonesia. *SpringerPlus* 3, 627. doi: 10.1186/2193-1801-3-627.

Antonelli, L., Quilichini, Y. and Marchand, B. (2010) *Sparicotyle chrysophrii* (Van Beneden and Hesse 1863) (Monogenea: Polyopisthocotylea) parasite of cultured gilthead sea bream *Sparus aurata* (Linnaeus

1758) (Pisces: Teleostei) from Corsica: ecological and morphological study. *Parasitology Research* 107(2), 389–398.

Apablaza, P., Frisch, K., Brevik, Ø.J., Småge, S.B., Vallestad, C., *et al.* (2017) Primary isolation and characterization of *Tenacibaculum maritimum* from Chilean Atlantic salmon mortalities associated with a *Pseudochattonella* spp. algal bloom. *Journal of Aquatic Animal Health* 29(3), 143–149. doi: 10.1080/08997659.2017.1339643.

Athanassopoulou, F. (1998) A case report of *Pleistophora* sp. infection in cultured sea bream (*Sparus aurata* L.) in Greece. *Bulletin of the European Association of Fish Pathologists* 18, 19–21.

Athanassopoulou, F., Karagouni, E., Dotsika, E., Ragias, V., Tavla, J., *et al.* (2004) Efficacy and toxicity of orally administrated anti-coccidial drugs for innovative treatments of *Myxobolus* sp. infection in *Puntazzo puntazzo*. *Diseases of Aquatic Organisms* 62(3), 217–226.

Avendaño-Herrera, R., Irgang, R., Sandoval, C., Moreno-Lira, P., Houel, A., *et al.* (2016) Isolation, characterization and virulence potential of *Tenacibaculum dicentrarchi* in salmonid cultures in Chile. *Transboundary and Emerging Diseases* 63(2), 121–126. doi: 10.1111/tbed.12464.

Bahri, S. (2012) Protozoan and myxozoan infections in wild gilthead seabream (*Sparus aurata* L.) from North Lake of Tunis, Tunisia. *Acta Parasitologica* 57(2), 114–121.

Bandín, I. and Souto, S. (2020) Betanodavirus and VER disease: a 30-year research review. *Pathogens* 9(2), 106. doi: 10.3390/pathogens9020106.

Bannister, J., Sievers, M., Bush, F. and Bloecher, N. (2019) Biofouling in marine aquaculture: a review of recent research and developments. *Biofouling* 35(6), 631–648.

Baseggio, L., Silayeva, O., Buller, N., Landos, M., Englestädter, J., *et al.* (2021) Complete, closed and curated genome sequences of *Photobacterium damselae* subsp. *piscicida* isolates from Australia indicate mobilome-driven localized evolution and novel pathogenicity determinants. *Microbial Genomics* 7(4), 000562. doi: 10.1099/mgen.0.000562.

Baums, C.G., Hermeyer, K., Leimbach, S., Adamek, M., Czerny, C.P., *et al.* (2013) Establishment of a model of *Streptococcus iniae* meningoencephalitis in Nile tilapia (*Oreochromis niloticus*). *Journal of Comparative Pathology* 149(1), 94–102. doi: 10.1016/j.jcpa.2012.10.003.

Blazhekovikj-Dimovska, D. and Stojanovski, S. (2020) Ectoparasitic species of the genus *Trichodina* (Ciliophora: Peritrichida) parasitizing Macedonian freshwater fish. *Acta Biologica* 27, 11–20.

Boerlage, A.S., Ashby, A., Herrero, A., Reeves, A., Gunn, G.J., *et al.* (2020) Epidemiology of marine gill diseases in Atlantic salmon (*Salmo salar*) aquaculture: a review. *Reviews in Aquaculture* 12(4), 2140–2159. doi: 10.1111/raq.12426.

Borrego, J.J., Valverde, E.J., Labella, A.M. and Castro, D. (2017) Lymphocystis disease virus: its importance in aquaculture. *Reviews in Aquaculture* 9, 179-193. doi: 10.1111/raq.12131

Boxshall, G.A. (2005) Crustacean parasites (Copepoda). In: Rohde, K. (ed.) *Marine Parasitology*. CAB International, Wallingford, UK, pp. 123–138.

Bridel, S., Bourgeon, F., Marie, A., Saulnier, D., Pasek, S., *et al.* (2020) Genetic diversity and population structure of *Tenacibaculum maritimum*, a serious bacterial pathogen of marine fish: from genome comparisons to high throughput MALDI-TOF typing. *Veterinary Research* 51(1), 60. doi: 10.1186/s13567-020-00782-0.

Bromage, E.S. and Owens, L. (2002) Infection of barramundi *Lates calcarifer* with *Streptococcus iniae*: effects of different routes of exposure. *Diseases of Aquatic Organisms* 52(3), 199-205. doi: 10.3354/dao052199

Buchmann, K. and Bresciani, J. (2006) Monogenea (phylum Platyhelminthes). In: Woo, P.T.K. (ed.) *Fish Diseases and Disorders*. Vol. 1. *Protozoan and Metazoan Infections*, 2nd edn. CAB International, Wallingford, UK, pp. 297–344.

Buján, N., Toranzo, A.E. and Magariños, B. (2018) *Edwardsiella piscicida*: a significant bacterial pathogen of cultured fish. *Diseases of Aquatic Organisms* 131(1), 59–71. doi: 10.3354/dao03281.

Caffara, M., Marcer, F., Florio, D., Quaglio, F. and Fioravanti, M.L. (2003) Heart infection due to *Henneguya* sp. (Myxozoa, Myxosporea) in gilthead sea bream (*Sparus aurata*) cultured in Italy. *Bulletin of the European Association of Fish Pathologists* 23(3), 108–112.

Carbello, C., Ortiz-Delgado, J.B., Berbel, C., Castro, D., Borrego, J.J., *et al.* (2019) Feed and immersion challenges with lymphocystic diseae virus (LCDV) reveals specific mechanisms for horizontal transmission and immune response in Senegalese sole post-larvae. *Fish & Shellfish Immunology* 89, 710–718. doi: 10.1016/j.fsi.2019.04.049.

Carraro, R., Dalla Rovere, G., Ferraresso, S., Carraro, L., Franch, R., *et al.* (2018) Development of a real-time PCR assay for rapid detection and quantification of *Photobacterium damselae* subsp. *piscicida* in fish tissues. *Journal of Fish Diseases* 41(2), 247–254. doi: 10.1111/jfd.12703.

Cascarano, M.C., Stavrakidis-Zachou, O., Mladineo, I., Thompson, K.D., Papandroulakis, N., *et al.* (2021) Mediterranean aquaculture in a changing climate: temperature effects on pathogens and diseases of three farmed fish species. *Pathogens* 10(9), 1205. doi: 10.3390/pathogens/10091205.

Castillo, D., Rørbo, N., Jørgensen, J., Lange, J., Tan, D., *et al.* (2019) Phage defense mechanisms and their genomic and phenotypic implications in the fish pathogen *Vibrio anguillarum*. *FEMS Microbiology Ecology* 95(3), fiz004. doi: 10.1093/femsec/fiz004.

Certad, G., Dupouy-Camet, J., Gantois, N., Hammouma-Ghelboun, O., Pottier, M., *et al.* (2015) Identification of *Cryptosporidium* species in fish from Lake Geneva (lac Leman) in France. *PLoS ONE* 10(7), e0133047.

Chang, C.T., Benedict, S. and Whipps, C.M. (2019a) Transmission of *Mycobacterium chelonae* and *Mycobacterium marinum* in laboratory zebrafish through live feeds. *Journal of Fish Diseases* 42(10), 1425–1431. doi: 10.1111/jfd.13071.

Chang, C.T., Lewis, J. and Whipps, C.M. (2019b) Source or sink: examining the role of biofilms in transmission of *Mycobacterium* spp. in laboratory zebrafish. *Zebrafish* 16(2), 197–206. doi: 10.1089/zeb.2018.1689.

Chideroli, R.T., Amoroso, N., Mainardi, R.M., Suphoronski, S.A., de Padua, S.B., *et al.* (2017) Emergence of a new multidrug-resistant and highly virulent serotype of *Streptococcus agalactiae* in fish farms from Brazil. *Aquaculture* 479, 45–51. doi: 10.1016/j.aquaculture.2017.05.013.

Chu, K.B., Seng, L.T., Yong, W.S., Thye, C.T. and Veloo, P. (2001) The life span of theront stages of *Cryptocaryon irritans* in different media, temperature and salinity. *The Journal of Protozoology Research* 11(3–4), 66–73.

Colorni, A. and Diamant, A. (2014) Infectious diseases of warmwater fish in marine and brackish waters. In: Woo, P.T.K. and Bruno, D.W. (eds) *Diseases and Disorders of Finfish in Cage Culture*, 2nd edn. CAB International, Wallingford, UK, pp. 155–192.

Costa, F.A., Leal, C.G., Leite, R.C. and Figueiredo, H.C.P. (2014) Genotyping of *Streptococcus dysgalactiae* strains isolated from Nile tilapia, *Oreochromis niloticus* (L.). *Journal of Fish Diseases* 37(5), 463–469. doi: 10.1111/jfd.12125.

Diamant, A. (1997) Fish-to-fish transmission of a marine myxosporean. *Diseases of Aquatic Organisms* 30(2), 99–105.

Diamant, A. (1998) *Brooklynella hostilis* (Hartmannulidae), a pathogenic ciliate from the gills of maricultured sea bream. *Bulletin of the European Association of Fish Pathologists* 18, 33–36.

Diamant, A., Lom, J. and Dyková, I. (1994) *Myxidium leei* n. sp., a pathogenic myxosporean of cultured sea bream *Sparus aurata*. *Diseases of Aquatic Organisms* 20(2), 137–141.

Diamant, A., Ucko, M., Paperna, I., Colorni, A. and Lipshitz, A. (2005) *Kudoa iwatai* (Myxosporea: Multivalvulida) in wild and cultured fish in the Red Sea: redescription and molecular phylogeny. *Journal of Parasitology* 91(5), 1175–1189.

Doan, Q.K., Vandeputte, M., Chatain, B., Morin, T. and Allal, F. (2017) Viral encephalopathy and retinopathy in aquaculture: a review. *Journal of Fish Diseases* 40(5), 717–742. doi: 10.1111/jfd.12541.

Dohle, A., Schmahl, G., Raether, W., Schmidt, H. and Ritter, G. (2002) Effects of orally administered chemotherapeutics (quinine, salinomycin) against *Henneguya* sp. Thelohán, 1892 (Myxozoa: Myxobolidae), a gill parasite in the tapir fish *Gnathonemus petersii* Günther, 1862 (Teleostei). *Parasitology Research* 88(9), 861–867.

Domingos, J.A., Shen, X., Terence, C., Senapin, S., Dong, H.T., *et al.* (2021) Scale drop disease virus (SDDV) and *Lates calcarifer* herpes virus (LCHV) coinfection downregulate immune-relevant pathways and cause splenic and kidney necrosis in barramundi under commercial farming conditions. *Frontiers in Genetics* 12, 666897. doi: 10.3389/fgene.2021.666897.

Downes, J.K., Yatabe, T., Marcos-Lopez, M., Rodger, H.D., MacCarthy, E., *et al.* (2018) Investigation of co-infections with pathogens associated with gill disease in Atlantic salmon during an amoebic gill disease outbreak. *Journal of Fish Diseases* 41(8), 1217–1227. doi: 10.1111/jfd.12814.

Egusa, S. (1982) A microsporidian species from yellowtail juveniles, *Seriola quinqueradiata*, with 'Beko' disease. *Fish Pathology* 16(4), 187–192.

Essam, H.M., Abdellrazeq, G.S., Tayel, S.I., Torky, H.A. and Fadel, A.H. (2016) Pathogenesis of *Photobacterium damselae* subspecies infections in sea bass and sea bream. *Microbial Pathogenesis* 99, 41–50. doi: 10.1016/j.micpath.2016.08.003.

Ferguson, H.W., Christian, M.J.D., Hay, S., Nicolson, J., Sutherland, D., *et al.* (2010) Jellyfish as vectors of bacterial disease for farmed salmon (*Salmo salar*). *Journal of Veterinary Diagnostic Investigation* 22(3), 376–382. doi: 10.1177/104063871002200305.

Fernández-Álvarez, C. and Santos, Y. (2018) Identification and typing of fish pathogenic species of the genus *Tenacibaculum*. *Applied Microbiology and Biotechnology* 102(23), 9973–9989. doi: 10.1007/s00253-018-9370-1.

Ferrario, C., Ricci, G., Milani, C., Lugli, G.A., Ventura, M., *et al.* (2013) *Lactococcus garvieae*: where is it from? A first approach to explore the evolutionary history of this emerging pathogen. *PLoS ONE* 8(12), e84796. doi: 10.1371/journal.pone.0084796.

Fiala, I., Bartošová-Sojková, P. and Whipps, C.M. (2015) Classification and phylogenetics of Myxozoa. In: Okamura, B., Gruhl, A. and Bartholomew, J. (eds) *Myxozoan Evolution, Ecology and Development*. Springer, Cham, Switzerland, pp. 85–110. doi: 10.1007/978-3-319-14753-6_5.

Francis-Floyd, R. and Floyd, M. (2011) *Amyloodinium ocellatum*, an important parasite of cultured marine fish. SRAC Publication No. 4705, Southern Regional Aquaculture Center. Available at: https://fisheries.tamu.edu/files/2013/09/SRAC-Publication-No.-4705-Amyloodinium-ocellatum-an-Important-Parasite-of-Cultured-Marine-Fish.pdf (accessed 22 September 2022).

Freeman, M.A., Yanagida, T. and Kristmundsson, À. (2020) A novel histozoic myxosporean, *Enteromyxum caesio* n. sp., infecting the redbelly yellowtail fusilier, *Caesio cuning*, with the creation of the Enteromyxidae n. fam., to formally accommodate this commercially important genus. *PeerJ* 8, e9529. doi: 10.7717/peerj.9529.

Fu, Y., Li, Y., Fu, W., Su, H., Zhang, L., *et al.* (2021) Scale drop disease virus associated yellowfin seabream (*Acanthopagrus lates*) ascites diseases, Zhuhai, Guangdong, Southern China: the first description. *Viruses* 13(3), 1617. doi: 10.3390/v13081617.

Gabriel, M.W., Matsui, G.Y., Friedman, R. and Lovell, C.R. (2014) Optimization of multilocus sequence analysis for identification of species in the genus *Vibrio*. *Applied and Environmental Microbiology* 80(17), 5359–5365. doi: 10.1128/AEM.01206-14.

Gauthier, D.T. and Rhodes, M.W. (2017) *Mycobacterium* spp. In: Woo, P.T.K. and Cipriano, R.C. (eds) *Fish Viruses and Bacteria: Pathobiology and Protection*. CAB International, Wallingford, UK, pp. 245–257. doi: 10.1079/9781780647784.0245.

Genin, A., Levy, L., Sharon, G., Raitsos, D.E. and Diamant, A. (2020) Rapid onsets of warming events trigger mass mortality of coral reef fish. *Proceedings of the National Academy of Sciences USA* 117(41), 25378–25385. doi: 10.1073/pnas.2009748117.

Giulietti, L., Karlsbakk, E., Cipriani, P., Shayo, S.D., Storesund, J.E., *et al.* (2020) Molecular characterization of the myoliquefactive fish parasite *Kudoa mirabilis* (Cnidaria, Kudoidae) from SW Indian Ocean and its phylogenetic relationship with the *Kudoa thyrsites* species complex. *Microorganisms* 8(9), 1352.

Giulietti, L., Karlsbakk, E., Cipriani, P., Bao, M., Storesund, J.E., *et al.* (2022) Long-term investigation of the 'soft flesh' condition in Northeast Atlantic mackerel induced by the myxosporean parasite *Kudoa thyrsites* (Cnidaria, Myxozoa): temporal trends and new molecular epidemiological observations. *Fisheries Research* 248, e106221.

Godwin, S.C., Fast, M.D., Kuparinen, A., Medcalf, K.E. and Hutchings, J.A. (2020) Increasing temperatures accentuate negative fitness consequences of a marine parasite. *Scientific Reports* 10(1), e18467. doi: 10.1038/s41598-020-74948-3.

Godwin, S.C., Krkosek, M., Reynolds, J.D. and Bateman, A.W. (2021) Sea-louse abundance on salmon farms in relation to parasite-control policy and climate change. *ICES Journal of Marine Science* 78(1), 377–387. doi: 10.1093/icesjms/fsaa173.

Golomazou, E., Athanassopoulou, F., Karagouni, E., Vagianou, S. and Tsantilas, H. (2006) Efficacy and toxicity of orally administrated anti-coccidial drug treatment on *Enteromyxum leei* infections in sharpsnout seabream (*Diplodus puntazzo* C.). *Israeli Journal of Aquaculture – Bamidgeh* 58(3), 157–169. doi: 10.46989/001c.20444.

Gomez-Gil, B., Thompson, C.C., Matsumura, Y., Sawabe, T., Iida, T., *et al.* (2014) The family *Vibrionaceae*. In: Rosenberg, E., DeLong, E.F., Lory, S., Stackebrandt, E. and Thompson, F. (eds) *The Prokaryotes: Gammaproteobacteria*. Springer, Berlin, pp. 659–747. doi: 10.1007/978-3-642-38922-1_225.

Griffin, M.J., Wise, D.J. and Pote, L.M. (2009) Morphology and small-subunit ribosomal DNA sequence of *Henneguya* adiposa (Myxosporea) from *Ictalurus punctatus* (Siluriformes). *The Journal of Parasitology* 95(5), 1076–1085.

Griffin, M., Quiniou, S., Ware, C., Bogdanovic, L. and Soto, E. (2014) *Kudoa thunni* from blackfin tuna (*Thunnus atlanticus*) harvested off the island of St. Kitts, West Indies. *The Journal of Parasitology* 100(1), 110–116.

Griffin, M.J., Greenway, T.E. and Wise, D.J. (2017) *Edwardsiella* spp. In: Woo, P.T.K. and Cipriano, R.C. (eds) *Fish Viruses and Bacteria: Pathobiology and Protection*. CAB International, Wallingford, UK, pp. 190–210. doi: 10.1079/9781780647784.0190.

Griffin, M.J., Khoo, L.H., Reichley, S.R., Ware, C., Rosser, T.G., *et al.* (2018) Encapsulation of *Bolbophorus damnificus* (Digenea: Bolbophoridae) metacercariae in juvenile channel catfish, *Ictalurus punctatus*, is linked to delayed-onset mortality. *Journal of the World Aquaculture Society* 49(3), 601–611.

Griffin, M.J., Soto, E. and Wise, D.J. (2020) Edwardsiellosis. In: Woo, P.T.K., Leong, J. and Buchmann, K. (eds) *Climate Change and Infectious Fish Diseases*. CAB International, Wallingford, UK, pp. 235–264. doi: 10.1079/9781789243277.0235.

Groner, M.L., Hoenig, J.M., Pradel, R., Choquet, R., Vogelbein, W.K., *et al.* (2018) Dermal mycobacteriosis and warming sea surface temperatures are associated with elevated mortality of striped bass in Chesapeake Bay. *Ecology and Evolution* 8(18), 9384–9397. doi: 10.1002/ece3.4462.

Gunn, M.A., Allen, P.J., Rosser, T.G., Wise, D.J. and Griffin, M.J. (2021) Comparative mortality of juvenile channel and hybrid catfish exposed to *Bolbophorus damnificus* cercariae. *North American Journal of Aquaculture* 83(4), 346–353.

Gunter, N. and Adlard, R. (2010) The demise of *Leptotheca* Thélohan, 1895 (Myxozoa: Myxosporea: Ceratomyxidae) and assignment of its species to *Ceratomyxa* Thélohan, 1892 (Myxosporea: Ceratomyxidae), *Ellipsomyxa* Køie, 2003 (Myxosporea: Ceratomyxidae), *Myxobolus* Bütschli, 1882 and *Sphaerospora* Thélohan, 1892 (Myxosporea: Sphaerosporidae). *Systematic Parasitology* 75(2), 81–104.

Guo, J.J., Kuo, C.M., Chuang, Y.C., Hong, J.W., Chou, R.L., *et al.* (2012) The effects of garlic-supplemented diets on antibacterial activity against *Streptococcus iniae* and on growth in orange-spotted grouper, *Epinephelus coioides*. *Aquaculture* 364–365, 33–38. doi: 10.1016/j.aquaculture.2012.07.023.

Guo, J.J., Kuo, C.M., Hong, J.W., Chou, R.L., Lee, Y.H., *et al.* (2015) The effects of garlic-supplemented diets on antibacterial activities against *Photobacterium damselae* subsp. *piscicida* and *Streptococcus iniae* and on growth in cobia, *Rachycentron canadum*. *Aquaculture* 435, 111–115. doi: 10.1016/j.aquaculture.2014.09.029.

Habib, C., Houel, A., Lunazzi, A., Bernardet, J.F., Olsen, A.B., *et al.* (2014) Multilocus sequence analysis of the marine bacterial genus *Tenacibaculum* suggests parallel evolution of fish pathogenicity and endemic colonization of aquaculture systems. *Applied and Environmental Microbiology* 80(17), 5503–5514. doi: 10.1128/AEM.01177-14.

Hawke, J.P. (2017) *Photobacterium damselae*. In: Woo, P.T.K. and Cipriano, R.C. (eds) *Fish Viruses and Bacteria: Pathobiology and Protection*. CAB International, Wallingford, UK, pp. 258–271. doi: 10.1079/9781780647784.0258.

Heckman, T.I. (2021) Understanding and controlling the widespread piscine pathogen *Streptococcus iniae*. PhD dissertation, University of California, Davis, Davis, California.

Heckman, T.I. and Soto, E. (2021) *Streptococcus iniae* biofilm formation enhances environmental persistence and resistance to antimicrobials and disinfectants. *Aquaculture* 540, 736739. doi: 10.1016/j.aquaculture.2021.736739.

Heckman, T.I., Griffin, M.J., Camus, A.C., LaFrentz, B.R., Morick, D., *et al.* (2020) Multilocus sequence analysis of diverse *Streptococcus iniae* isolates indicates an underlying genetic basis for phenotypic heterogeneity. *Diseases of Aquatic Organisms* 141, 53–69. doi: 10.3354/dao03521.

Heckman, T.I., Shahin, K., Henderson, E.E., Griffin, M.J. and Soto, E. (2022) Development and efficacy of *Streptococcus iniae* live-attenuated vaccines in Nile tilapia, *Oreochromis niloticus*. *Fish & Shellfish Immunology* 121, 152–162. doi: 10.1016/j.fsi.2021.12.043.

Henning, S.S., Hoffman, L.C. and Manley, M. (2013) A review of *Kudoa*-induced myoliquefaction of marine fish species in South Africa and other countries. *South African Journal of Science* 109(11), 1–5.

Hickey, M.E. and Lee, J.L. (2018) A comprehensive review of *Vibrio* (*Listonella*) *anguillarum*: ecology, pathology and prevention. *Reviews in Aquaculture* 10(3), 585–610. doi: 10.1111/raq.12188.

Hickey, M.E., Richards, G.P. and Lee, J.L. (2015) Development of a two-step, non-probed multiplex real-time PCR for surveilling *Vibrio anguillarum* in seawater. *Journal of Fish Diseases* 38(6), 551–559. doi: 10.1111/jfd.12264.

Hossain, M., Ahamed, S., Hassan, M., Islam M., Baten, M., *et al.* (2021) Assessing cage culture potentiality of long whiskers catfish, *Mystus gulio* (Hamilton, 1822) in relation to climate change adaptation in Bangladesh coast. *Journal of Applied Aquaculture* 34(3), 658–673. doi: 10.1080/10454438.2021.1881683.

Hvas, M., Folkedal, O. and Oppedal, F. (2020) Fish welfare in offshore salmon aquaculture. *Reviews in Aquaculture* 13(2), 836–852. doi: 10.1111/raq.12501.

Ina-Salwany, M.Y., Al-Saari, N., Mohamad, A., Mursidi, F.A., Mohd-Aris, A., *et al.* (2019) Vibriosis in fish: a review on disease development and prevention. *Journal of Aquatic Animal Health* 31(1), 3–22. doi: 10.1002/aah.10045.

Ishimaru, K., Matsuura, T., Tsunemoto, K. and Shirakashi, S. (2014) Seasonal monitoring of *Kudoa yasunagai* from sea water and aquaculture water using quantitative PCR. *Diseases of Aquatic Organisms* 108(1), 45–52.

Izawa, K. and Choi, K.H. (2000) Redescription of *Caligus latigenitalis* Shiino, 1954 (Copepod, Siphonostomatoida, Caligidae), parasitic on Japanese black sea bream, *Acanthopagrus schlegeli* (Bleeker, 1854). *Crustaceana* 73, 995–1005.

Johan, C.A.C. and Zainathan, S.C. (2020) Megalocytiviruses in ornamental fish: a review. *Veterinary World* 13(11), 2565–2577. doi: 10.14202/vetworld.2020.2565-2577.

Jones, N., Bohnsack, J.F., Takahashi, S., Oliver, K.A., Chan, M.S., *et al.* (2003) Multilocus sequence typing system for group B streptococcus. *Journal of Clinical Microbiology* 41(6), 2530–2536. doi: 10.1128/JCM.41.6.2530-2536.2003.

Jung, S.J. and Woo, P.T.K. (2012) *Miamiensis avidus* and related species. In: Woo, P.T.K. and Buchmann, K. (eds) *Fish Parasites: Pathobiology and Protection*. CAB International, Wallingford, UK, pp. 73–91. doi: 10.1079/9781845938062.0073.

Kai, Y.H., Su, H.M., Tai, K.T. and Chi, S.C. (2010) Vaccination of grouper broodfish (*Epinephelus tukula*) reduces the risk of vertical transmission by nervous necrosis virus. *Vaccine* 28(4), 996–1001. doi: 10.1016/j.vaccine.2009.10.132.

Katharios, P., Garaffo, M., Sarter, K., Athanassopoulou, F. and Mylonas, C.C. (2007) A case of high mortality due to heavy infestation of *Ceratomyxa diplodae* in sharpsnout sea bream (*Diplodus puntazzo*) treated with reproductive steroids. *Bulletin of the European Association of Fish Pathologists* 27(1), 43–47.

Kent, M.L., Shaw, R.W. and Sanders, J.L. (2014) Microsporidia in fish. In: Weiss, L.M. and Becnel, J.J. (eds) *Microsporidia: Pathogens of Opportunity*. Wiley-Blackwell, Ames, Iowa, pp. 493–520. doi: 10.1002/9781118395264.ch20.

Khalil, R., Medhat, S., Hanan, G., Khallaf, M., Sherif, S., *et al.* (2018) Seasonal parasitic infestations and their close relationship to immune suppression in cultured sea bass (*Dicentrarchus labrax* L.) and sea bream (*Sparus auratus*). *Biomedical Journal of Science and Technology Research* 6, 5371–5379.

Kim, J.H., Gomez, D.K., Choresca, C.H. and Park, S.C. (2007) Detection of major bacterial and viral pathogens in trash fish used to feed cultured flounder in Korea. *Aquaculture* 272(1), 105–110. doi: 10.1016/j.aquaculture.2007.09.008.

LaFrentz, B.R., Lozano, C.A., Shoemaker, C.A., García, J.C., Xu, D.H., *et al.* (2016) Controlled challenge experiment demonstrates substantial additive genetic variation in resistance of Nile tilapia (*Oreochromis niloticus*) to *Streptococcus iniae*. *Aquaculture* 458, 134–139. doi: 10.1016/j.aquaculture.2016.02.034

Lee, S.J., Yokoyama, H. and Ogawa, K. (2004) Modes of transmission of *Glugea plecoglossi* (Microspora) via the skin and digestive tract in an experimental infection model using rainbow trout, *Oncorhynchus mykiss* (Walbaum). *Journal of Fish Diseases* 27(8), 435–444.

Leiva-Rebello, R., Labella, A.M., Valverde, E.J., Castro, D. and Borrego, J.J. (2020) Persistence of lymphocystis disease virus (LCDV) in seawater. *Food and Environmental Virology* 12, 174–179. doi: 10.1007/s12560-020-09420-6.

Leung, K.Y., Wang, Q., Zheng, X., Zhuang, M., Yang, Z., *et al.* (2022) Versatile lifestyles of *Edwardsiella*: free-living, pathogen, and core bacterium of the aquatic resistome. *Virulence* 13(1), 5–18. doi: 10.1080/21505594.2021.2006890.

Levipan, H.A., Tapia-Cammas, D., Molina, V., Irgang, R., Toranzo, A.E., *et al.* (2019) Biofilm development and cell viability: an undervalued mechanism in the persistence of the fish pathogen *Tenacibaculum maritimum*. *Aquaculture* 511, 734267. doi: 10.1016/j.aquaculture.2019.734267.

Li, Y., Jiang, B., Mo, Z., Li, A. and Dan, X. (2022) *Cryptocaryon irritans* (Brown, 1951) is a serious threat to aquaculture of marine fish. *Reviews in Aquaculture* 14(1), 218–236. doi: 10.1111/raq.12594.

Lin, C.L., Ho, J.S. and Chen, S.N. (1994) Two species of *Caligus* (Copepoda: Caligidae) parasitic on black sea bream (*Acanthopagrus schlegeli*) cultured in Taiwan. *Fish Pathology* 29(4), 253–264.

Liu, X.H., Xu, L.W., Luo, D., Zhao, Y.L., Liu, G.F., *et al.* (2018) *Henneguya ovata* n. sp. (Myxosporea: Bivalvulida), causing severe enteric henneguyosis of net-cage-cultured ovate pompano, *Trachinotus ovatus* in China. *Aquaculture* 483, 8–15.

Loch, T.P., Hawke, J.P., Reichley, S.R., Faisal, M., Del Piero, F., *et al.* (2017) Outbreaks of edwardsiellosis caused by *Edwardsiella piscicida* and *Edwardsiella tarda* in farmed barramundi (*Lates calcarifer*). *Aquaculture* 481, 202–210. doi: 10.1016/j.aquaculture.2017.09.005.

Lom, J. (2002) A catalogue of described genera and species of microsporidians parasitic in fish. *Systematic Parasitology* 53(2), 81–99.

Lom, J. and Dyková, I. (2005) Microsporidian xenomas in fish seen in wider perspective. *Folia Parasitologica* 52(1/2), 69–81.

Lom, J. and Dyková, I. (2006) Myxozoan genera: definition and notes on taxonomy, life-cycle terminology and pathogenic species. *Folia Parasitologica* 53(1), 1–36.

López-Porras, A., Morales, J.A., Alvarado, G., Koda, S.A., Camus, A., *et al.* (2018) Red seabream iridovirus associated with cultured Florida pompano *Trachinotus carolinus* mortality in Central America. *Diseases of Aquatic Organisms* 130(2), 109–115. doi: 10.3354/dao03267.

Ma, J., Bruce, T.J., Jones, E.M. and Cain, K.D. (2019) A review of fish vaccine development strategies: conventional methods and modern biotechnological approaches. *Microorganisms* 7(11), 569. doi: 10.3390/microorganisms7110569.

Ma, Y., Hao, L., Liang, Z., Ma, J., Ke, H., *et al.* (2020) Characterization of novel antigenic vaccine candidates for Nile tilapia (*Oreochromis niloticus*) against *Streptococcus agalactiae* infection. *Fish & Shellfish Immunology* 105, 405–414. doi: 10.1016/j.fsi.2020.07.024.

Mabrok, M., Machado, M., Serra, C.R., Afonso, A., Valente, L.M.P., *et al.* (2016) Tenacibaculosis induction in the Senegalese sole (*Solea senegalensis*) and studies of *Tenacibaculum maritimum* survival against host mucus and plasma. *Journal of Fish Diseases* 39(12), 1445–1455. doi: 10.1111/jfd.12483.

Maekawa, S., Wang, Y.T., Yoshida, T., Wang, P.C. and Chen, S.C. (2020) Group C *Streptococcus dysgalactiae* infection in fish. *Journal of Fish Diseases* 43(9), 963–970. doi: 10.1111/jfd.13211.

Mahmoud, N.E., Mahmoud, A.M. and Fahmy, M.M. (2014) Parasitological and comparative pathological studies on monogenean infestation of cultured sea bream (*Sparus aurata*, Spariidae) in Egypt. *Oceanography* 2, 129. doi: 10.4172/2332-2632.1000129.

Mainous, M.E. and Smith, S.A. (2005) Efficacy of common disinfectants against *Mycobacterium marinum*. *Journal of Aquatic Animal Health* 17(3), 284–288. doi: 10.1577/H04-051.1.

Martínez-Lara, P., Martínez-Porchas, M., Gollas-Galván, T., Hernández-López, J. and Robles-Porchas, G.R. (2021) Granulomatosis in fish aquaculture: a mini review. *Reviews in Aquaculture* 13(1), 259–268. doi: 10.1111/raq.12472.

Martos-Sitcha, J.A., Mancera, J.M., Prunet, P. and Magnoni, L.J. (2020) Welfare and stressors in fish: challenges facing aquaculture. *Frontiers in Physiology* 11, 162. doi: 10.3389/fphys.2020.00162.

Marzouk, M., Anter, O., Ali, M., Mahmaud, M. and Kenawy, A. (2010) Studies on epizootic microsporidiosis in wild dusky grouper (*Epinephelus guaza*) from Mediterranean Sea at Matrouh Governorate, Egypt. *Egyptian Journal of Aquatic Biology and Fisheries* 14(1), 1–12.

Mataragka, A., Tzimotoudis, N., Kolygas, M., Karavanis, E. and Ikonomopoulos, J. (2022) Diagnostic investigation for the detection of mycobacteria in samples of fish feeds and tissue from sea bream and sea bass with severe granulomatous lesions. *Aquaculture* 546, 737283. doi: 10.1016/j.aquaculture.2021.737283.

Meyburgh, C.M., Bragg, R.R. and Boucher, C.E. (2017) *Lactococcus garvieae*: an emerging bacterial pathogen of fish. *Diseases of Aquatic Organisms* 123(1), 67–79. doi: 10.3354/dao03083.

Miniero Davies, Y., Xavier de Oliveira, M.G., Paulo Vieira Cunha, M., Soares Franco, L., Pulecio Santos, S.L., *et al.* (2019) *Edwardsiella tarda* outbreak affecting fishes and aquatic birds in Brazil. *Veterinary Quarterly* 38(1), 99–105. doi: 10.1080/01652176.2018.1540070.

Mishra, A., Nam, G.H., Gim, J.A., Lee, H.E., Jo, A., *et al.* (2018) Current challenges of *Streptococcus* infection and effective molecular, cellular, and environmental control methods in aquaculture. *Molecules and Cells* 41(6), 495–505. doi: 10.14348/molcells.2018.2154.

Mitchell, S.O. and Rodger, H.D. (2011) A review of infectious gill disease in marine salmonid fish. *Journal of Fish Diseases* 34(6), 411–432.

Mladineo, I. (2003) Life cycle of *Ceratothoa oestroides*, a cymothoid isopod parasite from sea bass *Dicentrarchus labrax* and sea bream *Sparus aurata*. *Diseases of Aquatic Organisms* 57(1–2), 97–101.

Mladineo, I. (2005) Parasite communities of Adriatic cage-reared fish. *Diseases of Aquatic Organisms* 64(1), 77–83.

Mladineo, I. and Maršić-Lučić, J. (2007) Host switch of *Lamellodiscus elegans* (Monogenea: Monopisthocotylea) and *Sparicotyle chrysophrii* (Monogenea: Polyopisthocotylea) between cage-reared sparids. *Veterinary Research Communications* 31(2), 153–160.

Mohd Yazid, S.H., Mohd Daud, H., Azmai, M.N.A., Mohamad, N. and Mohd Nor, N. (2021) Estimating the economic loss due to vibriosis in net-cage cultured Asian seabass (*Lates calcarifer*): evidence from the east coast of peninsular Malaysia. *Frontiers in Veterinary Science* 8, 644009. doi: 10.3389/fvets.2021.644009.

Molnár, K. (2006) Phylum Apicomplexa. In: Woo, P.T.K. (ed.) *Fish Diseases and Disorders*. Vol. 1. *Protozoan and Metazoan Infections*, 2nd edn. CAB International, Wallingford, UK, pp. 183–204.

Molnár, K., Baska, F. and Székely, C. (1987) Fumagillin, an efficacious drug against renal sphaerosporosis of the common carp *Cyprinus carpio*. *Diseases of Aquatic Organisms* 2(3), 187–190.

Molnár, K., Ostoros, G., Dunams-Morel, D. and Rosenthal, B.M. (2012) Eimeria that infect fish are diverse and are related to, but distinct from, those that infect terrestrial vertebrates. *Infection, Genetics and Evolution* 12(8), 1810–1815.
Morsy, K., Abdel-Monem, S., Abdel-Ghaffar, F., Bashtar, A.R., Al-Ghamdi, A., *et al.* (2011) First record of *Benedenia sciaenae* (Monogenea: Capsalidae) infecting the brown-spotted grouper fish *Epinephelus chlorostigma* (family: Serranidae) from the Red Sea in Egypt. *Life Science Journal – Acta Zhengzhou University Overseas Edition* 8(4), 245–252.
Mugetti, D., Varello, K., Pastorino, P., Tomasoni, M., Menconi, V., *et al.* (2021) Investigation of potential reservoirs of non-tuberculous mycobacteria in a European sea bass (*Dicentrarchus labrax*) farm. *Pathogens* 10(8), 1014. doi: 10.3390/pathogens10081014.
Nakajima, K. (2021) Red seabream iridoviral disease. In: *Manual of Diagnostic Tests for Aquatic Animals 2021*. Available at: http://www.oie.int/en/what-we-do/standards/codes-and-manuals/aquatic-manual-online-access/ (accessed 15 April 2022).
Noga, E.J. (2010) *Fish Disease: Diagnosis and Treatment*, 2nd edn. Wiley-Blackwell, Ames, Iowa.
Nowak, B.F. (2007) Parasitic diseases in marine cage culture – an example of experimental evolution of parasites? *International Journal for Parasitology* 37(6), 581–588.
Nowlan, J.P., Lumsden, J.S. and Russell, S. (2020) Advancements in characterizing *Tenacibaculum* infections in Canada. *Pathogens* 9(12), 1029. doi: 10.3390/pathogens9121029.
Ogawa, K. (1994) *Anoplodiscus tai* sp. nov. (Monogenea: Anoplodiscidae) from cultured red sea bream *Pagrus major*. *Fish Pathology* 29(1), 5–10.
Ogawa, K. and Egusa, S. (1986) Two new species of *Paradeontacylix* McIntosh, 1934 (Trematoda: Sanguinicolidae) from the vascular system of a cultured marine fish, *Seriola purpurascens. Fish Pathology* 21(1), 15–19.
Ogawa, K. and Fukudome, M. (1994) Mass mortality caused by blood fluke (*Paradeontacylix*) among amberjack (*Seriola dumerili*) imported to Japan. *Fish Pathology* 29(4), 265–269.
Ogawa, K. and Yokoyama, H. (1998) Parasitic diseases of cultured marine fish in Japan. *Fish Pathology* 33(4), 303–309.
Olsen, A.B., Spilsberg, B., Nilsen, H.K., Lagesen, K., Gulla, S., *et al.* (2020) *Tenacibaculum piscium* sp. nov., isolated from skin ulcers of sea-farmed fish, and description of *Tenacibaculum finnmarkense* sp. nov. with subdivision into genomovars *finnmarkense* and *ulcerans*. *International Journal of Systematic and Evolutionary Microbiology* 70(12), 6079–6090. doi: 10.1099/ijsem.0.004501.
Osorio, C.R., Vences, A., Matanza, X.M. and Terceti, M.S. (2018) *Photobacterium damselae* subsp. *damselae*, a generalist pathogen with unique virulence factors and high genetic diversity. *Journal of Bacteriology* 200(15), e00002-18. doi: 10.1128/JB.00002-18.
Palenzuela, O. Sitjà-Bobadilla, A. and Álvarez-Pellitero, P. (1997) *Ceratomyxa sparusaurati* (Protozoa: Myxosporea) infections in cultured gilthead sea bream *Sparus aurata* (Pisces: Teleostei) from Spain: aspects of the host-parasite relationship. *Parasitology Research* 83(6), 539–548.
Palenzuela, O., Redondo, M.J. and Álvarez-Pellitero, P. (2002) Description of *Enteromyxum scophthalmi* gen. nov., sp. nov. (Myxozoa), an intestinal parasite of turbot (*Scophthalmus maximus* L.) using morphological and ribosomal RNA sequence data. *Parasitology* 124(4), 369–379.
Palenzuela, O., Redondo, M.J., Cali, A., Takvorian, P.M., Alonso-Naveiro, M., *et al.* (2014) A new intranuclear microsporidium, *Enterospora nucleophila* n. sp., causing an emaciative syndrome in a piscine host (*Sparus aurata*), prompts the redescription of the family Enterocytozoonidae. *International Journal for Parasitology* 44(3–4), 189–203.
Pang, L., Zhang, X.H., Zhong, Y., Chen, J., Li, Y. and Austin, B. (2006) Identification of *Vibrio harveyi* using PCR amplification of the *toxR* gene. *Letters in Applied Microbiology* 43(3), 249–255. doi: 10.1111/j.1472-765X.2006.01962.x.
Papapanagiotou, E.P. and Trilles, J.P. (2001) Cymothoid parasite *Ceratothoa parallela* inflicts great losses on cultured gilthead sea bream *Sparus aurata* in Greece. *Diseases of Aquatic Organisms* 45, 237–239.
Paperna, I. (1977) Preliminary data on the Gnathiidae (Isopoda) of the northern Red Sea, the Bitter Lakes and the Eastern Mediterranean and the biology of *Gnathia piscivora* n. sp. *Rapports et Proces-Verbaux des Reunions – Commission Internationale pour l'Exploration Scientifique de la Mer Mediterranee (CIESM)* 24, 195–197.
Pečur Kazazić, S., Topić Popović, N., Strunjak-Perović, I., Florio, D., Fioravanti, M., *et al.* (2019) Fish photobacteriosis – the importance of rapid and accurate identification of *Photobacterium damselae* subsp. *piscicida*. *Journal of Fish Diseases* 42(8), 1201–1209. doi: 10.1111/jfd.13022.

Pham, T.H., Cheng, T., Wang, P., and Chen, S. (2020) Genotypic diversity, and molecular and pathogenic characterization of *Photobacterium damselae* subsp. *piscida* isolated from different fish species in Taiwan. *Journal of Fish Diseases* 43(7), 757–774. doi: 10.1111/jfd.13173.

Piñeiro-Vidal, M., Gijón, D., Zarza, C. and Santos, Y. (2012) *Tenacibaculum dicentrarchi* sp. nov., a marine bacterium of the family Flavobacteriaceae isolated from European sea bass. *International Journal of Systematic and Evolutionary Microbiology* 62(2), 425–429. doi: 10.1099/ijs.0.025122-0.

Picard-Sánchez, A., Piazzon, M.C., Ahmed, N.H., Del Pozo, R., Sitjà-Bobadilla, A. and Palenzuela, O. (2020) *Enterospora nucleophila* (Microsporidia) in gilthead sea bream (*Sparus aurata*): pathological effects and cellular immune response in natural infections. *Veterinary Pathology* 57(4), 565–576.

Picard-Sánchez, A., Piazzon, M.C., Estensoro, I., Del Pozo, R., Hossameldin Ahmed, N., *et al.* (2021) Experimental horizontal transmission of *Enterospora nucleophila* (Microsporea: Enterocytozoonidae) in gilthead sea bream (*Sparus aurata*). *Animals* 11(2), 362.

Pilarski, F., Ferreira de Oliveira, C.A., Darpossolo de Souza, F.P.B. and Zanuzzo, F.S. (2017) Different β-glucans improve the growth performance and bacterial resistance in Nile tilapia. *Fish & Shellfish Immunology* 70, 25–29. doi: 10.1016/j.fsi.2017.06.059.

Poloczanska, E.S. and Butler, A.J. (2010) Biofouling and climate change. In: Dürr, S. and Thomason, J.C. (eds) *Biofouling*. Wiley-Blackwell, Oxford, UK, pp. 333–347.

Porcellato, D., Smistad, M., Skeie, S.B., Jørgensen, H.J., Austbø, L., *et al.* (2021) Whole genome sequencing reveals possible host species adaptation of *Streptococcus dysgalactiae*. *Scientific Reports* 11(1), 17350. doi: 10.1038/s41598-021-96710-z.

Pote, L.M.W., Khoo, L. and Griffin, M. (2012) *Henneguya ictaluri*. In: Woo, P.T.K. and Buchmann, K. (eds) *Fish Parasites: Pathobiology and Protection*. CAB International, Wallingford, UK, pp. 177–192.

Prunescu, C.C., Prunescu, P., Pucek, Z. and Lom, J. (2007) The first finding of myxosporean development from plasmodia to spores in terrestrial mammals: *Soricimyxum fegati* gen. et sp. n. (Myxozoa) from *Sorex araneus* (Soricomorpha). *Folia Parasitologica* 54(3), 159–164.

Redondo, M.J., Palenzuela, O., Riaza, A., Macías, Á. and Álvarez-Pellitero, P. (2002) Experimental transmission of *Enteromyxum scophthalmi* (Myxozoa), an anteric parasite of turbot *Scophthalmus maximus*. *The Journal of Parasitology* 88(3), 482–488. doi: 10.1645/0022-3395(2002)088[0482:ETOESM]2.0.CO;2.

Reichley, S.R., Ware, C., Steadman, J., Gaunt, P.S., García, J.C., *et al.* (2017) Comparative phenotypic and genotypic analysis of *Edwardsiella* isolates from different hosts and geographic origins, with emphasis on isolates formerly classified as *E. tarda*, and evaluation of diagnostic methods. *Journal of Clinical Microbiology* 55(12), 3466–3491. doi: 10.1128/JCM.00970-17.

Remuzgo-Martínez, S., Lázaro-Díez, M., Padilla, D., Vega, B., El Aamri, F., *et al.* (2014) New aspects in the biology of *Photobacterium damselae* subsp. *piscicida*: pili, motility and adherence to solid surfaces. *Veterinary Microbiology* 174(1–2), 247–254. doi: 10.1016/j.vetmic.2014.08.005.

Rigos, G. and Katharios, P. (2010) Pathological obstacles of newly-introduced fish species in Mediterranean mariculture: a review. *Reviews in Fish Biology and Fisheries* 20(1), 47–70.

Rigos, G., Grigorakis, K., Christophilogannis, P., Nengas, I. and Alexis, M. (1997) *Ceratomyxa* spp. (Myxosporea) infection in cultured common dentex from Greece. *Bulletin of the European Association of Fish Pathologists* 17, 174–176.

Roberts, L.S. and Janovy, J. (2005) Phylum Apicomplexa: gregarines, coccidia and related organisms. In: Roberts, L.S. and Janovy, J. (eds) *Foundations in Parasitology*, 7th edn. McGraw-Hill, New York, pp. 123–146.

Rodriguez-Tovar, L.E., Speare, D.J. and Markham, R.F. (2011) Fish microsporidia: immune response, immunomodulation and vaccination. *Fish & Shellfish Immunology* 30(4–5), 999–1006.

Rosser, T.G., Griffin, M.J., Quiniou, S.M., Greenway, T.E., Khoo, L.H., *et al.* (2014) Molecular and morphological characterization of myxozoan actinospore types from a commercial catfish pond in the Mississippi Delta. *The Journal of Parasitology* 100(6), 828–839.

Roubal, F.R. (1995) Changes in monogenean and copepod infestation on captive *Acanthopagrus australis* (Sparidae). *Journal of Fish Biology* 46, 423–431.

Roux, F.L., Wegner, K.M., Baker-Austin, C., Vezzulli, L., Osorio, C.R., *et al.* (2015) The emergence of *Vibrio* pathogens in Europe: ecology, evolution, and pathogenesis. *Frontiers in Microbiology* 6, 830. doi: 10.3389/fmicb.2015.00830.

Senderovich, Y., Izhaki, I. and Halpern, M. (2010) Fish as reservoirs and vectors of *Vibrio cholerae*. *PLoS ONE* 5(1), e8607. doi: 10.1371/journal.pone.0008607.

Shahin, K., Subramaniam, K., Camus, A.C., Yazdi, Z., Yun, S., *et al.* (2021a) Isolation, identification and characterization of a novel *Megalocytivirus* from cultured tilapia (*Oreochromis* spp.) from Southern California, USA. *Animals* 11(12), 3524. doi: 10.3390/ani11123524.
Shahin, K., Veek, T., Heckman, T.I., Littman, E., Mukkatira, K., *et al.* (2021b) Isolation and characterization of *Lactococcus garvieae* from rainbow trout, *Onchorhyncus mykiss*, from California, USA. *Transboundary and Emerging Diseases* 69(4), 2326–2343. doi: 10.1111/tbed.14250.
Shao, S., Lai, Q., Liu, Q., Wu, H., Xiao, J., *et al.* (2015) Phylogenomics characterization of a highly virulent *Edwardsiella* strain ET080813^T encoding two distinct T3SS and three T6SS gene clusters: propose a novel species as *Edwardsiella anguillarum* sp. nov. *Systematic and Applied Microbiology*, 38(1), 36–47. doi: 10.1016/j.syapm.2014.10.008.
Shoemaker, C.A., Xu, D. and Soto, E. (2017) *Streptococcus iniae* and *S. agalactiae*. In: Woo, P.T.K. and Cipriano, R.C. (eds) *Fish Viruses and Bacteria: Pathobiology and Protection*. CAB International, Wallingford, UK, pp. 298–313. doi: 10.1079/9781780647784.0298.
Sitjà-Bobadilla, A. and Álvarez-Pellitero, P. (1990) *Sphaerospora testicularis* sp. nov. (Myxosporea: Sphaerosporidae) in wild and cultured sea bass, *Dicentrarchus labrax* (L.), from the Spanish Mediterranean area. *Journal of Fish Diseases* 13(3), 193–203.
Sitjà-Bobadilla, A. and Álvarez-Pellitero, P. (2001) *Leptotheca sparidarum* n. sp. (Myxosporea: Bivalvulida), a parasite from cultured common dentex (*Dentex dentex* L.) and gilthead sea bream (*Sparus aurata* L.). *Journal of Eukaryotic Microbiology* 48(6), 627–639.
Sitjà-Bobadilla, A. and Palenzuela, O. (2012) *Enteromyxum* species. In: Woo, P.T.K. and Buchmann, K. (eds) *Fish Parasites: Pathobiology and Protection*. CAB International, Wallingford, UK, pp. 163–176.
Sitjà-Bobadilla, A., de Felipe, M.C. and Álvarez-Pellitero, P. (2006) *In vivo* and *in vitro* treatments against *Sparicotyle chrysophrii* (Monogenea: Microcotylidae) parasitizing the gills of gilthead sea bream (*Sparus aurata* L.). *Aquaculture* 261(3), 856–864.
Småge, S.B., Brevik, Ø.J., Frisch, K., Watanabe, K., Duesund, H., *et al.* (2017) Concurrent jellyfish blooms and tenacibaculosis outbreaks in northern Norwegian Atlantic salmon (*Salmo salar*) farms. *PLoS ONE* 12(11), e0187476. doi: 10.1371/journal.pone.0187476.
Småge, S.B., Frisch, K., Vold, V., Duesund, H., Brevik, Ø.J., *et al.* (2018) Induction of tenacibaculosis in Atlantic salmon smolts using *Tenacibaculum finnmarkense* and the evaluation of a whole cell inactivated vaccine. *Aquaculture* 495, 858–864. doi: 10.1016/j.aquaculture.2018.06.063.
Soto, E., Wang, R., Wiles, J., Baumgartner, W., Green, C., *et al.* (2015) Characterization of isolates of *Streptococcus agalactiae* from diseased farmed and wild marine fish from the US Gulf Coast, Latin America, and Thailand. *Journal of Aquatic Animal Health* 27(2), 123–134. doi: 10.1080/08997659.2015.1032439.
Souto, S., Olveira, J.G. and Bandín, I. (2015) Influence of temperature on *Betanodavirus* infection in Senegalese sole (*Solea senegalensis*). *Veterinary Microbiology* 179, 162–167. doi: 10.1016/j.vetmic.2015.07.004.
Stentiford, G.D., Bass, D. and Williams, B.A. (2019) Ultimate opportunists – the emergent *Enterocytozoon* group microsporidia. *PLoS Pathogens* 15(5), e1007668.
Stilwell, J.M., Griffin, M.J., Waldbieser, G.C., Stanton, J.B., Leary, J.H., *et al.* (2022) Myxozoan community composition and diversity in clinical cases of proliferative gill disease in Mississippi catfish aquaculture. *The Journal of Parasitology* 108(2), 132–140.
Stoskopf, M.K. (2015) Biology and management of laboratory fishes. In: Fox, J.G., Anderson, L.C., Otto, G., Pritchett-Corning, K.R. and Whary, M.T. (eds) *Laboratory Animal Medicine*, 3rd edn. Academic Press, Cambridge, Massachusetts, pp. 1063–1086.
Sun, J., Fang, W., Ke, B., He, D., Liang, Y., *et al.* (2016) Inapparent *Streptococcus agalactiae* infection in adult/commercial tilapia. *Scientific Reports* 6(1), 26319. doi: 10.1038/srep26319.
Székely, C., Molnár, K. and Baska, F. (1988) Efficacy of fumagillin against *Myxidium giardi* Cepede, 1906 infection of the European eel (*Anguilla anguilla*): new observations on myxidiosis of imported glass eels. *Acta Veterinaria Hungarica* 36(3–4), 239–246.
Toranzo, A.E., Magariños, B. and Romalde, J.L. (2005) A review of the main bacterial fish diseases in mariculture systems. *Aquaculture* 246(1–4), 37–61. doi: 10.1016/j.aquaculture.2005.01.002.
Toranzo, A.E., Magariños, B. and Avendaño-Herrera, R. (2017) Vibriosis: *Vibrio anguillarum*, *V. ordalii* and *Aliivibrio salmonicida*. In: Woo, P.T.K. and Cipriano, R.C. (eds) *Fish Viruses and Bacteria: Pathobiology and Protection*. CAB International, Wallingford, UK, pp. 314–333. doi: 10.1079/9781780647784.0314.
Tsai, M.A., Wang, P.C., Liaw, L.L., Yoshida, T. and Chen, S.C. (2012) Comparison of genetic characteristics and pathogenicity of *Lactococcus garvieae* isolated from aquatic animals in Taiwan. *Diseases of Aquatic Organisms* 102(1), 43–51. doi: 10.3354/dao02516.

Tun, T., Yokoyama, H., Ogawa, K. and Wakabayashi, H. (2000) Myxosporeans and their hyperparasitic microsporeans in the intestine of emaciated tiger puffer. *Fish Pathology* 35(3), 145–156.

Vallecillos, A., Chaves-Pozo, E., Arizcun, M., Perez, R., Afonso, J.M., *et al.* (2021) Genetic parameters for *Photobacterium damselae* subsp. *piscicida* resistance, immunological markers and body weight in gilthead seabream (*Sparus aurata*). *Aquaculture* 543, 736892. doi: 10.1016/j.aquaculture.2021.736892.

Valverde, E.J., Labella, A.M., Borrego, J.J. and Castro, D. (2019) *Artemia* spp., a susceptible host and vector for lymphocystis disease virus. *Viruses* 11(6), 506. doi: 10.3390/v11060506.

van Samkar, A., Brouwer, M.C., van der Ende, A. and van de Beek, D. (2016) Zoonotic bacterial meningitis in human adults. *Neurology* 87(11), 1171–1179. doi: 10.1212/WNL0000000000003101

Vendrell, D., Balcázar, J.L., Ruiz-Zarzuela, I., de Blas, I., Gironés, O., *et al.* (2006) *Lactococcus garvieae* in fish: a review. *Comparative Immunology, Microbiology and Infectious Diseases* 29(4), 177–198. doi: 10.1016/j.cimid.2006.06.003.

Vezzulli, L., Pezzati, E., Brettar, I., Höfle, M. and Pruzzo, C. (2015) Effects of global warming on *Vibrio* ecology. *Microbiology Spectrum* 3(3). doi: 10.1128/microbiolspec.VE-0004-2014.

Weinstein, M.R., Litt, M., Kertesz, D.A., Wyper, P., Rose, D., *et al.* (1997) Invasive infections due to a fish pathogen, *Streptococcus iniae*. *New England Journal of Medicine* 337(9), 589–594. doi: 10.1056/NEJM199708283370902.

Whipps, C.M. and Kent, M.L. (2006) Phylogeography of the cosmopolitan marine parasite *Kudoa thyrsites* (Myxozoa: Myxosporea). *Journal of Eukaryotic Microbiology* 53(5), 364–373.

Whipps, C.M., Gauthier, D.T. and Kent, M.L. (2020) Fish mycobacteriosis. In: Woo, P.T.K., Leong, J. and Buchmann, K. (eds) *Climate Change and Infectious Fish Diseases*. CAB International, Wallingford, UK, pp. 265–279. doi: 10.1079/9781789243277.0265.

Whittington, I.D. (2004) The Capsalidae (Monogenea: Monopisthocotylea): a review of diversity, classification and phylogeny with a note about species complexes. *Folia Parasitologica* 51(2/3), 109–122.

Wise, D.J., Hanson, T.R. and Tucker, C.S. (2008) Farm-level economic impacts of *Bolbophorus* infections of channel catfish. *North American Journal of Aquaculture* 70(4), 382–387.

Yamashita, K. (1980) Parasitic Copepoda (*Ergasilus* sp.) found on the epidermis of larval fish of red sea bream (*Pagrus major*). *Fish Pathology* 15, 91–94.

Yanagida, T., Sameshima, M., Nasu, H., Yokoyama, H. and Ogawa, K. (2006) Temperature effects on the development of *Enteromyxum* spp. (Myxozoa) in experimentally infected tiger puffer, *Takifugu rubripes* (Temminck & Schlegel). *Journal of Fish Diseases* 29(9), 561–567.

Yang, R., Palermo, C., Chen, L., Edwards, A., Paparini, A., *et al.* (2015) Genetic diversity of *Cryptosporidium* in fish at the 18S and actin loci and high levels of mixed infections. *Veterinary Parasitology* 214(3–4), 255–263.

Yanong, R.P.E. and Terrell, S.P. (2003) Iridoviral-associated disease in oscars (*Astronotus ocellatus*). In: *Proceedings of the 34th Annual Meeting of International Association of Aquatic Animal Medicine, Waikoloa, Hawaii*. International Association of Aquatic Animal Medicine, Stamford, Connecticut, pp. 81–83.

Yanong, R.P.E. and Waltzek, T.B. (2019) *Megalocytivirus infections in fish, with emphasis on ornamental species*. University of Florida Institute of Food and Agriculture Science, Gainesville, Florida. Available at: http://edis.ifas.ufl.edu/pdf/FA/FA18200.pdf (accessed 15 January 2022).

Yasuda, H., Ooyama, T., Iwata, K., Tun, T., Yokoyama, H., *et al.* (2002) Fish-to-fish transmission of *Myxidium* spp. (Myxozoa) in cultured tiger puffer suffering from emaciation disease. *Fish Pathology* 37(1), 29–33.

Yilmaz, S., Yilmaz, E., Dawood, M.A.O., Ringø, E., Ahmadifar, E., *et al.* (2022) Probiotics, prebiotics and synbiotics used to control vibriosis in fish: a review. *Aquaculture* 547, 737514. doi: 10.1016/j.aquaculture.2021.737514.

Yokoyama, H., Liyanage, Y.S., Sugai, A. and Wakabayashi, H. (1999) Efficacy of fumagillin against haemorrhagic thelohanellosis caused by *Thelohanellus hovorkai* (Myxosporea: Myxozoa) in coloured carp, *Cyprinus carpio* L. *Journal of Fish Diseases* 22(3), 243–245.

Yokoyama, H., Kawakami, H., Yasuda, H. and Tanaka, S. (2003) *Henneguya lateolabracis* sp. n. (Myxozoa: Myxosporea), the causative agent of cardiac henneguyosis in Chinese sea bass *Lateolabrax* sp. *Fisheries Science* 69(6), 1116–1120.

Yokoyama, H., Whipps, C.M., Kent, M.L., Mizuno, K. and Kawakami, H. (2004) *Kudoa thyrsites* from Japanese flounder and *Kudoa lateolabracis* n. sp. from Chinese sea bass: causative myxozoans of post-mortem myoliquefaction. *Fish Pathology* 39(2), 79–85.

Yokoyama, H., Itoh, N. and Tanaka, S. (2005) *Henneguya pagri* n. sp. (Myxozoa: Myxosporea) causing cardiac henneguyosis in red sea bream, *Pagrus major* (Temminck & Schlegel). *Journal of Fish Diseases* 28(8), 479–487.

Yokoyama, H., Ayado, D., Miyahara, J., Matsukura, K., Takami, I., *et al.* (2011) Infection dynamics of *Microsporidium seriolae* (Microspora) causing the Beko disease of *Seriola* spp. *Fish Pathology* 46(2), 51–58.

Young, E.J., Bannister, J., Buller, N.B., Vaughan-Higgins, R.J., Stephens, N.S., *et al.* (2020) *Streptococcus iniae* associated mass marine fish kill off Western Australia. *Diseases of Aquatic Organisms* 142, 197–201. doi: 10.3354/dao03545.

Zlotkin, A., Hershko, H. and Eldar, A. (1998) Possible transmission of *Streptococcus iniae* from wild fish to cultured marine fish. *Applied and Environmental Microbiology* 64(10), 4065–4067. doi: 10.1128/AEM.64.10.4065-4067.1998.

6 Infectious Diseases of Warmwater Fish in Fresh Water

Andrew P. Shinn[1*†], Ha Thanh Dong[2†], Nguyen Tien Vinh[2], Ratchakorn Wongwaradechkul[1] and Gilda D. Lio-Po[3]
[1]*INVE (Thailand), Nonthaburi, Thailand;* [2]*Asian Institute of Technology, Pathum Thani, Thailand;* [3]*Southville Subd., Molo, Iloilo City, Philippines*

6.1 Introduction

Eurythermal fish species are those that can tolerate a wide range of temperatures (i.e. 2–32°C) (Molnár *et al.*, 2019). There is, however, no consensus on the definition of a warmwater species. Pope *et al.* (2009) consider these as species preferring temperatures >15°C, while Howard (2019) defines them as species preferring water temperatures around 27°C. In 2020, freshwater aquaculture accounted for 53,494,889 tonnes (i.e. 43.64% of total aquaculture production); a map of global finfish aquaculture is provided in Fig. 6.1. Finfish accounted for 90.04% of freshwater production in 2020 with 197 species or categories of fish (FishStatJ, 2022) farmed across 179 territories. Freshwater finfish production is increasing at 2.95% year on year (2016–2020), against an annual growth of 3.48% for the entire freshwater aquaculture sector.

Global freshwater cultured finfish production in 2020 is provided in Table 6.1. The cyprinids have major representation, while the top ten species, not including the general category 'freshwater fishes', account for 71.89% of the total tonnage. A summary of country-level cage and pen culture is presented in Tables 6.2 and 6.3.

For some species, cage culture is growing (e.g. *Siniperca* spp. in China) (Gui *et al.*, 2018), while elsewhere new policies linked to environmental protection have resulted in cage culture being discouraged (Nogueira *et al.*, 2020) or banned (Hu and Lv, 2016). In China, the demolition of cages used for rearing *Ictalurus punctatus* led to a 10.64% decrease in production (i.e. down 236,786 tonnes in 2016) and a consequential change to farming practices. At the same time, the Ministry of Fisheries and Aquaculture, Brazil has explored the creation of 100 aquaculture parks within 11 reservoirs for cage culture that could result in annual production of 226,934 tonnes of fish. The assessment and development of additional aquaculture parks could result in Brazil emerging as one of the largest producers of cage aquaculture in the world (Bueno *et al.*, 2015).

While there is a wealth of information on diseases of feral and cultured fish in warm fresh water (Lio-Po and Lim, 2014; Paladini *et al.*, 2017; Jahangiri *et al.*, 2021; Shinn *et al.*, 2022), there is a comparative scarcity of information on freshwater, cage culture-reared stocks. For many disease reports, the culture system is not clearly specified. Infections appear to occur

*a.shinn@inveaquaculture.com
†Joint senior authors.

 Climate Change on Diseases and Disorders of Finfish in Cage Culture, 3rd Edition (eds P.T.K. Woo and R.P. Subasinghe)

DOI: 10.1079/9781800621640.0006

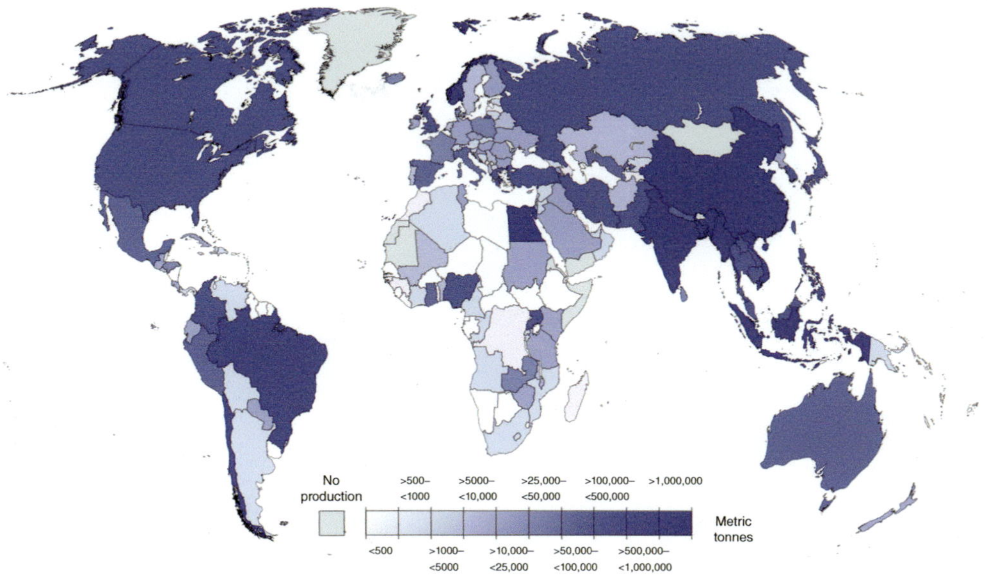

Fig. 6.1. Global finfish aquaculture by production (metric tonnes) in 2019. (Data drawn from FishStatJ, 2022.)

more frequently in cages than in ponds, where the captive-held stocks are exposed to pathogens of feral fish and to a greater number of intermediate hosts. Disease reports of cage culture in river systems and in urban waters exemplify this, where the biosecurity threats are greater and more challenging to control. There are also commensurate health risks to humans from stocks reared in unsanitary waters, polluted waters and in areas where fish-borne zoonotic diseases are prevalent. Diseases afflicting pond-reared and cage-cultured fish are in many cases similar; those that are significant in pond aquaculture are treated as potential problems of finfish reared in net enclosures and are the focus of this chapter. The potential impacts of climate change on the frequency and severity of infections of finfish reared in cage culture are also commented upon.

6.2 Viral Infections

Viral infections often cause mass mortality among fry, fingerlings and juveniles, although older fish may either develop pathological signs and die or may appear unaffected but are carriers. Most viral infections in fish occur at cool water temperatures of 15–25°C (Dishon *et al.*, 2007). Stress from handling, poor water quality, water temperature, age of fish, high stocking density and poor nutrition are factors that facilitate the development of viral diseases (Plumb, 1999). Viral infections of freshwater fish in tropical Asia are summarized in Table 6.4.

6.2.1 Tilapia lake virus (TiLV) disease

Tilapia lake virus disease (TiLVD), or syncytial hepatitis of tilapia (SHT), is an important viral disease of tilapia. Although first recorded in Israel and Ecuador from 2011, suspicious mortality events predating 2009 were known (Ferguson *et al.*, 2014). Reports from 16 countries rapidly followed, suggesting that the translocation of infected fish prior to the discovery of the pathogen may have contributed to its transboundary spread (Dong *et al.*, 2017a; Jansen *et al.*, 2019; Tang *et al.*, 2021).

The causative agent was initially termed tilapia lake virus (TiLV) upon the first discovery in Israel (Eyngor *et al.*, 2014). Further genome characterization suggested that TiLV is a novel

Table 6.1. The top freshwater finfish species/species groups with tonnages over 50,000 tonnes landed from aquaculture activities in 2020 with details regarding their distribution and production compared with tonnages landed from capture fisheries. (As categorized by FishStatJ, 2022; additional data are drawn from Froese and Pauly, 2022.)

Species (ranked by tonnage)	Latin name	Temperature range (°C)	No. of countries/islands reported from			Tonnage from aquaculture	No. of aquaculture producing countries	Tonnage from capture fisheries (no. of countries)	
			Native/endemic	Introduced	Not established			Native/endemic	Introduced
Grass carp (= white Amur)	*Ctenopharyngodon idella* (Valenciennes, 1844)	0–35	2	68	23	5,791,541	42	215 (1)	24,681 (17)
Silver carp	*Hypophthalmichthys molitrix* (Valenciennes, 1844)	6–28	4	58	19	4,896,612	35	1,016 (1)	33,182 (15)
Common carp	*Cyprinus carpio* (L., 1758)	3–35	17	116	6	4,236,326	80	29,048 (193)	100,421 (34)
Nile tilapia	*Oreochromis niloticus* (L., 1758)	14–33	22	77	6	3,561,316	74	208,307 (6)	73,337 (7)
Catla	*Gibelion catla* (Hamilton, 1822)	25–32	6	7 (8)[a]	4	3,540,312	8	nr	nr
Bighead carp	*Hypophthalmichthys nobilis* (Richardson, 1845)	4–26	1	44	23	3,187,236	20	nr	2,762 (8)
Carassius spp.	e.g. *Carassius carassius* (L., 1758)	2–22	42	15 (16)[a]	3	2,753,342	14	–	–
Freshwater fishes nei	Not specified					2,567,220	69	–	–
Striped catfish	*Pangasianodon hypophthalmus* (Sauvage, 1878)	22–26	4	4 (5)[a]	1	2,520,422	11	nr	nr
Roho labeo	*Labeo rohita* (Hamilton, 1822)	22–31	6	9	4	2,484,817	11	nr	nr
Torpedo-shaped catfishes nei	e.g. *Clarias batrachus* (L., 1758)	8–35	1	7	12[b]	1,249,012	10	–	–
	e.g. *Clarias gariepinus* (Burchell, 1822)		40	23 (24)[a]	5				

Tilapias nei	Various *Coptodon*, *Oreochromis*, *Sarotherodon*, tilapia spp.	8–36				1,146,316	69	–	–
	e.g. *Coptodon rendalli* (Boulenger, 1897)		11	26 (29)[a]	0				
	e.g. *Coptodon zillii* (Gervais, 1848)		28	23 (26)[a]	3				
	e.g. *Oreochromis aureus* (Steindachner, 1864)		10	35	4				
	e.g. *Oreochromis mossambicus* (Peters, 1852)	17–35	7 (9)[a]	93 (95)[a]	8 (9)[b]		10	0	21,450 (2)
	e.g. *Sarotherodon galilaeus* (L., 1758)		29	3 (4)[a]	1				
Wuchang bream	*Megalobrama amblycephala* (Yih, 1955)	10–20	1	2	0	781,737	1	nr	nr
Rainbow trout	*Oncorhynchus mykiss* (Walbaum, 1792)	12–21	6	84	14	733,999	75	140 (3)	1,216 (13)
Black carp	*Mylopharyngodon piceus* (Richardson, 1846)	0–40	4	17 (18)[a]	7	695,541	3	nr	nr
Snakeheads									
e.g. Indonesian snakehead	*Channa micropeltes* (Cuvier, 1831)	23–27	6	1 (2)[a]	1	648,363	10	15,335 (1)	0
e.g. striped snakehead	*Channa striata* (Bloch, 1793)		14	10	0		6	55,280 (2)	8,439 (2)
Largemouth black bass	*Micropterus salmoides* (Lacepède, 1802)	10–32	3	61 (63)[a]	11	621,327	8	863 (1)	180 (1)

Continued

Table 6.1. Continued

Species (ranked by tonnage)	Latin name	Temperature range (°C)	No. of countries/islands reported from			Tonnage from aquaculture	No. of aquaculture producing countries	Tonnage from capture fisheries (no. of countries)	
			Native/endemic	Introduced	Not established			Native/endemic	Introduced
Cyprinids nei	Various *Abramis*, *Chanodichthys*, *Cirrhinus*, *Rutilus*, *Tinca* spp.	4–26	–	–	–	578,041	26	–	–
Mrigal carp	*Cirrhinus cirrhosus* (Bloch, 1795)	14–31	4	8 (9)[a]	2	576,513	9	nr	nr
Yellow catfish	*Tachysurus fulvidraco* (Richardson, 1846)	16–25	5	0	0	565,477	1	nr	nr
Pangas catfishes nei	Various *Pangasius* spp.	26–28				462,216	4	–	–
	e.g. *Pangasius bocourti* (Sauvage, 1880)		4	0	0				
	e.g. *Pangasius pangasius* (Hamilton, 1822)		5	2 (4)[a]	2				
Channel catfish	*Ictalurus punctatus* (Rafinesque, 1818)	10–32	3	23	7	453,986	6	966 (1)	0
Blue-Nile tilapia, hybrid	*Oreochromis aureus* × *O. niloticus*	8–33	na	na	na	414,042	2	–	–
Mandarin fish	*Siniperca chuatsi* (Basilewsky, 1855)	4–22	3	1	0	376,986	1	nr	nr
Pond loach	*Misgurnus anguillicaudatus* (Cantor, 1842)	5–25	13	9	0	368,406	3	nr	nr
Amur catfish	*Silurus asotus* (L., 1758)	5–25	6	0 (1)[a]	0	351,468	3	nr	nr

Silver barb	*Barbonymus gonionotus* (Bleeker, 1849)	22–28	5	9	0	335,863	7	36,350 (2)	0
Asian swamp eel	*Monopterus albus* (Zuiew, 1793)	25–28	15	2	0	307,333	2	800 (1)	0
Japanese eel	*Anguilla japonica* (Temminck et Schlegel, 1846)	4–27	9	2 (3)[a]	2	279,011	4	121 (3)	0
North African catfish	*C. gariepinus* (Burchell, 1822)	8–35	39 (40)[a]	24	5	222,996	47	74,573 (7)	10 (1)
Giant gourami	*Osphronemus goramy* (Lacepède, 1801)	20–30	8	12	1	149,216	3	6,635 (1)	0
Sturgeons nei	Various *Acipenser*, *Huso* spp.	5–22	na	na	na	123,451	38	–	–
Pirapatinga (= piaractus)	*Myletes brachypomus* (G. Cuvier, 1817)	23–28	6 (7)[a]	2	3	122,021	10	2,772 (3)	0
Milkfish	*Chanos chanos* (Forsskål, 1775)	15–43	88	2	0	116,151	5	6,472 (1)	0
Cachama (= pacu or tambaqui)	*Colossoma macropomum* (G. Cuvier, 1818)	22–28	4	12	3	104,915	9	4,104 (3)	0
Africa–bighead catfish hybrid	*C. gariepinus* (Burchell, 1822) × *Clarias macrocephalus* (Günther, 1864)	8–35	na	na	na	99,344	1	–	–
Silver, bighead carps nei	*Hypophthalmichthys harmandi* (Sauvage, 1884), etc.	0–43	na	na	na	68,974	4	–	–
Climbing perch	*Anabas testudineus* (Bloch, 1792)	22–30	17	3	1	59,914	6	33,216 (2)	1,500 (1)
Orangefin labeo	*Labeo calbasu* (F. Hamilton, 1822)	19–33	7	1	0	52,808	2	–	–

nei, not included elsewhere; na, not applicable; nr, not reported.

[a] Includes unconfirmed/questionable reports

[b] Reports are based on misidentifications.

Table 6.2. A summary of freshwater cage and pen aquaculture production in selected countries ('est.' indicates an estimate).

Country	Cage/pen details	No.	Hectares	No. of culturists	Tonnage	Reference
Bangladesh	Cage		17.9		4,590	Department of Fisheries, Bangladesh (2021)
	Pen culture		7,263.0		13,337	
Brazil		100 parks	1,920.0		est. 284,778	Roubach *et al.* (2015)
China	Cage culture (given for 2020)		1,350.9		320,905	Ministry of Agriculture, Rural Affairs Fishery Administration (2021)
	Cage culture (given for 2015)		14,737.6		1,379,086	
India	Cages/pens in reservoirs	15,454	138.7			Department of Fisheries, India (2020)
Indonesia	Floating net cages (Karamba)				183,509	Indonesian Ministry of Marine Affairs and Fisheries (2020)
	Floating net (Jaring Apung Tawar)				432,698	
	Jaring Tancap Tawar				19,287	
Malaysia	Freshwater cages	30,416	55.0	1,963	15,736	Department of Fisheries Malaysia (2020)
	Pen culture	317	8.0	155	678	
Philippines					80,182	Philippine Statistics Authority (2022)
Thailand	Cage sites	7,471	220.3		52,335	Department of Fisheries, Ministry of Agriculture and Cooperatives, Thailand (2020); Shinn *et al.* (2022)
Total			25,711.4		2,502,343	

orthomyxo-like virus (Bacharach *et al.*, 2016). The virus was then officially named *Tilapinevirus tilapiae* (previously known as *Tilapia tilapinevirus*), a member of the family Amnoonviridae. TiLV is an enveloped single-stranded RNA (ssRNA) virus with a diameter of 55–100 nm.

TiLV affects several species in the family Cichlidae, including *Oreochromis niloticus*; a list of natural and susceptible hosts is summarized elsewhere (Surachetpong *et al.*, 2020; Tang *et al.*, 2021).

TiLV infects all stages of tilapia but is more severe in young fish (<50 g) than in older fish. Disease outbreaks usually occur in the first few weeks after stocking and result in a mortality of 20–90% leading to significant economic loss (Eyngor *et al.*, 2014; Ferguson *et al.*, 2014).

Table 6.3. Finfish species farmed in net and cage aquaculture. For each, where available, the production is provided in metric tonnes ('est.' indicates an estimate). Figures in brackets are production figures specifically from pen culture and are in addition to the figures provided. (Data sources: a = Department of Fisheries, Bangladesh, 2021; b = FAO, 2022; c = Lio-Po and Lim, 2014; d = Gui *et al.*, 2018; e = Radhakrishnan *et al.*, 2019; f = Indonesian Ministry of Marine Affairs and Fisheries, 2020; g = LARReC, 2001; h = Department of Fisheries, Malaysia, 2020; i = Griffiths *et al.*, 2019; j = Gurung *et al.*, 2010; k = Philippine Statistics Authority, 2022; l = Department of Fisheries, Ministry of Agriculture and Cooperatives, Thailand, 2020.)

Common name	Latin name	Bangladesh[a]	Brazil[b]	Cambodia[c]	China[d]	India[e]	Indonesia[f]	Lao PDR[g]	Malaysia[h]	Myanmar[i]	Nepal[j]	Philippines[k]	Thailand[l]
Basa	*Pangasius bocourti*							yes					
Bata	*Labeo bata*	(319)											
Bighead carp	*Hypophthalmichthys nobilis*				yes						yes		
Black carp	*Mylopharyngodon piceus*				yes								
Black ear catfish	*Pangasius larnaudii*							yes					157
Blunt-snout bream	*Megalobrama amblycephala*				yes								
Boggut labeo ('ghonia')	*Labeo boggut*	(177)											
Carp spp.	Not known								47.7			419.9	
Catla	*Gibelion catla*	(1,288)		yes (*Labeo* spp.)		yes				yes			
Channel catfish ('baung')	*Hemibagrus nemurus*								792.6				
Channel catfish	*Ictalurus punctatus*				yes								
Common carp	*Cyprinus carpio*	(292)					148,542.0	yes	0.2	yes	yes		
Common climbing perch ('koi')	*Anabas testudineus*	(18)											10

Continued

Table 6.3. Continued.

Common name	Latin name	Bangladesh[a]	Brazil[b]	Cambodia[c]	China[d]	India[e]	Indonesia[f]	Lao PDR[g]	Malaysia[h]	Myanmar[i]	Nepal[j]	Philippines[k]	Thailand[l]
Common silver barb/ Java barb	*Barbonymus gonionotus*	(1,440)					289.1	yes	53	yes			30
Crucian carp	*Carassius carassius*				yes								
Freshwater catfish ('keli')	*Clarias* sp.			yes			18,754.6		1,093.6 (101.3)	yes			769
Giant gourami	*Osphronemus goramy*						529.2	yes	0.1				96
Giant/ Indonesian snakehead	*Channa micropeltes*			yes									
Grass carp/ Chinese carp	*Ctenopharyngodon idella*	(327)			yes			yes		yes	yes		
Kalibaus	*Labeo calbasu*	(146)				yes							
Kuria labeo	*Labeo gonius*					yes							
Mandarin fish	*Siniperca* spp.				yes								
Milkfish	*Chanos chanos*											8,265.20	
Mozambique tilapia	*Oreochromis mossambicus*						2,145.2		7.3				2
Mrigal	*Cirrhinus cirrhosus*	(1,179)				yes							
Mud carp	*Cirrhinus molitorella*							yes					
Nile tilapia (incl. GIFT and red)	*Oreochromis niloticus*	4,590 (3,261)	est. >186,448		yes	yes	318,515.8	yes	6,108.5 (0.8)	yes		71,488.40	50,079
Pacu	*Piaractus mesopotamicus*		est. 10,112										
Pirapitinga	*Piaractus brachypomus*		est. 8,518										
Pomfret	Various pomfret species						25,044.7						

River carp ('lampam sungai')	*Barbonymus schwanenfeldii*								13.6		
River carp ('jelawut')	*Leptobarbus hoevenii*			yes			1,508.1		80.1		
Rohu	*Labeo rohita*	(1,849)				yes					
Sand/marble goby ('ketutu')	*Oxyeleotris marmoratus*			yes (spp.)			143.6		0.8		1
Silver carp	*Hypophthalmichthys molitrix*	(1,003)			yes			yes		yes	
Silver shark minnow	*Osteochilus vittatus*						63.5				
Small scale mud carp	*Cirrhinus microlepis*			yes							
Snakehead, various and hybrids	*Channa* spp.	(47)	yes		yes		42,637.3	yes			4
South American catfish/ jundiá	*Rhamdia quelen*		yes								
Streaked prochilod/ curimbatá	*Prochilodus lineatus*										
Striped catfish	*Pangasianodon hypophthalmus*	(345)		yes (spp.)		yes	72,437.6	yes	7,525.1 (576)		66
Sturgeons	Various species				yes						
Swamp eels	*Monopterus* and *Ophichthys* spp.	(13)			yes						
Tambaqui	*Colossoma macropomum*		est. 21,856								
Tambaqui × pacu hybrids	Various hybrids		yes								

Continued

Table 6.3. Continued.

Common name	Latin name	Bangladesh[a]	Brazil[b]	Cambodia[c]	China[d]	India[e]	Indonesia[f]	Lao PDR[g]	Malaysia[h]	Myanmar[i]	Nepal[j]	Philippines[k]	Thailand[l]
Walking catfish	*Clarias batrachus*	(9)											
Wallago catfish ('boal')	*Wallago attu*	(51)											
Yellow mystus	*Mystus nemurus*				yes		4,867.9						
Other freshwater fish	Other freshwater fish	(1,573)					15		13.8			8.5	835
Total		17,927	226,934				635,493.60		16,414.50			80,182	52,049

Table 6.4. Taxonomic classification, viral properties, susceptible host(s) and geographical distribution of major viruses causing diseases in tropical finfish commonly cultured in fresh waters.

Disease	Virus agent	Genus/Family	Nucleic acid/ Genome	Morphology/Size (nm)	Enveloped	Susceptible host(s)	Geographical distribution
Tilapia lake virus (TiLV) disease or syncytial hepatitis of tilapia (SHT)	*Tilapinevirus tilapiae*	*Tilapinevirus* Amnoonviridae	ssRNA 10.323 kb 10 RNA segments	Icosahedron 55–100	yes	Tilapia, giant gourami, African cichlids	17 countries in Asia, Africa and American continents
Tilapia parvovirus (TiPV) disease	Tilapia parvovirus (TiPV)	*Parvovirus Parvoviridae*	ssDNA >4269 bp	Spherical ~30	no	Tilapia	China, Thailand
Carp edema virus disease (CEVD) or koi sleepy disease (KSD)	*Cyprinid herpesvirus 1*	*Poxvirus Chordopoxvirinae*	dsDNA	Spherical to pleomorphic shape ~416–450	yes	Carps	Europe, USA, Japan, China, India, Korea, Thailand
Spring viraemia of carp (SVC)	Spring viraemia of carp virus (SVCV)	*Vesiculovirus* Rhabdoviridae	ssRNA	Bullet shape 60–90 wide by 90–180 long	yes	Carps	Europe, Asia
Koi herpesvirus disease (KHVD)	*Cyprinid herpesvirus 3* (CyHV-3) or koi herpesvirus (KHV)	*Herpesvirus* Alloherpesviridae	dsDNA	Icosahedron ~100–110	yes	Carps, tilapia	Worldwide
Grass carp reovirus disease (GCRVD)	Grass carp reovirus (GCRV) or Chinese grass carp reovirus (CGRV) or golden shiner virus (GSV)	*Aquareovirus C* Reoviridae	dsRNA	Icosahedron ~60–70	no	Carps	Asia
Channel catfish virus disease (CCVD)	Channel catfish virus (CCV) or *Ictalurid herpesvirus 1*	*Ictalurivirus* Alloherpesviridae	dsDNA	Icosahedron ~90–100	yes	Channel catfish, blue catfish, striped catfish	USA, Malaysia
Epizootic ulcerative syndrome (EUS)	EUS rhabdovirus	*Vesiculovirus* Rhabdoviridae	ssRNA	Bullet shape ~65–175	yes	Snakehead, catfish, goby, gourami, climbing perch, barbs	South-East Asia, East Asia and Africa

Diagnosis

Affected fish may present lethargy, loss of appetite, abnormal swimming, cessation of schooling, skin erosion, scale protrusion, gill pallor, abdominal distension, anaemia and exophthalmia (Ferguson *et al.*, 2014; Dong *et al.*, 2017b). In diseased hybrid red tilapia, a pale body and reddish opercula are usually observed, while diseased Nile tilapia usually present a darkened body (Fig. 6.2A). Internally, abnormalities include pale, necrotic and watery livers, with bile diffusion giving it a dark or an unusual green colour. An enlarged spleen and gall bladder, and fluid accumulation in the intestine and abdominal cavity, may also be observed (Dinh-Hung *et al.*, 2021; Tang *et al.*, 2021).

Pathognomonic lesions for the histopathological diagnosis are the presence of syncytial cells containing multiple nuclei in the liver (syncytial hepatitis) (Ferguson *et al.*, 2014; Del-Pozo *et al.*, 2017; Dong *et al.*, 2017b) (Fig. 6.2B). Stereotypical histopathological changes were also reported in the liver including hepatic vacuolization, focal necrosis, losing storage fats, cell dissociation, and a foamy cytoplasm to hepatocytes containing lipoprotein-like droplets or eosinophilic inclusions. Pancreatic degeneration and infiltration of inflammatory cells were also described (Jansen *et al.*, 2019). Other histopathological alterations in the spleen, kidney, brain, gills, gastrointestinal tract and reproductive organs are not pathognomonic.

Molecular methods for the early screening of TiLV from asymptomatic fish or as a confirmatory diagnosis of clinically sick fish include conventional reverse transcription polymerase chain reaction (RT-PCR), nested RT-PCR, semi-nested RT-PCR, SYBR Green- and TaqMan probe-based quantitative RT-PCR (RT-qPCR) and reverse transcription loop-mediated isothermal amplification (RT-LAMP) methods for pond-side diagnosis of TiLV (see Taengphu *et al.*, 2022). Taengphu *et al.* (2022) published a validated RT-qPCR method that can detect and quantify TiLV from both fish and environmental samples. Delamare-Deboutteville *et al.* (2021) integrated rapid amplicon sequencing, called Oxford Nanopore Technologies (ONT), with a PCR assay that serves as a promising platform for backpack, pond-side diagnosis and genotyping of TiLV in a single day. Other molecular methods for the localization of TiLV from infected tissues include *in situ* hybridization (ISH) using TiLV-specific probes (Bacharach *et al.*, 2016; Dong *et al.*, 2017b; Dinh-Hung *et al.*, 2021) and immunohistochemistry (IHC) using TiLV-binding antibodies (Piewbang *et al.*, 2021).

Several cell lines are reported to propagate TiLV including E-11, a clone of the cell line SSN-1 derived from *Channa striata*, tilapia primary brain cells OmB and TmB derived from *Oreochromis mossambicus*, CFF derived from *Pristolepis fasciatus*, OnlB and OnlL derived from *O. niloticus* liver and brain, TiB cell line established from tilapia brain, and CAMB originated from hybrid snakehead brain (see Wang *et al.*, 2020). Cytopathic effects (CPEs) typically appear 3–12 days after inoculating cells at 25°C with filtered homogenate containing TiLV virions.

Transmission

TiLV can be transmitted both horizontally and vertically. Cohabitation of infected and naïve fish induced disease and resulted in 56–80% mortality in a couple of weeks (Eyngor *et al.*, 2014; Liamnimitr *et al.*, 2018). Using environmental RNA (eRNA) to quantify viral loads in water, Taengphu *et al.* (2022) found ~10^3–10^5 copies per litre of pond water during disease outbreaks, confirming its waterborne transmission. Vertical transmission of TiLV from fertilized eggs and fry was confirmed through experimental infection (Yamkasem *et al.*, 2019; Dong *et al.*, 2020).

Control and prevention

Biosecurity, selective breeding and vaccination are effective in managing viral diseases in aquaculture. Using TiLV-free broodstock for seed production, early screening for TiLV-negative stocks by PCR before stocking, reducing stress and improving fish immunity in the early days of stocking are recommended to reduce the risk of disease outbreaks (Jansen *et al.*, 2019; Tang *et al.*, 2021).

Disinfection of the pond, water and equipment could prevent contamination. Jaemwimol *et al.* (2019) reported that 2.5 ppm iodine, 10 ppm sodium hypochlorite, 300 ppm hydrogen peroxide, 80 ppm formalin and 0.5% Virkon™ can inhibit the *in vitro* replication of TiLV; while Soto *et al.* (2019) found that 50 ppm buffered

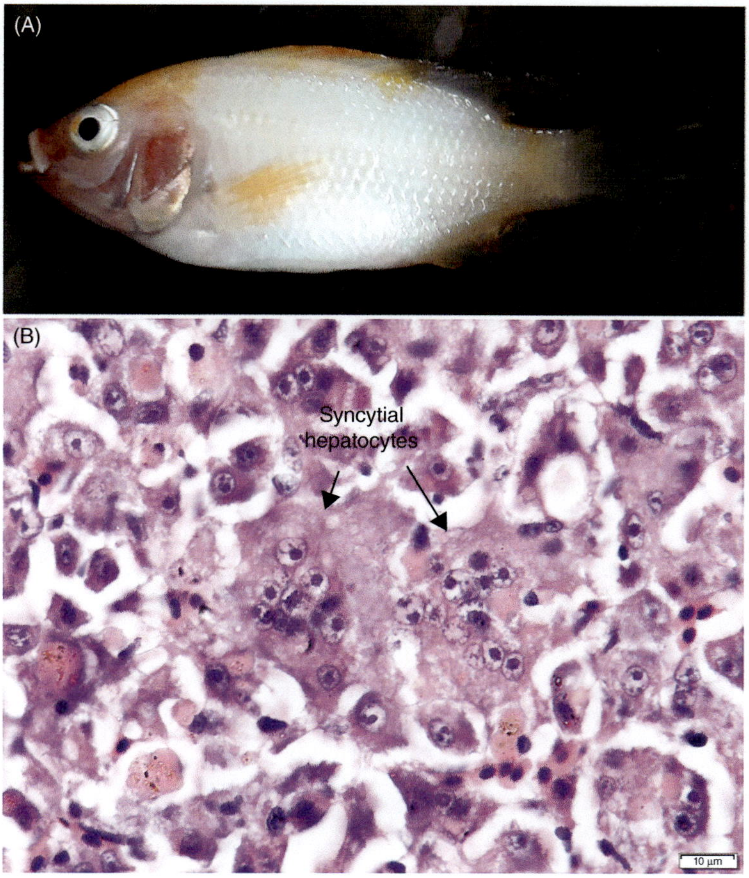

Fig. 6.2. (A) Juvenile hybrid red tilapia naturally infected with TiLV showing pale colour, scale protrusion, reddish colour around the mouth and on the operculum. (B) Pathognomonic lesions (syncytial hepatocytes) found in the fish liver are useful for histopathological diagnosis of TiLV infection. (Images by H.T. Dong.)

povidone iodine and household bleach (20 ppm free chlorine) can inactivate TiLV in 30 min.

Several studies report vaccination efficacy in preventing TiLVD. A live attenuated vaccine developed in Israel had a relative percentage survival (RPS) of 56–58% (Bacharach and Eldar, 2016). In China, a β-propiolactone-inactivated vaccine using the adjuvant Montanide IMS 1312 gave protection with RPS of 85.7% (Zeng *et al.*, 2021). Segment 8 (VP20)-based DNA and recombinant protein vaccines have also been investigated. DNA vaccine prime-recombinant VP20 (rVP20) boost, resulted in 72.5% protection (Zeng *et al.*, 2021). In Thailand, Mai *et al.* (2021) developed heat- and formalin-killed vaccines which gave good protection in tilapia juveniles with RPS of 71.3 and 79.6 %, respectively. These vaccines stimulated both branches of the adaptive immune systems (B-cell and T-cell responses) and elicited both systemic and mucosal anti-TiLV immunoglobulin M (IgM). Interestingly, the protective antibodies raised in vaccinated broodstock can be transferred maternally to the fertilized eggs and these persist in 1- to 3-day-old larvae. Thus, vaccination of tilapia broodstock was suggested to be a promising strategy for preventing TiLV infection in tilapia fertilized eggs and larvae (Mai *et al.*, 2022). Up-scaling TiLV vaccine production at an affordable cost for tilapia farmers remains a bottleneck for the tilapia industry.

6.2.2 Tilapia parvovirus (TiPV) disease

The first mass mortality report of adult *O. niloticus* was from China in 2020. The causative agent was identified as a novel parvovirus, called tilapia parvovirus (TiPV), a new member of the genus *Chapparvovirus* (family Parvoviridae) (Liu *et al.*, 2020). TiPV has been detected in tilapia samples collected from multiple provinces in China during 2015–2019 with relatively high prevalence (23.1 to 64.6%) (Du *et al.*, 2019; Liu *et al.*, 2020). Recently, a co-infection of TiLV and TiPV resulted in mortalities of 50–70% in juvenile hybrid red tilapia (*Oreochromis* sp.) in Thailand (Yamkasem *et al.*, 2021; Piewbang *et al.*, 2022). Co-infection of TiPV and *Streptococcus agalactiae* in adult Nile tilapia was also reported most recently in Thailand where <1% of the affected fish population showed signs of illness, and only minor mortalities were recorded because of handling stress (Dong *et al.*, 2022).

TiPV is a non-enveloped, spherical-shaped virus that is ~30 nm in diameter. Its genome is linear single-stranded DNA (ssDNA).

Relatively high mortality (60–70%) was estimated in adult Nile tilapia during disease events in China. Experimental infection using tissue homogenates from diseased fish resulted in 90% mortality within a couple of weeks post-infection (Liu *et al.*, 2020).

Diagnosis

Currently, no pathognomonic gross signs are proposed for the diagnosis of TiPV disease. Sick adult *O. niloticus* displayed lethargy, anorexia, abnormal corkscrew swimming movements, skin haemorrhages, exophthalmia and pronounced ocular lesions (Liu *et al.*, 2020) (Fig. 6.3A). TiLV and TiPV co-infected hybrid red tilapia displayed lethargy, haemorrhages and focal ulceration on the body surface, exophthalmos and abnormal swimming (Yamkasem *et al.*, 2021; Piewbang *et al.*, 2022). On the other hand, TiPV and *S. agalactiae* co-infected fish showed skin erosions, cloudy eye lenses and corneal injury. Most of the internal organs appeared normal, except for a few cases of microhepatica, intestinal pallor and necrotic areas in the ovary (Dong *et al.*, 2022).

Cowdry type A inclusion bodies (CAIB) within pancreatic acinar cells were proposed as the pathognomonic lesions that are diagnostic of TiPV infection (Dong *et al.*, 2022) (Fig. 6.3B).

Conventional PCR and SYBR Green-based quantitative PCR (qPCR) methods for the diagnosis of TiPV were published by Liu *et al.* (2020).

Liu *et al.* (2020) successfully isolated TiPV from infected fish using the tilapia brain cell line (TiB) originating from Nile tilapia. After inoculation with tissue homogenate, CPEs were initially observed at 3 days post-infection (dpi) and were clearly observed at 6 dpi, with completely detached cells at 8 dpi. Observed CPEs were cell shrinkage, rounding, and cell fusion with cytoplasmic vacuolization.

Transmission

The TiPV transmission route is still unknown. TiPV, like other aquatic viruses, is likely to spread vertically and horizontally. Dong *et al.* (2022) found TiPV within the oocytes of infected Nile tilapia.

Control and prevention

TiPV is currently reported in China and Thailand; wider distribution of the virus is likely due to the long history of active movement of live fish for aquaculture. Most disease cases were co-infections of TiPV and other pathogens, suggesting that further investigation to assess the true impact of TiPV is required. Apart from disease surveillance, increasing awareness and biosecurity measures are recommended to prevent the transboundary spread of this virus.

6.2.3 Carp edema virus (CEV) disease

Carp edema virus disease (CEVD) is caused by carp edema virus (CEV), a pox-like virus belonging to the Poxviridae (Adamek *et al.*, 2017a). As infected fish become lethargic and unresponsive at the bottom of their culture system, the disease is therefore also known as 'koi sleeping disease' (KSD). It was first described in juvenile koi carp in Japan in the 1970s with a mortality rate of up to 80–100%. Later, it was reported sporadically in adult koi and common carp (Oyamatsu *et al.*, 1997a; Miyazaki *et al.*, 2005). Due to the popularity of koi and common carp and their international trade, the virus has been introduced to

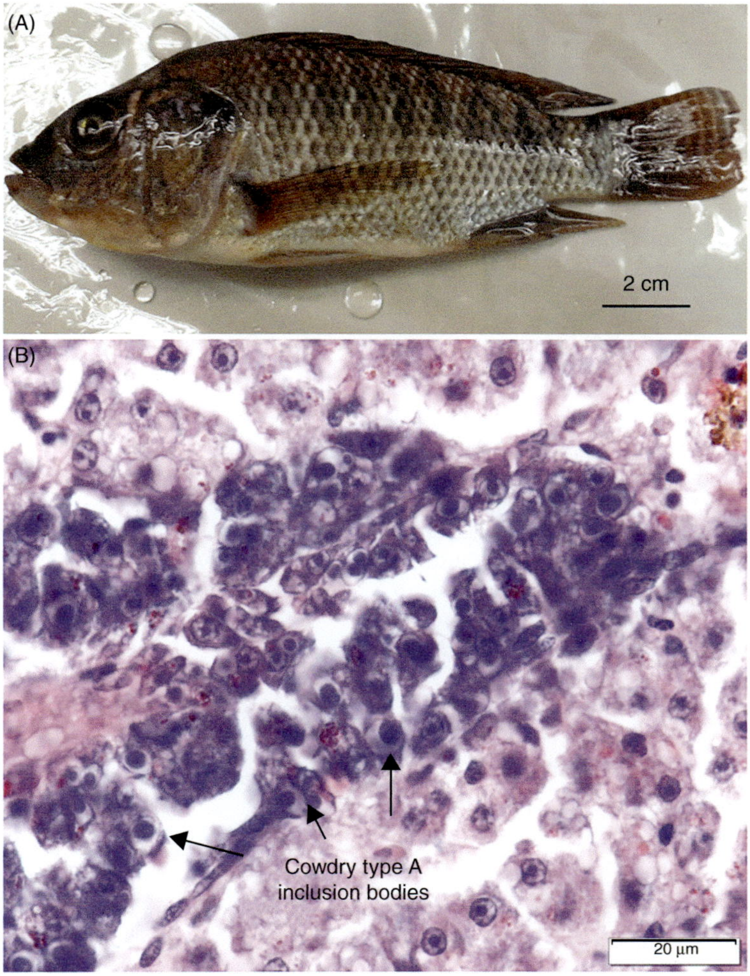

Fig. 6.3. (A) Adult Nile tilapia naturally infected with TiPV showing ocular damage. (B) Haematoxylin and eosin-stained section revealing pathognomonic lesions (Cowdry type A inclusion bodies) in the pancreatic tissue. (Images by H.T. Dong.)

many other nations (Marsella *et al.*, 2021). These reports serve as a warning of a new threat to the carp industry, which constitutes about 8.3% (in 2018) of the aquaculture food production sector globally (FAO, 2020).

The virus is described as enveloped with a 'mulberry-like' morphology. Two types of virions are described: immature ones with a rounded shape ranging from 416 to 450 nm in diameter, and mature ovoid virions measuring 300–400 nm × 250–400 nm. Mature virions have capsomeres arranged around a kidney-shaped nucleoid (Miyazaki *et al.*, 2005). The genome of CEV consists of double-stranded linear DNA; its complete genome has been published recently (Mekata *et al.*, 2021). Based on the P4a nucleotide sequence encoding the P4a core protein, at least three genogroups have been confirmed (Soliman *et al.*, 2019).

Diagnosis

In addition to the 'sleeping' behaviours of fish, signs of infection include gill necrosis, swelling of the primary filaments, sunken eyes, an over-secretion of mucus, and haemorrhages at the mouth and the bases of fins (Miyazaki *et al.*, 2005; Way *et al.*, 2017). The virus mainly attacks the gills, causing congestion, interfering with respiration and rendering fish lethargic (Oyamatsu *et al.*, 1997a).

Over 80% of the virus was found in the gills (Oyamatsu *et al.*, 1997a), making it a suitable tissue for diagnosis. Histopathology of the gills may show vacuolization in the primary gill lamellae; clubbing and fusion of the secondary gill lamellae; blockage of the interlamellar spaces by cellular debris; and epithelial cell enlargement (Adamek *et al.*, 2017a).

Molecular detection is based on the gene that encodes the P4a core protein. The first PCR protocols were developed in Japan using single and nested PCR (Oyamatsu *et al.*, 1997b), thereafter nested PCR and qPCR methods (Adamek *et al.*, 2016; Matras *et al.*, 2017) were developed. More recently, a novel droplet digital PCR (ddPCR) assay was developed that can detect about 2.2 ± 0.26 copies of CEV DNA per reaction and serves as a valuable tool for the diagnosis of this emerging disease (Wang *et al.*, 2021).

To date, no susceptible cell line has been reported for CEV, making it difficult to isolate the virus *in vitro* as part of disease diagnostics.

Transmission

Outbreaks occur more frequently in low temperatures (15–25°C) (Miyazaki *et al.*, 2005) and lower (6–15°C) (Way and Stone, 2013; Soliman *et al.*, 2019). Although temperature is the key factor, stressors such as poor water conditions and transportation may lead to manifestation of clinical signs (Lewisch *et al.*, 2015). Following lethargy, mortality occurs within 4–5 days, with a peak of mortality around 2 weeks when 75–100% of fish die (Miyazaki *et al.*, 2005). Co-infections of CEV with other pathogens such as *Cyprinid herpesvirus 3* (CyHV-3) (Adamek *et al.*, 2016; Kim *et al.*, 2020), flavobacteria (Adamek *et al.*, 2018) and *Aeromonas* spp. (Jung-Schroers *et al.*, 2015; Lewisch *et al.*, 2015) have been casually reported. The findings suggest that CEV acts as the primary agent that injures the gill structure or as an immunosuppressant paving the way for secondary infection (Jung-Schroers *et al.*, 2015; Way *et al.*, 2017).

To date, there is no solid evidence regarding the mechanism of transmission, including effective vectors of CEV. Water, however, is suggested to be the main cause of contagion by carrying epithelial cells that have been shed from the gill filaments of infected fish (Hedrick *et al.*, 1997).

Control and prevention

Currently there is no effective treatment for CEVD, hence preventive measures are the best resort for this disease. Translocation of fish remains a high-risk factor that can introduce the pathogen, so an effective diagnosis regime is imperative. As temperature constitutes a factor in virus proliferation, the harvesting and handling of fish in low temperatures of 15–25°C should be avoided. In Japan, keeping fish in 0.5% saline water prior to stressful events such as translocation, grading and handling is a common practice that helps delay the growth of CEV and helps rejuvenate the physiological condition of the host. It should be noted though that bathing fish in saline does not eradicate the virus. CEVD appears to cause more serious clinical signs in some strains of carp than in others, so in production it is judicious to select varieties that can resist the virus such as wild Amur carp (Adamek *et al.*, 2017b).

6.2.4 Koi herpesvirus (KHV) disease

Outbreaks of koi herpesvirus disease (KHVD), also known as carp nephritis and gill necrosis virus (CNGV) disease (Pikarsky *et al.*, 2004), have been reported extensively in Europe and North America. In Asia, epizootic levels were reported in Israel, Japan, Indonesia, Taiwan, Thailand, Singapore and Korea (Lio-Po and Lim, 2014). A three-year surveillance in 2004 to 2006 did not detect koi herpesvirus (KHV) in Cambodia, Lao PDR, Myanmar, the Philippines and Vietnam (Lio-Po *et al.*, 2009). Economic losses can be significant (Sunarto *et al.*, 2005; Yuasa and Sano, 2009).

KHV or CyHV-3 belongs to the family Herpesviridae (see Waltzek *et al.*, 2005). This double-stranded DNA (dsDNA) virus has an icosahedral nucleocapsid measuring 101.9 ± 10.3 nm and a total genome length of approximately 295 kbp (Aoki *et al.*, 2007). The KHV strains found in the USA, Israel and the Netherlands were closely related and of genotype E, while those found in Japan, Indonesia,

Taiwan and China were also closely related and of genotype A (Kurita *et al.*, 2009).

KHV infects common carp, koi carp and hybrid ghost carp although there are conflicting reports on the susceptibility of other fish species to KHV.

Natural and experimental transmission of the infection from common carp to *Carassius auratus*, *Ctenopharyngodon idella*, *Hypophthalmichthys molitrix*, *O. niloticus*, *Pangasianodon hypophthalmus*, *Plecoglossus altivelis*, etc. were unsuccessful (Perelberg *et al.*, 2003; Ito *et al.*, 2007; Yuasa and Sano, 2009). Hedrick *et al.* (2006), however, reported that goldfish × common carp hybrids were moderately sensitive to KHV.

Natural infections of susceptible fish result in 80–90% mortality within a week after the onset of clinical signs (Lio-Po and Lim, 2014). The virus can be transmitted from infected fish to naïve koi or common carp via intraperitoneal injection, by bath or by cohabitation.

Experimental exposure of healthy common carp to KHV at 22°C can cause up to 82% mortality within 15 days (Ronen *et al.*, 2003). The incubation period of the disease is 5–7 days, characterized by onset of clinical signs, and spread to koi and common carp was rapid when water temperatures were 15–25°C (Gilad *et al.*, 2003; Ronen *et al.*, 2003).

Diagnosis

Infected fish typically have white, necrotic patches on the gill filaments, sunken eyes, haemorrhages on the body surface, and excessive mucus production with rough pale patches on the skin. Affected fish often swim at the surface and exhibit respiratory distress.

The gills develop lamellar epithelial degeneration, focal areas of necrosis and exfoliation. There is hyperplasia and fusion of the secondary gill lamellae (Tu *et al.*, 2004). Intranuclear inclusion bodies (Fig. 6.4), a pathognomonic lesion for histopathological diagnosis of KHVD, can be observed from multiple organs of the KHV-infected fish such as the branchial epithelium, liver, kidney and spleen. The virus can be visualized by transmission electron microscopy (TEM). Necrotic tissues are also seen in the liver, spleen, and kidney parenchymal cells (Hedrick *et al.*, 2000). Non-specific secondary infections may be associated with KHV infections (e.g. *Flavobacterium columnare* and *Aeromonas* spp.) (Sunarto *et al.*, 2005).

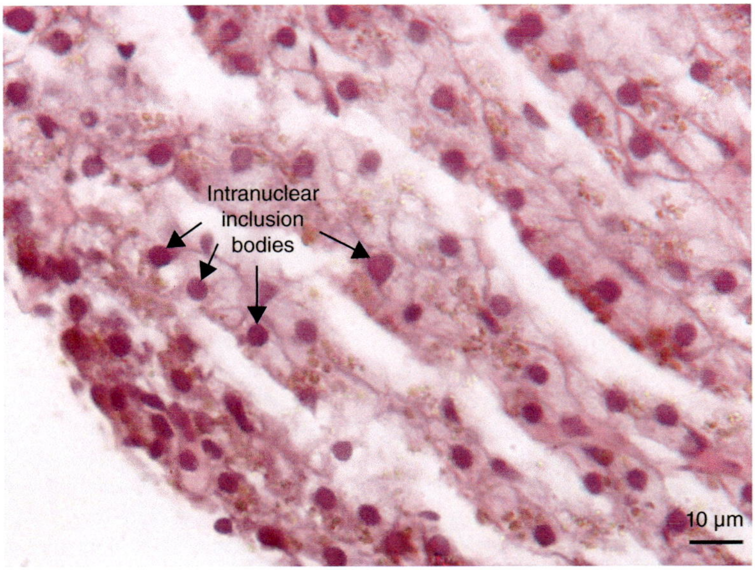

Fig. 6.4. Naturally KHV-diseased koi carp showed pathognomonic lesions (intranuclear inclusion bodies) in the liver (haematoxylin and eosin stain). (Image by H.T. Dong.)

To date, PCR is one of the most sensitive, specific and rapid tests for KHV detection (Gray *et al.*, 2002; Gilad *et al.*, 2004; Bercovier *et al.*, 2005; El-Matbouli *et al.*, 2007). A ring trial confirmed that the Bercovier-TK (1) and modified Gray SpH (2) primer sets were the most robust for detection of KHV DNA (Way *et al.*, 2008). A real-time PCR for the quantitative estimation of the KHV virus has also been developed using TaqMan real-time PCR (Gilad *et al.*, 2004).

Despite high sensitivity, the detection of latent and persistent stages, or during the onset of the disease when viral loads are low, may result in false negative results (Gilad *et al.*, 2003; Bergmann *et al.*, 2010). Subsequently, Monaghan *et al.* (2015a) assessed the sensitivity of seven PCRs for the early detection of KHV and found all PCRs could detect higher positivity from mucus samples (gill, skin, fin base swabs) compared with fish organs (gill, spleen, kidney, gut) in early infection (<4 days post-challenge). Viral DNA may be more concentrated in the mucus during the first days of transmission. The World Organisation for Animal Health (Office International des Epizooties, OIE), however, recommends diagnosis from internal organs of late-stage infected fish (OIE, 2019) as the virus colonizes these organs in abundance.

The development of a rapid LAMP test proved equally sensitive in its detection (Yoshino *et al.*, 2006). The analysis of mucus serves as an appropriate non-destructive sampling of suspected fish, particularly given the value of koi specimens. The recently developed fluorescence real-time LAMP assays to simultaneously detect CyHV-3 and CEV genogroup II with high sensitivity serve as an appropriate tool for the screening of transboundary fish (Cano *et al.*, 2021).

Other RT-PCR methods targeting the mRNA terminase (Yuasa *et al.*, 2012) and ISH (Haenen *et al.*, 2004) can also detect KHV in paraffin-embedded tissues from infected common carp. Strong ISH-positive responses were detected within epithelial cells of the secondary gill lamellae and in mucus cells of the basal area (Lee *et al.*, 2012). ISH demonstrated its utility in detecting viral DNA in gill and gut sections at 1 h post-infection (Monaghan *et al.*, 2015b).

Enzyme-linked immunosorbent assay (ELISA) allows for the detection of antibodies to KHV in the serum of fish previously exposed to the virus (St-Hilaire *et al.*, 2009). This method confirms the presence of anti-KHV antibodies found in the sera of survivors from KHV outbreaks that become carriers of the virus. The combination of ELISA and qPCR is recommended in the diagnosis of KHV-suspected carrier fish (Soto *et al.*, 2020).

The virus may be isolated in susceptible fish cell lines such as koi fin (KF-1), koi fin (KFC), koi tail (KT-2), common carp brain (CCB), koi snout (KS) or head kidney of common carp (KoK) (Eckart *et al.*, 2019). The KF-1 cells inoculated with KHV exhibit typical vacuolations at 20–25°C in 7–14 days. In contrast, survival of KHV in CCB cells was maintained for 30 days at 30°C (Dishon *et al.*, 2007). After 3 dpi, CPE is characterized by giant syncytia. Current recommended cell lines may not be highly sensitive to KHV (OIE, 2019a); newer cell lines such as KoK are expected to improve detection of KHV (Eckart *et al.*, 2019).

Transmission

Temperature is a critical factor in the pathogenesis of KHV infection. Cell cultures inoculated with KHV and incubated at optimum temperature develop typical vacuolation. However, the vacuolated cells may revert to normal, and plaques may disappear when the cells are shifted to non-permissive temperatures and can reappear after transfer to permissive temperatures (Dishon *et al.*, 2007). *In vivo*, the virus induces infection/mortalities at 18–25°C; if the fish are held at 13 or 30°C, no infection or mortality develops (Goodwin, 2005).

Some studies reported that virulence and pathogenesis of KHV are strain dependent. The virus can remain latent in the host for long periods without manifesting clinical signs of KHVD, becoming active only at permissive temperatures (Lio-Po and Lim, 2014).

Other species may be potential vectors for KHV transmission. In an experimental set-up, Matras *et al.* (2019) confirmed several potential vectors with the detection of viral DNA at 49 dpi. Fabian *et al.* (2013) found that with naturally occurring infections, many wild cyprinid and non-cyprinid species cohabitating with carp can become carriers of the virus, regardless of season, temperature variation, age and infection status of the carp stock. In addition, Minamoto *et al.* (2011) found that several rotifers from natural lakes tested positive for KHV DNA.

Prevention and control

A live attenuated KHV commercial vaccine, developed in Israel, is available for the prevention of KHV (Ronen *et al.*, 2003; Perelberg *et al.*, 2005). Vaccinated carp develop high antibody titres resulting in RPS of 80–95%. Protective immunity after vaccination lasts for at least 8 months. Using two formalin-inactivated vaccines entrapped within the liposomal membrane experimentally sprayed on to dry pellets and fed to common carp, Yasumoto *et al.* (2006) produced RPS scores of 74.4 and 65% when challenged with its homologous virus 22 days after vaccination. By an immersion method, the recombinant plasmid of the ORF149 gene of KHV can be delivered by single-walled carbon nanotubes (SWCNTs) proved to confer 56% protection (Hu *et al.*, 2021). Other recombinant DNA vaccines using partial viral genes (such as ORF25) are in research.

The virus is persistent, and this is related to a cycle of persistent infection and reactivation in hosts given that the virus was found to be present not only in water and pond sediment (Honjo *et al.*, 2012), but also in tank standpipes up to 6 days after removing fish (Tolo *et al.*, 2021). Effective disinfection is crucial to reduce the spread of the disease. The virus can be killed by iodophor (200 mg/l), benzalkonium chloride (60 mg/l) or 30% ethyl alcohol for 20 min (Kasai *et al.*, 2005). Other disinfectants such as peracetic acid (0.1% v/v), quicklime (0.015 M) and a protease, Neutrase® (8 mU/ml), were able to reduce the viral load of KHV and viral haemorrhagic septicaemia virus (VHSV) by at least four orders of magnitude within 24 h (Amtmann *et al.*, 2020).

The transboundary introduction of KHV into non-endemic countries should be closely monitored. Somga *et al.* (2010) intercepted an illegal importation of koi carp found positive for KHV in Manila. Thus, adequate biosecurity measures on importation of koi and common carp should be in place. Sensitive and quick detection of KHV prior to translocation undoubtedly is necessary.

6.2.5 Grass carp reovirus (GCRV) disease

Grass carp reovirus (GCRV) disease was previously known as haemorrhage of grass carp (Nie and Pan, 1985), haemorrhagic disease of grass carp (Jiang, 2009) and Chinese grass carp reovirus (CGRV) disease in China. This is the most serious viral infection reported in grass carp in China, causing ~80% mortalities (Jiang, 2009). Annual outbreaks with losses in central, southern and eastern China and northern Vietnam occurred at water temperatures of 24–30°C (Jiang, 2009). In the USA, golden shiner virus (GSV) disease, a similar disease, was first detected in cultured *Notemigonus crysoleucas* in 1977 (Plumb *et al.*, 1979). Outbreaks of a haemorrhagic GCRV-like disease also occurs in *I. punctatus* called channel catfish reovirus (CCRV) disease. Fish were introduced into China in 1984, and infection at 28°C results in mortalities of up to 60% in cultured fingerlings (Xu, J. *et al.*, 2013).

The *Aquareovirus* causing GCRV disease is a 60–80 nm virus belonging to the family Reoviridae. The virus is characterized by a non-enveloped icosahedron with a 5:3:2 three-dimensional symmetry. Its genetic material consists of 11 segmented double-stranded RNAs (dsRNAs) encoding seven structural proteins (VP1–VP7), the remainder are non-structural proteins. Proteins VP1–VP4 and VP6 build the core layer, while VP5 and VP7 form the outer capsid composed of 200 trimers (Cheng *et al.*, 2008; Fan *et al.*, 2010). Two serotypes with distinct antigenicities were reported in China, namely GV-87/3 from Hunan Province and GV-90/14 from Hubei Province (Jiang, 2009). CCRV-730 from *I. punctatus* measures 60–70 nm in diameter and has 99–100% nucleotide sequence similarity to the GCRV-873 strain. Strains are based on structural proteins V4, V6 and V7; among GCRV strains, GCRV-873, GCRV-HZ08 and Hubei GCRV are the three representative genotypes (Rao and Su, 2015).

In China, epizootics occur in grass carp *C. idella*, *Mylopharyngodon piceus*, *Pseudorasbora parva* and *Gobiocypris rarus* (see Wang *et al.*, 1994); however, the virus can replicate subclinically in *H. molitrix* and *Hemiculter bleekeri* (see Ding *et al.*, 1991). Other cyprinids such as bighead carp, common carp and golden carp can be asymptomatic carriers (Jiang, 2009).

In the USA, GSV infects golden shiner and grass carp (McEntire *et al.*, 2003); the virus induces low-grade mortality (Plumb *et al.*, 1979) but under crowded conditions and at high water temperatures, acute epizootics result in 50–75% mortality (Schwedler and Plumb, 1982).

Diagnosis

Signs of infection in grass carp include exophthalmia, dark body coloration, haemorrhages in the mouth cavity, and haemorrhagic or pale gills, fin bases and opercula. Affected fish are listless, swimming near the surface, have petechial haemorrhages in the cornea and dorsal musculature (reddish back), petechiae on the ventral body surfaces, in visceral fat and in the intestinal mucosa. Internally, haemorrhages develop in the musculature, intestine, liver, spleen and kidney. Degeneration and necrosis of the liver cells, hyperaemia and haemorrhagic vessels of the liver and spleen occur (Guo and Jiang, 1993). The virus replicates rapidly in the head kidney of infected fish causing an acute infection and significant mortalities of up to 80% among fingerlings and sometimes yearlings.

CCRV-infected channel catfish exhibit pale gills, localized haemorrhages in the liver, a dark and enlarged spleen, an empty gut, and haemorrhages on the intestinal wall with a clear yellow to blood-tinged fluid in the abdomen (Xu, J. *et al.*, 2013). Experimental infection yielded initial deaths at 2 dpi, reaching up to 50% mortalities at 7 dpi; clinical signs were like those observed in natural infections. Diseased channel catfish fingerlings become lethargic, float at the surface, display spiral swimming and die within 24 h manifesting severe haemorrhages of the operculum, lower jaw and fin bases, body colour depigmentation, bilateral exophthalmia and abdominal distension.

A neutralization test, ELISA or RT-PCR is used for virus detection (Li *et al.*, 1997). Seng *et al.* (2004) developed an RT-PCR for the detection of GCRV and other aquareoviruses. Later, a multiplex PCR was designed to detect GCRV and to distinguish between the three genotypes (Zeng *et al.*, 2013).

Isolation of GCRV is possible in cell cultures of grass carp kidney (CK or CIK), swim bladder cells of grass carp (GSB) and proboscis snout into fibres (PSF) cells of grass carp at 25°C (Li *et al.*, 2016). CPEs in CCRV-exposed channel catfish kidney (CCK) cells occurs 48 h post-inoculation (Xu, J. *et al.*, 2013).

Transmission

Experimental transmission of the virus by immersion in water containing GCRV resulted in disease and mortality within 1–2 weeks at 25–28°C. The serotype GV-90/14 showed increased virulence, inducing 100% mortality in 3–5 days in experimentally exposed grass carp fingerlings, while only 38% mortalities in 5–10 days were attributed to the serotype GV-87/3 (Jiang, 2009).

Prevention and control

In China, vaccination of naïve fingerlings with formalin-inactivated GCRV containing lipopolysaccharide induced immunity of up to 90%, a week after vaccination, when held at >20°C (Jiang *et al.*, 1991). An inactivated vaccine prepared from infected fish was effective when given to fish from the same region but less effective when given to fish in another region (Yang *et al.*, 1989). Vaccine administration by immersion yielded lower protection rate; however, the rate of protection can be increased using hyperosmotic infiltration (Ye *et al.*, 1992). An attenuated vaccine was developed but gave poor protection (Jiang, 1995). The vaccination of grass carp fingerlings in China has reduced mortality from >80 to <30% (Jiang, 2009). Subunit vaccines focusing on partial structural proteins show promise. The V35 subunit vaccine resulted in a 60% survival rate post-challenge (Gao *et al.*, 2018), and the VP56 subunit vaccine gave 71–75% protection (Pei *et al.*, 2019). Laboratory-scale trials, using DNA vaccines based on S6 and S10, gave moderate (59.9%) to low (23.1%) protection against GCRV-II (Chen *et al.*, 2018).

6.2.6 Channel catfish virus (CCV) disease

Channel catfish virus (CCV) is a viral pathogen of fry and fingerling channel catfish that can result in significant losses in commercial production. The first outbreak of channel catfish viral disease (CCVD) occurred in the USA in 1968 and it now occurs wherever channel catfish are reared (Plumb, 1999). *Ictalurus furcatus* and *P. hypophthalmus* are naturally susceptible to CCV, while common carp, koi carp and *Carassius carassius* are experimentally susceptible to CCV; other catfish species such as *Clarias gariepinus*, *Clarias batrachus* and *Silurus glanis* were resistant to an experimental challenge with CCV. CCV or *Ictalurid herpesvirus 1*, measuring 90–100 nm, is a double-stranded DNA virus belonging to the family Alloherpesviridae.

Diagnosis

Clinical signs are abdominal distension, exophthalmia, pale or haemorrhagic gills, and petechial haemorrhages at fin bases and the ventrum. In 20–50% of epizootics, affected fish swim in an upright or hanging position. Weak fish sink to the bottom and respire weakly, which increases prior to death (Plumb, 1999).

Acute infection occurs primarily in <10 g fish following waterborne exposure to CCV (Plumb, 1999). Infection is a haemorrhagic viraemia, initially replicating in the kidney and then in the spleen. Thereafter, the virus is transported to the intestine, liver, heart and brain. After experimental immersion, CCV was detected at 2–8 dpi in the skin and blood, at 10 dpi in the posterior kidney and gills (Kancharia and Hanson, 1996). Necrosis of the renal haematopoietic tissue and tubules; oedema, necrosis and congestion of the liver; intestinal oedema; and congestion and haemorrhage in the spleen are characteristic histopathological findings (Wolf, 1988). In experimentally infected fish, skeletal muscle haemorrhages are seen. Severe mortality approaching 100% occurs in <1-year-old fish in water temperatures of >25°C within 7–10 days. CCV causes moderate mortalities at 21–24°C and almost no mortalities at <18°C.

A rapid, sensitive and specific test for CCV detection is by PCR, which can detect the virus in brain, blood, intestine, kidney and liver of acutely infected fish (Gray *et al.*, 1999). Likewise, a nested PCR assay can detect <10 copies of CCV DNA in 10^8 times as much DNA of asymptomatic broodstock carriers (Baek and Boyle, 1996).

CCV replicates in channel catfish ovary (CCO) and in brown bullhead (BB) cells at 25–33°C (Wolf and Darlington, 1971; Bowser and Plumb, 1980). The virus can also be detected by TEM, a serum neutralization test and an indirect fluorescent antibody technique (IFAT). In addition, ELISA can detect antibodies against CCV in the serum of previously CCV-exposed fish (Crawford *et al.*, 1999).

Transmission

Transmission is horizontal from fish to fish or through infected water. Experimental infection is possible by injection (intramuscular or intraperitoneal), incorporation of the virus into the feed, swabbing the gills with a virus suspension and by immersion in virus-inoculated water. Juveniles (<10 g) can be infected by CCV via waterborne transmission while injection is needed for larger fish. Infection of 1-year-old fish results in low mortality which is usually associated with transportation or handling stress (Plumb, 1999).

Fingerlings surviving a primary epizootic become carriers of the virus. Survivors were 33% shorter and 85% lighter than their naïve control counterparts (McGlamery and Gratzek, 1974).

The portal of virus entry is via the gills and the gut. Naturally exposed fry die within 3 dpi, while experimentally infected fry die within 7–10 days following the death of the first fish used for the challenge. The virus also persists in apparently healthy adult broodfish in a latent state. CCV can occur in conjunction with a secondary bacterial infection with *F. columnare* that prolongs mortality (Plumb, 1999).

Prevention and control

Experimental studies have shown that attenuated CCV can induce protective immunity to channel catfish after injection or bath exposures (Walczak *et al.*, 1981). Passive transfer of adult channel catfish sera with anti-CCV neutralizing antibody activity also protected juvenile channel catfish from challenge with CCV (Hedrick and McDowell, 1987). Subsequent reports, likewise, indicated that vaccination of channel catfish with a combination vaccine pair (ORF59 and ORF6) provided strong protection against CCV challenge (Nusbaum *et al.*, 2002). These results showed that virus neutralizing antibodies were elicited with an anamnestic response upon viral challenge.

6.3 Bacterial Diseases

Intensive culture with high stocking density of fish leads to increased feed ration and waste with concomitant increases in ammonia and nitrite toxicity (Mitchell, 1997). This increases the organic load of the rearing water and permits rapid bacterial multiplication. Stress and trauma from handling are also predisposing factors.

In addition, most bacterial pathogens release enzymes that facilitate their entry/invasion into the fish host tissues. Although they may cause primary infection, they may also act as secondary disease agents to a primary virus or parasite. The major bacterial infections among warm freshwater fish are motile *Aeromonas* septicaemia, *Pseudomonas* septicaemia, edwardsiellosis, francisellosis, enteric septicaemia of catfish, columnaris disease and streptococcal septicaemia/meningoencephalitis (Table 6.5).

6.3.1 Motile *Aeromonas* septicaemia (MAS)

Motile *Aeromonas* septicaemia (MAS) is the most common bacterial infection in fish, also known as haemorrhagic septicaemia, infectious dropsy, infectious abdominal dropsy, red pest, red disease or red sore. MAS affects tropical freshwater fish worldwide and occasionally brackish-water fish. MAS infections are frequently reported in farmed milkfish, common carp, grass carp, Nile tilapia, hybrid red tilapia, snakehead fish, giant gourami, channel catfish and striped catfish (Lio-Po and Lim, 2014; Dong *et al.*, 2017c; Hoai *et al.*, 2019; Ngo *et al.*, 2022). Co-infection of motile *Aeromonas* and other pathogens is common during disease outbreaks (Dong *et al.*, 2015a).

Aeromonas hydrophila is known as the aetiological agent of MAS. Recent studies, however, revealed an increasing number of pathogenic motile *Aeromonas* which can also be attributed to outbreaks of MAS including *Aeromonas veronii*, *Aeromonas jandaei*, *Aeromonas caviae*, *Aeromonas sobria*, *Aeromonas schubertii* and *Aeromonas dhakensis* (Dong *et al.*, 2017c; Hoai *et al.*, 2019; Azzam-Sayuti *et al.*, 2021; Mursalim *et al.*, 2022). A recent investigation on the prevalence of *Aeromonas* isolated from cultured freshwater fish in Malaysia revealed that *A. dhakensis* was the predominant species (43%), followed by *A. veronii* (22%), *A. hydrophila* (20%), *A. caviae* (8%) and *A. jandaei* (7%) (Azzam-Sayuti *et al.*, 2021). In Thailand, *A. veronii* was the most dominant (72.1%) motile *Aeromonas* recovered from disease outbreaks from freshwater fish, while *A. hydrophila* was occasionally found (2.3%) (Mursalim *et al.*, 2022). *Aeromonas* spp. are a free-living, mesophilic bacteria found in the soil, freshwater lakes, ponds, streams, bottom mud, domestic tap water and sewage. They are often associated with the normal flora of fish and have been isolated from both healthy and diseased fish (Lio-Po and Lim, 2014). Motile *Aeromonas* species cause infections not only in aquatic animals but also in avian hosts, cows and sometimes humans.

Immersion of scarified milkfish fingerlings in *A. hydrophila* resulted in up to 100% mortality at 2 dpi but not in fish with intact skin, while intraperitoneal injection of milkfish with the bacterium caused mortalities within 12 h of injection (Lio-Po and Duremdez-Fernandez, 1986). In walking catfish and snakeheads, *A. hydrophila* induced dermal lesions after intramuscular injection of at least 10^5 cells per fish that eventually ulcerated (Lio-Po *et al.*, 1992). Likewise, immersion of Nile tilapia fingerlings yielded an LD_{50} of 1.5×10^6 colony-forming units (cfu)/ml, with 100% mortality at 10^8 cfu/ml and no mortality at 10^3 cfu/ml (Yambot, 1997). In channel catfish with mechanically abraded skin, *A. hydrophila* experimentally induced systemic infection in 80% of exposed fish while cutaneous lesions developed in the remaining fish (Matsche and Grizzle, 1999). Challenge of *P. hypophthalmus* with *A. hydrophila* induced external signs of haemorrhagic infection (Ly *et al.*, 2009).

Experimental infections with *A. veronii* or *A. jandaei* resulted in 100% mortality in Nile tilapia within 24 h when high doses ($3.7–8.9 \times 10^6$ cfu/fish) were administered intraperitoneally; a tenfold reduction in dose resulted in 50–70% mortality (Dong *et al.*, 2017c). An *A. veronii* dose of 10^5 cfu/fish given to *I. punctatus* induced 100% mortality in 10 days (Hoai *et al.*, 2019). In another study, experimental infection of hybrid red tilapia with *A. dhakensis* (1.86×10^5 cfu/g fish) resulted in over 80% mortality in 24 h (Soto-Rodriguez *et al.*, 2018).

Diagnosis

Infected fish lose appetite, become lethargic and swim near the surface. External signs are typical of bacterial septicaemia, namely exophthalmia, distended abdomen, fin and tail erosion, skin haemorrhage, ulceration and loss of scales. Septicaemia in acute MAS can be fatal with no obvious clinical signs. Internally, infected fish present ascites, swollen internal organs and

Table 6.5. Important bacterial diseases, causative agents, susceptible host(s), distribution and zoonotic potential.

Bacterial pathogen	Disease name	Susceptible host(s)	Geographical distribution	Zoonotic potential
Aeromonas hydrophila *Aeromonas veronii* *Aeromonas jandaei* *Aeromonas caviae* *Aeromonas sobria* *Aeromonas dhakensis* *Aeromonas schubertii*	Motile *Aeromonas* septicaemia (MAS) Aeromonasis	Freshwater fish, e.g., carp, tilapia, catfish, snakehead fish, etc.	Wide distribution in tropical countries	Yes (*A. hydrophila*, *A. veronii*, *A. jandaei*, *A. dhakensis*)
Pseudomonas fluorescens *Pseudomonas anguilliseptica* *Pseudomonas aeruginosa* *Pseudomonas alcaligenes* *Pseudomonas otitidis* *Pseudomonas plecoglossicida*	*Pseudomonas* septicaemia	Eel, tilapia	Wide distribution in tropical countries	No
Francisella orientalis	Francisellosis	Tilapia	Wide distribution in tropical countries	No
Edwardsiella ictaluri	Enteric septicaemia of catfish (ESC) Bacillary necrosis of pangasius (BNP) Edwardsiellosis of tilapia (EOT)	44 fish species including channel catfish, striped catfish and tilapia	Wide distribution in the USA, Japan, South-East Asia	No
Edwardsiella tarda *Edwardsiella anguillarum* *Edwardsiella piscicida*	Edwardsiellosis	Tilapia, catfish	Wide distribution in tropical countries	Yes (*E. tarda*)
Flavobacterium columnare *Flavobacterium covae* *Flavobacterium davisii* *Flavobacterium oreochromis*	Columnaris (or saddleback disease)	Carp, tilapia, catfish and various freshwater fish	Worldwide	No
Streptococcus agalactiae *Streptococcus iniae* *Streptococcus dysgalactiae* *Streptococcus ictaluri* *Streptococcus suis*	Streptococcal septicaemia (= meningoencephalitis or streptococcosis)	Freshwater fish, e.g. tilapia, striped bass, channel catfish, rainbow trout, barramundi, snakeskin gourami, etc.	Wide distribution in tropical countries	Yes (*S. agalactiae* ST283)

haemorrhages, and an accumulation of yellowish liquid and gas in the intestine (Dong *et al.*, 2017c; Hoai *et al.*, 2019). Tilapia infected with *A. schubertii* exhibited visceral white spots in the liver, spleen and kidney (Liu *et al.*, 2018).

The infection elicits an intense inflammatory response, with massive infiltration of monocytic and granulocytic cells into infected tissues. Infected goldfish are anaemic and there is a shift in the differential counts of lymphocytes to a predominance of neutrophils. Nile tilapia infected with *A. veronii* or *A. jandaei* present severe blood congestion, tissue degeneration with hemosiderin accumulation in the liver, hyperaemia and haemorrhages within the spleen, and there is severe necrosis of microvilli and sloughing of epithelial cells within the intestine (Dong *et al.*, 2017c). In tilapia infected with *A. schubertii*, histopathological manifestation includes vacuolization in the liver, splenic haemorrhages and swelling of capillaries in the brain. Necrotic lesions in the spleen, liver and kidney filled with many short, rod-shaped bacteria are seen (Liu *et al.*, 2018; Ren *et al.*, 2019).

Motile *Aeromonas* spp. are flagellated, Gram-negative, short rods. The bacteria grow at between 18 and 39°C. In tryptic soy agar (TSA) or in brain heart infusion agar (BHIA) at 25–30°C incubation for 24–48 h, *Aeromonas* spp. produce white to creamy, convex, moist colonies. In Rimler–Shotts' medium, the bacterium forms orange-yellow colonies at 35°C (Lio-Po and Lim, 2014). Identification of motile *Aeromonas*-complex is typically a combination of biochemical testing together with sequencing of 16S rRNA and/or *gyrB* for phylogenetic analysis. Recently, matrix-assisted laser desorption/ionization time-of-flight mass spectrometry (MALDI-TOF-MS) has been used as a rapid diagnostic tool for *Aeromonas* species (Dong *et al.*, 2017c; Hoai *et al.*, 2019; Mursalim *et al.*, 2022; Ngo *et al.*, 2022). A multiplex PCR assay has also been developed for the detection of multiple *Aeromonas* species (Persson *et al.*, 2015).

Transmission

Motile aeromonads are mostly isolated from mucosal surfaces and intestine of healthy fish (Harikrishnan and Balasundaram, 2005). Outbreaks of these opportunistic pathogens in farmed freshwater fish are usually related to transport and handling stress, adverse environmental conditions of low oxygen concentration, low pH, and increased levels of ammonia and carbon dioxide (Dong *et al.*, 2017c). These predisposing conditions are possibly immunodepressive and, along with the virulence of the *Aeromonas* strain, are important factors in the development of MAS epizootics (Thune *et al.*, 1993). Dong *et al.* (2015a) reported concurrent infections involving three to five pathogens in each individual tilapia where *A. veronii* was found in all diseased fish. Horizontal transmission through waterborne cohabitation is confirmed but vertical transmission is not yet elucidated.

Prevention and control

Prevention of stressful conditions in cultured fish will minimize MAS outbreaks. Inactivated vaccines for injection and immersion have been developed and tested in various freshwater fish species, but the focus has been on *A. hydrophila* (see Lio-Po and Lim, 2014). There have been a limited number of vaccine studies for other *Aeromonas* species (e.g. *A. veronii*, *A. jandaei* and *A. dhakensis*).

Using antibiotics (novobiocin and rifampicin) to induce mutation in pathogenic *A. hydrophila* strains, several live attenuated vaccines were obtained that showed impressive protection against pathogenic strains in tilapia and channel catfish (Pridgeon *et al.*, 2013, 2014). Bivalent live or inactivated vaccines were effective in laboratory trials (Pridgeon *et al.*, 2013; Pasaribu *et al.*, 2018). Nevertheless, further effort would be required to produce cost-affordable vaccines on a large scale.

Prophylactic bath treatments with 1–3% NaCl can help reduce post-handling infections. Likewise, bath treatments with potassium permanganate 2–4 mg/l are also effective for external lesions. Medicated feed with 2–4 g oxytetracycline/kg feed (50–100 mg/kg fish) for 14 days is recommended (Plumb, 1994); however, drug-resistant strains of *A. hydrophila* may evolve (Aoki, 1999). Administration of *A. hydrophila*-specific bacteriophages *per os* or by immersion was reportedly effective in protecting tilapia and striped catfish from *A. hydrophila* challenge (Dang *et al.*, 2021; Dien *et al.*, 2021a). A recent report describing ozone nanobubble technology as a non-chemical method for reducing the pathogenic bacterial load in water, thereby enhancing the survivability of Nile tilapia when challenged with a multidrug-resistant *A. hydrophila*, suggests this may serve as a promising technology for modern aquaculture (Dien *et al.*, 2021b).

6.3.2 *Pseudomonas* septicaemia

Pseudomonas (order Pseudomonadales) are Gram-negative, aerobic, motile, rod-shaped bacteria with one to three polar flagella that are found in diverse aquatic and terrestrial environments. Most *Pseudomonas* species are non-pathogenic, but a few cause fish diseases. In freshwater culture systems, *Pseudomonas* septicaemia has been associated with epizootic outbreaks in Nile tilapia, grass, silver and bighead carp, walking catfish, and Japanese and European eels (Thune *et al.*, 1993; Haenen and Davidse, 2001). *Pseudomonas fluorescens*, *Pseudomonas anguilliseptica* and *Pseudomonas aeruginosa* are the aetiological agents of the disease. Other species including *Pseudomonas alcaligenes* and *Pseudomonas otitidis* are usually associated with unusual problems of tilapia eggs and fry in the Philippines and Thailand (Duremdez and Lio-Po, 1985; S. Senapin, Thailand, 2022, personal communication). Most recently, *Pseudomonas plecoglossicida* was reported as an emerging pathogen causing visceral white nodule disease in cultured *Larimichthys crocea* in China (Li *et al.*, 2020).

Diagnosis

The clinical signs in fish affected with *Pseudomonas* septicaemia are very similar to those of MAS. Gross signs include ascites, exophthalmia, septicaemia and ulcers. The infection may be acute or chronic with the latter commonly associated with skin lesions. The pathogen has been associated with fin rot. Histopathological findings in Nile tilapia include focal necrosis, abscesses and granulomas in the eyes, gills, liver, swim bladder, kidney and spleen (Miyashita *et al.*, 1984).

Isolation and identification of the pathogen are required, and it can be cultured on nutrient agar, *Pseudomonas* F agar and blood agar (Austin and Austin, 2007). For strains pathogenic to fish, the optimum growth temperature is 20–25°C. These secrete oxidase, catalase and gelatinase, but not amylase, galactosidase, urease or hydrogen sulfide. They are citrate-positive, oxidative for glucose and produce a fluorescent pigment (Plumb, 1994). Molecular identification of *Pseudomonas* species is usually successful with amplification and sequencing of the 16S rRNA gene.

Prevention and control

Stress from low dissolved oxygen, high stocking density, physical trauma and poor nutrition are predisposing factors in the development of *Pseudomonas* septicaemia (Post, 1983). Bath treatments during the early stage of the disease include 1–2 mg benzalkonium chloride/l for 1 h and 0.5–1 mg furanace/l for 5–10 min (Austin and Austin, 2007). Swain *et al.* (2007) showed that vaccination using formalin-killed whole-cell antigens of *P. fluorescens* singly or as a polyvalent vaccine with *A. hydrophila*, *E. tarda* and *P. fluorescens* provided protection to *Labeo rohita*, yielding RPS of 80 and 70% in single and polyvalent vaccines, respectively.

6.3.3 Enteric septicaemia of catfish (ESC)

Enteric septicaemia of catfish (ESC) caused by *Edwardsiella ictaluri* is also known as bacillary necrosis of pangasius (BNP) in striped catfish, as red head disease in yellow catfish and as edwardsiellosis of tilapia (EOT) in tilapia (Machimbirike *et al.*, 2022). In American channel catfish culture, infection accounts for significant losses when temperatures are between 22 and 28°C (Peterman *et al.*, 2019). Infections are reported from Asia, Australia, the Caribbean, Europe and North America; to date, 44 fish species are reported as being susceptible to *E. ictaluri* infection (31 naturally susceptible and 13 experimentally susceptible) (Machimbirike *et al.*, 2022).

The causative agent *E. ictaluri* is a Gram-negative, pleomorphic, rod-shaped bacterium belonging to the family *Enterobacteriaceae*. The genomic sizes of *E. ictaluri* isolates range between 3.6 and 3.9 Mbp. Only 11 *E. ictaluri* genomes are currently available in GenBank (Machimbirike *et al.*, 2022).

E. ictaluri is pathogenic to channel catfish but only slightly pathogenic to *I. furcatus*. Recently, outbreaks in tilapia resulting in 30–65% mortality are considered an emerging problem for the Vietnamese tilapia industry and possibly wider. The causative agent is a highly pathogenic, multidrug-resistant *E. ictaluri* strain (Dong *et al.*, 2019; Nhinh *et al.*, 2022). The origin of this isolate remains unclear; however, most disease outbreaks originated from imported stocks.

Diagnosis

In acute infections, ingested bacteria enter the bloodstream through the intestine and colonize various organs, causing necrosis and ulceration (Noga, 2010). There is haemorrhaging, liver necrosis, splenic and renal hypertrophy; the peritoneal cavity secretes a bloody or clear fluid. Fish exhibit abdominal distension, exophthalmia and pale gills. With chronic infections, bacteria invade the olfactory organ through the nasal openings and spread from the meninges to the skull and skin, forming a hole-in-the-head lesion (Shotts *et al.*, 1986).

Infected striped catfish and tilapia are pale, having a swollen abdomen, a protruding bloodshot anus and cloudy eyes. Internally, ascites and numerous white nodules in the spleen, kidney, liver, gills, and rarely in the intestine, are associated with infection. Histopathologically, moderate to severe multifocal necrosis and pyogranulomatous lesions are observed in in the spleen (Fig. 6.5), kidney, liver and other organs, with the presence of white nodules. Other changes include an infiltration of inflammatory cells, cell pyknosis and karyorrhexis, and the presence of basophilic rod-shaped bacterial clusters in the necrotic tissues (Ferguson *et al.*, 2001; Crumlish *et al.*, 2002; Soto *et al.*, 2013a; Dong *et al.*, 2015b, 2019; Nhinh *et al.*, 2022).

E. ictaluri produces pinpoint to small-sized, off-white, translucent, irregular-surfaced colonies (0.14 ± 0.13 mm). After growth for 48 h on trypticase soy agar (TSA) at 28°C, the biochemical profile can be determined using conventional biochemical tests, the API 20E kit, combined with genus- or species-specific PCR and sequencing of the 16S rRNA. qPCR can be used to detect *E. ictaluri* in the brain, gill, kidney and liver of channel catfish (Shoemaker *et al.*, 2012).

Transmission

Horizontal transmission of *E. ictaluri* from infected channel catfish to naïve fish was confirmed though cohabitation experiments. *E. ictaluri* shed from dead fish, or cannibalism of them, are recognized mechanisms of horizontal transmission in channel catfish (Klesius, 1994). Infected channel catfish that recover from the disease can be asymptomatic carriers and may serve as a reservoir for disease transmission (Klesius, 1992). Detection of bacteria in the gonads indicates the possibility of vertical transmission (Mqolomba and Plumb, 1992).

Fish exposed to *E. ictaluri* at 25°C had the highest death rate, while those grown at 20 and 30°C had moderate death rates; infected groups kept at 15 and 35°C showed no or minor mortality (Baxa-Antonio *et al.*, 2011). Soto *et al.* (2013a) suggested the cutaneous and oral routes as the main portals of entry of the bacteria with the spleen and head kidney as major targets of infection.

Co-infection with either *Ichthyophthirius multifiliis* or *F. columnare* increased the disease severity, resulting in higher rates of mortality when compared with single infections (Shoemaker *et al.*, 2012; Dong *et al.*, 2015b).

Prevention and control

Current approaches to disease prevention and control are systematically reviewed by Machimbirike *et al.* (2022). Only a few commercial vaccines are available, including a live attenuated vaccine (AQUAVAC-ESC) for channel catfish in the USA and two oil-based inactivated vaccines (Alpha Ject Panga 1, Alpha Ject Panga 2) for striped catfish in Vietnam (Machimbirike *et al.*, 2022). In the USA, immersion vaccination of catfish against ESC is recommended at the fry stage (Bebak and Wagner, 2012). In Vietnam, the vaccines are injectable for juvenile striped catfish (>15 g). In Asia, the use of antibiotics for disease control in the catfish and tilapia industries has led to an alarming rise in the prevalence of multidrug-resistant strains (Machimbirike *et al.*, 2022; Nhinh *et al.*, 2022). Barnes *et al.* (2022) stated that autogenous vaccination in aquaculture is a locally feasible solution to tackle the global antimicrobial resistance crisis. Selective breeding for selection of *E. ictaluri*-resistant strains is

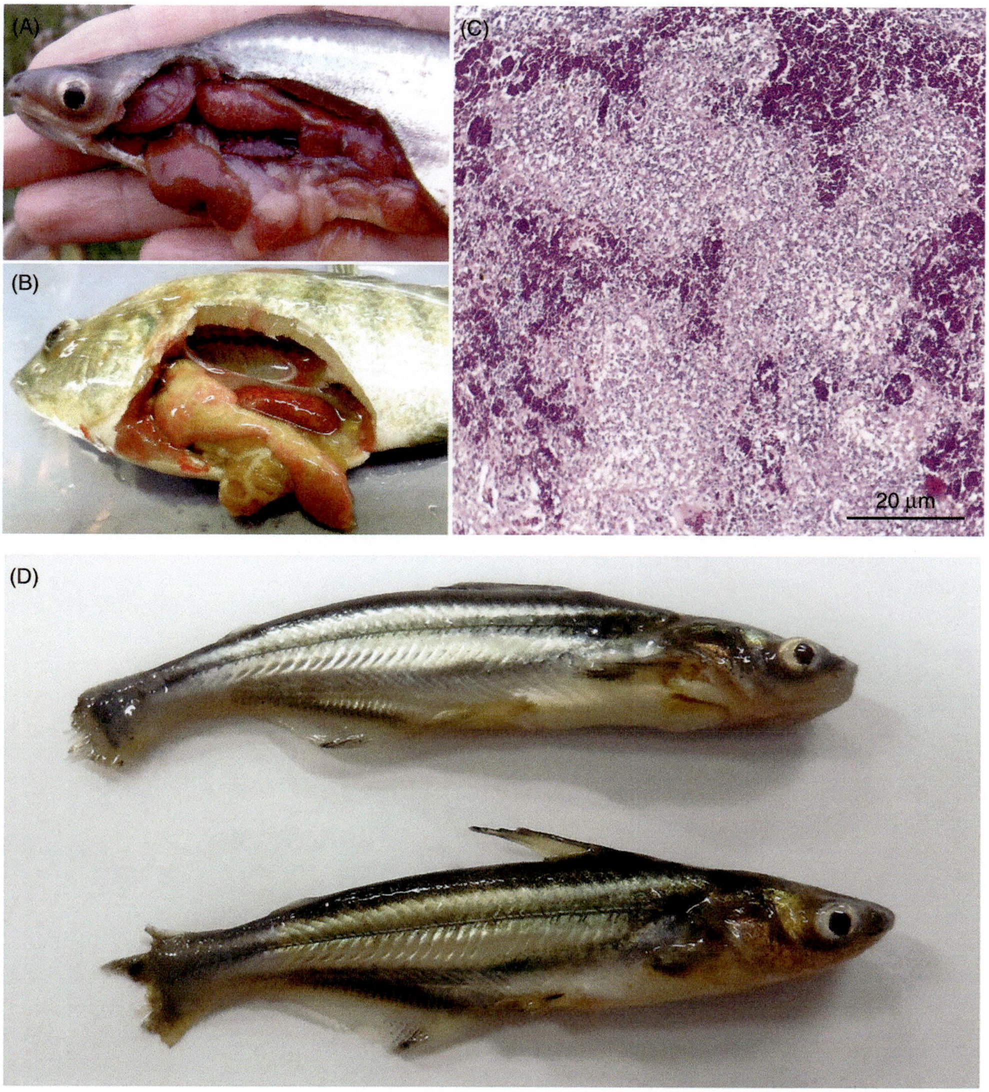

Fig. 6.5. (A) Striped catfish infected with *Edwardsiella ictaluri* showing numerous white spots in the head kidney, spleen and liver. (B) Nile tilapia infected with *E. ictaluri* exhibiting white spots in the swollen spleen, pale liver and ascites. (C) Histopathology of *E. ictaluri*-infected Nile tilapia spleen showing multifocal necrosis with multiple pyogranulomas. (D) Clinical signs of striped catfish naturally infected with *Flavobacterium columnare* showing severe fin erosion and skin necrosis. (Images by H.T. Dong.)

being developed for both channel catfish and striped catfish. In another study, macrophage chemotaxis in channel catfish in response to an *E. ictaluri* exoantigen was significantly low among unfed fish vis-à-vis the fish group that were fed daily. Moreover, mortality from ESC was significantly lower for fish fed daily or every other day than for fish that were not fed before and after challenge (Lim and Klesius, 2003).

6.3.4 Edwardsiellosis

The causative bacteria of edwardsiellosis are *Edwardsiella tarda*, *Edwardsiella piscicida* and *Edwardsiella anguillarum*, members of the family *Enterobacteriaceae* (Katharios *et al*., 2019; Leung *et al*., 2019; Oh *et al*., 2020). *E. ictaluri* is considered separately (see Section 6.3.3 above). Infections are reported worldwide (Austin and

Austin, 2007), affecting eels, mullet, channel catfish, tilapia, carp, striped bass and olive flounder (see Han *et al.*, 2006; Lio-Po and Lim, 2014; Leung *et al.*, 2019; Oh *et al.*, 2020).

Diagnosis

Infected fish may exhibit haemorrhagic areas on the body and malodorous lesions on the flank or caudal peduncle, and a swollen abdomen. Larger fish (>40 cm) are commonly infected. Internally, white nodules may develop in the liver. Histopathology of edwardsiellosis in tilapia consists of liquefactive necrosis of infected tissues with bacterial infiltration of macrophages engorging bacterial cells that eventually leads to granuloma formation in the liver, spleen and kidney (Miyazaki and Kaige, 1985). Virulence factors of *E. tarda* include haemolysin, dermatotoxin, siderophore, superoxide dismutase and catalase (Kokubo *et al.*, 1990; Han *et al.*, 2006). Infection is usually prevalent in channel catfish at 30°C. *E. tarda* is considered a zoonotic pathogen.

Edwardsiella species can be diagnosed based on typical clinical signs and isolation/identification of the pathogen *in vitro*. This Gram-negative, motile bacillus can be isolated on common bacterial culture media. On TSA at 25–30°C, *Edwardsiella* produces small, grey, circular transparent colonies. Species identification is based on either conventional biochemical tests or determined using the API 20E kit combined with molecular analysis, particularly sequencing of *16S rRNA* and DNA gyrase subunit B (*gyrB*) for phylogenetic analysis (Oh *et al.*, 2020).

Prevention and control

Vaccination by hyperosmotic infiltration of *O. niloticus* did not provide protection against *E. tarda* infection (Lio-Po and Wakabayashi, 1986). In contrast, vaccination by intraperitoneal injection using formalin-killed whole-cell antigens of *E. tarda* singly or as a polyvalent vaccine consisting of *A. hydrophila*, *E. tarda* and *P. fluorescens* provided protection to *L. rohita*. An RPS of 80% in either single or polyvalent vaccine was observed (Swain *et al.*, 2007). Castro *et al.* (2008) developed an *E. tarda* adjuvanted vaccine for cultured turbot which showed RPS values of >90% at least 6 months post-vaccination. Vaccination with the outer membrane recombinant protein OmpA was effective against *E. anguillarum* in Japanese eels, *Anguilla japonica*, with RPS of 77.7% (LiHua *et al.*, 2019).

6.3.5 Francisellosis

Francisellosis caused by *Francisella orientalis* (formerly known as *Francisella noatunensis* subsp. *orientalis*, *Rickettsia*-like or *Piscirickettsia*-like organism) is a significant bacterial disease of farmed *Oreochromis* spp. Francisellosis infections are reported worldwide and are summarized in Nguyen *et al.* (2016) and Ramirez-Paredes *et al.* (2020).

Francisellosis outbreaks usually occur in water temperatures of 23–26°C resulting in a 50–60% mortality in cultured tilapia (Nguyen *et al.*, 2016). A co-infection of *F. orientalis* and *I. multifiliis* resulted in higher mortality and more severe clinical signs of infection (Nguyen *et al.*, 2020).

The causative agent, *F. orientalis*, is a Gram-negative, non-motile coccobacillus to pleomorphic spherical bacterium measuring 0.1−1.5 µm in size (Colquhoun *et al.*, 2014). Intracellular, strictly aerobic and fastidious, *Francisella* requires specific media for its cultivation (Colquhoun *et al.*, 2014; Nguyen *et al.*, 2016).

Diagnosis

Presumptive diagnosis of francisellosis can be made through observation of gross signs. Clinically sick fish typically present whitish nodules (granulomas) on the gills, spleen, head kidney and sometimes in other organs (i.e. liver, intestine, muscle). Other non-specific signs such as a pale body, lethargy and anaemia are also seen during disease outbreaks (Soto *et al.*, 2009; Nguyen *et al.*, 2016).

Rapid tissue stamp smears of the head kidney or spleen stained with Giemsa can be used to visualize numerous coccobacilli, intracellular bacteria in the macrophages under a light microscope. Histopathologically, infected fish exhibit typical granulomas in multiple organs, notably the spleen and head kidney. Isolation of *F. orientalis* from infected fish can be achieved using enriched blood agar plates supplemented with 0.1% cysteine and 1% glucose, cysteine heart agar (CHA) with 5% sheep's blood, CHA with 1% haemoglobin (CHAH) or Thayer–Martin medium (Colquhoun *et al.*, 2014). Supplement of polymyxin B (100 units/ml) with or without ampicillin (50 μg/ml) added to the culture medium increases the chance of success; the optimal temperature for isolation is 28–30°C (Soto *et al.*, 2009).

Molecular diagnostic methods for the detection and quantification of *F. orientalis* include conventional PCR (Nguyen *et al.*, 2016), real-time qPCR (Rodrigues *et al.*, 2018), recombinase polymerase amplification (RPA) (Shahin *et al.*, 2018), LAMP (Pradeep *et al.*, 2017) and ISH (Dong *et al.*, 2016). Amplification of *16S rRNA* combined with sequencing and phylogenetic analysis is a common approach for the identification of *Francisella* from new geographical locations.

Transmission

The waterborne transmission of *F. orientalis* in *O. niloticus* fingerlings has been confirmed by laboratory trials (Soto *et al.*, 2013c). Vertical transmission of *F. orientalis* is also possible; asymptomatic tilapia broodstock can transmit *F. orientalis* to their reproductive organs, fertilized eggs and larvae (Pradeep *et al.*, 2017; Nguyen *et al.*, 2019).

Prevention and control

Using tilapia broodstock that test negative for *F. orientalis* is recommended to prevent vertical transmission and to produce specific pathogen-free seed. The onset of francisellosis in tilapia is significantly influenced by temperature, but not by salinity. Infected fish cultured at a high temperature (i.e. 30°C) displayed suppressed development of clinical signs and mortality while at lower temperatures (i.e. 25°C) the fish developed francisellosis with high mortality. Vaccination is effective in disease prevention. Pulpipat *et al.* (2020) reported that an injectable formalin-killed vaccine administered to cultured tilapia via intraperitoneal injection elicited protective antibodies and protection with an RPS of 71–76%. The presence of *F. orientalis*-specific IgM antibodies in the serum and mucus of tilapia that received an oral vaccine dose suggests that oral immunization has the potential to protect tilapia from francisellosis (Hoare *et al.*, 2021). Effective treatment using antibiotics incorporated with feed is also reported (Soto *et al.*, 2013c). Removing infected fish and disinfecting culture facilities is recommended to inactivate planktonic and biofilm forms of *F. orientalis* (Soto *et al.*, 2015).

6.3.6 Columnaris disease

Columnaris disease is a common infection of freshwater fish worldwide and can result in significant economic loss (Bernardet and Bowman, 2006). Natural outbreaks in tilapia hatcheries and grow-out farms result in a 10–70% mortality (Dong *et al.*, 2015c).

F. columnare, formerly known as *Flexibacter columnaris*, is the causative agent. It is a slender, Gram-negative, non-flagellated rod (about 0.5 μm × 4–12 μm) with gliding motility that forms 'haystacks' or columns. Recently, three new species were split out from *F. columnare*, namely *Flavobacterium covae*, *Flavobacterium davisii* and *Flavobacterium oreochromis* (see LaFrentz *et al.*, 2022).

Diagnosis

Columnaris disease is easy to recognize based on external clinical signs. Diseased fish usually show fin erosion with greyish to white margins; depigmented, necrotic skin lesions with yellowish or pale margins which can develop into shallow ulcers; yellowish mucoid material at the mouth; and light to dark brown gill discoloration. Infection primarily begins at the mouth, fins and gills. Gills or dermal/muscular capillaries of infected fish become congested and degenerated (Declercq *et al.*, 2013; Dong *et al.*, 2015c). Figure 6.5D shows an example of striped catfish with typical clinical signs of columnaris disease.

Gill lesions initiate at the distal end of the filaments and extend to the base. Epithelial vacuolation, necrosis, congestion, oedema, fusion and degeneration of the secondary lamellae subsequently follow. Colonization of filamentous bacteria on the necrotic areas of the gills and skin can be observed in both wet-mount and histological sections (Declercq *et al.*, 2013). Acute mortality is usually associated with gill lesions. Internal pathology or host inflammatory response may occur, and the pathogen may be isolated from internal tissues (Shotts and Starliper, 1999).

Primary isolation of the pathogen can be achieved on selective Anacker and Ordal's medium (or Cytophaga agar), modified Shield agar (MSA) or tryptone yeast extract salts (TYES) agar supplemented with neomycin at 5 μg/ml and polymyxin B at 200 IU/ml (Bernardet and Bowman, 2006). Colonies are yellow to orange, and rhizoid. This aerobic organism cannot tolerate more than 0.5% NaCl and grows between 4 and 36°C, producing gelatinase, caseinase, catalase, oxidase and chondroitin sulfatase (Bernardet and Bowman, 2006; Declercq *et al.*, 2013; Dong *et al.*, 2015c).

Rapid diagnosis is by specific PCR (Mabrok *et al.* 2020), LAMP (Suebsing *et al.*, 2015) and monoclonal antibodies (Ponpukdee *et al.* 2021). Amplification and sequencing of *16S rRNA* and multilocus sequence analysis (MLSA) are commonly used for identifying and characterizing this bacterium (Dong *et al.*, 2015a; LaFrentz *et al.*, 2018).

Transmission

Bacterial transmission is via water. The disease is most associated with stress from high temperatures, elevated organic loads, high stocking density, low dissolved oxygen and trauma from excessive handling. In channel catfish, it occurs more often at 25–32°C with significant mortality. Young fish are more susceptible than older fish. It may occur as a primary infection or as a mixed infection with another bacterium, *E. ictaluri* or *A. hydrophila*, or in association with a parasite, such as *Henneguya* sp. or *Ichthyobodo* sp. (Duarte *et al.*, 1993; Plumb, 1994). Columnaris disease appears to follow outbreaks of other diseases (Duarte *et al.*, 1993).

Survivors of columnaris disease release the pathogen into the water at rates of up to 5×10^3 cells/ml per h (Fujihara and Nakatani, 1971), and surviving fish may release the bacterium up to 140 dpi. The severity of the lesion depends on the virulence of the strain and the ability of the pathogen to elaborate proteolytic enzymes (Lio-Po and Lim, 2014).

The bacterium can survive for up to 16 days at 25°C in hard, alkaline water with high organic load; survival decreases at pH 7 or less and in waters with <50 mg $CaCO_3$/l and with low organic matter. In sterile mud at 25°C, the organism survives for 16 days (Becker and Fujihara, 1978).

F. columnare was found in the reproductive organs of apparently healthy tilapia broodstock, fertilized eggs and newly hatched fry, implying possible maternal transmission (Suebsing *et al.*, 2015).

Prevention and control

Maintenance of fish under optimal environmental conditions, proper handling of fish, prophylactic treatment and good health management practices are recommended for disease prevention (Declercq *et al.*, 2013). Fish exposed to *F. columnare* and maintained at 3–9 ppt salinity did not develop clinical signs or mortality, whereas fish cultured at lower salinity or in fresh water died with clear clinical symptoms of columnaris disease (Altinok and Grizzle, 2001).

Daily oral vaccination with heat-killed *F. columnare* for 4 weeks reportedly reduced mortality of rainbow trout from 48 to 8%, with protection correlated with antibody levels (Fujihara and Nakatani, 1971). Similarly, tilapia also mounts a significant humoral response in plasma and cutaneous mucus to *F. columnare* after intraperitoneal immunization with formalin-killed sonicated cells of *F. columnare* in Freund's complete adjuvant within 2 weeks. The mean titre remained significantly elevated above controls even at 10 weeks post-immunization (Grabowski *et al.*, 2004). The development of mucoadhesive immersion nanovaccines, based on chitosan, for the prevention of columnaris in tilapia resulted in an RPS of 58–81% (Kitiyodom *et al.*, 2019a,b).

Experimental application of FCP1 phage therapy via the intramuscular route in walking catfish, *C. batrachus*, provided protection against *F. columnare* infection. After treatment, gross signs disappeared and all experimental fish survived (Prasad *et al.*, 2011).

Therapeutic agents such as potassium permanganate at 5 mg/l (depending on the organic load of the rearing water) in combination with oxytetracycline added to feed at 50 mg/kg fish per day for 10 days is effective in controlling outbreaks in cages (Duarte *et al.*, 1993).

6.3.7 Streptococcal septicaemia/ meningoencephalitis

Streptococcosis caused by *S. agalactiae* is the most important bacterial disease of farmed tilapia worldwide; infections result in significant economic loss (Shinn *et al.*, 2018). Species that are pathogenic to freshwater fish include *S. agalactiae* (syn. *Streptococcus difficile*), *Streptococcus iniae* (syn. *Streptococcus shiloi*), *Streptococcus dysgalactiae*, *Streptococcus ictaluri* and *Streptococcus suis*. *Streptococcus* spp. are facultative, anaerobic, Gram-positive bacteria that are non-motile and non-spore-forming, exhibiting varying degrees of haemolysis according to species and strain (Buller, 2004; Noga, 2010). To date, *S. agalactiae* is the most dominant and significant of the reported species causing disease in freshwater fish. *S. agalactiae* (GBS) can be divided into ten serotypes (Ia, Ib, II–IX) based on capsular polysaccharides. Five serotypes have been identified in aquatic animals, namely serotypes Ia, Ib, II, III and IX. GBS sequence type 283 (GBS ST283) is a reported zoonotic type which is widely distributed in South-East Asia (Barkham *et al.*, 2019).

Diagnosis

Clinical signs include unilateral and bilateral exophthalmia with or without conjunctival haemorrhaging and corneal opacities. Petechiae occur beneath the opercula, perinal, on the caudal and pectoral fins, and around the mouth. There may be nodular or abscess formation, darkening and discoloration of the body. Abdominal swelling with ascites is common. Affected fish are anorexic, swim sluggishly in a circle, turn laterally and eventually die (Zamri-Saad *et al.*, 2010).

Internal signs include petechiae and haemorrhaging of the intestinal tract, liver and pyloric caeca. Systemic infection with evidence of bacterial dissemination in the heart, liver, kidney, stomach, small intestine, brain, eyes and musculature has been recorded. Multiple necrosis with granulomas, increasing melanomacrophage centres with an overload of melanophores, occur in the hepatic parenchyma. The spleen develops hyperplasia of the reticuloendothelial cells with necrotic foci. Degenerative changes in the renal tubules, catarrhal enteritis in the small intestine and stomach, bacterial meningitis and abscess formation in the muscle have been recorded (Buller, 2004; Rodkhum *et al.*, 2011).

Microscopic examination of Gram- or Giemsa-stained tissue smears from head kidney of infected fish is useful for a rapid presumptive diagnosis of streptococcosis. Diseased fish usually show numerous extra- and intracellular Gram-positive cocci. *Streptococcus* spp. can be isolated from the brain, kidney, heart, spleen, blood and exophthalmic eyes using nutrient agar (NA) supplemented with sheep's or goat's blood, BHIA or TSA. After incubation for 24–48 h at 28–30°C, *Streptococcus* spp. usually form pinpoint colonies on culture media. Modified Hucker's Gram staining showing small Gram-positive cocci, approximately 0.3–0.5 μm in diameter, most often occurring in chains, is a presumptive diagnosis (Kitao *et al.*, 1981; Buller, 2004). These organisms are non-motile and encapsulated. The pathogen does not grow in 40% bile, 6.5% saline, 0.1% methylene blue milk, or at 10 and 45°C (Kusuda and Salati, 1999; Buller, 2004). Details on the classification of *Streptococcus* spp. based on biochemical and serological tests are provided in Kitao *et al.* (1981) and Plumb (1994). Isolates from freshwater fish are usually β-haemolytic (Buller, 2004). Molecular methods for identification include specific duplex PCR for *S. agalactiae* and *S. iniae* (see Rodkhum *et al.*, 2012), qPCR for *S. agalactiae* (see Leigh *et al.*, 2020), and LAMP for *S. agalactiae* and *S. iniae* (see Suebsing *et al.*, 2013). Amplification of *16S rRNA* for sequencing is usually used for identification of *Streptococcus* species recovered from diseased fish.

Transmission

Experimental transmission occurs by immersion, injection, via the oral route or cohabitation and is enhanced by injury to the skin or stressful environment. Infection via the nares is a potential route of infection in Nile tilapia and hybrid striped bass (Evans *et al.*, 2000). Sources of infection are water, mud, contaminated feed or carrier fish (Plumb, 1994). Hatchery-to-grow-out farm transmission of *S. agalactiae* has also been reported (Amal *et al.*, 2013). Streptococcosis has been shown to be transmitted from parents to offspring (Pradeep *et al.*, 2017). *Streptococcus* is also more pathogenic to Nile tilapia than to channel catfish (Chang and Plumb, 1996).

Streptococcal outbreaks are triggered by several environmental factors, one of which is high temperature during the summer. Several studies revealed that tilapia infected with *S. agalactiae* and maintained at 33–35°C exhibited significantly higher rates of mortality compared with those maintained at 25–28°C (Rodkhum *et al.*, 2012; Kayansamruaj *et al.*, 2014).

Prevention and control

Avoidance of stress due to adverse or poor water quality, rough handling, high stocking density or overfeeding should be followed, and infected or dead fish removed.

Formalin-killed *S. difficile* vaccine injected intraperitoneally protects tilapia (Eldar *et al.*, 1995). In addition, Klesius *et al.* (2000) showed that intramuscular injection of a combined vaccine prepared from two strains of *S. iniae* obtained from Nile tilapia provided RPS of 63.1 and 87.3% when challenged with its homologous pathogens. Anti-*S. iniae* whole sera provided immunity to *O. niloticus* against *S. iniae* (see Shelby *et al.*, 2002). Subsequent studies showed that intraperitoneal injection of *O. niloticus* with *S. iniae* bacterin vaccine yielded RPS of 79–100% upon challenge with heterologous *S. iniae* derived from diverse geographical locations (Shoemaker *et al.*, 2010). In Malaysia, Ismail *et al.* (2016) developed a formalin-killed *S. agalactiae* feed-based vaccination regimen for prevention of streptococcosis in hybrid red tilapia with 45–70% survival after bacterial challenge. In the USA, novel attenuated *S. iniae* and *S. agalactiae* vaccines developed through selection for antibiotic resistance offered relatively good protection in tilapia (Pridgeon and Klesius, 2011; Liu *et al.*, 2019). Recently, Linh *et al.* (2022) reported that the pre-treatment of Nile tilapia with ozone nanobubbles for 10 min enhanced uptake of immersion heat-killed *S. agalactiae* vaccine and resulted in better protection compared with control without nanobubble treatment.

Selective breeding for *S. agalactiae*-resistance has gained promising results for Nile tilapia in Thailand. Selection response revealed that the risk of death decreased to 54% and survival rate increased to 21%, suggesting possible genetic improvements for the fish population (Suebsong *et al.*, 2019).

6.4 Diseases Caused by Oomycetes

Commonly known as water moulds, oomycetes are fungus-like eukaryotes which have cell walls that do not contain chitin but instead possess glycans and small amounts of cellulose (Yanong, 2003; Walker and van West, 2007). Most oomycetes are aquatic and produce characteristic heterokont types of zoospores with two flagella, one whiplash and one tinsel type. Most oomycetes that are pathogenic to fish belong to the subclass Saprolegniomycetidae and to the orders Saprolegniales (e.g. *Achlya*, *Branchiomyces*, *Dictyuchus*, *Saprolegnia*, etc.), Leptomitales (e.g. *Aphanomyces*) and Pythiales (e.g. *Pythium*) (Phillips *et al.*, 2008) (Fig. 6.6).

Diagnosis

Accurate identification is based on morphology; the sexual reproductive stages are typically required, although their general lack can make speciation difficult. Direct histochemical staining and/or fluorescent microscopy can assist in visualizing the hyphae and in determining cell-wall composition (see Paladini *et al.*, 2017). Molecular-based methods are crucial for identification (e.g. Diéguez-Uribeondo *et al.*, 2009).

6.4.1 Saprolegniosis

Most species in the *Saprolegniaceae* family are relatively intolerant of brackish water but thrive in fresh water, affecting a wide range of warm-water fish species. Some species of *Achlya* and *Saprolegnia* are important parasites of fish and their eggs (Panchai *et al.*, 2007). *Achlya* do not possess motile primary zoospores; instead they encyst as a hollow ball at the mouth of the zoosporangium. The infective secondary zoospores then emerge from the zoosporangium. *Achlya* spp. are reported from cage-reared *O. niloticus* (see Shinn *et al.*, 2022) and from *Gibelion catla* and *C. idella* from Pakistan (Iqbal *et al.*, 2012). The primary zoospores of *Aphanomyces* also encyst in a ball at the mouth of the sporangium, while in *Dictyuchus* the zoospores encyst inside the sporangium. *Saprolegnia* releases motile primary zoospores that possess two apical flagella. After a brief period of swimming, they encyst then release another motile secondary zoospore

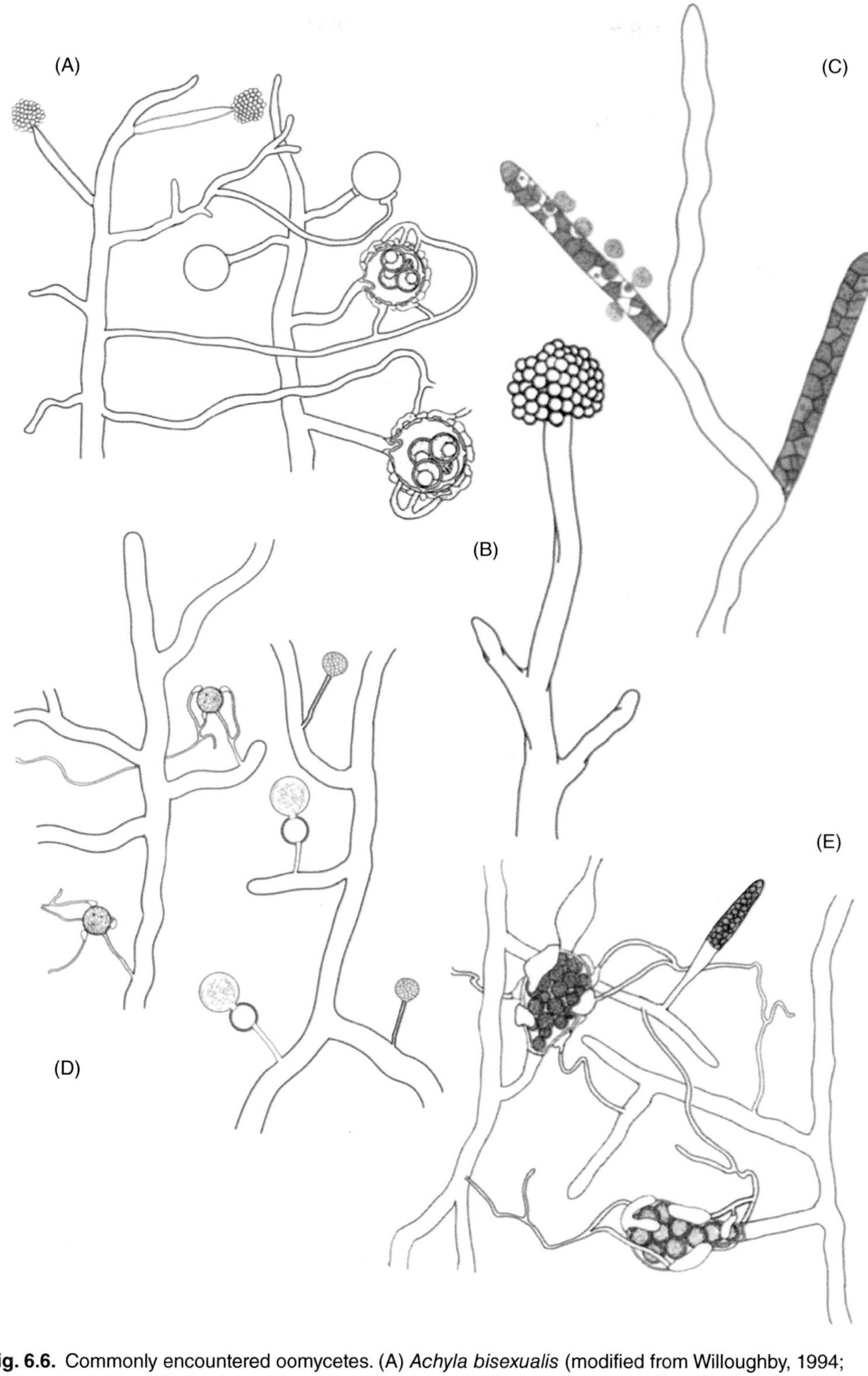

Fig. 6.6. Commonly encountered oomycetes. (A) *Achyla bisexualis* (modified from Willoughby, 1994; Markovskaja, 2004). (B) *Aphanomyces invadans* (modified from Noga, 2010). (C) *Dictyuchus sterilis* (modified from Webster, 1986). (D) *Pythium debaryanum* (modified from Webster, 1986; Trigiano *et al.*, 2004). (E) *Saprolegnia parasitica* (modified from Coker, 1923; Willoughby, 1994).

with two flagella. These secondary zoospores encyst; the cyst is covered by hooked hairs facilitating their attachment. Infections of *Saprolegnia parasitica* have resulted in significant mortalities of *P. hypophthalmus* in farm sites in India (Ravindra *et al.*, 2022). Infections are also reported from *G. catla*, *Cirrhinus mrigala* and *L. rohita* in India (Chauhan *et al.*, 2012); *G. catla* and *C. idella* in Pakistan (Iqbal *et al.*, 2012); *Channa punctata* from India (Mastan, 2015); and *Cyprinus carpio* from India (Magray *et al.*, 2021). Lio-Po and Lim (2014) and Chauhan (2014) provide a summary of oomycete infections various catfish, cyprinids, oreochromids and other fish species.

Pathology

Following their release from zoosporangia, the zoospores exhibit positive chemotactic responses to amino acids released from the tissues of susceptible hosts (El-Feki *et al.*, 2003; Sarowar *et al.*, 2014). Host contact triggers encystment of the zoospore; germination and host infection follow. Subsequent secretion of effector proteins by the pathogen can modulate or inhibit the host's immune responses, facilitating infection (Pradhan *et al.*, 2020). With the progression of infection there may be swelling, erosion and ulceration which can result in impaired osmoregulation, haemodilution, respiratory failure in the case of infections within the gills, and mortality. With *Aphanomyces invadans*, the hyphae invade the fish skin and skeletal muscles causing ulceration, commonly resulting in death (Lilley *et al.*, 1998). Infections of *Aphanomyces*, *Achlya*, *Allomyces* and *Saprolegnia* spp. are also reported in association with epizootic ulcerative syndrome in snakeheads (see Lio-Po and Lim, 2014). Experimental infection with *Aphanomyces* has been shown to induce lesions in naive snakeheads (Lilley and Roberts, 1997).

6.4.2 Branchiomycosis

Branchiomycosis or gill rot is due to infection either by *Branchiomyces demigrans* resulting in the penetration of gill tissue or by *Branchiomyces sanguinis* in the blood vessels of the gills. Infections result in respiratory distress and are associated with high rates of mortality, notably in waters >20°C. While infections are commonly reported in cyprinids (Post, 1983), significant losses are also documented for various oreochromids (Shinn *et al.*, 2022).

Pathology

Infected fish may be lethargic and present gills that are ragged, corroded or have evident areas of infarctive necrosis due to the intravascular growth of *Branchiomyces*. Infection can result in hyperplasia, fusion of the gill lamellae, reduction in blood supply to the gills, thrombosis, vascular necrosis and tissue sloughing (Post, 1983; Roberts, 2012). Spores are released from the necrotic gills; secondary bacterial invasion of the gill filament edges follows. Under favourable conditions, the disease may develop in 2–4 days; *in vitro* culture of the pathogen produced spores on day 14 of culture (Post, 1983). The hyphae of *B. sanguinis* are branched, thin-walled (0.2 μm) and non-septate, measuring 8–30 μm in diameter, with spores measuring 5–9 μm, that infect the gill filaments and lamellar capillaries. By comparison, *B. demigrans* has a thicker hyphal wall (0.5–0.7 μm), spores that are 12–17 μm in diameter and hyphae that infect the gill parenchyma (Post, 1983).

Prevention and control

The presence of organic matter, algal blooms, low pH <6.5, low dissolved oxygen, high stocking density and temperatures between 25 and 32°C are considered predisposing factors. Affected fish should be burned and/or buried. Survivors of an epizootic are carriers of the pathogen and should not be cultured with naïve fish or transported into *Branchiomyces*-free geographical areas.

6.4.3 Epizootic ulcerative syndrome (EUS) attributed to the EUS rhabdovirus, *Aeromonas hydrophila* and *Aphanomyces*

The epizootic ulcerative syndrome (EUS) is characterized by severe, ulcerative, dermal necrosis with extensive erosion/sloughing of the underlying musculature. Externally, fish may have red spots and/or large red or grey, shallow ulcers which are subject to secondary infections. Lesions on snakehead can be extensive leading to complete erosion of the posterior; necrosis to both the soft and hard tissues of the cranium can result in the brain being exposed.

The EUS caused serious economic losses in a wide range of wild and cultured fish, such as snakeheads (*Ophicephalus striatus*; *Channa* spp.), catfish (*Clarias* spp.), climbing perch (*Anabas testudineus*), barbs (*Puntius* spp.), goby (*Oxyeleotris marmoratus*, *Glossogobius giuris*), gourami (*Trichogaster pectoralis*, *Trichogaster trichopterus*, *Trichopsis vittatus*), Siamese fighting fish (*Betta splendens*) and eels (*Fluta alba*, *Mastacembelus armatus*, *Monopterus albus*) (Lilley *et al.*, 1998; Lio-Po, 1998). In the Philippines, cage-cultured snakeheads are very susceptible to the disease while carps (*C. carpio*), tilapia (*O. niloticus*) and milkfish (*Chanos chanos*) are considered naturally resistant. In Laguna de Bay, Philippines, the EUS morbidity rate among snakeheads was approximately 59% in January 1986 (Lio-Po and Lim, 2014).

By and large, EUS outbreaks show a seasonal pattern occurring from September to March when fish-rearing water temperatures in the region are at their lowest range of below 25°C. First outbreaks were reported in Vietnam in 1973, Singapore in 1977, Malaysia in 1979, Indonesia and Thailand in 1980, Cambodia, Myanmar and Lao PDR in 1984, the Philippines in 1985, Sri Lanka in 1987, Bangladesh and India in 1988, and Bhutan and Nepal in 1989. This spreading pattern of EUS outbreaks in South-East and East Asia strongly indicates the infectious nature of the aetiological agent/s. EUS was also reported in *C. gariepinus* in South Africa and Zambia (Huchzermeyer *et al.*, 2018; Malherbe *et al.*, 2019). To date, EUS rhabdovirus, *A. hydrophila* and *A. invadans* have been associated with EUS-affected fish (Frerichs *et al.*, 1986; Llobrera and Gacutan, 1987; Boonyaratpalin, 1989; Costa and Wejeyaratne, 1989; Kasornchandra *et al.*, 1992; Lio-Po *et al.*, 1992; Pathiratne *et al.*, 1994; Chinabut *et al.*, 1995; Karunasagar *et al.*, 1995; Thanpuran *et al.*, 1995; Lilley and Roberts, 1997; Lilley *et al.*, 1998; Lio-Po, 1998; Kanchanakhan *et al.*, 1999; Lio-Po *et al.*, 2000).

The EUS rhabdovirus was isolated from EUS-affected fishes in Thailand, Sri Lanka, Myanmar and the Philippines (Frerichs *et al.*, 1986; Kasornchandra *et al.*, 1992; Kanchanakhan *et al.*, 1999; Lio-Po *et al.*, 2000). Positive isolations in catfish spleen (CFS) and snakehead spleen (SHS) cell cultures were predominantly obtained from slightly lesioned fishes (Lio-Po, 1998). CPEs were also induced in monolayer cells of blue gill fry (BF2), snakehead (SSN-1) and CCO, resulting in a virus titre of 10^6 $TCID_{50}$/ml at 25°C in 2–3 days with optimum replication at 15°C (Lilley and Frerichs, 1994; Frerichs *et al.*, 1989; Kasornchandra *et al.*, 1992; Kanchanakhan *et al.*, 1999; Lio-Po *et al.*, 1999, 2000). Electron micrographs of the virus showed its bullet shape, typical of the family Rhabdoviridae, with an estimated size of 65 nm × 175 nm (Lio-Po *et al.*, 2000) (Table 6.4). Characterization and serological comparison of the virus with other fish rhabdoviruses associated with EUS-affected fish in Thailand showed that the Philippine virus isolate is morphologically similar and slightly antigenically related to the ulcerative dermal rhabdovirus (UDRV) isolated in Thailand.

The bacterium, *A. hydrophila*, has been consistently isolated from lesions of EUS-affected fish (Llobrera and Gacutan, 1987; Boonyaratpalin, 1989; Costa and Wejeyaratne, 1989; Pal and Pradhan, 1990; Torres, 1990; Lio-Po *et al.*, 1992; Pathiratne *et al.*, 1994; Angka *et al.*, 1995; Karunasagar *et al.*, 1995; Thanpuran *et al.*, 1995; Rahman *et al.*, 1999). Related studies experimentally demonstrated the dermonecrotic effects of this pathogen to naïve fish.

Mycotic isolations and identification of *Aphanomyces* were reported in EUS-affected snakeheads (Roberts *et al.*, 1993; Willoughby *et al.*, 1995). In Africa, only *Aphanomyces* was tested in sampled EUS-affected fish (Huchzermeyer *et al.*, 2018; Malherbe *et al.*, 2019). Fungal hyphae invade the fish skin and skeletal muscles causing ulceration, commonly resulting in death (Lilley *et al.*, 1998). Experimental infection with *Aphanomyces* induced lesions in naïve snakeheads and sand whiting (Roberts *et al.*, 1993; Chinabut *et al.*, 1995; Lilley and Roberts, 1997; Catap and Munday, 1998). The pseudofungi grow invasively through the fish skin and skeletal muscles causing severe myonecrosis and ulceration, commonly resulting in death (Callinan *et al.*, 1995; Chinabut *et al.*, 1995; Lilley and Roberts, 1997; Lilley *et al.*, 1998). However, granuloma development was observed at 26°C or above, while at lower temperatures only acute inflammation developed (Chinabut *et al.*, 1995). In addition, Catap and Munday (1998) observed that sand whiting injected with zoospores of *Aphanomyces* at 26°C developed highly inflamed, haemorrhagic external lesions while similarly treated fish held at 17°C had slightly inflamed injection sites. Temperature-related growth rate of this pathogen correlates with the findings that *Aphanomyces* from EUS-affected fish thrives better at 26–30°C than at lower temperatures (Lilley and Roberts, 1997).

Temperature is a significant factor in the pathogenesis of EUS. Replication of the EUS virus was observed at 15–25°C and the virus induced dermal lesions in fish reared at 20–22.5°C but not at 28–32°C (Lio-Po *et al.*, 2001). On the other hand, *A. hydrophila* and *Aphanomyces* thrive at 18–39°C and 26–30°C, respectively, with the latter pathogen producing 1.3–8 mm larger colonies at the high temperature. As such, among these three EUS-associated pathogens, only the EUS virus is temperature sensitive in terms of replication and dermal lesion induction in EUS-susceptible naïve fish viz-à-viz the EUS-season prevailing temperature of below 25°C. Moreover, studies on the interaction of these three pathogens in inducing EUS-like lesions indicate that the EUS virus is the primary pathogen and superimposed by a change to secondary/tertiary infections with either *A. hydrophila* and/or *Aphanomyces* (Lio-Po, 1998).

Diagnosis

The EUS-associated virus can be isolated from organ tissues of slightly lesioned catfish and snakeheads in cell monolayers of CFS, SHS, snakehead liver (SHL), CCO, BF2 and SSN-1 (Frerichs *et al.*, 1989; Kasornchandra *et al.*, 1992; Kanchanakhan *et al.*, 1999; Lio-Po *et al.*, 1999, 2000). Tissue filtrates derived from the visceral organs of EUS-affected fish induce CPEs when inoculated on to susceptible cells.

Bacterial isolation and identification of *A. hydrophila* from EUS-affected fish are detailed in Section 6.3.1 of this chapter on MAS. Bacterial colonies can also be histologically demonstrated in EUS-affected snakeheads (Cruz-Lacierda, 1995).

The pseudofungus, *Aphanomyces*, is isolated from ulcerated muscles of EUS-affected fish in Czapek Dox medium (Callinan *et al.*, 1995; Willoughby *et al.*, 1995). The hyphae of *Aphanomyces* are aseptate, 12–25 μm in diameter. Histological sections of muscular lesions of EUS-affected fish show the development of a necrotic granulomatous mycosis that may invade the abdominal viscera (Lilley *et al.*, 1998). Moreover, *Aphanomyces* can also be detected by PCR (Phadee *et al.*, 2004; Vandersea *et al.*, 2006; OIE, 2019b).

Transmission

Intramuscular injection of cell-cultured EUS rhabdovirus induced dermal lesion development and mortality of naïve snakehead (*Ophicephalus striatus*) juveniles reared at 20–22.5°C but not at 28–32°C (Lio-Po *et al.*, 2001). Slight dermal lesions developed 3 to 10 days following intramuscular inoculation which progressed to moderate lesions at 10–12 dpi but not deep ulcers in 5 to 12 days. Moreover, naïve snakehead fry and fingerlings exposed to cell-cultured EUS rhabdovirus by bath manifested significant mortalities ($P < 0.01$) of 100% at 5 and 9 dpi, respectively, with no apparent lesions (Lio-Po *et al.*, 2001; Kanchanakhan *et al.*, 2002). The lower temperature throughout the infection experiments simulated the water temperature during natural EUS outbreaks in EUS-affected countries. Moreover, cohabitation of naïve snakeheads with EUS fish or with apparently healthy snakeheads in lake water led to EUS-like lesion development attaining 100% morbidity by day 12 and 14, respectively, at 23–26°C (Lio-Po *et al.*, 2003), while naïve snakeheads stocked with EUS fish in aquifer water did not manifest EUS-like lesions. That study demonstrated the horizontal transmission of the virus from EUS fish and apparently healthy snakeheads held in endemic areas to naïve fish.

A. hydrophila has been associated with EUS-affected *C. striata* (syn. *O. striatus*) and *C. batrachus* in the wild as well as in ponds and cages (Lio-Po and Lim, 2014). Intramuscular injection of pure cultures of *A. hydrophila* induced dermonecrotic lesions in healthy catfish and snakeheads (Lio-Po *et al.*, 1992, 1996, 1998; Pathiratne *et al.*, 1994; Angka *et al.*, 1995; Karunasagar *et al.*, 1995). This bacterium thrives at 18–39°C and secretes a dermonecrotic factor at 10 and 30°C (Olivier *et al.*, 1981). Moreover, extracellular proteins such as protease are secreted by *A. hydrophila* associated with EUS-affected fish (Yadav *et al.*, 1992; Leaño *et al.*, 1996; Uddin *et al.*, 1997).

The hyphae of *Aphanomyces* develop primary zoospores within the sporangium which are then released to the tip where they form a spore cluster (Fig. 6.6B). Here the primary zoospores quickly transform into reniform, laterally biflagellated, motile secondary zoospores. On attaching to the fish skin, the spore germinates, and the hyphae invade the skin, musculature and internal organs resulting in the development of mycotic granulomas. Refer to Section 6.4.1 on saprolegniosis for further details.

Prevention and control

Quarantine and restricted movement of EUS-susceptible fish from endemic areas to non-endemic sites should be practised. Outbreaks are common when water temperatures drop to <25°C and the immunity of the fish is at its lowest (Catap and Munday, 1998). The transmission of the EUS virus/*A. hydrophila*/*A. invadans* is horizontal, with the presence/release of the virus, bacteria and zoospores from infected fish or through infected water supplies (Lio-Po *et al.*, 2003). Raising the salinity of the water to >2 ppt can help stop the spread of the pathogen/s, but on detection of infection it is advised that water exchanges are stopped, stock movements are restricted, and ponds are limed (Lilley *et al.*, 1998). Proper sun-drying and liming of ponds, coupled with thorough disinfection of all farm materials used for cage production, can destroy the pathogen. Currently, there is no vaccine available and there are no effective chemotherapeutant treatments. Dietary inclusion (2 g/kg) of the immunostimulant Salar-Bec, fed to satiation for 14 days prior to injection with *A. invadans* zoospores, resulted in a 59.2% higher survival in the treatment group at 40 days post-challenge (Miles *et al.*, 2001). Prophylactic treatment with 5 ppm Coptrol (a chelated copper compound) was reported to prevent induction of EUS lesions, while a proprietary mixture, CIFAX, may be curative (Lilley *et al.*, 1998).

6.5 Parasitic Diseases

While there is a wealth of information on the parasite fauna and diseases of warm, freshwater finfish aquaculture (Lio-Po and Lim, 2014; Paladini *et al.*, 2017; Jahangiri *et al.*, 2021; Shinn *et al.*, 2022), there is comparatively little information dealing with parasitic diseases in cage-pen culture systems. In many instances, the specific identity of the parasites is infrequently provided and at best only the genera are recorded (Paperna, 1991). Rarely are details regarding patterns of infection, pathology and prevailing factors predisposing fish to disease provided.

6.5.1 Diseases caused by protistans

Protistan parasites of importance include those belonging to the Ciliophora, Myxozoa, Microspora, Sarcomastigophora and Apicomplexa (Canning and Lom, 1986; Woo, 1995). The pathogenic protistans most reported include the myxosporeans, trichodinids and the dinoflagellates. The general under-reporting of protistan infections may result from infections going undetected, undiagnosed, not recorded at the farm level, or ignored because of their common occurrence. The translocation of fish stocks for aquaculture and fisheries has contributed to the worldwide distribution of many parasites, notably parasitic protistans (Shinn *et al.*, 2022). The earliest recorded translocations of tilapia, for example, were to South-East Asia in the late 1930s (Atz, 1954). Shinn *et al.* (2022) summarizes >820 translocations of tilapia worldwide, many of which occurred prior to mandatory health certification as part of national strategies in health management and aquaculture biosecurity.

6.5.1.i Diseases caused by ciliates

The ciliates (phylum Ciliophora) are common ectoparasites of fish, especially in hatcheries and on young fish in grow-out ponds. The vast majority of protistan problems in aquaculture are due to poor husbandry. Under normal culture conditions, *Apiosoma* spp., *Epistylis* spp. and *Trichodina* spp. are commonly encountered, as harmless ectocommmensals. However, if culture conditions deteriorate (e.g. due to overstocking, increased organic loading of the water due to overfeeding or poor waste management, low water exchange) then under these favourable, organic-rich conditions, the size of the parasite populations can increase quickly. Without the need for an intermediate host and reproducing by binary fission, parasite infections can spread rapidly, leading to morbidity and mortality. While notable pathogenic ciliates such as *I. multifiliis*, *Chilodonella* spp. and *Trichodina* spp. are readily identifiable, the specific identities of most ciliate infections in tropical aquaculture remain undetermined. For infections of *Chilodonella* spp., refer to the works of Lom and Dyková (1992), Basson and Van As (2006) and Bastos Gomes *et al.* (2016).

6.5.1.ii Trichodinid diseases

Pathogenic trichodinids include members of the genera *Trichodina*, *Tripartiella* and *Trichodinella*. Many trichodinids are associated with *C. idella*, *C. carpio*, *H. molitrix* and *H. nobilis*, and these were introduced into Israel and South-East Asia from China (Chen, 1955; Paperna, 1991). Various trichodinids from African cichlids have also been introduced into South-East Asia (Albaladejo and Arthur, 1989; Shinn *et al.*, 2022). Trichodinids commonly cause mortality in hatcheries, and these may continue to be a problem after fish are transferred to cage-culture systems. Trichodinids are prevalent on young clariid hybrids of *C. gariepinus* and *Clarias* sp. in cages. They are also found on *C. idella*, *H. molitrix* and *H. nobilis* in hatcheries in China and Vietnam, as well as on pangasiids and *Catla* spp. in cage culture. Although there are many species of trichodinids, only a few are known to be pathogenic (Lom, 1995). For a review of trichodinids infecting commercial fish species in South America, see the review of Maciel *et al.* (2018). For a review of infections on commercially important species of tilapia, see Shinn *et al.* (2022).

Pathology

Pathological effects are dependent on the host's response, the intensity of infections and environmental conditions, since stressful conditions can compromise the host's ability to counteract infections (Paperna, 1996). Infections on the fish's exterior may be at preferred sites (e.g. gill or fin bases). Parasite activity results in abrasion, cell damage, erosion and desquamation of the epidermis. The host response is through increased mucus secretion and epithelial hyperplasia, cellular destruction and inflammation. Infections may lead to respiratory distress, emaciation and mortalities, if left untreated (Paperna, 1996). It is important to emphasize that ciliate infections frequently facilitate the establishment of secondary microbial species; a review of their interactions and control is provided in Jahangiri *et al.* (2021) and is the subject of several case studies that investigate various bacterial and ciliate pathogens (Xu, D.H. *et al.*, 2013; Xu *et al.*, 2019).

Diagnosis

Trichodinids are easily observed microscopically from skin and gill scrapings (Paperna, 1996). Taxonomy of the trichodinids is based on the structure of the buccal ciliature, the morphology of the adhesive disc, and the number and size of its components (Lom, 1995). Trichodinids are essentially flat discs, with somatic ciliature consisting of three or four ciliary wreaths around the aboral surface of the body which is transformed into an adhesive disc. The disc is a proteinaceous skeleton, composed of a ring of hollow conical denticles. The denticles consist of blades (centrifugal flat projections) and horns (rod-like centripetal projections), connected to each other by radial pins (Fig. 6.7D). *Trichodina* is characterized by denticles with massive central conical parts, flat semi-circular blades, straight thorns and a diameter of 50–100 μm. Comprehensive reviews are provided by Van As and Basson (1989), Lom and Dyková (1992) and Basson and Van As (2006).

Prevention and control

In most cases an outbreak of *Trichodina* results from deteriorating environmental conditions, which are common in intensive culture systems. If *Trichodina* numbers rise, then fouled nets should be changed to ensure good water exchange through the net pen, pond aerators repositioned if necessary to facilitate water flow through the cages, feed management reviewed to ensure the fish are not being overfed, stocking density reduced by splitting stock out into clean cages, and waste management issues reviewed and addressed. To eliminate trichodinids, a saline solution (0.1–0.2% as a dip for 1–2 days) or formalin (150–250 ppm as a dip for 30–60 min) can be applied (Lom, 1995). The efficacy of formalin in controlling trichodinids depends on water quality (pH, salinity and ambient temperature) and the species of fish to be treated.

6.5.1.iii Ich or white spot disease

I. multifiliis, ich or white spot, is a pathogenic ciliate infecting freshwater fish and causing ichthyophthiriosis. This pathogen was first reported from China (Dickerson and Dawe, 1995), but it is a cosmopolitan pathogen commonly encountered in temperate and tropical warmwater fish in temperatures typically 2–27°C (Matthews, 1994). Development is temperature dependent, making them a potential danger to cage-culture systems in tropical warm waters. This parasite is maintained within the fish as a low subclinical (enzootic) infection and as encysted tomonts. It persists in the environment, becoming epizootic clinical infections when fish are stressed because of poor management practices (e.g. poor feed, overcrowding and poor sanitation). The pathogen is not host specific and recovery from the disease confers resistance to reinfection (Paperna, 1996).

Pathology

The parasitic trophont stage located within the host's epidermis feeds on the basal layer of the epidermis. Aggregations of trophonts can arise within galleries because of multiple entry at single sites, from parasite tunnelling and/or possibly from reproduction (Matthews, 1994) (see Fig. 6.7C). Trophonts need to reach a critical size before they can exit their hosts and their departure is an active process which may involve the discharge of contractile vacuoles (Ewing and Kocan, 1987). The exiting of mature trophonts damages the epidermis, causing detachment. The synchronized exiting of numerous trophonts, notably in small fish or from the gills, can be traumatic, resulting in osmotic shock, haemodilution and high rates of mortality.

Diagnosis

Clinical signs include anorexia, lethargy and respiratory distress; infections are characterized by white spots on the skin and gills (Dickerson and Dawe, 1995). Skin and gill scrapes examined under a compound microscope reveal ciliates (up to 1 mm in diameter) with a small cytostome. *Ichthyophthirius* fixed and stained with Giemsa or haematoxylin reveals a large crescent-shaped macronucleus and a small micronucleus.

Prevention and control

The management and control of *I. multifiliis* lies in having a comprehensive understanding of its biology, life cycle, and the factors contributing to its establishment and pathogenicity. In warm waters, the life cycle can complete in *c.*8 days (i.e. 7 days for development of the parasitic trophont; 1 day for the external phase leading to the release of infective theronts). Infection dynamics can be affected by several factors including the speed of water moving through cages, stocking density, cage/system design, water depth, aeration and cleaning frequency. The chemotherapeutants that have been assessed *in vivo* have typically been administered as a short (i.e. <60 min) high-dose application, delivered either as a single dose or as a regimen of daily treatments over a period of *c.*7 days (Picón-Camacho *et al.*, 2012a,b). Most treatments have had limited efficacy because the chemical has been unable to treat the parasite situated beneath the epithelium. The narrow treatment window serves only to treat the external stages of the parasite in the water column at the time of the treatment. By increasing the window of chemical exposure, protomonts can be treated as they are continuously released from the host and before they encyst, and tomites/theronts treated as they are released and before they infect. There are, however, practical challenges and welfare considerations in delivering long-duration bath treatments and these need to be carefully factored into any treatment. The efficacy of potassium permanganate, widely used throughout South-East Asia, is affected by the organic load of the water; it can lower the dissolved oxygen concentration of the water and its toxicity increases in waters with a high pH. The long duration of 1–5 ppt NaCl as a permanent bath can reduce the number of trophonts establishing on fish and may prove beneficial, helping the host recover the osmotic imbalance and loss of salts created by exiting trophonts.

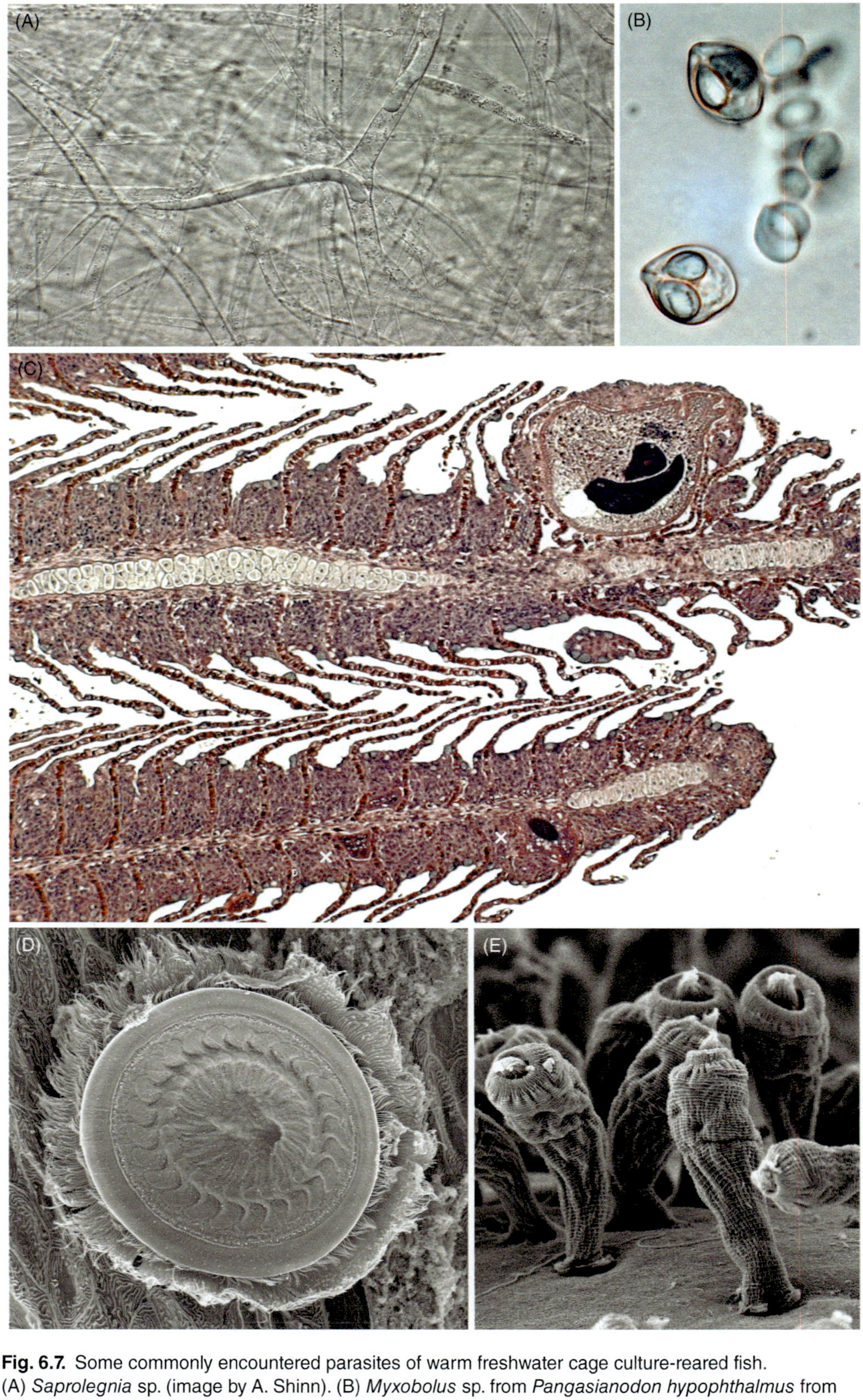

Fig. 6.7. Some commonly encountered parasites of warm freshwater cage culture-reared fish. (A) *Saprolegnia* sp. (image by A. Shinn). (B) *Myxobolus* sp. from *Pangasianodon hypophthalmus* from Vietnam (image by A. Shinn). (C) Developing trophonts (indicated as ×) of *Ichthyophthirius multifiliis* in the

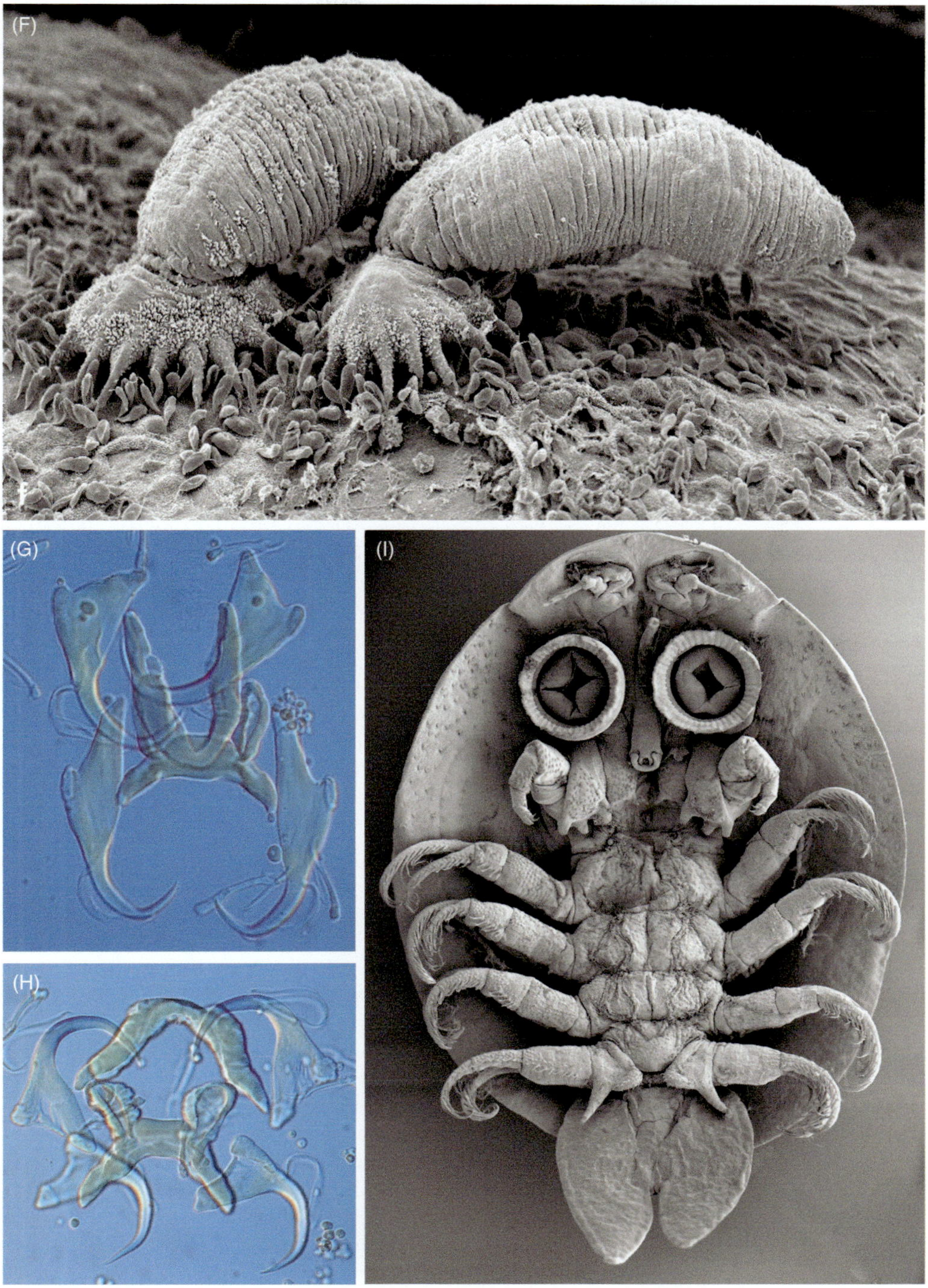

Fig 6.7. Continued.

gills of a rainbow trout, *Oncorhynchus mykiss* (image by A. Shinn). (D) *Trichodina* sp. from the skin of Nile tilapia, *Oreochromis niloticus* (image courtesy of G. Hanako Rosas Saito and A. García-Vásquez). (E) *Apiosoma* sp. (image courtesy of G. Paladini). (F) *Gyrodactylus* sp. and *Ichthyobodo necator* (image courtesy of G. Paladini). (G) *Cichlidogyrus tilapiae* from *O. niloticus* (image courtesy of A. García-Vásquez and A. Shinn). (H) *Cichlidogyrus sclerosus* from *O. niloticus* (image courtesy of A. García-Vásquez and A. Shinn). (I) *Argulus* sp. from *O. niloticus* reared in Mymensingh, Bangladesh (image by A. Shinn).

At present there are no commercial vaccines for *I. multifiliis*. However, fish can acquire a protective immunity against *I. multifiliis* (see Jørgensen *et al.*, 2008). Different vaccines using live, killed, parasite homogenate and DNA have been assessed for the protection they confer against *I. multifiliis* (see review of Shivam *et al.*, 2021). Studies by Cross and Matthews (1992) and Gurunathan *et al.* (2000) demonstrated that non-lethal infections or intraperitoneal injections of live theronts can confer immunity (Matthews, 2005). A DNA vaccine encoding immobilization antigens administered as an intramuscular injection in *I. punctatus* offered a lower level of protection than that of a live vaccine (Xu *et al.*, 2019).

6.5.2 Diseases caused by dinoflagellates

There are five genera of parasitic oodinid dinoflagellates on fish: *Amyloodinium*, *Piscinoodinium*, *Crepidoodinium*, *Ochthyodinium* and *Oodinioides* (Noga and Levy, 2006). The ichthyotoxins produced by dinoflagellates cause massive mortality in cultured and feral fish (Steindinger and Baden, 1984). In fresh water, the most important pathogenic dinoflagellate in fish is *Piscinoodinium*, which is closely related to the marine dinoflagellate pathogen, *Amyloodinium*. *Piscinoodinium* is not host specific and has been reported on feral, aquarium and cultured food-fish species from diverse families in warm waters (see Lio-Po and Lim, 2014).

6.5.2.i Velvet or rust disease

Fish with excessive mucus covering the body together with a rust-coloured appearance of the skin are infected with *Piscinoodinium pillulare*, the causative agent for velvet rust diseases, gold dust disease, pillularis disease and freshwater *Oodinium* disease (see Lio-Po and Lim, 2014 and references therein). *Piscinoodinium*, like its marine relative *Amyloodinium*, is found on a wide range of host species and is known to cause mortality in warmwater fish. Infections of *P. pillulare* in Malaysia have been reported to cause mortality by inducing hyperplastic gills and obstruction of the respiratory surface. Outbreaks with up to 100% mortality of farmed *Piaractus mesopotamicus* have been reported from Brazil (Sant'Ana *et al.*, 2012). A 30% mortality of farmed juvenile *C. macropomum* is reported from Peru (Arbildo-Ortiz *et al.*, 2020). Infections from farmed populations of tambacu hybrids, *P. mesopotamicus*, *C. macropomum*, *Leporinus macrocephalus*, *O. niloticus* and *Prochilodus lineatus* are reported from Brazil (Martins *et al.*, 2001). There is also a record of 8-year-old *Piaractus brachypomus* at a semi-intensive farm dying during the dry season in Brazil when temperatures fluctuate between 11 and 35°C (Ramos *et al.*, 2020). Infections of *Piscinoodinium* sp. on farmed *O. niloticus* are summarized in Shinn *et al.* (2022).

Pathology

Histopathological changes to gill structures occur with a massive proliferation of the gill epithelium, fusion of adjacent lamellae and separation of the gill respiratory epithelium resulting in a severe hyperplasia of the entire gill filament. The trophonts of *P. pillulare* penetrate the host cells by nail-like extensions resulting in degeneration and collapse of the cells, leading to focal erosion and proliferation of the epithelium and obliteration of the gill lamellae. The inner strata of the epithelium become spongious and may undergo complete lysis.

Diagnosis

Initial diagnosis can be based on clinical signs and confirmed by microscopic examination of the trophont stage (Gomes *et al.*, 2018). *Piscinoodinium* infects skin and gills with clinical signs like amyloodiniosis. Infected fish have a yellow- to rust-coloured (velvety) skin, a dense covering of mucus resulting in darkening of the skin, dyspnoea, anorexia and skin ulcers. All oodinids have a parasitic trophont stage and a sessile, stalked, sac-like trophozoite stage which feeds on the skin and gill epithelia. The trophont has a prominent stalk, which anchors the parasite to the host. It probably uses the stalk to absorb nutrients. After feeding, the trophont detaches, withdraws the stalk and forms an encysted tomont (reproductive cysts). The tomont divides asexually forming dinospores, the mobile infective stages. The trophonts and tomonts are important for definitive diagnosis, and microscopic identification of these stages is necessary. Trophonts are oval with smooth walls, usually visible to the naked eye as white spots (80–100 μm) and turn dark blue in Lugol's iodine. *Piscinoodinium* is distinguished from other oodinid dinoflagellates based on the morphology of the trophont, especially the type of host attachment and mode of nutrition. Fish should be examined live or immediately after death, and biopsies of the gills can be removed from live or recently dead fish and examined. Trophonts are removed by brushing the fish gently in a dish of water and the sediment is examined under the microscope. The trophont of *Piscinoodinium* is a yellow-green, pyriform or sac-like cell, almost round, 12 μm × 29 μm, with a rudimentary sulcus, and having a short stalk with an attachment disc extending from its base and thin holdfasts (rhizocysts) radiating from the stalk. Head parts of the rhizocysts are inverted in separate compartments (rhizothecas) in the sole of the disc, while their shafts are firmly embedded in the host cell cytoplasm. The theca covers the entire cell except for the area of the attachment disc.

Prevention and control

Outbreaks of oodinid infections result from stress due to poor environmental conditions. Hence, environmental manipulation is probably a viable approach to control outbreaks of *Piscinoodinium*. Long-duration treatments may result in better efficacies and should cover a period that allows for the emergence of dinospores from trophonts and the period over which dinospores remain infective (i.e. *c.*2 days). Salt (1.5 g/l) remains the safest method as a permanent bath; however, for-life threatening infections, Noga (2010) suggests that high concentrations of salt (e.g. up to 35 ppt) for 1–3 min dislodges trophonts. The latter treatment though will require establishing a treatment station where there is no possibility of the dislodged dinospores finding their way back into the main culture system.

6.5.3 Myxosporean diseases

Myxozoans are specialized cnidarian parasites which typically use a vertebrate and invertebrate host as part of their life cycle. Myxosporean cysts can be found infecting the skin and subcutaneous layer, muscle, gills, central nervous system and visceral organs, causing extensive lesions as cysts break and mortality (Lom and Dyková, 1995). Myxozoans are traditionally classified using morphological features of the spores (number and arrangement of polar capsules, features of spore walls, etc), morphometrics and tissue tropism. Notable problematic species in warmwater fish include *Myxobolus lentisuturalis* in *C. gibelio* (see Dyková *et al.*, 2002) and *Thelohanellus hovorkai* in *C. carpio* (see Yokoyama *et al.*, 1998). High rates of myxosporean-related mortality have been reported in *Colossoma macropomum* fry (see Videira *et al.*, 2016) and in cage-cultured *P. hypophthalmus* (see Molnár *et al.*, 2006).

Pathology

Myxozoan infections can result in a range of pathologies. *Myxobolus drjagini* infecting the skin of the head, the olfactory and oculomotor nerves in the cranial cavity and the intrafilamental epithelium of the gills results in destruction of the nerves and a condition known as 'silver carp twist disease' with consequential mass mortalities (Wu *et al.*, 1975; Chen and Ma, 1998). Infections are widely reported in *H. molitrix* and *Hypophthalmichthys nobilis*. There is, however, uncertainty regarding the identity of the myxosporeans recovered from different tissues; those from the gills may represent another morphologically similar species, *Myxobolus paratypicus* (see Xi *et al.*, 2019). Heavy infections of *Myxobolus pavlovskii* in the interlamellar epithelium and secondary lamellae in fingerling *H. molitrix* and *H. nobilis* results in impaired respiratory function and impacts survival (Molnár, 1979). Infection of *Myxobolus koi* in the gills of *C. carpio* and *C. auratus* results in numerous cysts forming within the gills (Paperna, 1991; Egusa, 1992). The formation of large cysts (*c.*5 mm in diameter) enveloped by vacularized host connective tissue within the gills restricts opercula movements; increased mucus production and epithelial proliferation leads to congestion and impacts on respiratory function (Hoshina, 1952). Spores of *M. koi* were also observed in the heart, liver, kidney and intestine. Heavy infections of *Myxobolus dispar* in *C. carpio* also result in impaired respiration (Lom and Dyková,

1992). *Myxobolus dahomeyensis* infection in the ovaries of *O. niloticus*, which results in a suppurating thick liquid replacing mature oocytes, can result in sterilization (Okaeme *et al.*, 1988; Gbankoto *et al.*, 2001). *O. niloticus* with ocular infections of *Myxobolus sarigi* present exophthalmos (Okaeme *et al.*, 1988). The elongated plasmodia of *Myxobolus encephalicus* in the brain blood vessels of *C. carpio* are in contact with the vessel walls and are coated with a layer of endothelial cells to prevent blood coagulation, which results in impaired blood flow, dilation of blood vessels and localized oedema. In the brain, mature spores released from plasmodia may provoke a heavy granulomatous inflammation. Heavily infected fry and fingerlings are emaciated, have shrunken eyes, have a loss of balance and swim in a circular motion (Lom and Dyková, 1992). The mass infection of *Sphaerospora molnari* in the gills of *C. carpio* results in heavy losses. Infections can elicit epithelial hyperplasia, with parasites replacing tissue consequentially leading to muscular dystrophy, circulatory disorder, and the necrotic destruction of the tissue with the release of spores into the water (Lom and Dyková, 1992). The plasmodia of *Thelohanellus nikolskii* in close association with the fin cartilage of *C. carpio* results in cartilage fragmentation and fin rays breaking off, impairing swimming performance (Molnár, 1982). In Indonesia, *Thelohanellus pyriformis* forms large plasmodia in the subcutaneous tissue and muscle of cyprinids causing fatal epizootics (Lom and Dyková, 1995). A co-infection of *Zschokkella auratis* together with various bacterial pathogens and *I. multifiliis* resulted in the complete loss of *C. striata* stock at a farm unit in India (Paul *et al.*, 2020). For a summary of myxozoans infecting oreochromids, where typically infections result in hyperplasia or hypertrophy of host tissues, see the recent review of Shinn *et al.* (2022).

Diagnosis

The gross signs of infection may be varied, but histozoic (i.e. those within tissues; cf. coelozoic, which are those occupying cavities) species frequently form cysts of various colours (i.e. usually white, or yellow), shapes and sizes (i.e. round, oval or branched). Heavy gill infections may hamper opercular movement and impair respiratory function, while infections of the brain, nervous system and fins may result in abnormal swimming behaviours and performance. Identification is assisted by spore morphology and based upon: the general spore shape; the number of polar capsules; the orientation of the polar filament within the polar capsule and the number of coils; the number of shell valves and whether these are smooth, ridged or possess protrusions; details of the infective sporoplasm (i.e. one binucleate or two uninucleated cells) with or without an iodophilic vacuole; use of morphometric data collected from key structures; and by the presence/absence of a mucus envelope. To identify the spores, cysts can be carefully excised, transferred to a glass slide and then the cyst broken to release the spores. For helpful guides to the Myxozoa, refer to the works of Dyková and Lom (1988), Lom and Dyková (1992, 1995, 2006) and Feist and Longshaw (2006).

Prevention and control

Currently, there are no effective treatments for myxosporeans and the best method is to remove and destroy heavily infected fish from cages. Management and control lie in the practice of good biosecurity to minimize the likelihood and potential impact of infections. New stocks should be held under quarantine while the relevant health checks are conducted to ensure their disease-free status prior to stocking, and the health of stock fish should be assessed regularly throughout production. For stocks reared in earth ponds, ponds should be drained, the sediment removed on a regular basis, the pond thoroughly dried between production cycles, and hydrated lime applied to ensure that any intermediate hosts (e.g. oligochaetes) are killed. Removing pond sediment reduces the amount of detritus available to oligochaetes. Using high-density polyethylene to completely line culture ponds may help in sediment management and its effective removal between production cycles, but the same disinfection practices of comprehensive cleaning of the liner between cycles is recommended. Where possible, the incoming water should be filtered and treated with ozone or by ultraviolet irradiation to minimize the likelihood of actinosporean stages and infected oligochaetes entering culture systems. Producers should be aware of the risks of feeding live oligochaetes which can harbour infective myxospores (Hallett *et al.*, 2005). Using specific pathogen-free oligochaetes sourced from a reputable supplier is therefore recommended.

6.5.4 Diseases caused by monogeneans

Monogeneans are among the most reported parasitic agents of fish (Scholz *et al.*, 2018; Shinn *et al.*, 2022). They are mainly ectoparasitic on the gills, buccal cavity, body surface and fins of freshwater fish, although some are endoparasitic (e.g. Gussev and Fernando, 1973; Euzet and Combes, 1998; Mendoza-Franco *et al.*, 2018). Monogeneans are oviparous except for the viviparous gyrodactylids. Although there are some species that cause epizootics in wild populations of fish (Bakke *et al.*, 2007; Paladini *et al.*, 2021), they are important pathogens in intensive fish culture (Ogawa *et al.*, 1995; Lopez *et al.*, 2002; García-Vásquez *et al.*, 2007; Grano-Maldonado *et al.*, 2011). Their direct life cycle results in rapid and continuous recruitment, especially in warm waters; this makes monogeneans especially dangerous in intensive culture (Deveney *et al.*, 2001). Disease caused by monogeneans is normally more debilitating than fatal, and subsequent mortality is usually attributed to viral or bacterial infection. Monogeneans stress the fish hosts by destroying the epidermal integrity of the fish, thus predisposing their hosts to other pathogens as mechanical vectors of bacterial and viral diseases (Xu *et al.*, 2007; Abdel-Latif and Khafaga, 2020).

In intensive culture systems, where the intensity of infection can be high on the gills, monogeneans can cause death directly by inhibiting respiration through physical damage to the gills that destroys the gill architecture. Fish mortality from monogenean infections may result from damage to gill tissues and skin caused by attachment organs, and by feeding on the integument, which stimulates cell proliferation and secretion of copious amounts of mucus (Paperna, 1991). Cage culture in tropical areas is usually conducive to the perpetuation of parasitic diseases with high stocking density. The nets trap eggs, infective larvae and food debris around the cages, which attract carrier/reservoir feral fish. Most monogenean genera are specific to a group of related host species; for example, *Dactylogyrus* is found on cyprinids, *Cichlidogyrus* on cichlids, while catfish harbour *Thaparocleidus* (see Shaharom, 1985). Although at the species level most monogeneans are specific to a particular host, some, like *Thaparocleidus caecus*, are found on several pangasiids (Lim *et al.*, 2001; Šimková *et al.*, 2013).

Many of the monogenean species on warm freshwater-cultured fish have not been identified or are incorrectly classified. For example, *Dactylogyrus* spp. have also been incorrectly implicated as being pathogenic to snakeheads, tilapia and clariids cultured in South-East Asia (Kabata, 1985). *Trianchoratus* is found on snakeheads other than giant snakehead, which is infected by *Sundanochus* spp., while *Cichlidogyrus* spp. infect the tilapias and *Quadriacanthus* spp. and *Bychowskyella* spp. infect the South-East Asian clariids. The gyrodactylids, on the other hand, are ubiquitous, although at species level they might be host specific. The most reported monogeneans on warm freshwater-cultured fish are the *Dactylogyrus* spp. on carp; *Cichlidogyrus* spp., *Enterogyrus* spp. and *Scutogyrus* spp. on cichlids; *Bychowskyella* spp. and *Quadriacanthus* spp. on clariids; *Trianchoratus* spp. on snakehead; *Pseudodactylogyroides* spp. on *Oxyeleotris marmorata*; *Thaparocleidus* spp. on catfish other than clariids; and the *Gyrodactylus* spp. (see Section 6.5.4.iv below) (Fig. 6.8).

In many cases, the specific identity of the pathogenic monogeneans, signs and pathology of the infection, disease mechanism, and control and preventive measures have not been specifically elucidated and documented. For instance, it is known that *Thaparocleidus siamensis* occurs in greater intensity than *T. caecus* on cultured *P. hypophthalmus* in Peninsular Malaysia and Thailand (Lim, 1990, 1996; Thuy and Buchmann, 2008), but it is not known which of the two species is pathogenic. The translocation of monogeneans, along with their hosts, has been well documented for the various *Dactylogyrus* spp. on imported Chinese carp, for *Cichlidogyrus* spp. on tilapia (Shinn *et al.*, 2022) and for *Quadriacanthus clariadis* on the *C. gariepinus* imported into Thailand (Paperna, 1991; Lerssutthichawal, 1999). There is information on the signs, pathology and control measures for some species of *Dactylogyrus*, *Gyrodactylus* and *Cichlidogyrus*, but not for other monogenean pathogens. In most cases, the information is derived from

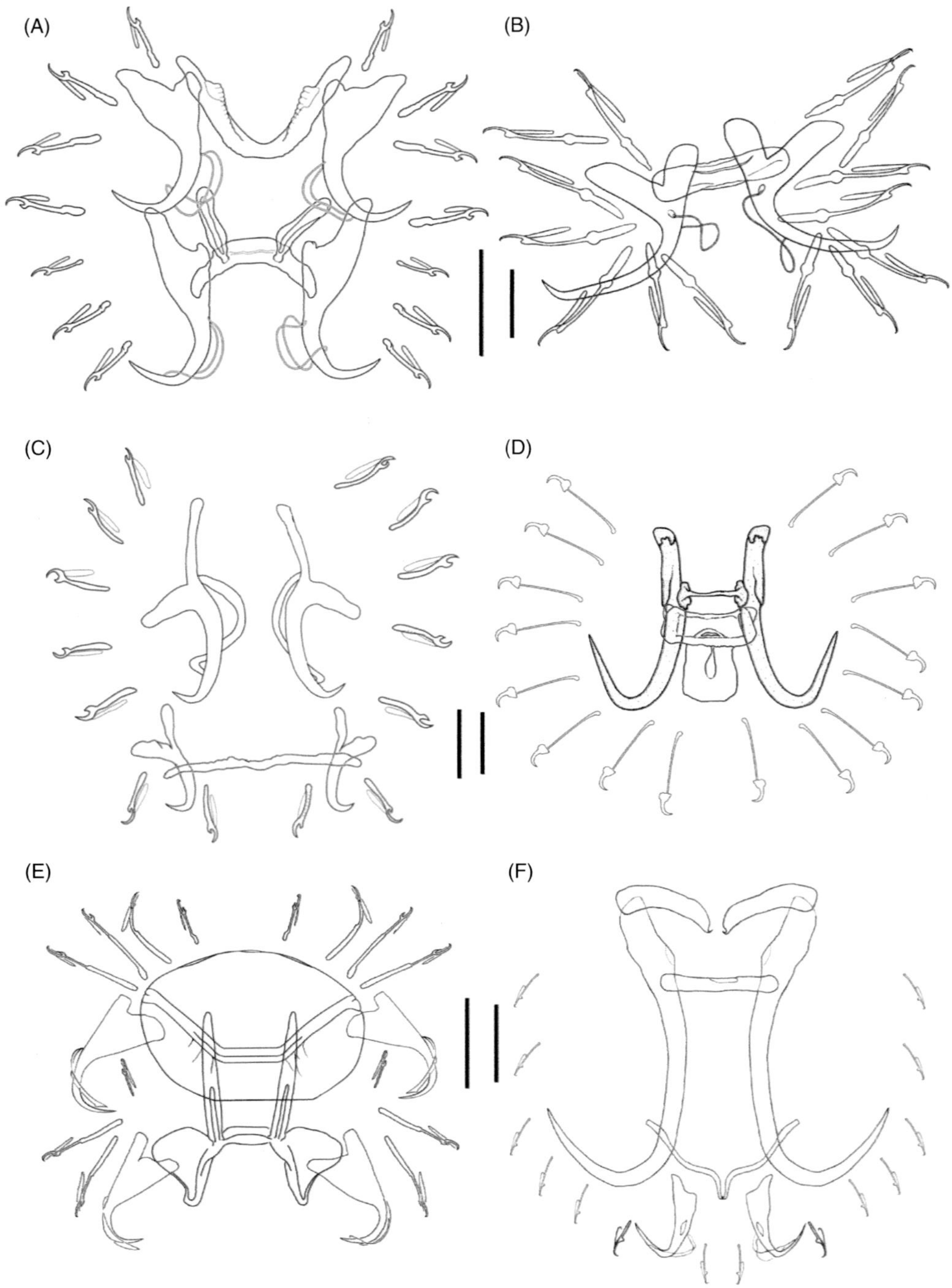

Fig. 6.8. Arrangement of the attachment hooks in six key species of monogenean. (A) *Cichlidogyrus tilapiae* Paperna, 1960 (modified from Šimková *et al.*, 2019). (B) *Dactylogyrus vastator* Nybelin, 1924 (modified from Gussev, 1985). (C) *Enterogyrus malmbergi* Bilong Bilong, 1988 (modified from Zhang *et al.*, 2019). (D) *Gyrodactylus cichlidarum* (adapted from García-Vásquez *et al.*, 2007). (E) *Scutogyrus longicornis* (adapted from Pariselle and Euzet, 1995). (F) *Tharoparocleidus* gen. sp. (adapted from Pariselle *et al.*, 2006). Scale bars: (A–D) 20 μm; (E, F) 30 μm.

pond-culture systems and not from cage-culture systems. It should be also noted that habitat can affect parasitic infections, as indicated by the infection of tilapia by *Neobenedenia* spp. instead of *Cichlidogyrus* spp. when farmed in cages in estuarine waters (see Shinn *et al.*, 2022).

6.5.4.i Diseases caused by *Dactylogyrus* species

Dactylogyrus species are specific to the Cyprinidae although they are also found on Hemiramphidae and one species on a catfish (Gussev, 1976). This genus is frequently listed as a disease-causing agent since cyprinids are the most cultured fish group. The genus *Dactylogyrus* includes over 900 species (Gibson *et al.*, 1996) and approximately 50 species are known from the seven top cultured species of cyprinid listed in Table 6.1. The four important species of *Dactylogyrus* that cause disease in cultured common carp in Israel are *Dactylogyrus anchoratus*, *Dactylogyrus extensus*, *Dactylogyrus minutus* and *Dactylogyrus vastator* (see Paperna, 1991). These have different temperature preferences: for example, *D. extensus* flourishes at low water temperatures (optimum temperatures of 16–17°C), while *D. vastator* prefers warmer waters (20–24°C). *Dactylogyrus* has also been shown to cause mass mortality of fry, small fish and broodfish (Paperna, 1991; Kritsky and Heckmann, 2002).

Pathology

The pathology caused by *Dactylogyrus* spp. on exotic carp has been reported in studies done in Europe, but not for the species infecting the indigenous cyprinids of South-East Asia. Feeding on epithelial cells and anchorage (attachment) by the monogeneans cause severe destruction of the gills, resulting in haemorrhage and metaplasia of the gill tissue. Secondary bacterial infections usually occur and result in death of the fish. The pathologies caused by *D. vastator* and *Dactylogyrus lamellatus* are similar (Molnár, 1972; Paperna, 1991). *D. vastator* infections cause severe hyperplasia of the epithelium of gill filaments. Extensive proliferation of the respiratory epithelium of the gills interferes with respiratory functions and may be a direct cause of death. The sites of proliferation are dependent on the preferred sites of the monogenean species. *D. vastator* prefers the tips of the gill filaments and causes mass mortality in young fish but seldom on fish greater than 32–35 mm since the functions of the remaining gill filaments are not affected. Massive infections of *D. extensus* can cause mortality in 4–7 kg broodfish (Paperna, 1991).

Diagnosis

Fish infected with *Dactylogyrus* spp. are lethargic and are seen swimming at the water surface. Fish heavily infected with *Dactylogyrus* have pale to greyish gills, swollen at the edges, and the opercula appear to open wider than normal and secrete an excessive amount of mucus. Heavily infected fish may be anorexic and seen gasping for air and exhibiting abnormal behaviour such as jumping out of the water. *Dactylogyrus* spp. are usually found on the gills, although in massive infections they can also be found on the buccal cavity. *Dactylogyrus* spp. can kill directly by damaging gill structures and affecting respiration. In warm eutrophic waters with low oxygen concentrations, this becomes serious (Molnár, 1994). *Dactylogyrus* infections usually result in secondary bacterial infections with subsequent mortality.

At present, *Dactylogyrus* infections are confirmed by examination of the gills and infected fins for presence of the monogeneans. The signs and pathology of monogenean infections are neither generic nor species specific. Hence, the diagnosis of a monogenean infection is based on the identification of the pathogen itself. Correct diagnosis of monogeneans requires proper preparation of the parasite specimens. Gills can either be completely removed or gill biopsies can be taken from the infected fish. Each parasite is removed carefully from the gills under a dissecting microscope, placed on a slide and covered with a coverslip. Excess water is removed, and the corners of the coverslip sealed with nail polish to prevent it from moving. Ammonium picrate is added underneath the coverslip to clear and fix the specimens, which are examined using a phase contrast microscope. Monogenean species are usually identified based on the hard reproductive and haptoral armaments (i.e. attachment hooks) on the cleared and flattened specimens. *Dactylogyrus* are oviparous monogeneans with or without four eye spots, 14 marginal hooks, two anchors, one or two connective bars and two needle-like structures, and spindle-shaped dactylogyrid-type seminal vesicles (Fig. 6.8B). For identification keys to the relevant species, see the works of Mizelle and Price (1964), Paperna (1991), Chinabut and Lim (1993), Wu *et al.* (2000), Pandey and Agrawal (2008) and Galli *et al.* (2010).

6.5.4.ii Diseases caused by *Cichlidogyrus* species

Cichlids are cultured in cages in warm fresh water as well as warm estuarine water. Tilapias cultured in fresh waters are affected by *Cichlidogyrus* spp., while in marine waters they are infected by the marine monogeneans, *Neobenedenia girellae* and *Neobenedenia melleni* (see Shinn *et al.*, 2022). In 2019, 11 commercially important species of tilapia (not including hybrids) were cultured in 124 countries worldwide resulting in an aquaculture production of >6.2 million tonnes (Shinn *et al.*, 2022). Cichlids are hosts to species of *Cichlidogyrus*, *Onchobdella* and *Enterogyrus* (an endoparasitic monogenean that lives within the stomach of its host). Shinn *et al.* (2022) documents >820 translocations of tilapia and comments on the parasites that have been moved with them. The most notable species that have been translocated to several countries include *Cichlidogyrus halli*, *Cichlidogyrus longicornis*, *Cichlidogyrus sclerosus*, *Cichlidogyrus thurstonae*, *Cichlidogyrus tilapiae* and *Scutogyrus longicornis* (see Zhang *et al.*, 2019; Shinn *et al.*, 2022). No accounts of mortality due to *Cichlidogyrus* spp. have been recorded, but *C. sclerosus* is reported to cause severe gill damage in tilapias cultured in the Philippines (Kabata, 1985) as does *Cichlidogyrus philander* on *Pseudocrenilabrus philander* in South Africa (Igeh and Avenant-Oldewage, 2020).

Diagnosis

The behaviour of the fish can indicate the presence of parasites, and this is like infections of *Dactylogyrus*. An accurate diagnosis requires removing the gills or taking gill biopsies; the monogeneans are collected and prepared as stated for *Dactylogyrus* (see Section 6.5.4.i above). *Cichlidogyrus* can be distinguished from other monogeneans by having a haptor with four anchors, with two bars, one of which is V-shaped and the other made up of three parts (see Fig. 6.8A).

6.5.4.iii Diseases caused by *Thaparocleidus* species

Species belonging to this monogenean genus are found on cultured pangasiids and bagrids in South-East Asia (see Pariselle *et al.*, 2002; Thuy and Buchmann, 2008; Šimková *et al.*, 2013; Lio-Po and Lim, 2014) (Fig. 6.8F). As with *Pseudodactylogyroides*, little is known about the pathology caused by this group of monogeneans, or the level of pathogenicity.

Diagnosis

Monogenean species from pangasiids are collected and prepared as for *Dactylogyrus* infection (see Section 6.5.4.i above). *Thaparocleidus* spp. (Fig. 6.8F) have four anchors, two connective bars, one of which may be whole or separated into two, 14 marginal hooks and a sac-like seminal vesicle (Lim, 1996).

6.5.4.iv Diseases caused by *Gyrodactylus* species

The gyrodactylids can be readily differentiated from the dactylogyrids – they are colourless, lack eye spots, viviparous with most specimens carrying a developing embryo within their uterus, epithelial grazers rather than blood feeders, and possess 16 rather than 14 marginal hooks. Mothers give birth to live young which immediately attach to the surface of their host. Following the first birth, gyrodactylids may increase their activity on their hosts, moving to the fin edges where, on close contact with another suitable host, they can loop across. The research findings of Soleng *et al.* (1999) show that gyrodactylids that become detached from their hosts and are drifting in the water column can reattach on contacting a new host. Detached specimens may also attach to the substrate and wait until another host passes over making immediate contact. Of the *Gyrodactylus* species infecting warm freshwater fish, some have wide host specificity and can cause fish mortalities in juvenile populations of fish (García-Vásquez *et al.*, 2007, 2011). For example, *Gyrodactylus cichlidarum* is reported from six species of commercially farmed cichlids and from at least 19 countries (Shinn *et al.*, 2022). *Gyrodactylus* and *Macrogyrodactylus* infections are also common on *Clarias* spp. with at least seven species of *Gyrodactylus* and three species of *Macrogyrodactylus* described from *C. gariepinus* (see Paperna, 1973; Přikrylová *et al.*, 2012; Maduenyane *et al.*, 2021).

Pathology

Gyrodactylus spp. are usually found on the skin, fins and gills and commonly in conjunction with protozoan and bacterial infections (Xu *et al.*, 2007; Abdel-Latif and Khafaga, 2020). Mucus secretion is increased during heavy infections, fins become frayed, skin ulcerated, swimming performance is impaired, and gills may be damaged by the feeding and attachment processes of the worm. Fish infected with *Gyrodactylus* exhibit abnormal behaviour, namely rubbing against the net, anorexia, hyperproduction of skin mucus, haemorrhagic ulcers on the body sides, fin rot (mainly anal and caudal fins), and thickening and opacity of the eye cornea. Infections can result in the mass mortality of juvenile fish (García-Vásquez *et al.*, 2007, 2011; Abdel-Latif and Khafaga, 2020).

Diagnosis

Initial diagnosis can be based on clinical signs with confirmation by examination of the parasites. The monogeneans are collected from skin and gill scrapings and prepared as for *Dactylogyrus* spp. (see Section 6.5.4.i above). Two sets of head glands are used as a means of temporary attachment during locomotion. Its haptor is armed with 16 marginal hooks, two anchors and two connective bars, one of which is a simple dorsal bar, the other is a triangular-shaped ventral bar (Fig. 6.8D). There are >400 species of *Gyrodactylus* (see Harris *et al.*, 2004); morphologically, species are separated on subtle differences in the shape of the attachment hooks, particularly the sickles of the marginal hooks (Malmberg, 1970; Gussev, 1985; Shinn *et al.*, 2001; Galli *et al.*, 2010). *Gyrodactylus* spp. in the tropical regions are poorly studied and more investigations are required.

6.5.4.v Other important monogenean infections of cage-culture species

For information regarding infections of *Pseudodactylogyroides marmorata* found on cage-cultured *O. marmorata*, *Sundanonchus* spp. primarily found on *Channa micropeltes* and *Trianchoratus* spp. commonly found on other channids and anabantids, refer to the review of Lio-Po and Lim (2014).

Prevention and control of monogenean infections

The main method for the control of small monogenean species is by a salt or formalin treatment. The recommended doses and concentrations to be used will vary according to host, size, stocking density, existing condition of the gills and parasite species, as well as the physicochemical properties of the water, notably the temperature (Shinn and Bron, 2012). Tancredo *et al.* (2019) recommend a short bath (1 h) using formalin at 75 mg/l, and not exceeding 100 mg/l, for the treatment of *C. carpio* fingerlings (average weight 1.7 g; 4.7 cm total length) infected with *D. minutus* in 28°C water. In all situations where a treatment may be required, it is advisable to seek veterinary advice and support. Eradication of feral reservoir fish from ponds is possible but not when the cages are in rivers or large lakes. The best alternative management strategy includes good husbandry based on knowledge of the reproductive biology and ecological requirements of the parasites such as temperature dependency. Using healthy fish fry from reliable hatcheries, limiting stocking density of fish, providing good-quality feed and sanitation of nets will help to keep infection at a low level.

6.5.5 Diseases caused by other helminths

While many digeneans, nematodes, cestodes and acanthocephalans have been recorded from tropical aquaculture species (Paperna, 1996), details regarding the pathogenic species causing disease in cage-culture systems are limited (Lio-Po and Lim, 2014). Important species include infections of the blood fluke *Sanguinicola* which has been recorded on exotic cultured grass carp and bighead carp; the nematodes *Anguillicola crassus* and *Philometroides cyprini* in common carp; the adult Asian tapeworm, *Schizocotyle* (syn. *Bothriocephalus*) *acheilognathi*, which causes mortality in heavily infected grass carp in Europe; and the cestodes *Lytocestus* spp. found in cultured and wild *C. batrachus* and *Senga* spp. found in cultured and wild snakeheads, *Channa* spp. Fish infected with intestinal (adult) cestodes have retarded growth, erratic swimming behaviour and distended abdomen, become emaciated, cease to feed, develop a haemorrhagic enteritis caused by destruction of the intestinal epithelium, and heavily infected fish have varying degrees of aseptic dropsy (Paperna, 1996). A cyclopoid copepod is the intermediate host, and the cestodes could be an important pathogen in cage-culture systems since the fish are in intimate contact with the environment. The helminths infecting a range of farmed tilapia species are provided in the review of Shinn *et al.* (2022).

6.5.6 Diseases caused by parasitic arthropods

Lernaea and *Ergasilus* spp. (Copepoda), *Argulus* (Branchiura) and *Alitropus* (Isopoda) have been recorded on a wide range of cultured fish species (Kabata, 1985; Shariff and Sommerville, 1986). For further details regarding infections on warmwater fish species, pathology and diagnostics, readers are signposted to the reviews provided by Lester and Roubal (1995) and Lio-Po and Lim (2014).

Prevention and control

Many of the treatments that were once commonly used for the treatment of parasitic copepods are no longer appropriate or permitted for use. For cages in rivers or large lake systems, however, the use of chemicals is ineffective and dipping fish in chemicals was not efficient in removing all the copepodid stages (Lester and Roubal, 1995). Raja *et al.* (2022) used emamectin benzoate (EB) given as a dose of 50 μg EB/kg per day for ten consecutive days to control infections of *Lernaea cyprinacea* on caged *Lates calcarifer* and on pond-reared *C. auratus* and *Cyprinus rubrofuscus*.

6.5.6.i *Argulus* infections

Most branchiurans are freshwater parasites with a few estuarine and marine species (Kabata, 1985). *Argulus* or the fish louse (Fig. 6.7I) is macroscopic and easily observable on the fish's exterior or in the oral cavity. These ectoparasitic crustaceans feed on the mucus layer, flesh and blood. The prolonged feeding and strong attachment of *Argulus* by its suckers on to the host result in direct mechanical damage to the skin and disruption of epithelial structure, resulting in lesions and subsequent invasions by opportunistic pathogens. The life cycle of the parasite is direct. Argulid eggs (ovoid, average size 0.2 mm × 0.3 mm) are covered by a gelatinous material and are laid in rows, typically on hard structures with rough rather than smooth surfaces. Development of the eggs is temperature dependent and is *c.*17 days at 23°C (Rahman, 1995). The eggs hatch into free-swimming larvae which must find a host within 2–3 days. Species such as *Argulus japonicus*, *Argulus foliaceus*, *Argulus indicus* and *Argulus siamensis* are economically important parasites of fish in warm freshwater aquaculture (see reviews of Lio-Po and Lim, 2014; Shinn *et al.*, 2022).

Pathology

This parasite is not host specific and is found on a wide range of fish species from cyprinids to siluriforms and perciforms, and infections can result in significant losses (Kumari and Nomani, 2021). Heavily infected fish are lethargic, listless, cease to feed and rub themselves on the substrate to dislodge attached parasites. The parasite is also common in India, affecting the major Indian carp, especially rohu species. *Argulus* feed via a mouth tube, a proboscis, that encloses heavily armed mandibles and two syphons for the release of digestive enzymes. There is also a slender, partially sheathed pre-oral stylet which repeatedly punctures the skin to induce blood flow to the feeding site (Kabata, 1970). Products released through the stylet and mouth tube result in the breakdown of host epithelial cells, facilitating blood feeding, which creates a localized inflammatory response with subcutaneous haemorrhaging (Walker *et al.*, 2011). AmbuAli *et al.* (2020) identified 27 proteins within the secretory products that play a role in digestion and immunomodulation. Haemorrhagic responses from adult feeding were observed within 24–48 h post-infection (Walker *et al.*, 2011). The lesion or wound made by the feeding *Argulus* may be restricted to the epidermis or may penetrate through to the stratum spongiosum of the dermis and even the stratum compactum, turning the dermis oedematous (Lester and Roubal, 1995). The area may become necrotic with secondary bacterial and fungal infections. Mortality may be associated with changes in the ionic and osmotic homeostasis, anorexia and secondary infections. *Argulus* feeding on blood causes fish to become anaemic and its piercing proboscis stylet causes haemorrhagic spots on the epidermis. The spots are formed by epidermal hyperplasia. Bacterial infections occur around the site of infection. *Argulus* may also serve as a vector of viral infections, such as spring viraemia of carp (SVC) (Pfeil-Putzien, 1977; Ahne, 1985) and carp pox (carp papilloma) which occurred in conjunction with an infection of *A. japonicus* infection in Israel (Sarig, 1971). *Argulus* also acts as an intermediate host for various skrjabillanid and daniconematid nematodes (Moravec, 1998) and argulids have been implicated in facilitating the establishment of several bacterial infections, for example *A. hydrophila* and *F. columnare* (see Bandilla *et al.*, 2006; Shameena *et al.*, 2021).

Diagnosis

The parasite is oval to round, dorsoventrally flattened (about 4–8 mm in diameter), with a pair of modified sucker-like first maxillae. Its proboscis or feeding organ is for inserting into the epidermis and the underlying tissue of the fish hosts to feed on blood (Fig. 6.7I).

Prevention and control

Historically, a range of noxious chemicals have been recommended for the control of *Argulus* infections; all are now either banned or undesirable (Kabata, 1985; Egusa, 1992). Current chemotherapeutic methods for the control of *Argulus* include the use of EB. Hakalahti *et al.* (2004) used a dose of 50 μg EB/kg per day for seven consecutive days to control infections of *Argulus coregoni* on rainbow trout. Raja *et al.* (2022) applied the same dose for seven to ten consecutive days to control *A. siamensis* on Indian major carps and *L. rohita*, and in removing *Argulus* spp. on caged *P. hypophthalmus* and on pond-reared Indian major carps. The fallowing of sites – that is, the drying of ponds or not stocking ponds until the larval stages have died – is an effective approach in the control of crustacean pathogens. Other non-chemical strategies include the use of removable substrates on to which *Argulus* lay their eggs, the substrates are then periodically removed, rotated, sun-dried or cleaned, and redeployed (Gault *et al.*, 2002).

6.6 Impacts from Climate Change

Since 2000, sea levels have risen by 61.74 mm (Church and White, 2006; Sweet *et al.*, 2022) and Arctic polar ice is decreasing at a rate of 3.5–4.1% per decade (IPCC, 2014). Ocean acidification has seen a mean global pH drop from 8.11 in 1989 to 8.05 in 2020 (Our World in Data, 2022a). Global atmospheric carbon dioxide has risen from 339.6 ppm in 1980 to 418.28 ppm in 2020 (Our World in Data, 2022b); atmospheric methane levels have risen from 1625.9 ppb in 1983 to 1906.8 ppb in 2022 (Our World in Data, 2022c). Global temperatures have risen by 0.354°C over the period 1970–2021; sea temperatures by 0.286°C over the same time frame (Morice *et al.*, 2021). Eckstein *et al.* (2021) predict that the frequency of severe tropical storm events will increase with every 0.1°C rise in global temperature. Changing climatic conditions resulting in flood events, altered patterns of precipitation, salt intrusion, siltation, oxygen depletion, heatwaves, drought, changes to water flow and velocity, can all exert effects with direct impacts on aquaculture production, systems, stocks, and the frequency of disease events. Damage to aquaculture structures can result in the loss of stock. Escaped fish may acquire new infections and, as reservoirs of infections, their activity around caged stocks can result in the transfer of infections to caged stock or in extending the window of infection as the treatment of fish falls outside the health management of cages. Rising water temperatures though may have the greatest impact on the frequency and severity of disease events, favouring pathogens like *Streptococcus* spp. which thrive in temperatures above 30°C. Without management interventions, warmer waters lower dissolved oxygen concentrations imposing greater degrees of stress on fish species, some of which may already be farmed at the upper extreme of their temperature tolerance (see Table 6.1). Warmer waters may also result in a general acceleration in parasite development times or may permit certain pathogens/parasites and/or their intermediate hosts to extend their geographic range.

The temporal, decadal change in global surface temperature is presented in Fig. 6.9. When the average climate risk index for each country is considered (i.e. countries most exposed and vulnerable to extreme weather events; Fig. 6.10A) and combined with current aquaculture production (Fig. 6.10B), then analysis reveals that many aquaculture industries within the tropical zone are at most risk (i.e. Myanmar, Haiti, Philippines, Mozambique, Bangladesh, Pakistan, Thailand, Nepal, Madagascar and Vietnam). Safeguarding those aquaculture industries at most risk from climate change will require national coordinated strategic plans of area management to protect water resources and the aquaculture industries within them. At the farm level, this will require a move away from abstracting and discharging waters from common water sources to closed farm systems, with on-site disinfection and recycling of water, as a biosecurity measure, to minimize the likelihood of waterborne pathogens entering farm sites. For cage-culture operations, protecting stocks from climate change and disease threats may lie in genetic breeding programmes selecting for strains that are specific-pathogen resistant (SPR) or specific-pathogen tolerant (SPT), or can tolerate increases in various environmental factors (salt, extremes in temperature, toxic nitrogenous compounds, etc.), or have faster growth reducing the 'in pond' culture time. There also need to be parallel programmes of vaccine development and vaccination against those pathogens where there is currently no protection or for new emerging diseases in warmwater aquaculture. The last decade has also seen radical changes to the chemotherapy arsenal – many of the products that

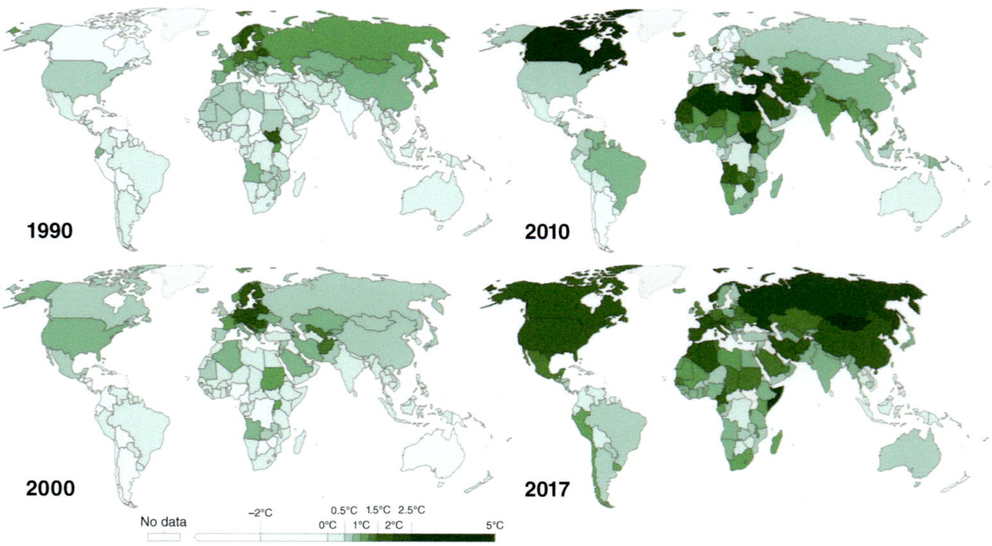

Fig. 6.9. Global surface temperature change, measured in °C, for the last four decades. (Data are based on HadCRUT analysis from the Climatic Research Unit of the University of East Anglia, together with that produced by the Hadley Centre, UK Meteorological Office.)

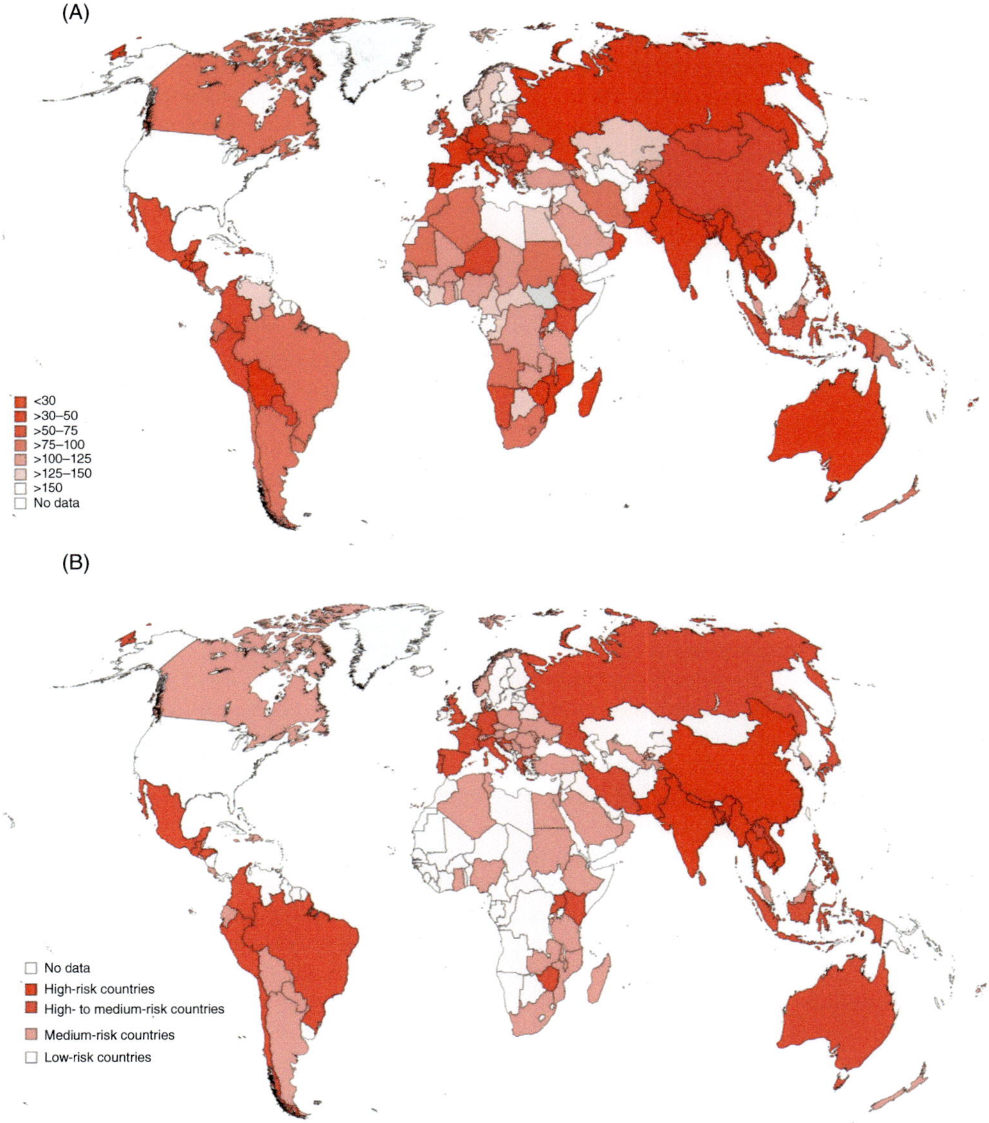

Fig. 6.10. (A) Average climate risk index (CRI) for 2000–2019 using scores calculated by Eckstein *et al.* (2021). Countries are colour-coded according to their exposure and vulnerability to extreme weather events and their capacity to cope with these – countries at the greatest risk are those with low CRI scores. The map indicates that the top ten countries currently at the greatest risk of changes to climatic conditions are Myanmar (most affected), Haiti, Philippines, Mozambique, Bangladesh, Pakistan, Thailand, Nepal, Madagascar and Vietnam. Please note that data for some countries, such as the USA, were not scored by the Eckstein *et al.* (2021) report. (B) Aquaculture-producing countries at most risk from the impacts of climate change. Countries are ranked by their average CRI score (2000–2019) and by the size of national finfish production. (Data on CRI scores are drawn from Eckstein *et al.*, 2021; while finfish production data are taken from FishStatJ, 2022.)

were once commonly used in tropical aquaculture as parasite treatments are no longer permitted or suitable for the treatment of cages and large water bodies. The limited availability of licensed efficacious treatments, therefore, remains a barrier to effective parasite control in aquaculture. While new and alternative regimens are researched, good aquaculture biosecurity and husbandry practices (e.g. stocking of disease-free stocks, recognizing the early signs of disease, conducting regular health surveillance, training in treatment administration) and access to good veterinary/health support can help minimize the frequency and magnitude of disease events.

6.7 Conclusions and Recommendations for Future Research

Microorganisms and parasites are part of a fish's normal flora and fauna and under normal conditions, many of these do not induce disease in their host. Changing environmental conditions from pollution, overstocking and/or climate change may be stressful to fish. Bacterial multiplication, for instance, is enhanced with increasing organic matter from uneaten feeds. Stress predisposes fish to invasion by opportunistic pathogens, with subsequent morbidity and mortality. Stress is also associated with handling, stocking, grading and shipping of fish. Often fish mortality can be attributed to several factors (e.g. fish condition, pathogens, poor husbandry, environment), and frequently it is difficult to determine the significance of any one of these factors (Mitchell, 1997). Some parasites may not result in direct mortality, but they debilitate fish, predisposing them to other pathogens. Parasite activity can create portals of entry for viral, bacterial and oomycete pathogens of fish. Other parasites may serve as reservoirs of viral pathogens. Despite the long history of aquaculture in the tropics and the importance of disease in aquaculture, large knowledge gaps in the health of cage-reared aquaculture species remain. This may be due to lack of trained manpower and institutional support. The usual approach to disease management is to use chemicals (usually indiscriminately) or, when this fails, to discard the fish and start afresh. In many countries there is no mandatory requirement to report fish deaths, and until recently fish were usually imported and stocked without quarantine. With the cross-boundary movement of fish there are risks of pathogen transfer (Hedrick, 1996) and of fish escapes. For commercial species of tilapia, Shinn *et al.* (2022) summarizes >820 translocations and comments on how some have facilitated the spread of parasites into new regions. In this regard, the provisions of the OIE International Aquatic Animal Health Code should be adhered to (OIE, 2021). The lack of institutional support in many countries results in reduced research on pathogens and consequently an inability to control and prevent diseases. There is also a lack of trained personnel in disease management and little reliable information on the specific identity of pathogens. A related issue is the lack of legislation and guidelines pertaining to the use of drugs and chemicals in aquaculture. In many instances, control agents are unregulated, used indiscriminately (usually the aetiological agents are not identified) and without a specific withdrawal period prior to the sale of the fish. Trained competent fish disease managers, who can diagnose pathogens and are capable of dispensing proper prevention and control measures, are important to sustain aquaculture.

References

Abdel-Latif, H.M.R. and Khafaga, A.F. (2020) Natural co-infection of cultured Nile tilapia *Oreochromis niloticus* with *Aeromonas hydrophila* and *Gyrodactylus cichlidarum* experiencing high mortality during summer. *Aquaculture Research* 51, 1880–1892.

Adamek, M., Jung-Schroers, V., Hellmann, J., Teitge, F., Bergmann, S.M., *et al.* (2016) Concentration of carp edema virus (CEV) DNA in koi tissues affected by koi sleepy disease (KSD). *Diseases of Aquatic Organisms* 119(3), 245–251.

Adamek, M., Matras, M., Jung-Schroers, V., Teitge, F., Heling, M., *et al.* (2017a) Comparison of PCR methods for the detection of genetic variants of carp edema virus. *Diseases of Aquatic Organisms* 126(1), 75–81.

Adamek, M., Oschilewski, A., Wohlsein, P., Jung-Schroers, V., Teitge, F., *et al.* (2017b) Experimental infections of different carp strains with the carp edema virus (CEV) give insights into the infection biology of the virus and indicate possible solutions to problems caused by koi sleepy disease (KSD) in carp aquaculture. *Veterinary Research* 48(1), 12.

Adamek, M., Teitge, F., Jung-Schroers, V., Heling, M., Gela, D., *et al.* (2018) Flavobacteria as secondary pathogens in carp suffering from koi sleepy disease. *Journal of Fish Diseases* 41(11), 1631–1642.

Ahne, W. (1985) *Argulus foliaceus* L. and *Piscicola geometra* L. as mechanical vectors of spring viraemia of carp virus (SVCV). *Journal of Fish Diseases* 8, 241–242.

Albaladejo, J.D. and Arthur, J.R. (1989) Some trichodinids (Protozoa: Ciliophora: Peritrichida) from freshwater fishes imported into the Philippines. *Asian Fisheries Science* 3, 1–25.

Altinok, I. and Grizzle, J.M. (2001) Effects of low salinities on *Flavobacterium columnare* infection of euryhaline and freshwater stenohaline fish. *Journal of Fish Diseases* 24(6), 361–367.

Angka, S.L., Lam, T.J. and Sin, Y.M. (1995) Some virulence characteristics of *Aeromonas hydrophila* in walking catfish (*Clarias gariepinus*). *Aquaculture* 130, 103–112.

Amal, M.N.A., Zamri-Saad, M., Siti-Zahrah, A. and Zulkafli, A.R. (2013) Transmission of *Streptococcus agalactiae* from a hatchery into a newly established red hybrid tilapia, *Oreochromis niloticus* (L.) × *Oreochromis mossambicus* (Peters), farm. *Journal of Fish Diseases* 36(8), 735–739.

AmbuAli, A., Monaghan, S., McLean, K., Inglis, N.F., Bekaert, M., *et al.* (2020) Identification of proteins from the secretory/excretory products (SEPs) of the branchiuran ectoparasite *Argulus foliaceus* (Linnaeus, 1758) reveals unique secreted proteins amongst haematophagous ecdysozoa. *Parasites & Vectors* 13, 88.

Amtmann, A., Ahmed, I., Zahner-Rimmel, P., Mletzko, A., Jordan, L.K., *et al.* (2020) Virucidal effects of various agents – including protease – against koi herpesvirus and viral haemorrhagic septicaemia virus. *Journal of Fish Diseases* 43(2), 185–195.

Aoki, T. (1999) Motile aeromonads (*Aeromonas hydrophila*). In: Woo, P.T.K. and Bruno, D.W. (eds) *Fish Diseases and Disorders*. Vol. 3. *Viral, Bacterial and Fungal Infections*, 1st edn. CAB International, Wallingford, UK, pp. 427–453.

Aoki, T., Hirono, I., Kurokawa, K., Fukuda, H., Nahary, R., *et al.* (2007) Genome sequences of three koi herpesvirus isolates representing the expanding distribution of an emerging disease threatening koi and common carp worldwide. *Journal of Virology* 81, 5058–5065.

Arbildo-Ortiz, H., Alvez-Robledo, J., Chuquipiondo Guardia, C. and Silva de Souza, A.K. (2020) Primer registro de infestación de *Piscinoodinium pillulare* (Dinoflagellida) en juveniles de *Colossoma macropomum* (Characiformes: Serrasalmidae) encultivo semi-intensivoen Loreto, Perú. *Revista de Investigaciones Veterinarias del Perú* 31(3), e16662.

Atz, J.W. (1954) The peregrinating tilapia. *Animal Kingdom* 57(5), 148–155.

Austin, B. and Austin, D.A. (2007) *Bacterial Fish Pathogens: Disease of Farmed and Wild Fish*, 4th edn. Springer-Praxis Publishing, Chichester, UK.

Azzam-Sayuti, M., Ina-Salwany, M.Y., Zamri-Saad, M., Yusof, M.T., Annas, S., *et al.* (2021) The prevalence, putative virulence genes and antibiotic resistance profiles of *Aeromonas* spp. isolated from cultured freshwater fishes in peninsular Malaysia. *Aquaculture* 540, 736719.

Bacharach, E. and Eldar, A. (2016) Tilapia lake virus vaccines. *US Patent* no. US2016/0354458A1.

Bacharach, E., Mishra, N., Briese, T., Zody, M.C., Kembou Tsofack, J.E., *et al.* (2016) Characterization of a novel orthomyxo-like virus causing mass die-offs of tilapia. *MBio* 7(2), e00431-16.

Baek, Y.S. and Boyle, J.A. (1996) Detection of channel catfish virus in adult channel catfish by use of a nested polymerase chain reaction. *Journal of Aquatic Animal Health* 8, 97–103.

Bakke, T.A., Cable, J. and Harris, P.D. (2007) The biology of gyrodactylid monogeneans: the 'Russian doll-killers'. *Advances in Parasitology* 64, 161–376.

Bandilla, M., Valtonen, E.T., Suomalainen, L.-R., Aphalo, P.J. and Hakalahti, T. (2006) A link between ectoparasite infection and susceptibility to bacterial disease in rainbow trout. *International Journal for Parasitology* 36(9), 987–999.

Barkham, T., Zadoks, R.N., Azmai, M.N.A., Baker, S., Bich, V.T.N., *et al.* (2019) One hypervirulent clone, sequence type 283, accounts for a large proportion of invasive *Streptococcus agalactiae* isolated from humans and diseased tilapia in Southeast Asia. *PLoS Neglected Tropical Diseases* 13(6), e0007421.

Barnes, A.C., Silayeva, O., Landos, M., Dong, H.T., Lusiastuti, A., *et al.* (2022) Autogenous vaccination in aquaculture: a locally enabled solution towards reduction of the global antimicrobial resistance problem. *Reviews in Aquaculture* 14(2), 907–918.
Basson, L. and Van As, J. (2006) Trichodinidae and other ciliophorans (Phylum Ciliophora). In: Woo, P.T.K. (ed.) *Fish Diseases and Disorders*. Vol. 1. *Protozoan and Metazoan Infections*, 2nd edn. CAB International, Wallingford, UK. pp. 154–182.
Bastos Gomes, G., Jerry, D.R., Miller, T.L. and Huston, K.S. (2016) Current status of parasitic ciliates *Chilodonella* spp. (Phyllopharyngea: Chilodonellidae) in freshwater fish aquaculture. *Journal of Fish Diseases* 405, 703–715.
Baxa-Antonio, D., Groff, J.M. and Hedrick, R.P. (1992) Effect of water temperature on experimental *Edwardsiella ictaluris* infections in immersion-exposed channel catfish. *Journal of Aquatic Animal Health* 4(2), 148–151.
Bebak, J. and Wagner, B. (2012) Use of vaccination against enteric septicemia of catfish and columnaris disease by the US catfish industry. *Journal of Aquatic Animal Health* 24, 30–36.
Becker, C.D. and Fujihara, M.P. (1978) *The Bacterial Pathogen* Flexibacter columnaris *and Its Epizootiology among Columbia River Fish*. American Fisheries Society, Washington, DC.
Bercovier, H., Fishman, Y., Nahary, R., Sinai, S., Zlotkin, A., *et al.* (2005) Cloning of the koi herpesvirus (KHV) gene encoding thymidine kinase and its use for a highly sensitive PCR based diagnosis. *BMC Microbiology* 5, 13.
Bergmann, S.M., Riechardt, M., Fichtner, D., Lee, P. and Kempter, J. (2010) Investigation on the diagnostic sensitivity of molecular tools used for detection of koi herpesvirus. *Journal of Virological Methods* 163(2), 229–233.
Bernardet, J.F. and Bowman, J.P. (2006) The genus *Flavobacterium.The Prokaryotes* 7, 481–531.
Boonyaratpalin, S. (1989) Bacterial pathogens involved in the epizootic ulcerative syndrome of fish in Southeast Asia. *Journal of Aquatic Animal Health* 1, 272–276.
Bowser, P.R. and Plumb, J.A. (1980) Channel catfish virus: comparative replication and sensitivity of cell lines from channel catfish ovary and brown bullhead. *Journal of Wildlife Diseases* 16, 451–454.
Bueno, G.W., Ostrensky, A., Canzi, C., de Matos, F.T. and Roubach, R. (2015) Implementation of aquaculture parks in Federal Government waters in Brazil. *Reviews in Aquaculture* 7, 1–12.
Buller, N.B. (2004) *Bacteria from Fish and Other Aquatic Animals: A Practical Identification Manual*. CAB International, Wallingford, UK.
Callinan, R.B., Paclibare, J.O., Bondad-Reantaso, M., Chin, J.C. and Gogolewski, R.P. (1995) *Aphanomyces* species associated with epizootic ulcerative syndrome (EUS) in the Philippines and red spot disease (RSD) in Australia: preliminary comparative studies. *Diseases of Aquatic Organisms* 21, 233–238.
Canning, E.U. and Lom, J. (1986) *The Microsporidia of Vertebrates*. Academic Press, London.
Cano, I., Worswick, J., Mulhearn, B., Stone, D., Wood, G., *et al.* (2021) A seasonal study of koi herpesvirus and koi sleepy disease outbreaks in the United Kingdom in 2018 using a pond-side test. *Animals* 11(2), 459.
Castro, N., Toranzo, A.E., Núñez, S. and Magariños, B. (2008) Development of an effective *Edwardsiella tarda* vaccine for cultured turbot (*Scophthalmus maximus*). *Fish & Shellfish Immunology* 25(3), 208–212.
Catap, E.S. and Munday, B.L. (1998) Effects of variations of water temperature and dietary lipids on the expression of experimental epizootic ulcerative syndrome (EUS) in sand whiting, *Sillago ciliata*. *Fish Pathology* 33, 327–335.
Chang, P.H. and Plumb, J.A. (1996) Histopathology of experimental *Streptococcus* sp. in tilapia, *Oreochromis niloticus* (L.) and channel catfish, *Ictalurus punctatus* (Rafinesque). *Journal of Fish Diseases* 19, 235–241.
Chauhan, R. (2014) Fungal attack on *Tilapia mossambicus* in culture pond, leading to mass mortality of fishes. *International Journal of Pharmaceutical Sciences and Research* 5(07), 425–428.
Chauhan, R., Kaur, P. and Sharma, S. (2012) Pathogenicity of some species of *Achlya* and *Saprolegnia* on Indian major carps viz *Catla catla*, *Cirrihinus mrigala* and *Labeo rohita*. *Journal of Environmental Science, Computer Science and Engineering & Technology* 1, 422–428.
Chen, C.L. (1955) The protozoan parasites from four species of Chinese pond fishes: *Ctenopharyngodon idellus*, *Mylopharyngodon aethiops*, *Aristichthys nobilis* and *Hypophthalmychthys molitrix*. I. The protozoan parasites of *Ctenopharyngodon idellus*. *Acta Hydrobiologica Sinica* 2, 123–164.
Chen, D.-D., Yao, Y.-Y., Cui, Z.-W., Zhang, X.-Y., Peng, K.-S., *et al.* (2018) Comparative study of the immunoprotective effect of two DNA vaccines against grass carp reovirus. *Fish & Shellfish Immunology* 75, 66–73.

Chen, Q.L. and Ma, C.L. (1998) *Fauna Sinica. Invertebrata*. Vol. 15. *Myxozoa, Myxosporea*. Science Press, Beijing.

Cheng, L., Fang, Q., Shah, S., Atanasov, I.C. and Zhou, Z.H. (2008) Subnanometer-resolution structures of the grass carp reovirus core and virion. *Journal of Molecular Biology* 382(1), 213–222.

Chinabut, S. and Lim, L.H.S. (1993) Seven new species of *Dactylogyrus* Diesing, 1850 (Monogenea) from *Puntius* Hamilton (Cyprinidae) of Thailand. *The Raffles Bulletin of Zoology* 41, 47–59.

Chinabut, S., Roberts, R.J., Willoughby, L.G. and Pearson, M.D. (1995) Histopathology of snakehead, *Channa striatus* (Bloch), experimentally infected with the specific *Aphanomyces* fungus associated with epizootic ulcerative syndrome (EUS) at different temperatures. *Journal of Fish Diseases* 18, 41–47.

Church, J.A. and White, N.J. (2006) A 20th century acceleration in global sea-level rise. *Geophysical Research Letters* 33(1), 1–4.

Coker, W.C. (1923) *The Saprolegniaceae, with Notes on Other Water Molds*. The University of North Carolina Press, Chapel Hill, North Carolina.

Colquhoun, D.J., Larsson P., Duodu, S. and Forsman, M. (2014) The family Francisellaceae. In: Rosenberg, E., DeLong, E.F., Lory, S., Stackebrandt, E. and Thompson, F. (eds) *The Prokaryotes*. Springer, Berlin/Heidelberg, Germany, pp. 287–314.

Costa, H.H. and Wejeyaratne, J.S. (1989) Epidemiology of the epizootic ulcerative syndrome occurring for the first time among fish in Sri Lanka. *Journal of Applied Ichthyology* 1, 48–52.

Crawford, S.A., Gardner, I.A. and Hedrick, R.P. (1999) An enzyme-linked immunosorbent assay (ELISA) for detection of antibodies to channel catfish virus (CCV) in channel catfish. *Journal of Aquatic Animal Health* 11, 148–153.

Cross, M.L. and Matthews, R.A. (1992) Ichthyophthiriasis in carp, *Cyprinus carpio* L.: fate of parasites in immunized fish. *Journal of Fish Diseases* 15, 497–505.

Crumlish, M., Dung, T.T., Turnbull, J.F., Ngoc, N.T.N. and Ferguson, H.W. (2002) Identification of *Edwardsiella ictaluri* from diseased freshwater catfish, *Pangasius hypophthalmus* (Sauvage), cultured in the Mekong Delta, Vietnam. *Journal of Fish Diseases* 25(12), 733–736.

Cruz-Lacierda, E. (1995) *Histopathology and hematology of epizootic ulcerative syndrome (EUS)-positive snakehead* (Ophicephalus striatus). PhD thesis, Universiti Pertanian Malaysia, Serdang, Malaysia.

Dang, T.H.O., Xuan, T.T.T., Duyen, L.T.M., Le, N.P. and Hoang, H.A. (2021) Protective efficacy of phage PVN02 against haemorrhagic septicaemia in striped catfish *Pangasianodon hypophthalmus* via oral administration. *Journal of Fish Diseases* 44(8), 1255–1263.

Declercq, A.M., Haesebrouck, F., Van den Broeck, W., Bossier, P. and Decostere, A. (2013) Columnaris disease in fish: a review with emphasis on bacterium–host interactions. *Veterinary Research* 44(1), 27.

Delamare-Deboutteville, J., Taengphu, S., Gan, H.M., Kayansamruaj, P., Debnath, P.P., *et al.* (2021) Rapid genotyping of tilapia lake virus (TiLV) using Nanopore sequencing. *Journal of Fish Diseases* 44(10), 1491–1502.

Del-Pozo, J., Mishra, N., Kabuusu, R., Cheetham, S., Eldar, A., *et al.* (2017) Syncytial hepatitis of tilapia (*Oreochromis niloticus* L.) is associated with *Orthomyxovirus*-like virions in hepatocytes. *Veterinary Pathology* 54(1), 164–170.

Department of Fisheries, Bangladesh (2021) *Yearbook of Fisheries Statistics of Bangladesh 2019–20*. Department of Fisheries Bangladesh, Ministry of Fisheries and Livestock, Government of the People's Republic of Bangladesh, Dhaka.

Department of Fisheries, India (2020) *Handbook on Fisheries Statistics 2020*. Fisheries Statistics Division, Department of Fisheries, Ministry of Fisheries, Animal Husbandry & Dairying, Government of India, New Delhi.

Department of Fisheries Malaysia (2020) *Fisheries Statistics 1. Annual Fisheries Statistics – Volume 1. Perangkaan Perikanan Tahunan 2020*. Department of Fisheries Malaysia, Ministry of Agriculture and Food Industries, Putrajaya, Malaysia. Available at: https://www.dof.gov.my/en/resources/fisheries-statistics-i/ (accessed 11 August 2022).

Department of Fisheries, Ministry of Agriculture and Cooperatives, Thailand (2020) *Fisheries Statistics of Thailand 2020*. Information and Communication Technology Center, Department of Fisheries, Ministry of Agriculture and Cooperatives, Bangkok. Available at: https://www4.fisheries.go.th/local/index.php/main/view_qr_group/1408/7527 (accessed 17 June 2022).

Deveney, M.R., Chisholm, L.A. and Whittington, I.D. (2001) First published record of the pathogenic monogenean parasite *Neobenedenia melleni* (Capsalidae) from Australia. *Diseases of Aquatic Organisms* 46, 79–82.

Dickerson, H.W. and Dawe, D.L. (1995) *Ichthyophthirius multifiliis* and *Cryptocaryon irritans* (Phylum Ciliophora). In: Woo, P.T.K. (ed.) *Fish Diseases and Disorders*. Vol. 1. *Protozoan and Metazoan Infections*, 1st edn. CAB International, Wallingford, UK, pp. 181–227.

Diéguez-Uribeondo, J., García, M.A., Cerenius, L., Kozubíková, E., Ballesteros, I., *et al.* (2009) Phylogenetic relationships among plant and animal parasites, and saprotrophs in *Aphanomyces* (Oomycetes). *Fungal Genetics and Biology* 46, 365–376.

Dien, L.T., Ky, L.B., Huy, B.T., Mursalim, M.F., Kayansamruaj, P., *et al.* (2021a) Characterization and protective effects of lytic bacteriophage pAh6.2TG against a pathogenic multidrug-resistant *Aeromonas hydrophila* in Nile tilapia (*Oreochromis niloticus*). *Transboundary and Emerging Diseases* 69(4), e435–e450.

Dien, L.T., Linh, N.V., Sangpo, P., Senapin, S., St-Hilaire, S., *et al.* (2021b) Ozone nanobubble treatments improve survivability of Nile tilapia (*Oreochromis niloticus*) challenged with a pathogenic multi-drug-resistant *Aeromonas hydrophila*. *Journal of Fish Diseases* 44(9), 1435–1447.

Ding, Q., Yu, L., Wang, X. and Ke, L. (1991) Study on infecting other fishes with grass carp hemorrhagic virus. *Chinese Journal of Virology* 6, 371–373.

Dinh-Hung, N., Sangpo, P., Kruangkum, T., Kayansamruaj, P., Rung-Ruangkijkrai, T., *et al.* (2021) Dissecting the localization of *Tilapia tilapinevirus* in the brain of the experimentally infected Nile tilapia, *Oreochromis niloticus* (L.). *Journal of Fish Diseases* 44(8), 1053–1064.

Dishon, A., Davidovich, M., Ilouze, M. and Kotler, M. (2007) Persistence of cyprinid herpesvirus 3 in infected cultured carp cells. *Journal of Virology* 81, 4828–4836.

Dong, H.T., Nguyen, V.V., Le, H.D., Sangsuriy, P., Jitrakorn, S., *et al.* (2015a) Naturally concurrent infections of bacterial and viral pathogens in disease outbreaks in cultured Nile tilapia (*Oreochromis niloticus*) farms. *Aquaculture* 448, 427–435.

Dong, H.T., Nguyen, V.V., Phiwsaiya, K., Gangnonngiw, W., Withyachumnarnkul, B., *et al.* (2015b) Concurrent infections of *Flavobacterium columnare* and *Edwardsiella ictaluri* in striped catfish, *Pangasianodon hypophthalmus* in Thailand. *Aquaculture* 448, 142–150.

Dong, H.T., LaFrentz, B., Pirarat, N. and Rodkhum, C. (2015c) Phenotypic characterization and genetic diversity of *Flavobacterium columnare* isolated from red tilapia, *Oreochromis* sp., in Thailand. *Journal of Fish Diseases* 38(10), 901–913.

Dong, H.T., Gangnonngiw, W., Phiwsaiya, K., Charoensapsri, W., Nguyen, V.V., *et al.* (2016) Duplex PCR assay and *in situ* hybridization for detection of *Francisella* spp. and *Francisella noatunensis* subsp. *orientalis* in red tilapia. *Diseases of Aquatic Organisms* 120(1), 39–47.

Dong, H.T., Ataguba, G.A., Khunrae, P., Rattanarojponga, T. and Senapin, S. (2017a) Evidence of TiLV infection in tilapia hatcheries from 2012 to 2017 reveals probable global spread of the disease. *Aquaculture* 479, 579–583.

Dong, H.T., Siriroob, S., Meemetta, W., Santimanawong, W., Gangnonngiw, W., *et al.* (2017b) Emergence of tilapia lake virus in Thailand and an alternative semi-nested RT-PCR for detection. *Aquaculture* 476, 111–118.

Dong, H.T., Techatanakitarnan, C., Jindakittikul, P., Thaiprayoon, A., Taengphu, S., *et al.* (2017c) *Aeromonas jandaei* and *Aeromonas veronii* caused disease and mortality in Nile tilapia, *Oreochromis niloticus* (L.). *Journal of Fish Diseases* 40(10), 1395–1403.

Dong, H.T., Senapin, S., Jeamkunakorn, C., Nguyen, V.V., Nguyen, N.T., *et al.* (2019) Natural occurrence of edwardsiellosis caused by *Edwardsiella ictaluri* in farmed hybrid red tilapia (*Oreochromis* sp.) in Southeast Asia. *Aquaculture* 499, 17–23.

Dong, H.T., Senapin, S., Gangnonngiw, W., Nguyen, V.V., Rodkhum, C., *et al.* (2020) Experimental infection reveals transmission of tilapia lake virus (TiLV) from tilapia broodstock to their reproductive organs and fertilized eggs. *Aquaculture* 515, 734541.

Dong, H.T., Sangpo, P., Dien, L.T., Mai, T.T., Linh, N.V., *et al.* (2022) Usefulness of the pancreas as a prime target for histopathological diagnosis of Tilapia parvovirus (TiPV) infection in Nile tilapia, *Oreochromis niloticus*. *Journal of Fish Diseases* 45(9), 1323–1331.

Du, J., Wang, W., Chan, J.F.W., Wang, G., Huang, Y., *et al.* (2019) Identification of a novel ichthyic parvovirus in marine species in Hainan Island, *China. Frontiers in Microbiology* 10, 2815.

Duarte, S.A., Masser, M.P. and Plumb, J.A. (1993) Seasonal occurrence of diseases in cage-reared channel catfish, 1987–1991. *Journal of Aquatic Animal Health* 5, 223–229.

Duremdez, R.C. and Lio-Po, G.D. (1985) Studies on the causative organism of *Sarotherodon niloticus* (Linnaeus) fry mortalities – II. Identification and characterization of the physiological properties of *Pseudomonas fluorescens*. *Fish Pathology* 20, 115–123.

Dyková, I. and Lom, J. (1988) Review of pathogenic myxosporeans in intensive culture of carp (*Cyprinus carpio*) in Europe. *Folia Parasitologica* 35, 289–307.
Dyková, I., Fiala, I. and Nie, P. (2002) *Myxobolus lentisuturalis* sp. n. (Myxozoa: Myxobolidae), a new muscle-infecting species from the Prussian carp, *Carassius gibelio from China. Folia Parasitologica* 49, 253–258.
Eckart, V., Yamaguchi, T., Franzke, K., Bergmann, S.M., Boudinot, P., *et al.* (2019) New cell lines for efficient propagation of koi herpesvirus and infectious salmon anaemia virus. *Journal of Fish Diseases* 42(2), 181–187.
Eckstein, D., Künzel, V. and Schäfer, L. (2021) *Global Climate Risk Index 2021. Who Suffers Most from Extreme Weather Events? Weather-Related Loss Events in 2019 and 2000–2019*. Germanwatch eV, Bonn, Germany.
Egusa, S. (1992) *Infectious Diseases of Fish*. A.A. Balkema, Rotterdam, The Netherlands.
Eldar, A., Shapiro, O., Bejerano, Y. and Bercovier, H. (1995) Vaccination with whole-cell vaccine and bacterial protein extract protects tilapia against *Streptococcus difficile* meningoencephalitis. *Vaccine* 13, 867–870.
El-Feki, M., Hatai, K. and Hussein, M.A. (2003) Chemotactic and chemokinetic activities of *Saprolegnia parasitica* toward different metabolites and fish tissue extracts. *Mycoscience* 44, 159–162.
El-Matbouli, M., Ruchker, U. and Soliman, H. (2007) Detection of Cyprinid herpesvirus-3 (CyHV-3) DNA in infected fish tissues by nested polymerase chain reaction. *Diseases of Aquatic Organisms* 78, 23–28.
Euzet, L. and Combes, C. (1998) The selection of habitats among the Monogenea. *International Journal for Parasitology* 28, 1645–165.
Evans, J.J., Shoemaker, C.A. and Klesius, P.H. (2000) Experimental *Streptococcus iniae* infection of hybrid striped bass (*Morone chrysops* × *Morone saxatilis*) and tilapia (*Oreochromis niloticus*) by nares inoculation. *Aquaculture* 189, 197–210.
Ewing, M.S. and Kocan, K.M. (1987) *Ichthyophthirius multifiliis* (Ciliophora) exit from gill epithelium. *Journal of Protozoology* 34, 309–312.
Eyngor, M., Zamostiano, R., Tsofack, J.E.K., Berkowitz, A., Bercovier, H., *et al.* (2014) Identification of a novel RNA virus lethal to tilapia. *Journal of Clinical Microbiology* 52(12), 4137–4146.
Fabian, M., Baumer, A. and Steinhagen, D. (2013) Do wild fish species contribute to the transmission of koi herpesvirus to carp in hatchery ponds? *Journal of Fish Diseases* 36(5), 505–514.
Fan, C., Shao, L. and Fang, Q. (2010) Characterization of the nonstructural protein NS80 of grass carp reovirus. *Archives of Virology* 155(11), 1755–1763.
FAO (Food and Agriculture Organization of the United Nations) (2020) *The State of World Fisheries and Aquaculture 2020: Sustainability in Action*. FAO, Rome.
FAO (Food and Agriculture Organization of the United Nations) (2022) Fisheries and Aquaculture. National Aquaculture Sector Overview. Brazil. Available at: https://www.fao.org/fishery/en/countrysector/br/en (accessed 18 June 2022).
Feist, S.W. and Longshaw, M. (2006) Phylum Myxozoa. In: Woo, P.T.K. (ed.) *Fish Diseases and Disorders*. Vol. 1. *Protozoan and Metazoan Infections*, 2nd edn. CABI, Wallingford, UK. pp. 230-296.
Ferguson, H.W., Kabuusu, R., Beltran, S., Reyes, E., Lince, J.A. and del Pozo, J. (2014) Syncytial hepatitis of farmed tilapia, *Oreochromis niloticus* (L.): a case report. *Journal of Fish Diseases* 37(6), 583–589.
Ferguson, H.W., Turnbull, J.F., Shinn, A., Thompson, K., Dung, T.T. and Crumlish, M. (2001) Bacillary necrosis in farmed *Pangasius hypophthalmus* (Sauvage) from the Mekong Delta, Vietnam. *Journal of Fish Diseases* 24(9), 509–513.
FishStatJ (2022) *FishStatJ, A Tool for Fishery Statistics Analysis. Release: 4.02.04* (Berger, T., Sibeni, F. and Calderini, F., eds). Fishery and Aquaculture Division (NFI), Food and Agriculture Organization of the United Nations, Rome.
Frerichs, G.N., Millar, S.D. and Roberts, R.J. (1986) Ulcerative rhabdovirus in fish in South-East Asia. *Nature* 322, 216.
Frerichs, G.N., Hill, B.J. and Way, K. (1989) Ulcerative disease rhabdovirus; cell line susceptibility and serological comparison with other fish rhabdoviruses. *Journal of Fish Diseases* 12, 51–56.
Froese, R. and Pauly, D. (2022) FishBase, Version 02/2022. Available at: www.fishbase.org (accessed 29 June 2022).
Fujihara, M.P. and Nakatani, R.E. (1971) Antibody production and immune responses of rainbow trout and coho salmon to *Chondrococcus columnaris*. *Journal of the Fisheries Board of Canada* 28, 1253–1258.
Galli, P., Pugachev, O.N. and Kritsky, D.C. (2010) *Guide to Monogenoidea of Freshwater Fish of Palaearctic and Amur Regions*. Ledizioni Ledi Publishing, Milan, Italy.

Gao, Y., Pei, C., Sun, X., Zhang, C., Li, L. and Kong, X. (2018) Novel subunit vaccine based on grass carp reovirus VP35 protein provides protective immunity against grass carp hemorrhagic disease. *Fish & Shellfish Immunology* 75, 91–98.

García-Vásquez, A., Hansen, H. and Shinn, A.P. (2007) A revised description of *Gyrodactylus cichlidarum* Paperna, 1968 (Gyrodactylidae) from the Nile tilapia, *Oreochromis niloticus niloticus* (Cichlidae), and its synonymy with *G. niloticus* Cone, Arthur et Bondad-Reantaso, 1995. *Folia Parasitologica* 54, 129–140.

García-Vásquez, A., Hansen, H., Christison, K.W., Bron, J.E. and Shinn, A.P. (2011) Description of three new species of *Gyrodactylus* Nordmann, 1832 (Monogenea) from oreochromids (*Oreochromis*, Cichlidae). *Acta Parasitologica* 56(1), 20–33.

Gault, N.F.S., Kilpatrick, D.J. and Stewart, M.T. (2002) Biological control of the fish louse in a rainbow trout fishery. *Journal of Fish Biology* 60, 226–237.

Gbankoto, A., Pampoulie, C., Marques, A. and Sakiti, G.N. (2001) *Myxobolus dahomeyensis* infection in ovaries of tilapia species from Benin (West Africa). *Journal of Fish Biology* 58, 883–886.

Gibson, D.I., Timofeeva, T.A. and Gerasev, P.I. (1996) A catalogue of the nominal species of the monogenean genus *Dactylogyrus* Diesing, 1850 and their host genera. *Systematic Parasitology* 35, 3–48.

Gilad, O., Yun, S., Adkison, M.A., Way, K., Willits, N.H., *et al.* (2003) Molecular comparison of isolates of an emerging fish pathogen, koi herpesvirus, and the effect of water temperature on mortality of experimentally infected koi. *Journal of General Virology* 84(10), 2661–2668.

Gilad, O., Yun, S., Zagmutt-Vergara, F.J., Leutenegger, C.M., Bercovier, H. and Hedrick, R.P. (2004) Concentrations of a koi herpesvirus (KHV) in tissues of experimentally infected *Cyprinus carpio* koi as assessed by real-time TaqMan PCR. *Diseases of Aquatic Organisms* 60, 179–187.

Gomes, A.L.S., Costa, J.I., Benetton, M.L.F.N., Bernardino, G. and Belem-Costa, A. (2018) A fast and practical method for initial diagnosis of *Piscinoodinium pillulare* outbreaks: piscinootest. *Ciência Rural* 48(7), e20170680.

Goodwin, A.E. (2005) The biology and politics of SVCV and KHV. *World Aquaculture* 36, 9–10.

Grabowski, L.D., LaPatra, S.E. and Cain, K.D. (2004) Systemic and mucosal antibody response in tilapia, *Oreochromis niloticus* (L.) following immunization with *Flavobacterium columnare*. *Journal of Fish Diseases* 27, 573–581.

Grano-Maldonado, M.I., Gisbert, E., Hirt-Chabbert, J., Paladini, G., Roque, A., *et al.* (2011) Infection by *Gyrodactylus anguillae* Ergens, 1960 (Monogenea) associated with the mortality of glass eels (*Anguilla anguilla* L.) on the north-western Mediterranean Sea board of Spain. *Veterinary Parasitology* 180, 323–331.

Gray, M., LaPatra, S., Groff, J.M. and Goodwin, A. (2002) Detection of koi herpesvirus DNA in tissues of infected fish. *Journal of Fish Diseases* 25, 171–178.

Gray, W.L., Williams, R.J. and Griffin, B.R. (1999) Detection of channel catfish virus DNA in acutely infected channel catfish, *Ictalurus punctatus* (Rafinesque), using the polymerase chain reaction. *Journal of Fish Diseases* 22, 111–116.

Griffiths, D., Ali, M.S., Karim, M. and Khin, M. (2019) *Nursing fish in a hapas (Burmese version)*. Inland MYSAP – Myanmar Sustainable Aquaculture Programme implemented by Deutsche Gesellschaft für Internationale Zusammenarbeit (GIZ) GmbH. WorldFish brochure. Available at: https://digitalarchive.worldfishcenter.org/handle/20.500.12348/3854 (accessed 18 June 2022).

Gui, J.F., Tang, Q., Li, Z., Liu, Z. and De Silva, S.S. (2018) *Aquaculture in China: Success Stories and Modern Trends*. Wiley, Chichester, UK.

Guo, Q.L. and Jiang, Y. (1993) Histopathological studies of the hemorrhage disease of grass carp infected by two types of grass carp reovirus. *Transactions of Research on Fish Diseases* 1, 7–11.

Gurunathan, S., Klinman, D.M. and Seder, R.A. (2000) DNA vaccines: immunology, application, and optimization. *Annual Review of Immunology* 18, 927–974.

Gurung, T.B., Mulmi, R.M., Kalyan, K.C., Wagle, G., Pradhan, G.B., *et al.* (2010) Cage fish culture: an alternative livelihood option for communities displaced by reservoir impoundment in Kulekhani, Nepal. In: De Silva, S.S. and Davy, F.B. (eds) *Success Stories in Asian Aquaculture*. Springer, Berlin, pp. 85–102.

Gussev, A.V. (1976) Freshwater Indian Monogenoidea: principles of systematics, analysis of world faunas and their evolution. *Indian Journal of Helminthology* 25–26 (1973–1974), 1–241.

Gussev, A.V. (1985) Monogenea. In: Bauer, O.N. (ed.) *Key to the Parasites of the Freshwater Fish Fauna of the USSR*, Vol. 2. Izdat 'Nauka', Leningrad, Russia, pp. 269–344 (in Russian).

Gussev, A.V. and Fernando, C.H. (1973) Dactylogyridae (Monogenoidea) from the stomach of fishes. *Folia Parasitologica* 20, 207–212.

Haenen, O.L.M. and Davidse, A. (2001) First isolation and pathogenicity studies with *Pseudomonas anguilliseptica* from diseased European eel *Anguilla anguilla* (L.) in the Netherlands. *Aquaculture* 196(1–2), 27–36.

Haenen, O.L.M., Way, K., Bergmann, S.M. and Ariel, E. (2004) The emergence of koi herpesvirus and its significance to European aquaculture. *Bulletin of the European Association of Fish Pathologists* 24, 293–307.

Hakalahti, T., Lankinen, Y. and Valtonen, E.T. (2004) Efficacy of emamectin benzoate in the control of *Argulus coregoni* (Crustacea: Branchiura) on rainbow trout *Oncorhynchus mykiss*. *Diseases of Aquatic Organisms* 60, 197–204.

Hallett, S.L., Atkinson, S.D., Erséus, C. and El-Matbouli, M. (2005) Dissemination of triactinomyxons (Myxozoa) via oligochaetes used as live food for aquarium fishes. *Diseases of Aquatic Organisms* 65, 137–152.

Han, H.-J., Kim, D.-H., Lee, D.-C., Kim, S.-M. and Park, S.-I. (2006) Pathogenicity of *Edwardsiella tarda* to olive flounder, *Paralichthys olivaceus* (Temminck & Schlegel). *Journal of Fish Diseases* 29, 601–609.

Harikrishnan, R. and Balasundaram, C. (2005) Modern trends in *Aeromonas hydrophila* disease management with fish. *Reviews in Fisheries Science* 13, 281–320.

Harris, P.D., Shinn, A.P., Cable, J. and Bakke, T.A. (2004) Nominal species of the genus *Gyrodactylus* von Nordmann 1832 (Monogenea: Gyrodactylidae), with a list of principal host species. *Systematic Parasitology* 59, 1–27.

Hedrick, R.P. (1996) Movement of pathogens with the international trade of live fish: problems and solutions. *Revue Scientifique et Technique, Office International des Epizooties* 15, 523–531.

Hedrick, R.P. and McDowell, T. (1987) Passive transfer of sera with antivirus neutralizing activity from adult channel catfish protects juveniles from channel catfish virus disease. *Transactions of the American Fisheries Society* 116, 277–281.

Hedrick, R.P., Antonio, D.B. and Munn, R.J. (1997) Poxvirus like agent associated with epizootic mortality in juvenile koi (*Cyprinus carpio*). *FHS Newsletter* 25(3), 1–2.

Hedrick, R.P., Gilad, O., Yun, S., Spangenberg, J.V., Marty, G.D., *et al.* (2000) A herpesvirus associated with mass mortality of juvenile and adult koi, a strain of common carp. *Journal of Aquatic Animal Health* 12, 44–57.

Hedrick, R.P., Thomas, B. and McDowell, T.S. (2006) Susceptibility of koi carp, common carp, goldfish, and goldfish × common carp hybrids to Cyprinid herpesvirus-2 and herpesvirus-3. *Journal of Aquatic Animal Health* 18, 26–34.

Hoai, T.D., Trang, T.T., Van Tuyen, N., Giang, N.T.H. and Van, K.V. (2019) *Aeromonas veronii* caused disease and mortality in channel catfish in Vietnam. *Aquaculture* 513, 734425.

Hoare, R., Leigh, W., Limakom, T., Wongwaradechkul, R., Metselaar, M., *et al.* (2021) Oral vaccination of Nile tilapia (*Oreochromis niloticus*) against francisellosis elevates specific antibody titres in serum and mucus. *Fish & Shellfish Immunology* 113, 86–88.

Honjo, M.N., Minamoto, T. and Kawabata, Z. (2012) Reservoirs of Cyprinid herpesvirus 3 (CyHV-3) DNA in sediments of natural lakes and ponds. *Veterinary Microbiology* 155(2–4), 183–190.

Hoshina, T. (1952) Notes on some myxosporidian parasites on fishes of Japan. *Journal of Tokyo University of Fisheries* 3, 69–89.

Howard, J. (2019) *Fish Biology and Fisheries*. ED-Tech Press, Waltham Abbey, UK.

Hu, F., Li, Y., Wang, Q., Zhu, B., Wu, S., *et al.* (2021) Immersion immunization of koi (*Cyprinus carpio*) against cyprinid herpesvirus 3 (CyHV-3) with carbon nanotube-loaded DNA vaccine. *Aquaculture* 539, 736644.

Hu, Q.Y. and Lv, J.L. (2016) Hubei promotes the demolition of fences and cages of rivers, lakes and reservoirs to be completed by the end of December next year. *Hubei Daily*, 11 November. Available at: http://www.hubei.gov.cn/hbfb/bmdt/201611/t20161111_1512354.shtml (accessed 14 April 2022).

Huchzermeyer, C.F., Huchzermeyer, K.D.A., Christison, K.W., Macey, B.M., Colly, P.A., *et al.* (2018) First record of epizootic ulcerative syndrome from the Upper Congo catchment: an outbreak in the Bangweulu swamps, Zambia. *Journal of Fish Diseases* 41, 87–94.

Igeh, P.C. and Avenant-Oldewage, A. (2020) Pathological effects of *Cichlidogyrus philander* Douëllou, 1993 (Monogenea, Ancyrocephalidae) on the gills of *Pseudocrenilabrus philander* (Weber, 1897) (Cichlidae). *Journal of Fish Diseases* 43, 177–184.

Indonesian Ministry of Marine Affairs and Fisheries (2020) Aquaculture production statistics. Indonesian Ministry of Marine Affairs and Fisheries, Jakarta. Available at: https://statistik.kkp.go.id/ (accessed 27 June 2022).

IPCC (Intergovernmental Panel on Climate Change) (2014) *Climate Change 2014: Synthesis Report. Contribution of Working Groups I, II and III to the Fifth Assessment Report of the Intergovernmental Panel on Climate Change* (Core Writing Team, Pachauri, R.K. and Meyer, L.A., eds). IPCC, Geneva, Switzerland.

Iqbal, Z., Asghar, M. and Rubaba (2012) Saprolegniasis in two commercially important carps. *Pakistan Journal of Zoology* 44(2), 591–595.

Ismail, M.S., Siti-Zahrah, A., Syafiq, M.R.M., Amal, M.N.A., Firdaus-Nawi, M. and Zamri-Saad, M. (2016) Feed-based vaccination regime against streptococcosis in red tilapia, *Oreochromis niloticus* × *Oreochromis mossambicus*. *BMC Veterinary Research* 12(1), 194.

Ito, T., Sano, M., Kurita, J., Yuasa, K. and Iida, T. (2007) Carp larvae are not susceptible to koi herpesvirus. *Fish Pathology* 42(2), 107–109.

Jaemwimol, P., Sirikanchana, K., Tattiyapong, P., Mongkolsuk, S. and Surachetpong, W. (2019) Virucidal effects of common disinfectants against tilapia lake virus. *Journal of Fish Diseases* 42(10), 1383–1389.

Jahangiri, L., Shinn, A.P., Pratoomyot, J. and Bastos Gomes, G. (2021) Unveiling associations between ciliate parasites and bacterial microbiomes under warm-water fish farm conditions – a review. *Reviews in Aquaculture* 13, 1097–1118.

Jansen, M.D., Dong, H.T. and Mohan, C.V. (2019) Tilapia lake virus: a threat to the global tilapia industry? *Reviews in Aquaculture* 11(3), 725–739.

Jiang, Y. (1995) Advances in fish virology research in China. In: Shariff, M., Arthur, J.R. and Subasinghe, R.P. (eds) *Diseases in Asian Aquaculture II*. Fish Health Section, Asian Fisheries Society, Manila, pp. 211–225.

Jiang, Y.L. (2009) Hemorrhagic disease of grass carp: status of outbreaks, diagnosis, surveillance and research. *Israeli Journal of Aquaculture – Bamidgeh* 61, 199–197.

Jiang, Y.L., Li, Y. and Yu, P. (1991) A preliminary study on immune response of grass carp. *Acta Hydrobiologia Sinica* 15, 321–326.

Jørgensen, L.V.G., Nemli, E., Heinecke, R.D., Raida, M.K. and Buchmann, K. (2008) Immune-relevant genes expressed in rainbow trout following immunisation with a live vaccine against *Ichthyophthirius multifiliis*. *Diseases of Aquatic Organisms* 80, 189–197.

Jung-Schroers, V., Adamek, M., Teitge, F., Hellmann, J., Bergmann, S.M., *et al.* (2015) Another potential carp killer?: carp edema virus disease in Germany. *BMC Veterinary Research* 11(1), 114.

Kabata, Z. (1970) *Diseases of Fishes. Book 1: Crustacea as Enemies of Fishes*. TFH Publications, Neptune City, New Jersey.

Kabata, Z. (1985) *Parasites and Diseases of Fish Cultured in the Tropics*. Taylor & Francis, London.

Kanchanakhan, S., Saduakdee, U. and Areerat, S. (1999) Virus isolation from epizootic ulcerative syndrome-diseased fishes. *Asian Fisheries Science* 12, 327–335.

Kanchanakhan, S., Chinabut, S., Tonguthai, K. and Richards, R.H. (2002) Epizootic ulcerative syndrome of fishes: rhabdovirus infection and EUS induction experiments in snakehead fish. In: Lavilla-Pitogo, C.R. and Cruz-Lacierda, E.R. (eds) *Diseases in Asian Aquaculture II*. Fish Health Section, Asian Fisheries Society, Manila, pp. 383–394.

Kancharia, S.R. and Hanson, L.A. (1996) Production and shedding of channel catfish virus (CCV) and thymidine kinase negative CCV in immersion exposed channel catfish fingerlings. *Diseases of Aquatic Organisms* 27, 25–34.

Karunasagar, I., Sugumar, G. and Karunasagar, I. (1995) Virulence characters of *Aeromonas* spp. isolated from fish affected by epizootic ulcerative syndrome. In: Shariff, M., Arthur, J.R. and Subasinghe, R.P. (eds) *Diseases in Asian Aquaculture II*. Fish Health Section, Asian Fisheries Society, Manila, pp. 307–314.

Kasai, H., Muto, Y. and Yoshimizu, M. (2005) Virucidal effects of ultraviolet, heat treatment and disinfectants against koi herpesvirus (KHV). *Fish Pathology* 40, 137–138.

Kasornchandra, J., Engelking, H.M., Lannan, C.N., Rohovec, J.S. and Fryer, J.L. (1992) Characteristics of three rhabdoviruses from snakehead fish *Ophicephalus striatus*. *Diseases of Aquatic Organisms* 13, 89–94.

Katharios, P., Kalatzis, P.G., Kokkari, C., Pavlidis, M. and Wang, Q. (2019) Characterization of a highly virulent *Edwardsiella anguillarum* strain isolated from Greek aquaculture, and a spontaneously induced prophage therein. *Frontiers in Microbiology* 10, 141.

Kayansamruaj, P., Pirarat, N., Hirono, I. and Rodkhum, C. (2014) Increasing of temperature induces pathogenicity of *Streptococcus agalactiae* and the up-regulation of inflammatory related genes in infected Nile tilapia (*Oreochromis niloticus*). *Veterinary Microbiology* 172(1–2), 265–271.

Kim, S.W., Giri, S.S., Kim, S.G., Kwon, J., Oh, W.T. and Park, S.C. (2020) Carp edema virus and cyprinid herpesvirus-3 coinfection is associated with mass mortality of koi (*Cyprinus carpio haematopterus*) in the Republic of Korea. *Pathogens* 9(3), 222.

Kitao, T., Aoki, A. and Sakoh, R. (1981) Epizootic caused by β-haemolytic *Streptococcus* species in cultured freshwater fish. *Fish Pathology* 15, 301–307.

Kitiyodom, S., Kaewmalun, S., Nittayasut, N., Suktham, K. and Surassmo, S. (2019a) The potential of mucoadhesive polymer in enhancing efficacy of direct immersion vaccination against *Flavobacterium columnare* infection in tilapia. *Fish & Shellfish Immunology* 86, 635–640.

Kitiyodom, S., Yata, T., Yostawornkul, J., Kaewmalun, S., Nittayasut, N., *et al.* (2019b) Enhanced efficacy of immersion vaccination in tilapia against columnaris disease by chitosan-coated 'pathogen-like' mucoadhesive nanovaccines. *Fish & Shellfish Immunology* 95, 213–219.

Klesius, P. (1992) Carrier state of channel catfish infected with *Edwardsiella ictaluri*. *Journal of Aquatic Animal Health* 4(3), 227–230.

Klesius, P. (1994) Transmission of *Edwardsiella ictaluri* from infected, dead to noninfected channel catfish. *Journal of Aquatic Animal Health* 6(2), 180–182.

Klesius, P.H., Shoemaker, C.A. and Evans, J.J. (2000) Efficacy of single and combined *Streptococcus iniae* isolate vaccine administered by intraperitoneal and intramuscular routes in tilapia (*Oreochromis niloticus*). *Aquaculture* 188, 237–246.

Kokubo, T., Iida, T. and Wakabayashi, H. (1990) Production of siderophore by *Edwardsiella tarda*. *Fish Pathology* 25, 237–241.

Kritsky, D.C. and Heckmann, R. (2002) Species of *Dactylogyrus* (Monogenoidea: Dactylogyridae) and *Trichodina mutabilis* (Ciliata) infesting koi carp, *Cyprinus carpio*, during mass mortality at a commercial rearing facility in Utah, USA. *Comparative Parasitology* 69(2), 217–218.

Kumari, P. and Nomani, M.M.R. (2021) Study of parasitic effect of *Argulus* sp. on Indian major carps in pond culture. *International Journal of Fisheries and Aquatic Studies* 9(3), 373–377.

Kurita, J., Yuasa, K., Ito, T., Sano, M., Hedrick, R.P., *et al.* (2009) Molecular epidemiology of koi herpesvirus (KHV). *Fish Pathology* 44, 59–66.

Kusuda, R. and Salati, F. (1999) *Enterococcus seriolicida* and *Streptococcus iniae*. In: Woo, P.T.K. and Bruno, D.W. (eds) *Fish Diseases and Disorders*. Vol. 3. *Viral, Bacterial, and Fungal Infections*, 1st edn. CAB International, Wallingford, UK, pp. 303–317.

LaFrentz, B.R., García, J.C., Waldbieser, G.C., Evenhuis, J.P., Loch, T.P., *et al.* (2018) Identification of four distinct phylogenetic groups in *Flavobacterium columnare* with fish host associations. *Frontiers in Microbiology* 9, 452.

LaFrentz, B.R., Králová, S., Burbick, C.R., Alexander, T.L., Phillips, C.W., *et al.* (2022) The fish pathogen *Flavobacterium columnare* represents four distinct species: *Flavobacterium columnare*, *Flavobacterium covae* sp. nov., *davisii* sp. nov. and *Flavobacterium oreochromis* sp. nov., and emended description of *Flavobacterium columnare*. *Systematic and Applied Microbiology* 45(2), 126293.

LARReC (Living Aquatic Resources Research Centre) (2001) *A Report on Cage Culture in Lao PDR. Prepared by the Living Aquatic Resources Research Centre (LARReC) and the Network of Aquaculture Centres in Asia-Pacific (NACA)*. LARReC Research Report No. 0002. LARReC, Vientiane, Lao PDR.

Leaño, E.M., Lio-Po, G.D. and Dureza, L.A. (1996) Virulence and production of extracellular proteins (ECP) of *Aeromonas hydrophila* associated with the epizootic ulcerative syndrome (EUS) of freshwater fish. *UPV Journal of Natural Sciences* 1, 30–38.

Lee, N.S., Jung, S.H., Park, J.W. and Do, J.W. (2012) *In situ* hybridization detection of koi herpesvirus in paraffin-embedded tissues of common carp *Cyprinus carpio* collected in 1998 in Korea. *Fish Pathology* 47, 100–103.

Leigh, W.J., Zadoks, R.N., Costa, J.Z., Jaglarz, A. and Thompson, K.D. (2020) Development and evaluation of a quantitative polymerase chain reaction for aquatic *Streptococcus agalactiae* based on the *groEL* gene. *Journal of Applied Microbiology* 129(1), 63–74.

Lerssutthichawal, T. (1999) *Monogeneans of the freshwater siluriform fishes of Thailand*. PhD thesis, University of Malaya, Kuala Lumpur.

Lester, R.J.G. and Roubal, F.R. (1995) Phylum Arthropoda. In: Woo, P.T.K. (ed.) *Fish Diseases and Disorders*. Vol. 1. *Protozoan and Metazoan Infections*, 1st edn. CAB International, Wallingford, UK, pp. 475–598.

Leung, K.Y., Wang, Q., Yang, Z. and Siame, B.A. (2019) *Edwardsiella piscicida*: a versatile emerging pathogen of fish. *Virulence* 10(1), 555.

Lewisch, E., Gorgoglione, B., Way, K. and El-Matbouli, M. (2015) Carp edema virus/koi sleepy disease: an emerging disease in Central-East Europe. *Transboundary and Emerging Diseases* 62(1), 6–12.

Li, C., Wang, S., Ren, Q., He, T. and Chen, X. (2020) An outbreak of visceral white nodules disease caused by *Pseudomonas plecoglossicida* at a water temperature of 12°C in cultured large yellow croaker (*Larimichthys crocea*) in China. *Journal of Fish Diseases* 43(11), 1353–1361.

Li, J., Wang, T.H., Yi, Y.L., Liu, H.Q., Lu, R.H. and Chen, H.X. (1997) A detection method for grass carp hemorrhagic virus (GCHV) based on a reverse transcription-polymerase chain reaction. *Diseases of Aquatic Organisms* 29, 7–12.

Li, X., Zeng, W.W., Wang, Q., Wang, Y.Y., Li, Y.Y., *et al.* (2016) The study on the proliferation of GCRV II in different fish cell lines. *Journal of Fisheries of China* 40, 1249–1257.

Liamnimitr, P., Thammatorn, W., U-thoomporn, S., Tattiyapong, P. and Surachetpong, W. (2018) Non-lethal sampling for tilapia lake virus detection by RT-qPCR and cell culture. *Aquaculture* 486, 75–80.

LiHua, D., JianJun, F., Peng, L., SongLin, G., Le, H. and YiQun, X. (2019) Evaluation of an outer membrane protein as a vaccine against *Edwardsiella anguillarum* in Japanese eels (*Anguilla japonica*). *Aquaculture* 498, 143–150.

Lilley, J.H. and Frerichs, G.N. (1994) Comparison of rhabdoviruses associated with epizootic ulcerative syndrome (EUS) with respect to their structural proteins, cytopathology and serology. *Journal of Fish Diseases* 17, 513–522.

Lilley, J.H. and Roberts, R.J. (1997) Pathogenicity and culture studies comparing the *Aphanomyces* involved in epizootic ulcerative syndrome (EUS) with other similar fungi. *Journal of Fish Diseases* 20, 135–144.

Lilley, J.H., Callinan, R.B., Chinabut, S., Kanchanakhan, S., MacRae, I.H. and Phillips, M.J. (1998) *Epizootic Ulcerative Syndrome (EUS) Technical Handbook*. The Aquatic Animal Health Research Institute, Bangkok.

Lim, C. and Klesius, P.H. (2003) Influence of feed deprivation on hematology, macrophage chemotaxis, and resistance to *Edwardsiella ictaluri* challenge of channel catfish. *Journal of Aquatic Animal Health* 15, 13–20.

Lim, L.H.S. (1990) *Silurodiscoides* Gussev, 1961 (Monogenea) from *Pangasius sutchi* Fowler, 1931 (Pangasiidae) cultured in Peninsular Malaysia. *The Raffles Bulletin of Zoology* 38, 55–63.

Lim, L.H.S. (1996) *Thaparocleidus* Jain 1952, the senior synonym of *Silurodiscoides* Gussev, 1976 (Monogenea: Ancylodiscoidinae Gussev, 1961). *Systematic Parasitology* 35, 207–218.

Lim, L.H.S., Timofeeva, T.A. and Gibson, D.I. (2001) Dactylogyridean monogeneans of the siluriform fishes of the Old World. *Systematic Parasitology* 50, 159–197.

Linh, N.V., Dien, L.T., Sangpo, P., Senapin, S., Thapinta, A., *et al.* (2022) Pre-treatment of Nile tilapia (*Oreochromis niloticus*) with ozone nanobubbles improve efficacy of heat-killed *Streptococcus agalactiae* immersion vaccine. *Fish & Shellfish Immunology* 123, 229–237.

Lio-Po, G. (1998) Studies on several viruses, bacteria and a fungus associated with epizootic ulcerative syndrome (EUS) of several fishes in the Philippines. PhD thesis, Simon Fraser University (SFU), Burnaby, Canada.

Lio-Po, G.D. and Duremdez-Fernandez, R. (1986) The pathogenicity of bacteria associated with transport-stressed *Chanos chanos* fingerlings. In: Maclean, J.L., Dizon, L.B. and Hosillos, L.V. (eds) *The First Asian Fisheries Forum*. Asian Fisheries Society, Manila, pp. 223–226.

Lio-Po, G.D. and Lim, L.H.S. (2014) Infectious diseases of warmwater fish in fresh water. In: Woo, P.T.K. and Bruno, D.W. (eds) *Diseases and Disorders of Finfish in Cage Culture*, 2nd edn. CABI Publishing, Wallingford, UK, pp. 193–253.

Lio-Po, G.D. and Wakabayashi, H. (1986) Immuno-response in tilapia *Sarotherodon niloticus* vaccinated with *Edwardsiella tarda* by the hyperosmotic infiltration method. *Veterinary Immunology and Immunopathology* 12, 351–357.

Lio-Po, G.D., Albright, L.J. and Alapide-Tendencia, E.V. (1992) *Aeromonas hydrophila* in the epizootic ulcerative syndrome (EUS) of snakehead, (*Ophicephalus striatus*) and catfish (*Clarias batrachus*): quantitative estimation in natural infection and experimental induction of dermo-muscular necrotic lesion. In: Shariff, M., Subasinghe, R.P. and Arthur, J.R. (eds) *Diseases in Asian Aquaculture I*. Fish Health Section, Asian Fisheries Society, Manila, pp. 461–474.

Lio-Po, G.D., Albright, L.J. and Leaño, E.M. (1996) Experiments on virulence dose and portals of entry of *Aeromonas hydrophila* in the walking catfish (*Clarias batrachus*). *Journal of Aquatic Animal Health* 8, 340–343.

Lio-Po, G.D., Albright, L.J., Michel, C. and Leaño, E.M. (1998) Experimental induction of lesions in snakeheads (*Ophicephalus striatus*) and catfish (*Clarias batrachus*) with *Aeromonas hydrophila*, *Aquaspirillum* sp., *Pseudomonas* sp. and *Streptococcus* sp. *Journal of Applied Ichthyology* 14, 75–79.

Lio-Po, G.D., Traxler, G.S. and Albright, L.J. (1999) Establishment of cell lines from catfish (*Clarias batrachus*) and snakeheads (*Ophicephalus striatus*). *Asian Fisheries Science* 12, 345–349.

Lio-Po, G.D., Traxler, G.S., Albright, L.J. and Leaño, E.M. (2000) Characterization of a virus obtained from the epizootic ulcerative syndrome (EUS) in snakeheads (*Ophicephalus striatus*) in the Philippines. *Diseases of Aquatic Organisms* 43, 191–198.

Lio-Po, G.D., Albright, L.J., Traxler, G.S. and Leaño, E.M. (2001) Pathogenicity of the epizootic ulcerative syndrome (EUS)-associated rhabdovirus to snakehead *Ophicephalus striatus*. *Fish Pathology* 36, 57–66.

Lio-Po, G.D., Albright, A., Traxler, G., Leaño, E.M. (2003) Horizontal transmission of the EUS rhabdovirus in simulated natural conditions. *Diseases of Aquatic Organisms* 57, 213–220.

Lio-Po, G.D., Amar, E., de la Peña, L., Orozco, Z.G.A., Faisan, J.P., *et al.* (2009) Surveillance of emerging fish viral pathogens in some Southeast Asian countries. *Israeli Journal of Aquaculture – Bamidgeh* 61, 208–214.

Liu, C., Chang, O.Q., Zhang, D.F., Li, K.B., Wang, F., *et al.* (2018) *Aeromonas shuberti* as a cause of multi-organ necrosis in internal organs of Nile tilapia, *Oreochromis niloticus*. *Journal of Fish Diseases* 41(10), 1529–1538.

Liu, L., Lu, D.-Q., Xu, J., Luo, H.-L. and Li, A.-X. (2019) Development of attenuated erythromycin-resistant *Streptococcus agalactiae* vaccine for tilapia (*Oreochromis niloticus*) culture. *Journal of Fish Diseases* 42(5), 693–701.

Liu, W., Zhang, Y., Ma, J., Jiang, N., Fan, Y., *et al.* (2020) Determination of a novel parvovirus pathogen associated with massive mortality in adult tilapia. *PLoS Pathogens* 16(9), e1008765.

Llobrera, A.T. and Gacutan, R.Q. (1987) *Aeromonas hydrophila* associated with ulcerative disease epizootic in Laguna de Bay, Philippines. *Aquaculture* 67, 273–278.

Lom, J. (1995) Trichodinidae and other ciliates (Phylum Ciliophora). In: Woo, P.T.K. (ed.) *Fish Diseases and Disorders*. Vol. 1. *Protozoan and Metazoan Infections*, 1st edn. CAB International, Wallingford, UK, pp. 229–262.

Lom, J. and Dyková, I. (1992) *Protozoan Parasites of Fishes. Developments in Aquaculture and Fisheries Science*, Vol. 26. Elsevier Science Publishers BV, Amsterdam.

Lom, J. and Dyková, I. (1995) Myxosporea (Phylum Myxozoa). In: Woo, P.T.K. (ed.) *Fish Diseases and Disorders*. Vol. 1. *Protozoan and Metazoan Infections*, 1st edn. CAB International, Wallingford, UK, pp. 97–148.

Lom, J. and Dyková, I. (2006) Myxozoan genera: definition and notes on taxonomy, life-cycle terminology and pathogenic species. *Folia Parasitologica* 53, 1–36.

Lopez, C., Rajan, P.R., Lin, J.H.Y., Kuo, T.Y. and Yang, H.L. (2002) Disease outbreak in seafarmed cobia (*Rachycentron canadum*) associated with *Vibrio* spp., *Photobacterium damselae* ssp. *piscicida*, monogenean and myxosporean parasites. *Bulletin of the European Association of Fish Pathologists* 22, 206–211.

Ly, L.T.T., Nguyen, D.N., Vo, P.H. and Doan, C.V. (2009) Hemorrhage disease of cultured tra catfish (*Pangasianodon hypophthalmus*) in Mekong Delta (Vietnam). *Israeli Journal of Aquaculture – Bamidgeh* 61, 215–224.

Mabrok, M., Chokmangmeepisarn, P., LaFrentz, B.R., Kayansamruaj, P., Dong, H.T. and Rodkhum, C. (2020) Development of a species-specific polymerase chain reaction for highly sensitive detection of *Flavobacterium columnare* targeting chondroitin AC lyase gene. *Aquaculture* 521, 734597.

Machimbirike, V.I., Crumlish, M., Dong, H.T., Santander, J., Khunrae, P. and Rattanarojpong, T. (2022) *Edwardsiella ictaluri*: a systemic review and future perspectives on disease management. *Reviews in Aquaculture* 14(3), 1613–1636.

Maciel, P.O., Garcia, F., Chagas, E.C., Fujimoto, R.Y. and Tavares-Dias, M. (2018) Trichodinidae in commercial fish in South America. *Reviews in Fish Biology and Fisheries* 28, 33–56.

Maduenyane, M., Dos Santos, Q.M. and Avenant-Oldewage, A. (2021) Light and scanning electron microscopy of the effects of *Macrogyrodactylus congolensis* (Prudhoe, 1957) on the skin of the African sharptooth catfish *Clarias gariepinus* (Burchell, 1822). *Journal of Fish Diseases* 45, 595–602.

Magray, A.R., Hafeez, S., Ganai, B.A., Lone, S.A., Dar, G.J., *et al.* (2021) Study on pathogenicity and characterization of disease causing fungal community associated with cultured fish of Kashmir valley, *India. Microbial Pathogenesis* 151, 104715.

Mai, T.T., Kayansamruaj, P., Taengphu, S., Senapin, S., Costa, J.Z., *et al.* (2021) Efficacy of heat-killed and formalin-killed vaccines against *Tilapia tilapinevirus* in juvenile Nile tilapia (*Oreochromis niloticus*). *Journal of Fish Diseases* 44(12), 2097–2109.

Mai, T.T., Kayansamruaj, P., Soontara, C., Kerddee, P., Nguyen, D.H., *et al.* (2022) Immunization of Nile tilapia (*Oreochromis niloticus*) broodstock with tilapia lake virus (TiLV) inactivated vaccines elicits protective antibody and passive maternal antibody transfer. *Vaccines* 10(2), 167.

Malherbe, W., Christison, K.W., Wepener, V. and Smit, N.J. (2010) Epizootic ulcerative syndrome – first report of evidence from South Africa's largest and premier conservation area, the Kruger National Park. *International Journal for Parasitology: Parasites and Wildlife* 10, 207–210.

Malmberg, G. (1970) The excretory systems and the marginal hooks as a basis for the systematics of *Gyrodactylus* (Trematoda, Monogenea). *Arkiv för Zoologi* 2, 1–235.

Markovskaja, S. (2004) Saprolegniaceae (Peronosporomycetes) in Lithuania. 1. The genera *Achyla* and *Newbya*. *Botanica Lithuanica* 10(2), 141–160.

Marsella, A., Pretto, T., Abbadi, M., Quartesan, R., Cortinovis, L., *et al.* (2021) Carp edema virus-related mortality in wild adult common carp (*Cyprinus carpio*) in Italy. *Journal of Fish Diseases* 44(7), 939–947.

Martins, M.L., Moraes, J.R.E., Andrade, P.M., Schalch, S.H. and Moraes, F.R. (2001) *Piscinoodinium pillulare* (Schäperclaus, 1954) Lom, 1981 (Dinoflagellida) infection in cultivated freshwater fish from the northeast region of São Paulo State, Brazil: parasitological and pathological aspects. *Brazilian Journal of Biology* 61, 639–644.

Mastan, S.A. (2015) Fungal infection in freshwater fishes of Andhra Pradesh, India. *African Journal of Biotechnology* 14, 530–534.

Matras, M., Borzym, E., Stone, D., Way, K., Stachnik, M., *et al.* (2017) Carp edema virus in Polish aquaculture – evidence of significant sequence divergence and a new lineage in common carp *Cyprinus carpio* (L.). *Journal of Fish Diseases* 40(3), 319–325.

Matras, M., Stachnik, M., Borzym, E., Maj-Paluch, J. and Reichert, M. (2019) Potential vector species of carp edema virus (CEV). *Journal of Fish Diseases* 42(7), 959–964.

Matsche, M.A. and Grizzle, J.M. (1999) Early changes in pigmented macrophages in head kidney of channel catfish infected with *Aeromonas hydrophila*. *Journal of Aquatic Animal Health* 11, 253–261.

Matthews, R.A. (1994) *Ichthyophthirius multifiliis* Fouquet, 1876: infection and protective response within the fish host. In: Pike, A.W. and Lewis, J.W. (eds) *Parasitic Diseases of Fish*. Samara Publishing Limited, Dyfed, UK, pp. 17–42.

Matthews, R.A. (2005) *Ichthyophthirius multifiliis* Fouquet and ichthyophthiriosis in freshwater teleosts. *Advances in Parasitology* 59, 159–241.

McEntire, M.E., Iwanowicz, L.R. and Goodwin, A.E. (2003) Molecular, physical, and clinical evidence that golden shiner virus and grass carp reovirus are variants of the same virus. *Journal of Aquatic Animal Health* 15, 257–263.

McGlamery, M.H. and Gratzek, J.B. (1974) Stunting syndrome associated with young channel catfish that survived exposure to channel catfish virus. *The Progressive Fish-Culturist* 36, 38–41.

Mekata, T., Kawato, Y. and Ito, T. (2021) Complete genome sequence of carp edema virus isolated from koi carp. *Microbiology Resource Announcements* 10(16), e00239-21.

Mendoza-Franco, E.F., Caspeta-Mandujano, J.M. and Osorio, M.T. (2018) Ecto- and endo-parasitic monogeneans (Platyhelminthes) on cultured freshwater exotic fish species in the state of Morelos, south-central Mexico. *ZooKeys* 776, 1–12.

Miles, D.J.C., Polchana, J., Lilley, J.H., Kanchanakhan, S., Thompson, K.D. and Adams, A. (2001) Immunostimulation of striped snakehead *Channa striata* against epizootic ulcerative syndrome. *Aquaculture* 195, 1–15.

Minamoto, T., Honjo, M.N., Yamanaka, H., Tanaka, N., Itayama, T. and Kawabata, Z. (2011) Detection of cyprinid herpesvirus-3 DNA in lake plankton. *Research in Veterinary Science* 90(3), 530–532.

Ministry of Agriculture, Rural Affairs Fishery Administration (2021) *China Fishery Statistical Yearbook*. China Agriculture Press, Beijing.

Mitchell, A.J. (1997) Fish disease summaries for the Southeastern United States from 1976–1995. *Aquaculture Magazine* 23, 87–93.

Miyashita, T., Kubota, S.S. and Miyashita, T. (1984) A histological study of *Pseudomonas fluorescens* infection in tilapia. *Fish Pathology* 19, 161–166.

Miyazaki, T. and Kaige, N. (1985) Comparative histopathology of Edwardsiellosis in fishes. *Fish Pathology* 20, 219–227.

Miyazaki, T., Isshiki, T. and Katsuyuki, H. (2005) Histopathological and electron microscopy studies on sleepy disease of koi *Cyprinus carpio* koi in Japan. *Diseases of Aquatic Organisms* 65(3), 197–207.
Mizelle, J.D. and Price, C.E. (1964) Studies on monogenetic trematodes. XXVII. Dactylogyrid species with the proposal of *Urocleidoides* gen. n. *Journal of Parasitology* 50, 579–584.
Molnár, K. (1972) Studies on gill parasitosis of the grass carps (*Ctenopharyngodon idella*) caused by *Dactylogyrus lamellatus* Achmerov, 1952. IV. Histopathological changes. *Acta Veterinaria Academiae Scientiarum Hungaricae* 22, 9–24.
Molnár, K. (1979) *Myxobolus pavlovskii* (Akhmerov, 1954) (Myxosporidia) infection in the silver carp and bighead. *Acta Veterinaria Academiae Scientiarum Hungaricae* 27, 207–216.
Molnár, K. (1982) Biology and histopathology of *Thelohanellus nikolskii* Akhmerov, 1955 (Myxosporea, Myxozoa), a protozoan parasite of the common carp (*Cyprinus carpio*). *Zeitschrift für Parasitenkunde* 68, 269–277.
Molnár, K. (1994) Effect of decreased water oxygen content on common carp fry with *Dactylogyrus vastator* (Monogenea) infection of varying severity. *Diseases of Aquatic Organisms* 20, 153–157.
Molnár, K., Székely, C., Mohamed, K. and Shaharom-Harrison, F. (2006) Myxozoan pathogens in cultured Malaysian fishes. I. Myxozoan infections of the sutchi catfish *Pangasius hypophthalmus* in freshwater cage cultures. *Diseases of Aquatic Organisms* 68, 209–218.
Molnár, K., Székely, C. and Láng, M. (2019) *Field Guide to the Control of Warmwater Fish Diseases in Central and Eastern Europe, the Caucasus and Central Asia*. FAO Fisheries and Aquaculture Circular No. 1182. Food and Agriculture Organization of the United Nations, Ankara.
Monaghan, S.J., Thompson, K.D., Adams, A. and Bergmann, S.M. (2015a) Sensitivity of seven PCRs for early detection of koi herpesvirus in experimentally infected carp, *Cyprinus carpio* L., by lethal and non-lethal sampling methods. *Journal of Fish Diseases* 38(3), 303–319.
Monaghan, S.J., Thompson, K.D., Adams, A., Kempter, J. and Bergmann, S.M. (2015b) Examination of the early infection stages of koi herpesvirus (KHV) in experimentally infected carp, *Cyprinus carpio* L. using *in situ* hybridization. *Journal of Fish Diseases* 38(5), 477–489.
Moravec, F. (1998) *Nematodes of Freshwater Fishes of the Neotropical Region*. Academia, Prague.
Morice, C.P., Kennedy, J.J., Rayner, N.A., Winn, J.P., Hogan, E., *et al.* (2021) An updated assessment of near-surface temperature change from 1850: the HadCRUT5 data set. *Journal of Geophysical Research: Atmospheres* 126(3), e2019JD032361.
Mqolomba, T.N. and Plumb, J.A. (1992) Longevity of *Edwardsiella ictaluri* in the organs of experimentally infected channel catfish, *Ictalurus punctatus*. *Aquaculture* 101(1–2), 1–6.
Mursalim, M.F., Budiyansah, H., Raharjo, H.M., Debnath, P.P., Sakulworakan, R., *et al.* (2022) Diversity and antimicrobial susceptibility profiles of *Aeromonas* spp. isolated from diseased freshwater fishes in Thailand. *Journal of Fish Diseases* 45(8), 1149–1163.
Ngo, T.P.H., Vu, H.T.T., Le, T.T.T., Bui, H.C.N., Liles, M.R. and Rodkhum, C. (2022) Comparative genomic analysis of hypervirulent *Aeromonas hydrophila* strains from striped catfish (*Pangasianodon hypophthalmus*) in Vietnam. *Aquaculture* 558, 738364.
Nguyen, V.V., Dong, H.T., Senapin, S., Pirarat, N. and Rodkhum, C. (2016) *Francisella noatunensis* subsp. *orientalis*, an emerging bacterial pathogen affecting cultured red tilapia (*Oreochromis* sp.) in Thailand. *Aquaculture Research,* 47(11), 3697–3702.
Nguyen, V.V., Dong, H.T., Senapin, S., Gangnonngiw, W., Pirarat, N., *et al.* (2019) Transmission of *Francisella noatuensis* subsp. *orientalis* from subclinically infected hybrid red tilapia broodstock (*Oreochromis* sp.) to their offspring. *Microbial Pathogenesis* 136, 103670.
Nguyen, V.V., Dong, H.T., Senapin, S., Kayansamruaj, P., Pirarat, N., *et al.* (2020) Synergistic infection of *Ichthyophthirius multifiliis* and *Francisella noatunensis* subsp. *orientalis* in hybrid red tilapia (*Oreochromis* sp.). *Microbial Pathogenesis* 147, 104369.
Nhinh, D.T., Giang, N.T.H., Van, K.V., Dang, L.T., Dong, H.T. and Hoai, T.D. (2022) Widespread presence of a highly virulent *Edwardsiella ictaluri* strain in farmed tilapia, *Oreochromis* spp. *Transboundary and Emerging Diseases* 69(5), e2276–e2290.
Nie, D.A. and Pan, J.P. (1985) Diseases of grass carp (*Ctenopharyngodon idellus* Valenciennes, 1844) in China, a review from 1953 to 1983. *Fish Pathology* 20, 323–330.
Noga, E.J. (2010) *Fish Disease: Diagnosis and Treatment*, 2nd edn. Wiley-Blackwell, Ames, Iowa.
Noga, E.J. and Levy, M.G. (2006) Phylum Dinoflagellata. In: Woo, P.T.K. (ed.) *Fish Diseases and Disorders*. Vol. 1. *Protozoan and Metazoan Infections*, 2nd edn. CAB International, Wallingford, UK, pp. 16–45.
Nogueira, S.M.S., Santos, M.A., Lordelo, S.A.V. and Farias Filho, J.R. (2020) Brazilian continental aquaculture: a model for the development of its regularization. *Acta Scientiarum Technology* 42, e47064.

Nusbaum, K.E., Smith, B.F., de Innocentes, P. and Bird, R.C. (2002) Protective immunity induced by DNA vaccination of channel catfish with early and late transcripts of the channel catfish herpesvirus (IHV-1). *Veterinary Immunology and Immunopathology* 84, 151–168.

Ogawa, K., Bondad-Reantaso, M.G., Fukudome, M. and Wakabayashi, H. (1995) *Neobenedenia girellae* (Hargis, 1955) Yamaguti, 1963 (Monogenea: Capsalidae) from cultured marine fishes of Japan. *Journal of Parasitology* 81, 223–227.

Oh, W.T., Jun, J.W., Kim, H.J., Giri, S.S., Yun, S., *et al.* (2020) Characterization and pathological analysis of a virulent *Edwardsiella anguillarum* strain isolated from Nile tilapia (*Oreochromis niloticus*) in Korea. *Frontiers in Veterinary Science* 7, 14.

OIE (World Organisation for Animal Health) (2019a) Infection with koi herpesvirus. In: *Manual of Diagnostic Tests for Aquatic Animals*, updated 2019 version. OIE, Paris.

OIE (World Organisation for Animal Health) (2019b) *Manual of Diagnostic Tests for Aquatic Animals* OIE, *Paris*.

OIE (World Organisation for Animal Health) (2021) *OIE Aquatic Animal Health Strategy 2021–2025*. OIE, Paris.

Okaeme, A.N., Obiekezie, A.I., Lehman, J., Antai, E.E. and Madu, C.T. (1988) Parasites and diseases of cultured fish of Lake Kainji area, Nigeria. *Journal of Fish Biology* 32, 479–481.

Olivier, G., Lallier, R. and Lariviere, S. (1981) A toxigenic profile of *Aeromonas hydrophila* and *A. sobria* isolated from fish. *Canadian Journal of Microbiology* 27, 330–333.

Our World in Data (2022a) Ocean acidification: mean seawater pH. Available at: https://ourworldindata.org/explorers/climate-change?facet=none&pickerSort=desc&pickerMetric=sea_level_rise_cw2011&Metric=Ocean+pH&Long-run+series%3F=false&country=OWID_WRL~ATA~Gulkana+Glacier~-Lemon+Creek+Glacier~North+America~South+Cascade+Glacier~Wolverine+Glacier (accessed 9 March 2022).

Our World in Data (2022b) Global atmospheric CO_2 concentration. Available at: https://ourworldindata.org/explorers/climate-change?facet=none&pickerSort=desc&pickerMetric=sea_level_rise_cw2011&Metric=CO%E2%82%82+concentrations&Long-run+series%3F=false&country=OWID_WRL~ATA~Gulkana+Glacier~Lemon+Creek+Glacier~North+America~South+Cascade+Glacier~Wolverine+Glacier (accessed 9 March 2022).

Our World in Data (2022c) Global atmospheric methane concentration. Available atL https://ourworldindata.org/explorers/climate-change?facet=none&pickerSort=desc&pickerMetric=sea_level_rise_cw2011&-Metric=CH%E2%82%84+concentrations&Long-run+series%3F=false&country=OWID_WRL~ATA~Gulkana+Glacier~Lemon+Creek+Glacier~North+America~South+Cascade+Glacier~Wolverine+Glacier (accessed 9 March 2022).

Oyamatsu, T., Hata, N., Yamada, K., Sano, T. and Fukuda, H. *et al.* (1997a) An etiological study on mass mortality of cultured colorcarp juveniles showing edema. *Fish Pathology* 32(2), 81–88.

Oyamatsu, T., Matoyama, H., Yamamoto, K. and Fukuda, H. (1997b) A trial for the detection of carp edema virus by using polymerase chain reaction. *Aquaculture Science* 45(2), 247–251.

Paladini, G., Longshaw, M., Gustinelli, A. and Shinn, A.P. (2017) Parasitic diseases in aquaculture: their biology, diagnosis and control. In: Austin, B.A. and Newaj-Fyzul, A. (eds) *Diagnosis and Control of Diseases of Fish and Shellfish*. Wiley, Chichester, UK, pp. 37–107.

Paladini, G., Shinn, A.P., Taylor, N.G.H., Bron, J.E. and Hansen, H. (2021) Geographical distribution of *Gyrodactylus salaris* Malmberg, 1957 (Monogenea, Gyrodactylidae). *Parasites & Vectors* 14, 34.

Pal, J. and Pradhan, K. (1990) Bacterial involvement in ulcerative condition of air-breathing fish from India. *Journal of Fish Biology* 36, 833–839.

Panchai, K., Hanjavanit, C. and Kitacharoen, N. (2007) Characteristics of *Achlya bisexualis* isolated from eggs of Nile tilapia (*Oreochromis niloticus*). *Khon Kaen University Research Journal* 12, 195–202.

Pandey, K.C. and Agrawal, N. (2008) *An Encyclopaedia of Indian Monogenoidea*. Vitasta Publishing Pvt. Ltd, New Delhi.

Paperna, I. (1973) New species of Monogenea (Vermes) from African freshwater fish. A preliminary report. *Revue de Zoologie et de Botanique Africaines* 87(3), 505–518.

Paperna, I. (1991) Diseases caused by parasites in the aquaculture of warm water fish. *Annual Review of Fish Diseases* 1, 155–194.

Paperna, I. (1996) *Parasite Infections and Diseases of Fish in Africa. An Update*. CIFA Technical Paper No. 31. Food and Agriculture Organization of the United Nations, Rome.

Pariselle, A. and Euzet, L. (1995) *Scutogyrus* gen. n. (Monogenea: Ancyrocephalidae) for *Cichlidogyrus longicornis minus* Dossou, 1982, *C. l. longicornis*, and *C. l. gravivaginus* Paperna and Thurston, 1969, with description of three new species parasitic on African cichlids. *Proceedings of the Helminthological Society of Washington* 62, 157–173.

Pariselle, A., Lim, L.H.S. and Lambert, A. (2002) Monogeneans from Pangasiidae (Siluriformes) in Southeast Asia: V. Five new species of *Thaparocleidus* Jain, 1952 (Ancylodiscoididae) from *Pangasius bocourti*, *P. djambal* and *P. hypophthalmus*. *Parasite* 9, 207–217.

Pariselle, A., Lim, L.H.S. and Lambert, A. (2006) Monogeneans from Pangasiidae (Siluriformes) in Southeast Asia: X. Six new species of *Tharoparocleidus* Jain, 1952 (Ancylodiscoididae) from *Pangasius micronema*. *Parasite* 13, 283–290.

Pasaribu, W., Sukenda, S. and Nuryati, S. (2018) The efficacy of Nile tilapia (*Oreochromis niloticus*) broodstock and larval immunization against *Streptococcus agalactiae* and *Aeromonas hydrophila*. *Fishes* 3(1), 16.

Pathiratne, A., Widanapathirana, G.S. and Chandrakanthi, W.H.S. (1994) Association of *Aeromonas hydrophila* with epizootic ulcerative syndrome (EUS) of freshwater fish in Sri Lanka. *Journal of Applied Ichthyology* 10, 204–208.

Paul, A., Pattanayak, S., Sahoo, M.K., Kumar, P.R., Kumar R. and Sahoo, P.K. (2020) Co-infection of bacterial and parasitic pathogens including the myxosporean *Zschokkella auratis* infecting brain in farmed striped murrel *Channa striata* (Bloch, 1793) causing large scale mortality. *Indian Journal of Fisheries* 67(4), 105-115.

Pei, C., Gao, Y., Sun, X., Li, L. and Kong, X. (2019) A developed subunit vaccine based on fiber protein VP56 of grass carp reovirus providing immune protection against grass carp hemorrhagic disease. *Fish & Shellfish Immunology* 90, 12–19.

Perelberg, A., Ronen, A., Hutoran, M., Smith, Y. and Kotler, M. (2005) Protection of cultured *Cyprinus carpio* against a lethal viral disease by an attenuated virus vaccine. *Vaccine* 23, 3396–3403.

Perelberg, A., Smirnov, M., Hutoran, M., Diamant, A., Bejerano, Y. and Kotler, M. (2003) Epidemiological description of a new viral disease afflicting cultured *Cyprinus carpio* in Israel. *Israeli Journal of Aquaculture – Bamidgeh* 55, 5–12.

Persson, S., Al-Shuweli, S., Yapici, S., Jensen, J.N. and Olsen, K.E. (2015) Identification of clinical *Aeromonas* species by *rpoB* and *gyrB* sequencing and development of a multiplex PCR method for detection of *Aeromonas hydrophila*, *A. caviae*, *A. veronii*, and *A. media*. *Journal of Clinical Microbiology* 53(2), 653–656.

Peterman, M.A. and Posadas, B.C. (2019) Direct economic impact of fish diseases on the east Mississippi catfish industry. *North American Journal of Aquaculture* 81(3), 222–229.

Pfeil-Putzien, C. (1977) New results in the diagnosis of spring viraemia of carp caused by experimental transmission of *Rhabdovirus carpio* with the carp louse (*Argulus foliaceus*). *Bulletin Office International des Épizooties* 87(5–6), 457.

Phadee, P., Kurata, O., Hatai, K., Hirono, I. and Aoki, T. (2004) Detection and identification of fish-pathogenic *Aphanomyces piscida* using polymerase chain reaction (PCR) with species-specific primers. *Journal of Aquatic Animal Health* 16, 220–230.

Philippine Statistics Authority (2022) *Freshwater Cage Production in the Philippines for 2021*. Philippine Fisheries Profile, Bureau of Fisheries and Aquatic Resources, Quezon City, The Philippines.

Phillips, A.J., Anderson, V.L., Robertson, E.J., Secombes, C.J. and van West, P. (2008) New insights into animal pathogenic oomycetes. *Trends in Microbiology* 16, 13–19.

Picón-Camacho, S.M., Marcos-Lopez, M., Bron, J.E. and Shinn, A.P. (2012a) An assessment of the use of chemotherapeutants and physical interventions in the treatment of *Ichthyophthirius multifiliis* Fouquet, 1876, a protozoan parasite of freshwater fish. *Parasitology* 138, 1–42.

Picón-Camacho, S.M., Taylor, N.G.H., Bron, J.E., Guo, F.C. and Shinn, A.P. (2012b) The effect of low dose, long exposure bronopol treatments on the infection dynamics of *Ichthyophthirius multifiliis* (Ciliophora), parasitising *Oncorhynchus mykiss*. *Veterinary Parasitology* 186, 237–244.

Piewbang, C., Tattiyapong, P., Khemthong, M., Lachroje, S., Boonrungsiman, S., *et al.* (2022) Dual infections of tilapia parvovirus (TiPV) and tilapia lake virus (TiLV) in multiple tilapia farms: their impacts, genetic diversity, viral tropism, and pathological effects. *Aquaculture* 550, 737887.

Piewbang, C., Tattiyapong, P., Techangamsuwan, S. and Surachetpong, W. (2021) Tilapia lake virus immunoglobulin G (TiLV IgG) antibody: immunohistochemistry application reveals cellular tropism of TiLV infection. *Fish & Shellfish Immunology* 116, 115–123.

Pikarsky, E., Ronen, A., Abramowitz, J., Levavi-Sivan, B., Hutoran, M., *et al.* (2004) Pathogenesis of acute viral disease induced in fish by carp interstitial nephritis and gill necrosis virus. *Journal of Virology* 78, 9544–9551.

Plumb, J.A. (1994) *Health Maintenance of Cultured Fishes: Principal Microbial Diseases*. CRC Press, Boca Raton, Florida.

Plumb, J.A. (1999) *Health Maintenance and Principal Microbial Diseases of Cultured Fishes*. Iowa State University Press, Ames, Iowa.

Plumb, J.A., Bowser, P., Grizzle, J. and Mitchell, A. (1979) Fish viruses: a double-stranded RNA icosahedral virus from a North American cyprinid. *Journal of the Fisheries Research Board of Canada* 36, 1390–1394.

Ponpukdee, N., Wangman, P., Rodkhum, C., Pengsuk, C., Chaivisuthangkura, P., *et al.* (2021) Detection and identification of a fish pathogen *Flavobacterium columnare* using specific monoclonal antibodies. *Aquaculture* 545, 737231.

Pope, K.L., Neumann, R.M. and Bryan, S.D. (2009) Warmwater fish in small standing waters. In: Bonar, S.A., Hubert, W.A. and Willis, D.W. (eds) *Standard Methods for Sampling North American Freshwater Fishes*. American Fisheries Society, Bethesda, Maryland, pp. 13–27.

Post, G. (1983) *Textbook of Fish Health*. TFH Publications, Neptune City, New Jersey.

Pradeep, P.J., Suebsing, R., Sirithammajak, S., Kampeera, J., Turner, W., *et al.* (2017) Vertical transmission and concurrent infection of multiple bacterial pathogens in naturally infected red tilapia (*Oreochromis* spp.). *Aquaculture Research* 48(6), 2706–2717.

Pradhan, P.K., Verma, D.K., Peruzza, L., Gupta, S., Haq, S.A., *et al.* (2020) Molecular insights into the mechanisms of susceptibility of *Labeo rohita* against oomycete *Aphanomyces invadans*. *Scientific Reports* 10(1), 19531.

Prasad, Y., Arpana and Kumar, D. (2011) Lytic bacteriophages specific to *Flavobacterium columnare* rescue catfish, *Clarias batrachus* (Linn.) from columnaris disease. *Journal of Environmental Biology* 32, 161–168.

Pridgeon, J.W. and Klesius, P.H. (2011) Development and efficacy of novobiocin and rifampicin-resistant *Aeromonas hydrophila* as novel vaccines in channel catfish and Nile tilapia. *Vaccine* 29(45), 7896–7904.

Pridgeon, J.W., Klesius, P.H. and Yildirim-Aksoy, M. (2013) Attempt to develop live attenuated bacterial vaccines by selecting resistance to gossypol, proflavine hemisulfate, novobiocin, or ciprofloxacin. *Vaccine* 31(18), 2222–2230.

Pridgeon, J.W., Klesius, P.H., Lewbart, G.A., Daniels, H.V. and Jacob, M. (2014) *Edwardsiella tarda* and *Aeromonas hydrophila* isolated from diseased Southern flounder (*Paralichthys lethostigma*) are virulent to channel catfish and Nile tilapia. *Journal of Coastal Life Medicine* 2(5), 337–343.

Přikrylová, I., Blažek, R. and Vanhove, M.P.M. (2012) An overview of the *Gyrodactylus* (Monogenea: Gyrodactylidae) species parasitizing African catfishes, and their morphological and molecular diversity. *Parasitology Research* 110, 1185–1200.

Pulpipat, T., Maekawa, S., Wang, P.C. and Chen, S.C. (2020) Immune responses and protective efficacy of a formalin-killed *Francisella noatunensis* subsp. *orientalis* vaccine evaluated through intraperitoneal and immersion challenge methods in *Oreochromis niloticus*. *Vaccines* 8(2), 163.

Radhakrishnan, K., Aanand, S., Padmavathy, P. and Biswas, I. (2019) Current status of freshwater cage aquaculture in India: towards blue revolution. *Aquaculture Asia Magazine* 23(1), 3–10.

Rahman, M.H., Kar, D. and Barman, N.N. (1999) Bacterial pathogens associated with epizootic ulcerative syndrome in fishes in Assam. *Indian Journal of Animal Sciences* 69, 1083–1085.

Rahman, M.M. (1995) Some aspects of the biology of a freshwater fish parasite, *Argulus foliaceus* (L.) (Argulidae, Branchiura, Crustacea). *Bangladesh Journal of Zoology* 23, 77–86.

Raja, R.A., Patil, P.K., Avunje, S., Kumaran, M., Solanki, H.G., *et al.* (2022) Efficacy of emamectin benzoate in controlling natural infestations of ectoparasites in economically important fish species of India. *Aquaculture* 551, 737940.

Ramirez-Paredes, J.G., Larsson, P., Thompson, K.D., Penman, D.J., Busse, H.-J., *et al.* (2020) Reclassification of *Francisella noatunensis* subsp. *orientalis* Ottem *et al.* 2009 as *Francisella orientalis* sp. nov., *Francisella noatunensis* subsp. *chilensis* subsp. nov. and emended description of *Francisella noatunensis*. *International Journal of Systematic and Evolutionary Microbiology* 70(3), 2034–2048.

Ramos, D.G.S., Moraes, R.S., Moreira, C.N., Ferraz, H.T., Lopes, D.T., *et al.* (2020) Infection outbreak for *Piscinoodinium pillulare* in *Piaractus brachypomus* from aquaculture farm in Brazilian cerrado. *International Journal of Hydrology* 4(2), 97–98.

Rao, Y. and Su, J. (2015) Insights into the antiviral immunity against grass carp (*Ctenopharyngodon idella*) reovirus (GCRV) in grass carp. *Journal of Immunology Research* 2015, 670437.

Ravindra, Kumar, R., Ahmad, N., Verma, D.K., Kantharajan, G., *et al.* (2022) Mortalities in cultured *Pangasianodon hypophthalmus* due to oomycete *Saprolegnia parasitica* infection in Uttar Pradesh, India. *Aquaculture Reports* 23, 101047.

Ren, Z., Cai, Y., Wang, S., Liu, S., Li, A., *et al.* (2019) First case of *Aeromonas schubertii* infection in brackish water wild Nile tilapia, *Oreochromis niloticus*, in China. *Aquaculture* 501, 247–254.

Roberts, R.J. (2012) *Fish Pathology*, 4th edn. Wiley, Chichester, UK.

Roberts, R.J., Willoughby, L.G. and Chinabut, S. (1993) Mycotic aspects of epizootic ulcerative syndrome (EUS) of Asian fishes. *Journal of Fish Diseases* 16, 169–183.

Rodkhum, C., Kayansamruaj, P. and Pirarat, N. (2011) Effect of water temperature on susceptibility to *Streptococcus agalactiae* serotype Ia infection in Nile tilapia (*Oreochromis niloticus*). *Thai Journal of Veterinary Medicine* 41(3), 309–314.

Rodkhum, C., Kayansamruaj, P., Pirarat, N. and Wongtawatchai, J. (2012) Duplex PCR for simultaneous and unambiguous detection of *Streptococcus iniae* and *Streptococcus agalactiae* associated with streptococcosis of cultured tilapia in Thailand. *Thai Journal of Veterinary Medicine* 42(2), 153–158.

Rodrigues, M.V., Francisco, C.J., David, G.S., da Silva, R.J., Falcone-Dias, M.F. and Júnior, J.P.A. (2018) Monitoring of *Francisella noatunensis* subsp. *orientalis* in farmed Nile tilapia (*Oreochromis niloticus*) in Brazil. *Aquaculture International* 26(1), 127–138.

Ronen, A., Perelberg, A., Abramowitz, J., Hutoran, M., Tinman, S., *et al.* (2003) Efficient vaccine against the virus causing a lethal disease in cultured *Cyprinus carpio*. *Vaccine* 21, 4677–4684.

Roubach, R., Ostrensky, A., Schulter, E.P. and Bueno, G.W. (2015) Aquaculture planning development in Brazilian federal waters. *Global Aquaculture Advocate* 18(1), 40–43.

Sant'Ana, F.J.F., Oliveira, S.L., Rabelo, R.E., Vulcani, V.A.S., da Silva, S.M.G. and Júnior, J.A.F. (2012) Surtos de infecção por *Piscinoodinium pillulare* e *Henneguya* spp. empacus (*Piaractus mesopotamicus*) criados intensivamente no Sudoeste de Goiás [Outbreaks of *Piscinoodinium pillulare* and *Henneguya* spp. infection in intensively raised *Piaractus mesopotamicus* in Southwestern Goiás, Brazil]. *Pesquisa Veterinária Brasileira* 32(2), 121–125.

Sarig, S. (1971) *Diseases of Fishes. Book 3: The Prevention and Treatment of Diseases of Warm Water Fishes under Sub-Tropical Conditions, With Special Emphasis on Intensive Farming*. TFH Publications, Neptune City, New Jersey.

Sarowar, M., Saraiva, M., Jessop, C.N., Lilje, O., Gleason, F.H. and van West, P. (2014) Infection strategies of pathogenic oomycetes in fish. In: Jones, E.B.G., Hyde, K.D. and Pang, K.L. (eds) *Freshwater Fungi and Fungal-like Organisms*. De Gruyter, Berlin, pp. 217–243.

Scholz, T., Vanhove, M.P.M., Smit, N., Jayasundera, Z. and Gelnar, M. (2018) A guide to the parasites of African freshwater fishes. *Abc Taxa* 18, 1–424.

Schwedler, T.C. and Plumb, J.A. (1982) Golden shiner virus and effects of stocking density on the incidence of viral infection. *The Progressive Fish-Culturist* 44, 151–152.

Seng, E.K., Fang, Q., Lam, T.J. and Sin, Y.M. (2004) Development of a rapid, sensitive and specific diagnostic assay for fish *Aquareovirus* based on RT-PCR. *Journal of Virological Methods* 118, 111–122.

Shaharom, M.F. (1985) The site-specificity of *Cichlidogyrus sclerosus*, its juveniles, and *Cichlidogyrus tubicirrus minutus* on the gills of tilapia species. *Tropical Biomedicine* 2, 107–112.

Shahin, K., Ramirez-Paredes, J.G., Harold, G., Lopez-Jimena, B., Adams, A. and Weidmann, M. (2018) Development of a recombinase polymerase amplification assay for rapid detection of *Francisella noatunensis* subsp. *orientalis*. *PLoS ONE* 13(2), e0192979.

Shameena, S.S., Kumar, K., Kumar, S., Kumari, P., Krishnan, R., *et al.* (2021) Dose-dependent co-infection of *Argulus* sp. and *Aeromonas hydrophila* in goldfish (*Carassius auratus*) modulates innate immune response and antioxidative stress enzymes. *Fish & Shellfish Immunology* 114, 199–206.

Shariff, M. and Sommerville, C. (1986) Identification and distribution of *Lernaea* spp. in Peninsular Malaysia. In: Maclean, J.L., Dizon, L.B. and Hosillos, L.V. (eds) *The First Asian Fisheries Forum*. Asian Fisheries Society, Manila, pp. 269–272.

Shelby, R.A., Klesius, P.H., Shoemaker, C.A. and Evans, J.J. (2002) Passive immunization of tilapia, *Oreochromis niloticus* (L.), with anti-*Streptococcus iniae* whole sera. *Journal of Fish Diseases* 25, 1–6.

Shinn, A.P. and Bron, J.E. (2012) Considerations for the use of anti-parasitic drugs in aquaculture. In: Austin, B. (ed.) *Infectious Disease in Aquaculture*. Woodhead Publishing Ltd, Cambridge, pp. 190–217.

Shinn, A.P., Gibson, D.I. and Sommerville, C. (2001) Morphometric discrimination of *Gyrodactylus salaris* Malmberg (Monogenea) from species of *Gyrodactylus* parasitising British salmonids using novel parameters. *Journal of Fish Diseases* 24, 83–97.

Shinn, A.P., Pratoomyot, J., Metselaar, M. and Bastos Gomes, G. (2018) Diseases in aquaculture – counting the costs of the top 100. In: Binder, E.M. (ed.) *Biomin World Nutrition Forum: SCOPE 'Scientific Challenges and Opportunities in the Protein Economy', Cape Town, South Africa, 3–5 October*. Erber AG, Austria, pp. 227–262.

Shinn, A.P., Avenant-Oldewage, A., Bondad-Reantaso, M.G., Cruz-Laufer, A., García-Vásquez, A., *et al.* (2022) A global review of problematic and pathogenic parasites of farmed tilapia. *Reviews in Aquaculture* 2022, 1–62. doi:10.1111/raq.12742.

Shivam, S., El-Matbouli, M. and Kumar, G. (2021) Development of fish parasite vaccines in the OMICs era: progress and opportunities. *Vaccines* 9, 179.

Shoemaker, C.A., LaFrentz, B.R., Klesius, P.H. and Evans, J.J. (2010) Protection against heterologous *Streptococcus iniae* isolates using a modified bacterin vaccine in Nile tilapia, *Oreochromis niloticus*. *Journal of Fish Diseases* 33, 537–544.

Shoemaker, C.A., Martins, M.I., Xu, D.H. and Klesius, P.H. (2012) Effect of *Ichthyophthirius multifiliis* parasitism on the survival, hematology and bacterial load in channel catfish previously exposed to *Edwardsiella ictaluri*. *Parasitology Research* 111, 2223–2228.

Shotts, E.B. and Starliper, C.E. (1999) Flavobacterial diseases: columnaris disease, cold-water disease and bacterial gill disease. In: Woo, T.K. and Bruno, D.W. (eds) *Fish Diseases and Disorders*. Vol. 3. *Viral, Bacterial, and Fungal Infections*, 1st edn. CAB International, Wallingford, UK, pp. 559–576.

Shotts, E.B., Blazer, V.S. and Waltman, W.D. (1986) Pathogenesis of experimental *Edwardsiella ictaluri* infections in channel catfish (*Ictalurus punctatus*). *Canadian Journal of Fisheries and Aquatic Sciences* 43, 36–42.

Šimková, A., Serbielle, C., Pariselle, A., Vanhove, M.P.M. and Morand, S. (2013) Speciation in *Thaparocleidus* (Monogenea: Dactylogyridae) parasitizing Asian pangasiid catfishes. *BioMed Research International* 2013, 353956.

Šimková, A., Řehulková, E., Rasoloariniaina, J.R., Jorissen, M.W.P., Scholz, T., *et al.* (2019) Transmission of parasites from introduced tilapias: a new threat to endemic Malagasy ichthyofauna. *Biological Invasions* 21, 803–819.

Soleng, A., Jansen, P.A. and Bakke, T.A. (1999) Transmission of the monogenean *Gyrodactylus salaris*. *Folia Parasitologica* 46, 179–184.

Soliman, H., Lewisch, E. and El-Matbouli, M. (2019) Identification of new genogroups in Austrian carp edema virus isolates. *Diseases of Aquatic Organisms* 136(2), 193–197.

Somga, J.R., de la Peña, L.D., Sombito, C.D., Paner, M.G., Suarnaba, V.S., *et al.* (2010) Koi herpesvirus-associated mortalities in quarantined koi carp in the Philippines. *Bulletin of the European Association of Fish Pathologists* 30, 2–7.

Soto, E., Hawke, J.P., Fernandez, D. and Morales, J.A. (2009) *Francisella* sp., an emerging pathogen of tilapia, *Oreochromis niloticus* (L.), in Costa Rica. *Journal of Fish Diseases* 32(8), 713–722.

Soto, E., Illanes, O., Revan, F., Griffin, M. and Riofrio, A. (2013a) Bacterial distribution and tissue targets following experimental *Edwardsiella ictaluri* infection in Nile tilapia *Oreochromis niloticus*. *Diseases of Aquatic Organisms* 104(2), 105–112.

Soto, E., Kidd, S., Gaunt, P.S. and Endris, R. (2013b) Efficacy of florfenicol for control of mortality associated with *Francisella noatunensis* subsp. *orientalis* in Nile tilapia, *Oreochromis niloticus* (L.). *Journal of Fish Diseases* 36(4), 411–418.

Soto, E., Kidd, S., Mendez, S., Marancik, D., Revan, F., *et al.* (2013c) *Francisella noatunensis* subsp. *orientalis* pathogenesis analyzed by experimental immersion challenge in Nile tilapia, *Oreochromis niloticus* (L.). *Veterinary Microbiology* 164(1–2), 77–84.

Soto, E., Halliday-Simmonds, I., Francis, S., Kearney, M.T. and Hansen, J.D. (2015) Biofilm formation of *Francisella noatunensis* subsp. *orientalis*. *Veterinary Microbiology* 181(3–4), 313–317.

Soto, E., Yun, S. and Surachetpong, W. (2019) Susceptibility of tilapia lake virus to buffered povidone–iodine complex and chlorine. *Aquaculture* 512, 734342.

Soto, E., Tamez-Trevino, E., Yazdi, Z., Stevens, B.N., Yun, S., *et al.* (2020) Non-lethal diagnostic methods for koi herpesvirus in koi *Cyprinus carpio*. *Diseases of Aquatic Organisms* 138, 195–205.

Soto-Rodriguez, S.A., Lozano-Olvera, R., Garcia-Gasca, M.T., Abad-Rosales, S.M., Gomez-Gil, B. and Ayala-Arellano, J. (2018) Virulence of the fish pathogen *Aeromonas dhakensis*: genes involved, characterization and histopathology of experimentally infected hybrid tilapia. *Diseases of Aquatic Organisms* 129(2), 107–116.

Steindinger, K.A. and Baden, D.G. (1984) Toxic marine dinoflagellates. In: Spector, D.L. (ed.) *Dinoflagellates*. Academic Press, New York, pp. 201–261.

St-Hilaire, S., Beevers, N., Joiner, C., Hedrick, R.P. and Way, K. (2009) Antibody response of two populations of common carp, *Cyprinus carpio* L., exposed to koi herpesvirus. *Journal of Fish Diseases* 32, 311–320.

Suebsing, R., Kampeera, J., Tookdee, B., Withyachumnarnkul, B., Turner, W. and Kiatpathomchai, W. (2013) Evaluation of colorimetric loop-mediated isothermal amplification assay for visual detection of *Streptococcus agalactiae* and *Streptococcus iniae* in tilapia. *Letters in Applied Microbiology* 57(4), 317–324.

Suebsing, R., Suebsing, R., Kampeera, J., Withyachumnarnkul, B., Turner, W. and Kiatpathomchai, W. (2015) Colorimetric method of loop-mediated isothermal amplification with the pre-addition of calcein for detecting *Flavobacterium columnare* and its assessment in tilapia farms. *Journal of Aquatic Animal Health* 27(1), 38–44.

Suebsong, W., Poompuang, S., Srisapoome, P., Koonawootrittriron, S., Luengnaruemitchai, A., *et al.* (2019) Selection response for *Streptococcus agalactiae* resistance in Nile tilapia *Oreochromis niloticus*. *Journal of Fish Diseases* 42(11), 1553–1562.

Sunarto, A., Rukyani, A. and Itami, T. (2005) Indonesian experience on the outbreaks of koi herpesvirus in koi and carp (*Cyprinus carpio*). *Bulletin of the Fisheries Research Agency* S2, 15–21.

Surachetpong, W., Roy, S.R.K. and Nicholson, P. (2020) Tilapia lake virus: the story so far. *Journal of Fish Diseases* 43(10), 1115–1132.

Swain, P., Behura, A., Dash, S. and Nayak, S.K. (2007) Serum antibody response of Indian major carp, *Labeo rohita* to three species of pathogenic bacteria; *Aeromonas hydrophila*, *Edwardsiella tarda*, and *Pseudomonas fluorescens*. *Veterinary Immunology and Immunopathology* 117, 137–141.

Sweet, W.V., Hamlington, B.D., Kopp, R.E., Weaver, C.P., Barnard, P.L., *et al.* (2022) *Global and Regional Sea Level Rise Scenarios for the United States: Updated Mean Projections and Extreme Water Level Probabilities along US Coastlines*. NOAA Technical Report NOS 01. National Oceanic and Atmospheric Administration, National Ocean Service, Silver Spring, Maryland. Available at: https://oceanservice.noaa.gov/hazards/sealevelrise/noaa-nos-techrpt01-global-regional-SLR-scenarios-US.pdf (accessed 29 September 2022).

Taengphu, S., Kayansamruaj, P., Kawato, Y., Delamare-Deboutteville, J., Mohan, C.V., *et al.* (2022) Concentration and quantification of *Tilapia tilapinevirus* from water using a simple iron flocculation coupled with probe-based RT-qPCR. *PeerJ* 10, e13157.

Tancredo, K.R., Marchiori, N.C., Pereira, S.A. and Martins, M.L. (2019) Toxicity of formalin for fingerlings of *Cyprinus carpio* var. koi and *in vitro* efficacy against *Dactylogyrus minutus* Kulwièc, 1927 (Monogenea: Dactylogyridae). *Journal of Parasitic Diseases* 43(1), 46–53.

Tang, K.F.J., Bondad-Reantaso, M.G., Surachetpong, W., Dong, H.T., Fejzic, N., *et al.* (2021) *Tilapia Lake Virus Disease Strategy Manual*. FAO Fisheries and Aquaculture Circular No. 1220. Food and Agriculture Organization of the United Nations, Rome.

Thanpuran, N., Surendran, P.K., Kukundan, M.K. and Gopakumar, K. (1995) Bacteriological studies on fish affected by epizootic ulcerative syndrome (EUS) in Kerala, India. *Asian Fisheries Science* 8, 103–111.

Thune, R.L., Stanley, L.A. and Cooper, R.K. (1993) Pathogenesis of gram-negative bacterial infections in warm water fish. *Annual Review of Fish Diseases* 3, 7–68.

Thuy, D.T. and Buchmann, K. (2008) Infections with gill parasitic monogeneans *Thaparocleidus siamensis* and *T. caecus* in cultured catfish *Pangasius hypophthalmus* in southern Vietnam. *Bulletin of the European Association of Fish Pathologists* 28(1), 10–15.

Tolo, I.E., Padhi, S.K., Williams, K., Singh, V., Halvorson, S., *et al.* (2021) Susceptibility of *Pimephales promelas* and *Carassius auratus* to a strain of koi herpesvirus isolated from wild *Cyprinus carpio* in North America. *Scientific Reports* 11(1), 1985.

Torres, J.L., Shariff, M. and Law, A.T. (1990) Identification and virulence screening of *Aeromonas* spp. isolated from healthy and epizootic ulcerative syndrome (EUS)-infected fish. In: Hirano, R. and Hanyu, I. (eds) *The Second Asian Fisheries Forum*. Asian Fisheries Society, Manila, pp. 663–666.

Tu, C., Weng, M.C., Shiau, J.R. and Lin, S.Y. (2004) Detection of koi herpesvirus in koi *Cyprinus carpio* in Taiwan. *Fish Pathology* 39, 109–110.

Uddin, M.N., Chowdhury, M.B.R. and Wakabayashi, H. (1997) Optimum temperatures for the growth and protease production of *Aeromonas hydrophila*. *Fish Pathology* 32, 117–120.

Van As, J.G. and Basson, L. (1989) A further contribution to the taxonomy of the Trichodinidae (Ciliophora: Peritrichia) and a review of the taxonomic status of some fish ectoparasitic trichodinids. *Systematic Parasitology* 14, 157–179.

Vandersea, N.W., Litaker, R.W., Yonnish, B., Sosa, E., Landsberg, J.H., *et al.* (2006) Molecular assays for detecting *Aphanomyces invadans* in ulcerative mycotic fish lesions. *Applied and Environmental Microbiology* 72, 1551–1557.

Videira, M., Velasco, M., Malcher, C.S., Santos, P., Matos, P. and Matos, E. (2016) An outbreak of myxozoan parasites in farmed freshwater fish *Colossoma macropomum* (Cuvier, 1818) (Characidae, Serrasalminae) in the Amazon region, Brazil. *Aquaculture Reports* 3, 31–34.

Walczak, E.M., Noga, E.J. and Hartmann, J.X. (1981) Properties of a vaccine for channel catfish virus disease and a method of administration. *Developments in Biological Standardization* 49, 410–429.

Walker, C.A. and van West, P. (2007) Zoospore development in the oomycetes. *Fungal Biology Reviews* 21, 10–18.

Walker, P.D., Russon, I.J., Haond, C., van der Velde, G. and Wendelaar-Bonga, S.E. (2011) Feeding in adult *Argulus japonicus* Thiele, 1900 (Maxillipoda, Branchiura), an ectoparasite of fish. *Crustaceana* 84(3), 307–318.

Waltzek, T.B., Kelly, G.O., Stone, D.M., Way, K., Hanson, L., *et al.* (2005) Koi herpesvirus represents a third cyprinid herpesvirus (CyHV-3) in the family Herpesviridae. *Journal of General Virology* 86, 1659–1667.

Wang, N., Zhang, Z., Jing, H., Zhang, M., Wu, S. and Lin, X. (2021) Development of a novel droplet digital PCR assay for the sensitive detection of carp edema virus. *Aquaculture* 545, 737162.

Wang, T., Chen, H. and Liu, H. (1994) Preliminary studies on the susceptibility of *Gobiocypris rarus* to hemorrhagic virus of grass carp. *Acta Hydrobiologia Sinica* 18, 144–149.

Wang, Y., Li, Z., Wang, Q., Zeng, W., Li, Y., *et al.* (2020) Establishment of a brain cell line obtained from hybrids of *Channa argus* × *Channa maculata* for the detection of tilapia lake virus. *Microbial Pathogenesis* 138, 103810.

Way, K. and Stone, D. (2013) Emergence of carp edema virus-like (CEV-like) disease in the UK. *CEFAS Finfish News* 15, 32–34.

Way, K., Paley, R. and Stone, S. (2008) Development and use of PCR assay protocols for Koi herpesvirus at CEFAS and the results of laboratory ring trials in 2006 and 2007. Abstract at the International Workshop on CyHV-3 (KHV) Cyprinid Herpes Viruses: Basic and Applied Aspects, held in Israel, 17–18 February 2008.

Way, K., Haenen, O., Stone, D., Adamek, M., Bergmann, S.M., *et al.* (2017) Emergence of carp edema virus (CEV) and its significance to European common carp and koi *Cyprinus carpio*. *Diseases of Aquatic Organisms* 126(2), 155–166.

Webster, J. (1986) *Introduction to Fungi*, 2nd edn. Cambridge University Press, Cambridge, UK.

Willoughby, L.G. (1994) *Fungi and Fish Diseases*. Pisces Press, Stirling, UK.

Willoughby, L.G., Roberts, R.J. and Chinabut, S. (1995) *Aphanomyces invaderis* sp. nov. the fungal pathogen of freshwater tropical fish affected by epizootic ulcerative syndrome. *Journal of Fish Diseases* 18, 273–275.

Wolf, K. (1988) *Fish Viruses and Fish Viral Diseases*. Cornell University Press, Ithaca, New York.

Wolf, K. and Darlington, R.W. (1971) Channel catfish virus: a new herpesvirus of ictalurid fish. *Journal of Virology* 8, 525–533.

Woo, P.T.K. (ed.) (1995) *Fish Diseases and Disorders*. Vol. 1. *Protozoan and Metazoan Infections*, 1st edn. CAB International, Wallingford, UK.

Wu, B., Lang, S. and Wang, W.J. (2000) *Fauna Sinica: Platyhelminthes: Monogenea*. Science Press, Beijing.

Wu, P.H., Chang, Z.H., Chang, J. and Chen, Y.S. (1975) Twist disease of *Hypophthalmichthys molitrix* in Hangchow region of Chekiang Province. *Acta ZoologicaSinica* 21(2), 190–196 (in Chinese).

Xi, B.W., Zhao, X., Li, P. and Xie, J. (2019) Morphological variation in *Myxobolus drjagini* (Akhmerov, 1954) from silver carp and description of *Myxobolus paratypicus* n. sp. (Cnidaria: Myxozoa). *Parasitology Research* 118, 2149–2157.

Xu, D.H., Shoemaker, C. and Klesius, P. (2007) Evaluation of the link between gyrodactylosis and streptococcosis of Nile tilapia, *Oreochromis niloticus* (L.). *Journal of Fish Diseases* 30(4), 233–238.

Xu, D.H., Shoemaker, C.A., Zhang, Q. and Klesius, P.H. (2013) Naturally infected channel catfish (*Ictalurus punctatus*) concurrently transmit *Ichthyophthirius multifiliis* and *Edwardsiella ictaluri* to naïve channel catfish. *Aquaculture* 376–379, 133–136.

Xu, D.H., Zhang, D., Shoemaker, C. and Beck, B. (2019) Immune response of channel catfish (*Ictalurus punctatus*) against *Ichthyophthirius multifiliis* post vaccination using DNA vaccines encoding immobilization antigens. *Fish & Shellfish Immunology* 94, 308–317.

Xu, J., Zeng, L., Luo, S., Wang, Y., Fan, Y. and Gong, S. (2013) Reovirus infection emerged in cultured channel catfish, *Ictalurus punctatus*, in China. *Aquaculture* 372–375, 39–44.

Yadav, M., Indira, G. and Ansary, A. (1992) Cytotoxin elaboration by *Aeromonas hydrophila* isolated from fish with epizootic ulcerative syndrome. *Journal of Fish Diseases* 15, 183–189.

Yambot, A.V. (1997) Isolation of *Aeromonas hydrophila* from *Oreochromis niloticus* during fish disease outbreaks in the Philippines. *Asian Fisheries Science* 10, 347–354.

Yamkasem, J., Tattiyapong, P., Kamlangdee, A. and Surachetpong, W. (2019) Evidence of potential vertical transmission of tilapia lake virus. *Journal of Fish Diseases* 42(9), 1293–1300.

Yamkasem, J., Tattiyapong, P., Gorgoglione, B. and Surachetpong, W. (2021) Uncovering the first occurrence of tilapia parvovirus in Thailand in tilapia during co-infection with *Tilapia tilapinevirus*. *Transboundary and Emerging Diseases*, 68(6), 3136–3144.

Yang, X.L., Xia, C. and Zuo, W.G. (1989) Inactivated vaccine for hemorrhage of grass carp: comparison of immunogenicity and immunizing dose between two strains. *Journal of Fisheries in China* 13, 138–144.

Yanong, R.P.E. (2003) Fungal diseases of fish. *Veterinary Clinics of North America: Exotic Animal Practice* 6, 377–400.

Yasumoto, S., Kazuya, Y., Yasuda, M., Yoshimura, T. and Miyazaki, T. (2006) Oral immunization of common carp with a liposome vaccine fusing koi herpesvirus antigen. *Fish Pathology* 41, 141–145.

Ye, X.P., Yang, G.Z., Luo, Y.Z., Chen, Y.L. and Chen, Z.H. (1992) Large-scale culture of grass carp cell and virus by using bioreactor. *Journal of Fisheries in China* 16, 1–6.

Yokoyama H., Liyanage, S.A., Sugai, A. and Wakabayashi, H. (1998) Hemorrhagic thelohanellosis of color carp caused by *Thelohanellus hovorkai* (Myxozoa: Myxosporea). *Fish Pathology* 33, 85–89.

Yoshino, M., Watari, H., Kojima, T. and Ikedo, M. (2006) Sensitive and rapid detection of koi herpesvirus by LAMP method. *Fish Pathology* 41, 19–27.

Yuasa, K., and Sano, M. (2009) Koi herpesvirus: status of outbreaks, diagnosis, surveillance and research. *Israeli Journal of Aquaculture – Bamidgeh* 62, 169–179.

Yuasa, K., Kurita, J., Kawana, M., *et al.* (2012) Development of mRNA-specific RT-PCR for the detection of koi herpesvirus (KHV) replication stage. *Diseases of Aquatic Organisms* 100(1), 11–18.

Zamri-Saad, M., Amal, M.N.A. and Siti-Zahrah, A. (2010) Pathological changes in red tilapias (*Oreochromis* spp.) naturally infected by *Streptococcus agalactiae*. *Journal of Comparative Pathology* 143(2–3), 227–229.

Zeng, W., Wang, Q., Wang, Y., Zhang, L., Liu, B., *et al.* (2013) Establishment of multiplex PCR for detection of grass carp reovirus and its application. *Journal of Fishery Sciences of China* 20(2), 419–426.

Zeng, W., Wang, Y., Chen, X., Wang, Q., Bergmann, S.M., *et al.* (2021) Potency and efficacy of VP20-based vaccine against tilapia lake virus using different prime-boost vaccination regimens in tilapia. *Aquaculture* 539, 736654.

Zhang, S., Zhi, T., Xu, X., Zheng, Y., Bilong Bilong, C.F., *et al.* (2019) Monogenean fauna of alien tilapias (Cichlidae) in south China. *Parasite* 26, 4.

7 Non-Infectious Disorders of Warmwater Fish

Florbela Soares[1,2]*, Benjamín Costas[3,4], Cátia Lourenço-Marques[1,2], Ignacio Fernández[5] and Paulo Gavaia[6,7]

[1]*IPMA/EPPO National Institute for the Ocean and Atmosphere/Aquaculture Research Station, Olhão, Portugal;* [2]*S2AQUA – Collaborative Laboratory, Association for a Sustainable and Smart Aquaculture, Olhão, Portugal;* [3]*CIIMAR – Centro Interdisciplinar de Investigação Marinha e Ambiental, Matosinhos, Portugal;* [4]*ICBAS-UP – Instituto de Ciências Biomédicas Abel Salazar, Universidade do Porto, Porto, Portugal;* [5]*Centro Oceanográfico de Vigo, Instituto Español de Oceanografía (IEO), Vigo, Spain;* [6]*CCMAR – Centre of Marine Sciences, University of Algarve, Faro, Portugal;* [7]*FMCB-UALG – Faculty of Medicine and Biomedical Sciences/ University of Algarve, Faro, Portugal*

7.1 Introduction

Research and developments in aquaculture have been remarkable, especially the improvements in the efficiency of production systems and the quality of the fish produced, while mitigating environmental impacts. New technologies include up-scaling of recirculation systems; development of cages and nets that can be used in higher-energy locations; and integrated multitrophic production systems. These technological advances have allowed diversification with several new species in Mediterranean aquaculture like meagre (*Argyrosomus regius*), tuna (*Thunnus thynnus*), white sea bream (*Diplodus sargus*), sharp-snout sea bream (*Diplodus puntazzo*) and greater amberjack (*Seriola dumerili*), for which the industry is adapting, and production is increasing. Also, a few established species have reached production volumes sufficient for stable markets to develop, such as European sea bass (*Dicentrarchus labrax*), Senegalese sole (*Solea senegalensis*) and gilthead sea bream (*Sparus aurata*).

Warmwater fish species, mainly European sea bass and gilthead sea bream, are reared in intensive on-growing systems in floating sea cages in lagoons, sheltered bays, or in semi-exposed and offshore conditions (EFSA, 2008). Juveniles (2.5–5.0 g) are transferred from the pre-growing tanks into the sea cages. However, it is now possible to stock smaller juvenile (0.3–2 g) gilthead sea bream or larger size fish (8.0–10.0 g). Farmers prefer to stock younger fish (2.5–150 g) at a density of 5 to 10 kg/m^3 and older fish (>150 g) at a density of 10 to 20 kg/m^3. Originally, cages used to be placed in well-protected, largely enclosed sites. Issues related to oxygen deficits and cage fouling during summer months, coupled with the scarcity of suitable sites and questions related to management of the coastal zone, resulted in the development and implementation of offshore cage technology. Control of offshore

*fsoares@ipma.pt

 Climate Change on Diseases and Disorders of Finfish in Cage Culture, 3rd Edition (eds P.T.K. Woo and R.P. Subasinghe)
DOI: 10.1079/9781800621640.0007

environmental conditions is still very limited, with offshore sites being widely exposed to extreme events like thermal and saline alterations, currents and hydrodynamics. Additionally, interventions related to routine management during the intensive grow-out phase in sea cages (e.g. antifouling treatments and noise), the development of morpho-anatomical abnormalities and other anthropogenic factors (such as pollutants and contaminants) have further significant impacts on farm revenues. European sea bass and gilthead sea bream are exposed to disturbances that may cause stress during ordinary farming procedures. Stressors include handling and manipulation, cleaning routines, grading, crowding and confinement, transportation between units, prophylactic measures and use of chemicals. Also, predators, boats and divers are factors of disturbance for on-growing fish kept in sea cages. All these factors raise concerns about fish farmers, since tertiary stress responses may take place. These responses to chronic stress extend to the level of the organism and the population, and refer to aspects of whole-animal performance such as changes in growth.

Commercial feeds have been greatly improved and are now produced by major manufacturers in highly sophisticated facilities under controlled conditions. Although nutrient deficiency conditions and toxic contaminations are nowadays much less frequent, they still occur. Nutritional imbalances due to deficiency or excess of specific nutrients can lead to undetected subclinical insufficiencies, possibly contributing to inefficient fish growth, losses to disease and baffling problems encountered with attempts to culture new fish species. The dependence on fishmeal has been reduced for several species. Alternatives such as soybean, corn meal and many other protein sources have not yet been perfectly adjusted to meet fish requirements and could have adverse consequences due to imbalances and/or the presence of antinutritional factors such as phytosterols and protease inhibitors.

Other issues such as genetic inbreeding, non-infectious disorders and climate change are also matters of concern. Aquaculture in temperate zones will be more affected by water warming, exceeding the tolerance of many farmed species. The increase in extreme weather events may affect aquaculture through physical destruction of facilities, loss of stock and spread of diseases, and the risks will be larger in farms located in more open and exposed sites. Moreover, little is known about the influence of fish domestication and/or selection on behaviour and adaptation within the context of fish culture. The European sea bass and gilthead sea bream industry has been based on empirical criteria for genetic selection, and systematic genetic improvement programmes have only recently been implemented.

The present chapter aims to give an overview on how different aspects of current finfish cage culture, such as environmental and anthropogenic-related problems, stress and nutritional imbalances, and the development of morpho-anatomical abnormalities may affect growth performance and increase disease susceptibility. Due to space limitations, citations in this chapter are not exhaustive. In addition, this chapter also specifies the clinical signs and histopathological lesions associated with these conditions and evaluates the current possibilities for prevention and control of these non-infectious disorders in warmwater fish.

7.2 Environmental and Anthropogenic-Related Problems

Since the marine aquaculture grow-out phase in warmwater fish species is mainly done in offshore locations, fish farmers have no control over environmental conditions, particularly in a climate change scenario. The freshwater aquaculture grow-out phase is still mainly done in (earth) ponds, with low environmental control, but an increasing effort is being made to develop recirculation aquaculture systems (RAS). This contrasts with the hatchery and nursery phases where farmers have control over most of the environmental factors. Consequently, there is a bidirectional relationship between cage/pond aquaculture and the environment, which might limit the growth of this economic activity. In one direction, certain environmental factors may impact fish physiology; while some aquaculture activities may lead to environmental impacts that can affect aquaculture. For example, the feeding activity (rates and regimens) might increase water eutrophication through the increase of water phosphorus and nitrogen content. In addition, several anthropogenic activities such as vessel traffic or wastewater flows from urban communities, intensive agriculture industry or antifouling treatments can affect aquaculture suitability.

This section briefly describes the environmental factors (water temperature, dissolved oxygen, ammonia concentration and algal blooms) and anthropogenic activities modifying environmental conditions (presence and concentration of several pollutants, underwater noise, fouling process and antifouling treatments) that were reported to have some impact in finfish cage production.

7.2.1 Ammonia toxicity

In teleost fish, ammonia is the main end product of nitrogen metabolism. In high production densities, ammonia levels rise and are extremely toxic to aquatic animals, causing several adverse effects (Egnew *et al.*, 2019; Zhang *et al.*, 2021). This problem is intensified as a result of global climate change, since it is predicted that variations in temperature, salinity and pH will lead to an increase in ammonia concentrations (Portz *et al.*, 2006). Kir and Sunar (2018) showed that, in gilthead sea bream, the toxic effect of ammonia and nitrite increases with decreasing salinity. Thus, increased precipitation predicted under a climate change scenario could therefore lead to increased ammonia and nitrite toxicity in gilthead sea bream. Comparing species cultured in the Mediterranean and eastern Atlantic coasts, since European sea bass is less tolerant to ammonia compared with gilthead sea bream (Person-Le Ruyet *et al.*, 1995), fish species-specific differences might be expected due to climate change impacts in different regions.

Increased levels of ammonia under intensive culture conditions have an impact on fish survival and growth (Person-Le Ruyet *et al.*, 2003). Briefly, ammonia is highly toxic in recirculating systems, especially when the pH is high, which is not the case under cage-culture conditions, where the pH is almost constant (7.8–8.2) and a naturally sustained water circulation occurs throughout the year. Unionized ammonia concentration of 0.5 mg/l can be harmful to finfish and crustaceans (Santacruz-Reyes and Chien, 2009). Average values of NH_4-N in Mediterranean regions with little anthropogenic impact are low (between 0.60 ± 0.56 and 0.73 ± 0.60 μM) (Faragallah *et al.*, 2009). Thus, seawater ammonia content should not be considered a major environmental factor impacting cage culture. Indeed, since the ammonia excretion from fish depends mainly on the amount of protein in feeds and the metabolic efficiency of the species (Vatsos and Angelidis, 2010), fish farmers should pay attention to local intensive agriculture outputs and water current regimes to monitor the potential toxicity risk by ammonia levels.

7.2.2 Fouling and antifouling treatments

Offshore aquaculture requires infrastructures that consist of a complex assortment of submerged components: cages, nets, floats and ropes. The presence of such large and diverse surfaces provides for a broad diversity of epibiotic organisms to settle and grow. This natural event is termed biofouling and is defined as the undesirable accumulation of micro- and macroorganisms on submerged artificial surfaces (Fig. 7.1). The direct economic cost to control biofouling has been estimated to be 5–10% of production cost, which can be US$1.5 billion to US$3 billion per year (Lane and Willemsen, 2004; Fitridge *et al.*, 2012). Indirect effects on fish (e.g. net occlusion, fish escapes, decrease in weight gain and risk of disease) (Bannister *et al.*, 2019; Sen *et al.*, 2020) might be expected when biofouling is not managed properly.

Water exchange restriction is a key impact factor due to net occlusion through the growth of fouling organisms (Madin *et al.*, 2010) leading to poor water quality with low dissolved oxygen levels and low removal of feed and waste excess, which impacts fish physiology. Therefore, fish could be subjected to stressful conditions due to environmental factors (hypoxia and high ammonia levels). Excess of biofouling can also compromise cage structure, causing deformation and structural fatigue (Swift *et al.*, 2006), and leading to fish escapes due to net breaks. A third negative effect is the risk of disease, with the fouling communities acting as reservoirs for pathogenic microorganisms harboured by macro or microbial fouling species, or due to increased fish vulnerability to diseases through exposure to lower dissolved oxygen and high ammonia

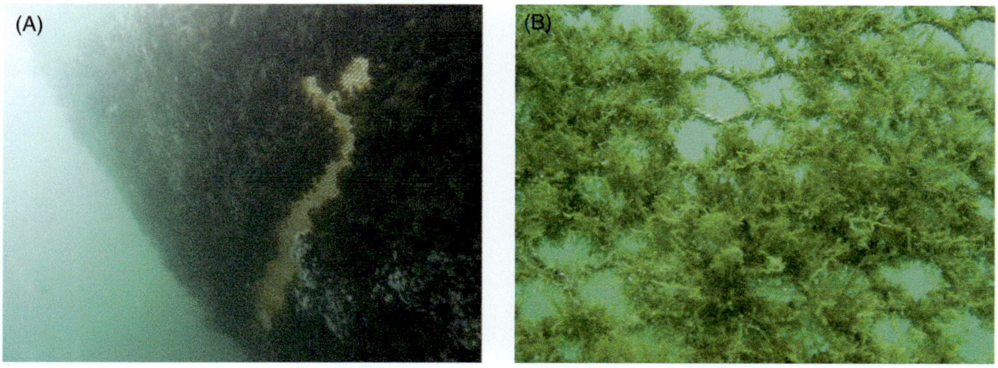

Fig. 7.1. Fouling by heavy growth of mussels (A) and by macroalgae (B) in sea cages (southern Portugal). (Photographs courtesy of Pedro Pousão-Ferreira.)

levels that lead to increased stress levels and mortality.

The operations performed in a commercial fish farm to minimize biofouling usually employ a multifaceted approach of: net changing and cleaning; chemical treatment of structures to prevent the recruitment by fouling organisms; and biological control to graze biofouling (Fitridge *et al.*, 2012). Frequent net changing damages the infrastructure, increases the risk of fish escape and disturbs the feeding regimens of fish, inducing lower growth rates. Nets can be cleaned *in situ*, primarily with cleaning discs on remotely operated vehicles, or manually by divers, in procedures that are likely to cause disturbances in fish behaviour. In addition, *in situ* cleaning, through brushing, increases fouling problems because it increases surface rugosity and loose filaments that are ideal settlement substrata for some fouling species (Carl *et al.*, 2011). Use of chemical antifoulants mainly consists of copper coatings, which are toxic for most marine invertebrates. Its application should be done each year in temperate regions (Braithwaite *et al.*, 2007). Copper released from antifouling-treated nets could be a health risk factor for European sea bass (Cotou *et al.*, 2012), although low concentrations were detected in muscle (1.5 μg/g) and liver (117 μg/g). Moreover, although copper levels did not induce oxidative stress, it affected the immediate immune defence mechanism of European sea bass, making them more vulnerable to diseases. In a way similar to other antifoulants (zinc and different biocides), copper also causes negative impacts in non-target species (Mochida *et al.*, 2006). The European Commission proposed to classify copper as a dangerous substance based on the 67/548/EEC directive. Consequently, new candidates to supplement or replace copper as an antifoulant have been proposed (Fitridge *et al.*, 2012). Accordingly, organic booster biocides were introduced (Guardiola *et al.*, 2012) as alternatives to the tributyltin (TBT)-related compounds found in previous antifouling products, due to the restrictions on their use caused by their negative impact in the marine environment. The third alternative to reduce biofouling is the use of herbivorous fish or invertebrates, and some attempts to do so have been done experimentally and on a small scale. However, this biological control of fouling is constrained by the fact that the fish or invertebrates used should have a broad dietary range due to the high diversity of biofouling organisms. More recently, Sen *et al.* (2020) reported an eco-friendly, copper-free antifouling paint (AP), effective in preventing biofouling on nets. Furthermore, it was as effective as the copper-based commercial APs. Also, a study with sea urchins used as biocontrol has shown a 74% reduction of fouling on infrastructures (Lodeiros and Garcia, 2004). Future research on the potential use of biocontrol techniques for fouling control is required.

Antifouling strategies should be non-specific, environmentally friendly, capable of withstanding onshore handling and cleaning, and economically viable. Fish farmers should consider the potential range of species that could be fouling their nets and the mechanisms to control and remove fouling, in order to manage this problem without affecting their fish.

7.2.3 Physical factors

Stocking density and handling

High stocking density is considered a stress factor for fish (Carbonara *et al.*, 2019) and values in production can reach 100 kg/m^3 (e.g. red sea bream in cages, China). Stocking density depends on the carrying capacity of cages, species requirements and size of the fish, among other factors. Stocking density affects the health, welfare and productivity of fish and although it does not necessarily cause direct harm or damage, detrimental effects may occur through social interactions and decline in water quality (MacIntyre *et al.*, 2008). Social interactions include aggression, resulting in physical damage or reduced access to feed, and water quality including lowered oxygen levels or raised ammonia levels in some parts of the cage. Despite the complexity and uncertainty, increasing stock densities will lead to the occurrence of problems and to reduced growth and/or increased mortalities due to infectious and/or non-infectious origins. Another important and related aspect is to have uniform sized fish in cages to limit inequal social interactions, thus obtaining better growth which will help to avoid cannibalism.

Handling of fish causes stress and consequently may contribute to disease outbreaks. Fish sampling, fishing and net changing, among other handling activities, may also cause stress or injury and can make fish vulnerable to diseases. It is important to monitor and recognize behavioural changes that are usually associated with stress (eating less, starvation, swimming erratically, skin discolorations and any other unusual behaviours).

Noise

Sound plays a crucial role in aquatic animals as means of communication and its role in echolocation, predator avoidance or perception of changes in the environment is of utmost importance. In fish, sound perception is based on two interconnected systems: the inner ears and the lateral line (Popper *et al.*, 2019). In offshore culture systems, fish are exposed to a wide range of sounds, such as sea ambient noise, noise generated by cage and feeding machinery, and noise from marine traffic. The anthropogenic sources of noise may induce several types of responses (Slabbekoorn *et al.*, 2010). Santulli *et al.* (1999) demonstrated typical primary and secondary stress responses (variation of cortisol, glucose, lactate, AMP, ADP, ATP and cAMP levels), in different tissues of European sea bass, to air-gun detonations. Adverse effects on oxidative status and on some immune parameters in gilthead sea bream juveniles were registered when fish were exposed to sea soundscape, determining a mild stress condition that could affect welfare (Filiciotto *et al.*, 2017) and consequently growth parameters and survival.

Under climate change, wind speed is expected to increase and for that reason, also the sounds associated with the overall offshore environment. Nevertheless, the acoustic pollution expected to have major impact in finfish cage-rearing sites is that related to shipping activities. There is an increase in low-frequency (100–500 Hz) ambient noise related to international transportation and recreational shipping (Ross, 2005) which can be detected by many fish species (Popper *et al.*, 2003). However, physiological responses to underwater noise are poorly understood. European sea bass and gilthead sea bream exposed to a 0.1–1 kHz acoustic stimulus had a significant increase in swimming activity, as well as in lactate and haematocrit levels (Buscaino *et al.*, 2010). These responses reflected an intense metabolic activity in white muscle anaerobic fibres during the acoustic exposure. Furthermore, the increase of metabolic muscle activity implies a higher demand for oxygen and thus an increase in respiratory rate. Sarà *et al.* (2007) reported altered fish behaviour in tuna (*T. thynnus*) due to the sounds generated by hydrofoil ferries, small boats and large car ferries. In the absence of boat noise, tuna assumed a concentrated co-ordinated school structure with unidirectional swimming and without a precise shape. Under noise from hydrofoils, tuna changed swimming direction and increased their vertical movement. This perturbation in fish behaviour was more evident when the sound came from outboard motors of small boats. Another impact on fish behaviour was reported by Picciulin *et al.* (2012), who observed that noise from boats caused variations in brown meagre (*Sciaena umbra*) vocalizations; in fact, the mean pulse rate increased over multiple boat passages.

Since fish are affected by noises, anthropogenic noise deserves greater attention by fish farmers than it has previously been given. Consequently, to reduce potential stress in fish caused by anthropogenic activities, limiting vessel traffic around fish cages and using newly developed engines (producing less noise) in activities related to fish production should be considered.

7.3 Emerging Problems

7.3.1 Dissolved oxygen

Normal levels of oxygen (normoxia) for marine fish are 4–8 mg/l, and lower or higher values of dissolved oxygen are defined as hypoxia or hyperoxia, respectively (Shultz *et al.*, 2011). Aquatic hypoxia is a frequent event that could take place: when water temperatures rise (midday/summer season); on cloudy days; early in the morning (due to the consumption of oxygen by algae during the night) (Noga, 2010); with excess feeding; with low seawater circulation; when there is increased net fouling; and/or with organic pollution, eutrophication and algal blooms. Hypoxia is a major source of stress (Karim *et al.*, 2003) and can also induce skeletal deformities (Castro *et al.*, 2011), resulting in considerable economic losses to aquaculture. Moreover, hypoxia is reported as one of the greatest risks in tuna farming (Nowak, 2004). This problem is likely to be exacerbated in the coming years, caused by increasing human activities such as coastal construction that changes the current and mixing patterns of water. Low-oxygen conditions result in hyperventilation (Shultz *et al.*, 2011); consequently fish under hypoxia have ionic disturbances and an accumulation of lactate (Vanlandeghem *et al.*, 2010), which leads to a reduction of the fish's ability to burst swim (Kieffer, 2000). Furthermore, adult European sea bass exposed to hypoxic conditions showed significantly lower ascorbate and α-tocopherol levels (Di Marco *et al.*, 2008). Hypoxia can also affect growth, feed consumption and feed conversion efficiency (Pichavant *et al.*, 2001), as well as influence the immune system (Henrique *et al.*, 2002), making fish more susceptible to opportunistic pathogens. In contrast, hyperoxia reduces ventilation rates (Shultz *et al.*, 2011), from which the most serious consequence is an inability of fish to excrete wastes as carbon dioxide (CO_2). High levels of CO_2 in the water cause respiratory acidosis and nephrocalcinosis (Gómez, 2000). Hyperoxia and high CO_2 levels could happen in tanks under intensive fish-rearing conditions, but none of those conditions has been reported so far in cage culture.

Aquatic hypoxia conditions in cage culture could be easily prevented by monitoring dissolved oxygen regularly, selecting proper cage location with minimum requirements of water currents (Burt *et al.*, 2012) and lowering anthropogenic impacts through agriculture and urban activities (eutrophication and organic pollution). Nevertheless, since climate change might alter temperature regimes (reducing dissolved oxygen content) and water circulation (Scavia *et al.*, 2002; IPCC, 2020), fish farm locations should be reconsidered to avoid problems in the future. However, reducing stocking densities (Pichavant *et al.*, 2001), feeding ratio and timing, as well as preventing fouling in net cages, could be good procedures to minimize the impacts on fish related to hypoxia, particularly those associated with climate change.

7.3.2 Pollutants

In general, a substance introduced into the environment that has undesired effects that adversely affects the usefulness of a resource, is called a pollutant. Diverse human land and coastal activities have influenced the existence and distribution of pollutants, and undoubtedly increased and spread them in aquatic systems. When pollutants enter the aquatic environment, either accidently or deliberately, they may lead to large-scale and sudden mortality of organisms. Even low quantities can have important impacts on marine organisms and may result in immunosuppression, physical damage to gills and epithelia, and adverse effects on metabolism, as well as increasing susceptibility to various infectious and non-infectious diseases. Several diseases such as carcinomas, epidermal papilloma, fin/tail erosion, gill disease/hyperplasia, liver disease, neoplasia, parasitic diseases, skeletal malformations, skin disease/ulceration and viral infections have been linked to pollution (Austin, 2007; Fernández *et al.*, 2018a; Kahlon *et al.*, 2018). More recently, reports on the presence of microplastics in fish feeds and pollutants adsorbed on these plastics (e.g. polycyclic aromatic hydrocarbons (PAHs) and benzo[a]pyrene (BAP)) raise new concerns in finfish production (Thiele *et al.*, 2021).

The adverse effects of pollutants in farmed fish depend on fish size and species, exposure time and source of the pollutants, as well as on environmental conditions (e.g. dissolved oxygen or pH). Some species accumulate heavy metals regardless of whether the habitat is contaminated or not, as seems to be the case with copper or mercury accumulation in mullet (*Mugil cephalus*) or tuna, respectively (Austin, 2007). In addition, adverse effects also depend on the type of pollutant (Austin, 2007). Metals, pesticides and hydrocarbons taken up by fish through the gills, digestive tract and body surface are important inducers of oxidative stress in aquatic organisms. Therefore, components of the antioxidant defence (such as metallothioneins or superoxide dismutase, catalase, glutathione peroxidase and glutathione *S*-transferase enzymes) could be used as biomarkers for the oxidative stress produced by pollutants. Although more than 90% of the heavy metal loads received by coastal aquatic systems are bound to suspended particulate matter and sediments, and thus not toxic to pelagic fish, they may effectively be recycled back into the overlying water phase through a variety of biological (e.g. activities of bottom-dwelling fish) and physicochemical processes (e.g. pH changes, sediment oxidation, heavy metal complexion by anions like chloride) (Sylaios *et al.*, 2012).

Little is known about contamination pathways in aquaculture systems. The presence of pollutants in farmed fish could be due to contaminants present in the rearing water, coming from local sources such as the aquaculture activity itself, for example the antifouling treatments (Varvarigos, 2007), or from neighbouring pollution sources such as industrial, agriculture or urban wastewater discharges. In some cases, the contaminant concentration is increased with the trophic level (biomagnification), a process which occurs at different rates depending on the physicochemical properties of the different contaminants (Harmelin-Vivien *et al.*, 2012). Recently, it was reported by Vaz *et al.* (2019) that higher concentrations of copper sulfate (0.5 mg/l) might be toxic to fish, showing histological alterations and hepatic copper accumulation.

Interestingly, some unexpected pollution comes from fish feeds. Different persistent organic pollutants, such as polychlorinated biphenyls (PCBs), organochlorine pesticides and polybrominated diphenyl ethers (PBDEs), have been found in high and moderate concentrations in salmon and in feeds (Jacobs *et al.*, 2002). Similarly, organochlorine pollutants were found in farmed and wild gilthead sea bream from the western Mediterranean (Spain) and their diets (Serrano *et al.*, 2008). White sea bream and European sea bass also presented low levels of organochlorine content (Ferreira *et al.*, 2008; Schnitzler *et al.*, 2008). Concentrations of organochlorine compounds in gilthead sea bream tissues were found to be strongly correlated with seasonal changes and with the biological cycle of fishes (Blanes *et al.*, 2009). Nacher-Mestre *et al.* (2010) showed that changes in diets led to different pollutant contents in fish and that PAH levels decreased after replacing fish oil by vegetable oil. This mechanism of wash-out of pollutants was further confirmed in European sea bass by water-bath exposure to light cycle oil, a refined product of heavy fuel oil. Importantly, fish showed severe external lesions (tissue necrosis, suppurative exudates and haemorrhagic areas) three days after the beginning of the recovery period and reduced phagocytic activity and lysozyme concentration suggested some degree of immunosuppression (Bado-Nilles *et al.*, 2011).

A careful design of the facilities, selection of the cage site, use of specialized equipment, and feed selection and composition can minimize or even eliminate the effects of contaminants in fish physiology (Vatsos and Angelidis, 2010). As an example, selecting the proper antifouling treatment in sea cages will reduce or eliminate the already reported exceeded risk for irgarol concerning seawater organisms (Muñoz *et al.*, 2010). However, it is important to note that until now, all residues and contaminants detected in farmed fish species are in fact low and always below the guidelines recommended for human consumption (Ferreira *et al.*, 2008; Blanes *et al.*, 2009; Muñoz *et al.*, 2010; Padula *et al.*, 2012; Varol and RaşitSünbül, 2019).

7.3.3 Climate change

Climate change is one of the most serious environmental threats for global ecosystems and extreme weather conditions are becoming more intense and frequent. The potential effects of climate change on aquaculture have been reviewed by Rosa *et al.* (2012) and Cascarano *et al.* (2021). Also reviews on infectious diseases of warmwater fish in marine and brackish waters (Parker-Graham *et al.*, Chapter 5, this volume, 2023) and in fresh water (Shinn *et al.*, Chapter 6, this volume, 2023) are recommended for further reading.

Global climate change may impose severe risks for aquatic animal health, since increasing water temperature leads to an increase in the incidence of parasitic and bacterial diseases (Harvell *et al.*, 2002; Pounds *et al.*, 2006; Lafferty, 2009) because fish will be subject to different stresses and physiological effects which, consequently, will impact growth and may further increase their susceptibility to diseases. This could take place through a temperature-driven effect on the epidemiology of the disease. Higher temperatures may boost the rate of disease spread through positive effects on parasite fitness in a weakened host due to the thermal stress (Harvell *et al.*, 2002). Increased temperature may also extend the transmission season, leading to higher total prevalence of infection and more widespread epidemics. However, the direction and magnitude of the changes in the incidence of extreme high temperatures are also known to vary between different parts of the world (Easterling *et al.*, 2000), and this should be considered. Temperature variation has in fact decreased in some areas due to an increase in the minimum temperatures rather than the maximum (Easterling *et al.*, 1997), emphasizing the asymmetric nature of climate warming. The effects of global warming will also depend on the age of fish, their health condition, ecology, farming practices and immunity (Karvonen *et al.*, 2010; Cascarano *et al.*, 2021). Another example could be the spread of *Ciona intestinalis* and other invasive biofouling organisms. This increases the complexity of strategies against biofouling in sea cages (Fitridge *et al.*, 2012). Global warming has nowadays many unpredictable effects on the marine environment and, consequently, on offshore aquaculture. Side effects of the global climate change also include the acidification of seawater and the rise of sea levels. Recent evidence suggests that both increased water temperature and elevated levels of dissolved CO_2 can change the behaviour of fishes in ways that reduce individual fitness, since the interaction of increased temperature and CO_2 could have significant effects on the growth and survival of fishes, as was reported in juvenile reef fish (*Amphiprion melanopus*) (Nowicki *et al.*, 2012). The impacts of ocean acidification on finfish species are less well known than those on shellfish. Although fish contain calcified otoliths which can be impacted by ocean acidification, becoming larger (Bignami *et al.*, 2013), the behaviour of marine fish under more acidified conditions has been well studied and several teleost behaviours are known to be impacted. Fish may experience impaired growth and development, tissue damage, respiration problems and decreased RNA viability under more acidified conditions (Munday *et al.*, 2009; Franke and Clemmesen 2011; Frommel *et al.*, 2012). Since climate change is already a reality, it becomes crucial to understand the complexity of its effects on fish health and to develop timely tools and management practices to limit economic damages in the aquaculture industry worldwide (Cascarano *et al.*, 2021). New and emerging diseases are expected, and selective breeding of more robust fish (e.g. disease resistance) is strongly recommended.

7.4 Nutritional Imbalance-Related Problems

Nutritional disorders of fish could be the result of a deficiency, excess or imbalance of nutrients. Deficiency or imbalance is an insufficiency of either macronutrients (e.g. protein, carbohydrate, lipid, fibre) or micronutrients (e.g. vitamins, minerals) in the diet. Lipid deficiency is among the most serious problems. Among the micronutrients, any of a wide range of components can exert an effect, especially in fast-growing young fish (Roberts, 2002). Nutritional imbalance also includes the excess of particular nutrients. As an example, excessive carbohydrate or lipid levels may result in hepatocyte degeneration and increased mortality (associated with fatty livers), respectively (Roberts, 2012).

7.4.1 Impacts on fish production

Nutritional imbalances are likely to occur with undetected subclinical deficiencies and possibly contribute to inefficient fish growth, increased disease susceptibility and unsolved problems encountered with attempts to culture new fish species. These include dorsal fin erosion, which has been associated with lysine deficiency, and cataracts associated with methionine and tryptophan deficiency (Roberts, 2012). However, dietary disease problems associated with the lipid fraction of the diet appear to be more serious. Three long-chain polyunsaturated fatty acids (PUFAs), namely docosahexaenoic acid (DHA; 22:6*n*-3), eicosapentaenoic acid (EPA; 20:5*n*-3) and arachidonic acid (ARA; 20:4*n*-6), have a variety of relevant functions in fish. Inadequate contents of these dietary essential fatty acids (EFAs) give rise to alterations such as reduction in erythrocyte volume and increase in erythrocyte fragility, haemoglobin content and erythrocyte counts. Additionally, renal morphology was also seen to be affected, showing extreme dilation of capillaries and occlusion of Bowman's capsule of the glomeruli (Montero *et al.*, 2004). Although the requirements for vitamins are small, deficiencies of these micronutrients can cause symptoms ranging from poor appetite to severe tissue deformities (Lovell, 1998). Moreover, the combination of high dietary content of vitamin K and antioxidant vitamins E and C (i.e. 23, 450 and 230 mg/kg, respectively) reduced the incidence of granulomatosis in meagre, further suggesting that this pathology could be mediated by nutritional factors (Ruiz *et al.*, 2019).

7.4.2 Diagnosis

A wide range of specific clinical features due to dietary deficiencies has been described in experimental studies (Table 7.1). Additionally, starvation may occur in farmed fish due to complete deprivation of food, inadequate feeding levels of a diet that is completely satisfactory in itself,

Table 7.1. Disorders associated with nutrient deficiencies in warmwater fish species.

Nutrient	Disorder	Species	Reference
Amino acid deficiencies			
Arginine, lysine and threonine	Decreased growth and feed efficiency	European sea bass Gilthead sea bream	Tibaldi and Lanari (1991); Tibaldi *et al.* (1994); Tibaldi and Tulli (1999); Fournier *et al.* (2002)
Fatty acid deficiencies			
ARA	Lower hepatosomatic index	European sea bass	Geay *et al.* (2011)
EPA	Growth and immune deficiencies Decreased antioxidant capacity Decreased fat retention and anaemia Higher incidence of hepatic granulomas	European sea bass Gilthead sea bream Meagre	Montero *et al.* (1998, 2008); Saera-Vila *et al.* (2009); Geay *et al.* (2011); Ballester-Lozano *et al.* (2015); Carvalho *et al.* (2019)
DHA	Growth and immune deficiencies Decreased fat retention and anaemia Decreased swimming activity and antioxidant capacity Higher incidence of hepatic granulomas	European sea bass Gilthead sea bream Meagre	Montero *et al.* (1998, 2008); Saera-Vila *et al.* (2009); Geay *et al.* (2011); Ballester-Lozano *et al.* (2015); Carvalho *et al.* (2019)
Vitamin deficiencies			
Vitamin B	Reduced haematocrit Atrophic pancreas	Gilthead sea bream	Morris *et al.* (1995)
Vitamin E	Decreased complement activity	Gilthead sea bream	Montero *et al.* (1998)

or behavioural, physiological or mechanical prevention of food intake.

7.4.3 Clinical signs, gross and histopathological lesions

The most frequent clinical description associated with nutritional deficiency disorders in European sea bass is loss of weight and decrease in condition factor, and is associated with darkening of the skin, lethargy and poor growth (Echevarría *et al.*, 1997).

EFA deficiencies have been reported to produce alterations in the oxygen-carrying capacity of gilthead sea bream blood, provoking anaemia, immunosuppression (of both cellular and humoral systems) and renal injury, including renal tube degeneration and systemic glomerulonephritis (Montero *et al.*, 2004; Ballester-Lozano *et al.*, 2015). Moreover, low dietary *n*-3 long-chain PUFA content (i.e. 0.8% of diet) can increase the incidence of hepatic granulomas in meagre (Carvalho *et al.*, 2019). Reported cases of vitamin B deficiency have been associated with significant reductions in haematocrit levels in gilthead sea bream. These vitamin-deficient specimens also revealed that the pancreas became atrophic with an accumulation of pigmented granules around the organ while the normal homogeneity of the liver parenchyma was lost (Morris *et al.*, 1995).

Lipid peroxidation, specifically PUFA oxidation, is highly deleterious and results in damage of cellular biomembranes, which contain large amounts of PUFAs. This process is possibly the most extensively studied aspect of oxidative damage in biological systems. In fact, feeding diets with high PUFA content resulted in signs of increased peroxidative stress in juvenile gilthead sea bream, as evidenced by increased levels of tissue lipid peroxidation products (Tocher *et al.*, 2002).

7.4.4 Prevention and control

Careful management of fish nutrition and health are two critical factors in today's intensive aquaculture systems. Consequences of amino acid deficiencies are not catastrophic, usually slower growth and higher diet conversion ratios, rather than fish mortality. Similarly, EFA imbalances are rare in commercial fish production, excluding spawning marine fish (Hardy, 2001). Vitamin and mineral deficiencies, although the easiest problems to avoid in fish diets, are the most common category of deficiencies observed in commercial aquaculture. A constant supply of essential water-soluble vitamins is required to prevent deficiency signs in fish, since these vitamins are not stored for long periods in body tissues (Lovell, 1998).

7.5 Winter Disease

The so-called 'winter syndrome' (WS) or 'winter disease' mainly affects farmed gilthead sea bream reared at low temperatures and causes production losses in France, Italy, Portugal, Greece and Spain, especially in the northern areas.

WS was first detected in cultured gilthead sea bream in the Mediterranean area in 1991 and was reported for the western and central Mediterranean. Although gilthead sea bream is apparently the most affected, WS has also been described rarely in greater amberjack reared in the Mediterranean, during the cold season. Conversely, other species like European sea bass and meagre, that are frequently reared in the same facilities, do not seem to be affected by this disease (Gallardo *et al.*, 2003; Ibarz *et al.*, 2010).

WS-affected fish showed a pathological condition associated with long-term exposure to low ambient temperatures during the winter months, when water temperature dropped below 13°C. The causes for WS are not clear and involve several factors such as chemical stressors (water pollutants), biological stressors (parasites) (Lemly, 1996), depressed immune status, nutritional imbalance and osmoregulatory problems (Tort *et al.*, 1998b), factors that increase fish susceptibility to opportunistic pathogens and decrease their ability to fight infections (Pickering and Pottinger, 1985).

Pseudomonas anguilliseptica, *Pseudomonas* spp., *Aeromonas* spp., *Vibrio* spp. and *Staphylococcus* spp. as well as unidentified enterobacteria have been detected in different outbreaks of WS in gilthead sea bream during winter months (Berthe *et al.*, 1995; Bovo *et al.*, 1995; Doimi, 1996; Ibarz *et al.*, 2010; Birincioğlu *et al.*, 2013). Similarly, different viral agents, namely picorna-like virus, parvo-like virus and reovirus (Bovo *et al.*, 1995;

Doimi, 1996), were detected in sea bream diagnosed with WS. Parasites such as monogenean parasites and eggs of *Sanguinicolid* were reported in the gills of affected fish and *Myxidium* spp. in the intestine (Birincioğlu *et al.*, 2013).

7.5.1 Impacts on fish production

WS mainly affects 1-year-old gilthead sea bream specimens, causing mortalities ranging from 5 to 80%, usually about 5–10%, but the effects can be severe and account for up to 30–50% (Coutteau *et al.*, 2001) or as high as 80% of stock losses (Padrós *et al.*, 1998; Ibarz *et al.*, 2010). Mortalities can occur as acute peaks or constant mortalities. WS affects individual fish weighing 100–300 g by reducing their feed uptake.

7.5.2 Diagnosis

When water temperature decreases below 13°C, fish reduce ingestion and do not show external clinical signs. Fish become lethargic, present disturbed swimming movements, develop anaemia, darker skin colour, distended and haemorrhagic abdomen and corneal cloudiness, and are marked by typical 'stress bands'.

External clinical signs are:

- swimming belly-up or on the side;
- distended abdomen;
- slightly protruded red anus; and
- occasionally, small haemorrhages around the base of the fins.

Internal clinical signs are:

- distended intestine filled with clear fluid;
- ascitic fluid in abdominal cavity;
- distended gall bladder;
- pale liver with occasional bloodshot areas; and
- enlarged spleen.

7.5.3 Clinical, physiological and histopathological changes

Tissue lesions include: (i) granular degeneration and necrosis in white muscle fibres, caused by starvation that can induce changes in muscle related to catabolic reactions (Gallardo *et al.*, 2003); (ii) severe liver alteration characterized by a fatty degeneration in hepatocytes (Galeotti *et al.*, 1998; Padrós *et al.*, 1998; Tort *et al.*, 1998b; Contessi *et al.*, 2000); (iii) severe distension of the intestine, which appears distended and filled with a clear liquid and mucous casts, indicating a potentially reduced nutrient absorption; and (iv) presence of lesions in the pancreas.

Fish with WS showed increased plasma cortisol and total proteins and, on the other hand, decreased levels of haematocrit, glycaemia, plasma calcium and magnesium, a decrease in complement and lysozyme activities, and reduced circulating lymphocytes (Tort *et al.*, 1998a, Galeotti *et al.*, 1999; Gallardo *et al.*, 2003) and erythrocytes (Padrós *et al.*, 1999), suggesting a severe immunosuppression in fish during winter months, allowing infection by opportunistic pathogens.

According to Schrama *et al.* (2017), the use of winter feed for gilthead sea bream induced an upregulation of some proteins involved in the immune system and cell protection compared with a control diet, suggesting that the addition of vitamins and higher-quality lipids and proteins to the diet benefits fish health condition and can help to prevent WS.

Histopathologically, WS is characterized by a substantial increase in the number of mucous cells in the gut epithelium, while the liver of most of the affected fish exhibits fatty infiltrations (Fig. 7.2).

As for bacteriology, in different studies related to WS no bacterial growth was observed. Nevertheless, there are records of bacteria being detected, namely *P. anguilliseptica* (particularly important in brain tissue) (Doménech *et al.*, 1997), enterobacteria (Doimi, 1996), *Aeromonas* spp., *Vibrio* spp. and *Staphylococcus* spp. (Birincioğlu *et al.*, 2013).

7.5.4 Prevention and control

WS has a complex aetiology that involves several factors such as environmental (low temperature), physiological (high stress, immune suppression, metabolic function depression) and occasional opportunistic pathogens (bacteria, viruses, parasites). However, cold conditions may be the trigger factor that allows all the other factors

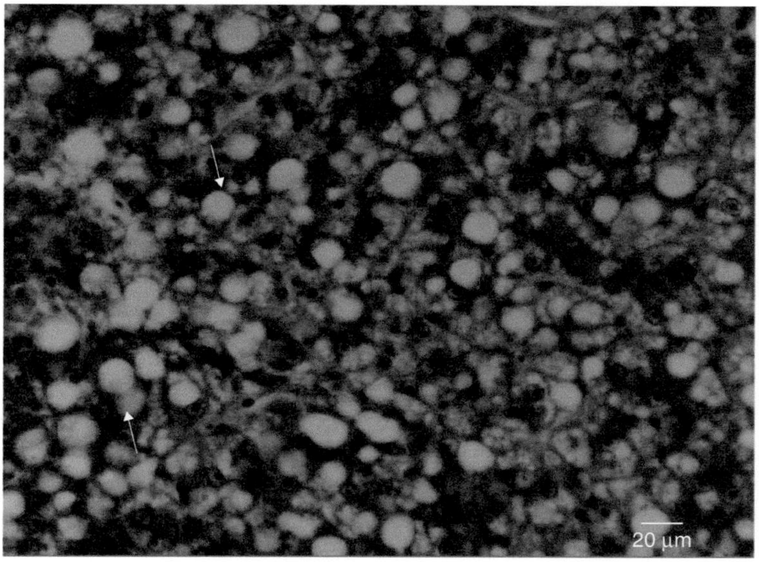

Fig. 7.2. Haematoxylin and eosin-stained liver section of gilthead sea bream (*Sparus aurata*) affected by winter disease. Notice the highly vacuolated hepatocytes (arrows).

to compromise the physiology of gilthead sea bream and start the disease. To prevent WS, some measures are suggested: (i) restrict feeding throughout the winter season and restart feeding the fish only when water temperature exceeds 13°C; (ii) use lower dietary lipid content (or appropriate diets for winter) in fish feeds before the winter months, especially at the end of summer and the approaching winter season; (iii) improve fish immunological status and provide optimal feeding; (iv) avoid handling and stressful episodes during the winter period; and (v) avoid high stocking density conditions, considered an additional stressor that may impair the physiological condition of the fish.

7.6 Stress in Warmwater Fish

Stress responses have been extensively studied in fish in the last decades, presenting the key role of the endocrine system in the process (Herrera *et al.*, 2019). The primary stress response is based on hormonal cascades which promote secondary stress responses to stressors (Schreck and Tort, 2016). The hypothalamus–pituitary–interrenal and hypothalamus–sympathetic–chromaffin axes are activated during the primary response, releasing corticosteroids and catecholamines (e.g. adrenaline, noradrenaline) into the bloodstream. Thereafter, energy metabolic pathways increase (secondary responses) and, if stress still remains, severe failures at organism level (e.g. pathologies, decreasing growth, immunosuppression) may appear as tertiary responses (Iwama *et al.*, 2006).

7.6.1 Impacts on fish production

Water quality is one of the most important contributors to fish health and stress level. Temperature, dissolved oxygen, ammonia, nitrite, nitrate, salinity, pH and CO_2 are the most common water quality parameters affecting physiological stress (Portz *et al.*, 2006). Therefore, a chronic exposure to high ammonia and low dissolved oxygen will lead to decreased growth and higher feed conversion ratio, and may increase mortality (MacIntyre *et al.*, 2008). Studies have suggested that fin damage intensifies with increasing stocking density and high temperature (25 and 29°C) in European sea bass (Person-Le Ruyet and Le Bayon, 2009). Moreover, water temperature can also change the degree of acute stress responses in meagre and European sea bass reared in net-pen sea cages (Samaras *et al.*, 2016). In gilthead sea bream cage culture, a possible impairment of the corticosteroid stress response from rearing at high densities could result in a reduction in the physiological ability to cope with social stressors from conspecifics or abiotic changes in their confined environment (Barton *et al.*, 2005).

7.6.2 Diagnosis

While classic diagnoses of the physiological and health status are provided by haematological and clinical chemical analyses, some skin mucus metabolites and hormones have been recently shown to respond to stressful events (Fernández-Alacid *et al.*, 2019), rendering this matrix a novel, low-invasive approach to measure stress in fish. Recent methodologies such as proteomics, genomics and metabolomics give better insights into the mechanisms involved in stress-related processes in fish. These facilitate the identification of stress and/or welfare indicators and help in proposing different biomarkers for fish stress (Prunet *et al.*, 2012; Rodrigues *et al.*, 2012). Studies on liver transcriptome have reported significant changes in various functions including inflammation (acute-phase proteins), immune response, gluconeogenesis and glycogenolysis, energy metabolism and protein degradation (Prunet *et al.*, 2012). Analysis of hepatic RNA from gilthead sea bream exposed to confinement led to the characterization of four major temporally defined gene expression profiles that comprise rapid metabolic readjustment followed by tissue repair and remodelling processes, and finally re-establishment of cellular homeostasis and regulation of the immune system (Calduch-Giner *et al.*, 2010). A comparative proteomics analysis was employed to discover a set of liver proteins involved in the adaptive processes that tune the physiological response of gilthead sea bream to different suboptimal rearing conditions and physical challenges, leading to 71 differentialy abundant proteins distributed among the trials (Magalhães *et al.*, 2021). The latter study reported that prolonged exposure to stress in sea bream appears to induce extensive changes in amino acid, carbohydrate and lipid metabolisms, antioxidant response, and protein folding, sorting and degradation processes.

Adverse weather conditions

Environmental changes in cage-culture systems can be rapid and immediate, and therefore can affect fish physiology since farmed fish are incapable of evading environmental changes. For instance, water salinity can decrease in estuaries or enclosed basins during rainy seasons, inducing osmotic and, consequently, metabolic problems for fish (Vargas-Chacoff *et al.*, 2009). As another example, temperature reduction during winter induces a cessation in feed intake, which alters both osmoregulatory and metabolic responses to salinity acclimatization, compromising the osmoregulatory capacity of fish (Polakof *et al.*, 2006). Moreover, low plasma cortisol levels were observed in European sea bass and meagre reared in net-pen sea cages during spring compared with fish reared in winter (Samaras *et al.*, 2016).

Suboptimal culture conditions

There is a general tendency to maximize stocking density for profit. However, this may exert adverse effects on fish health, depending on the acquired or inherited characteristics in a given population and on the capacity of the producer to maintain water quality. Inappropriate stocking densities are known to lead to poor welfare and compromise health conditions of farmed fish, as well as the profitability of the aquaculture industry (Pottinger, 2008). Moreover, several studies have addressed the effects of handling and crowding conditions on oxidative stress in fish. For instance, an increase in both cortisol and glucose in skin mucus and plasma was observed in meagre submitted to acute stress (Fernández-Alacid *et al.*, 2019). Furthermore, lower growth and higher plasma cortisol levels were reported for meagre reared at low density (3 kg/m^3) compared with those at high density (13 kg/m^3) (Millán-Cubillo *et al.*, 2016). A microarray analysis of genes expressed during the time course of stress response in gilthead sea bream after an acute confinement exposure (100 kg/m^3) highlighted a vast array of metabolic adjustments, with an increased reactive oxygen species (ROS) scavenging accompanied by a general decline of ROS production (Calduch-Giner *et al.*, 2010). The complement component C3 was downregulated in gilthead sea bream exposed to low oxygen levels (Magalhães *et al.*, 2020). Additionally, heat-shock cognate protein 70 (chaperoning) was downregulated in gilthead sea bream under different chronic stressful conditions (Alves *et al.*, 2010), which might indicate immunosuppressive effects of stressors related to rearing conditions. The risk of spreading infectious diseases and triggering disease outbreaks is exacerbated by the fact that stocks at risk may already be compromised by handling/transport stress, suboptimal water quality, higher-than-usual stocking densities and by any injuries that fish may have suffered. It is therefore imperative that appropriate biosecurity measures be taken to minimize the risk of transmitting disease and that pre-transport health checks are performed to ensure that only healthy individuals are transported (MacIntyre *et al.*, 2008).

7.6.3 Clinical signs, gross and histopathological lesions

Gross signs of a tertiary-stage stress response include reduction or cessation of growth, decrease in condition factor and increased immune impairment. An important effect of high stocking density can be haemoconcentration, affecting haematocrit, haemoglobin, erythrocyte count and total plasma proteins, an effect that has been described as a strategy for increasing the oxygen-carrying capacity of blood during periods of high energy demand (Montero *et al.*, 1999a). Several histopathological lesions and immune parameters in chronically stressed fish are compiled in Table 7.2.

7.6.4 Prevention and control

The European Food Safety Authority (EFSA) has identified some hazards and risk factors potentially affecting the welfare of European sea bass and gilthead sea bream (EFSA, 2008). An obvious approach to limiting the stress associated with a particular regime is to reduce the frequency, duration and severity of stressors; however, an interesting approach for reducing the stress experienced by fish under intensive rearing

Table 7.2. Gross signs, histopathological lesions and immune parameters indicative of a tertiary-stage stress response in warmwater fish.

Stressor	Disorders	Species	Reference
High stocking density	Fin damage Chronic erosive dermatopathy Decreased complement activity and circulating lymphocytes Increase of melano-macrophage centres in spleen	European sea bass Gilthead sea bream Sharp-snout sea bream Meagre	Tort *et al.* (1996); Montero *et al.* (1999a,b); Person-Le Ruyet and Le Bayon (2009); Rigos and Katharios (2010)
Low temperature	Mucous hyperplasia in the intestinal mucosa Decreased circulating lymphocytes and plasma lysozyme	Gilthead sea bream	Tort *et al.* (1998a, 2004)
High temperature	Fin damage	European sea bass	Person-Le Ruyet and Le Bayon (2009)
High CO_2 level	Increased mortality Gas bubble disease Nephrocalcinosis	European sea bass	Grøttum and Sigholt (1996); Athanassopoulou *et al.* (2004); Vandeputte *et al.* (2009)
Osmotic stress	Cataracts	European sea bass Gilthead sea bream	Bjerkås *et al.* (2000)

conditions is to accelerate the process of domestication through selective breeding programmes. Despite considerable research in this area, the implications for the reduced endocrine stress response in fish used in aquaculture are still not clear. While it may be advantageous to select fish exhibiting low stress responses for intensive commercial aquaculture, those being reared for stock enhancement could conceivably be at a disadvantage when released into a natural environment (Barton *et al.*, 2005).

7.7 Skeletal Abnormalities

Disorders associated with skeletal abnormalities represent one of the major constraints to the development of some marine fish for intensive production. The development of morpho-anatomical abnormalities constitutes an important problem for the fish farming industry. However, despite being widely reported in cultivated fish and several causative factors identified, a definitive solution has not been found. Special attention should be given to the dietary regimens applied during early stages since these can have a significant influence on the development of skeletal abnormalities (Djellata *et al.*, 2021). Since most of the problems related to deformities occur during the early life stages, it is essential to have an effective selection of the fry in order to cull out all the specimens that show an altered phenotype (Boglione and Costa, 2011).

7.7.1 Axial deformities

The terminology and the types of vertebral deformities range from vertebral malformations, platyspondyly caused by fusions or compressions, deformed arches/spines to vertebral column curvatures (Favaloro and Mazzola, 2000, 2003, 2006; Fraser *et al.*, 2004; Lewis *et al.*, 2004; Boglione *et al.*, 2013a,b), classified according to the affected vertebral region (Matsuoka, 1987, 1997; Boglione *et al.*, 2001). Macroscopically, most vertebral deformations can be distributed into: (i) scoliosis, an abnormal lateral curvature; (ii) lordosis, an abnormal V-shaped dorsal curvature; and (iii) kyphosis, an abnormal inverted V-shaped ventral curvature (Koumoundouros, 2010; Cardeira *et al.*, 2012, 2015). These deformities frequently co-occur in the same fish, causing lordosis–scoliosis–kyphosis (LSK) and saddleback syndromes with various severity levels (Afonso *et al.*, 2000; Sfakianakis *et al.*, 2003). Lordosis and kyphosis are very frequent in farmed pelagic fish species like gilthead sea bream (Chatain, 1994; Fernández *et al.*, 2008), European sea bass (Chatain, 1994; Koumoundouros *et al.*, 2002), white sea bream (Fig. 7.3) and meagre (Fig. 7.4), also being present in

Fig. 7.3. Juvenile specimens of white sea bream (*Diplodus sargus*). (A) Radiography of a normal fish (top) and a lordotic fish (bottom). (B) External appearance of a fish displaying saddleback syndrome (arrow). (C) Radiography of the same fish revealing internal vertebral compressions (white arrow). Radiography performed on a Carestream Kodak DSX-4000 apparatus (35 KVP, 45 s).

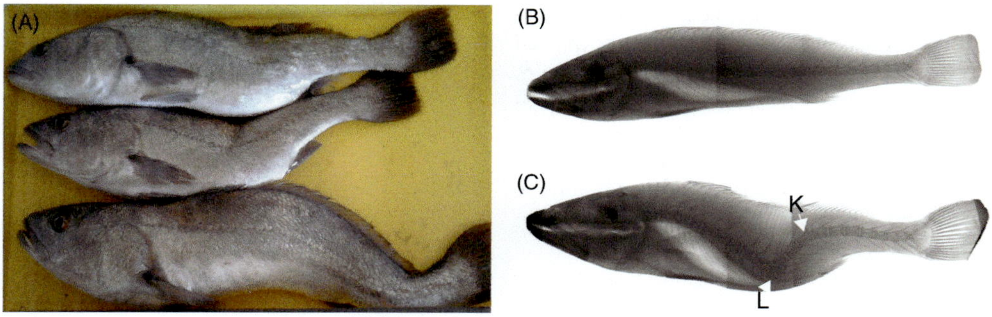

Fig. 7.4. (A) Some lordo-kyphotic specimens of meagre (*Argyrosomus regius*) captured at the end of the production cycle. (B) Composite radiography of a normal fish. (C) Composite radiography of the top fish in (A) displaying a severe deformation with a haemal lordosis (L) angle of 90° and a compensatory kyphosis (K). Radiography performed on a Carestream Kodak DSX-4000 apparatus (35 KVP, 30 s).

flatfish such as Atlantic halibut (*Hippoglossus hippoglossus*) (Lewis *et al.*, 2004) and Senegalese sole (Cardeira *et al.*, 2012; Fernández *et al.*, 2009, 2018b).

The occurrence of skeletal anomalies in cage-cultured fish has been known since 1984 (Backiel *et al.*, 1984) where a high incidence of vertebral column anomalies (up to 94% incidence) developed in large-scale cultivation of carp (*Cyprinus carpio*) in floating cages. A strong correlation with high current speed was found to be the causative factor. In European sea bass exposed to high current velocities, 77% of the individuals developed a high number of deformities with an incidence of 20% of severely deformed fish with body curvatures like lordosis and kyphosis (Divanach *et al.*, 1997).

In the case of European sea bass, Loy *et al.* (2000) assessed the effects on body morphology of different rearing conditions during larval and post-larval stages followed by rearing in cages until 15 months. Differences in shape were evident in juveniles and in adults the differences were smaller but still significant, which indicated the crucial importance of larval rearing conditions in determining adult European sea bass shape. Başaran *et al.* (2007, 2009) showed that juveniles presenting vertebral deformities at the age of transference to the rearing cages had a significantly reduced swimming speed and an affected capacity to grow and convert food.

The incidence of vertebral column deformities in European sea bass gradually decreases with growth since there is a high mortality of the affected fish during metamorphosis (Koumoundouros *et al.*, 2002). The surviving fish with deformities like lordosis or kyphosis will have a significantly altered body shape, with bent posterior abdominal region (Sfakianakis *et al.*, 2006). However, among fish affected by haemal lordosis at transfer, Fragkoulis *et al.* (2019, 2022) described the recovery from lordosis in 50% of gilthead sea bream and in 60% of European sea bass during on-growth in sea cages. These studies show that it is possible to revert vertebral curvatures in fish in cages. Nevertheless, the transference of batches of fry carrying high incidences of these type of deformities into cages may lead to production losses by feeding deformed fish; owing to their increased mortality and poor growth performance, these fish, if not removed before the end of the production cycle, cannot be commercialized. In European sea bass, fry can be moved from weaning tanks to on-growing facilities and are usually graded two or three times for size separation and evaluation of deformities until they reach a size of 1–2.5 g. Juveniles produced for sea farming are kept in flow-through tanks until they reach approximately 10 g size before transfer to sea cages (Moretti *et al.*, 1999).

7.7.2 Gill-cover and jaw anomalies

Gill-cover anomalies are common in intensively reared species such as gilthead sea bream, European sea bass and some candidate species, with incidences of up to 80%. These anomalies noticeably alter not only the external morphology but also the biological performances, with deformed fish showing lower resistance to low oxygen levels than undeformed ones (Paperna, 1978; Barahona-Fernandes, 1982; Chatain, 1994; Francescon *et al.*, 1988; Verhaegen *et al.*, 2007; Koumoundouros, 2010). Opercula deformities are unilateral (81.4%) with similar right/left frequency (fluctuant asymmetry) (Koumoundouros *et al.*, 1997).

These anomalies develop in larvae during the preflexion and flexion stages (Koumoundouros *et al.*, 1997; Koumoundouros, 2010). In gilthead sea bream, two types of opercula deformities were reported and characterized (Ortiz-Delgado *et al.*, 2014). The inwards folding of the opercular plate into the gill chamber is detectable in larvae at 25 days post-hatch (Beraldo *et al.*, 2003). However, during the on-growing stage in sea cages, fish display monolateral inside folding of the gill cover, a partial recovery of the opercula, but only if the degree of the anomaly was low (Beraldo and Canavese, 2011). Further, in gilthead sea bream juveniles, opercular-plate anomaly develops mostly in intensive conditions: siblings reared in semi-intensive conditions showed no deformed opercular plates (Boglione *et al.*, 2013b), an effect that can be due to a higher selective mortality of deformed fish. Recent studies suggested that opercular deformities may be reversible. Regeneration, however, proceeds in different ways in relation to the anatomical structure involved, and is not yet complete after 9 months of rearing (Beraldo *et al.*, 2003).

Mouth deformities can severely affect the fitness and health of affected fish. From a functional viewpoint, deformities of the mouth result in an altered capacity to properly open and close the mouth, leading to reduced feeding ability (Fraser and de Nys, 2005). Up to 35.7% of larval barramundi (*Lates calcarifer*) have been reported to exhibit jaw deformities (Fraser and de Nys, 2005), but lower-jaw deformity and pug-headedness have also been observed in other fish species (Branson and Turnbull, 2008). Another major problem affecting fish with mouth deformity results from the reduced capacity to use buccal-opercular pumping to properly ventilate their gills, causing breathing impairment, and these fish have to swim more to create proper water flow over their gills (Lijalad and Powell, 2009). Under high stocking densities, like in cages, that may cause reduced swimming capacity, these deformities can potentially compromise the welfare of these fish (Noble *et al.*, 2012).

The causes for mouth deformities have been attributed to dietary and zootechnical factors. In greater amberjack these deformities have been associated with inadequate dietary levels of DHA and EPA during early development stages (Roo *et al.*, 2019), but they may also be related to walling behaviour that is induced by positive phototaxis, leading to collisions with clear colour tank walls and causing severe deformations on the mouth (Gavaia *et al.*, 2021). This walling behaviour can be partially prevented by rearing fish in low-brightness tanks (Sawada *et al.*, 2020). An inability to express normal behaviours can help explain these injuries to the mouth of fish in culture due to mechanical damage (Noble *et al.*, 2012). Ishibashi *et al.* (2009) have described that during the juvenile and early adult stages of bluefin tuna, night-time collisions with tank walls and cage netting are common, therefore inducing damage to the mouth. It is therefore crucial to adapt culture conditions and cage sizes and colour to the specificities of the species under culture to prevent deformities that ultimately may lead to low-quality fish and mortalities.

7.7.3 Prevention and control

Grading is a particularly important exercise for intensively farmed fish that is performed two or three times for European sea bass and gilthead sea bream during the weaning and before the on-growing phase, combined with an inspection for functional swim bladder and to check for skeletal deformities.

European sea bass affected by axial malformations will show lethargic behaviour and low swimming performances (Başaran *et al.*, 2007). An obvious bending of the body is indicative of a vertebral curvature deformity that can be due to lordo-kyphosis or scoliosis, sometimes accompanied by dorsal fin alterations such as saddleback syndrome. Such fish are not fit for the market and should be removed. Small internal deformities that do not affect the external morphology can only be detected by radiographic examination or by histological preparation. This type of deformity normally does not impair growth performance and will not affect the marketability of the fish.

A proper management of water current velocities and implementing feeding plans that ensure a proper vitamin supply in the diet will help prevent vertebral and opercular anomalies. Juveniles from hatcheries with deformity-free certification for stocking cages will ensure a low incidence of deformed fish at the end of the production cycle.

7.8 New Problems Arising: Cellular and Metabolic Responses

7.8.1 Neoplasia

Offshore fish production is susceptible to different biotic and abiotic factors that can cause diseases and/or pathological conditions in fish which decrease profits. The physico-chemical parameters controlled for water quality (which usually are more stable in offshore production) and some environmental contaminants can be the cause of neoplasia in fish. Tumours or neoplasms are tissue growths of abnormal cells that proliferate uncontrollably and are generally a rare event in fish culture.

Impacts on fish production

Neoplasms are known to cause problems in commercial fish culture. A large number of tumours in fish are caused by infectious agents. Plasmacytoid leukaemia, caused by a retroviral agent, was proven to cause high mortality in net-pen-reared chinook salmon (*Oncorhynchus tshawytscha*) (Kent, 1997; Grizzle and Goodwin, 1998). Neoplasms that are not caused by a virus are normally not related to high mortalities in fish. However, any cutaneous abnormalities could affect the marketability of fish.

Diagnosis

Fish neoplasms can appear in every organ, tissue and cell type (Rocha *et al.*, 2017). Neoplasms are rarely observed in cultured fish, possibly due to the limited lifetime of the fish under production. The cause of tumours is still poorly understood, although known and suspected factors identified as oncogenic agents in fish include viruses, chemical compounds (pesticides, heavy metals, plastics ingestion), age, parasites, genetic background, ultraviolet (UV) radiation and physical trauma-related husbandry practices (Hinton, 1989; Grizzle and Goodwin, 1998; Sweet *et al.*, 2012; Rochman *et al.*, 2013). The mutagenesis and carcinogenesis mechanisms appear to be intrinsically linked and often influenced by biological characteristics and chemical or physical agents in the environment.

There are four principal categories of neoplasms (Sinderman, 1990). Besides the rare occurrences of unusual tumours, the tumour types are those affecting epithelia, mesenchyme, pigment and neural cells.

Among mesenchymal types, fibrosarcoma has been described in teleosts since the early 1940s (Lucke and Schlumberger, 1941). Heavy metal pollution was the primary causative agent in black sea bream (*Spondyliosoma cantharus*) (Marino *et al.*, 2010). In Sparidae, a case of ameloblastic fibro-odontoma of the lips was described by Paperna *et al.* (1977), a case of branchial osteochondroma was described by Nash and Porter (1985), a case of schwannoma affecting the dorsolateral part of the fish body was described by Marino *et al.* (2008) and a case of tumour in the mouth was described by Gutierrez *et al.* (1977), all in farmed gilthead sea bream. The first two cases were from the Red Sea (Israel), the third from Spain and the last one from the South Tyrrhenian Sea (Italy). In fish the neoplasia of haematopoietic origin has most frequently been observed from lymphoid tissues (Harshbarger, 1977). Ramos and Peleteiro (2003) reported the presence of a tumour inserted in the right operculum in a European sea bass breeder under experimental conditions. In addition, juvenile meagre maintained in intensive production developed a thymus sarcoma (Soares *et al.*, 2012a) (Fig. 7.5), but the cause of that remains unknown. In general, the origin of these neoplastic lesions is often unknown.

Clinical signs, gross and histopathological lesions

Neoplasm typically is apparent as an external or internal abnormal tissue growth. The diagnosis is made by histological observation of the abnormal cells. Benign tumours are normally well differentiated and circumscribed, without invading surrounding normal tissue and do not metastasize. Malignant tumours are not well differentiated,

Fig. 7.5. Thymus sarcoma in a meagre juvenile (*Argyrosomus regius*).

may grow rapidly, infiltrating normal tissues and tending to metastasize (sarcoma or carcinoma).

Juvenile meagre maintained in experimental conditions developed lateral and/or bilateral circular-shaped sarcoma within the opercular cavity. The sarcoma was dense, reddish, and its growth from the branchial arch exerted pressure on the operculum, forcing it to open. Histologically, the neoplasm exhibited marked proliferation of mesenchymal connective tissue composed of fusiform cells, which developed in a solid pattern accompanied by abundant mononuclear cell types. Multifocal areas of discrete necrosis were also observed, compatible with a sarcomatous proliferation (Soares *et al.*, 2012b).

Prevention and control

Horizontal transmission of neoplasia does not occur, except in the case of neoplasia caused by infectious viruses or parasites. Avoiding exposure to unfavourable biotic and abiotic factors that are potential causes of neoplasia in fish (see above) constitutes the main prevention measure. When neoplasms appear a general screening of potential causative factors is recommended and when they occur during the production cycle, appropriate control measures should be designed.

7.8.2 Antinutritional problems

The expansion of aquaculture has been accompanied by a rapid growth of fish feed production, which for carnivorous species has been largely based on fishmeal and fish oil. As the production of fishmeal and fish oil is stagnating, a shortage of supply on the world market occurs concomitant with the increase in price of these raw materials (Tacon and Metain, 2008). Thus, replacement of fishmeal with alternative protein sources is necessary if further growth in aquaculture production is to be pursued. Plant-derived products, mainly soybean meal (SBM), are still nowadays the most widely implemented alternative to fishmeal (Nie and Hallerman, 2021). The use of vegetable products is recognized to have several disadvantages, particularly related to their protein contents, amino acid profiles and unsaturated fatty acid imbalances, but they also include endogenous antinutritional factors. These so-called antinutritional factors (natural or synthetic substances found in feeds) may have possible harmful effects such as reducing palatability, decreasing efficient nutrient utilization for growth, altering nutrient balances and inhibiting growth, implying intestinal dysfunction, altering gut microflora, immune modulation, and inducing pancreatic hypertrophy, hypoglycaemia or liver damage (Krogdahl *et al.*, 2010).

Impacts on fish production

Fishmeal replacement with plant-protein ingredients has been accomplished at different levels (50–95%) without jeopardizing growth and feed utilization in European sea bass, gilthead sea bream, sharp-snout sea bream, black sea bream and meagre (Kaushik *et al.*, 2004; Dias *et al.*, 2009; Mérida *et al.*, 2010; Estévez *et al.*, 2011; Ngandzali *et al.*, 2011; Guardiola *et al.*, 2018). However, a high or total substitution of fish oil with rapeseed, linseed and soybean oils for several months induced decreases in growth rate or lower haematological parameters in gilthead sea bream, European sea bass and meagre (Izquierdo *et al.*, 2005; Montero *et al.*, 2008, 2010; Geay *et al.*, 2010; Emre *et al.*, 2015). In contrast, dietary fish oil can be totally replaced by camelina oil in diets for red sea bream (*Pagrus major*) with no detrimental effects as long as *n*-3 PUFAs are incorporated at the recommended level (Mzengereza *et al.*, 2021). Other metabolic consequences were also observed in fish fed high levels of vegetable oil mixtures (Benedito-Palos *et al.*, 2008). European sea bass fed a vegetable-based diet exhibited lower expression of genes related to immune response compared with specimens fed a fish-based diet (Geay *et al.*, 2011). Moreover, Sitjà-Bobadilla *et al.* (2005) observed in gilthead sea bream that a fishmeal replacement at 75% by a mixture of plant-protein sources decreased complement values. Since long-chain PUFAs are important regulators of the inflammatory response in fish, their reduction in diets containing vegetable oils as a single lipid source may affect inflammatory processes, with potential consequences on the fishes' response to pathogen infection. Most of the antinutrients due to alternative protein sources in fish feeds do not lead to mortality but could produce adverse effects such as immune impairment and decreased productivity.

Antinutritional factors due to fishmeal and fish oil replacement

Antinutrients can be broadly divided into four groups: (i) factors affecting protein utilization and digestion, such as protease inhibitors, tannins and lectins; (ii) factors affecting mineral utilization, which include phytates, gossypol pigments, oxalates and glucosinolates; (iii) antivitamins; and (iv) miscellaneous substances such as mycotoxins, mimosine, cyanogens, nitrate, alkaloids, photosensitizing agents, phyto-oestrogens and saponins (Francis *et al.*, 2001). Interactions between the effects of antinutritional factors seem to be very important; however, the picture is complicated as the gut microbiota may modify the antinutrients and therefore their interactions and biological effects (Krogdahl *et al.*, 2010). At levels likely to be present in fish diets containing commercially available plant-derived protein sources, protease inhibitors, phytates and antigenic compounds are unlikely to affect fish growth performance. Several studies in gilthead sea bream and European sea bass have shown that partial replacement of fishmeal by plant proteins is possible (Robaina *et al.*, 1995; Kaushik *et al.*, 2004). However, the use of plant-based ingredients increases the amounts of non-starch polysaccharides that need to be fermented by microorganisms in the intestine and will therefore induce modifications not only in the intestinal microbiota, but also in the expression of immune and inflammation regulatory genes (Estruch *et al.*, 2015, 2018; Porcino and Genovese, 2022). It has been shown by Piazzon *et al.* (2017) that vegetable-based diets can induce high levels of parasite infection, compromising growth performance, decreasing the intestinal microbiota diversity and altering the gut mucosa proteome, thus suggesting detrimental effects on intestinal function. Nevertheless, addition of sodium butyrate could reverse these effects. It is therefore crucial to perform further research and evaluate the effects on health of fish fed diets with replacement of fish base ingredients, to achieve a desirable substitution by more sustainable plant-based diets without compromising fish growth and health.

Clinical signs, gross and histopathological lesions

Histological studies showed an increased deposition of lipid and decreased glycogen deposits in the liver with increased levels of fishmeal replacement (Robaina *et al.*, 1995; Sitjà-Bobadilla *et al.*, 2005). The lipid accumulation in enterocytes also led to desquamation and degeneration of the epithelium in gilthead sea bream fed lipids from different sources (Caballero *et al.*, 2003). Moreover, a marked dilation of the gut submucosa (frequently with infiltration of eosinophilic granular cells) and liver steatosis were found in fish fed total fishmeal replacement (Sitjà-Bobadilla *et al.*, 2005).

Prevention and control

Some antinutritional factors are easy to eliminate by processing, while others are more difficult to eliminate. Depending on the chemical nature of the compound, it is sometimes possible to decrease the antinutritional effect using an appropriate treatment. Since organic acids are particularly heat stable, a beneficial effect can be obtained from heat-treating compounds which are proteic in nature (Guillaume and Métailler, 2001). In addition, oligosaccharides become somewhat more digestible after heat treatment, whereas the effect of thermal treatment on substances such as tannins remains uncertain (Francis *et al.*, 2001). For all antinutritional factors, fermentation or enzyme treatments directly focusing on inactivation of a specific antinutritional factor may reduce content or activity in the feedstuff (Krogdahl *et al.*, 2010). Selective breeding and genetic modification may also alter the content in antinutritional factors. It is therefore theoretically possible to create genetically engineered ingredients with low concentrations of antinutritional factors. Obtaining varieties which are free or practically free from antinutritional factors, when possible, has provided a radical solution to the problems encountered by nutritionists (Guillaume and Métailler, 2001; Krogdahl *et al.*, 2010).

7.8.3 Organic discharge

The bidirectional interaction between offshore aquaculture and the environment has been pointed out (Edwards, 2015) and this might limit the success and growth of fish farming in offshore cages. In addition, most environmental conditions are not under the fish farmer's control. However, this is not the case for eutrophication and destruction of the natural ecosystem, which partly depends on husbandry and management practices. For example, excessive feeding could lead to the release of organic compounds that can promote water eutrophication. Another example could be the increased release of pollutants associated with antifouling treatments such as copper or irgarol. Therefore, a controlled waste production strategy is necessary to maintain sustainable aquaculture growth, and this has been reviewed by Amirkolaie (2012). As feed is the major source of waste, the management of aquaculture waste should be approached from feed producers through diet formulation and through feeding strategies by fish farmers. Highly digestible diets have been introduced as a solution to reduce solid waste excretion. Further reductions in solid waste can be achieved through careful selection of feed ingredients and feed processing to improve nutrient availability. A reduction in dissolved nitrogen waste can be achieved by ensuring a balance between protein and energy that promotes fish to use non-protein sources as an energy source. Phosphorus waste can be decreased through careful ingredient selection and proper processing to improve diet digestibility. A proper feed ration and feeding method for each species should be adopted because feed waste constitutes a large part of waste production. However, organic waste output from aquaculture feeding operations cannot be eliminated since fish cannot retain all the food they consume, and part of the feed always remains uneaten. Furthermore, research is needed to reduce the use of chemicals in antifouling and of antibiotics that will be harmful to the environment surrounding the offshore cages.

7.9 Molecular Tools to Characterize Non-Infectious Disorders

Great advances have been made in the last decades, mainly in the application of large-scale genomic studies to farmed species of relevance and in the development of high-throughput next-generation sequencing (NGS) technologies at affordable cost (Yue and Wang, 2017). Although still used considerably more in genetic selection to improve broodstock (Janssen *et al.*, 2017; Robledo *et al.*, 2018; You *et al.*, 2020), to characterize and prevent diseases and/or identify their causative infectious agents (Stärk *et al.*, 2019; Natnan *et al.*, 2021; Ramos-Vivas *et al.*, 2021), and more recently in studying the interaction of microbial communities with fish hosts (Perry *et al.*, 2020; Diwan *et al.*, 2022), the application of molecular methodologies to non-infectious diseases and disorders is promising in other fields such as inbreeding management, genetic environmental adaptation, and the monitoring and control of stress-related disorders.

7.9.1 Inbreeding management

Inbreeding occurs in nature, especially within small populations. It is an issue to be considered in animal farming, including aquaculture, as it may result in increased disease susceptibility, decreased ability to respond to and cope with environmental changes, and higher incidence of skeletal deformities (Swatdipong *et al.*, 2010; Lorenzo-Felipe *et al.*, 2021).

Molecular methodologies to estimate the genetic diversity of population stocks, using allozymes, microsatellites and random amplified polymorphic DNA (RAPD) markers (Desvignes *et al.*, 2001; Nguyen *et al.*, 2005; Verspoor *et al.*, 2005), have been complemented over the years with more sophisticated techniques based on high-throughput genotyping approaches like paired-end restriction site-associated DNA sequencing (RAD-seq) (Plough, 2016).

New methodologies such as optimum contribution selection (OCS) and alternative genomic relationship matrices should be applied to improve inbreeding management in economically relevant aquaculture species (Gebregiwergis *et al.*, 2020). This is of particular importance in the implementation of genetic breeding programmes, to ensure that genetic selection is done without losing variation and to guarantee the adaptation of the population to the breeding objectives (D'Ambrosio *et al.*, 2019). In this sense, DNA microsatellite analysis could guarantee successful retrospective parental assignments in more than 99% of cases. Moreover, it allows comparisons of offspring viability with different parental origins after the common rearing and mass spawning in the culture of gilthead sea bream species (Borrell *et al.*, 2007). This tool was used to improve gilthead sea bream production, resulting in better growth and carcass traits at harvesting time (Navarro *et al.*, 2009) or fast growth of offspring (Borrell *et al.*, 2011). Nevertheless, uncontrolled or not well-designed genetic drift could also lead to undesired side effects in fish farming. This could be the case for higher deformity incidences which seem to be related to a specific genetic background (Afonso *et al.*, 2000; Castro *et al.*, 2008). In particular, broodstock with a genetic background for the presence of deformities will have a negative influence on reproduction traits and quality of progeny (Lorenzo-Felipe *et al.*, 2021). Moreover, it has been shown that European sea bass could have differential growth depending on genetic background when fed diets with partial replacement by vegetable oil (Le Boucher *et al.*, 2010). In this sense, holistic approaches are recommended in breeding programmes, considering different fish traits. In addition, most of the genetically improved strains reaching the aquaculture industry were developed through traditional selective breeding (selection, crossbreeding and hybridization) (Hulata, 2001) that is well accepted by consumers. However, emerging and more modern technologies for genetic manipulations (ploidy manipulation, sex manipulation and transgenesis) are not so well perceived by the public and could hamper consumer perception of a responsible, natural and suitable aquaculture production industry.

7.9.2 Genetic environmental adaptation

The adaptation of populations to specific conditions (particularly important in farmed fish) can lead to a decrease in genetic variation. This was reported for turbot (*Scophthalmus maximus*) between Baltic and Atlantic populations, where the latter was better adapted to higher ranges of salinity (do Prado *et al.*, 2018); for barramundi with regard to thermal tolerance (Newton *et al.*, 2010); and where single-nucleotide polymorphisms (SNPs) associated with hypoxia tolerance were recently identified (Yang *et al.*, 2020). In this context, epigenetic mechanisms that mediate phenotypic variation through genotype–environment interaction are of great importance in the sense that they may regulate the way the organisms cope with environmental changes and can significantly impact aquaculture productivity and sustainability (Eirin-Lopez and Putnam, 2019). The most important epigenetics findings in an aquaculture context are related to altered methylation levels and histone activity, which have a direct impact on masculinization of females, thermal and salinity tolerance, and disease resistance in several fish species. In addition, nutritional epigenetic studies reference the potential impact of early nutrition programming in the performance of broodstock and their progeny (Eirin-Lopez and Putnam, 2019).

As the pressure for a more sustainable aquaculture industry grows, the need to decrease the dependency on fishmeal and fish oils rises, increasing the need for sustainable alternatives for fish diets. This may have an impact on digestion capacity and nutrient absorption and, consequently, on fish growth and health, particularly in carnivorous species. The ability to adapt to a plant-based diet was demonstrated for rainbow trout after a single generation of selection (Eirin-Lopez and Putnam, 2019) and a study using grass carp (*Ctenopharyngodon idella*) cells showed a higher expression of the sweet taste receptor gene (*t1r2s*) in response to plant-specific fructose, suggesting that this gene might play a role in the food habit transition from carnivory to herbivory in grass carp (Eirin-Lopez and Putnam, 2019).

Information on these environmental adaptations may be particularly relevant for producers when implementing breeding programmes or to find broodstock at and for different geographical locations.

7.9.3 Molecular markers for stress monitoring

Stress in farmed fish is a response to physical or physiological changes that occur in their environment and can be due to climate change, transportation, density, handling, variations in physicochemical parameters and other management procedures. Upon a trigger, several biomarkers in stress response mechanisms are activated. Advances in -omic tools make possible the early identification of these biomarkers, which can be used to anticipate and prevent stressful situations in aquaculture, mainly due to temperature, low oxygen levels and/or high ammonia content in the water.

Heat-shock proteins (HSPs) are markers for thermal stress (with HSP70 being the most suitable molecular biomarker of thermal stress) and their expression was found to be upregulated in several marine organisms (Eissa and Wang, 2016; Liu *et al.*, 2018; Ju-Ngam *et al.*, 2021; Zarantoniello *et al.*, 2021; Ma *et al.*, 2022). HSPs are not exclusive for thermal stress and were found in other stressful situations, such as confinement, infection and osmoregulatory stress (Dini *et al.*, 2006; Şükrü, 2019; Song *et al.*, 2022). Therefore, although it might be a good biomarker of fish stress, its evaluation is not specific to stress type. Water quality factors like temperature, salinity and ammonia negatively influence expression of the *insulin growth factor 1* gene (*igf1*) through the interaction with other genes, such as expression of the *myostatin* gene (*mstn*), a member of the *transforming growth factor* β (TGF-β) family. In gilthead sea bream, *mstn* gene expression was altered when the fish were exposed to thermal stress, with a significant increase in the first 30 min after stress followed by a significant decrease after 60 min. This indicates the *myostatin* gene is a candidate gene to evaluate acute response to thermal stress (Zarantoniello *et al.*, 2021).

Hypoxia is a common stress in aquaculture produced species, especially when the farming occurs in intensive conditions. A transcriptomic analysis of the oriental river prawn (*Macrobrachium nipponense*) revealed that genes involved in signalling pathways associated with stress response and adaptation to extreme environments were significantly upregulated, with 18 genes involved in hypoxia regulation being differentially expressed (Sun *et al.*, 2015). Studies on cobia (*Rachycentron canadum*) under hypoxic stress revealed that *hypoxia inducible factor-1a* (*hif-1a*) and *vascular endothelial growth factor* (*vegf*) gene expression were upregulated in gills, liver and muscle, indicating that these genes play a critical role in response to hypoxia (Huang *et al.*, 2022).

Continuous exposure to high levels of ammonia can cause oxidative damage to fish tissues and organs. However, the mechanism(s) remain to be clarified and the associated genes/proteins remain largely unknown. A transcriptomic analysis in Pacific white shrimp (*Litopenaeus vannamei*) evidenced 94 differentially expressed genes related to immune function, apoptosis, growth, moulting and osmoregulation upon acute ammonia stress (Lu *et al.*, 2016). Moreover, transcriptomic analysis in the gills and liver of large-scale loach (*Paramisgurnus dabryanus*) exposed to high ammonia concentrations revealed that amino acid metabolism pathways were highly enriched in liver and ammonia transporters were upregulated in gills at 48 h after environmental ammonia exposure. These results suggested that this species responds to high levels of ammonia through the enhancement of amino acid metabolism and by increasing the expression of Rhesus glycoproteins and aquaporins (Zhang *et al.*, 2021).

The implementation of biomarker panels, adapted to each stressor, may help to clarify the stress response of farmed fish while enabling a better understanding of the mechanism of action. In the future, diagnostic markers of animal physiology should also be incorporated towards improvement of the welfare of farmed animals.

7.10 Conclusions

A diverse set of factors has an impact on fish physiology that might be translated into different non-infectious disorders, reducing fish growth performance and resistance to stress and/or increasing susceptibility to diseases. How variations in some environmental factors (e.g. water temperature, dissolved oxygen, ammonia concentration and algal blooms) and how anthropogenic activities modifying environmental conditions (e.g. presence and concentration of several pollutants, underwater noise, biofouling process and antifouling treatments) and rearing practices affect fish physiology have been reviewed in this chapter. The impact of these factors on non-infectious disorders seems to be specific for the species reared and the location of sea cages. Although the potential effect of climate change on these environmental factors has been evidenced, the potential synergistic effects of new and extreme weather conditions interacting with these factors are still largely unknown. Considering that climate change is different geographically, a massive effort is urged to predict and adapt fish-rearing systems to mitigate and/or avoid the effects of climate change for the species of interest. In this sense, modelling how fish physiology and rearing operations in cage systems might be impacted by climate change depending on the fish species and the geographical location using holistic approaches (including -omic technologies) might provide new insights on which monitoring procedures should be developed and which strategies need to be implemented at industrial levels. For instance, predicting how climate change affects biofouling and reared species allows more effective control and may help to schedule operations for changing nets during the year and/or the search for more suitable/resistant materials. Spatial and temporal monitoring of oxygen levels and water temperature might allow the most suitable locations for siting of sea cages to be identified and specific locations with increased variability of temperature/oxygen under a climate change scenario to be avoided. Also, adaptation to changes in environmental temperature may increase some nutritional requirements, leading to problems related to nutrient imbalances in the long term. This fact could be aggravated in modern dietary formulations with the presence of antinutritional actors. Furthermore, not only is the identification of biomarkers for specific stressful conditions required, but also the implementation of new feasible methodologies for less-invasive (blood plasma, skin mucus and/or water) sampling might boost their implementation at farm level.

7.11 Acknowledgements

This work was partially funded by the MET2VI project (Ref. RTI2018-099029-A-I00) funded by the European Regional Development Fund (ERDF) – Ministerio de Ciencia, Innovación y Universidades (MICIU) – Agencia Estatal de Investigación of the Spanish Government, European Union's Horizon 2020 research and innovation programme under the Marie Skłodowska-Curie grant agreement No. 766347-BIOMEDAQU and by Portuguese national funds from the Foundation for Science and Technology (FCT) through projects UIDB/04423/2020, UIDB/04326/2020, UID/00350/2020, UIDP/04423/2020, UIDP/04326/2020 and LA/P/0101/2020; and by Fundo Europeu dos Assuntos Marítimos e das Pescas (FEAMP) through the project SAÚDE&AQUA (MAR-02.05.01-FEAMP-0009). I.F. acknowledges the funding from the MICIU and the European Social Fund, 'The European Social Fund invests in your future', through the Ramón y Cajal (Ref. RYC2018-025337-I) contract from the Plan Estatal de Investigación Científica y Técnica e Innovación 2017–2020.

References

Afonso, J.M., Montero, D., Robaina, L., Astorga, N., Izquierdo, M.S. and Gines, R. (2000) Association of a lordosis–scoliosis–kyphosis deformity in gilthead seabream (*Sparus aurata*) with family structure. *Fish Physiology and Biochemistry* 22, 159–163.

Alves, R.N., Cordeiro, O., Silva, T.S., Richard, N., de Vareilles, M., *et al.* (2010) Metabolic molecular indicators of chronic stress in gilthead seabream (*Sparus aurata*) using comparative proteomics. *Aquaculture* 299, 57–66.

Amirkolaie, A.K. (2012) Reduction in the environmental impact of waste discharged by fish farms through feed and feeding. *Reviews in Aquaculture* 3, 19–26.

Athanassopoulou, F., Billinis, C. and Prapas, T. (2004) Important disease conditions of newly cultured species in intensive freshwater farms in Greece: first incidence of nodavirus infection in *Acipenser* sp. *Diseases of Aquatic Organisms* 40, 247–252.

Austin, B. (2007) The involvement of pollution with fish health. In: Mothersill, C., Mosse, I. and Seymour, C. (eds) *Multiple Stressors: A Challenge for the Future*. Springer, Dordrecht, The Netherlands, pp. 13–30.

Backiel, T., Kokurewicz, B. and Ogorzalek. A. (1984) High incidence of skeletal anomalies in carp, *Cyprinus carpio*, reared in cages in flowing water. *Aquaculture* 43, 369–380.

Bado-Nilles, A., Quentel, C., Mazurais, D., Zambonino-Infante, J.L., Auffret, M., *et al.* (2011) *In vivo* effects of the soluble fraction of light cycle oil on immune functions in the European sea bass, *Dicentrarchus labrax* (Linné). *Ecotoxicology and Environmental Safety* 74, 1896–1904.

Ballester-Lozano, G., Benedito-Palos, L., Estensoro, I., Sitjà-Bobadilla, A., Kaushik, S. and Pérez-Sánchez, J. (2015) Comprehensive biometric, biochemical and histopathological assessment of nutrient deficiencies in gilthead sea bream fed semi-purified diets. *British Journal of Nutrition* 114, 713–726.

Bannister, J., Sievers, M., Bush, F. and Bloecher, N. (2019) Biofouling in marine aquaculture: a review of recent research and developments. *Biofouling* 35, 631–648.

Barahona-Fernandes, M.H. (1982) Body deformation in hatchery reared European sea bass *Dicentrarchus labrax* (L). Types, prevalence and effect on fish survival. *Journal of Fish Biology* 21, 239–249.

Barton, B.A., Ribas, L., Acerete, L. and Tort, L. (2005) Effects of chronic confinement on physiological responses of juvenile gilthead sea bream, *Sparus aurata* L., to acute handling. *Aquaculture Research* 36, 172–179.

Başaran, F., Özbilgin, H. and Özbilgin, Y.D. (2007) Effect of lordosis on the swimming performance of juvenile sea bass (*Dicentrarchus labrax* L.). *Aquaculture Research* 38, 870–876.

Başaran, F., Özbilgin, H., Doğanyilmaz Özbilgin, Y., Paruğ, Ş.Ş. and Özden, O. (2009) The effect of lordosis severity on juvenile sea bass (*Dicentrarchus labrax* L., 1758) swimming performance. *Turkish Journal of Zoology* 33, 413–419.

Benedito-Palos, L., Navarro, J.C., Sitjà-Bobadilla, A., Bell, J.G., Kaushik, S. and Perez-Sanchez, J. (2008) High levels of vegetable oils in plant protein-rich diets fed to gilthead sea bream (*Sparus aurata* L.): growth performance, muscle fatty acid profiles and histological alterations of target tissues. *British Journal of Nutrition* 100, 992–1003.

Beraldo, P. and Canavese, B. (2011) Recovery of opercular anomalies in gilthead sea bream, *Sparus aurata* L.: morphological and morphometric analysis. *Journal of Fish Diseases* 34, 21–30.

Beraldo, P., Pinosa, M., Tibaldi, E. and Canavese, B. (2003) Abnormalities of the operculum in gilthead seabream (*Sparus aurata*): morphological description. *Aquaculture* 220, 89–99.

Berthe, F.C.J., Michel, C. and Bernardet, F. (1995) Identification of *Pseudomonas anguilliseptica* isolated from several fish species in France. *Diseases of Aquatic Organisms* 21, 151–155.

Bignami, S., Enrochs, I.C., Manzello, D.P., Sponagule, S. and Cowen, R.K. (2013) Ocean acidification alters the otoliths of a pantropical fish species with implications for sensory function. *Proceedings of the National Academy of Sciences USA* 110, 7366–7370.

Birincioğlu, S.S., Aydoğan, A. and Avci, H. (2013) Pathological investigations of winter-like disease in cultured sea breams (*Sparus aurata*). *Kafkas Üniversitesi Veteriner Fakültesi Dergisi* 19(2), 343–346.

Bjerkås, E., Wall, A.E. and Prapas, A. (2000) Screening of farmed sea bass (*Dicentrarchus labrax* L.) and sea bream (*Sparus aurata* L.) for cataract. *Bulletin of the European Association of Fish Pathologists* 20, 180–185.

Blanes, M.A., Serrano, R. and Lopez, F.J. (2009) Seasonal trends and tissue distribution of organochlorine pollutants in wild and farmed gilthead sea bream (*Sparus aurata*) from the western Mediterranean Sea and their relationship with environmental and biological factors. *Archives of Environmental Contamination and Toxicology* 57, 133–144.

Boglione, C. and Costa, C. (2011) Skeletal deformities and juvenile quality. In: Pavlidis, M. and Mylonas, C. (eds) *Sparidae: Biology and Aquaculture of Gilthead Sea Bream and Other Species*. Wiley-Blackwell, Oxford, UK, pp. 233–294.

Boglione, C., Gagliardi, F., Scardi, M. and Cataudella, S. (2001) Skeletal descriptors and quality assessment in larvae and post-larvae of wild-caught and hatchery-reared gilthead sea bream (*Sparus aurata* L. 1758). *Aquaculture* 192, 1–22.

Boglione, C., Gavaia, P., Koumoundouros, G., Gisbert, E., Moren, M., *et al.* (2013a) Skeletal anomalies in reared European fish larvae and juveniles. Part 1: Normal and anomalous skeletogenic processes. *Reviews in Aquaculture* 5(1), S99–S120.

Boglione, C., Gisbert, E., Gavaia, P., Witten, P.E., Moren, M., *et al.* (2013b) Skeletal anomalies in reared European larvae and juveniles. Part 2: Main typologies, occurrences and causative factors. *Reviews in Aquaculture* 5(1), S121–S167.

Borrell, Y.J., Carleos, C.E., Asturiano, J.F., Bernardo, D., Vázquez, E., et al. (2007) Use of microsatellites and a combinatorial optimization approach in the acquisition of gilthead seabream (*Sparus aurata* L.) broodstocks for hatcheries. *Aquaculture* 269, 200–210.

Borrell, Y.J., Gallego, V., García-Fernández, C., Mazzeo, I., Pérez, L., *et al.* (2011) Assessment of parental contributions to fast- and slow-growing progenies in the sea bream *Sparus aurata* L. using a new multiplex PCR. *Aquaculture* 314, 58–65.
Bovo, G., Borghesan, F., Comuzzi, M., Ceschia, G. and Giorgetti, G. (1995) 'Winter disease' in orata di allevamento: osservazionin preliminari. *Bolletino della Societa Italiana di Patologia Ittica* 17, 2–11.
Braithwaite, R.A., Carrascosa, M.C.C. and McEvoy, L.A. (2007) Biofouling of salmon cage netting and the efficacy of a typical copper-based antifoulant. *Aquaculture* 262, 219–226.
Branson, E.J. and Turnbull, T. (2008) Welfare and deformities in fish. In: Branson, E.J. (ed.) *Fish Welfare*. Blackwell, Oxford, UK, pp. 202–216.
Burt, K., Hamoutene, D., Mabrouk, G., Lang, C., Puestow, T., *et al.* (2012) Environmental conditions and occurrence of hypoxia within production cages of Atlantic salmon on the south coast of Newfoundland. *Aquaculture Research* 43, 607–620.
Buscaino, G., Filiciotto, F., Buffa, G., Bellante, A., Di Stefano, V., *et al.* (2010) Impact of an acoustic stimulus on the motility and blood parameters of European sea bass (*Dicentrarchus labrax* L.) and gilthead sea bream (*Sparus aurata* L.). *Marine Environmental Research* 69, 136–142.
Caballero, M.J., Izquierdo, M.S., Kjorsvik, E., Montero, D., Socorro, J., *et al.* (2003) Morphological aspects of intestinal cells from gilthead sea bream (*Sparus aurata*) fed diets containing different lipid sources. *Aquaculture* 225, 325–340.
Calduch-Giner, J.A., Davey, G., Saera-Vila, A., Houeix, B., Talbot, A., *et al.* (2010) Use of microarray technology to assess the time course of liver stress response after confinement exposure in gilthead sea bream (*Sparus aurata* L.). *BMC Genomics* 11, 193.
Carbonara, P., Alfonso, S., Zupa, W., Manfrin, A., Fiocchi, E., *et al.* (2019) Behavioral and physiological responses to stocking density in sea bream (*Sparus aurata*): do coping styles matter? *Physiology & Behavior* 212, 112698.
Cardeira, J., Bensimon-Brito, A., Pousão-Ferreira, P., Cancela, M.L. and Gavaia, P.J. (2012) Lordotic-kyphotic vertebrae develop ectopic cartilage-like tissue in Senegalese sole (*Solea senegalensis*). *Journal of Applied Ichthyology* 28, 460–463.
Cardeira, J., Mendes, A.C., Pousão-Ferreira, P., Cancela, M.L. and Gavaia, P.J. (2015) Micro-anatomical characterization of vertebral curvatures in Senegalese sole *Solea senegalensis*. *Journal of Fish Biology* 86(6), 1796–1810.
Carl, C., Guenther, J. and Sunde, L.M. (2011) Larval release and attachment modes of the hydroid *Ectopleura larynx* on aquaculture nets in Norway. *Aquaculture Research* 42, 1056–1060.
Carvalho, M., Castro, P., Montero, D., Peres, H., Acosta, F., *et al.* (2019) Essential fatty acid deficiency increases hepatic non-infectious granulomatosis incidence in meagre (*Argyrosomus regius*, Asso 1801) fingerlings. *Aquaculture* 505, 393–404.
Cascarano, M.C., Stavrakidis-Zachou, O., Mladineo, I., Thompson, K.D., Papandroulakis, N. and Katharios, P. (2021) Mediterranean aquaculture in a changing climate: temperature effects on pathogens and diseases of three farmed fish species. *Pathogens* 10, 1205.
Castro, J., Pino-Querido, A., Hermida, M., Chavarrias, D., Romero, R., *et al.* (2008) Heritability of skeleton abnormalities (lordosis, lack of operculum) in gilthead seabream (*Sparus aurata*) supported by microsatellite family data. *Aquaculture* 279, 18–22.
Castro Sánchez, R., Bustos Obregón, E. and Rojas Rauco, M. (2011) Hypoxia is like an ethiological factor in vertebral column deformity of salmon (*Salmo salar*). *Aquaculture* 316, 13–19.
Chatain, B. (1994) Abnormal swimbladder development and lordosis in sea bass (*Dicentrarchus labrax*) and sea bream (*Sparus aurata*). *Aquaculture* 119, 371–379.
Contessi, B., Votpatti, D., Gusmani, L., Bovo, G., Maltese, C., *et al.* (2000) 'Winter disease' in farmed gilthead sea bream (*S. aurata*): immunological aspects. *Bolletino della Societa Italiana di Patologia Ittica* 12, 2–16.
Cotou, E., Henry, M., Zeri, C., Rigos, G., Torreblanca, A. and Catsiki, V.A. (2012) Short-term exposure of the European sea bass *Dicentrarchus labrax* to copper-based antifouling treated nets: copper bioavailability and biomarkers responses. *Chemosphere* 89, 1091–1097.
Coutteau, P., Robles, R., De Nigris, G., Cirillo, A., Verstraete, P. and Tort, L. (2001) Nutritional solutions to winter syndrome in gilthead seabream, *Sparus aurata*: verification at a land-based farm. *Aquafeed International* 2, 30–33.
D'Ambrosio, J., Phocas, F., Haffray, P., Bestin, A., Brard-Fudulea, S., *et al.* (2019) Genome-wide estimates of genetic diversity, inbreeding and effective size of experimental and commercial rainbow trout lines undergoing selective breeding. *Genetics Selection Evolution* 51(1), 26.

Desvignes, J.F., Laroche, J., Durand, D. and Bouvet, Y. (2001) Genetic variability in reared stocks of common carp (*Cyprinus carpio* L.) based on allozymes and microsatellites. *Aquaculture* 194(3–4), 291–301.

Dias, J., Conceição, L.E.C., Ribeiro, A.R., Borges, P., Valente, L.M.P. and Dinis, M.T. (2009) Practical diet with low fish-derived protein is able to sustain growth performance in gilthead seabream (*Sparus aurata*) during the grow-out phase. *Aquaculture* 293, 255–262.

Di Marco, P., Priori, A., Finoia, M.G., Petochi, T., Marino, G., *et al.* (2008) Plasma total oxidant/antioxidant status in *Dicentrarchus labrax* after exposure to experimental hypoxia, hyperoxia and hypercapnia. *Comparative Biochemistry and Physiology Part A: Molecular & Integrative Physiology* 151, S15.

Dini, L., Lanubile, R., Tarantino, P., Mandich, A. and Cataldi, E. (2006) Expression of stress proteins 70 in tilapia (*Oreochromis mossambicus*) during confinement and crowding stress. *Italian Journal of Zoology* 73(2), 117–124.

Divanach, P., Papandroulakis, N., Anastasiadis, P., Koumoundouros, G. and Kentouri, M. (1997) Effect of water currents during postlarval and nursery phase on the development of skeletal deformities in sea bass (*Dicentrarchus labrax* L.) with functional swimbladder. *Aquaculture* 156, 145–155.

Diwan, A.D., Harke, S.N., Gopalkrishna and Panche, A.N. (2022) Aquaculture industry prospective from gut microbiome of fish and shellfish: an overview. *Journal of Animal Physiology and Animal Nutrition* 106(2), 441–469.

Djellata, A., Sarih, S., Hernández-Cruz, C.M., Martínez-Rodríguez, G., Gilannejad, N. and Roo, J. (2021) The effect of different co-feeding protocols on greater amberjack (*Seriola dumerili*, Risso 1810) larvae. *Aquaculture Nutrition* 27, 1761–1776.

Doimi, M. (1996) A new winter disease in sea bream (*Sparus aurata*): a preliminary report. *Bulletin of the European Association of Fish Pathologists* 16, 17–18.

Doménech, A., Fernandez-Garayzabal, J.F., Lawson, P., García, J.A., Cutuli, M.T., *et al.* (1997) Winter disease outbreak in sea-bream (*Sparus aurata*) associated with *Pseudomonas anguilliseptica* infection. *Aquaculture* 156, 317–326.

do Prado, F.D., Vera, M., Hermida, M., Bouza, C., Pardo, B.G., *et al.* (2018) Parallel evolution and adaptation to environmental factors in a marine flatfish: Implications for fisheries and aquaculture management of the turbot (*Scophthalmus maximus*). *Evolutionary Applications* 11(8), 1322–1341.

Easterling, D.R., Horton, B., Jones, P.D., Peterson, T.C., Karl, T.R., *et al.* (1997) Maximum and minimum temperature trends for the globe. *Science* 277, 364–367.

Easterling, D.R., Meehl, G.A., Parmesan, C., Changnon, S.A., Karl, T.R. and Mearns, L.O. (2000) Climate extremes: observations, modelling, and impacts. *Science* 289, 2068–2074.

Echevarría, G., Martínez-Bebiá, M. and Zamora, S. (1997) Evolution of biometric indices and plasma metabolites during prolonged starvation in European sea bass (*Dicentrarchus labrax*, L.). *Comparative Biochemistry and Physiology Part A: Physiology* 118, 111–123.

Edwards, P. (2015) Aquaculture environment interactions: past, present and likely future trends. *Aquaculture* 447, 2–14.

EFSA (European Food Safety Authority) (2008) Scientific opinion of the Panel on Animal Health and Welfare on a request from the European Commission on animal welfare aspects of husbandry systems for farmed European seabass and gilthead seabream. *The EFSA Journal* 844, 1–21.

Egnew, N., Renukdas, N., Ramena, Y., Yadav, A.K., Kelly, A.M., *et al.* (2019) Physiological insights into largemouth bass (*Micropterus salmoides*) survival during long-term exposure to high environmental ammonia. *Aquatic Toxicology* 207, 72–82.

Eirin-Lopez, J.M. and Putnam, H.M. (2019) Marine environmental epigenetics. *Annual Review of Marine Science* 11(1), 335–368.

Eissa, N. and Wang, H. (2016) Transcriptional stress responses to environmental and husbandry stressors in aquaculture species. *Reviews in Aquaculture* 8(1), 61–88.

Emre, Y., Kurtoğlu, A., Emre, N., Güroy, B. and Güroy, D. (2015) Effect of replacing dietary fish oil with soybean oil on growth performance, fatty acid composition and haematological parameters of juvenile meagre, *Argyrosomus regius*. *Aquaculture Research* 47, 2256–2265.

Estévez, A., Treviño, L., Kotzamanis, Y., Karacostas, I., Tort, L. and Gisbert, E. (2011) Effects of different levels of plant proteins on the ongrowing of meagre (*Argyrosomus regius*) juveniles at low temperatures. *Aquaculture Nutrition* 17, e572–e582.

Estruch, G., Collado, M.C., Peñaranda, D.S., Tomás Vidal, A., Jover Cerdá, M., *et al.* (2015) Impact of fishmeal replacement in diets for gilthead sea bream (*Sparus aurata*) on the gastrointestinal microbiota determined by pyrosequencing the 16S rRNA gene. *PLoS ONE* 10, e0136389.

Estruch, G., Collado, M.C., Monge-Ortiz, R., Tomás-Vidal, A., Jover-Cerdá, M., *et al.* (2018) Long-term feeding with high plant protein based diets in gilthead seabream (*Sparus aurata*, L.) leads to changes in the inflammatory and immune related gene expression at intestinal level. *BMC Veterinary Research* 14, 302.

Faragallah, H.M., Askar, A.I., Okbah, M.A. and Moustafa, H.M. (2009) Physico-chemical characteristics of the open Mediterranean sea water far about 60 Km from Damietta harbor, Egypt. *Journal of Ecology and the Natural Environment* 1, 106–119.

Favaloro, E. and Mazzola, A. (2000) Meristic character analysis and skeletal anomalies during growth in reared sharpsnout seabream. *Aquaculture International* 8, 417–430.

Favaloro, E. and Mazzola, A. (2003) Meristic variation and skeletal anomalies of wild and reared sharpsnout seabream juveniles (*Diplodus puntazzo*, Cetti 1777) off coastal Sicily, Mediterranean Sea. *Aquaculture Research* 34, 575–579.

Favaloro, E. and Mazzola, A. (2006) Meristic character counts and incidence of skeletal anomalies in the wild *Diplodus puntazzo* (Cetti, 1777) of an area of the south-eastern Mediterranean Sea. *Fish Physiology and Biochemistry* 32, 159–166.

Fernández, I., Hontoria, F., Ortiz-Delgado, J.B., Kotzamanis, J., Estevez, A., *et al.*. (2008) Larval performance and skeletal deformities in farmed gilthead sea bream (*Sparus aurata*) fed with graded levels of vitamin A enriched rotifers (*Brachionus plicatilis*). *Aquaculture* 283, 102–115.

Fernández, I., Pimentel, M.S., Ortiz-Delgado, J.B., Hontoria, F., Sarasquete, C., *et al.* (2009) Effect of dietary vitamin A on Senegalese sole (*Solea senegalensis*) skeletogenesis and larval quality. *Aquaculture* 295, 250–265.

Fernández, I., Gavaia, P.J., Laizé, V. and Cancela, M.L. (2018a) Fish as a model to assess chemical toxicity in bone. *Aquatic Toxicology* 194, 208–226.

Fernández, I., Granadeiro, L., Darias, M., Gavaia, P.J., Andree, K.B. and Gisbert, E. (2018b) *Solea senegalensis* skeletal ossification and gene expression patterns during metamorphosis: new clues on the onset of skeletal deformities. *Aquaculture* 496, 153–165.

Fernández-Alacid, L., Sanahuja, I., Ordóñez-Grande, B., Sánchez-Nuño, S., Herrera, M. and Ibarz, A. (2019) Skin mucus metabolites and cortisol in meagre fed acute stress-attenuating diets: correlations between plasma and mucus. *Aquaculture* 499, 185–194.

Ferreira, M., Antunes, P., Costa, J., Amado, J., Gil, O., *et al.* (2008) Organochlorine bioaccumulation and biomarkers levels in culture and wild white seabream (*Diplodus sargus*). *Chemosphere* 73, 1669–1674.

Filiciotto, F., Cecchini, S., Buscaino, G., Maccarrone, V., Piccione, G. and Fazio, F. (2017) Impact of aquatic acoustic noise on oxidative status and some immune parameters in gilthead sea bream *Sparus aurata* (Linnaeus, 1758) juveniles. *Aquaculture Research* 48(4), 1895–1903.

Fitridge, I., Dempster, T., Guenther, J. and de Nys, R. (2012) The impact and control of biofouling in marine aquaculture: a review. *Biofouling* 28, 649–669.

Fournier, V., Gouillou-Coustans, M.F., Métailler, R., Vachot, C., Guedes, M.J., *et al.* (2002) Protein and arginine requirements for maintenance and nitrogen gain in four teleosts. *British Journal of Nutrition* 87, 459–469.

Fragkoulis, S., Printzi, A., Geladakis, G., Katribouzas, N. and Koumoundouros, G. (2019) Recovery of haemal lordosis in gilthead seabream (*Sparus aurata* L.). *Scientific Reports* 9, 9832.

Fragkoulis, S., Kourkouta, C., Geladakis, G., Printzi, A., Glaropoulos, A. and Koumoundouros, G. (2022) Recovery of haemal lordosis in European seabass *Dicentrarchus labrax* (Linnaeus 1758). *Aquaculture Journal* 2, 1–12.

Francescon, A., Freddi, A., Barbaro, A. and Giavenni, R. (1988) Daurade *S. aurata* L. reproduite artificiellement et daurade sauvage. Expériences paralleles en diverses conditions d'elevage. *Aquaculture* 72, 273–285.

Francis, G., Makkar, H.P.S. and Becker, K. (2001) Antinutritional factors present in plant-derived alternate fish feed ingredients and their effects in fish. *Aquaculture* 199, 197–227.

Franke, A. and Clemmesen, C. (2011) Effect of ocean acidification on early life stages of Atlantic herring (*Clupea harengus* L.). *Biogeosciences* 8, 3697–3707.

Fraser, M.R. and de Nys, R. (2005) The morphology and occurrence of jaw and operculum deformities in cultured barramundi (*Lates calcarifer*) larvae. *Aquaculture* 250, 496–503.

Fraser, M.R., Anderson, T.A. and de Nys, R. (2004) Ontogenic development of the spine and spinal deformities in larval barramundi (*Lates calcarifer*) culture. *Aquaculture* 242, 697–711.

Frommel, A.Y., Maneja, R., Lowe, D., Malzahn, A.M., Geffen, A.J., *et al.* (2012) Severe tissue damage in Atlantic cod larvae under increasing ocean acidification. *Nature Climate Change* 2, 42–46.

Galeotti, M., Contessi, B., Volpatti, L., Fusmani, L., Maltese, C., *et al.* (1999) 'Winter disease' in farmed gilthead sea bream (*Sparus aurata*): etiopathological and immunological aspects. *Presented at the 9th International Conference of the European Association of Fish Pathologists on 'Diseases of Fish and Shellfish', Rhodes, Greece, 19–24 September 1999.*

Gallardo, M.A., Sala-Rabanal, M., Ibarz, A., Padrós, F., Blasco, J., *et al.* (2003) Functional alterations associates with 'winter syndrome' in gilthead sea bream (*Sparus aurata*). *Aquaculture* 223, 15–27.

Gavaia, P.J., Barata, M., Oliveira, C., Mendes, A.C., Soares, F., *et al.* (2021) Skeletal deformities in aquaculture produced greater amberjack *Seriola dumerili*. *Presented at Aquaculture Europe 21, Funchal, Madeira, Portugal, 4–7 October 2021.*

Geay, F., Santigosa, I., Culi, E., Corporeau, C., Boudry, P., *et al.*. (2010) Regulation of FADS2 expression and activity in European sea bass (*Dicentrarchus labrax*, L.) fed a vegetable diet. *Comparative Biochemistry and Physiology Part B: Biochemistry and Molecular Biology* 156, 237–243.

Geay, F., Ferraresso, S., Zambonino-Infante, J.L., Bargelloni, L., Quentel, C., *et al.* (2011) Effects of the total replacement of fish-based diet with plant-based diet on the hepatic transcriptome of two European sea bass (*Dicentrarchus labrax*) half-sib families showing different growth rates with the plant-based diet. *BMC Genomics* 12, 522.

Gebregiwergis, G.T., Sørensen, A.C., Henryon, M. and Meuwissen, T. (2020) Controlling coancestry and thereby future inbreeding by optimum-contribution selection using alternative genomic-relationship matrices. *Frontiers in Genetics* 11, 345.

Gómez, S. (2000) Nephrocalcinosis in Mediterranean cultured white bream (*Diplodus sargus* L.). *Bulletin of the European Association of Fish Pathologists* 20, 40–42.

Grizzle, J.M. and Goodwin, A.E. (1998) Neoplasms and related lesions. In: Leatherland, J.F. and Woo, P.T.K. (eds) *Fish Diseases and Disorders*. Vol. 2. *Non-infectious Disorders*, 1st edn. CAB International, Wallingford, UK, pp. 37–104.

Grøttum, J.A. and Sigholt, T. (1996) Acute toxicity of carbon dioxide on European seabass (*Dicentrarchus labrax*): mortality and effects on plasma ions. *Comparative Biochemistry and Physiology Part A: Physiology* 115, 323–327.

Guardiola, F.A., Cuesta, A., Meseguer, J. and Esteban, M.A. (2012) Risks of using antifouling biocides in aquaculture. *International Journal of Molecular Sciences* 13, 1541–1560.

Guardiola, F.A., Barroso, C., Enes, P., Couto, A., Díaz-Rosales, P., *et al.* (2018) Humoral and mucosal immune responses in meagre (*Argyrosomus regius*) juveniles fed diets with varying inclusion levels of carob seed germ meal. *Fish & Shellfish Immunology* 79, 209–217.

Guillaume, J. and Métailler, R. (2001) Antinutritional factors. In: Guillaume, J., Kaushik, S., Bergot, P. and Metailler, R. (eds) *Nutrition and Feeding of Fish and Crustaceans*. Praxis Publishing, Chichester, UK, pp. 297–307.

Gutierrez, M., Perez Crespo, J. and Arias, A. (1977) Particulas virus-like en un tumor en boca de dorada, *Sparus aurata* L. *Investigation Pesquera* 41, 331–336.

Hardy, R.W. (2001) Nutritional deficiencies in commercial aquaculture: likelihood, onset, and identification. In: Lim, C. and Webster, C.D. (eds) *Nutrition and Fish Health*. Food Products Press, Binghamton, New York, pp. 131–147.

Harmelin-Vivien, M., Bodiguel, X., Charmasson, S., Loizeau, V., Mellon-Duval, C., *et al.* (2012) Differential biomagnification of PCB, PBDE, Hg and radiocesium in the food web of the European hake from the NW Mediterranean. *Marine Pollution Bulletin* 64, 974–983.

Harshbarger, J.C. (1977) Role of the registry of tumors in lower animals in the study of environmental carcinogenesis in aquatic animals. *Annals of the New York Academy of Sciences* 298, 280–289.

Harvell, C.D., Mitchell, C.E., Ward, J.R., Altizer, S., Dobson, A.P., *et al.* (2002) Climate warming and disease risks for terrestrial and marine biota. *Science* 296, 2158–2162.

Henrique, M.M.F., Gouillou-Coustans, M.F. and Gomes, E. (2002) Effect of dietary ascorbic acid supplementation and chronic hypoxia on sea bream growth and vitamin C status. *Journal of Fish Biology* 60, 442–452.

Herrera, M., Mancera, J.M. and Costas, B. (2019) The use of dietary additives in fish stress mitigation: comparative endocrine and physiological responses. *Frontiers in Endocrinology* 10, 447.

Hinton, D. (1989) Environmental contamination and cancer in fish *Marine Environmental Research* 28(1–4), 411–416.

Huang, J.S., Amenyogbe, E., Yang, L.T., Wang, Z., Chen, G., *et al.* (2022) Cloning and expression analysis of *hif-1α* and downstream genes during hypoxic stress in cobia (*Rachycentron canadum*). *Aquaculture International* 30, 803–824.

Hulata, G. (2001) Genetic manipulations in aquaculture: a review of stock improvement by classical and modern technologies. *Genetica* 111, 155–173.
Ibarz, A., Padrós, F., Gallardo, M., Fernández-Borràs, J., Blasco, J. and Tort, L. (2010) Low-temperature challenges to gilthead sea bream culture: review of cold-induced alterations and 'winter syndrome'. *Reviews in Fish Biology and Fisheries* 20, 539–556.
IPCC (Intergovernmental Panel on Climate Change) (2020) *Climate Change (2022): Impacts, Adaptation and Vulnerability. Contribution of Working Group II to the Sixth Assessment Report of the Intergovernmental Panel on Climate Change* (Pörtner, H.-O., Roberts, D.C., Tignor, M., Poloczanska, E.S., Mintenbeck, K., *et al.*, eds). Cambridge University Press, Cambridge, UK. Available at: https://www.ipcc.ch/report/ar6/wg2/ (accessed 9 October 2022).
Ishibashi, Y., Honryo, T., Saida, K., Hagiwara, A., Miyashita, S., *et al.* (2009) Artificial lighting prevents high night-time mortality of juvenile Pacific bluefin tuna *Thunnus orientalis* caused by poor scotopic vision. *Aquaculture* 293, 157–163.
Iwama, G.K., Afonso, L.O.B. and Vijayan, M.M. (2006) Stress in fishes. In: Evans, D.H. and Clairbone J.B. (eds) *The Physiology of Fishes*. CRC Press, Boca Raton, Florida, pp. 319–342.
Izquierdo, M.S., Montero, D., Robaina, L., Caballero, M.J., Roselund, G. and Gines, R. (2005) Alterations in fillet fatty acid profile and flesh quality in gilthead seabream (*Sparus aurata*) fed vegetable oils for a long term period. Recovery of fatty acid profiles by fish oil feeding. *Aquaculture* 250, 431–444.
Jacobs, M.N., Covaci, A. and Schepens, P. (2002) Investigation of selected persistent organic pollutants in farmed Atlantic salmon (*Salmo salar*), salmon aquaculture feed, and fish oil components of the feed. *Environmental Science and Technology* 36, 2797–2805.
Janssen, K., Chavanne, H., Berentsen, P. and Komen, H. (2017) Impact of selective breeding on European aquaculture. *Aquaculture* 472, 8–16.
Ju-Ngam, T., McMillan, N., Yoshimizu, M., Kasai, H., Wongpanya, R. and Srisapoome, P. (2021) Functional and stress response analysis of heat shock proteins 40 and 90 of giant river prawn (*Macrobrachium rosenbergii*) under temperature and pathogenic bacterial exposure stimuli. *Biomolecules* 11(7), 1034.
Kahlon, S.K., Sharma, G., Julka, J.M., Kumar, A., Sharma, S. and Stadler, F.J. (2018) Impact of heavy metals and nanoparticles on aquatic biota. *Environmental Chemistry Letters* 16, 919–946.
Karim, M.R., Sekine, M. and Ukita, M. (2003) A model of fish preference and mortality under hypoxic water in the coastal environment. *Marine Pollution Bulletin* 47, 25–29.
Karvonen, A., Rintamäki, P., Jokela, J. and Valtonen, E.T. (2010) Increasing water temperature and disease risks in aquatic systems: climate change increases the risk of some, but not all, diseases. *International Journal for Parasitology* 40, 1483–1488.
Kaushik, S.J., Covès, D., Dutto, G. and Blanc, D. (2004) Almost total replacement of fish meal by plant protein sources in the diet of a marine teleost, the European seabass, *Dicentrarchus labrax*. *Aquaculture* 230, 391–404.
Kent, M.L. (1997) Neoplastic diseases and related disorders. In: Kent, M.L. and Poppe, T.T. (eds) *Diseases of Seawater Netpen-Reared Salmonid Fishes*. Fisheries and Oceans Canada, Nanaimo, British Columbia, Canada, pp. 106–113.
Kieffer, J.D. (2000) Limits to exhaustive exercise in fish. *Comparative Biochemistry and Physiology Part A: Molecular & Integrative Physiology* 126, 161–179.
Kir, M. and Sunar, M.C. (2018) Acute toxicity of ammonia and nitrite to sea bream, *Sparus aurata* (Linnaeus, 1758), in relation to salinity. *Journal of the World Aquaculture Society* 49(3), 516–522.
Koumoundouros, G. (2010) Morpho-anatomical abnormalities in Mediterranean marine aquaculture. In: Koumoundouros, G. (ed.) *Recent Advances in Aquaculture Research*. Transworld Research Network, Kerala, India, pp. 125–148.
Koumoundouros, G., Gagliardi, F. and Divanach, P. (1997) Normal and abnormal osteological development of caudal fin in *Sparus aurata* L. fry. *Aquaculture* 149, 215–226.
Koumoundouros, G., Maingot, E., Divanach, P. and Kentouri, M. (2002) Kyphosis in reared sea bass (*Dicentrarchus labrax* L.): ontogeny and effects on mortality. *Aquaculture* 209, 49–58.
Krogdahl, Å., Penn, M., Thorsen, J., Refstie, S. and Bakke, A.M. (2010) Important antinutrients in plant feedstuffs for aquaculture: an update on recent findings regarding responses in salmonids. *Aquaculture Research* 41, 333–344.
Lafferty, K.D. (2009) The ecology of climate change and infectious diseases. *Ecology* 90, 888–900.
Lane, A. and Willemsen, P.R. (2004) Collaborative effort looks into biofouling. *Fish Farming International* 34–35.

Le Boucher, R., Vandeputte, M., Dupont-Nivet, M., Quillet, E., Mazurais, D., *et al.* (2010) A first insight into genotype × diet interactions in European sea bass (*Dicentrarchus labrax* L. 1756) in the context of plant-based diet use. *Aquaculture Research* 42, 583–592.

Lemly, A.D. (1996) Winter stress syndrome: an important consideration for hazard assessment of aquatic pollutants. *Ecotoxicology and Environmental Safety* 34(3), 223–227.

Lewis, L.M., Lall, S.P. and Witten, P.E. (2004) Morphological descriptors of the early stages of spine and vertebral development in hatchery-reared larval and juvenile Atlantic halibut (*Hippoglossus hippoglossus*). *Aquaculture* 241, 47–59.

Lijalad, M. and Powell, M.D. (2009) Effects of lower jaw deformity on swimming performance and recovery from exhaustive exercise in triploid and diploid Atlantic salmon *Salmo salar* L. *Aquaculture* 290, 145–154.

Liu, Z.M., Zhu, X., Lu, J., Cai, W., Ye, Y.P. and Lv, Y.P. (2018) Effect of high temperature stress on heat shock protein expression and antioxidant enzyme activity of two morphs of the mud crab *Scylla paramamosain*. *Comparative Biochemistry and Physiology Part A: Molecular & Integrative Physiology* 223, 10–17.

Lodeiros, C. and Garcia, N. (2004) The use of sea urchins to control fouling during suspended culture of bivalves. *Aquaculture* 231, 293–298.

Lorenzo-Felipe, Á., Shin, H.S., León-Bernabeu, S., Pérez-García, C., Zamorano, M.J., *et al.* (2021) The effect of the deformity genetic background of the breeders on the spawning quality of gilthead seabream (*Sparus aurata* L.). *Frontiers in Marine Science* 8, 656901.

Lovell, T. (1998) *Nutrition and Feeding of Fish*, 2nd edn. Kluwer Academic Publishers, Dordrecht, The Netherlands.

Loy, A., Boglione, C., Gagliardi, F., Ferrucci, L. and Cataudella, S. (2000) Geometric morphometries and internal anatomy in sea bass shape analysis (*Dicentrarchus labrax* L., Moronidae). *Aquaculture* 186, 33–44.

Lu, X., Kong, J., Luan, S., Dai, P., Meng, X., *et al.* (2016) Transcriptome analysis of the hepatopancreas in the Pacific white shrimp (*Litopenaeus vannamei*) under acute ammonia stress. *PLoS ONE* 11(10), e0164396.

Lucke, B. and Schlumberger, H.G. (1941) Transplantable epitheliomas of the lip and mouth of catfish: I. Pathology. Transplantation to anterior chamber of eye and into cornea. *Journal of Experimental Medicine* 74, 397–408.

Ma, J., Bruce, T.J., Jones, E.M., Cain, K.D. (2019) A review of fish vaccine development strategies: conventional methods and modern biotechnological approaches. *Microorganisms* 7, 569.

MacIntyre, C.M., Ellis, T., North, B.P. and Turnbull, J.F. (2008) The influences of water quality on the welfare of farmed rainbow trout: a review. In: Branson, E.J. (ed.) *Fish Welfare*. Blackwell, Oxford, UK, pp. 150–184.

Madin, J., Chong, V.C. and Basri, B. (2010) Development and short-term dynamics of macrofouling assemblages on fish-cage nettings in a tropical estuary. *Estuarine, Coastal Shelf Science* 83, 19–29.

Magalhães, C., Schrama, D., Farinha, A.P., Revets, D., Kuehn, A., *et al.* (2020) Protein changes as robust signatures of fish chronic stress: a proteomics approach to fish welfare research. *BMC Genomics* 21, 309.

Magalhães, C., Schrama, D., Nakharuthai, C., Boonanuntanasarn, S., Revets, D., *et al.* (2021) Metabolic plasticity of gilthead seabream under different stressors: analysis of the stress responsive hepatic proteome and gene expression. *Frontiers in Marine Science* 8, 676189.

Marino, F., Germanà, A. and Panebianco, A. (2008) A case of schwannoma in farmed seabream *Sparus aurata*. *Diseases of Aquatic Organisms* 82, 249–252.

Marino, F., Mazzullo, G., Manganaro, M., De Vico, G. and Macrìv, B. (2010) Metastatic fibrosarcoma in black seabream (*Spondyliosoma cantharus*). *Bulletin of the European Association of Fish Pathologists* 30(4), 150–153.

Matsuoka, M. (1987) Development of the skeletal tissues and skeletal muscles in the red sea bream. *Bulletin of Seikai Regional Fisheries Research Laboratory* 5, 1–114.

Matsuoka, M. (1997) Osteological development in the Japanese sardine, *Sardinops melanostictus*. *Ichthyological Research* 44(3), 275–295.

Mérida, S.N., Tomás-Vidal, A., Martínez-Llorens, S. and Cerdá, M.J. (2010) Sunflower meal as a partial substitute in juvenile sharpsnout sea bream (*Diplodus puntazzo*) diets: amino acid retention, gut and liver histology. *Aquaculture* 298, 275–281.

Millán-Cubillo, A.F., Martos-Sitcha, J.A., Ruiz-Jarabo, I., Cárdenas, S. and Mancera, J.M. (2016) Low stocking density negatively affects growth, metabolism and stress pathways in juvenile specimens of meagre (*Argyrosomus regius*, Asso 1801). *Aquaculture* 451, 87–92.

Mochida, K., Ito, K., Harino, H., Kakuno, A. and Fuji, K. (2006) Acute toxicity of pyrithione antifouling biocides and joint toxicity with copper to red sea bream (*Pagrus major*) and toy shrimp (*Heptacarpus futilirostris*). *Environmental Toxicology and Chemistry* 25, 3058–3064.
Montero, D., Tort, L., Izquierdo, M.S., Robaina, L. and Vergara, J.M. (1998) Depletion of serum alternative complement pathway activity in gilthead seabream caused by α-tocopherol and *n*-3 HUFA dietary deficiencies. *Fish Physiology and Biochemistry* 18, 399–407.
Montero, D., Izquierdo, M.S., Tort, L., Robaina, L. and Vergara, J.M. (1999a) High stocking density produces crowding stress altering some physiological and biochemical parameters in gilthead seabream, *Sparus aurata*, juveniles. *Fish Physiology and Biochemistry* 20, 53–60.
Montero, D., Blazer, V.S., Socorro, J., Izquierdo, M.S. and Tort, L. (1999b) Dietary and culture influences on macrophage aggregate parameters in gilthead seabream (*Sparus aurata*) juveniles. *Aquaculture* 179, 523–534.
Montero, D., Socorro, J., Tort, L., Caballero, M.J., Robaina, L.E., *et al.* (2004) Glomerulonephritis and immunosuppression associated with dietary essential fatty acid deficiency in gilthead sea bream, *Sparus aurata* L., juveniles. *Journal of Fish Diseases* 27, 297–306.
Montero, D., Grasso, V., Izquierdo, M.S., Ganga, R., Real, F., *et al.* (2008) Total substitution of fish oil by vegetable oils in gilthead sea bream (*Sparus aurata*) diets: effects on hepatic Mx expression and some immune parameters. *Fish & Shellfish Immunology* 24, 147–155.
Montero, D., Mathlouthi, F., Tort, L., Afonso, J.M., Torrecillas, S., *et al.* (2010) Replacement of dietary fish oil by vegetable oils affects humoral immunity and expression of proinflammatory cytokines genes in gilthead sea bream *Sparus aurata*. *Fish & Shellfish Immunology* 29, 1073–1081.
Moretti, A., Pedini Fernandez-Criado, M., Cittolin, G. and Guidastri, R. (1999) *Manual on Hatchery Production of Seabass and Gilthead Seabream*, Vol. 1. Food and Agriculture Organization of the United Nations, Rome.
Morris, P.C., Davies, S.J. and Lowe, D.M. (1995) Qualitative requirement for vitamin-B in diets for the gilthead seabream (*Sparus aurata* L). *Animal Science* 61, 419–426.
Munday, P.L., Crawley, N.E. and Nilsson, G.E. (2009) Interacting effects of elevated temperature and ocean acidification on the aerobic performance of coral reef fishes. *Marine Ecology Progress Series* 388, 235–242.
Muñoz, I., Martínez Bueno, M.J., Agüera, A. and Fernández-Alba, A.R. (2010) Environmental and human health risk assessment of organic micro-pollutants occurring in a Spanish marine fish farm. *Environmental Pollution* 158, 1809–1816.
Mzengereza, K., Ishikawa, M., Koshio, S., Yokoyama, S., Yukun, Z., *et al.* (2021) Effect of substituting fish oil with camelina oil on growth performance, fatty acid profile, digestibility, liver histology, and antioxidative status of red seabream (*Pagrus major*). *Animals* 11, 1990.
Nacher-Mestre, J., Serrano, R., Benedito-Palos, L., Navarro, J., Lopez, F., *et al.* (2010) Bioaccumulation of polycyclic aromatic hydrocarbons in gilthead sea bream (*Sparus aurata* L.) exposed to long term feeding trials with different experimental diets. *Archives of Environmental Contamination and Toxicology* 59, 137–146.
Nash, G. and Porter, C. (1985). Branchial osteochondroma in a gilthead sea bream, *Sparus aurata* L., cultured in the Gulf of Aqaba. *Journal of Fish Diseases* 8, 333–336.
Natnan, M.E., Mayalvanan, Y., Jazamuddin, F.M., Aizat, W.A., Low, C., *et al.* (2021) Omics strategies in current advancements of infectious fish disease management. *Biology* 10, 1086.
Navarro, A., Zamorano, M.J., Hildebrandt, S., Ginés, R., Aguilera, C. and Afonso, J.M. (2009) Estimates of heritabilities and genetic correlations for growth and carcass traits in gilthead seabream (*Sparus auratus* L.), under industrial conditions. *Aquaculture* 289, 225–230.
Newton, J.R., Smith-Keune, C. and Jerry, D.R. (2010) Thermal tolerance varies in tropical and sub-tropical populations of barramundi (*Lates calcarifer*) consistent with local adaptation. *Aquaculture* 308(Suppl. 1), S128–S132.
Ngandzali, B.O., Zhou, F., Xiong, W., Shao, Q.J. and Xu, J.Z. (2011) Effect of dietary replacement of fish meal by soybean protein concentrate on growth performance and phosphorus discharging of juvenile black sea bream, *Acanthopagrus schlegelii*. *Aquaculture Nutrition* 17, 526–535.
Nguyen, T.T.T., Burridge, C.P. and Austin, C.M. (2005) Population genetic studies on the Australian freshwater crayfish, *Cherax destructor* (Crustacea: Parastacidae) using allozyme and RAPD markers. *Aquatic Living Resources* 18(1), 55–64.
Nie, P. and Hallerman, E. (2021) Advancing the sustainability of aquaculture. *Reviews in Aquaculture* 13, 781–782.

Noble, C., Cañon Jones, H.A., Damsgård, B., Flood, M.J., Midling, K.O., *et al.* (2012) Injuries and deformities in fish: their potential impacts upon aquacultural production and welfare. *Fish Physiology and Biochemistry* 38, 61–83.
Noga, E.J. (2010) *Fish Disease: Diagnosis and Treatment*, 2nd edn. Wiley-Blackwell, Ames, Iowa.
Nowak, B.F. (2004) Assessment of health risks to southern bluefin tuna under current culture conditions. *Bulletin of the European Association of Fish Pathologists* 24, 45–51.
Nowicki, J.P., Miller, G.M. and Munday, P.L. (2012) Interactive effects of elevated temperature and CO_2 on foraging behavior of juvenile coral reef fish. *Journal of Experimental Marine Biology and Ecology* 412, 46–51.
Ortiz-Delgado, J.B., Fernández, I., Sarasquete, C. and Gisbert, E. (2014) Normal and histopathological organization of the opercular bone and vertebrae in gilthead sea bream *Sparus aurata*. *Aquatic Biology* 21, 67–84.
Padrós, F., Crespo, S., Sala, S., Sánchez, J., Blasco, J., *et al.* (1998) Winter syndrome in gilthead seabream (*Sparus aurata*). Structural and functional alterations and the influence of stress. *Presented at the International Congress on the Biology of Fish, Baltimore, Maryland, 26–30 July 1998*.
Padrós, F., Hernández, A., Rotllant, J., Puigcerver, M., Sala, R., *et al.* (1999) La enfermedad de invierno en la dorada (*Sparus aurata* L.). Características del síndrome, disfunciones observadas y metodología de análisis y prevención. *Presented at the VII Congreso Nacional de Acuicultura, Las Palmas de Gran Canaria, Spain, 19–22 May 1999*.
Padula, D., Madigan, T. and Nowak, B.F. (2012) Australian farmed yellowtail kingfish (*Seriola lalandi*) and mulloway (*Argyrosomus hololepidotus*): residues of metallic, agricultural and veterinary chemicals, dioxins and polychlorinated biphenyls. *Chemosphere* 86, 709–717.
Paperna, I. (1978) Swimbladder and skeletal deformations in hatchery bred *Sparus aurata*. *Journal of Fish Biology* 12, 109–114.
Paperna, I., Colorni, A., Gordin, H. and Kissil, G.W. (1977). Diseases of *Sparus aurata* in marine culture at Elat. *Aquaculture* 10, 195–213.
Parker-Graham, C.A., Heckman, T.I., Griffin, M.J. and Soto, E. (2023) Infectious diseases of warmwater fish in marine and brackish waters. In: Woo, P.T.K. and Subasinghe, R.P. (eds) *Climate Change on Diseases and Disorders of Finfish in Cage Culture*, 3rd edn. CAB International, Wallingford, UK, pp. 163–201.
Perry, W.B., Lindsay, E., Payne, C.J., Brodie, C. and Kazlauskaite, R. (2020) The role of the gut microbiome in sustainable teleost aquaculture. *Proceedings of the Royal Society B: Biological Sciences* 287(1926), 20200184.
Person-Le Ruyet, J. and Le Bayon, N. (2009) Effects of temperature, stocking density and farming conditions on fin damage in European sea bass (*Dicentrarchus labrax*). *Aquatic Living Resources* 22, 349–362.
Person-Le Ruyet, J., Chartois, H. and Quemener, L. (1995) Comparative acute ammonia toxicity in marine fish and plasma ammonia response. *Aquaculture* 136, 181–194.
Person-Le Ruyet, J., Lamers, A., Le Roux, A., Severe, A., Boeuf, G. and Mayer-Gostan, N. (2003) Long-term ammonia exposure of turbot: effects on plasma parameters. *Journal of Fish Biology* 62, 879–894.
Piazzon, M.C., Calduch-Giner, J.A., Fouz, B., Estensoro, I., Simó-Mirabet, P., *et al.* (2017) Under control: how a dietary additive can restore the gut microbiome and proteomic profile, and improve disease resilience in a marine teleostean fish fed vegetable diets. *Microbiome* 5, 164.
Picciulin, M., Sebastianutto, L., Codarin, A., Calcagno, G. and Ferrero, E.A. (2012) Brown meagre vocalization rate increases during repetitive boat noise exposures: a possible case of vocal compensation. *Journal of the Acoustical Society of America* 132, 3118.
Pichavant, K., Person-Le-Ruyet, J., Le Bayon, N., Severe, A., Le Roux, A. and Boeuf, G. (2001) Comparative effects of long-term hypoxia on growth, feeding and oxygen consumption in juvenile turbot and European sea bass. *Journal of Fish Biology* 59, 875–883.
Pickering, A.D. and Pottinger, T.G. (1985) Cortisol can increase the susceptibility of brown trout, *Salmo trutta* L., to disease without reducing the white blood cell count. *Journal of Fish Biology* 27, 611–619.
Plough, L.V. (2016) Genetic load in marine animals: a review. *Current Zoology* 62(6), 567–579.
Polakof, S., Arjona, F.J., Sangiao-Alvarellos, S., Martín del Río, M.P., Mancera, J.M. and Soengas, J.L. (2006) Food deprivation alters osmoregulatory and metabolic responses to salinity acclimation in gilthead sea bream *Sparus auratus*. *Journal of Comparative Physiology B* 176, 441–452.
Popper, A.N., Fay, R.R., Platt, C. and Sand, O. (2003) Sound detection mechanism and capabilities of teleost fishes. In: Collin, S.P. and Marshall, N.J. (eds) *Sensory Processing in Aquatic Environments*. Springer, New York, pp. 3–38.

Popper, A.N., Hawkins, A.D., Sand, O. and Sisneros, J.A. (2019) Examining the hearing abilities of fishes. *Journal of the Acoustical Society of America* 146(2), 948–955.
Porcino, N. and Genovese, L. (2022) Review on alternative meals for gilthead seabream, *Sparus aurata*. *Aquaculture Research* 53, 2109–2145.
Portz, D.E., Woodley, C.M. and Cech, J.J. (2006) Stress-associated impacts of short-term holding of fishes. *Reviews in Fish Biology and Fisheries* 16, 125–170.
Pottinger, T.G. (2008) The stress response in fish – mechanisms, effects and measurements. In: Branson, E.J. (ed.) *Fish Welfare*. Blackwell, Oxford, UK, pp. 32–48.
Pounds, J.A., Bustamante, M.R., Coloma, L.A., Consuegra, J.A., Fogden, M.P.L., *et al.* (2006) Widespread amphibian extinctions from epidemic disease driven by global warming. *Nature* 439, 161–167.
Prunet, P., Øverli, Ø., Douxfils, J., Bernardini, G., Kestemont, P. and Baron, D. (2012) Fish welfare and genomics. *Fish Physiology and Biochemistry* 38, 43–60.
Ramos, P. and Peleteiro, M.C. (2003) Três casos de neoplasias espontâneas em peixes. *Revista Portuguêsa de Ciências Veterinarias* 98(546), 77–80.
Ramos-Vivas, J., Superio, J., Galindo-Villegas, J. and Acosta, F. (2021) Phage therapy as a focused management strategy in aquaculture. *International Journal of Molecular Sciences* 22(19), 10436.
Rigos, G. and Katharios P. (2010) Pathological obstacles of newly-introduced fish species in Mediterranean mariculture: a review. *Reviews in Fish Biology and Fisheries* 20, 47–70.
Robaina, L., Izquierdo, M.S., Moyano, F.J., Socorro, J., Vergara, J.M., *et al.* (1995) Soybean and lupin seed meals as protein sources in diets for gilthead seabream (*Sparus aurata*): nutritional and histological implications. *Aquaculture* 130, 219–233.
Roberts, R.J. (2002) Nutritional pathology. In: Halver, J.E. and Hardy, R.W. (eds) *Fish Nutrition*. Elsevier Science, San Diego, California, pp. 453–504.
Roberts, R.J. (2012) *Fish Pathology*, 4th edn. Wiley-Blackwell, Oxford, UK.
Robledo, D., Palaiokostas, C., Bargelloni, L., Martínez, P. and Houston, R. (2018) Applications of genotyping by sequencing in aquaculture breeding and genetics. *Reviews in Aquaculture* 10(3), 670–682.
Rocha, C., Moreira-Nunes, C., Rocha, S., Silva, M. and Burbano, R. (2017) A review on occurrence of neoplasia in fish. *Acta of Fisheries and Aquatic Resources* 5(2), 19–24.
Rochman, C.M., Hoh, E., Kurobe, T. and Teh, S.J. (2013) Ingested plastic transfers hazardous chemicals to fish and induces hepatic stress. *Scientific Reports* 3, 3263.
Rodrigues, P.M., Silva, T.S., Dias, J. and Jessen, F. (2012) Proteomics in aquaculture: applications and trends. *Journal of Proteomics* 75, 4325–4345.
Roo, J., Hernández-Cruz, C.M., Mesa-Rodríguez, A., Fernández-Palacios, H. and Izquierdo, M.S. (2019) Effect of increasing *n*-3 HUFA content in enriched *Artemia* on growth, survival and skeleton anomalies occurrence of greater amberjack *Seriola dumerili* larvae. *Aquaculture* 500, 651–659.
Rosa, R., Marques, A. and Nunes, M.L. (2012) Impact of climate change in Mediterranean aquaculture. *Reviews in Aquaculture* 4, 163–177.
Ross, D. (2005) Ship sources of ambient noise. *Journal of Oceanic Engineering* 30, 257–261.
Ruiz, M.A., Betancor, M.B., Robaina, L., Montero, D., Hernández-Cruz, C.M., *et al.* (2019) Dietary combination of vitamin E, C and K affects growth, antioxidant activity, and the incidence of systemic granulomatosis in meagre (*Argyrosomus regius*). *Aquaculture* 498, 606–620.
Saera-Vila, A., Benedito-Palos, L., Sitjà-Bobadilla, A., Nácher-Mestre, J., Serrano, R., *et al.* (2009) Assessment of the health and antioxidant trade-off in gilthead sea bream (*Sparus aurata* L.) fed alternative diets with low levels of contaminants. *Aquaculture* 296, 87–95.
Samaras, A., Papandroulakis, N., Costari, M. and Pavlidis, M. (2016) Stress and metabolic indicators in a relatively high (European sea bass, *Dicentrarchus labrax*) and a low (meagre, *Argyrosomus regius*) cortisol responsive species, in different water temperatures. *Aquaculture Research* 47, 3501–3515.
Santacruz-Reyes, R.A. and Chien, Y. (2009) Efficacy of *Yucca schidigera* extract for ammonia reduction in freshwater: effectiveness analysis and empirical modeling approach. *Aquaculture* 297, 106–111.
Santulli, A., Modica, A., Messina, C., Ceffa, L., Curatolo, A., *et al.* (1999) Biochemical responses of European sea bass (*Dicentrarchus labrax* L.) to the stress induced by offshore experimental seismic prospecting in the Mediterranean Sea. *Marine Pollution Bulletin* 38, 1105–1114.
Sarà, G., Dean, J.M., D'Amato, D., Buscaino, G., Oliveri, A., *et al.* (2007) Effect of boat noise on the behaviour of bluefin tuna *Thunnus thynnus* in the Mediterranean Sea. *Marine Ecology Progress Series* 331, 243–253.
Sawada, Y., Sasaki, T., Nishio, K., Kurata, M., Honryo, T. and Agawa, Y. (2020) Positive phototaxis as the cause of jaw malformations in larval greater amberjack, *Seriola dumerili* (Risso, 1810): mitigation by rearing in tanks with low-brightness walls. *Aquaculture Research* 51(6), 2261–2274.

Scavia, D., Field, J.C., Boesch, D.F., Buddemeier, R.W., Burkett, V., *et al.* (2002) Climate change impacts on US coastal and marine ecosystems. *Estuaries* 25, 149–164.

Schnitzler, J.G., Koutrakis, E., Siebert, U., Thomi, J.P. and Das, K. (2008) Effects of persistent organic pollutants on the thyroid function of the European sea bass (*Dicentrarchus labrax*) from the Aegean Sea, is it an endocrine disruption? *Marine Pollution Bulletin* 56, 1755–1764.

Schrama, D., Richard, N., Silva, T.S., Figueiredo, F.A., Conceição, L.E.C., *et al.* (2017) Enhanced dietary formulation to mitigate winter thermal stress in gilthead seabream (*Sparus aurata*): a 2D-DIGE plasma proteome study. *Fish Physiology and Biochemistry* 43(2), 603–617.

Schreck, C.B. and Tort, L. (2016) The concept of stress in fish. In: Schreck, C.B., Tort, L., Farrell, A.P. and Brauner, C.J. (eds) *Biology of Stress in Fish*. Academic Press, London, pp. 2–34.

Sen, K., Erdogan, U.H. and Cavas, L. (2020) Prevention of biofouling on aquaculture nets with eco-friendly antifouling paint formulation Korhan. *Coloration Technology* 136, 120–129.

Serrano, R., Blanes, M. and Lopez, F.J. (2008) Biomagnification of organochlorine pollutants in farmed and wild gilthead sea bream (*Sparus aurata*) and stable isotope characterization of the trophic chains. *Science of the Total Environment* 389, 340–349.

Sfakianakis, D.G., Koumoundouros, G., Anezaki, L., Divanach, P. and Kentouri, M. (2003) Development of a saddleback-like syndrome in reared white seabream *Diplodus sargus* (Linnaeus, 1758). *Aquaculture* 217, 673–676.

Sfakianakis, D.G., Georgakopoulou, E., Papadakis, I.E., Divanach, P., Kentouri, M. and Koumoundouros, G. (2006) Environmental determinants of haemal lordosis in European sea bass, *Dicentrarchus labrax* (Linnaeus, 1758). *Aquaculture* 254, 54–64.

Shinn, A.P., Dong, H.T., Vinh, N.T., Wongwaradechkul, R. and Lio-Po, G.D. (2023) Infectious diseases of warmwater fish in fresh water. In: Woo, P.T.K. and Subasinghe, R.P. (eds) *Climate Change on Diseases and Disorders of Finfish in Cage Culture*, 3rd edn. CAB International, Wallingford, UK, pp. 202–277.

Shultz, A.D., Murchie, K.J., Griffith, C., Cooke, S.J., Danylchuk, A.J., *et al.* (2011) Impacts of dissolved oxygen on the behavior and physiology of bonefish: implications for live-release angling tournaments. *Journal of Experimental Marine Biology and Ecology* 402, 19–26.

Sinderman, C.J. (1990) Neoplastic diseases. In: *Principal Diseases of Marine Fish and Shellfish*. Academic Press, New York, pp. 173–199.

Sitjà-Bobadilla, A., Peña-Llopis, S., Gómez-Requeni, P., Médale, F., Kaushik, S. and Pérez-Sánchez, J. (2005) Effect of fish meal replacement by plant protein sources on non-specific defence mechanisms and oxidative stress in gilthead sea bream (*Sparus aurata*). *Aquaculture* 249, 387–400.

Slabbekoorn, H., Bouton, N., van Opzeeland, L., Coers, A., ten Cate, C. and Popper, A.N. (2010) A noisy spring: the impact of globally rising underwater sound levels on fish. *Trends in Ecology and Evolution* 25, 419–427.

Soares, F., Leitão, A., Moreira, M., Teixeira de Sousa, J., Almeida, A.C., *et al.* (2012a) Sarcoma in the thymus of meagre juveniles (*Argyrosomus regius*). *Bulletin of the European Association of Fish Pathologists* 31(5), 189–192.

Soares, F., Leitão, A., Moreira, M., Teixeira de Sousa, J., Almeida, A.C., *et al.* (2012b) Presence of sarcoma in the thymus of meagre juveniles (*Argyrosomus regius*, Asso) reared in an intensive system. *Diseases of Aquatic Organisms* 102, 119–127.

Song, J.A., Choi, Y.J. and Choi, C.Y. (2022) Effects of salinity changes on the osmoregulatory and stress responses in the bay scallop *Argopecten irradians*. *Fisheries Science* 88(2), 275–283.

Stärk, K.D.C., Pękala, A. and Muellner, P. (2019) Use of molecular and genomic data for disease surveillance in aquaculture: towards improved evidence for decision making. *Preventive Veterinary Medicine* 167, 190–195.

Şükrü, Ö. (2019) Expression differences of stress and immunity genes in rainbow trout (*Oncorhynchus mykiss*, Walbaum 1792) with different bacterial fish diseases. *Israeli Journal of Aquaculture – Bamidgeh* 31(2), 196–201.

Sun, S., Xuan, F., Fu, H., Zhu, J., Ge, X. and Gu, Z. (2015) Transcriptomic and histological analysis of hepatopancreas, muscle and gill tissues of oriental river prawn (*Macrobrachium nipponense*) in response to chronic hypoxia. *BMC Genomics* 16, 491.

Swatdipong, A., Primmer, C.R. and Vasemägi, A. (2010) Historical and recent genetic bottlenecks in European grayling, *Thymallus thymallus*. *Conservation Genetics* 11(1), 279–292.

Sweet, M., Kirkham, N., Bendall, M., Currey, L., Bythell, J. and Heupel, M. (2012). Evidence of melanoma in wild marine fish populations. *PLoS ONE* 7(8), e41989.

Swift, M.R., Fredriksson, D.W., Unrein, A., Fullerton, B., Patursson, O. and Baldwin, K. (2006) Drag force acting on biofouled net panels. *Aquaculture Engineering* 35, 292–299.

Sylaios, G., Kamidis, N. and Stamatis, N. (2012) Assessment of trace metals contamination in the suspended matter and sediments of a semi-enclosed Mediterranean gulf. *Soil and Sediment Contamination* 21, 673–700.

Tacon, A.G.J. and Metain, M. (2008) Global overview on the use of fish meal and fish oil in industrially compounded aquafeeds: trends and future prospects. *Aquaculture* 285, 146–158.

Thiele, C.J., Hudson, M.D., Russell, A.E., Saluveer, M. and Sidaoui-Haddad, G. (2021) Microplastics in fish and fishmeal: an emerging environmental challenge? *Scientific Reports* 11, 2045.

Tibaldi, E. and Lanari, D. (1991) Optimal dietary lysine levels for growth and protein utilisation of fingerling sea bass (*Dicentrarchus labrax* L.) fed semipurified diets. *Aquaculture* 95, 297–304.

Tibaldi, E. and Tulli, F. (1999) Dietary threonine requirement of juvenile European sea bass (*Dicentrarchus labrax*). *Aquaculture* 175, 155–166.

Tibaldi, E., Tulli, F. and Lanari, D. (1994) Arginine requirement and effect of different dietary arginine and lysine levels for fingerling sea bass (*Dicentrarchus labrax*). *Aquaculture* 127, 207–228.

Tocher, D.R., Mourente, G., Van der Eecken, A., Evjemo, J.O., Diaz, E., *et al.* (2002) Effects of dietary vitamin E on antioxidant defence mechanisms of juvenile turbot (*Scophthalmus maximus* L.), halibut (*Hippoglossus hippoglossus* L.) and sea bream (*Sparus aurata* L.). *Aquaculture Nutrition* 8, 195–207.

Tort, L., Sunyer, J.O., Gómez, E. and Molinero, A. (1996) Crowding stress induced changes in serum haemolytic and agglutinating activity in the gilthead sea bream *Sparus aurata*. *Veterinary Immunology and Immunopathology* 51, 179–188.

Tort, L., Padrós, F., Rotllant, J. and Crespo, S. (1998a) Winter syndrome in the gilthead seabream *Sparus aurata*. Immunological and histopathological features. *Fish & Shellfish Immunology* 8, 37–47.

Tort, L., Rotllant, J. and Rovira, L. (1998b) Immunological suppression in gilthead sea bream *Sparus aurata* of the north-west Mediterranean at low temperatures. *Comparative Biochemistry and Physiology Part A: Molecular & Integrative Physiology* 120, 175–179.

Tort, L., Rotllant, J., Liarte, C., Acerete, L., Hernández, A., *et al.* (2004) Effects of temperature decrease on feeding rate, immune indicators and histopathological changes of gilthead sea bream *Sparus aurata* fed with an experimental diet. *Aquaculture* 229, 55–65.

Vandeputte, M., Dupont-Nivet, M., Haffray, P., Chavanne, H., Cenadelli, S., *et al.* (2009) Response to domestication and selection for growth in the European sea bass (*Dicentrarchus labrax*) in separate and mixed tanks. *Aquaculture* 286, 20–27.

Vanlandeghem, M.M., Wahl, D.H. and Suski, C.D. (2010) Physiological responses of largemouth bass to acute temperature and oxygen stressors. *Fisheries Management and Ecology* 17, 414–425.

Vargas-Chacoff, L., Arjona, F.J., Ruiz-Jarabo, I., Páscoa, I., Gonçalves, O., *et al.* (2009) Seasonal variation in osmoregulatory and metabolic parameters in earthen pond-cultured gilthead sea bream *Sparus auratus*. *Aquaculture Research* 40, 1279–1290.

Varol, M. and RaşitSünbül, M. (2019) Environmental contaminants in fish species from a large dam reservoir and their potential risks to human health. *Ecotoxicology and Environmental Safety* 169, 507–515.

Varvarigos, P. (2007) Chronic sub-lethal copper toxicity from net antifoulants causing anemia, gill degeneration and inducing bacterial and parasitic diseases to fish in coastal marine farms in Greece. In: *Proceedings of the 13th International EAFP Conference on Fish and Shellfish Diseases, Grado, Italy, 17–21 September 2007*, p. 159.

Vatsos, I.N. and Angelidis, P. (2010) Water quality and fish diseases. *Journal of the Hellenic Veterinary Medical Society* 61, 40–48.

Vaz, C., Afonso, F., Barata, M., Ribeiro, L., Pousão-Ferreira, P. and Soares, F. (2019) Effect of copper exposure and recovery period in reared *Diplodus sargus*. *Ecotoxicology* 28(9), 1075–1084.

Verhaegen, Y., Adriaens, D., de Wolf, T., Dhert, P. and Sorgeloos, P. (2007) Deformities in larval gilthead sea bream (*Sparus aurata*): a qualitative and quantitative analysis using geometric morphometrics. *Aquaculture* 268, 156–168.

Verspoor, E., Beardmore, J.A., Consuegra, S., Garcia, C., Hindar, K., *et al.* (2005) Population structure in the Atlantic salmon: insights from 40 years of research into genetic protein variation. *Journal of Fish Biology* 67(s1) 3–54.

Yang, Z., Wang, L., Wong, S.M. and Yue, G.H. (2020) The *HIF1αn* gene and its association with hypoxia tolerance in the Asian seabass. *Gene* 731, 144341.

You, X., Shan, X. and Shi, Q. (2020) Research advances in the genomics and applications for molecular breeding of aquaculture animals. *Aquaculture* 526, 735357.

Yue, G.H. and Wang, L. (2017) Current status of genome sequencing and its applications in aquaculture. *Aquaculture* 468, 337–347.

Zarantoniello, M., Bortoletti, M., Olivotto, I., Ratti, S., Poltronieri, C., *et al.* (2021) Salinity, temperature and ammonia acute stress response in seabream (*Sparus aurata*) juveniles: a multidisciplinary study. *Animals* 11(1), 97.

Zhang, I., Shang, Z., Wang, G., You, K. and Mi, D. (2021) High concentrations of environmental ammonia induced changes in large-scale loach (*Paramisgurnus dabryanus*) immunity. *Ecology and Evolution* 11(13), 8614–8622.

8 Sporadic Emerging Infectious and Non-Infectious Diseases and Disorders

Sophie N. St-Hilaire[1]*, Giana Bastos Gomes[2], Ana Rita Pinheiro Marques[1], Siow Foong Chang[3] and Edgar Brun[4]

[1]*City University, Hong Kong SAR, People's Republic of China;*
[2]*Temasek Life Sciences Laboratory, Singapore;* [3]*National Parks Board, Singapore;*
[4]*Norwegian Veterinary Institute, Ås, Norway*

8.1 Introduction

The previous chapters in this volume have highlighted infectious and non-infectious diseases in freshwater, brackish water and marine net-pen aquaculture that have been known for some time. However, as aquaculture grows and expands into new geographic areas, novel health issues arise; and as the industry intensifies and modernizes, health issues also emerge because of new husbandry practices. This chapter reviews some recent, emerging issues in finfish culture, highlighting some environmental pollutants, infectious diseases and problems associated with the use of antimicrobials that have surfaced in the last decade and are anticipated to worsen with climate change.

ENVIRONMENTAL POLLUTANTS

Pollution of aquatic ecosystems has been worsening progressively over the last few decades (Zeitoun and Mehana, 2014). The health implications of environmental pollutants have been raised by researchers, and more recently it has become a topic of interest in the mainstream media especially for marine species (Williamson, 2018). Endocrine disruptors and microplastics are pollutants that can accumulate through the trophic food chain and are associated with numerous health issues in both fish and humans (Diamanti-Kandarakis *et al.*, 2009). This part of the chapter discusses these compounds and their potential impact on fish health.

8.2 Endocrine Disruption in Aquatic Animals

8.2.1 Introduction

Endocrine disruptors are chemicals or metals that interfere with the hormonally controlled systems of an animal. These are exogenous compounds in the environment, mainly in water and food, that can 'mimic' important hormones such as oestrogens and androgens, causing agonistic or antagonistic effects on cellular pathways. They could disrupt the production of hormones, interrupt their transport, and/or enhance or disrupt the regulation of cellular pathways (Goksøyr, 2006; Kar *et al.*, 2021).

*ssthilai@cityu.edu.hk

 Climate Change on Diseases and Disorders of Finfish in Cage Culture, 3rd Edition (eds P.T.K. Woo and R.P. Subasinghe)
DOI: 10.1079/9781800621640.0008

Some of the common endocrine-disrupting chemicals (EDCs) include pesticides such as chlorpyrifos, dichlorodiphenyltrichloroethane (DDT), dioxins, phthalates, bisphenol A, polychlorinated and polybrominated biphenyls (PCBs and PBBs), and hormone-based pharmaceuticals (diethylstilbestrol) (Diamanti-Kandarakis *et al.*, 2009). A specific category of EDCs, the per- and polyfluoroalkyl substances (PFASs), do not degrade and are problematic as they have been used for more than 50 years in the commercial and industrial manufacturing of products such as Teflon, food packaging containers, lubricants, water-resistant coatings for textiles and firefighting foams, to list a few. These chemicals have recently been associated with negative reproductive and immunological effects in humans and animals and some are banned in the USA (Rogers Worley *et al.*, 2017). Many heavy metals (e.g. zinc, copper, iron, mercury, manganese, arsenic, cadmium, chromium and lead) can also be EDCs through several pathways including binding to oestrogen receptors (Georgescu *et al.*, 2011).

EDCs are present in water, sediment, soil and air, and they can transition between gas and particulate states under different environmental conditions (Simcik *et al.*, 1998). The rapid development and modernization of agriculture, industries and urban centres have led to an increase of these pollutants in the environment, of particular importance to aquaculture. These chemicals in aquatic systems are reported globally in oceans, rivers, estuaries and lakes, but are particularly prominent in coastal regions close to urban areas or large agricultural and/or industrial sites (Landos *et al.*, 2021). The consequence of years of EDCs accumulating in aquatic environments is anomalies in animals such as fish, amphibians, reptiles, aquatic mammals and seabirds. For example, there are increasing numbers of reports of intersex fish in lakes and rivers globally (Kar *et al.*, 2021). There are also reports of abnormalities in invertebrates and insects associated with EDCs; however, fish from wild capture fisheries and to a lesser extent from aquaculture enterprises are the animals in which the impact of EDCs is commonly reported (Landos *et al.*, 2021).

8.2.2 Impacts on fish production

EDCs can have individual- and population-level effects. There are many types of EDC and their modes of action differ widely (i.e. androgenic or estrogenic), some may act synergistically or have tangential effects (Diamanti-Kandarakis *et al.*, 2009). The predominant consequence of these compounds is on cellular pathways that are hormonally regulated, particularly those pertaining to the reproductive and immunological systems (Diamanti-Kandarakis *et al.*, 2009). The effects of EDCs on fish and other animals can range from developmental issues to disrupted reproductive, osmotic and metabolic functions, and/or immune system impairments. The latter could further result in increased susceptibility to infectious diseases (Landos *et al.*, 2021; Tarnecki *et al.*, 2022).

There are multiple mechanisms by which EDCs interfere in the natural physiological processes in fish. The most common mode of action is their binding to nuclear and/or cell-membrane hormone receptors, thereby disrupting the production, transport, metabolism, secretion and/or breakdown of natural endogenous hormones (Goksøyr, 2006; La Merrill *et al.*, 2020). Some EDCs directly affect the activity of gene receptors or disrupt the receptor-regulated feedback (autoregulation) activities, resulting in disruption of the activation and/or inhibition of cellular and molecular pathways (Rotchell and Ostrander, 2003; Goksøyr, 2006; Kar *et al.*, 2021).

Laboratory studies on specific EDCs have provided knowledge on effects on individual animals, but under field conditions single exposure to a high level of a single chemical is not common. Animals are usually exposed to chronic (prolonged) low doses of complex mixtures of compounds. Chronic low-grade exposure with different combinations of EDCs is difficult to mimic under a laboratory setting, so impacts are limited to those seen in observational field studies. To complicate matters, the age at which different animals are exposed to different chemical cocktails, as well as the sex of the animal, also influence the outcome(s) (Diamanti-Kandarakis *et al.*, 2009; Kar *et al.*, 2021; Landos *et al.*, 2021). Despite these limitations, there is a significant body of literature on the potential impacts of EDCs in fish stocks.

Developmental effects

Life stage of the animal at the time of exposure to EDCs is an important factor in determining the impact on endocrine disruption (Kar *et al.*, 2021; Landos *et al.*, 2021). Embryonic to larval stages, when fish undergo sexual maturation, are generally the most vulnerable to EDC exposure as this is when the most significant structural and physiological damage occurs (Ostrach *et al.*, 2008; Chen *et al.*, 2014; Landos *et al.*, 2021).

There is a 'window of susceptibility' in aquatic animals to the toxicology of endocrine disruptors (Kar *et al.*, 2021; Landos *et al.*, 2021). EDCs cause major structural and physiological defects in multiple organ systems in early life stages (Landos *et al.*, 2021). These compounds can bioaccumulate in the yolk of eggs, due to its high lipid content, and cause toxicity to early larval stages when fish start absorbing this source of nutrients. Fish developed from eggs and larvae in rivers in the eastern USA that contained high levels of EDCs (PCBs, polybrominated diphenyl ethers (PBDEs) and pesticides) showed abnormal brain and liver development, as well as poor growth (Ostrach *et al.*, 2008; Landos *et al.*, 2021). PCBs and other chemical pollutants also have a negative effect on the egg development of green turtle (*Chelonia mydas*) in Malaysia (Van de Merwe *et al.*, 2009). Hydrocarbons (e.g. oil spills and drilling) have affected heart development in larval stages of tuna, kingfish and billfishes (Muncaster *et al.*, 2016; Brette *et al.*, 2017; Landos *et al.*, 2021) and EDC exposure during older life stages appears to manifest predominantly as immune and/or reproductive problems.

Reproductive system effects

Exposure to androgen- and oestrogen-mimicking compounds has an effect on the reproduction and sex determination of fish, as aquaculture industries have been using high-dose exposure to testosterone and oestradiol to generate same-sex fish (all-male and all-female, respectively) for generations. The effect of low-dose exposure to EDCs will depend on what stage of development the fish are exposed at, and on what type of EDC(s) is involved. One of the most commonly reported impacts of EDCs on aquaculture production is the disruption of the reproductive processes of fish.

The impact of EDCs on the reproductive organs of fish can manifest as morphological gonadal anomalies, as well as gonadal dysfunction. The type and extent of the impact depend somewhat on the specific chemicals involved, as EDCs can cause antagonistic or agonistic effects on either oestrogen or androgen pathways (Kar *et al.*, 2021). However, common pathways disrupted include the hypothalamo–hypophyseal–gonadal (HHG) axis including interference with the gonadotrophin-releasing hormone pathway (mainly responsible for reproductive processes), thyroid metabolism (responsible for regulation of metabolic processes, growth and differentiation of tissues) and glucocorticogenesis (responsible for the development of glucocorticosteriods). The interruptions to these processes can manifest as a decrease in fertility, inhibition of gametogenesis, alteration in the gonadosomatic index (a measure of an animal's sexual maturity in relation to ovary and testis development) and intersex gonads in some species.

In female teleosts, 17β-oestradiol (E2) is responsible for the production of vitellogenin (Vtg) and eggshell zona radiata (zr) proteins by the liver. The presence of xenoestrogens, exogenous compounds that mimic E2, can confuse and disrupt the fish's physiological processes, resulting in the premature and/or abnormal development of eggs in females and the production of female characters in males (Goksøyr, 2006; Letcher *et al.*, 2010). Male fish exposed to EDCs have also shown reduced sperm fertility, motility, quality and density (Carnevali *et al.*, 2018; Kar *et al.*, 2021).

Exposure to EDCs may be more problematic for species that sexually differentiate during their development stages (gonochoristic species). An example is the benthopelagic cobia (*Rachycentron canadum*), which is an emerging aquaculture fish species in tropical and subtropical regions. Intersex gonads are becoming common in farmed and wild fish (Dutney *et al.*, 2017). While environmental factors, genetics and diet can play a role in the sex differentiation process (Dutney *et al.*, 2017), water pollution with EDCs dysregulates the synthesis and action of endogenous hormones resulting in reproductive defects in fish. Dutney *et al.* (2017) showed that after periods of heavy rain (300 mm within a 24 h period), captive cobia raised in Queensland, Australia had a high percentage of intersex gonads, which is problematic for reproduction. They speculated that this was associated with the urban and agricultural runoff during these weather events.

Another example of intersexed fish was reported in wild populations of roach (*Rutilus rutilus*) from two rivers in the UK, the Nene in Northamptonshire and the Aire in Yorkshire. These rivers received treated sewage effluent and feminizing chemicals (potential EDCs) in the water (Jobling *et al.*, 2002). In addition, a higher incidence of oocyte degeneration (uncommon in healthy females) was observed in roach exposed to the effluent from these rivers compared with control females (Jobling *et al.*, 2002).

EDCs can also have a negative effect on fish that are sequential hermaphrodites. These fish change sex at a specific time (i.e. protandrous – from male to female; protogynous – from female to male), with sex determination in many of these species being based on social cues and environmental changes such as increased temperature, salinity and photoperiod (Jenssen, 2006; Good and Davidson, 2016; Terence *et al.*, 2021). Exposure to EDCs skews the ratios of males to females or vice versa depending on the EDC. This could pose a challenge for breeding programmes as it may reduce the number of progeny and also potentially negatively influence the genetic diversity of the spawn (Terence *et al.*, 2021).

EDCs can also lead to precocious sexual maturation in fish (Terence *et al.*, 2021), which could be a problem for aquaculture industries. Premature sexual maturation results in energy being diverted from growth to the reproductive organs, which reduces feed efficiency in a grow-out operation. In some species it may also increase aggression (Medeiros *et al.*, 2007).

In summary, EDCs can have several impacts on the reproductive success of fish depending on the chemical, the dose and the life stage of the fish at the time of exposure (Kar *et al.*, 2021; Delbes *et al.*, 2022). The exposure to these compounds can have population-level impacts that are devastating to broodstock programmes and pose challenges for the growth of aquaculture industries.

Immune system effects

Most publications on EDCs and their impact on fish focus on their effects on the reproductive system; however, recent research has also shown these chemicals have negative and serious effects on the immune and metabolic pathways of fish. The mechanism of EDC action on the immune system depends on the chemical (see Milla *et al.*, 2011; Kar *et al.*, 2021). An indirect impact of EDCs on the immune system is the increased susceptibility to infectious agents (e.g. viruses, bacteria, parasites) (De Wit *et al.*, 2003; Landos *et al.*, 2021).

The immune system of teleosts is regulated by hormones including oestrogens and androgens. This is evident by the presence of receptors for these hormones in fish immune organs (liver, anterior kidney and spleen) which are responsible for the expression of immune genes (Lamková *et al.*, 2007; Milla *et al.*, 2011), particularly the genes responsible for innate immune proteins, acquired antibodies and the quantity of leucocytes (Steine *et al.*, 2001; Bowden *et al.*, 2007; Lamková *et al.*, 2007; Morgan *et al.*, 2008; Milla *et al.*, 2011). The relationship between oestrogen/androgen receptors and the immune system explains the sensitivity of fish to infectious diseases during gametogenesis as a consequence of immunosuppression due to sex steroids and their related EDCs (Milla *et al.*, 2011).

PCBs can influence the stress response of animals. Laboratory tests on Arctic charr showed that fish exposure to PCBs and a stressor, such as acute handling and challenges with *Aeromonas salmonicida*, had lower plasma cortisol than control animals (Jørgensen *et al.*, 2002; Letcher *et al.*, 2010). This suggests that EDCs may reduce a fish's ability to cope with secondary stressors (Vijayan *et al.*, 2005; Hontela and Vijayan, 2008; Letcher *et al.*, 2010). The mechanism for the reduction in cortisol production is hypothesized to be a decrease in inter-renal tissue's sensitivity to adrenocorticotrophic hormone stimulation, inter-renal exhaustion, or negative feedback on the pathway due to the presence of the EDC (Vijayan *et al.*, 1997; Hontela and Vijayan, 2008; Letcher *et al.*, 2010). As fish under farming conditions are regularly exposed to 'secondary stressors' (manual handling, transport, treatments, etc.), PCBs may impair their ability to fight infections.

8.2.3 Diagnosis/detection of endocrine-disrupting chemicals

The detection and diagnosis of endocrine disruptors and their impacts in fish are challenging. These chemicals may be responsible for production issues, and also may pose a food safety issue as they persist in adipose tissue for extended periods of time. In contrast to diagnosing infectious diseases, confirming exposure to EDCs can be costly and time-consuming. In most cases the clinical signs associated with EDCs have a gradual onset and are not pathognomonic. As previously mentioned, the impact of EDCs is varied; reproductive and immune pathways are usually affected, but these systems are influenced by many other factors which may make investigations difficult. To further complicate the issue there are numerous chemical compounds that have endocrine-disrupting properties and may act synergistically or antagonistically, so confirming their presence in low levels may be difficult.

Identifying issues with the reproductive system or other target organs for EDCs in fish is likely the initial step to warrant further investigation of EDCs in an aquaculture setting. Histology can be a valuable tool to identify cellular changes in different tissues, including intersex fish and/or other developmental anomalies. Biomarkers can also be used to indicate endocrine disruption. For example, plasma Vtg, a precursor of egg yolk which should only be present in adult female fish, may be elevated in males or in young females if these animals are exposed to oestrogen-mimicking compounds (Hutchinson *et al.*, 2006). Impacts of EDCs on the immune system may be more subtle and difficult to diagnose; however, if exposure to EDCs is suspected it may be possible to test fish tissues for specific chemicals of interest.

Traditionally, the detection of EDCs in fish tissues relies on specialized chromatographic assays (i.e. gas chromatography or liquid chromatography) with or without mass spectrometry (Azzouz *et al.*, 2019). For detecting and quantifying trace metal contamination, inductively coupled plasma–optical emission spectrometry (ICP-OES) can be used (Azzouz *et al.*, 2019). These analytical techniques are time-consuming due to the different extraction and sample preparation steps, expensive to conduct because of the specialized equipment and materials, and are not widely available to the aquaculture industry (Beyer and Biziuk, 2008; Kar *et al.*, 2021; Mwanza *et al.*, 2021).

Other more rapid tests are available to address the limitations of these conventional chemical analyses. Immunoassays have recently been developed for common pesticides and endocrine disruptors (Beyer and Biziuk, 2008; Ren *et al.*, 2017; Jarque *et al.*, 2019; Kar *et al.*, 2021). The principle behind these tests is based on enzyme-labelled specific antibodies binding to specific organic compounds. There are now commercially available (CALUX assays) to detect dioxins and other compounds that bind to oestrogen and androgen receptors (Beyer and Biziuk, 2008). These tests are live cell assays where the cells have been modified to include a luciferase gene linked to a receptor, which is activated by the binding of the chemical ligand in question (Chobtang *et al.*, 2011). Chemical sensor technology is also showing promise for the rapid and sensitive detection of some chemical contaminants in the environment (e.g. dioxins and PCBs) without the need for extraction and separation steps (Beyer and Biziuk, 2008; Jarque *et al.*, 2019); however, this technology is not yet commonly used.

Another approach to assess low concentrations of EDCs is a model organism assay with biomarkers specifically linked to endocrine disruption in fish. This multiplex method is based on zebrafish embryos and can be used to detect multiple EDCs from aquatic environments, including those mimicking oestrogen, androgen and thyroid hormones (Jarque *et al.*, 2018, 2019; Kar *et al.*, 2021). A detailed list of current assays for screening and assessing endocrine disruptors in fish is available elsewhere (Beyer and Biziuk, 2008; Kar *et al.*, 2021).

8.2.4 Sources of endocrine-disrupting chemicals and modes of exposure

Sources of EDCs including unintentional releases of agricultural chemicals (e.g. spray drift, stormwater runoff) and industrial pollutants (e.g. petrochemical and mining) into water bodies is probably one of the most common ways that EDCs end up in different aquatic environments. Wastewater treatment plants, runoffs and waste dumping from urban areas are also other common activities involved in the spread of these compounds into natural waterways (Landos *et al.*, 2021). The reduction of wetlands in coastal areas and freshwater lakes has decreased the buffering capacity of aquatic ecosystems to absorb and trap pollutants, which has likely exacerbated the levels of EDCs near open-pond and net-pen aquaculture industries.

Another major source of EDCs in waterways is atmospheric pollution. Deposition of volatilized air pollutants occurs at cold temperatures, by way of rain and snow deposition (Wania and Mackay, 1993). This phenomenon contributes significantly to ground- and surface-water contamination. Atmospheric deposition of air pollution is the reason for the high levels of volatile organic compounds in the Arctic and Antarctica, as well as in high mountain lakes, where there are no known sources of industrial or agricultural pollution for hundreds or thousands of miles (George and Frear, 1966; Oehme and Ottar, 1984; Wania and Mackay, 1993).

Absorption of EDCs by fish occurs predominantly through passive and facilitated diffusion of chemicals at the gill, gastrointestinal and skin epithelium interface (McKim *et al.*, 1985; Nichols *et al.*, 1996; Kelly *et al.*, 2004). Many EDCs are lipophilic, so they are relatively easily transported across epithelial cell membranes (La Merrill *et al.*, 2020). Once across the cell membrane they can be transported via the circulatory system to different cells and act on nuclear and membrane-bound hormone receptors, as well as directly on the synthesis, transport or breakdown of hormones in the body. The mechanism of action depends on the type of EDC (La Merrill *et al.*, 2020). These compounds are stored in adipose tissue so they can accumulate in animals at different trophic levels of the food chain.

Biomagnification through the consumption of contaminated feed is another important route by which EDCs and heavy metals find their way into fish. Biomagnification happens when chemicals are transferred from lower trophic levels to higher trophic levels within a food web, promoting a higher concentration in top predators (Goksøyr, 2006; Kar *et al.*, 2021). Because EDCs persist in animals, and humans are at the top of the food chain, biomagnification raises particular concerns for public health. There are several examples of this phenomenon with EDCs in higher vertebrates. For example, fish-consuming birds like herring gulls from the Great Lakes, USA have shown embryonic deformities due to contamination with PCBs and polycyclic aromatic hydrocarbons (PAHs) (Fox, 1993; Kar *et al.*, 2021) and reproductive issues due to organohalogens (Bosveld and van den Berg, 2002; Kar *et al.*, 2021). While EDC biomagnification may not pose a significant problem for aquaculture fish in net-pen systems as they are fed artificial feeds, the use of wild-caught fish (meal and oils) in diets may pose a problem for exposure to these fat-soluble compounds.

8.2.5 Anticipated effects of climate change on endocrine disruption

As countries become more industrialized, climate change and pollution are expected to worsen. Events and environmental changes associated with climate change will further exacerbate the contamination of aquatic systems by pollutants and potentially the availability of these contaminants for uptake by aquatic species. Changes to environmental water quality parameters, such as increases in temperature and reductions in dissolved oxygen, pH and salinity, may also directly impact aquatic animal health. Increased water temperature will have its own effect on gonadal development and sexual maturation in fish, potentially further exacerbating the effects of EDCs on reproduction. Changes in environmental factors may also impart additional stress on fish, and possibly change the diffusion of EDCs across epithelial cell membranes (Noyes and Lema, 2015; DeCourten and Brander, 2017; Kar *et al.*, 2021). All these factors suggest that the impact of EDCs on freshwater and saltwater net-pen aquaculture will become more problematic over the next decade.

Environmental changes will influence the global distribution of chemical pollutants including EDCs. Changes in water chemistry will impact chemicals differently depending on their composition. Recent work evaluating the effect of temperature changes on the volatility of PCBs suggests that as water temperature increases the volatilization of these EDCs will also increase (Zhang and Huo, 2019), possibly resulting in their global redistribution through long-range atmospheric deposition to colder regions and areas of high precipitation (Wania and Mackay, 1993). As storms become more frequent and stronger there will also be an increase in the atmospheric deposition of air pollutants including EDCs. Increases in water temperature will also result in changes in the metabolism of fish (increased feeding and respiratory rate), which in turn may increase the uptake of EDCs from the environment or feeds, thus increasing exposure to these compounds and exacerbating the health and reproductive effects previously described (Clarke and Fraser, 2004; Jin *et al.*, 2011; Noyes and Lema, 2015; DeCourten and Brander, 2017).

Researchers reported mixed temperature-dependent effects with exposures to two oestrogenic EDCs (bifenthrin and 17α-ethinyloestradiol (EE2)) in inland silversides (*Menidia beryllina*). The higher-temperature EE2-exposed group had reduced expression of the CYP19b (aromatase) gene, which plays a key role in fish sex determination by converting androgens into oestrogens (DeCourten *et al.*, 2019). A previous study by the same research team showed that the sex ratio of *M. beryllina* was more significantly influenced when these fish were exposed to EE2 at higher water temperature than at lower water temperature (DeCourten and Brander, 2017). Similar findings have been described in zebrafish, with higher temperatures compounding the effects of EE2 exposure by inhibiting masculinization of fish (Luzio *et al.*, 2016).

Changes in water pH may also affect leaching and uptake of EDCs from sediment-bound pollutants. This effect would vary depending on the chemical composition of the EDCs, but many of the heavy metals (e.g. cadmium, nickel and copper, which act as EDCs) are likely to be leached into the water from soil under acidic conditions (Zhang, Y. *et al.*, 2018).

Large changes in salinity are expected to happen between the dry and the monsoon seasons in tropical and subtropical areas. These changes may cause stress to some species of fish associated with osmoregulation adaptation. Coho salmon (*Oncorhynchus kisutch*) exposed to the organophosphate insecticide phorate presented higher mortalities at elevated salinity (32 g/l) than at lower salinity (0.5 g/l). The increase in salinity was assumed to magnify the toxicity of the compound by increasing the formation of phorate oxon and phorate oxon sulfoxide metabolites (Lavado *et al.*, 2011). Changes in carbamates at different salinities have also been observed (Noyes and Lema, 2015). Rainbow trout (*Oncorhynchus mykiss*) exposed to aldicarb for 96 h under high salinity demonstrated elevated cholinesterase inhibition and increased formation of the toxic metabolite alicarb sulfoxide. However, in the same experimental design with hybrid striped bass (*Morone saxatilis* × *Morone chrysops*), fish did not present toxicity or metabolic alterations. This suggests that EDCs and climate stressors can have different interactions and impacts depending on the fish species in question (Wang *et al.*, 2001; Noyes and Lema, 2015). Compounding the impact of salinity changes associated with heavy rains will be the increase in agricultural and industrial runoffs that commonly occurs during these events and the subsequent accumulation of sediments and release of pollutants into water systems.

While these studies can elucidate some of the effects that abiotic factors such as temperature, pH and salinity have on the impact of EDCs in aquatic animals, it is evident that much is unknown. If the aquaculture industry is to mitigate the effect of EDCs in climate change, it will be important to consider the interaction between EDCs and environmental factors. Determining the most vulnerable aquatic species affected by such interactions and how they disrupt reproductive, metabolic and immunological functions at different developmental stages will also be crucial for the success of new/alternative ways to control endocrine dysfunctions in aquatic animals.

8.2.6 Control and/or prevention

The control and prevention of endocrine disruption in fish are certainly not easy. The complexity of this issue, together with the limitations for its diagnosis and the pervasive nature of EDCs in aquatic ecosystems, contributes to the difficulty in preventing exposure to these compounds. Perhaps addressing exposure to EDCs in fish requires augmentation and stricter enforcement of existing regulations pertaining to the manufacturing and use of these products. Global agreements banning some of these compounds are in existence; however, it is difficult to address the use of all ECDs because of economic implications and the need for these products in many everyday activities (Kar *et al.*, 2021). Even with the strict banning of certain substances such as PCBs, these continue to persist in the environment. The volatile nature of many of these compounds also means they will continue to be released from sediments and disperse globally via the atmosphere (Wania and Mackay 1993; Simcik *et al.*, 1998).

Public education and awareness on the dangers of EDC exposure can help increase pressure on local governments to improve current industrial, agricultural and urban standards related to these compounds. Consumer awareness can also reduce the demand for products associated with EDC pollution.

It may be possible to relocate fish farms away from places where there are high levels of EDCs from point-source inputs such as urban areas, industrial or agricultural industries. Considering the pervasiveness of these chemicals, finding new locations where the aquatic ecosystem is not compromised may be challenging. Increasing wetlands around coastal areas, rivers and lakes will increase sequestration of pollutants and may reduce availability of compounds in the aquatic environment.

Alternative ways to control and prevent endocrine disruption in fish by EDCs may require adopting recirculating aquaculture systems (RAS) or closed-containment net-pen systems. These systems have been expanding in use worldwide for the production of multiple species but are still considered costly compared with open net-pen facilities. Using RAS or closed-containment systems permits the use of artificial seawater and/or filtration devices to clean water, which can reduce the presence of chemical pollutants.

Another potential source of EDCs is fish feed. There is increasing evidence that this source of EDCs is significant and likely associated with the fish meal and/or fish oil components in fish diets (Zhang, G.L. *et al.*, 2018; Oliveira and Vasconcelos, 2020). Reducing and/or replacing these components of the feed with plant-based, agricultural commodities such as soybeans, wheat or maize, or with land-based animal proteins and fats, should help reduce the level of some EDCs associated with wild fish; however, some plants may contain other types of EDCs. For example, if grains are farmed using pesticides or chemical fertilizers, or are contaminated with mycotoxins such as zearalenone, residues of these compounds may be strong oestrogen-active EDCs (Zhang, G.L. *et al.*, 2018). Therefore, feed companies should implement proactive and regular testing of feeds for the presence of zearalenone or other known EDCs as part of their quality control to prevent high levels of these active compounds in aquatic animals' feeds. This example reinforces the importance of developing more innovative and robust assays for identification and quantification of EDCs in multiple matrices (water, feed, tissues, organs, etc.), so that adequate preventive measures can be implemented (Kar *et al.*, 2021).

Another way to deal with EDCs in aquatic animals would be regenerative farming (Landos *et al.*, 2021). These systems employ a more ecological approach than traditional farming by rebuilding soil organic matter and restoring degraded soil biodiversity, which will reduce carbon emissions and improve water recycling. The use of pesticides in these farming systems is reduced. This type of farming is usually less intensive than industrial farming so the question of whether these can meet the ever-growing demand for agriculture products remains. The return on investment in these types of environmentally friendly farms may not be as high as with industrial farms, thus the incentive to switch to this system to protect a shared commodity from contamination that takes years to manifest may be difficult to sell. Also, this change in practice needs to occur on a large scale to have significant global impact.

While the multiple chemicals and heavy metals described in this section can cause endocrine disruption in aquatic animals, plastics have not been discussed specifically. This class of compounds often has endocrine-disrupting properties and is becoming more problematic in aquatic ecosystems. The awareness of microplastics and nanoplastics in food items has increased over the last 10 years, and therefore the following section is dedicated to this emerging issue as it is likely to worsen with climate change.

8.3 Microplastic Pollution

8.3.1 Introduction

Contamination of food with microplastics and their detection in humans have made it into the mainstream media. Consumers are concerned about food safety as microplastics can contain harmful chemicals and/or bind toxic pollutants that can biomagnify in the food chain. This is an emerging issue for both wild capture fisheries and the aquaculture industry.

Plastic production has increased exponentially, with approximately 300 million tonnes of plastic produced per year. In the last 2 years since the beginning of the COVID-19 outbreak, the situation has worsened. More than 8 million tonnes of pandemic-associated plastic waste was generated globally, mostly due to medical waste, personal protective equipment and packaging (Peng *et al.*, 2021). The production of plastics has outpaced their disposal, management and degradation. It is estimated that 14 million tonnes end up in our oceans, making up 80% of all marine debris (IUCN, 2021).

Plastic particles are classified as macroplastics (1 cm and larger), microplastics (1 to 5000 μm) or nanoplastics (from 1 to 100 nm) (Hartmann *et al.*, 2019). These originate from the physical, chemical, biological or photodegradation of larger plastic debris (Auta *et al.*, 2017) or from primary sources. The most common types of microplastic particles in seawater are composed of polypropylene (PP), polyethylene (PE), polystyrene (PS), polyvinylchloride (PVC) and polyethylene terephthalate (PET) (Rocha-Santos and Duarte, 2015). Low-density polyethylene (LDPE) is the second most common polymer made within the European Union, and the most commonly used plastic in packaging (IPEN, 2020). It is also the most reported microplastic in the marine environment, as well as inside the gastrointestinal tract, liver, skin and muscle of commercially caught fish species (Collard *et al.*, 2017; Compa *et al.*, 2018; Daniel *et al.*, 2020).

Fish are attracted to microplastics due to their coloration and buoyancy, which resembles food in aquatic environments (Jovanović, 2017). The first report of microplastic ingestion in fish was documented in 1972 (Carpenter *et al.*, 1972). Since then, plastic ingestion has been reported in as many as 386 marine fish species, including 210 species of commercial importance (Savoca *et al.*, 2021). Plastic products can accumulate in the gastrointestinal tract and on occasion lead to gastrointestinal blockage (Li *et al.*, 2016). Smaller nanoplastic particles, as well as chemical compounds leached from larger particles, can penetrate the circulatory system to reach other tissues, such as the liver and other organs (Lu *et al.*, 2016).

The toxicity and mechanical damage of microplastics in fish are related to the quantity, shape, size and chemical composition of the particles (Jeong *et al.*, 2016; McDevitt *et al.*, 2017). When chemical additives in plastics are absorbed by fish, this can result in a number of sequelae depending on their composition. Microplastics in the aquatic environment adsorb and transport persistent organic pollutants (POPs) such as PCBs. Once consumed these particles may function as mobile reservoirs for these highly toxic pollutants that bioaccumulate in the fatty tissues of fish (Jensen *et al.*, 2020).

Many compounds in plastics are endocrine disruptors, hepatoxic and/or neurotoxic (Rochman *et al.*, 2014; Barboza *et al.*, 2019; Iheanacho *et al.*, 2020; IPEN, 2020). Common additives include perfluorinated chemicals, phthalates, bisphenols and nonylphenols, among others. Some of these plastic chemicals are thought to produce reactive oxygen species, either directly or indirectly, which may negatively affect marine organisms (Anifowoshe *et al.*, 2022). Further, many of the plastic breakdown products act as EDCs, which can lead to physiological stress, decreased fecundity (Wang *et al.*, 2019), compromised immune regulation (Espinosa *et al.*, 2017), behavioural alterations (Limonta *et al.*, 2019) and metabolic abnormities (Wan *et al.*, 2019) (see Section 8.2 above on EDCs for more details).

8.3.2 Impacts on fish production

The ubiquitous nature of microplastics in the aquatic environment has led to their occurrence in many food products including aquaculture fish (Garcia *et al.*, 2021). This has had direct and indirect impacts on this industry. The impact of microplastics varies depending on the quantity of exposure and the nature of the adsorbed particles (Collard *et al.*, 2017). For the most part, farmed fish are not exposed to significant quantities of microplastics in a single dose, rather they bioaccumulate these compounds over time, which can lead to chronic problems.

Most studies are on individual fish with exposure levels that are significantly higher than natural exposures, but they reveal important effects on the host. Physiological responses to microplastics include intestinal distension, inflammation and weight loss (Chen, Q. *et al.*, 2020; Varó *et al.*, 2021). Fish exposed to microplastics also showed increased lipid peroxidation in gills and muscle tissues, compromising both respiratory function and movement (Barboza *et al.*, 2019). The breakdown of the epithelial wall in the gastrointestinal tract has also been reported from mechanical trauma and can lead to increased uptake of toxic chemicals leached from the plastic particles (Wardrop *et al.*, 2016). Often areas with high plastic concentrations also have high levels of other pollutants, aggravating the problem. Chemicals from plastics have the potential to cross the blood–brain barrier and affect fish physiology and behaviour (Mattsson *et al.*, 2017). The specific effect of chemical absorption depends on the plastic's composition. The translocation of microplastics and their associated chemicals to other organs is a health concern for both fish and humans (Collard *et al.*, 2017; Compa *et al.*, 2018; Daniel *et al.*, 2020; Leslie *et al.*, 2022).

Ingestion of microplastics may impair feeding in aquatic organisms as these particles can remain in the gut lumen for multiple days and give a false sense of satiation, which will lead to weight loss and other energy deficiency-related changes (de Sá *et al.*, 2015). The quantity of microplastics in the digestive tract of fish has been correlated to their type of habitat and feeding behaviour, with evidence of pelagic fish ingesting more microplastics compared with benthic fish species (Güven *et al.*, 2017). Microplastics in the gastrointestinal tract can reduce the diversity and alter the composition of the gut microbiome (Jin *et al.*, 2018).

In a study on Japanese medaka, microplastics lodged in the gills of fish increased mucus production and led to deformed opercula (Zhu *et al.*, 2020). Increased mucus production is a common response to aquatic pollution and an important part of the innate immune response of fish, protecting the epithelium from foreign substances (Shephard, 1994). Excessive mucus can potentially impair respiratory functions and decrease the oxygenation of blood in the gills. This can predispose animals to infectious agents.

Exposure to microplastics in humans through consumption of aquaculture fish is lower than exposure via consumption of wild fish (Garcia *et al.*, 2021), but the notion that there is any risk is a concern to consumers. Contamination of food is difficult to screen for, and as some of the additives or chemical components of microplastics are known to be carcinogens and endocrine disruptors (Lithner *et al.*, 2011), consumer perception of the risk associated with fish consumption is a foreseeable growing public concern.

8.3.3 Diagnosis and detection of exposure to microplastics

Identifying chronic low-level exposure to microplastics is very difficult in aquaculture. Chronic exposure to micro- and nanoplastics can result in trauma to tissues and other fish health issues that are not pathognomonic for microplastics. Changes associated with microplastics at the cellular level will vary depending on the shape and size of the particles, the chemical composition of the particles and the quantity of exposure. Histologically, physical damage to the villi and enterocytes in the epithelial layer of gastrointestinal tissues can suggest exposure to microplastics (Lei *et al.*, 2018). Clear intercellular spaces between myocytes in the muscularis layer of the gastrointestinal tract, as well as head kidney glomerulopathy with expansion and congestion of glomerular capillaries, glomerulomegaly and expansion of Bowman's space, are also sometimes reported (Chen, Q. *et al.*, 2020; Zhu *et al.*, 2020). Some plastic additives and stabilizers have been shown to be EDCs (Rochman *et al.*, 2014). Clinical signs of exposure to EDCs are outlined in Section 8.2.3 above. Once it is suspected that fish have been exposed to microplastic particles, confirmation of exposure can be done using infrared spectroscopy.

Consumers are increasingly concerned with the quality and safety of their food, including farmed fish. Microplastics detection in the flesh of fish can be identified through microscopic analysis coupled with Fourier transform infrared spectroscopy (FTIR), laser-based Raman microscopy or scanning electron microscopy. The latter is more commonly used for detection of smaller plastic particles (Kwon *et al.*, 2020).

Microplastics have been identified in fishmeal and so evaluating ingredients used in aquaculture feeds may also become standard practice as awareness about food safety around these products becomes more common (Thiele *et al.*, 2021). Similar spectroscopy methods applied to fish tissue can be used for this purpose (Thiele *et al.*, 2021). Recent identification of microplastics in human blood samples made use of a semi-quantitative technique based on the thermal degradation of plastic particles, namely pyrolysis double shot–gas chromatography/mass spectrometry (Py-GC/MS) (Leslie *et al.*, 2022); however, these techniques require specialized laboratories which are currently not easily or routinely available to the aquaculture industry.

Detection of specific chemicals associated with absorption of microplastics is achieved using traditional chemical analysis such as high-resolution gas chromatography coupled with high-resolution mass spectrometry (HRGC/HRMS) (Matsumoto *et al.*, 2014). These tests require highly specialized laboratories and are expensive to perform. For common chemicals such as PCBs, there may be commercially available immunoassays and/or bioassays for their detection.

8.3.4 Modes of exposure

Microplastics are widespread especially in aquatic systems, many of which tend to be sinks for pollutants (Martin *et al.*, 2020). Increased exposure of fish to microplastics occurs with increased proximity to water pollution sources. The increasing population in coastal urban cities means that increasing amounts of land-based pollutants, including microplastics, are transported to the coastal waters through wastewater discharges, rivers or urban runoff.

In the aquatic environment fish are likely exposed to microplastics directly through ingestion and absorption through the gill. Microplastics and harmful adsorbed chemicals associated with these particles have been known to accumulate in various marine organisms and move up the food chain (Mattsson *et al.*, 2015; Lusher *et al.*, 2017). Recently, polymers used in plastics were identified in human blood (Leslie *et al.*, 2022).

Although wild fish are more prone to accumulating microplastics than aquaculture fish, the latter are at risk of exposure because these compounds have been found in contaminated feeds (Garcia *et al.*, 2021; Thiele *et al.*, 2021). Aquaculture fish are often fed diets containing fishmeal and fish oil produced from wild caught fish, which are more prone to scavenge on microplastics than farmed species. Aquaculture fish can also come into direct contact with these particles in the aquatic environment as microplastics move passively through open net pens (Garcia *et al.*, 2021). There may also be the possibility of parental transfer of nanoplastics to progeny as reported by Pitt *et al.* (2018).

8.3.5 Anticipated effects of climate change on exposure to microplastics

The increase in frequency and magnitude of extreme weather events associated with climate change will contribute to the increased dispersion of plastic pollution in aquatic ecosystems. The fossil fuel usage associated with the manufacturing of plastics contributes to greenhouse gas emissions, which drives climate change, exacerbating the problem of microplastics. Additionally, the greenhouse gases methane and ethylene are produced during degradation of plastics via exposure to solar radiation, further contributing to global warming and climate change (Royer *et al.*, 2018).

The precise impact of climate change on the toxicity of microplastics and its effect on fish are not well established. Under some circumstances it is possible that increased temperatures may affect the decomposition and uptake of pollutants in and on microplastics (Min *et al.*, 2020; Wang *et al.*, 2022). It is also possible that some surface chemicals disassociate from particles under different pH and temperature conditions; however, predicting these changes depends on the volatility of the chemicals involved (Zhang and Huo, 2019).

8.3.6 Control and prevention

Reducing exposure to microplastics in open net-pen farms is difficult. Decreasing these pollutants requires a concerted effort by local, regional and global communities. Novel strategies to reduce or degrade plastics are needed to address this issue on a large scale. One potential approach has been the adoption of bio-based plastics; however, their greenhouse gas emissions are variable, making it preferable to shift to reducing our reliance on single-use products (Ford *et al.*, 2022). Marine plastic pollution would also benefit directly from the conservation and restoration of blue carbon coastal habitats that support high sediment accumulation rates and trap plastic pollution (Martin *et al.*, 2020).

On a more local farm level, including buffers around net pens in areas where there is plastic pollution may reduce their surface concentration within the net pens, decreasing the risk of microplastic consumption by fish and trapping of these particles within the gills. Certified plastic- or chemical-free fish feeds and/or fish-feed ingredients is another approach that can help to reduce the level of exposure via feed consumption and reassure customers about food safety. Lastly, improved testing of fish/seafood for different chemicals commonly found in or adsorbed on microplastics may increase customer confidence in farmed fish regarding pollutants.

INFECTIOUS DISEASES

Impacts of pollution on the immunity of fish will likely exacerbate the occurrence and severity of infectious diseases, which are also likely to be more problematic as the aquaculture industry grows and intensifies. In addition, new and complex health issues are predicted to emerge with changing husbandry practices and climate change.

8.4 Gill and Skin Health

8.4.1 Introduction

The gill and skin of fish are the first tissues to be exposed to the environment, as such they are vulnerable to changes to the aquatic surroundings. The epithelium and mucus layers of these organs are essential for protecting fish from infectious diseases and toxic insults, as well as for many basic physiological activities such as gas exchange, excretion of nitrogenous waste products, acid–base balance and osmoregulation (Alshammari *et al.*, 2019; Cabillon and Lazado, 2019). As net-pen aquaculture industries intensify and husbandry practices change to accommodate larger farms, there is the potential for increased health problems. In many salmon-farming operations gill and skin issues have become the most important underlying reasons for mortality and welfare issues (Shinn *et al.*, 2015; Sommerset *et al.*, 2021).

The skin consists of the epidermis and dermis. The latter is where mineralized scales originate; the former can be a few to several cells thick. The primary role of the skin is to create a barrier against the aquatic environment, which is essential to maintain homeostasis and protection against pathogens. However, gills consist of only one layer of epithelial cells embedded with specialized cells, separating the environment from the bloodstream. The gills have a surface area similar to the size of the entire body surface area of the fish to enable efficient gas exchange. The thin epithelial layer of the gills also permits exchange of ions to maintain homeostasis and acid–base balance.

The epithelial layers of both the gills and the skin have mucus-producing cells. Mucus contains immunoglobulins, enzymes and antibacterial peptides (Dash *et al.*, 2018; Reverter *et al.*, 2018). The mucus layer is a critical component of the innate immune systems. Additionally, the mucus layer of the skin has its own microbial flora which interacts with the environmental microbiota (Minniti, 2018) and reduces the risk of colonization by pathogenic bacteria. This protective barrier is essential to reducing bacterial invasion. When either the skin or the gill epithelium is damaged it provides an opportunity for infectious agents to colonize, as well as osmoregulatory issues to manifest.

Skin and gills are in direct contact with the aquatic environment, so they are vulnerable to water quality issues such as suboptimal water temperature, hypoxia, pH changes, ammonia and toxins. Chemical irritants from bath treatments against parasites or bacteria, or exogenous exposure to pollutants, can damage the structure of the integument (usually impacts the mucus layer) and this can provide an opportunity for secondary pathogens to manifest. Mechanical damage especially to the gills from algae, jellyfish and organic debris can also be significant in open net-pen fish farms (Foyle *et al.*, 2020).

Handling of fish in aquaculture is inevitable and as industries intensify these activities may increase. These activities can result in physical trauma to the skin, which can increase the risk of secondary infections and osmoregulatory problems.

Researchers have investigated the role of nutrition on the health of the skin and the mucosal layer of fish (Ángeles Esteban, 2012; Masso-Silva and Diamond, 2014; Sveen *et al.*, 2017; Minniti, 2018). If fish are not provided the adequate essential nutrients needed to produce the immunological components of the mucus layer, then this may impair its function and provide a route of entry for secondary pathogens. In a Norwegian study, zinc and *n*-3 fatty acids in feed were shown to affect the lipid composition of skin, gut and gills (Berge *et al.*, 2019). As diets evolve and move away from fishmeal and fish oil, the effect of the mucus barrier remains to be seen (Sweetman *et al.*, 2010; Torrecillas *et al.*, 2011; Camilleri *et al.*, 2019).

With the ongoing climate change, the negative impact on gill and skin health is anticipated to become more significant.

8.4.2 Impacts on fish production

Skin and gill issues in aquaculture occur in both freshwater and saltwater species. In this section the focus is on saltwater issues, with examples predominantly from the salmonid aquaculture industry; however, many of the principles are transferable to other species and industries.

Skin issues

One of the most common issues in farmed fish is skin ulcers or the loss of the epidermal and/or dermal layers of the skin (Fig. 8.1). Skin ulcers are not pathognomonic and can be a result of infection with viral, bacterial or parasitic pathogens, and can also be caused by exposure to non-infectious irritants including poor water quality or toxins. Often skin ulcers are due to multifactorial causes and several issues can contribute to skin damage. The impact of skin ulcers on fish depends on the extent of the lesion(s) and the causative factor(s).

Sea lice infestations are a major concern in the marine environment. Iversen *et al.* (2019) estimated the cost of salmon lice alone for the Norwegian salmon industry to be more than €500 million annually and this does not include effects on growth. These ectoparasites are temperature and host density dependent, so they can increase in numbers rapidly if they are left untreated on large farms. If sea lice are left untreated infections can lead to mortality; however, in most

Fig. 8.1. Natural infection of Atlantic salmon with *Moritella viscosa* showing typical ulcerative lesion. (Photograph courtesy of Anton Sæter, MarinHelse, Norway.)

industries sea lice are kept under control with various types of treatments. As industries increase in intensity and size, this group of parasites has increased in prevalence, and consequently this has led to more treatments (Kristoffersen *et al.*, 2013; Overton *et al.*, 2019). The latter includes in-feed treatments with emamectin benzoate or diflubenzuron, and bath treatments with pyrethroids, organophosphates and/or hydrogen peroxide (Overton *et al.*, 2019). Due to the high degree of resistance to chemicals, the salmon industry in Norway and other areas has increased the use of non-chemical treatments for sea lice control including cleaner fish, mechanical removal and thermal treatments (Overton *et al.*, 2019; Sommerset *et al.*, 2021). Many of these treatments, when administered repeatedly, result in damage to the gills and skin. These treatments sometimes cause mortality directly and/or predispose to osmoregulatory stress and secondary infections. Gill injuries are one of several side effects observed particularly with thermal sea lice treatments (Gismervik *et al.*, 2019; Sommerset *et al.*, 2021).

A common skin issue is tenacibaculosis, caused by various *Tenacibaculum* spp. (Zrnčić *et al.*, 2013; Avendaño-Herrera *et al.*, 2016; Småge *et al.*, 2016, 2018; Lopez *et al.*, 2022). This genus of bacteria is commonly associated with ulcers around the head, tail, fins and flank of numerous fish species in salt water, including salmonids (Fig. 8.2).

In salmonids, this condition usually occurs shortly after transfer into salt water and results in acute mortality (Klakegg *et al.*, 2020a) in smolts. Outbreaks in larger fish may develop more slowly and are often associated with handling and mechanical treatment for salmon lice. Lesions associated with these bacteria are driven by extracellular products that have strong proteolytic properties (Margarinos *et al.*, 1995; Pazos, 1997). This bacterial infection is prominent in most salmon-farming countries including Chile.

Fig. 8.2. Ulcerative lesion (arrow) from salmon naturally infected with *Tenacibaculum* spp. (Photograph courtesy of Duncan Colquhoun, Veterinærinstituttet, Norway.)

In the northern hemisphere where salmonids are the primary aquaculture species cultured in marine net pens, one of the most problematic skin issues besides tenacibaculosis is skin ulcers associated with *Moritella viscosa* and/or *Aliivibrio wodanis* (Karlsen *et al.*, 2014). The skin lesions associated with this infection are primarily on the flank of the fish that is protected from the water flow (Fig. 8.1). These ulcers are often referred to as winter ulcers because they occur more often in colder water temperatures. However, in some areas in Canada, ulcers from *M. viscosa* infections have been observed when water temperatures were above 10°C (MacKinnon *et al.*, 2019a). Recent work suggests the lesions associated with *M. viscosa* are driven by extracellular products produced by the bacteria, which may explain the rare occurrence of systemic infections in fish, the exclusive appearance of large ulcers on fish, the minimal horizontal transmission of the pathogen between pens and the acute mortality (MacKinnon *et al.*, 2019b, 2020). Mortality rates from *M. viscosa* infections in Atlantic salmon can be as high as 23% in Canada (MacKinnon *et al.*, 2019a). In addition to mortality, the stress on fish reduces growth and the scars from large ulcers lead to downgrading at processing plants. Lastly, the use of antibiotics to treat these conditions is costly and may result in increased risk of antimicrobial resistance (AMR).

In contrast, skin lesions in the southern hemisphere where the water temperatures are not as cold are dominated by *Tenacibaculum* spp. and *Piscirickettsia salmonis* infections, which favour warmer water temperatures. The latter form distinct small ulcers over the flanks of the affected fish. Regardless of the specific cause of the skin lesions, their impacts include high mortality, welfare issues and marketability problems for the salmon industries (MacKinnon *et al.*, 2019a; Sommerset *et al.*, 2021).

Gill health

Many aquaculture industries are also reporting a complex gill pathology impacting production (Shinn *et al.*, 2015; Sommerset *et al.*, 2021). Gill lesions may occur at any time, from fingerlings to market-size fish. There is very little knowledge on the impact of subclinical gill issues or on what proportion of the gill surface needs to be damaged before basic physiological processes are affected, but even minor issues may lead to physiological stress for fish, reduced growth and mortality. When smolts are transferred from fresh to salt water, subtle gill damage prior to saltwater introduction may predispose fish to more severe physiological stress in salt water. Because of multiple causes of gill issues it is not possible to estimate the overall cost of gill disease to aquaculture. However, it has been identified in several areas of Europe as a significant health challenge (Boerlage *et al.*, 2020). In a recent Norwegian questionnaire, fish health personnel identified complex gill problems during the grow-out phase among the diseases that rank highest as an underlying cause of death, reduced growth, early harvest and welfare concern (Sommerset *et al.*, 2021). One anonymous company in Norway estimated improved gill health of their fish increased their earnings by fourfold (Moore, 2020).

As with skin issues, gill pathology can result from multiple aetiologies and is often multifactorial, thus it is sometimes referred to as complex gill disease (CGD) (Gjessing *et al.*, 2019). Chronic gill pathology may accumulate over time as fish are exposed to multiple insults including chemicals, pollutants, toxins, pathogens, as well as mechanical and/or thermal stressors. The latter have been investigated more recently due to the thermal and/or mechanical delousing that is commonly used to control sea lice. Interestingly, damage to the gill structure from these treatments is not limited to physical damage but can also include changes in gill-associated microorganisms (Østevik *et al.*, 2022). Other non-infectious causes of gill damage include microplastics and other pollutants, toxins and organic material in the water column that gets trapped in the gills during respiration. High organic matter can also lead to blooms of various types of phytoplankton and jellyfish, which can cause gill damage either directly or indirectly via release of toxins (see Wahli *et al.*, Chapter 4, this volume, 2023). In addition, as farms increase in size, and husbandry practices such as net cleaning, feeding and fish transfers become more automated, this may lead to more issues with gill health. For example, there are reports that water hydroids released during *in situ* net cleaning can attach to gills and cause trauma to epithelium tissue, making fish more susceptible to pathogens (Bloecher *et al.*, 2018).

Gills are constantly exposed to infectious agents, some of which can cause disease while others require co-infections or trauma before they induce significant health issues. Mixed gill infections are not uncommon in many species of fish. Amoebic gill disease (AGD) (Fig. 8.3), caused by the ectoparasite *Paramoeba perurans*, has become an increasing problem in several marine fish species worldwide and in particular in salmonid aquaculture (Boerlage *et al.*, 2020). This parasite leads to hyperplasia of the gill lamella which reduces gill efficacy (Wiik-Nielsen *et al.*, 2016). The infection is highly seasonal, and in Europe it is most prominent in late summer/autumn when water temperature and salinity are highest (Rodger, 2014). Other emerging gill diseases/pathogens that on their own tend not to be deadly, but can contribute to CGD, include epitheliocystis (Mendoza *et al.*, 2013; Contador *et al.*, 2016) and salmon gill pox virus (SGPV) (Nylund *et al.*, 2008; Gjessing *et al.*, 2015, 2017; Boerlage *et al.*, 2020). SGPV may affect fish during all stages of the salmon production cycle, but the infection seems to be most devastating in fresh water. Experiences from the field suggest that stress may be a critical factor for induction of SGPV disease. Thoen *et al.* (2020) supported these assumptions by demonstrating experimentally how elevated plasma cortisol was a prerequisite for developing disease due to SGPV. Fish infected with this virus in fresh water may be more susceptible to other pathogens after marine transfer.

The impact of gill disease on production can be significantly affected by the environment. If water quality is good fish may be able to recover from gill disease with minimal impact, but if water quality is suboptimal then the impact can be severe. As climate change influences environmental parameters, gill disease may become more significant. Further, as aquaculture intensifies water quality around farms may worsen, which can result in reduced gas exchange,

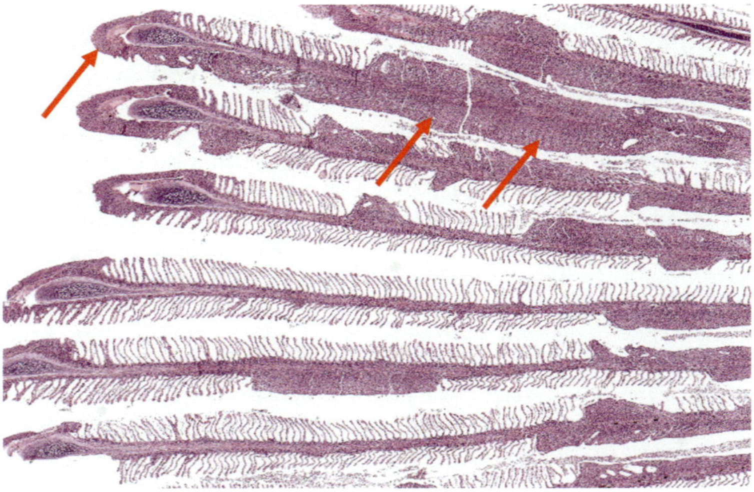

Fig. 8.3. Hyperplasia of the gill lamella (arrows) in salmon naturally infected with the ectoparasite *Paramoeba perurans*. (Image courtesy of Ole Bendik Dale, Veterinærinstituttet, Norway.)

increased metabolic demands on the gill (nitrogen waste removal), as well as changes in parasite and microbial agents that otherwise would not be found at harmful concentrations.

8.4.3 Diagnosis and surveillance of the condition(s)

Diagnosis of subclinical skin and gill lesions is difficult on commercial fish farms. Gill issues can develop slowly and unless there is a monitoring programme, most common gill problems will be observed and diagnosed only when clinical effects are apparent. At this stage chronic lesions may have developed, and a larger proportion of the population may be affected, making it more difficult to address the issue. A scoring system of gross pathology has been suggested for field-based macroscopic assessment of complex or multifactorial gill disease in farmed Atlantic salmon, to detect lesions earlier in the disease process and improve mitigation efforts (Fridman *et al.*, 2021). This scoring system has proved quite effective for managing AGD in many salmonid industries, as early detection and treatment can reduce mortality (Taylor *et al.*, 2016). This scoring system may be adopted as a first step in surveillance for gill disease.

Gill wet mounts are often used as a preliminary diagnostic tool for AGD and other parasitic infections, as well as for identification of algae and organic and inorganic debris trapped in gill filaments. For bacterial gill disease culture can be used; however, results are sometimes difficult to interpret as commensal bacteria and secondary pathogens can obscure the primary aetiology. Single and multiplex molecular tools such as polymerase chain reaction (PCR) can be used to identify single or multiple bacterial and viral pathogens associated with gill lesions; however, to fully understand the nature of the gill disease histopathology is often required. Gjessing *et al.* (2019) established a system for scoring histopathology of gills. They focused on a wide range of gill lesions and investigated potential associations between histopathological findings and infections with specific gill pathogens. This guide complements a gross anatomical scoring system that can help identify when gill health may be compromised. Taken together, these diagnostic tools can provide a reliable diagnostic picture and estimate the extent of the gill damage. More advanced techniques such as immunohistochemistry and *in situ* hybridization can also be used to visualize pathogens within the tissue to better understand the mechanism of disease.

Surveillance for early detection of skin lesions requires observation of animals submerged in water, which can be challenging. The use of cameras on salmon farms can improve the visualization of fish under water, but it is still difficult to identify early signs of skin disease (i.e. dulling of mucus sheen on the fish or excessive mucus production). For monitoring of sea lice, most salmon industries have very sophisticated weekly or biweekly monitoring programmes where fish are non-lethally sampled from multiple pens and the different life stages of parasites are counted and reported for industry analysis (Myksvoll, 2019). This surveillance programme is useful to understand the infection levels in an area, and the timing and efficacy of treatments.

Diagnostics for different bacterial causes of skin lesions usually starts with bacterial culture, but often it is difficult to separate primary from secondary bacterial issues. In the case of *M. viscosa*, this Gram-negative bacterium grows on marine agar and/or blood agar with 1.5% salt, but it can be difficult to isolate from lesions. In some advanced cases bacteria such as *M. viscosa* can also be isolated from internal organs (MacKinnon *et al.*, 2020); however, many fish that die from this bacterial disease are not systemically infected, which suggests osmoregulatory issues as the cause of death (Tørud and Håstein, 2008). *Tenacibaculum* spp., which are also Gram-negative bacteria that grow on marine agar, are also not typically found in internal organs. For differentiation of the various genotypes sequencing is necessary.

Histological evaluation of skin tissue may also help reveal the cellular characteristics of lesions and provide information on aetiology. Molecular tests supported by immunohistopathology are necessary to identify viruses and associated pathological lesions caused by these agents (e.g. SGPV).

8.4.4 Transmission

The multifactorial nature of many gill and skin issues makes it difficult to identify specific mechanisms of transmission. In the case of disease caused by sea lice and *P. perurans*, which are host dependent and transmitted horizontally, the higher the fish density the higher the infestation. Several studies have demonstrated farm-to-farm spread of sea lice (Kristoffersen *et al.*, 2013). Transmission of *P. perurans* is not as well studied. This parasite likely does not transmit between farms, but fish-to-fish transmission occurs within a pen and between pens (Hjeltnes *et al.*, 2014).

Transmission of *Tenacibaculum* spp. and *M. viscosa* seems to be more from environmental sources and is less dependent on host density compared with sea lice transmission. These pathogens are common in the marine environment where they have an important ecological function in breaking down organic material. The diversity of *Tenacibaculum* strains seen during an outbreak within the same farm suggests fish are infected from environmental sources and not from fish within the pens (Småge *et al.*, 2018). Further, in 2016, *Tenacibaculum maritimum* was isolated for the first time in Norway on a diseased lumpsucker (*Cyclopterus lumpus*) (Småge *et al.*, 2016), but the infection was not spread to the salmon in the same net pen despite the close proximity between the two fish species. This lack of transmission to the larger salmon in the net pen may also indicate that salmonids must be predisposed to infection for bacteria to attach and create skin ulcers.

A study on the distribution of ulcer disease outbreaks associated with *M. viscosa* in Canada also suggested that pen-to-pen and farm-to-farm spread of this pathogen was not prevalent as many sites only had a few pens affected despite their close proximity during outbreaks (MacKinnon *et al.*, 2019a). The pattern of infection in space and time was also not suggestive of farm-to-farm pathogen spread.

Skin ulcers appear to be increasing in prevalence in many salmonid industries; this suggests that either fish are becoming more susceptible to infection or pathogens are increasing in abundance and/or virulence, or both issues are occurring simultaneously. Given the increase in heat treatments and mechanical delousing in Norway (Sommeret *et al.*, 2021), as well as the increase in nutrient loading in fish-farming areas from intensification of the industry, which would promote bacterial growth, it is likely that both processes are at play.

8.4.5 Anticipated effects of climate change on gill and skin health

Skin and gill diseases are usually due to multiple factors, including interactions between hosts and pathogens that are modulated by the environment; therefore it is likely that climate change will influence these conditions. The precise effect of climate change will vary depending on the host and the pathogen. In general, gill and skin issues, with perhaps the exception of *M. viscosa* infections which are more prevalent in colder temperatures, are predicted to increase with climate change (Toranzo *et al.*, 2005; Cidad *et al.*, 2018). Several chapters in this volume have already discussed the effects of climate change on specific pathogens, so in the present chapter the focus is on the predicted impact of these changes on host gill and skin tissues. These tissues, particularly the gills, are responsible for gas exchange, acid–base regulation and osmoregulation, thus changes to the environment have the potential to affect these critical functions.

Increasing water temperature, especially if it is already at the upper limit of the normal range of the fish, may have direct detrimental effects on the physiology of the fish host, including its immune system. Higher water temperatures result in a higher metabolic rate, which requires higher gill perfusion for oxygen (O_2) and carbon dioxide (CO_2) exchange. Compounding this issue is that higher-temperature water also has a lower ability to hold dissolved O_2, thus requiring higher respiration and heart rates for fish to achieve the same gas exchange. Further, as farms intensify, the biomass and organic matter in an area increase the demand for O_2, which may result in suboptimal O_2 levels for aquaculture species. When fish have subclinical or chronic gill disease, the effect of high temperature with or without low dissolved O_2 may be exacerbated as the gills may not be able to cope with the increased physiological demands.

Increasing the body temperature of fish may also lead to a decline in plasma pH associated with the high CO_2 and a loss of bicarbonate (HCO_3^-) (Foyle *et al.*, 2020). Over time this could cause acid–base and osmoregulatory stress on fish. It may also result in an increase in nephrocalcinosis, which has been linked to high CO_2 levels in fish (Fivelstad *et al.*, 2018).

In addition to temperature changes, the acidification of the oceans will also stress the acid–base regulation of fish. Increased carbonic acid in seawater due to increased CO_2 levels will decrease the pH of water. The higher the ambient water CO_2 levels the more difficult it is for fish to excrete this gas; however, most fish are able to compensate for this by the uptake or retention of HCO_3^- (Foyle *et al.*, 2020). Ocean acidification also appears to increase the drinking rate of fish and thus the uptake of salts (Foyle *et al.*, 2020). Given that marine fish lose water to their saline environment, cells in the gills are continuously pushing out sodium and chloride ions, which requires energy. Ion exchange, both active transport and loss of Na^+ and Cl^- through diffusion across the gill epithelium, maintains the acid–base balance over time, but it may result in osmoregulatory stress (McDonald, 1983). To compensate, some species can remodel their gill structure and increase or decrease the surface area for gas and ion exchange in response to salinity, hypoxia or acidification (Foyle *et al.*, 2020). The ability to remodel the gills varies by species, making some animals more sensitive to climate change than others. The acidification of water may help fish with diffusion of ammonia (NH_3) as this will be facilitated under more acidic conditions. However, as mentioned earlier, other end-product metabolites such as ammonium (NH_4^+), hydrogen (H^+), CO_2 and HCO_3^- may be more difficult to regulate in acidic environments.

Increases in eutrophication are also predicted with climate change and this may have a significant impact on the pathogenicity and proliferation of many infectious agents associated with gill and skin conditions. The more nutrients in the water the more likely that bacteria, algae, jellyfish and biofouling will proliferate (Foyle *et al.*, 2020). This increases the demand for O_2, compounding the issues of reduced O_2 from increased water temperature. Further, these organisms can directly act as primary invaders or indirectly cause mechanical damage and increase the risk of secondary infections. NH_3 from high nutrients in the water can also be toxic to fish. Overall eutrophication will directly and indirectly exacerbate skin and gill conditions.

8.4.6 Control and/or prevention

Preventing or reducing exposure to infectious diseases, pollutants, toxins or sources of mechanical damage will help reduce the incidence of the problem. In addition, improving the immunity, particularly the mucosal immunity, of fish may also improve the outcome. Lastly, addressing environmental issues that exacerbate the proliferation of pathogens and increase the physiological demand on fish will also reduce losses.

Preventing exposure to pathogens is difficult in open net-pen systems; however, if the primary or secondary cause(s) of the skin or gill disease is transmitted horizontally from fish to fish or from farm to farm, early detection and treatment of the pathogen will likely reduce the magnitude of the problem and reduce spread to other animals. This is the principle that underpins the sea lice and AGD surveillance and control strategies used by salmon-farming industries to control these pathogens. However, when the frequency of treatments required to control pathogens is too great, this can lead to its own set of issues (i.e. damage to gills and skin from repeated sea lice treatments or handling; AMR from overuse of antibiotics). If horizontally transmitted infectious agents cannot be controlled without overusing chemical and/or pharmaceutical treatments, it may be time to assess the natural biocapacity of the area and reduce the biomass.

For bacterial pathogens that are transmitted from fish to fish (i.e. *P. salmonis*), it may be advisable to avoid practices that crowd animals and result in skin-to-skin contact. For example, feeding practices such as once-a-day feeding may lead to increased contact between fish and provide an opportunity for transmission of infectious agents. For other pathogens such as *Tenacibaculum* spp. and *M. viscosa* where there is likely limited horizontal transmission between fish (MacKinnon *et al.*, 2019a), early treatment of these pathogens may help reduce loss of infected fish, but may not reduce transmission between fish, as infection is mostly from the environment.

Reducing bacterial load in the environment is difficult, but as the concentration of bacteria may be linked to the nutrient level in the area, ensuring the biocapacity of the site is not exceeded may be important in reducing the ambient levels of these pathogens. As salmon aquaculture industries have intensified, many farmers have had to provide supplemental O_2 during the warmer periods. This may indicate that the biocapacity of an area has been exceeded. Increasing O_2 addresses one of the limiting factors; however, this is likely not the only water quality issue. Poor water quality (i.e. pH changes, high CO_2 and high NH_3) can also lead to gill and skin issues which will predispose fish to pathogens. Further, the water quality in an area (i.e. changes in temperature, low dissolved O_2 and acidification) may also alter the microbial ecology of an area, favouring different bacterial species under different conditions; however, this is not well understood.

In the case of some pathogens such as *M. viscosa* there are vaccines available to mitigate the severity of infection (Løvoll *et al.*, 2009; Karlsen *et al.*, 2014). For bacteria that produce extracellular proteolytic toxins and cause acute mortality, vaccination given systemically to fish may provide only limited protection against these organisms (Småge *et al.*, 2018). Recently, Klakegg *et al.* (2020b) demonstrated some success in the use of probiotic (baths) to treat ulcers in lumpfish (*C. lumpus*); however, the use of this type of treatment in net-pen facilities may be difficult. Similarly, these researchers also showed the addition of a probiotic consisting of *Aliivibrio* spp. to freshwater tanks with Atlantic salmon smolts increased their growth, reduced mortality and improved food conversion rates after they were transferred to seawater (Klakegg *et al.*, 2020a). The studies indicate the role naturally occurring bacterial species may play in inhibiting more pathogenic organisms. Further, probiotics added to feed activated selected innate immune defence molecules, suggesting that probiotics can be utilized to evoke favourable responses on the skin, gills and intestine (Nimalan *et al.*, 2022).

Bacteriophages are small viruses which infect bacteria. They are found in high numbers in aquatic environments. In a study at the Norwegian Veterinary Institute, phage therapy against *M. viscosa* infections showed promising results and may be a treatment option in the future (K. Sveinsson, Oslo, 2022, personal communication).

Improving water quality is always helpful in mitigating the impact of these types of conditions. In open net-pen farming there are limits to what can be done to change environmental conditions, but it may be possible to increase O_2 levels using nanobubblers or other supplemental O_2 devices. This could make a difference in survival of fish if gill disease is severe. If net cleaning is releasing debris, hydroids and/or particulate matter that is damaging gills, net washers with extraction devices may reduce this problem, but these devices are more expensive. Other water quality issues may be more difficult to address; yet if these are constantly a limiting factor for gill and skin health on a site, it may be advisable to consider area-level management strategies. Relocation to avoid pollution sources or to improve water flow on a farm may be an option. Assessing the source(s) of nutrients in the area, including aquaculture fish biomass, and reducing the density of fish in areas with high nutrient loads may help alleviate low dissolved O_2, high NH_3 and high CO_2, which may reduce the risk of local algal and jellyfish blooms as well.

If water quality cannot be improved in the open system, then closed-containment systems with water treatment may be a consideration; however, the cost of such systems may not be economical. Offshore farming may also be less influenced by coastal water quality issues; however, these structures require high capital investments and are logistically more difficult to operate than net-pen facilities in coastal watersheds. They are also vulnerable to storms and offshore weather events.

In the long term it may be important to select fish that are more tolerant to environmental stressors to improve their resistance to climate change. Perhaps future expansion of aquaculture areas can also consider the resilience of different species and the predicted environmental parameters when deciding on the biocapacity of fish-farming areas.

8.5 Scale Drop Disease in Asian Sea Bass

8.5.1 Introduction

Scale loss and skin lesions associated with high mortalities in Asian sea bass are known in the South-East Asian region but with no confirmed aetiology. Asian sea bass (*Lates calcarifer*, Bloch 1790), also known as barramundi, is an economically important warmwater fish species farmed throughout the year in the Asia Pacific region. Commercial farming is often carried out in brackish or freshwater ponds, but more recently production has shifted towards large-scale integrated farms with on-land hatcheries and nurseries, followed by net-cage grow-out operations, similar to salmon farming. This fish has high market value and also suits intensive culture in land-based RAS with high stocking densities (Schipp *et al.*, 2007).

The scale drop syndrome was attributed to various pathogens commonly found in fish with skin lesions. Laboratory findings often reported ubiquitous external parasites, *T. maritimum* and/or *Vibrio* spp. infections associated with lesions; however, these were not consistent and were suspected as secondary pathogens (Gibson-Kueh *et al.*, 2012).

Isolation of a novel virus from clinically affected fish, and fulfilment of Koch's postulate and virus identification, confirmed the causative agent, namely *Scale drop disease virus* (SDDV), as a member of the *Iridoviridae* family (Guelen *et al.*, 2014; de Groof *et al.*, 2015). The virus showed phylogenetic relatedness to viruses in the genus *Megalocytivirus*, which consists of many known fish pathogens of temperate and warmwater fish. Interestingly, SDDV forms a separate branch from *Infectious spleen and kidney necrosis virus* (ISKNV) and is a new species within this genus (Chinchar *et al.*, 2021).

Typical clinical signs of fish affected by SDDV include darkened dorsal region of the body, scale loss (not always observed), reddening and haemorrhage of the ventral part of the body, pallor of gills, tail and fin erosions, and exophthalmia (Fig. 8.4). Affected fish are lethargic, lack schooling behaviour, and sometimes exhibit neurological signs, including spiral swimming. The pathology of this condition is a result of generalized vasculitis in multiple organs (de Groof *et al.*, 2015).

Fig. 8.4. Adult farmed Asian sea bass showing scale loss, skin and fin lesions (arrows) due to infection with SDDV. (Photograph courtesy of Masato Miyata, Prime Aquaculture, Singapore.)

Another recently discovered virus, *Lates calcarifer* herpesvirus (LCHV) (Chang *et al.*, 2018), causes similar clinical signs and mortality pattern as SDDV infections. Infections with LCHV are often observed within 7 days of stocking in marine net cages and this virus causes a more acute infection with higher morbidity (Y.S. Lee, Singapore, 2022, personal communication). Differentiating the two viruses may be important as autogenous vaccines exist for SDDV but not for LCHV (D. Chee, Singapore, 2022, personal communication).

8.5.2 Impacts on fish production

Since the initial identification of SDDV and the sharing of diagnostic data, there has been subsequent reporting of this virus in several South-East Asian countries including Indonesia, Thailand, Malaysia, Vietnam and Singapore (Senapin *et al.*, 2019; Kerddee *et al.*, 2020; Nurliyana *et al.*, 2020; Sriisan *et al.*, 2020; OIE, 2022). These reports confirm SDDV as a key pathogen causing a significant impact on Asian sea bass culture in the region. While for many years SDDV was reported only in Asian sea bass, recently SDDV-like viruses have been reported in European chub, *Squalius cephalus* (Halaly *et al.*, 2019), in England and in yellowfin sea bream, *Acanthopagrus latus* (Fu *et al.*, 2021), in China. This highlights the potential wider host range for this virus.

SDDV is a primary pathogen of fish that causes chronic, persistent disease with mortalities of up to 50% (de Groof *et al.*, 2015). Farmers observe that subsequent secondary bacterial infections cause fish to be less productive, with some surviving fish exhibiting chronic abscesses that result in postharvest losses.

In contrast to infections with ISKNV, which mainly affect juvenile and younger fish, SDDV has the propensity to infect and cause disease in both juveniles and adult fish with infections being reported in fish greater than 3 kg (Y.S. Lee, Singapore, 2022, personal communication). These late-onset infections lead to significant economic losses for affected farms.

8.5.3 Diagnosis

In the field, skin lesions caused by SDDV are easily confused with those caused by *T. maritimum* and other marine bacterial species, as well as parasite infestation. SDDV is often observed within a month post-stocking of fingerlings from freshwater land-based nurseries to grow-out net cages in the sea. The first clinical signs are darkened coloration in a few fish with patches of skin necrosis and scale loss (Fig. 8.4). Affected fish usually look moribund and are in the corners of the nets (Fig. 8.5). Initial morbidity is low; however, the disease spreads progressively to other fish within the cage and net-pen system over a few weeks (S.F. Chang, Singapore, 2022, personal communication).

SDDV can be detected using PCR tests (de Groof *et al.*, 2015). Development of other test methodologies using semi-nested PCR, quantitative PCR (qPCR), cross-priming amplification and clustered regularly interspaced short palindromic repeats (CRISPR)-based nucleic acid detection have also yielded useful outcomes (Charoenwai *et al.*, 2019; Sriisan *et al.*, 2020; Prasitporn *et al.*, 2021; Sukonta *et al.*, 2022). In addition, SDDV grows in susceptible cell lines including sea bass kidney (SK) SK21 and Asian sea bass brain (SBB) cells (de Groof *et al.*, 2015) at

Fig. 8.5. Asian sea bass in an ocean net pen showing scale loss and skin lesions. Note: observable clinical signs are not pathognomonic for SDDV or LCHV infection, and confirmatory laboratory tests are critical. (Photograph courtesy of Siow Foong Chang, National Parks Board, Singapore.)

28°C. Typical cytopathogenic effects (CPEs) such as enlarged cells are noted at 2 days post-inoculation with advanced CPEs observed after 6 days.

8.5.4 Transmission

SDDV occurs mainly in open net cages but transmission to the nursery or tank systems has also been reported anecdotally (Y.S. Lee, Singapore, 2022, personal communication), with untreated seawater, contaminated fish or fomites being the likely sources of exposure for these land-based systems. In open-water production systems, the disease usually occurs at 1 to 2 months post-introduction to net cages, gradually progressing over a period of 1–2 months. Infections are also seen in older fish in endemic areas. The epidemiology of this disease has not been fully elucidated. While transmission occurs through horizontal transmission between fish in adjacent nets, more studies are needed to determine if vertical transmission also occurs.

The spread of viral diseases in aquaculture is often attributed to movement of live fish. The expansion of Asian sea bass production to new farming locations carries a risk of spreading this pathogen around South-East Asia. The transboundary transfer of fingerlings could introduce this disease into new regions (Ramírez-Paredes *et al.*, 2021). Once a pathogen is introduced to an open-water system it is difficult to prevent its spread. It is possible that wild or feral sea bass become infected and act as reservoirs to maintain the infection in an area and spread infection between farms. Further, long-distance transportation via ballast water is plausible. Other similar viruses such as red sea bream iridovirus (RSIV) have been transferred via ballast water as far as North America from Asia (Kim *et al.*, 2016). The wide host range of other megalocytiviruses suggests this family of viruses has the potential to establish in different fish species and in new geographic areas.

8.5.5 Anticipated effects of climate change

Early reports suggested that the severity of clinical disease associated with SDDV appears to follow a seasonal pattern, with the arrival of the south-west monsoon/inter-monsoon season starting around September bringing cooler temperatures and other environmental stressors, such as salinity shifts, particularly in tropical regions (de Groof *et al.*, 2015). Farmers report an increase in SDDV occurrence during colder seasons in South-East Asia associated with increased rainfall and temperatures below 30°C. The increase in water temperature related to climate change may reduce the clinical signs associated with warmwater megalocytiviruses, as they tend to affect fish at cooler temperatures (Sah Putra *et al.*, 2020); however, the increase in the number and severity of extreme weather events (IPCC, 2021) such as storms during the monsoon season in these regions could increase stress and predispose fish to infection with SDDV. As weather patterns change due to increased air and ocean temperatures, there is also a potential risk of SDDV spreading to new areas and new species. There have been recent discoveries of SDDV in novel host species (Halaly *et al.*, 2019; Fu *et al.*, 2021), which highlights the potential for SDDV to establish itself in other aquaculture systems worldwide.

8.5.6 Control and prevention

One of the key strategies for prevention of infectious diseases is vaccination. Experimental vaccines against SDDV are effective under laboratory conditions (de Groof *et al.*, 2015). Both inactivated SDDV and recombinant subunit vaccines comprising the major capsid protein are partially efficacious, but neither of these vaccines are licensed for commercial use. A commercially available megalocytivirus vaccine containing RSIV as the antigen did not offer heterologous cross-protection against SDDV. This was postulated to be due to the differences in genetic makeup as well as the different biological mechanisms of replication between RSIV and SDDV (de Groof *et al.*, 2015).

The successful use of commercial vaccines against other megalocytiviruses, such as ISKNV and RSIV, suggests that vaccination could be important for the control of SDDV. Production systems based on biosecure land-based nurseries could vaccinate fingerlings prior to transfer to saltwater net pens where exposure to SDDV predominantly occurs. Small-scale production of autogenous recombinant vaccines has been tested with mixed results (D. Chee, Singapore, 2022, personal communication) and research continues for a commercial vaccine against SDDV.

For farms within SDDV-unaffected areas, biosecurity is key to keeping the pathogen out. This requires coordinated efforts on the part of farmers within the area with shared water. The World Organisation for Animal Health (OIE) recently published a new chapter in the *Aquatic Animal Health Code* (OIE, 2021) on broad biosecurity guidelines which can help farmers manage the risk of introduction of non-endemic pathogens to their farms. A recent publication on sea-based net-pen farm and area-level biosecurity in Asia outlines specific practices that can be adopted to reduce the risk of introducing diseases of these types on to farms (L. Jahangiri, S. St-Hilaire and C.F. Leung, 2022, unpublished results). Some practices such as area- or farm-level fallowing that normally help reduce pathogens from the surrounding area may not be economically viable for small farm holders or ineffective if the pathogen is endemic in farmed and/or wild fish populations. If it is not possible to prevent the introduction of this virus in an area, farmers may be able to reduce losses by minimizing spread within their farm using different strategies to reduce stress, such as maintaining good water quality and nutrition, using immunostimulants, as well as detecting infections early to cull and/or apply treatments to reduce secondary bacterial infections.

There is also a new hyperthermic treatment approach (in contained systems) that is still experimental but shows promise in reducing the impact of viral diseases and is currently being evaluated by some farmers. This involves heating water up to 39°C for about 30 min to a few hours before allowing the temperature to cool gradually (Michel, 2018). Oxygenation of water must be increased and monitored closely, together with toxic ammonia build-up, to avoid fish mortality. Regular repetitive treatments are sometimes recommended depending on the epidemiology of the target pathogen. The empirical explanation for this heat treatment follows two lines of thought. First, it may be outside the optimum temperature range of SDDV, which is around 28°C. Second, expression of heat-shock proteins (HSPs) in response to hyperthermic exposure may induce disease resistance (Sung and MacRae, 2011; Sun *et al.*, 2021). The use of hyperthermic treatments offers an alternative option to farmers facing viral pathogens early in their fish production when vaccination is not an option (i.e. nurseries). When used in combination with good biosecurity principles, this treatment regime could enable recovery of fish infected with a number of potential viruses including viral nervous necrosis virus (VNNV), SDDV, ISKNV and LCHV (A. Michel, Paris, 2019, personal communication). This treatment regime is gaining widespread use in nurseries and hatcheries in Asia and Africa for multiple species of fish including sea bass, groupers and tilapia (A. Michel, Paris, 2019, personal communication). However, extensive research is still needed to understand the mechanisms behind heat treatment to control viruses in aquaculture.

Once the principle behind heat treatments is better understood and its limitations and risk to fish clarified, it may be an effective supplement to an integrated biosecurity strategy that includes early screening for pathogens and importing healthy fish, periodic fallowing of farms, and vaccination programmes to prevent and control diseases in aquaculture systems.

ANTIMICROBIALS

8.6 Use of Antimicrobials and Antimicrobial Resistance

8.6.1 Introduction

As aquaculture industries intensify and farms increase in size, infectious disease outbreaks are increasing in frequency, as are antibiotic treatments used to mitigate these events (Schar *et al.*, 2020). Schar *et al.* (2020) estimated an increase in antibiotic use of 33% (from approximately 10,259 to 13,600 tonnes) between 2017 and 2030 if industries continue their current practices and growth projections. Paralleling the current increase in use of antibiotics is an increase in AMR (antimicrobial resistance) (Watts *et al.*, 2017).

Two commonly accepted drivers of AMR are the overuse of one or similar classes of drugs in an area or system, and the underdosing of antimicrobials (Andersson and Hughes, 2014), both of which occur in aquaculture. The limited number of approved products for treating bacterial infections in aquatic animals in many countries (Chen, J. *et al.*, 2020) means the same product will be used repeatedly over time, contributing to AMR risk. The primary method of delivering antibiotic medication to aquaculture fish is through medicated feed, which means that all fish in a pen, and often all fish on a farm, will be given antibiotics when a treatment is initiated. The sheer size of the commercial aquaculture systems means that large numbers of fish will consume antibiotics during treatments. So, not only is there a limited number of products licensed for use, but when they are used, substantial quantities are dispensed.

Perhaps even more problematic than the overuse of antibiotics is the underdosing that occurs with medicated feed treatments. The difficulty in properly dosing fish in large aquaculture populations occurs because of several issues, including using poor-quality or substandard drugs (Leung *et al.*, 2020), leaching of the medication before fish have time to consume it (Daniel, 2009) and uneven consumption of medicated feed within the population (Talbot *et al.*, 1999; Price *et al.*, 2019). Treating fish with subtherapeutic levels of antibiotics is a known risk for AMR (Andersson and Hughes, 2014).

8.6.2 Impacts on fish production

The increased use of antibiotics in aquaculture has direct and indirect impacts on the industry. The more a drug is used in an area the greater the risk of AMR in the bacterial populations. As there are only a handful of approved antibiotics for use in aquaculture in most countries (Chen, J. *et al.*, 2020), resistance to any of these products reduces the treatment options for farmers. AMR also directly results in increased costs to farmers, as it leads to an inability to treat animals and/or requires higher dosages (more product) to achieve a successful treatment and/or necessitates more frequent repeated treatments.

An indirect cost of the increased use of antibiotics in aquaculture globally is the negative market perception of this practice and the public health implications that the use of these products has on food safety. Although all antibiotics licensed for food animals have labelled withdrawal periods, where animals should not be slaughtered for food consumption, the more these products are used the higher the probability that some fish with residues will reach the market. There are many examples of fish products that have failed antibiotic residue tissue testing in different parts of the world (Liu *et al.*, 2017). Consumption of low dosages of antibiotics through food is considered a public health concern, as it may result in AMR.

The other issue that has negatively impacted the public's perception of farmed fish with regard to antibiotic usage is the environmental contamination that occurs with its use (He *et al.*, 2012; Lulijwa *et al.*, 2020). The nature of open net-pen farming and the method of delivering medicated feeds mean there is very high likelihood that some product will be lost to the environment. The quantity of environmental contamination depends on the product, the quality of the top coating, the feeding strategy used, the behaviour of the animals and the amount that is excreted in the faeces. Different fish species have different feeding behaviours: the slower feed is consumed, the greater the environmental leaching. Also, uneaten pellets will end up under the cage system to be decomposed by bacteria or consumed by other aquatic animals. Even when the pellets are consumed, fish do not absorb all the medication and some active compounds may be excreted in the faeces. Some studies have found that, depending on the medication, the excretion of antibiotics within faeces can be up to 90% (Sarmah *et al.*, 2006). Environmental contamination of the aquatic ecosystem around fish farms has been well documented (He *et al.*, 2012; Miranda *et al.*, 2018; Lulijwa *et al.*, 2020) and may lead to reduced bacterial biodiversity and AMR issues in environmental bacterial populations, which may potentially spill over to fish pathogens as well as human pathogens.

8.6.3 Anticipated effect of climate change on antibiotics use and antimicrobial resistance

Proliferation of bacteria resulting from the increase in water temperature is relatively well documented, and more recent research also suggests that temperature may influence the expression of bacterial virulence factors (Cascarano *et al.*, 2021). In addition to the direct increase in exposure to pathogens associated with variation in weather patterns, increases in water turbidity from runoff, leading to high organic levels and/or low dissolved oxygen, and changes in water temperature can also directly stress fish, which can increase their susceptibility to disease and exacerbate mortalities. There is evidence that both warmer water temperatures and increased frequency and volume of rain events lead to increased use of antibiotics by fish farmers, presumably due to increases in bacterial disease manifestation (Nguyen *et al.*, 2015).

In addition to increasing the frequency of bacterial infection and therefore antibiotic treatments, an increase in the water temperature may also result in an increase in treatment failure if the higher metabolism of fish is not taken into consideration when dosing. Even a change of a few degrees Celsius can alter the pharmacokinetics of drugs, which impacts their efficacy and clearance time (Zanuzzo *et al.*, 2022). This phenomenon is likely to be more significant for antibiotics that have short half-lives such as florfenicol (Zanuzzo *et al.*, 2022). The net effect could be more animals being underdosed, resulting in higher treatment failures and potentially an increase in the risk of AMR.

The level of environmental leaching and breakdown of drugs in the aquatic ecosystem is also likely to be impacted by climate change via changes in pH and water temperature. The impact of pH on leaching and drug breakdown is difficult to predict as it will depend on the product. However, the effect of temperature is clear: the warmer the temperature the greater the leaching and the faster the degradation of medications (Daniel, 2009). Although the two phenomena could be viewed as counteracting each other, the likely result will be greater levels of drugs escaping into the ecosystem, thus potentially increasing the risk of AMR. Also, the greater the leaching, the more problematic it is to treat fish accurately and effectively.

The environmental changes documented with climate change are also likely to have a direct effect on AMR in bacterial populations (Burnham, 2021). Several researchers have recently found that temperature-stressed bacteria can increase their selection for AMR genes (Kong *et al.*, 2017; MacFadden *et al.*, 2018; Pepi and Focardi, 2021). In addition, the anticipated changes to other water quality parameters, such as changes in pH, increases in pollutants and agricultural runoff, may also stress bacterial populations and exacerbate AMR (Cabello *et al.*, 2013; Burnham, 2021).

8.6.4 Control and/or prevention

The best way to prevent AMR is to prevent the use of antibiotics, which means to prevent bacterial infections. Individual farms can implement specific strategies to reduce the likelihood of bacterial diseases (see chapters on bacterial diseases and their control, this volume). Area-level biosecurity practices can also help reduce the spread of infectious agents between farms within close proximity (Midtlyng *et al.*, 2011); however, no net-pen industry has been able to completely eliminate pathogens, due to the interactions with wild fisheries and incomplete farm and area-level biosecurity management practices. It is therefore imperative to improve treatments strategies, as well as control the misuse of antibiotics, to reduce the risk of AMR.

Limiting the use of antibiotics for treatment purposes only, and monitoring and/or adjusting dosages based on pharmacokinetic data to avoid underdosing fish, would reduce the risk of bacterial communities developing AMR. Ensuring prudent use of antibiotics requires good stewardship and governance. Stronger regulations on drug use and how these products are incorporated into feeds are required to avoid their misuse. Perhaps more importantly, enforcement of these regulations is necessary. The Norwegian government provides an example of good governance over antibiotics use in aquaculture (Norwegian Veterinary Institute, 2016). They have been monitoring use of antibiotics in aquaculture for several years and have been able to reduce usage over time (reviewed in Burridge *et al.*, 2008).

In addition to good governance, strategies to reduce underdosing of fish are needed as this problem is common that, as mentioned earlier, will only increase with the increasing water temperatures predicted under climate change. Using quality-assured medications will also help with accurate dosing. Finding methods to improve the top coating of feeds to reduce environmental leaching will help reduce both underdosing and environmental contamination. Products such as pre-gelatinized potato starch and pectin-based products exist and could reduce leaching, but they have been explored on a small scale only (Kankate *et al.*, 2020).

Ensuring even consumption of medicated feed across fish populations will also reduce the variation in dosing; however, this is sometimes difficult to achieve as there are many reasons for uneven consumption of medicated feed, including sick fish and feeding strategies. Early detection and treatment of bacterial diseases on farms can reduce the number of sick fish not feeding during an antibiotic treatment and help decrease treatment failures, thus the need to repeat treatments (Price *et al.*, 2019). The practice of delivering small amounts of feed frequently throughout the day appears to reduce the variation in feed consumption in saltwater salmon farms in Chile (R. Ibarra, California, 2021, personal communication). However, no publications are available that assess this practice for its efficacy at delivering medication to large populations of fish. Avoiding and minimizing the use of antibiotics are critical to reducing the risk of bacterial AMR, but this is likely to be difficult given the predicted growth of the aquaculture industry globally and the lack of enforcement of antimicrobial regulations.

8.7 Conclusion

This chapter highlights some of the potential and emerging issues facing aquaculture over the next 10 years. Mitigation for some of the problems is provided, but it is recognized that many of the topics covered are outside the control of individual farmers and the industry. The solutions to many of the aquatic ecosystem issues highlighted here require global changes across many industries. Further, there may be mitigating circumstances that require new innovations to circumvent or address problems until the underlying issues are addressed.

References

Alshammari, E., Patel, M., Sachidanandan, M., Kumar, P. and Adnan, M. (2019) Potential evaluation and health fostering intrinsic traits of novel probiotic strain *Enterococcus durans* F3 isolated from the gut of fresh water fish *Catla catla*. *Food Science of Animal Resources* 39(5), 844.

Andersson, D.I. and Hughes, D. (2014) Microbiological effects of sublethal levels of antibiotics. *Nature Reviews Microbiology* 12(7), 465–478.

Ángeles Esteban, M. (2012) An overview of the immunological defenses in fish skin. *International Scholarly Research Notices* 2012, 853470.

Anifowoshe, A.T., Roy, D., Dutta, S. and Nongthomba, U. (2022) Evaluation of cytogenotoxic potential and embryotoxicity of KRS–Cauvery River water in zebrafish (*Danio rerio*). *Ecotoxicology and Environmental Safety* 233, 113320.

Auta, H.S., Emenike, C.U. and Fauziah, S.H. (2017) Distribution and importance of microplastics in the marine environment: a review of the sources, fate, effects, and potential solutions. *Environment International* 102, 165–176.

Avendaño-Herrera, R., Irgang, R., Sandoval, C., Moreno-Lira, P., Houel, A., *et al.* (2016) Isolation, characterization and virulence potential of *Tenacibaculum dicentrarchi* in salmonid cultures in Chile. *Transboundary and Emerging Diseases* 63(2), 121–126.

Azzouz, A., Colón, L.P., Souhail, B. and Ballesteros, E. (2019) A multi-residue method for GC-MS determination of selected endocrine disrupting chemicals in fish and seafood from European and North African markets. *Environmental Research* 178, 108727.

Barboza, L.G.A., Lopes, C., Oliveira, P., Bessa, F., Otero, V., *et al.* (2019) Microplastics in wild fish from North East Atlantic Ocean and its potential for causing neurotoxic effects, lipid oxidative damage, and human health risks associated with ingestion exposure. *Science of the Total Environment* 717, 134625. https://doi.org/10.1016/j.scitotenv.2019.134625

Berge, G.M., Ytteborg, E., Østbye, T.K.K., Sundh, H., Rud, I., *et al.* (2019) *Ernæringens betydning for skinn-, tarm-, og gjellehelse hos laks. Faglig sluttrapport.* Nofima Rapport 17/2019. Available at: https://nofima.brage.unit.no/nofima-xmlui/bitstream/handle/11250/2600679/Rapport%2b17-2019%2bErn%25C3%

25A6ringens%2bbetydning%2bfor%2bskinn-%2btarm-%2bog%2bgjellehelse%2bhos%2blaks.pdf?sequence=1&isAllowed=y (accessed15 October 2022).
Beyer, A. and Biziuk, M. (2008) Methods for determining pesticides and polychlorinated biphenyls in food – problems and challenges. *Critical Reviews in Food Science and Nutrition* 48(10), 888–904.
Bloecher, N., Powell, M., Hytterød, S., Gjessing, M., Wiik-Nielsen, J., *et al*. (2018) Effects of cnidarian biofouling on salmon gill health and development of amoebic gill disease. *PLoS ONE* 13(7), e0199842.
Boerlage, A.S., Ashby, A., Herrero, A., Reeves, A., Gunn, G.J. and Rodger, H.D. (2020) Epidemiology of marine gill diseases in Atlantic salmon (*Salmo salar*) aquaculture: a review. *Reviews in Aquaculture* 12(4), 2140–2159.
Bosveld, A.T. and van den Berg, M. (2002) Reproductive failure and endocrine disruption by organohalogens in fish-eating birds. *Toxicology* 181, 155–159.
Bowden, T.J., Thompson, K.D., Morgan, A.L., Gratacap, R.M. and Nikoskelainen, S. (2007) Seasonal variation and the immune response: a fish perspective. *Fish & Shellfish Immunology* 22(6), 695–706.
Brette, F., Shiels, H., Galli, G., Cross, C., Incardona, J.P., *et al*. (2017) A Novel cardiotoxic mechanism for a pervasive global pollutant. *Scientific Reports* 7, 41476.
Burnham, J.P. (2021) Climate change and antibiotic resistance: a deadly combination. *Therapeutic Advances in Infectious Disease* 8, 2049936121991374.
Burridge, L., Weis, J.S., Cabello, F., Pizarro, J. and Bostick, K. (2008) Chemical use in salmon aquaculture: a review of current practices and possible environmental effects. *Aquaculture* 306, 7–23.
Cabello, F.C., Godfrey, H.P., Tomova, A., Ivanova, L., Dölz, H., *et al*. (2013) Antimicrobial use in aquaculture re-examined: its relevance to antimicrobial resistance and to animal and human health. *Environmental Microbiology* 15(7), 1917–1942.
Cabillon, N.A.R. and Lazado, C.C. (2019) Mucosal barrier functions of fish under changing environmental conditions. *Fishes* 4(1), 2.
Camilleri, M., Lyle, B.J., Madsen, K.L., Sonnenburg, J., Verbeke, K. and Wu, G.D. (2019) Role for diet in normal gut barrier function: developing guidance within the framework of food-labeling regulations. *American Journal of Physiology: Gastrointestinal and Liver Physiology* 317(1), G17–G39.
Carnevali, O., Santangeli, S., Forner-Piquer, I., Basili, D. and Maradonna, F. (2018) Endocrine-disrupting chemicals in aquatic environment: what are the risks for fish gametes? *Fish Physiology and Biochemistry* 44(6), 1561–1576.
Carpenter, E.J., Anderson, S.J., Harvey, G.R., Miklas, H.P. and Peck, B.B. (1972) Polystyrene spherules in coastal waters. *Science* 178(4062), 749–750.
Cascarano, M.C., Stavrakidis-Zachou, O., Mladineo, I., Thompson, K.D., Papandroulakis, N. and Katharios, P. (2021) Mediterranean aquaculture in a changing climate: temperature effects on pathogens and diseases of three farmed fish species. *Pathogens* 10(9), 1205.
Chang, S.F., Ng, K.S., Grisez, L., De Groof, A., Vogels, W., *et al*. (2018) Novel fish pathogenic virus. *International Patent Publication no. WO2018029301A1*.
Charoenwai, O., Meemetta, W., Sonthi, M., Dong, H.T. and Senapin, S. (2019) A validated semi-nested PCR for rapid detection of scale drop disease virus (SDDV) in Asian sea bass (*Lates calcarifer*). *Journal of Virological Methods* 268, 37–41.
Chen, J., Tanguay, R.L., Tal, T.L., Gai, Z., Ma, X., *et al*. (2014) Early life perfluorooctanesulphonic acid (PFOS) exposure impairs zebrafish organogenesis. *Aquatic Toxicology* 150, 124–132.
Chen, J., Sun, R., Pan, C., Sun, Y., Mai, B. and Li, Q.X. (2020) Antibiotics and food safety in aquaculture. *Journal of Agricultural and Food Chemistry* 68(43), 11908–11919.
Chen, Q., Lackmann, C., Wang, W., Seiler, T.B., Hollert, H. and Shi, H. (2020) Microplastics lead to hyperactive swimming behaviour in adult zebrafish. *Aquatic Toxicology* 224, 105521.
Chinchar, V.G., Duffus, A.L. and Brunner, J.L. (2021) Ecology of viruses infecting ectothermic vertebrates: the impact of *Ranavirus* infections on amphibians. In: Hurst, C.J. (ed.) *Studies in Viral Ecology*, 2nd edn. Wiley, Chichester, UK, pp. 405–447.
Chobtang, J., De Boer, I.J., Hoogenboom, R.L., Haasnoot, W., Kijlstra, A. and Meerburg, B.G. (2011) The need and potential of biosensors to detect dioxins and dioxin-like polychlorinated biphenyls along the milk, eggs and meat food chain. *Sensors* 11(12), 11692–11716.
Cidad, M., Peral, I., Ramos, S., Basurco, B., López-Francos, A., *et al*. (2018) *Assessment of Mediterranean Aquaculture Sustainability*. Deliverable 1.2 of the Horizon 2020 project MedAID (GA number 727315), published in the project web site on 21.12.2018. Available at: https://archimer.ifremer.fr/doc/00515/62630/ (accessed 15 October 2022).
Clarke, A. and Fraser, K.P.P. (2004) Why does metabolism scale with temperature? *Functional Ecology* 18(2), 243–251.

Collard, F., Gilbert, B., Compère, P., Eppe, G., Das, K., *et al.* (2017) Microplastics in livers of European anchovies (*Engraulis encrasicolus*, L.). *Environmental Pollution* 229, 1000–1005.

Compa, M., Ventero, A., Iglesias, M. and Deudero, S. (2018) Ingestion of microplastics and natural fibres in *Sardina pilchardus* (Walbaum, 1792) and *Engraulis encrasicolus* (Linnaeus, 1758) along the Spanish Mediterranean coast. *Marine Pollution Bulletin* 128, 89–96.

Contador, E., Methner, P., Ryerse, I., Huber, P., Lillie, B.N., *et al.* (2016) Epitheliocystis in lake trout *Salvelinus namaycush* (Walbaum) is associated with a β-proteobacteria. *Journal of Fish Diseases* 39(3), 353–366.

Daniel, D.B., Ashraf, P.M. and Thomas, S.N. (2020) Microplastics in the edible and inedible tissues of pelagic fishes sold for human consumption in Kerala, India. *Environmental Pollution* 266, 115365.

Daniel, P. (2009) Drugs and chemicals in aquafeeds: the problems and solutions. In: Rogers, C. and Basurco, B. (eds) *The Use of Veterinary Drugs and Vaccines in Mediterranean Aquaculture*. Options Méditerranéennes: Série A. Séminaires Méditerranéens No. 86. CIHEAM, Zaragoza, Spain, pp. 85–94.

Dash, S., Das, S.K., Samal, J. and Thatoi, H.N. (2018) Epidermal mucus, a major determinant in fish health: a review. *Iranian Journal of Veterinary Research* 19(2), 72–81.

DeCourten, B.M. and Brander, S.M. (2017) Combined effects of increased temperature and endocrine disrupting pollutants on sex determination, survival, and development across generations. *Scientific Reports* 7(1), 9310.

DeCourten, B.M., Connon, R.E. and Brander, S.M. (2019) Direct and indirect parental exposure to endocrine disruptors and elevated temperature influences gene expression across generations in a euryhaline model fish. *PeerJ* 7, e6156.

de Groof, A., Guelen, L., Deijs, M., van der Wal, Y., Miyata, M., *et al.* (2015) A novel virus causes scale drop disease in *Lates calcarifer*. *PLoS Pathogens* 11(8), e1005074.

Delbes, G., Blázquez, M., Fernandino, J.I., Grigorova, P., Hales, B.F., *et al.* (2022) Effects of endocrine disrupting chemicals on gonad development: mechanistic insights from fish and mammals. *Environmental Research* 204, 112040.

de Sá, L.C., Luís, L.G. and Guilhermino, L. (2015) Effects of microplastics on juveniles of the common goby (*Pomatoschistus microps*): confusion with prey, reduction of the predatory performance and efficiency, and possible influence of developmental conditions. *Environmental Pollution* 196, 359–362.

De Wit, C.A., Fisk, A.T., Hobbs, K.E. and Muir, D.C. (2003) Summary of effects of persistent organic pollutants (POPs) in Arctic wildlife. *Organohalogen Compounds* 61, 309–312.

Diamanti-Kandarakis, E., Bourguignon, J.P., Giudice, L.C., Hauser, R., Prins, G.S., *et al.* (2009) Endocrine-disrupting chemicals: an Endocrine Society scientific statement. *Endocrine Reviews* 30(4), 293–342.

Dutney, L., Elizur, A. and Lee, P. (2017) Analysis of sexually dimorphic growth in captive reared cobia (*Rachycentron canadum*) and the occurrence of intersex individuals. *Aquaculture* 468, 348–355.

Espinosa, C., Cuesta, A. and Esteban, M.Á. (2017) Effects of dietary polyvinylchloride microparticles on general health, immune status and expression of several genes related to stress in gilthead seabream (*Sparus aurata* L.). *Fish & Shellfish Immunology* 68, 251–259.

Fivelstad, S., Hosfeld, C.D., Medhus, R.A., Olsen, A.B. and Kvamme, K. (2018) Growth and nephrocalcinosis for Atlantic salmon (*Salmo salar* L.) post-smolt exposed to elevated carbon dioxide partial pressures. *Aquaculture* 482, 83–89.

Ford, H.V., Jones, N.H., Davies, A.J., Godley, B.J., Jambeck, J.R., *et al.* (2022) The fundamental links between climate change and marine plastic pollution. *Science of the Total Environment* 806, 150392.

Fox, G.A. (1993) What have biomarkers told us about the effects of contaminants on the health of fish-eating birds in the Great Lakes? The theory and a literature review. *Journal of Great Lakes Research* 19(4), 722–736.

Foyle, K.L., Hess, S., Powell, M.D. and Herbert, N.A. (2020) What is gill health and what is its role in marine finfish aquaculture in the face of a changing climate? *Frontiers in Marine Science* 7, 400.

Fridman, S., Tsairidou, S., Jayasuriya, N., Sobolewska, H., Hamilton, A., *et al.* (2021) Assessment of marine gill disease in farmed Atlantic salmon (*Salmo salar*) in Chile using a novel total gross gill scoring system: a case study. *Microorganisms* 9(12), 2605.

Fu, Y., Li, Y., Fu, W., Su, H., Zhang, L., *et al.* (2021) Scale drop disease virus associated with yellowfin seabream (*Acanthopagrus latus*) ascites diseases, Zhuhai, Guangdong, southern China: the first description. *Viruses* 13(8), 1617.

Garcia, A.G., Suárez, D.C., Li, J. and Rotchell, J.M. (2021) A comparison of microplastic contamination in freshwater fish from natural and farmed sources. *Environmental Science and Pollution Research* 28(12), 14488–14497.

George, J.L. and Frear, D.E.H. (1966) Pesticides in the Antarctic. *Journal of Applied Ecology* 3(Suppl.), 155–167.

Georgescu, B., Georgescu, C., Dărăban, S., Bouaru, A. and Paşcalău, S. (2011) Heavy metals acting as endocrine disrupters. *Scientific Papers Animal Science and Biotechnologies* 44(2), 89–93.

Gibson-Kueh, S., Chee, D., Chen, J., Wang, Y.H., Tay, S., *et al.* (2012) The pathology of 'scale drop syndrome' in Asian seabass, *Lates calcarifer* Bloch, a first description. *Journal of Fish Diseases* 35(1), 19–27.

Gismervik, K., Gåsnes, S.K., Gu, J., Stien, L.H., Madaro, A. and Nilsson, J. (2019) Thermal injuries in Atlantic salmon in a pilot laboratory trial. *Veterinary and Animal Science* 8, 100081.

Gjessing, M.C., Yutin, N., Tengs, T., Senkevich, T., Koonin, E., *et al.* (2015) Salmon gill poxvirus, the deepest representative of the *Chordopoxvirinae*. *Journal of Virology* 89(18), 9348–9367.

Gjessing, M.C., Thoen, E., Tengs, T., Skotheim, S.A. and Dale, O.B. (2017) Salmon gill poxvirus, a recently characterized infectious agent of multifactorial gill disease in freshwater-and seawater-reared Atlantic salmon. *Journal of Fish Diseases* 40(10), 1253–1265.

Gjessing, M.C., Steinum, T., Olsen, A.B., Lie, K.I., Tavornpanich, S., *et al.* (2019) Histopathological investigation of complex gill disease in sea farmed Atlantic salmon. *PLoS ONE* 14(10), e0222926.

Goksøyr, A. (2006) Endocrine disruptors in the marine environment: mechanisms of toxicity and their influence on reproductive processes in fish. *Journal of Toxicology and Environmental Health, Part A* 69(1–2), 175–184.

Good, C. and Davidson, J. (2016) A review of factors influencing maturation of Atlantic salmon, *Salmo salar*, with focus on water recirculation aquaculture system environments. *Journal of the World Aquaculture Society* 47(5), 605–632.

Guelen, L., De Groof, A.D., Schrier, C.C., Grisez, L., Chang, S.F., *et al.* (2014) Scale drop disease (SDD) causative virus and derivatives thereof. *International Patent Publication no. WO/2014/191445*. Available at: https://patentscope.wipo.int/search/en/detail.jsf?docId=WO2014191445 (accessed 28 March 2022).

Güven, O., Gökdağ, K., Jovanović, B. and Kıdeyş, A.E. (2017) Microplastic litter composition of the Turkish territorial waters of the Mediterranean Sea, and its occurrence in the gastrointestinal tract of fish. *Environmental Pollution* 223, 286–294.

Halaly, M.A., Subramaniam, K., Koda, S.A., Popov, V.L., Stone, D., *et al.* (2019) Characterization of a novel megalocytivirus isolated from European chub (*Squalius cephalus*). *Viruses* 11(5), 440.

Hartmann, N.B., Huffer, T., Thompson, R.C., Hassellöv, M., Verschoor, A., *et al.* (2019) Are we speaking the same language? Recommendations for a definition and categorization framework for plastic debris. *Environmental Science & Technology* 53(3), 1039–1047.

He, X., Wang, Z., Nie, X., Yang, Y., Pan, D., *et al.* (2012) Residues of fluoroquinolones in marine aquaculture environment of the Pearl River Delta, *South China. Environmental Geochemistry and Health* 34(3), 323–335.

Hjeltnes, B., Karlsbakk, E., Mo, T.A., Mortensen, S., Olsen, R.E. and Rimstad, E. (2014) *Risk Assessment of Amoebic Gill Disease: Opinion of the Panel on Animal Health and Welfare of the Norwegian Scientific Committee for Food Safety*. VKM Report 2014:11. Norwegian Scientific Committee for Food Safety (VKM), Oslo.

Hontela, A. and Vijayan, M.M. (2008) Adrenocortical toxicology in fishes. In: Harvey, P.W., Everett, D.J. and Springall, C.J. (eds) *Adrenal Toxicology*. CRC Press, Boca Raton, Florida, pp. 259–282.

Hutchinson, T.H., Ankley, G.T., Segner, H. and Tyler, C.R. (2006) Screening and testing for endocrine disruption in fish – biomarkers as 'signposts,' not 'traffic lights,' in risk assessment. *Environmental Health Perspectives* 114(Suppl. 1), 106–114.

Iheanacho, S.C., Igberi, C., Amadi-Eke, A., Chinonyerem, D., Iheanacho, A. and Avwemoya, F. (2020) Biomarkers of neurotoxicity, oxidative stress, hepatotoxicity and lipid peroxidation in *Clarias gariepinus* exposed to melamine and polyvinyl chloride. *Biomarkers* 25(7), 603–610.

IPCC (Intergovernmental Panel on Climate Change) (2021) Summary for policymakers. In: *Climate Change 2021: The Physical Science Basis. Contribution of Working Group I to the Sixth Assessment Report*. IPCC, Geneva, Switzerland. Available at: https://www.ipcc.ch/report/ar6/wg2/resources/spm-headline-statements/ (accessed 28 March 2022).

IPEN (International Pollutants Elimination Network) (2020) *Plastic's Toxic Additives and The Circular Economy*. Available at: https://ipen.org/sites/default/files/documents/plastics_and_additives_final-low-o-en.pdf (accessed 17 March 2022).

IUCN (International Union for the Conservation of Nature) (2021) Marine plastic pollution. *Issues Brief*, November 2021. Available at: https://www.iucn.org/sites/default/files/2022-04/marine_plastic_pollution_issues_brief_nov21.pdf (accessed 15 October 2022).

Iversen, A., Hermansen, Ø., Nystøyl, R., Hess, E.J., Rolland, K.H., *et al.* (2019) *Kostnadsutvikling og forståelse av drivkrefter i norsk lakseoppdrett. Faglig sluttrapport.* Nofima Rapport 35/2019. Available at: https://nofima.brage.unit.no/nofima-xmlui/bitstream/handle/11250/2632322/Rapport+35-2019+Kostnadsutvikling+og+forst%C3%A5else+av+drivkrefter+i+norsk+lakseoppdrett+-+faglig+sluttrapport.pdf?sequence=2 (accessed 15 October 2022).

Jarque, S., Fetter, E., Veneman, W.J., Spaink, H.P., Peravali, R., *et al.* (2018) An automated screening method for detecting compounds with goitrogenic activity using transgenic zebrafish embryos. *PLoS ONE* 13(8), e0203087.

Jarque, S., Ibarra, J., Rubio-Brotons, M., García-Fernández, J. and Terriente, J. (2019) Multiplex analysis platform for endocrine disruption prediction using zebrafish. *International Journal of Molecular Sciences* 20(7), 1739.

Jensen, I.J., Eilertsen, K.E., Otnæs, C.H.A., Mæhre, H.K. and Elvevoll, E.O. (2020) An update on the content of fatty acids, dioxins, PCBs and heavy metals in farmed, escaped and wild Atlantic salmon (*Salmo salar* L.) in Norway. *Foods* 9(12), 1901.

Jenssen, B.M. (2006) Endocrine-disrupting chemicals and climate change: a worst-case combination for Arctic marine mammals and seabirds? *Environmental Health Perspectives* 114(Suppl. 1), 76–80.

Jeong, C.B., Won, E.J., Kang, H.M., Lee, M.C., Hwang, D.S., *et al.* (2016) Microplastic size-dependent toxicity, oxidative stress induction, and p-JNK and p-p38 activation in the monogonont rotifer (*Brachionus koreanus*). *Environmental Science & Technology* 50(16), 8849–8857.

Jin, Y., Shu, L., Huang, F., Cao, L., Sun, L. and Fu, Z. (2011) Environmental cues influence EDC-mediated endocrine disruption effects in different developmental stages of Japanese medaka (*Oryzias latipes*). *Aquatic Toxicology* 101(1), 254–260.

Jin, Y., Xia, J., Pan, Z., Yang, J., Wang, W. and Fu, Z. (2018) Polystyrene microplastics induce microbiota dysbiosis and inflammation in the gut of adult zebrafish. *Environmental Pollution* 235, 322–329.

Jobling, S., Beresford, N., Nolan, M., Rodgers-Gray, T., Brighty, G.C., *et al.* (2002) Altered sexual maturation and gamete production in wild roach (*Rutilus rutilus*) living in rivers that receive treated sewage effluents. *Biology of Reproduction* 66(2), 272–281.

Jørgensen, E.H., Vijayan, M.M., Aluru, N. and Maule, A.G. (2002) Fasting modifies Aroclor 1254 impact on plasma cortisol, glucose and lactate responses to a handling disturbance in Arctic charr. *Comparative Biochemistry and Physiology, Part C: Toxicology & Pharmacology* 132(2), 235–245.

Jovanović, B. (2017) Ingestion of microplastics by fish and its potential consequences from a physical perspective. *Integrated Environmental Assessment and Management* 13(3), 510–515.

Kankate, D., Panpalia, S.G., Kumar, J.K., Kennedy, J.F. (2020) Studies to predict the effect of pregelatinization on excipient property of maize and potato starch blends. *International Journal of Biological Macromolecules* 164, 1206–1214. https://doi.org/10.1016/j.ijbiomac.2020.07.170

Kar, S., Sangem, P., Anusha, N. and Senthilkumaran, B. (2021) Endocrine disruptors in teleosts: evaluating environmental risks and biomarkers. *Aquaculture and Fisheries* 6(1), 1–26.

Karlsen, C., Vanberg, C., Mikkelsen, H. and Sørum, H. (2014) Co-infection of Atlantic salmon (*Salmo salar*), by *Moritella viscosa* and *Aliivibrio wodanis*, development of disease and host colonization. *Veterinary Microbiology* 171(1–2), 112–121.

Kelly, B.C., Gobas, F.A. and McLachlan, M.S. (2004) Intestinal absorption and biomagnification of organic contaminants in fish, wildlife, and humans. *Environmental Toxicology and Chemistry* 23(10), 2324–2336.

Kerddee, P., Dong, H.T., Chokmangmeepisarn, P., Rodkhum, C., Srisapoome, P., *et al.* (2020) Simultaneous detection of scale drop disease virus and *Flavobacterium columnare* from diseased freshwater-reared barramundi *Lates calcarifer*. *Diseases of Aquatic Organisms* 140, 119–128.

Kim, Y., Aw, T.G. and Rose, J.B. (2016) Transporting ocean viromes: invasion of the aquatic biosphere. *PLoS ONE* 11(4), e0152671.

Klakegg, Ø., Salonius, K., Nilsen, A., Fülberth, M. and Sørum, H. (2020a) Enhanced growth and decreased mortality in Atlantic salmon (*Salmo salar*) after probiotic bath. *Journal of Applied Microbiology* 129(1), 146–160.

Klakegg, Ø., Myhren, S., Juell, R.A., Aase, M., Salonius, K. and Sørum, H. (2020b) Improved health and better survival of farmed lumpfish (*Cyclopterus lumpus*) after a probiotic bath with two probiotic strains of *Aliivibrio*. *Aquaculture* 518, 734810.

Kong, H.K., Law, H.W., Liu, X., Law, C.O., Pan, Q., *et al.* (2017) Transcriptomic analysis of *Laribacter hongkongensis* reveals adaptive response coupled with temperature. *PLoS ONE* 12(1), e0169998.

Kristoffersen, A.B., Rees, E.E., Stryhn, H., Ibarra, R., Campisto, J.L., *et al.* (2013) Understanding sources of sea lice for salmon farms in Chile. *Preventive Veterinary Medicine* 111(1–2), 165–175.

Kwon, J.H., Kim, J.W., Pham, T.D., Tarafdar, A., Hong, S., *et al.* (2020) Microplastics in food: a review on analytical methods and challenges. *International Journal of Environmental Research and Public Health* 17(18), 6710.

La Merrill, M.A., Vandenberg, L.N., Smith, M.T., Goodson, W., Browne, P., *et al.* (2020) Consensus on the key characteristics of endocrine-disrupting chemicals as a basis for hazard identification. *Nature Reviews Endocrinology* 16(1), 45–57.

Lamková, K., Šimková, A., Palíková, M., Jurajda, P. and Lojek, A. (2007) Seasonal changes of immunocompetence and parasitism in chub (*Leuciscus cephalus*), a freshwater cyprinid fish. *Parasitology Research* 101(3), 775–789.

Landos, M., Smith, M.L. and Immig, J. (2021) *Aquatic Pollutants in Oceans and Fisheries*. International Pollutants Elimination Network (IPEN). Available at: https://ipen.org/sites/default/files/documents/ipen-fisheries-v1_6cw-en.pdf (accessed 15 October 2022).

Lavado, R., Maryoung, L.A. and Schlenk, D. (2011) Hypersalinity acclimation increases the toxicity of the insecticide phorate in coho salmon (*Oncorhynchus kisutch*). *Environmental Science & Technology* 45(10), 4623–4629.

Lei, L., Wu, S., Lu, S., Liu, M., Song, Y., *et al.* (2018) Microplastic particles cause intestinal damage and other adverse effects in zebrafish *Danio rerio* and nematode *Caenorhabditis elegans*. *Science of the Total Environment* 619–620, 1–8.

Leslie, H.A., van Velzen, M.J., Brandsma, S.H., Vethaak, D., Garcia-Vallejo, J.J. and Lamoree, M.H. (2022) Discovery and quantification of plastic particle pollution in human blood. *Environment International* 163, 107199.

Letcher, R.J., Bustnes, J.O., Dietz, R., Jenssen, B.M., Jørgensen, E.H., *et al.* (2010) Exposure and effects assessment of persistent organohalogen contaminants in arctic wildlife and fish. *Science of the Total Environment* 408, 2995–3043.

Leung, K.C., Huang, Q., St-Hilaire, S., Liu, H., Zheng, X., *et al.* (2020) Fraudulent antibiotic products on the market for aquaculture use. *Preventive Veterinary Medicine* 181, 105052.

Li, W.C., Tse, H.F. and Fok, L. (2016) Plastic waste in the marine environment: a review of sources, occurrence and effects. *Science of the Total Environment* 566, 333–349.

Limonta, G., Mancia, A., Benkhalqui, A., Bertolucci, C., Abelli, L., *et al.* (2019) Microplastics induce transcriptional changes, immune response and behavioral alterations in adult zebrafish. *Scientific Reports* 9(1), 15775.

Lithner, D., Larsson, Å. and Dave, G. (2011) Environmental and health hazard ranking and assessment of plastic polymers based on chemical composition. *Science of the Total Environment* 409, 3309–3324.

Liu, X., Steele, J.C. and Meng, X.Z. (2017) Usage, residue, and human health risk of antibiotics in Chinese aquaculture: a review. *Environmental Pollution* 223, 161–169.

Lopez, P., Saulnier, D., Swarup-Gaucher, S., David, R., Lau, C., *et al.* (2022) First isolation of virulent *Tenacibaculum maritimum* isolates from diseased orbicular batfish (*Platax orbicularis*) farmed in Tahiti Island. *Pathogens* 11(2), 131.

Løvoll, M., Wiik-Nielsen, C.R., Tunsjø, H.S., Colquhoun, D., Lunder, T., *et al.* (2009) Atlantic salmon bath challenged with *Moritella viscosa* – pathogen invasion and host response. *Fish & Shellfish Immunology* 26(6), 877–884.

Lu, Y., Zhang, Y., Deng, Y., Jiang, W., Zhao, Y., *et al.* (2016) Uptake and accumulation of polystyrene microplastics in zebrafish (*Danio rerio*) and toxic effects in liver. *Environmental Science & Technology* 50(7), 4054–4060.

Lulijwa, R., Rupia, E.J. and Alfaro, A.C. (2020) Antibiotic use in aquaculture, policies and regulation, health and environmental risks: a review of the top 15 major producers. *Reviews in Aquaculture* 12(2), 640–663.

Lusher, A., Hollman, P. and Mendoza-Hill, J. (2017) *Microplastics in Fisheries and Aquaculture: Status of Knowledge on Their Occurrence and Implications for Aquatic Organisms and Food Safety*. FAO Fisheries and Agriculture Technical Paper No. 615. Food and Agriculture Organization of the United Nations, Rome.

Luzio, A., Santos, D., Fontaínhas-Fernandes, A.A., Monteiro, S.M. and Coimbra, A.M. (2016) Effects of 17α-ethinylestradiol at different water temperatures on zebrafish sex differentiation and gonad development. *Aquatic Toxicology* 174, 22–35.

MacFadden, D.R., McGough, S.F., Fisman, D., Santillana, M. and Brownstein, J.S. (2018) Antibiotic resistance increases with local temperature. *Nature Climate Change* 8(6), 510–514.

MacKinnon, B., Groman, D., Fast, M.D., Manning, A.J., Jones, P. and St-Hilaire, S. (2019a) Atlantic salmon challenged with extracellular products from *Moritella viscosa*. *Diseases of Aquatic Organisms* 133(2), 119–125.

MacKinnon, B., Jones, P., Hawkins, L., Dohoo, I., Stryhn, H., *et al.* (2019b) The epidemiology of skin ulcers in saltwater reared Atlantic salmon (*Salmo salar*) in Atlantic Canada. *Aquaculture* 501, 230–238.

MacKinnon, B., Groman, D., Fast, M.D., Manning, A.J., Jones, P., *et al.* (2020) Transmission experiment in Atlantic salmon (*Salmo salar*) with an Atlantic Canadian isolate of *Moritella viscosa*. *Aquaculture* 516, 734547.

Margarinos, B., Pazos, F., Santos, Y., Romalde, J.L. and Toranzo, A.E. (1995) Response of *Pasteurella piscicida* and *Flexibacter maritimus* to skin mucus of marine fish. *Diseases of Aquatic Organisms* 21(2), 103–108.

Martin, C., Baalkhuyur, F., Valluzzi, L., Saderne, V., Cusack, M., *et al.* (2020) Exponential increase of plastic burial in mangrove sediments as a major plastic sink. *Science Advances* 6(44), eaaz5593.

Masso-Silva, J.A. and Diamond, G. (2014) Antimicrobial peptides from fish. *Pharmaceuticals* 7, 265–310.

Matsumoto, R., Tu, N.P.C., Haruta, S., Kawano, M. and Takeuchi, I. (2014) Polychlorinated biphenyl (PCB) concentrations and congener composition in masu salmon from Japan: a study of all 209 PCB congeners by high-resolution gas chromatography/high-resolution mass spectrometry (HRGC/HRMS). *Marine Pollution Bulletin* 85(2), 549–557.

Mattsson, K., Ekvall, M.T., Hansson, L.A., Linse, S., Malmendal, A. and Cedervall, T. (2015) Altered behavior, physiology, and metabolism in fish exposed to polystyrene nanoparticles. *Environmental Science & Technology* 49(1), 553–561.

Mattsson, K., Johnson, E.V., Malmendal, A., Linse, S., Hansson, L.A. and Cedervall, T. (2017) Brain damage and behavioural disorders in fish induced by plastic nanoparticles delivered through the food chain. *Scientific Reports* 7(1), 11452.

McDevitt, J.P., Criddle, C.S., Morse, M., Hale, R.C., Bott, C.B. and Rochman, C.M. (2017) Addressing the issue of microplastics in the wake of the Microbead-Free Waters Act – a new standard can facilitate improved policy. *Environmental Science & Technology* 51(12), 6611–6617.

McDonald, D.G. (1983) The effects of H^+ upon the gills of freshwater fish. *Canadian Journal of Zoology* 61(4), 691–703.

McKim, J., Schmieder, P. and Veith, G. (1985) Absorption dynamics of organic chemical transport across trout gills as related to octanol–water partition coefficient. *Toxicology and Applied Pharmacology* 77(1), 1–10.

Medeiros, A.P.T., Chellappa, S. and Yamamoto, M.E. (2007) Agonistic and reproductive behaviors in males of red hybrid tilapia, *Oreochromis niloticus* (Linnaeus, 1758) × *O. mossambicus* (Peters, 1852) (Osteichthyes: Cichlidae). *Brazilian Journal of Biology* 67(4), 701–706.

Mendoza, M., Güiza, L., Martinez, X., Caraballo, X., Rojas, J., *et al.* (2013) A novel agent (*Endozoicomonas elysicola*) responsible for epitheliocystis in cobia *Rachycentrum canadum* larvae. *Diseases of Aquatic Organisms* 106(1), 31–37.

Michel, A. (2018) A potential, unconventional approach to pathogen control in aquaculture. Available at: https://www.globalseafood.org/advocate/hyperthermia-boost-innate-immune-system-juvenile-fish/ (accessed 21 March 2022).

Midtlyng, P.J., Grave, K. and Horsberg, T.E. (2011) What has been done to minimize the use of antibacterial and antiparasitic drugs in Norwegian aquaculture? *Aquaculture Research* 42, 28–34.

Milla, S., Depiereux, S. and Kestemont, P. (2011) The effects of estrogenic and androgenic endocrine disruptors on the immune system of fish: a review. *Ecotoxicology* 20(2), 305–319.

Min, K., Cuiffi, J.D. and Mathers, R.T. (2020) Ranking environmental degradation trends of plastic marine debris based on physical properties and molecular structure. *Nature Communications* 11(1), 727.

Minniti, G. (2018) The farmed Atlantic salmon (*Salmo salar*) skin-mucus microbiome: bacterial diversity, mucus degrading potential and putative virulence factors. PhD thesis, Norwegian University of Life Sciences, Ås, Norway.

Miranda, C.D., Godoy, F.A. and Lee, M.R. (2018) Current status of the use of antibiotics and the antimicrobial resistance in the Chilean salmon farms. *Frontiers in Microbiology* 9, 1284.

Moore, G. (2020) Better fish health and EBIT for Scottish Sea Farms. *Fishfarmingexpert*, 11 November 2020. Available at: https://www.fishfarmingexpert.com/article/good-fish-health-and-earnings-for-scottish-sea-farms/ (accessed 20 April 2022).

Morgan, A.L., Thompson, K.D., Auchinachie, N.A. and Migaud, H. (2008) The effect of seasonality on normal haematological and innate immune parameters of rainbow trout *Oncorhynchus mykiss* L. *Fish & Shellfish Immunology* 25(6), 791–799.

Muncaster, S.P., Jacobson, G., Taiarui, M., King, S. and Bird, S. (2016) Effects of MV Rena heavy fuel oil and dispersed oil on yellowtail kingfish early life stages. *New Zealand Journal of Marine and Freshwater Research* 50(1), 131–143.

Mwanza, M.M., Ndunda, E.N., Bosire, G.O., Nyamori, V.O. and Martincigh, B.S. (2021) Advances in sample pretreatment and detection of PCBs in the environment. *Journal of Hazardous Materials Advances* 4, 100028.

Myksvoll, M.S. (2019) From physics to farming: an evaluation of an operational hydrodynamic and behavioral monitoring system. *Global Seafood Alliance*, September 2019. Available at: https://www.globalseafood.org/advocate/norwegian-researchers-develop-sea-lice-tracking-model/ (accessed 22 June 2022).

Nguyen, D.G.C., Sebesvari, Z., Renaud, F., Rosendahl, I., Hoang Minh, Q. and Amelung, W. (2015) Occurrence and dissipation of the antibiotics sulfamethoxazole, sulfadiazine, trimethoprim, and enrofloxacin in the Mekong Delta, Vietnam. *PLoS ONE* 10(7), e0131855.

Nichols, J.W., McKim, J.M., Lien, G.J., Hoffman, A.D., Bertelsen, S.L. and Elonen, C.M. (1996) A physiologically based toxicokinetic model for dermal absorption of organic chemicals by fish. *Fundamental and Applied Toxicology* 31(2), 229–242.

Nimalan, N., Sørensen, S.L., Fečkaninová, A., Koščová, J., Mudroňová, D., *et al.* (2022) Mucosal barrier status in Atlantic salmon fed marine or plant-based diets supplemented with probiotics. *Aquaculture* 547, 737516.

Norwegian Veterinary Institute (2016) *Use of Antibiotics in Norwegian Aquaculture*. Norwegian Veterinary Institute, Oslo. Available at: https://www.vetinst.no/rapporter-og-publikasjoner/rapporter/2016/use-of-antibiotics-in-norwegian-aquaculture (accessed 30 March 2022).

Noyes, P.D. and Lema, S.C. (2015) Forecasting the impacts of chemical pollution and climate change interactions on the health of wildlife. *Current Zoology* 61(4), 669–689.

Nurliyana, M., Lukman, B., Ina-Salwany, M.Y., Zamri-Saad, M., Annas, S., *et al.* (2020) First evidence of scale drop disease virus in farmed Asian seabass (*Lates calcarifer*) in Malaysia. *Aquaculture* 528, 735600.

Nylund, A., Watanabe, K., Nylund, S., Karlsen, M., Saether, P.A., *et al.* (2008) Morphogenesis of salmonid gill poxvirus associated with proliferative gill disease in farmed Atlantic salmon (*Salmo salar*) in Norway. *Archives of Virology* 153(7), 1299–1309.

Oehme, M. and Ottar, B. (1984) The long range transport of polychlorinated hydrocarbons to the Arctic. *Geophysical Research Letters* 11(11), 1133–1136.

OIE (World Organisation for Animal Health) (2021) Chapter 4.1. Biosecurity for aquaculture establishments. In: *Aquatic Animal Health Code*. OIE, Paris. Available at: https://www.oie.int/en/what-we-do/standards/codes-and-manuals/aquatic-code-online-access/ (accessed 21 March 2022).

OIE (World Organisation for Animal Health) (2022) Regional Aquatic Animal Disease Report. Region: Asia and the Pacific. Available at: https://rr-asia.oie.int/en/projects/qaad-reports/ (accessed 5 April 2022).

Oliveira, M. and Vasconcelos, V. (2020) Occurrence of mycotoxins in fish feed and its effects: a review. *Toxins* 12(3), 160.

Østevik, L., Stormoen, M., Evensen, Ø., Xu, C., Lie, K.I., *et al.* (2022) Effects of thermal and mechanical delousing on gill health of farmed Atlantic salmon (*Salmo salar* L.). *Aquaculture* 552, 738019.

Ostrach, D.J., Low-Marchelli, J.M., Eder, K.J., Whiteman, S.J. and Zinkl, J.G. (2008) Maternal transfer of xenobiotics and effects on larval striped bass in the San Francisco Estuary. *Proceedings of the National Academy of Sciences USA* 105(49), 1935–19359.

Overton, K., Dempster, T., Oppedal, F., Kristiansen, T.S., Gismervik, K. and Stien, L.H. (2019) Salmon lice treatments and salmon mortality in Norwegian aquaculture: a review. *Reviews in Aquaculture* 11(4), 1398–1417.

Pazos, F. (1997) *Flexibacter maritimus*: estudio fenotípico, inmunológico y molecular. Doctoral thesis, Universidad de Santiago de Compostela, Santiago de Compostela, Spain.

Peng, Y., Wu, P., Schartup, A.T. and Zhang, Y. (2021) Plastic waste release caused by COVID-19 and its fate in the global ocean. *Proceedings of the National Academy of Sciences USA* 118(47), e2111530118.

Pepi, M. and Focardi, S. (2021) Antibiotic-resistant bacteria in aquaculture and climate change: a challenge for health in the Mediterranean area. *International Journal of Environmental Research and Public Health* 18(11), 5723.

Pitt, J.A., Trevisan, R., Massarsky, A., Kozal, J.S., Levin, E.D. and Di Giulio, R.T. (2018) Maternal transfer of nanoplastics to offspring in zebrafish (*Danio rerio*): a case study with nanopolystyrene. *Science of the Total Environment* 643, 324–334.

Prasitporn, T., Senapin, S., Vaniksampanna, A., Longyant, S. and Chaivisuthangkura, P. (2021) Development of cross-priming amplification (CPA) combined with colorimetric and lateral flow dipstick visualization for scale drop disease virus (SDDV) detection. *Journal of Fish Diseases* 44(9), 1411–1422.
Price, D., Sánchez, J., Ibarra, R. and St-Hilaire, S. (2019) Variation in the concentration of antibiotics in tissue during oral antibiotic treatments in farmed salmonids. *Aquaculture* 498, 587–593.
Ramírez-Paredes, J.G., Paley, R.K., Hunt, W., Feist, S.W., Stone, D.M., *et al.* (2021) First detection of infectious spleen and kidney necrosis virus (ISKNV) associated with massive mortalities in farmed tilapia in Africa. *Transboundary and Emerging Diseases* 68(3), 1550–1563.
Ren, R., Zhang, Y., Nadappuram, B.P., Akpinar, B., Klenerman, D., *et al.* (2017) Nanopore extended field-effect transistor for selective single-molecule biosensing. *Nature Communications* 8(1), 586.
Reverter, M., Tapissier-Bontemps, N., Lecchini, D., Banaigs, B. and Sasal, P. (2018) Biological and ecological roles of external fish mucus: a review. *Fishes* 3(4), 41.
Rocha-Santos, T. and Duarte, A.C. (2015) A critical overview of the analytical approaches to the occurrence, the fate and the behavior of microplastics in the environment. *TrAC Trends in Analytical Chemistry* 65, 47–53.
Rochman, C.M., Kurobe, T., Flores, I. and Teh, S.J. (2014) Early warning signs of endocrine disruption in adult fish from the ingestion of polyethylene with and without sorbed chemical pollutants from the marine environment. *Science of the Total Environment* 493, 656–661.
Rodger, H.D. (2014) Amoebic gill disease (AGD) in farmed salmon (*Salmo salar*) in Europe. *Fish Veterinary Journal* 14(1), 16–26.
Rogers Worley, R., McAfee Moore, S., Tierney, B.C., Ye, X., Calafat, A.M., *et al.* (2017) Per- and polyfluoroalkyl substances in human serum and urine samples from a residentially exposed community. *Environment International* 106, 135–143. https://doi.org/10.1016/j.envint.2017.06.007
Rotchell, J.M. and Ostrander, G.K. (2003) Molecular markers of endocrine disruption in aquatic organisms. *Journal of Toxicology and Environmental Health, Part B: Critical Reviews* 6(5), 453–496.
Royer, S.J., Ferrón, S., Wilson, S.T. and Karl, D.M. (2018) Production of methane and ethylene from plastic in the environment. *PLoS ONE* 13(8), e0200574.
Sah Putra, B., Hick, P.M., Hall, E., Whittington, R.J., Khairul, R., *et al.* (2020) Prevalence of *Infectious spleen and kidney necrosis virus* (ISKNV), nervous necrosis virus (NNV) and ectoparasites in juvenile *Epinephelus* spp. farmed in Aceh, Indonesia. *Pathogens* 9(7), 578. https://doi.org/10.3390/pathogens9070578
Sarmah, A.K., Meyer, M.T. and Boxall, A.B. (2006) A global perspective on the use, sales, exposure pathways, occurrence, fate and effects of veterinary antibiotics (VAs) in the environment. *Chemosphere* 65(5), 725–759.
Savoca, M.S., McInturf, A.G. and Hazen, E.L. (2021) Plastic ingestion by marine fish is widespread and increasing. *Global Change Biology* 27(10), 2188–2199.
Schar, D., Klein, E.Y., Laxminarayan, R., Gilbert, M. and Van Boeckel, T.P. (2020) Global trends in antimicrobial use in aquaculture. *Scientific Reports* 10(1), 21878.
Schipp, G., Bosmans, J. and Humphrey, J. (2007) *Northern Territory Barramundi Farming Handbook*. Department of Primary Industry, Fisheries and Mines, Northern Territory Government, Darwin, Australia.
Senapin, S., Dong, H.T., Meemetta, W., Gangnonngiw, W., Sangsuriya, P., *et al.* (2019) Mortality from scale drop disease in farmed *Lates calcarifer* in Southeast Asia. *Journal of Fish Diseases* 42(1), 119–127.
Shephard, K.L. (1994) Functions for fish mucus. *Reviews in Fish Biology and Fisheries* 4(4), 401–429.
Shinn, A.J., Pratoomyot, J., Bron, J., Paladini, G., Brooker, E. and Brooker, A. (2015) Economic impacts of aquatic parasites on global finfish production. *Global Aquaculture Advocate* 2015, 58–61.
Simcik, M.F., Franz, T.P., Zhang, H. and Eisenreich, S.J. (1998) Gas-particle partitioning of PCBs and PAHs in the Chicago urban and adjacent coastal atmosphere: states of equilibrium. *Environmental Science & Technology* 32(2), 251–257.
Småge, S.B., Frisch, K., Brevik, Ø.J., Watanabe, K. and Nylund, A. (2016) First isolation, identification and characterisation of *Tenacibaculum maritimum* in Norway, isolated from diseased farmed sea lice cleaner fish *Cyclopterus lumpus* L. *Aquaculture* 464, 178–184.
Småge, S.B., Frisch, K., Vold, V., Duesund, H., Brevik, Ø.J., *et al.* (2018) Induction of tenacibaculosis in Atlantic salmon smolts using *Tenacibaculum finnmarkense* and the evaluation of a whole cell inactivated vaccine. *Aquaculture* 495, 858–864.
Sommerset, I., Jensen, B.B., Bornø, G., Haukaas, A. and Brun, E. (2021) *Fiskehelserapporten 2020*. Rapport 41a/2021. Norwegian Veterinary Institute, Ås, Norway.

Sriisan, S., Boonchird, C., Thitamadee, S., Sonthi, M., Dong, H.T. and Senapin, S. (2020) A sensitive and specific SYBR Green-based qPCR assay for detecting scale drop disease virus (SDDV) in Asian sea bass. *Diseases of Aquatic Organisms* 139, 131–137.

Steine, N.O., Melingen, G.O. and Wergeland, H.I. (2001) Antibodies against *Vibrio salmonicida* lipopolysaccharide (LPS) and whole bacteria in sera from Atlantic salmon (*Salmo salar* L.) vaccinated during the smolting and early post-smolt period. *Fish & Shellfish Immunology* 11(1), 39–52.

Sukonta, T., Senapin, S., Meemetta, W. and Chaijarasphong, T. (2022) CRISPR-based platform for rapid, sensitive and field-deployable detection of scale drop disease virus in Asian sea bass (*Lates calcarifer*). *Journal of Fish Diseases* 45(1), 107–120.

Sun, Y., Wen, H., Tian, Y., Mao, X., Li, X., *et al.* (2021) *HSP90* and *HSP70* families in *Lateolabrax maculatus*: genome-wide identification, molecular characterization, and expression profiles in response to various environmental stressors. *Frontiers in Physiology* 12, 784803.

Sung, Y.Y. and MacRae, T.H. (2011) Heat shock proteins and disease control in aquatic organisms. *Journal of Aquaculture Research & Development* S2, 006. https://doi.org/10.4172/2155-9546.S2-006

Sveen, L.R., Grammes, F.T., Ytteborg, E., Takle, H. and Jørgensen, S.M. (2017) Genome-wide analysis of Atlantic salmon (*Salmo salar*) mucin genes and their role as biomarkers. *PLoS ONE* 12(12), e0189103.

Sweetman, J.W., Torrecillas, S., Dimitroglou, A., Rider, S., Davies, S.J. and Izquierdo, M.S. (2010) Enhancing the natural defences and barrier protection of aquaculture species. *Aquaculture Research* 41(3), 345–355.

Talbot, C., Corneillie, S. and Korsøen, Ø. (1999) Pattern of feed intake in four species of fish under commercial farming conditions: implications for feeding management. *Aquaculture Research* 30(7), 509–518.

Tarnecki, A.M., Miller, C., Sherwood, T.A., Griffitt, R.J., Schloesser, R.W. and Wetzel, D.L. (2022) Dispersed crude oil induces dysbiosis in the red snapper *Lutjanus campechanus* external microbiota. *Microbiology Spectrum* 10(1), e00587-21.

Taylor, R.S., Huynh, C., Cameron, D., Evans, B., Cook, M. and Ritchie, G. (2016) *Gill Score Guide – Amoebic Gill Disease (AGD) Management Training Document*. Tassal Operations Pty, Hobart, Australia.

Terence, C., Domingos, J.A., Uichanco, J.A., Monteil, Y., Monteil, L., *et al.* (2021) Early sex change of barramundi *Lates calcarifer* farmed in Singapore. *Aquaculture Reports* 20, 100712.

Thiele, C.J., Hudson, M.D., Russell, A.E., Saluveer, M. and Sidaoui-Haddad, G. (2021) Microplastics in fish and fishmeal: an emerging environmental challenge? *Scientific Reports* 11(1), 2045.

Thoen, E., Tartor, H., Amundsen, M., Dale, O.B., Sveinsson, K., *et al.* (2020) First record of experimentally induced salmon gill poxvirus disease (SGPVD) in Atlantic salmon (*Salmo salar* L.). *Veterinary Research* 51(1), 63.

Toranzo, A.E., Magariños, B. and Romalde, J.L. (2005) A review of the main bacterial fish diseases in mariculture systems. *Aquaculture* 246(1–4), 37–61.

Torrecillas, S., Makol, A., Benítez-Santana, T., Caballero, M.J., Montero, D., *et al.* (2011) Reduced gut bacterial translocation in European sea bass (*Dicentrarchus labrax*) fed mannan oligosaccharides (MOS). *Fish & Shellfish Immunology* 30(2), 674–681.

Tørud, B. and Håstein, T. (2008) Skin lesions in fish: causes and solutions. *Acta Veterinaria Scandinavica* 50(1), 1–3.

Van de Merwe, J.P., Hodge, M., Olszowy, H.A., Whittier, J.M., Ibrahim, K. and Lee, S.Y. (2009) Chemical contamination of green turtle (*Chelonia mydas*) eggs in peninsular Malaysia: implications for conservation and public health. *Environmental Health Perspectives* 117(9), 1397–1401.

Varó, I., Osorio, K., Estensoro, I., Naya-Català, F., Sitjà-Bobadilla, A., *et al.* (2021) Effect of virgin low density polyethylene microplastic ingestion on intestinal histopathology and microbiota of gilthead sea bream. *Aquaculture* 545, 737245.

Vijayan, M.M., Feist, G., Otto, D.M., Schreck, C.B. and Moon, T.W. (1997) 3,3′,4,4′-Tetrachlorobiphenyl affects cortisol dynamics and hepatic function in rainbow trout. *Aquatic Toxicology* 37(2–3), 87–98.

Vijayan, M.M., Prunet, P. and Boone, A.N. (2005) Xenobiotic impact on corticosteroid signaling. In: Mommsen, T.P. and Moon, T.W. (eds) *Biochemistry and Molecular Biology of Fishes*. Vol. 6. *Environmental Toxicology*. Elsevier, Amsterdam, pp. 365–394.

Wahli, T., Schmidt-Posthaus, H. and Segner, H. (2023) Non-infectious disorders of coldwater fish. In: Woo, P.T.K. and Subasinghe, R.P. (eds) *Climate Change on Diseases and Disorders of Finfish in Cage Culture*, 3rd edn. CAB International, Wallingford, UK, pp. 125–162.

Wan, Z., Wang, C., Zhou, J., Shen, M., Wang, X., *et al.* (2019) Effects of polystyrene microplastics on the composition of the microbiome and metabolism in larval zebrafish. *Chemosphere* 217, 646–658.

Wang, J., Grisle, S. and Schlenk, D. (2001) Effects of salinity on aldicarb toxicity in juvenile rainbow trout (*Oncorhynchus mykiss*) and striped bass (*Morone saxatilis* × *chrysops*). *Toxicological Sciences* 64(2), 200–207.

Wang, J., Li, Y., Lu, L., Zheng, M., Zhang, X., *et al.* (2019) Polystyrene microplastics cause tissue damages, sex-specific reproductive disruption and transgenerational effects in marine medaka (*Oryzias melastigma*). *Environmental Pollution* 254, 113024.

Wang, X., Zhang, R., Li, Z. and Yan, B. (2022) Adsorption properties and influencing factors of Cu (II) on polystyrene and polyethylene terephthalate microplastics in seawater. *Science of the Total Environment* 812, 152573.

Wania, F. and Mackay, D. (1993) Global fractionation and cold condensation of low volatility organochlorine compounds in polar regions. *Ambio* 22(1), 10–18.

Wardrop, P., Shimeta, J., Nugegoda, D., Morrison, P.D., Miranda, A., *et al.* (2016) Chemical pollutants sorbed to ingested microbeads from personal care products accumulate in fish. *Environmental Science & Technology* 50(7), 4037–4044.

Watts, J.E., Schreier, H.J., Lanska, L. and Hale, M.S. (2017) The rising tide of antimicrobial resistance in aquaculture: sources, sinks and solutions. *Marine Drugs* 15(6), 158.

Wiik-Nielsen, J., Mo, T.A., Kolstad, H., Mohammad, S.N., Hytterød, S. and Powell, M.D. (2016) Morphological diversity of *Paramoeba perurans* trophozoites and their interaction with Atlantic salmon, *Salmo salar* L., gills. *Journal of Fish Diseases* 39(9), 1113–1123.

Williamson, S.H. (2018) What's in the water? How media coverage of corporate GenX pollution shapes local understanding of risk. *Critical Criminology* 26, 289–305. https://doi.org/10.1007/s10612-018-9389-8

Zanuzzo, F.S., Peroni, E.d.F.C., Sandrelli, R.M., St-Hilaire, S., O'Brien, N. and Gamperl, A.K. (2022) Temperature has considerable effects on plasma and muscle antibiotic concentrations in Atlantic salmon (*Salmo salar*). *Aquaculture* 546, 737372.

Zeitoun, M.M. and Mehana, E.E. (2014) Impact of water pollution with heavy metals on fish health: overview and updates. *Global Veterinaria* 12(2), 219–231.

Zhang, H. and Huo, S. (2019) Climate change impacts on polychlorinated biphenyls (PCBs) deposition in a shallow lake, China. *IOP Conference Series: Earth and Environmental Science* 227(5), 052074.

Zhang, G.L., Feng, Y.L., Song, J.L. and Zhou, X.S. (2018) Zearalenone: a mycotoxin with different toxic effect in domestic and laboratory animals' granulosa cells. *Frontiers in Genetics* 9, 667.

Zhang, Y., Zhang, H., Zhang, Z., Liu, C., Sun, C., *et al.* (2018) pH effect on heavy metal release from a polluted sediment. *Journal of Chemistry* 2018, 7597640.

Zhu, M., Chernick, M., Rittschof, D. and Hinton, D.E. (2020) Chronic dietary exposure to polystyrene microplastics in maturing Japanese medaka (*Oryzias latipes*). *Aquatic Toxicology* 220, 105396.

Zrnčić, S., Oraić, D., Račić, I. and Beck, R. (2013) *Tenacibaculum maritimum* isolated from sea bass (*Dicentrarchus labrax*) in Croatia. In: *Book of Abstracts of the 16th EAFP International Conference on Diseases of Fish and Shellfish, Tampere, Finland, 2–6 September 2013*. European Association of Fish Pathologists, p. 387.

9 Movement of Infectious Agents between Wild and Farmed Fish

Sonja M. Saksida[1]*, Mark D. Fast[1], Kyle A. Garver[2] and Stewart C. Johnson[2]
[1]*University of Prince Edward Island, Charlottetown, Prince Edward Island, Canada;*
[2]*Fisheries and Oceans Canada, Nanaimo, British Columbia, Canada*

9.1 Introduction

Aquaculture is a key sector in fulfilling the ever-increasing global food demand (Costello *et al.*, 2020). Significant technological advances and improved husbandry/production practices have permitted substantial growth of world aquaculture over the past three decades (FAO, 2020). However, there remains some public opposition to the expansion of aquaculture due to concerns related to sustainable resource use and environmental impacts. In particular, rearing fish in cage systems has garnered much criticism with valid concerns regarding potential ecological consequences. One such concern is whether fish held in such systems are a source and/or amplifier of infectious disease agents. Cage aquaculture, which allows for the exchange of water with its surrounding environment, has the potential risk of permitting movements of pathogens as well. As such, the extent of infectious agent movement and resulting consequences to wild and farmed populations are of critical interest and a major research topic of net-cage aquaculture systems (Håstein and Lindstad, 1991; McVicar, 1997; Holmer, 2010; Johansen *et al.*, 2011; Peeler and Taylor, 2011; Jones *et al.*, 2015; Krkošek, 2017).

This chapter examines the dynamic interaction of infectious agent movements between farmed and sympatric wild finfish populations. Whether pathogen exposure of farmed and wild host populations leads to infection and the development of disease is dependent on a myriad of environmental, infection agent and host factors that often interact in unforeseen ways. Those factors contributing to infection, amplification, spread and the occurrence of infectious disease as a consequence of wild and farmed fish interactions are highlighted in this chapter. Utilizing examples, the chapter illustrates the complexities of these interactions, their potential outcomes and uncertainties in the face of climate change. Lastly, methodologies to better define the magnitude and consequences of pathogen exchange associated with net-cage aquaculture are recommended and further research needs and recommendations for management and policy are suggested.

9.2 Transmission and Spread of Infectious Agents in the Aquatic Environment

Infectious agents spread rapidly in the aquatic environment, likely much faster than in terrestrial environments. For example, it is estimated that

*ssaksida@upei.ca

 Climate Change on Diseases and Disorders of Finfish in Cage Culture, 3rd Edition (eds P.T.K. Woo and R.P. Subasinghe)
DOI: 10.1079/9781800621640.0009

herpes virus spread at a rate of 10,000 km/year through pilchard populations in Australia (McCallum *et al.*, 2003). The most likely explanations for the relatively rapid spread of aquatic pathogens are the lack of physical barriers to dispersal and the potential for longer-term survival of infectious agents outside the host in aquatic environments. There are five main exposure pathways by which infectious agents can be transmitted (within a population) or spread (between populations) in the aquatic environment, namely waterborne, fish-to-fish, vector-borne, fomite and food (OIE, 2021). These pathways, however, are not mutually exclusive as fish can be exposed to many infectious agents by more than one route; for example, *Ichthyophonus* is transmitted through predation from small fish to large fish. While this mode of infection is unidirectional, the parasite has also evolved to direct infection via waterborne exposure, allowing for greater dispersion (Kocan, 2019).

9.2.1 Waterborne exposure

Waterborne exposure is the most common route by which infectious agents of fish transmit between individuals and spread among populations (Bricknell, 2017). Viral, bacterial and parasitic agents released into the aquatic environment from infected hosts are transported by water movements to expose other potential hosts. The distance over which an agent disperses while remaining infectious depends on the capacity of the agent to survive outside its host, which in turn is related to the physical and chemical characteristics of the aquatic environment. The dispersal of infectious agents can be facilitated through anthropogenic activities such as the use of ballast water in ships or through the discharge of untreated processing plant wastewaters. For example, ballast water movement is postulated to be the mechanism by which viral haemorrhagic septicaemia virus (VHSV) was introduced and spread in the Great Lakes of North America (Bain *et al.*, 2010; Sieracki *et al.*, 2014), while wastewater from fish slaughter and processing activities has been implicated in facilitating the spread of VHSV in Finland (Vennerström *et al.*, 2020) and infectious salmon anaemia virus (ISAV) in the coastal waters of Norway (Vågsholm *et al.*, 1994; Jarp and Karlsen, 1997).

9.2.2 Fish-to-fish exposure

Infectious agent exposure can occur horizontally from fish to fish through direct physical contact between an infected and naïve fish (Hammell *et al.*, 2009; Tolo *et al.*, 2022) or vertically from an infected parent to its offspring. In net-cage culture where high numbers of individuals are reared within a confined space, fish-to-fish transmission of pathogens is facilitated by the repeated physical contact between uninfected and infected individuals.

Faecal–oral transmission is initiated by ingestion of contaminated faecal material. Faecal–oral transmission is also an example of fish-to-fish contact due to the natural buoyancy of faeces in water allowing for maintained suspension of faeces and thus a longer exposure time for direct contact with fish in the water column. This is an important route of transmission for a range of infectious agents including *Renibacterium salmoninarum*, causative agent of bacterial kidney disease (BKD), which is transmitted horizontally in cage-cultured Atlantic salmon through the ingestion of contaminated faeces (Balfry *et al.*, 1996). Species within the genera *Aeromonas*, *Mycobacterium*, *Nocardia* and *Enterococcus* have also been reported to be transmitted by this route, as have viruses such as ISAV and infectious haematopoietic necrosis virus (IHNV) as well as parasites such as *Hexamita* (Romalde *et al.*, 1996; Hammell *et al.*, 2009; Meyers *et al.*, 2019).

9.2.3 Vector-borne exposure

Vectors are living organisms capable of spreading an infectious agent to a susceptible species but are not in themselves susceptible (OIE, 2021). Viral and bacterial agents of fish have on occasion been detected in invertebrates during disease outbreaks and these have been suggested as vectors (e.g. Gregory *et al.*, 2007; Kitamura *et al.*, 2007; Skar and Mortensen, 2007; Fitridge *et al.*, 2012; Molloy *et al.*, 2013, 2014; Kim *et al.*, 2017; Vennerström *et al.*, 2020). However, since direct horizontal transmission of these agents can readily be accomplished, vector-mediated transmission is not necessary in spreading these infectious agents.

Vectors should not be confused with intermediate hosts, with the difference being that intermediate hosts perform an essential role in the life cycle of the infectious agent. Many invertebrates serve as intermediate hosts for a whole range of parasites, such as snails serving as an intermediate host for diplostomoid digeneans that affect culture of catfish and trout (Overstreet and Curran, 2004) and polychaetes being an intermediate life history stage for the blood fluke, *Cardicola forsteri*, which causes serious disease in cage-reared southern bluefin tuna (Cribb *et al.*, 2011). Leeches are intermediate hosts for parasitic pathogens such as *Cryptobia salmositica*, a pathogenic haemoflagellate of salmonid fishes (Woo, 2003).

9.2.4 Fomite exposure

Fomites are inanimate objects or materials (e.g. cage structures, ropes, anchors) or equipment (e.g. boats, nets, seines) which can retain and transmit infectious agents. Transmission/spread can occur when a susceptible fish comes in contact with contaminated farm equipment and structures. Often, the deposition of viral and bacterial agents on to inanimate objects is facilitated through biofilms that can adhere to cage structures and serve as reservoirs of infectious agents (Carballo *et al.*, 2000; Levipan *et al.*, 2019). For example, *Yersinia ruckeri*, a freshwater bacterial pathogen, forms biofilms on hard surfaces which are a source of recurrent infections in freshwater rainbow trout farms (Coquet *et al.*, 2002). Contaminated equipment has been suspected in the transmission of epizootic epitheliotropic disease virus (EEDV) within hatchery populations of lake trout (Purbayu *et al.*, 2021); and fishing boats and equipment were suggested as the source of spring viraemia of carp virus (SVCV) in common carp in Minnesota, USA (Phelps *et al.*, 2012).

9.2.5 Food-related pathogen exposure

Infectious agents can be spread through the consumption of contaminated foods. For example, *Ichthyophonus* is spread to fish through the consumption of infected prey. Infectious agents can spread to cage-cultured fish through feeding them contaminated feeds. However, in many fish-culture systems this route of transmission has been eliminated with the switch from the use of raw fish or fish by-product to manufactured feeds in which infectious agents in the feed components have been eliminated (e.g. through pasteurization). In situations where raw feeds are still used, there is the risk that wild fish attracted to and feeding on wastes from cage sites can be exposed via this route.

9.2.6 Air-borne exposure

Air-borne transmission has not been well studied in aquatic infectious diseases. Bioaerosols are a subcategory of particles released from terrestrial and marine ecosystems into the atmosphere. These biological particles can include living (e.g. viral, bacterial, fungal) organisms, as well as non-living components (Xie *et al.*, 2021). Depending on the size and type of particle and the physical/chemical properties of the atmosphere, bioaerosols can carry particles over large distances (Xie *et al.*, 2021). There is some evidence for transmission of viral, bacterial and parasitic agents in aquatic systems via droplets and/or bioaerosols, especially over short distances, under laboratory and intensive culture conditions (Wooster and Bowser, 1996; Bishop *et al.*, 2003; Snow *et al.*, 2003; Troyer and Kurath, 2003; Roberts-Thomson *et al.*, 2006; Ramsay *et al.*, 2009; Hick *et al.*, 2011).

9.3 Outcomes of Exposure to Infectious Agents

The outcome of exposure to a pathogen is determined by the characteristics of the infectious agent and the susceptibility of the host, both of which are highly influenced by the environment. For disease to occur, the host must be susceptible, the agent must be infectious and virulent, and environmental conditions must be conducive to the development of disease (Fig. 9.1). A shift in any one of these parameters could shift the outcome of the exposure from a no-disease to a disease state.

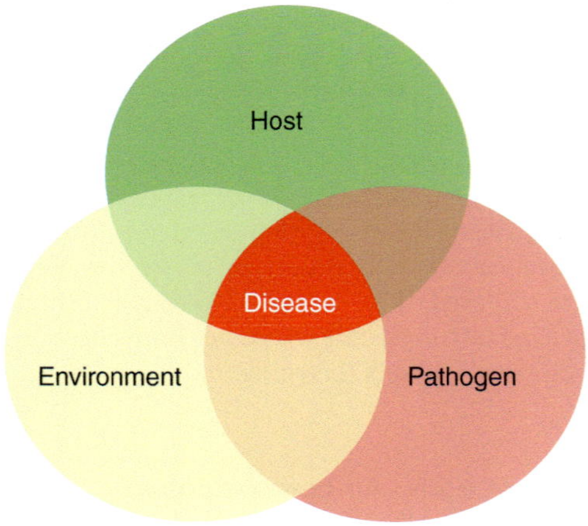

Fig. 9.1. The interaction of the host, pathogen (infectious agent) and environment in causing disease, where there has to be an infectious agent exposed to a susceptible host and both must be present in a conducive environment.

With respect to ectotherms such as fish, the water temperature at the time of exposure plays a large role in determining the outcome of exposure due to its role as a primary regulator of physiological and immunological responses. The outcome of exposure to an infectious agent can be unpredictable as it depends on a myriad of environmental, infectious agent and host factors, which often interact in unforeseen ways (commonly referred to as risk factors). Examples of risk factors that have been associated with increasing the chance of developing disease are identified in Table 9.1. The occurrence of an infectious agent within the environment or its presence in the host may not necessarily lead to the development of disease in the absence of risk factors. Similarly, the disease may not be triggered by the same risk factors in all cases because each exposure and the conditions under which it occurs are unique.

9.3.1 Host

Exposure to an infectious agent can result in disease if the host is susceptible to infection and lacks an adequate immune response to limit disease development. Host factors which are important to its susceptibility and capacity to respond to infections include genetics, age/reproductive stage, immune competency and pre-existing conditions such as co-infections (for recent reviews see Anacleto *et al.*, 2019; Abdel-Latif *et al.*, 2020; Wise *et al.*, 2021). For example, juvenile salmonids appear to be more susceptible to the disease infectious haematopoietic necrosis (IHN) than adults (Saksida, 2006; Garver and Wade, 2017). Different stocks of chinook salmon have different susceptibility to BKD (Purcell *et al.*, 2014).

9.3.2 Infectious agent

The capability of an agent to cause disease depends on its ability to infect and evade or outcompete the host's immune system. This can be influenced by the strain of agent, its concentration and the duration of host exposure. For example, the disease infectious salmon anaemia (ISA) occurs only when fish are infected with ISAV variants that have deletions within the highly polymorphic region (ISAV-HPR) of the haemagglutinin esterase gene, while variants that do not have these deletions (ISAV-HPR0) are highly infectious but not causative of disease (Markussen *et al.*, 2008; Ritchie *et al.*, 2009).

Exposure time and infective dose (number of viral particles or bacteria required to establish an infection) can also influence disease development and have to be considered in relation to the biology and ecology of the host. For instance, a host may be exposed to high concentrations of an infectious agent, but exposure time is too short (i.e. fish actively migrating versus resident population) to initiate an infection. Transmission/spread of infectious agents to susceptible finfish in cage culture is further restricted in their life cycle if the agent requires an intermediate host.

9.3.3 Environment

There are many abiotic and biotic factors (Table 9.1) that affect both the pathogen and the host. The capacity of aquatic infectious agents to survive in a viable and infectious state outside a host is highly species/agent specific. However,

Table 9.1. Examples of host and pathogen factors influencing the outcome of exposure and the environmental factors which influence the host (h), the pathogen (p) or both (h/p).

		Environment factors	
Host factors	Pathogen or infectious agent factors	Abiotic	Biotic
Species	Infectivity	Physical (h/p)	Intermediate hosts (p)
Genetics (e.g. stock)	Pathogenicity	Temperature (h/p)	Vectors (p)
Sex	Virulence	Turbidity (h/p)	Carrier populations (h)
Age and/or reproductive state	Level of exposure (infective dose and exposure time)	Water movements (p)	Water column
Life history (e.g. migratory, resident)	Route of exposure	UV light (p)	Microbial community (h/p)
Thermal tolerance range	Duration of exposure	Chemical (h/p)	Toxic algae (h)
Immunocompetency	Life cycle	Oxygen (h)	Predators/grazers (h/p)
Stress		Salinity (h/p)	Anthropogenic activities (h/p)
Co-infections		pH (h/p)	
Co-morbidities		Pollutants (h)	
Nutritional status		Anthropogenic activities (h/p)	

regardless of the agent, physiochemical and biological factors in the environment affect its capacity to survive and remain infectious and thereby limit dispersal. The effects of water temperature are most studied, with temperature recognized as having major effects on survival, inactivation and infectivity rates of viral and bacterial agents (Oidtmann *et al.*, 2018). In general higher temperatures increase the rates of viral inactivation. As an example, the average time for a 3-log reduction in VHSV infectivity in natural seawater (31 ppt) held in the dark at 4, 10, 15 and 20°C was 12, 7, 4 and 2 days, respectively (Hawley and Garver, 2008). Likewise, *Piscirickettsia salmonis*, the causative agent of piscirickettsiosis, remained viable in natural seawater for up to 3 weeks when held at 5 and 10°C but was completely inactivated after 1 week at 20°C (Jones, 2017).

Salinity also impacts survival of viral or bacterial agents. IHNV, VHSV and *Y. ruckeri* appear to remain infectious longer in lower salinities (Tobback *et al.*, 2007; Hawley and Garver, 2008; Garver and Hawley, 2021) while salmonid alphavirus (SAV) has higher survival at higher salinities (Graham *et al.*, 2007). For parasites, water temperature and salinity greatly affect development across multiple life history stages. For the salmon louse, *Lepeophtheirus salmonis*, temperature and salinity affect the number of eggs produced as well as the rate at which they develop and hatch. Similarly, the free-swimming naupliar stages, the infectious copepodid stage and the period of time over which copepodids remain infective are impacted by temperature and salinity (Groner *et al.*, 2016; Samsing *et al.*, 2018). Additionally, the hatching rates and larval development of monogenean parasites are also variable and depend upon temperature and salinity of the aquatic environment (Hoai, 2020). With respect to marine cage-cultured fish, members of the monogenean parasite *Neobenedenia* are harmful as ectoparasites in tropical and subtropical waters. Brazenor and Hutson (2015) examined the effects of temperature and salinity on the embryonation, hatching success, longevity of the free-swimming infectious oncomiracidia stage, infection success and time to sexual maturity of *Neobenedenia* sp. Within their preferred temperature ranges the dispersal capacity of both the non-feeding stages of caligid copepods and *Neobenedenia* sp. are higher at lower temperatures due to their increased longevity.

Sunlight (ultraviolet (UV) light) inactivates bacteria and viruses in shallow waters, with the time to inactivation dependent on the intensity and wavelength of the radiation, depth of the pathogen in the water column, turbidity, exposure period and pathogen species (Gardner *et al.*, 2014). For example, it has been demonstrated that IHNV is very sensitive to exposure to solar radiation, with infectious virus concentration reduced by six orders of magnitude compared with that of dark controls over a 3 h experiment (Garver *et al.*, 2013b). Both UV radiation and biological activity of natural seawater are reported to limit the survival time of ISAV under normal conditions (Vike *et al.*, 2014). It is known that other viral and bacterial agents are sensitive to exposure to UV light in the laboratory, suggesting they may have a similar sensitivity to solar radiation exposure in the water column.

The presence of organic material in fresh water and seawater can affect survival and viability, especially of viral agents (Oidtmann *et al.*, 2018). For example, Kocan *et al.* (2001) reported that the presence of ovarian fluids in natural seawater, such as would occur during spawning events, significantly increases survival of VHSV. However, in the case of SAV the presence of organic matter resulted in reduced survival (Graham *et al.*, 2007).

It is well recognized that the presence of microbial biota affects both viral and bacterial survival with survival times reduced and/or inactivation rates increased in natural fresh water and natural seawater when compared with sterile waters (Oidtmann *et al.*, 2018). For example, IHNV and VHSV can remain infectious for up to a year at 4°C in sterile waters in the dark (Hawley and Garver, 2008; Garver and Wade, 2017; Garver and Hawley, 2020), whereas survival in non-sterile waters may be limited to hours or days (Hawley and Garver, 2008; Garver *et al.*, 2013b). However, it is not known to what extent seasonal or other changes in the composition and abundance of microbes in seawater would affect viral and bacterial survival rates under natural conditions.

9.3.4 Consequence of exposure to an infectious agent

Figure 9.2 illustrates the potential consequences to the host following exposure to an infectious agent, showing that infectious agent exposure does not necessarily lead to infection and infection does not automatically lead to disease. A carrier state is an infectious individual who is capable of transmitting disease but exhibits no clinical evidence of disease. Chronic or latent infections are characterized by the continued presence of infectious viruses or bacteria following the primary infection. Rhyan and Spraker (2010) propose that, in general, there are four possible outcomes from pathogen exposure in a new or novel individual/population (spillover). The individual/population may be: (i) resistant to the infection, therefore it incurs no effect; (ii) a 'dead-end' host or sink that is unable to maintain the infection without external hosts; (iii) a spillover host that is able to transiently maintain the infection for a time but requires periodic exposure from another source, or is a maintenance host; or (iv) capable of maintaining the infection without further transmission from another species (Rhyan and Spraker, 2010).

It must be stressed that, within any population, the progression or response to exposure to an infectious agent may vary considerably among its individuals (i.e. from no effect to serious disease). Sublethal infection or parasitic infestation may lead to behavioural or physiological changes that increase the risk of being preyed upon (Brassard *et al.*, 1982; Binning *et al.*, 2017). Ultimately, each exposure is unique and for any individual host or agent the outcome of infection may differ under different scenarios.

9.4 Cage Culture and Potential Effects on the Natural Ecology

It is well recognized that pathogen introductions into farmed fish from the natural aquatic environment are common events. This is evidenced when fish known to be free of a specific pathogen before stocking acquire infection or disease after introduction to the cage system. Yet the occurrence and frequency at which farmed fish acquire infections of naturally occurring (endemic) pathogens are often unpredictable due to the complexity of interactions between the host, infectious agent and environment.

In instances where farmed fish are highly susceptible to an infectious agent that is present in the environment, spillover infections (movement of an infectious agent from its natural host

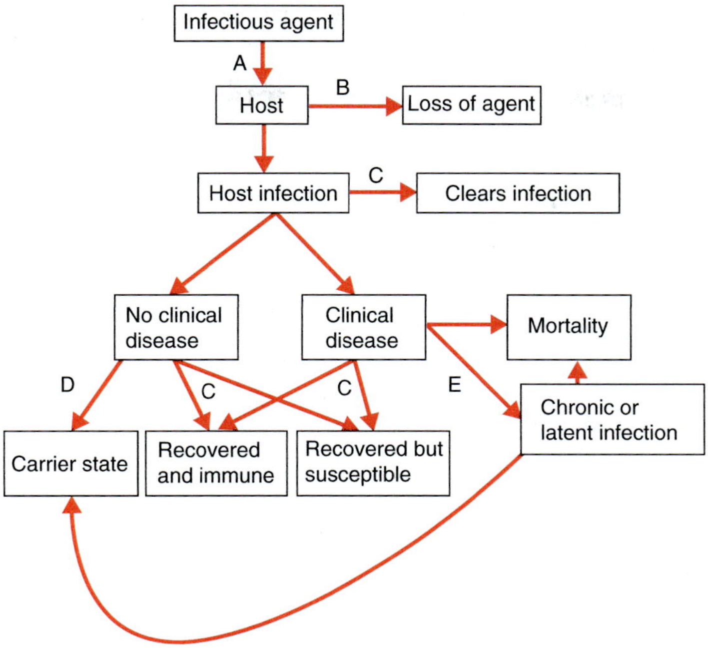

Fig. 9.2. The potential consequences to the host following exposure to an infectious agent: A, exposure; B, no entry into the host; C, effective immune response and the agent is cleared from the host; D, effective immune response but the agent is not cleared from the host; E, primary infection is not cleared by the immune response and/or the capacity of the agent to kill host cells is reduced.

to an alternative) of the agent into farmed populations can be frequent. Such a scenario has been observed with farmed Atlantic salmon reared in the coastal waters of British Columbia, where nearly all farmed salmon become infected with *Piscine orthoreovirus 1* (PRV-1) following seawater entry into open net cages (Polinski *et al.*, 2020). In fact, the time at sea was a significant predictor for the detection of PRV-1 in Atlantic salmon with the probability increasing up to 18 months after seawater entry (Laurin *et al.*, 2019). However, if the species being cultured is of low susceptibility to the endemic infectious agents and/or exposure to agents in the surrounding environment is uncommon, then spillover infection in farmed animals will be infrequent.

Limiting wild and cultured fish interactions is important to mitigating risk of infectious agent exchange; however, open cage culture presents a unique concern in that, similar to other man-made structures such wharfs, jetties, marinas, etc., cages are attractants to wild fish as potential sources of food and shelter. In the case of cage-culture farms this attraction brings wild and farmed fish into close contact, thereby increasing the opportunity for and risk of transfer of endemic agents. One example is the occurrence of VHSV in farmed Atlantic salmon of the north-eastern Pacific Ocean. Farmed Atlantic salmon are of low to moderate susceptibility to VHSV, however Pacific herring and other forage species native to the north-eastern Pacific Ocean are exceptionally susceptible to VHSV and are a natural reservoir of the virus (Hershberger *et al.*, 2021). Attracted to Atlantic salmon cage systems as a refuge from predators, the forage species carrying VHSV are placed in proximity with the farm fish presenting an optimal situation for the transmission of the virus to the less susceptible farmed Atlantic salmon. Shedding copious amounts of virus (Hershberger *et al.*, 2010), the cohabitating forage fish expose the farmed fish to high concentrations of virus. This results in the successful spillover of VHSV from the wild marine reservoir to the farmed Atlantic salmon (Garver *et al.*, 2013a).

The assemblages of wild fish and the nature of their interactions with cage aquaculture are extremely diverse, nevertheless understanding the biology, behaviours and ecological traits of the wild fish species provides insights into the potential interactions that can occur. Non-migratory reef fish reside in high numbers around tropical sea-cage culture (Sudirman *et al.*, 2009), while migratory fish species, such as sockeye and Atlantic salmon, may be around farms for only minutes while en route to their feeding grounds (Vollset *et al.*, 2016; Rechisky *et al.*, 2021). Those species which reside in and around farms for longer durations are expected to have an increased risk of exposure when compared with migratory species. For example, wild sea trout and Atlantic salmon have similar timing of juvenile seawater entry but have different spatial distributions once in the ocean (Harvey *et al.*, 2020). Atlantic salmon undertake extensive feeding migrations moving relatively rapidly as juveniles from coastal waters to offshore feeding grounds in the North Atlantic and adjacent waters, whereas sea trout remain in fjords and coastal waters, often in the vicinity of cage-farmed salmon (Eldoy *et al.*, 2015; Flaten *et al.*, 2016; Bøhn *et al.*, 2022). This places them at a higher risk than wild Atlantic salmon to infestations with sea lice, *L. salmonis*, originating from salmon farms (see Bøhn *et al.*, 2022 and references therein). Likewise, exposure of wild fish to farm-derived infectious agents is only possible if wild fish are in the vicinity of the farms when infectious agents are indeed present in farmed fish. Winter ulcer disease, caused by *Moritella viscosa* infection, in farmed Atlantic salmon in British Columbia is limited to December, January and February when water temperatures are below 10°C. During these months, neither juvenile nor adult wild sockeye salmon are present and therefore are extremely unlikely to be exposed to farm-origin *M. viscosa* (Mimeault *et al.*, 2020).

Another route by which pathogens from cage-cultured fish can be spread to wild fish is through escapes. Farmed fish, whose behaviours upon escape bring them into contact with susceptible wild species, and who are infected at the time of escape, or if they become infected after escaping, are a potential route of transmission. Escaped farmed Atlantic salmon, gilthead sea bream (*Sparus aurata*) and sea bass (*Dicentrarchus labrax*) have been identified by numerous authors as potential sources of pathogens for transfer to wild populations; however, such transfers have not been demonstrated (reviewed in Johansen *et al.*, 2011; Arechavala-Lopez *et al.*, 2013, 2018).

The origins of aquaculture go back thousands of years (see Chua and Tech, 2002) and with increasingly better technologies, both the quantity and the diversity of species being farmed are expanding continually. Of the species cultured, they can be classified as either indigenous or non-indigenous, with each scenario presenting different challenges and risks in the context of fish health and infectious agents. Non-indigenous species are often selected for cage culture when native species do not have desirable characteristics such as fast growth, good table quality, ease of breeding, tolerance to a wild range of environment conditions, early maturation and efficient food conversion ratios. This means that fish such as carp, tilapia, catfish and salmonids, which possess many of these quantities, represent the most highly produced species in cage aquaculture. This results in these species often being cultured outside their native origin, such as the production of Atlantic salmon in both the North and South Pacific Oceans or the farming of tilapia outside Africa. While the use of non-indigenous animals in aquaculture has proven successful, the introduction of species into a novel environment can be challenging from a fish health perspective. In particular, aquaculture of non-native species has the potential to represent a novel host for the endemic agents such that these fish might have a heightened susceptibility to endemic diseases. This is evident with farming of Atlantic salmon in the North Pacific where outbreaks of IHN have caused devasting losses in cage culture. Laboratory studies have demonstrated that Atlantic salmon have greater susceptibility to IHN than indigenous Pacific salmon species, possibly a consequence of the fact that Atlantic salmon are exotic to the Pacific Ocean and likely have less innate resistance to the IHNV (Traxler *et al.*, 1993; Garver *et al.*, 2013b; Long *et al.*, 2017).

Another consideration in utilizing non-indigenous species is the potential for introducing an exotic pathogen with the translocation or introduction of the non-native aquaculture stock (Peeler *et al.*, 2011). Examples include the introduction of ISAV into Chile likely from Norwegian exports of Atlantic salmon embryos (Vike *et al.*, 2009); IHNV to Asia and Europe likely through the importation of rainbow trout eggs from western North America (Bovo *et al.*, 1987); and PRV into Washington state likely through Atlantic salmon eggs imported from Iceland (Warheit, 2018).

The alternative practice of using indigenous species in aquaculture, while typically associated with a better understanding of how the farmed species will respond to endemic agents, is not devoid of risks concerning fish health and infectious agent transmission. Although native to the region, the farmed species is often a domesticated strain or of a genetic stock that is not identical to the truly native population. In this scenario, while the cultured and indigenous populations may be considered a single species, they undoubtedly have genetic differences that potentially lead them to respond to endemic agents differently. Furthermore, movement of a geographically isolated strain of species to another population of the same species may carry a similar risk of infectious agent introduction as observed for non-indigenous aquaculture. The emergence of *Gyrodactylus salaris*, an important disease affecting wild Atlantic salmon, resulted from the movement of a Baltic strain of Atlantic salmon from Sweden to Norway for aquaculture purposes (Mo, 1994). Historically in its original host, the Baltic strain of Atlantic salmon, *G. salaris* was not associated with disease; however, upon the host switch to an Atlantic strain of Atlantic salmon in Norway, the emergence of disease resulted. Consequently, implementation of pathogen-free broodstock, highly sensitive diagnostics, strict biosecurity measures and proper legislative controls are essential to safeguard against the risks of pathogen introductions associated with the use of native and non-native species in aquaculture.

Infectious agents and disease concerns originating in wild populations can spill over into farmed populations (McVicar, 1997; Murray 2009). As noted previously, the frequency of these spillover events is dependent upon the prevalence of the agent in the natural environment, the susceptibility of the farmed animals and their likelihood of interacting with the natural host. Nevertheless, when cage-cultured animals do become infected with an endemic infectious agent, concerns arise as to whether the agent may be amplified in the farm population, leading to disease outbreaks in the farms that then spill back (movement of an infectious agent from an alternative host back to the natural host) into the wild population and/or transfer to other farm populations. Often reared at high densities, cage-cultured fish have the potential to increase pathogen loads in the environment in which they reside, particularly if disease ensues and is left unmanaged. For instance, the prevalence of IPNV in wild marine fish in Scotland, albeit low overall, was found to be significantly higher within 5 km of virus-positive Atlantic salmon farms (Wallace *et al.*, 2008). As reviewed in Section 9.6 on the consequences of pathogen exchange, determining the effects of pathogen release from open net-cage farms on sympatric wild populations is often difficult to assess due to knowledge gaps in our understanding of background infectious agent prevalence and diseases in wild fish. Nevertheless, with estimates of transmission parameters associated with these endemic agents coupled to physical hydrodynamic models, accurate geospatial predictions of infectious agent spread can be simulated from aquaculture sites (Salama and Rabe, 2013). Model outputs from simulated infectious agent dispersions can be used to construct connectivity maps and quantify agent spread. Modelling IHNV dispersion from Atlantic salmon marine net-cage aquaculture using empirically derived estimates of infectious dose, viral shedding and decay rates (Garver *et al.*, 2013b) coupled to a water circulation model (Foreman *et al.*, 2012) illustrated that vaccination and disease management actions mitigate the risk of transmitting an infectious dose of IHNV to wild salmon (Mimeault *et al.*, 2017). However, if farms are without vaccination and/or appropriate disease management procedures are not in use, diseased farms were predicted to amplify sufficient virus to permit transmission of an infectious dose to nearby farms and wild salmonids (Foreman *et al.*, 2015).

In addition to the potential for amplification of endemic agents, infection and disease of farmed animals raise concerns regarding the natural evolution of pathogens and parasites. Generally, the selection pressures on pathogens and parasites infecting wild hosts differ from those on their farmed counterparts (Mennerat *et al.*, 2010; Kurath and Winton, 2011). Aquaculture, like all farming activities, can create conditions that may alter the evolutionary trajectory of the naturally occurring agent, which in some cases may favour the development of highly virulent pathogens and parasites (Nowak, 2007; Kurath and Winton, 2011; Kennedy *et al.*, 2016). In fact, features of aquaculture such as high rearing densities, limited genetic diversity and the constant production of overlapping age classes present conditions akin to laboratory serial passage studies designed to induce experimental evolution to investigate agent adaptation (Nowak, 2007). Although many disease burdens of aquaculture are likely a result of epidemiological changes, increasing evidence suggests that agent evolution, including evolution of virulence, is also playing a role in the emergence of some diseases in aquaculture (Walker and Winton, 2010). Some laboratory studies where intensive aquaculture practices have been suggested in the increased occurrence and/or severity of the pathogen or parasite include a bacterial fish disease caused by *Flavobacterium columnare* in salmon fingerlings at fish farms in Finland (Pulkkinen *et al.*, 2010; Sundberg *et al.*, 2016); the emergence of virulent ISAV strains in Norwegian Atlantic salmon farms (Mjaaland *et al.*, 2002; Nylund *et al.*, 2003); and the observation that sea lice were more harmful if originated from a farmed versus unfarmed host (Ugelvik *et al.*, 2017). In light of the continued expansion and intensification of aquaculture, fish health professionals need to consider the potential risks associated with evolution of virulence as a consequence of farm practices. Additional research is required on this topic, specifically examining how aquaculture practices might drive virulence evolution and what mechanism could be employed in the interest of preventing pathogen adaptation to aquaculture environments.

9.5 Infectious Agent/Disease Management in Net-Cage Culture

Cage-culture systems are most often situated in locations and designed in ways to facilitate water exchange between the water masses within and outside the cages. The openness of these systems allows for the potential of bidirectional infectious agent exchange between cultured and wild populations. There are a variety of tools which can be applied to farm populations to mitigate these exchanges and their potential consequences to both cultured and wild stocks. For example, the selection of sites with water conditions that meet the physiological requirements for optimal growth and health for the species under culture is critical, as this reduces the risk of infection and disease outbreaks within the cultured population. In addition, sites should be selected to limit exposure to known risks or stressors, as well as to avoid areas in which wild populations reside.

The selection of species and more specifically strains/stocks that have the physiology capacity to thrive in the water conditions of the site, as well as in a captive culture situation, is also extremely important to maintain optimal health and prevent infection and disease outbreaks. For any particular species, it is quite evident that not all stocks or strains are capable of being cultured and similarly not all of the stocks/strains that are suitable for culture are productive across all regions. For example, Atlantic salmon strains originating from Atlantic Canada do not perform as well as strains originating from Europe when cultured on the Pacific coast of Canada, leading to different strains being cultured in the Pacific and Atlantic regions. So just as in livestock farming, the characteristics of the environment and the culture species must be evaluated together to ensure operational success. Further, optimizing the husbandry and production needs of the cultured species is essential to maintaining the general health and welfare of the population.

Even with appropriate site and species/strain selection and optimal husbandry and rearing conditions, the implementation of appropriate biosecurity measures is essential to disease prevention/management and the economic success of a farming operation (Bennett, 2012) (for more details on biosecurity, see Subasinghe and Shinn, Chapter 11, this volume, 2023). Fish, water, vectors, fomites and feed are all potential entry portals for infectious agents that need to be considered when developing biosecurity mitigation measures in aquaculture. Eliminating all risk in cage culture through biosecurity measures may not be possible, however, due to the open nature of the system. In the case of areas where multiple farms operate, consideration should be given to coordinating activities to improve biosecurity (e.g. coordinating fallowing, stocking, movement, harvesting, treatment)

(Chang *et al.*, 2014; Murray and Salama, 2016). For example, Gustafson *et al.* (2014) suggested that issues with biosecurity at farms, on harvest vessels and in fish-processing plants contributed to the spread of ISAV in Chile. Poor biosecurity such as cohabiting naïve fish with survivors of IHN, and movement of equipment and fish, likely accounted for some farm outbreaks and led to secondary spread of the IHNV between farms in two major IHN outbreaks in farmed Atlantic salmon that occurred in British Columbia, Canada in 1992 and 2001 (St-Hilaire *et al.*, 2002; Saksida, 2006). Alternatively, for an occurrence of IHN in 2012, rapid industry-wide implementation of good biosecurity practices, such as effective use of vaccination and early culling of farms with IHNV, quickly controlled and prevented a widescale outbreak as occurred historically (Romero *et al.*, 2022). In addition to protecting the farmed fish, good biosecurity practices also lessen the risk of amplification and spillback of infectious agents into the environment, thereby limiting the risk of pathogen exchange. Off-farm biosecurity measures (e.g. import and movement regulations) play a significant role in limiting the introduction of new or novel infectious agents into an ecosystem (see Chapter 11, this volume).

Disease prevention is a cornerstone of biosecurity which in turn reduces the risks of amplification and spillback to the wild populations. In cage culture, pathogen and disease screening is crucial in disease prevention and typically is conducted prior to the introduction of fish into cages. In addition, some vertically transmitted diseases are controlled through broodstock screening programmes. For example, transmission of vertically transmitted infectious agents such as *R. salmoninarum* and IHNV are often less frequent in regions that have extensive broodstock segregation, screening and egg disinfection procedures within aquaculture operations (Elliot *et al.*, 1995).

Vaccination has also become one of the most important, easy and effective approaches to prevent and control infectious diseases in cultured fish. Commercial fish vaccines are available for many economically important infectious bacterial and viral diseases of fish. There are no successful vaccines for the parasitic or fungal diseases (Brudeseth *et al.*, 2013; Shefat, 2018). However, the majority of commercial vaccines that have been developed are for salmonids, leaving other cage-culture sectors without this important tool to effectively prevent/control pathogens. Development of fish vaccines is a challenging task, due to the variety of pathogens and host species, as well as the uniqueness of host susceptibility to each pathogen. Cost-effectiveness is an essential limitation to commercial fish vaccine development; consequently, the availability and use of licensed vaccines for disease prevention are still limited in many regions where cage-culture farms operate (e.g. regions in Africa, South America and Asia) (Brudeseth *et al.*, 2013). In order to increase availability of vaccines, production of autogenous vaccines (vaccines produced on small to medium scales from pathogens isolated directly from the farm on which they are deployed) has been proposed as a solution that could effectively limit disease on farms (Brudeseth *et al.*, 2013; Subasinghe *et al.*, 2019) and the risk of spillback to the environment.

Cage culture currently has the ability to manage/control diseases caused by infectious agents (in particular, bacterial and parasitic agents) through the use of therapeutants (e.g. antibiotics, antiparasitics) which reduce disease, pathogen amplification and potential risk to wild populations. However, as is being seen in other food-animal production sectors, there is an increasing awareness of issues regarding the use of therapeutants in aquaculture (particularly cage culture) (Rico *et al.*, 2013; Henriksson *et al.*, 2018). This has driven the push towards the concept of prudent use of therapeutants (Okocha *et al.*, 2018) and research towards alternatives. Examples include functional feeds (Encarnação, 2015; Olmos Soto *et al.*, 2015; Legrand *et al.*, 2020), as well as mechanical and biological treatments for the management of sea lice in salmon (for more information on sea lice, see Leong *et al.*, Chapter 2, this volume, 2023; St-Hilaire *et al.*, Chapter 8, this volume, 2023; Chapter 11, this volume; Huntingford *et al.*, Chapter 12, this volume, 2023).

Disease surveillance/monitoring is an essential component of a good biosecurity plan (Bondad-Reantaso *et al.*, 2021) (see Chapter 11, this volume). However, because of the importance the environment plays in aquatic animal health, environmental monitoring activities (e.g. measurement of temperature, dissolved oxygen, salinity, toxic phytoplankton abundance, predator activity) are important to incorporate as part of regular fish health monitoring activities on cage farms. The goal of any surveillance/monitoring programme is the early detection of a threat, in this case an infectious agent or disease, and this is accomplished through careful routine monitoring for signs of disease, prompt removal of dead fish from cages and examination/diagnostic testing of moribund/dead individuals (Hoinville *et al.*, 2013). Frequent removal of dead fish reduces potential pathogen load within the cage since shedding of pathogens may continue after the fish is dead (Kunttu *et al.*, 2009; Pulkkinen *et al.*, 2010; Johansen *et al.*, 2011) and reduces the likelihood for predators to be attracted to the cages that stresses the surviving population and potentially makes them further susceptible to disease.

An appreciation of normal versus not normal is developed through repeated and regular monitoring. Regular surveillance/monitoring can be very effective in detecting emerging diseases since the new or emerging disease is likely a spillover event from an unknown wild source, particularly in areas with established aquaculture activities (i.e. no imports) (Rhyan and Spraker, 2010; Hoinville *et al.*, 2013; Brugere *et al.*, 2017). This could become even more important with changing climate (see Section 9.7 below). It has also been suggested that farms regularly monitored could be used as sentinels to evaluate pathogen exposure in the aquatic environment (St-Hilaire *et al.*, 2001; Pert *et al.*, 2014). However, this approach has utility only if farmed fish are susceptible to the pathogen and have been introduced into the cage systems free of the pathogen of interest. In addition, occurrence of disease in farmed fish populations does not necessarily imply occurrence of the same disease in wild populations (Gardner *et al.*, 2014). Infectious agents that are considered to be serious in wild populations may not be pathogenic to farmed fish or the farmed fish may act as vectors with no evidence of disease.

Diagnostic assays are critical in the process of disease detection and diagnoses, and they are routinely used to support health management decisions in cage culture and other aquaculture activities, particularly as so many of the infectious diseases seen in cage culture share similar clinical presentation (are not pathognomonic). For example, many bacterial and viral diseases present with petechial haemorrhage on external and internal organs; therefore petechial haemorrhage is a commonly reported sign that without diagnostic analysis does not inform health management decisions and requires additional diagnostic analysis to confirm a diagnosis. Rapid and accurate diagnosis is essential in appropriately managing diseases to avoid amplification and the risk of spillback. Low availability of quality diagnostic laboratories and trained aquatic animal health specialists in a region can limit disease detection and ultimately timely control (Brugere *et al.*, 2017). This is a concern in many regions where cage culture occurs, particularly within developing countries where the consequences of pathogen exchange may be more severe due to an inability to rapidly diagnose and thereby effectively control infectious agents and disease.

9.6 Consequences and Assessment of Infectious Agent Exchange

9.6.1 In cage culture

The impact of infectious disease in cage culture consists of three economic components: (i) impact on the individual farmer including loss of capital (mortality) and reduction in productivity or quality and marketability (Lafferty *et al.*, 2015); (ii) wider impacts such as costs of surveillance and control, trade restrictions and animal welfare; and (iii) impact on rural economies and tourism (Bennett, 2012; Brooks-Pollock *et al.*, 2015). However, when it comes to cage culture, for a number of reasons including the high connectivity in the aquatic environment, the operations occurring in public rather than private waterways and the increased awareness of the environment, there is the added component of social license or public trust (Bennett, 2012). Social licence (such as reputational risk, regulatory risk or political risk such as market closure, etc.) could impact the success or failure of a given venture (Mather and Fanning, 2019). Recent (since 2020) events in British Columbia, particularly the threat of closure of salmon aquaculture activities in one region (consists of 20% of all farms), have highlighted the significant impact that public opinion, civil society actors (advocacy organizations, charities and think tanks) and other governmental actors (indigenous and municipal) can have on the aquaculture industry. Risks related to public trust are particularly important, as lack of social licence may be partly responsible for recent slow growth of aquaculture in other regions (Mather and Fanning, 2019). All these parameters impact the cost of production and ultimately the sustainability of a farm, or even an industry. The amount of effort or costs applied for disease management is often considered based on the resulting benefits: cost–benefit analysis of disease management.

A feature of cage culture is its close connection to the environment, which means fish health management is a global, transboundary concern (Asche and Smith, 2010). However, the motivation for controlling disease or infectious agents in cage culture is often a complex (and potentially poorly defined) mix that includes economic constraints bound by trade agreements and regulations, animal health and welfare issues, and political positioning, often based on historic norms, practicalities of control and public opinion, even if control is epidemiologically well defined (Brooks-Pollock *et al.*, 2015). Today the cost–benefit equation cannot simply focus on the benefits to the individual farmers, it must include the value towards the public good and social licence and the ecosystem.

In some regions/countries there are extensive rules and regulatory oversight surrounding disease management at aquaculture operations beyond those (e.g. control reportable diseases, import controls) recommended by the World Organization for Animal Health (WOAH formally OIE). In some countries such as Norway regulatory requirements and oversight are enforced nationally while in other countries, such as Canada, regulatory requirements differ among the regions. These additional oversight activities are costly, however, and as a result often restricted to countries that can afford it, thereby causing an imbalance in aquatic animal health oversight and the potential for transboundary effects. Over the last decade, a number of international certification standards have been developed for aquaculture. Certification standards such as the ASC (Aquaculture Stewardship Council) and BAP (Best Aquaculture Practices) include specific criteria on aquatic animal health that need to be maintained to receive certification. The application of certification in aquaculture is viewed as a potential market-based tool for minimizing potential negative impacts and increasing societal and consumer benefits and confidence in aquaculture (FAO, 2011; Henriksson *et al.*, 2018; Naylor *et al.*, 2021).

9.6.2 On wild populations

Pathogens and parasites are a natural and abundant component of our aquatic ecosystems that typically reside in balance with host and environmental conditions such that an endemic state is reached within populations. However, if this balance is disrupted through changes to host, pathogen or environmental conditions, a transition can occur from an endemic to a disease state and potentially be of risk to wild populations. Infectious diseases can impact individuals directly through mortality or indirectly through changes in various performance parameters including, but not limited to, swimming performance, growth and reproduction. Yet quantification of impacts due to disease are often hindered as mortality can go unnoticed or underestimated in wild populations or the disease contribution is masked by other factors (i.e. infectious disease increases susceptibility to predation due to reduced swimming performance).

Numerous factors, including large- and small-scale climate and ocean changes as well as shifts in predator and prey abundance and distribution, all interact with each other to regulate populations of wild fish in complex and poorly understood ways. Due to the enormity of factors involved in structuring wild population levels, determining the relative contributions of infectious diseases towards the overall mortality of a population is often difficult. For instance, while pathogen-induced mortality directly reduces population abundance, there are likely subsequent unpredictable (positive or negative) consequences incurred upon the population as a result of altered density-dependent interactions, as observed with an outbreak of IHN in wild sockeye salmon where despite catastrophic losses at the egg to fry stage, slightly higher than average survival was subsequently recorded at the fry to smolt stage, likely as a consequence of reduced in-lake competition for resources (Williams and Amend, 1976). Consequently, with high uncertainty in quantifying natural disease-derived mortality rates in wild populations, assessing the risk of whether aquaculture activities alter this rate with any significant precision is extremely challenging.

To assess the impact of farm-origin infectious agents on wild populations, it is necessary to know the biology of the agent; for example, whether the agent is infectious, virulent and causative of disease in particular host species. Defining the disease-causing potential of an infectious agent is best performed under controlled laboratory studies where host, pathogen and environmental conditions can be made to simulate natural exposure conditions or independently altered to assess the role of specific factors in the development of a disease state. These laboratory studies are typically the 'gold standard' in establishing disease causation and are extremely useful in defining the pathology and disease progression by examining infection, dissemination, persistence, recovery and immunity (Meyers and Hickey, 2022). These studies also assist with evaluating host susceptibilities, infectious dose and the physiological consequences of infection and disease. While they are instrumental in defining the relationship of pathogens and diseases in individuals, laboratory environments may be lacking necessary risk factors in order to replicate diseases seen in the natural environment. One must exercise caution in the application of the disease effects observed in individuals held in the laboratory to populations residing in the natural aquatic environment. In order to understand the mechanisms and scales in which infectious disease could impact wild populations, it is necessary to rely not only upon disease information derived in laboratory settings but also data obtained through field-based disease ecology studies.

A significant knowledge gap in our understanding of infectious diseases in wild populations is a lack of adequate historical pathogen/disease data to assess the general role of infectious disease in aquatic animal population extinction and endangerment (Smith *et al.*, 2006). In other words, if a disease event occurs, it is nearly impossible to determine if the event is unique or common, particularly when previous disease prevalence rates are unknown for the affected population. This information is often lacking as surveillance programmes on wild populations are laborious and expensive, often requiring vessels, special sampling equipment and knowledge of the fishery targeted for sampling. Furthermore, surveillance activities typically need to be performed over several generations and encompass large areas in an effort to capture potentially high spatiotemporal variability and to account for other factors that could drive pathogen prevalence. Importantly, the surveillance efforts need to be conducted in concert with ongoing fisheries research programmes so that pathogen and disease prevalence rates can be placed into context with fisheries biometrics. Moreover, to specifically investigate whether disease occurrence in wild fish is influenced by the vicinity of farmed fish, surveillance activities need to be conducted in areas both with and without aquaculture, keeping in mind that environmental/ecological conditions may not be similar between sites. Additionally, it is imperative to know the status of disease and the pathogen prevalence in the aquaculture settings so that it can be determined whether an association exists between disease in wild and farmed fish populations.

There are numerous diagnostic methodologies suitable for surveillance activities, but the suite of assays employed should be tailored to the objectives of the study. Molecular-based diagnostics have revolutionized the throughput, sensitivity and specificity of infectious agent detection, yet the identification and quantification of a nucleic acid component of an agent are not necessarily an accurate proxy for disease. In order to more reliably assess whether the infection is causing a meaningful disease, necropsy and histopathology are required whereby tissue damage and severity of the disease are visually assessed. Nevertheless, as aetiological agents of disease are better characterized through controlled laboratory investigations and pathology assessments, molecular-based assays, when properly validated to the specific agents of interest, provide an extremely powerful tool to detect minute traces of the agents that is particularly advantageous when analysing dilute samples such as water around farm sites (i.e. environmental DNA (eDNA)). When combined with automation, molecular-based assays afford researchers the capacity to process and analyse large sample sizes very quickly. Furthermore, as the molecular-based assays require small sample input, they are ideally suited for non-lethal testing where small biopsies can be performed without injury to the fish.

Barrett *et al.* (2019) reviewed studies looking at aquaculture impacts on wild populations and for disease/pathogen exchange, reporting that of 11 studies investigating infection levels (mostly on sea lice), all found higher levels in the presence of active fish farms. This is not too surprising as farms have the potential to amplify infectious agents. This was clearly acknowledged in the chapter on wild–farmed disease interaction in the previous edition of this book, where Saksida *et al.* (2014) stated: 'The question therefore, is not whether there is the likelihood of exposure to an infectious agent; as it is obvious that this is very likely, but rather the key question is whether this interaction leads to higher disease risks and rates in wild and cultured fish as a consequence of this interaction.'

Predicting and assessing the potential risks from these infectious agents are a substantial challenge that requires considerable understanding of the biology of the agent and the host as well as their interaction(s) with the physical environment, including the transport of agents from sources to potential host populations. The development of coupled biological–physical models has made investigating such interactions possible. Models run diverse scenarios within a system, act as tools to reproduce interlinked processes under different conditions (Salama and Rabe, 2013) and help to understand problems and identify questions relating to the system of interest. However, the strength of a prediction is only as good as the data that go into it (Kitching *et al.*, 2005); for diseases in wild fish such data are currently extremely sparse, hence accounting for the limited predictive value of the models. This uncertainty is further corroborated in the conflicting predictions between models. Some models predict stock extinctions due to aquaculture while others suggest aquaculture is not detrimental to wild fish (Teixeira Alves and Taylor, 2020). Models suggest pathogen risks to wild fish can be mitigated by acquired immunity in freshwater aquaculture systems (Teixeira Alves and Taylor, 2020). It is likely one model will not be universal to all aquaculture scenarios and each one will have to be built with data specific to the situation.

Quantifying pathogen-induced mortality, and specifically that which can be attributable to infectious agents originating from cage culture, is challenging due to the aforementioned reasons. However, through collaboration among fisheries scientists and aquatic animal health professionals and researchers, fisheries models can better incorporate disease mortality estimates to ultimately strengthen the predictive capacity of the models.

9.7 Climate Change Effects on Wild–Farm Population Pathogen Exchange

The idea that human activities lead to changes in climate is not new, with the proposed connection between atmospheric carbon dioxide and global warming being first identified in the late 19th century (Jackson, 2020). However, in recent years it has become obvious that climate-related human activities over the long term are gradually shifting, and over the shorter term are changing, the timing, magnitude and duration of naturally fluctuating environmental conditions of terrestrial and aquatic systems (Fig. 9.3). Climate-related changes in environmental conditions are now also recognized to play an important role in shaping the patterns of wild fish activities and behaviours, such as patterns of seasonal occupation and migration.

There are a number of recent reviews exploring the effects of climate change on aquaculture and wild capture fisheries and it is widely accepted that there will be impacts to both; the ways this will occur are less clear, however, but will likely include a series of direct and indirect inputs (McIlgorm *et al.*, 2010; Reid *et al.*, 2019; Reverter *et al.*, 2020; Wang *et al.*, 2020; Maulu *et al.*, 2021; Mengual *et al.*, 2021). Of these, only a few have addressed how climate change may impact diseases and disease management of cage-cultured fish or the occurrence and impacts of diseases in wild populations (Marcogliese, 2008; Karvonen *et al.*, 2010; Marcos-López *et al.*, 2010; Callaway *et al.*, 2012; Doubleday *et al.*, 2013).

The following subsections use specific examples to explore how climate-driven changes may change the risk of infectious agent transmission between cage-cultured and wild fish. Potential impacts on infectious agents, cage-cultured and wild hosts, frequency and severity of disease, transmission and contact patterns between cage-cultured and wild fish, and cur-

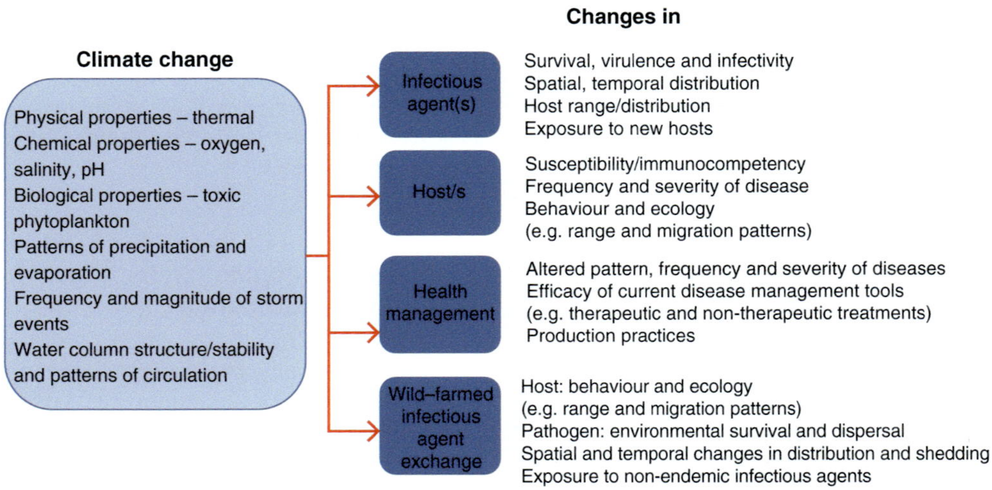

Fig. 9.3. Summary of the effects of climate change on the aquatic environment and possible outcomes in infection/disease dynamics.

rent health management practices are discussed (Fig. 9.3). However, it is important to remember that the complexity and interrelatedness of climate-driven changes, and the high amount of environmental variability between areas in which cage culture occurs, require that conditions at local scales must be considered as climate impacts will vary with site location (Falconer *et al.*, 2020).

9.7.1 Climate change impacts on infectious agents

Each infectious agent has a range of specific environmental parameters outside which their capacity to survive in the environment, spread and infect new hosts, and cause disease is reduced. Section 9.2 above reviewed the physiochemical and biological factors in the environment that affect the capacity of infectious agents to survive and remain infectious outside the host. For the majority of infectious agents our knowledge is limited and often based on laboratory studies on a single parameter, not considering how other parameters may interact to influence outcomes. In many cases, the results of laboratory studies will not accurately reflect what happens in the field. For this reason, laboratory results should be interpreted cautiously especially when using them to predict effects of climate change on infectious agents and their transmission potential between cage-cultured and wild fish.

Climate change may directly affect infectious agents leading to changes in virulence, infectivity and viability, as reflected in their capacity to survive and remain infectious outside a host (Callaway *et al.*, 2012). For example, the role of temperature and other physical and chemical properties of water on adhesion, virulence and interaction of fish pathogenic bacteria with host cells is reviewed in Hamed *et al.* (2018). The virulence of some aquatic infectious agents and parasites can be exacerbated under elevated temperatures (see Marcogliese, 2008).

Climate-related changes are also expected to have effects on the spatial and temporal distribution of infectious agents and diseases. For example, winter ulcer disease caused by *M. viscosa* in cage-reared Atlantic salmon occurs primarily during the winter, when water temperatures drop below 7–10°C (Wade and Weber, 2020). In this instance increased winter water temperatures could further restrict the time period over which this disease occurs, thereby further reducing the risk of disease in cage-cultured fish and the risk of disease transmission to wild fish. However, in areas where climate-related changes result in colder water temperatures this agent may become more of a problem and a greater risk to wild fish.

Climate change is expected to have a large impact on infectious agents that require intermediate hosts, as environmental changes may alter the interaction (positively or negatively) between the infectious agent and existing intermediate host and the infected intermediate host and the final host. Contact patterns, which determine individuals' interactions with others and their infection risk, may also be altered/expanded in situations where intermediate hosts experience an expanded range temporally and spatially. An example of this would be in the myxozoan parasite *Tetracapsuloides bryosalmonae*, which is the causative agent of proliferative kidney disease (PKD), a serious disease of salmon. It is believed that climate-related changes have resulted in an increased incidence of PKD across an expanded geographical range (Marcos-López *et al.*, 2010; Gorgoglione *et al.*, 2020). These increases may be related to changes in the distribution, abundance and/or biology of the intermediate bryozoan hosts and/or the capacity of fish to tolerate subclinical/chronic infections (Ros *et al.*, 2022). An extensive study of another myxozoan parasite of salmonids, *Ceratonova shasta*, has also shown that warm, dry years are associated with high disease-related mortality in hosts; and that cooler years with greater winter discharge are modelled to produce smaller numbers of polychaetes (intermediate host), thereby producing fewer actinospores and reducing infection rate in fish (True *et al.*, 2011; Adam Ray *et al.*, 2015).

Based on an analysis of a 20-year disease data set from two farms, Karvonen *et al.* (2010) found that the prevalence of infection for some agents (*Ichthyophthirius multifiliis*, *F. columnare*) increased with temperature, whereas other agents showed an opposite trend (*Ichthyobodo necator*) or no trend (*Chilodonella* spp.). Reverter *et al.* (2020) analysed the influence of temperature on the mortality of temperate and warmwater fish species infected with *Aeromonas* spp., *Edwardsiella* spp., *Flavobacterium* spp., *Streptococcus* spp., *Lactococcus* spp., *Vibrio* spp. and *Yersinia* spp. using published data from 270 laboratory infection trials. They found no significant increase in mortality in temperate fish infected with *Edwardsiella* spp. and *Y. ruckeri*, or in warmwater fish infected with *F. columnare*, with increasing temperature. For the other agents their analysis predicted that a temperature increase of 1°C could lead to a small increase in mortality of 3.9–6.0% and 2.8–4.1% in temperate and warmwater fishes, respectively.

The intracellular pathogen of salmonids, *R. salmoninarum*, has been shown to have a lethal dose tenfold lower at 15°C compared with 9°C in some salmonid species (Jones *et al.*, 2007), but also may exhibit biphasic control with an optimal temperature of 15°C, reduced growth at 18°C and again induced growth at 20°C (Perreira *et al.*, 2021). While the bacteria may have enhanced growth and impacts on some salmonids at these higher temperatures, in chinook salmon (*Oncorhynchus tshawytscha*) Purcell *et al.* (2016) found significantly higher *R. salmoninarum*-specific mortality, bacterial load and shedding rates at 8°C relative to the fish held at 12 or 15°C, and a trend of suppressed shedding in the group held at 15°C. Therefore, pathogen survival, infectivity and pathogenicity towards the host will not always have a linear or unidirectional correlation with temperature changes that may be seen with climate change.

Parasitic infections like amoebic gill disease (AGD), common in cage-cultured salmon worldwide (i.e. Australia, Europe, Chile and North America), are also positively associated with temperature. In fact, high salinity and temperature are two major risk factors for AGD in all areas (Douglas-Helders *et al.*, 2001; Rozas *et al.*, 2012; Oldham *et al.*, 2016). Elevated seawater temperature and infection with *Neoparamoeba perurans* have been reported to exacerbate complex gill disease (CGD) in farmed Atlantic salmon (*Salmo salar*) in British Columbia (Jones and Price, 2022).

It is clear from these examples that individual infectious agents differ in how they respond to changes in their environment and hosts. This, along with the uncertainty on how climate-related changes will be expressed in different areas, emphasizes the need to consider agents and the conditions they are experiencing as an individual case, when predicting the direction and magnitude of responses to climate-induced changes.

9.7.2 Climate change impacts on host(s)

Most finfish are ectotherms/poikilotherms and rely on behaviour to regulate body temperature within a physiological range. Cellular metabolism and the nervous system do not function properly at temperatures below the physiological range, while temperatures above can lead to protein and cell-membrane instability and death (Haesemeyer, 2020). Each fish species also has optimal ranges for other physical and chemical environmental parameters (e.g. dissolved gases, pH and salinity) within which they can maintain normal behavioural, physiological and immunological functions. Therefore short- and longer-term climate-related changes to the environment will have an effect on the physiological and immunological status of fish, possibly leading to changes in their susceptibility to infection and disease (Alfonso *et al.*, 2021; Mugwanya *et al.*, 2022).

Knowledge of the ranges and optima for environmental parameters for different fish species is necessary for predicting how and to what extent climate-related changes may impact them. Also important is an understanding of the extent to which different fish species have the capacity to deal with short-term environmental fluctuations that fall outside their optimal ranges (short-term stress response), as well as their capacity to adapt to longer-term (chronic) exposures to suboptimal conditions. Of these climate-related environmental parameters the effects of temperature have been the most extensively studied, especially the effects on physiological/stress and immunological responses. For example, enhanced/peak cellular and humoral capacity in rainbow trout (*Oncorhynchus mykiss*) has been observed at 15 and 20°C compared with 5°C (Nikoskelainen *et al.*, 2004), whereas in sockeye salmon (*Oncorhynchus nerka*) improved complement has actually been observed at lower temperatures (8 versus 12°C) (Alcorn *et al.*, 2002). These differences within such closely related species, the salmonids, are also paralleled by variation in local adaptation within wild salmon populations with respect to temperature tolerance (Chen *et al.*, 2013).

Temperature impact on antibody production is well described in the common carp. The magnitude of primary antibody production is not affected by lower temperatures while the secondary response (memory component) of the specific response is gradually lost as temperature drops, suggesting that immunological memory is relatively temperature sensitive (Rijkers *et al.*, 1980). Further, the relationship between temperature and peak day of the humoral immune response in carp closely matches the relationship between temperature and cellular immune response. For instance, time to peak antibody production is significantly shorter at 25°C compared with 16°C and both are shorter than at 12°C.

Exposure to acute or chronic thermal stress has the potential to stimulate innate and/or specific immune responses in farmed or wild fish, while also exerting an inhibitory impact on leucocyte/lymphocyte populations through the induction of cortisol (the predominant corticosteroid) through chronic stress. A main finding of Pérez-Casanova *et al.* (2008) was that Atlantic cod under chronic exposure to elevated temperatures (increases from 10 to 19.1°C) exhibited strong biphasic changes to immune-related gene expression either initiated by cortisol modulation/inhibition or through the presence of an optimal temperature for immune function in this species. Related to this, Larsen *et al.* (2018) showed Atlantic cod cleared infection with the intracellular pathogen *Brucella pinnipedialis* more rapidly at high water temperature (15°C) compared with its more optimal water temperature (6°C). This suggests a trade-off between resistance/tolerance to survive infection at suboptimal temperatures and impacts of increased water temperatures on the energetic costs of immune system activation.

Host thermal stress also intersects with the viability of the potential pathogen to induce disease and morbidity in a specific host species. In Atlantic salmon intraperitoneally injected with a virulent ISAV-HPR4 isolate, post-injection temperature increase from 12 to 20°C had no impact on mortality rate or total mortality (85% by 21 days post-injection). However, cohabitation exposure resulted in nearly two times higher mortality at 10°C than at 20°C; although mortality occurred earlier at the higher temperature, clearance of the virus was also more complete at the higher temperature, with a greater number of carriers and higher relative abundance of ISAV transcripts in survivors at 10°C as compared with 20°C (Groves *et al.*, 2021). Therefore, induced antiviral responses were greater at 10°C compared with 20°C, but the inverse was true for antigen presentation and differential T-lymphocyte activation. At the opposite end of the spectrum, cold stress has been shown in rainbow trout cells to negatively impact antigen-presenting molecules at 2°C but not 5°C, as well as to impact antigen presentation when cells are infected with VHSV at 2°C but not 14°C (reviewed in Abram *et al.*, 2017). Using an Arctic charr cell line, Semple *et al.* (2017) were able to show these same antigen presentation proteins were not impacted at 1, 4 or 14°C. Collectively, these works show a diversity of antiviral and antigen presentation pathway responses at both ends of the temperature spectrum (low and high), which have adapted differently, likely based on life history trajectories, in salmonids. This highlights the importance of understanding how species with different optimal temperature ranges will respond under extreme temperature shifts at both ends of the spectrum, which can lead to suboptimal immune responsiveness to bacterial or viral infections they may otherwise be able to resist.

These same temperature impacts also apply to parasitic infections. Godwin *et al.* (2020) showed a positive relationship between sea louse (*L. salmonis*) load, temperature and mortality rate in host Atlantic salmon at temperatures between 10 and 20°C. This coincided with reduced wound-healing capacity in these fish at the highest temperatures (M.D. Fast, 2022, unpublished results). Of particular concern in this host–parasite relationship is not only the potential for enhanced susceptibility of farmed Atlantic salmon, but also the impacts global climate change may have on infected juvenile wild salmon (Fast and Dalvin, 2020).

Increases in temperature and the associated carbon dioxide with climate change will contribute to ocean and freshwater acidification. Both wild and hatchery populations of salmon are more sensitive to lower pH during the parr–smolt transformation than at other stages (Rosseland and Skogheim, 1984; Staurnes *et al.*, 1993) and chronic exposure to low pH has been shown in Atlantic and sockeye salmon to impair seawater readiness and survival (Staurnes *et al.*, 1993; Kennedy and Picard, 2012). Impacts of these stressors, at early life stages, can increase susceptibility to subsequent pathogen challenge, either directly through non-specific infections from reduced seawater readiness (i.e. *Tenacibaculum* spp., CGD, etc.) or indirectly through lack of growth and protection against predation and parasitism.

The red sea bream iridovirus (RSIV), which causes significant mortality in marine-cultured sea bream (*Pagrus major*) and over 30 other cultured marine fish (Kawakami and Nakajima, 2002), can infect rock bream (*Oplegnathus fasciatus*) at temperatures as low as 13°C without subsequent mortality. However, the virus persists in these fish, with significant (up to 100%) mortality when temperatures shift up to 18–25°C (Jun *et al.*, 2009). These temperatures completely overlap with the optimal physiological temperatures for this host species, thereby potentially increasing the frequency of infections for this and other cultured marine species as coastal ocean temperatures stay at 20–28°C for extended periods. Given the low host specificity of RSIV it has the potential to impact numerous wild and other cultured fish that inhabit the same waters as infected red sea bream and rock bream. Moreover, the risk of RSIV to juvenile marine fish is greater than for adults; severity of disease in wild juveniles is enhanced where large bream are cultured in areas overlapping with wild, juvenile, susceptible marine fishes. Differential susceptibility of sympatric host species can also lead to greater impacts on wild populations in the case of RSIV, as shown for the myxozoan endoparasite *T. bryosalmonae* that causes temperature-dependent PKD in salmonids. Lauringson *et al.* (2021) identified that initial proliferation rate of the parasite in brown trout, which was associated with temperature, was followed by greater severity of PKD in this species compared with the sympatric Atlantic salmon.

Other ways in which the wild host can be impacted by climate change are via: (i) migration and residency patterns of wild populations, which in the case of salmonids are influenced by temperature and salinity linked to timing of the snow melt (Nekouei *et al.*, 2018); and (ii) temperature shifts, which change the abundance of prey/food items and feeding grounds, such as has been described in adult (higher survival in colder years) and juvenile (better survival in colder years) herring, *Clupea harengus*, which feed on the planktonic copepod *Calanus finmarchicus* in the Norwegian Sea (Engelhard and Heino, 2006). Increased tissue damage and reduced early-life growth and survival have all been described for larval Atlantic cod in response to carbon dioxide loading and the subsequent exposure to chronic acidified water (Baumann *et al.*, 2012; Frommel *et al.*, 2012). Effects on growth, development and immunological robustness thereby could reduce the ability of wild populations to withstand endemic infectious agents, resulting in increased disease that could spill over into farmed populations.

Coldwater species such as salmon and cod may be good examples to demonstrate the impacts of projected temperature increases on wild and farmed fish populations and their interactions. But according to Cheung *et al.* (2013) and Oyinlola *et al.* (2020) it is the tropics that will observe major losses in mariculture species richness potential (MSRP), which is defined as the total number of species that can be farmed given a set of specific environmental conditions. Tropical and subtropical species adapted to stable environments are less resilient to even small shifts in environmental temperature and pH, and for this reason they are considered to be highly at risk due to climate-derived changes. As these fish are reduced, they are replaced by others with greater environmental optima. In salmon and cod, 10°C shifts in temperature are stressful and can lead to some morbidity depending on the species and even strain. Fish such as gilthead sea bream, *S. aurata*, can show equally detrimental effects even with minor temperature changes. Mateus *et al.* (2017) showed that just a 4°C difference in temperature during early development (egg and larvae) of sea bream caused thermal imprinting with long-term consequences that resulted in significant suppression of innate immune mechanisms in adults.

The balance of the host–pathogen interaction in farmed and wild fish can be tipped in either direction depending on the type of thermal stress and the adaptations of the species of fish. There are at least two potentially linked ways, however, farmed and wild fish will differ in these responses, regardless of their basal immunological and physiological statuses. The first is in their ability to behaviourally alter these responses. In cage culture, the farmed fish have a limited ability to thermoregulate due to confinement, whereas wild populations have the ability to behaviourally thermoregulate without these constraints. Both coldwater and warmwater species choose the best thermal conditions to mount a successful immune response, otherwise known as behavioural fever (Boltana *et al.*, 2013, 2018). Behavioural fever impacts positively upon lymphocyte proliferation, inflammatory cytokine expression and other immune functions. For example, in salmon, individuals able to access different temperature regimes throughout the day (15 ± 7.4°C) had reduced viral load (IPNV), enhanced Th1 responses and greater survival than individuals maintained at a constant 15°C temperature (Boltana *et al.*, 2018). This thermal preference reduced clinical signs of disease and mortality. Behavioural thermoregulation allows the host to both boost its response to the pathogen while also avoiding the temperature most permissive for viral/pathogen replication. As identified by Boltana *et al.* (2018), behavioural fever can act as an integrative signal that also promotes specific epigenetic modifications. Epigenetic modifications in regulating the immune response likely lead to warm-seeking initiation. With respect to climate change, Miller *et al.* (2012) also described epigenetic impacts of carbon dioxide exposure from current parental environments leading to impacts on coral reef fish progenies' responses to elevated carbon dioxide in their future environment. Epigenetics would therefore be a second mechanism by which wild populations may adapt quicker to the changing environment, that may not be seen in the farmed populations.

9.7.3 Climate change impacts on health and disease management practices

The frequency, severity and outcomes of disease outbreaks are expected to change in response to climate-related changes. This may be due to effects on the infectious agents such as change in virulence, changes in rates of transmission and/or stress-related changes in host resistance. Increased frequency and severity of disease in cultured fish ultimately increase the risk of transmission of infectious agents to wild fish. Marcos-López *et al.* (2010) developed a risk framework to examine the influence of climate change on disease emergence in freshwater systems in the UK. These authors state that 'many of the endemic diseases of salmonids (e.g. enteric red mouth, furunculosis, proliferative kidney disease and white spot) will become more prevalent and difficult to control as water temperatures increase.' In marine waters it is expected that the outbreak frequency for some, but not all, fish diseases will increase due to climate-driven changes in the environment and hosts. However, as noted previously, outbreaks of some coldwater diseases may decrease. For example, ISAV has enhanced virulence/cytopathogenic effect at higher temperature (22°C > 18°C > 15°C) in cell culture (Atlantic salmon kidney cells) but reduced horizontal transmission *in vivo*, likely due to the speed at which the virus overtakes the host response at 20°C compared with 10°C. Despite the possibility of greater acute mortality at higher temperature as demonstrated in laboratory trials, risk factor analysis of field data actually found a strong latitude and lower temperature (between 4 and 12°C) association with the likelihood of ISA outbreaks in Norway (Lyngstad *et al.*, 2018). In a recent review of Mediterranean aquaculture, the absence of data on the occurrence of fish diseases and on the frequency of pathologies as they relate to increasing water temperatures was identified (Cascarano *et al.*, 2021) and a similar situation occurs for other regions where cage culture takes place (Collins *et al.*, 2020). This may make it difficult to track any deviation that may occur as a consequence of climate-related changes in these regions.

Climate change will likely require modifications in production and health management practices to maintain healthy populations in cages (Islam *et al.*, 2022). With respect to meeting the physiological/health requirements of the species under culture, climate-related changes in the physical, chemical and biological properties of the water may require the relocation of cage operations to areas where the magnitude and frequency of these changes do not exceed the adaptive capacity of the cage-reared fish. Changes to production schedules (e.g. timing of fish introduction), husbandry practices (e.g. reduced rearing densities) and/or transition to the rearing of species that are more tolerant of environmental conditions at the site may also be required. For example, in Atlantic salmon farming, changes in oceanic conditions can be partially mitigated by delaying the transition to seawater from freshwater hatcheries, thereby ensuring that larger and potentially more robust fish enter the changed environment (Collins *et al.*, 2020).

In many instances climate change will result in the amplification of environmental stressors, including changes in the magnitude and frequency of events that may exceed the physiological and immunological capacity of cage-cultured fish. As discussed previously, stressful conditions are often reflected as an increase in the frequency and severity of infectious disease outbreaks in cage culture, as well as in reduced effectiveness of some health management practices such as vaccination (reviewed in Alfonso *et al.*, 2021; Mugwanya *et al.*, 2022).

Increased temperature may lead to increased severity of bacterial diseases and thus to increases in antibiotics use in aquaculture, especially in lower- to middle-income countries (Reverter *et al.*, 2020; Pepi and Focardi, 2021; Mugwanya *et al.*, 2022). The combination of drug-resistant bacteria from land-derived sources making their way into water courses and increased antimicrobials use in aquaculture will likely increase antimicrobial resistance (AMR) in these systems. Selection and emergence of AMR may be further exacerbated by warmer temperatures reducing generation times and leading to higher AMR development rates for these bacteria (Pepi and Focardi, 2021). The loss of therapeutic-value antimicrobials through the development of AMR poses a significant risk in aquaculture, agriculture and human health.

Climate-related changes will likely cause changes in the spatial and temporal distribution of diseases, which may change the risk of infectious agent transmission between cage-cultured and wild fish (Marcos-López *et al.*, 2010; Rosa *et al.*, 2012; Cascarano *et al.*, 2021). Changes in the environment will likely alter the frequency and methods by which some infectious agents are managed, in particular where temperature variation had previously played a role in their control. For example, currently abundance of the salmon louse, *L. salmonis*, on net-cage-cultured Atlantic salmon declines in the North Atlantic Ocean starting in autumn/winter (November/December) and continues almost until June (see Chang *et al.*, 2011; Fast and Dalvin, 2020). Furthermore, recruitment and development of the parasite are hampered through these low-temperature months (<4°C) and are highest from June to September/October. However, over the last decade water temperatures >8°C have extended into December with minimum winter temperatures at or above 4°C.

These extended warm periods have resulted in increasing the annual reproductive output of the salmon louse and increasing the infection pressure on farmed and wild species sharing these coastal waters. With increased infection pressure and parasite loads on farmed fish, so too has there been a need to increase the use of chemotherapeutants and more recently, due to issues with therapeutant resistance, non-medical management practices to control the lice.

Increased frequency and intensity of storms may require better cage systems to reduce the risk of fish escapes due to cage system damage. This not only has an economic impact on the farmer but also potential and ecological impacts (e.g. increase in wild–farmed fish interaction, resource competition).

9.7.4 Climate change impacts on wild–farmed infectious agent exchange

This section illustrates the complexity in assessing the consequences of climate change on the infectious agent and host. Basically, different infectious agents and hosts will react differently to specific scenarios. In some situations, endemic fish hosts will be become exposed to exotic or emerging infectious agents that have moved into a region; while in others it will be the infectious agents that are endemic and the potential hosts exotic (having migrated from elsewhere). Climate change will alter the infectivity, pathogenicity and virulence of some infectious agents and susceptibility of the hosts. There is a growing body of evidence that fish and other aquatic species are shifting their ranges (geographical areas and depth distribution), migration patterns and seasonal activities in response to short- and longer-term climate-related changes in environmental conditions (reviewed in Kaimuddin *et al.*, 2016; Chaikin *et al.*, 2022; Galappaththi *et al.*, 2022). In addition to changing the risk of exposure of wild fish to cage culture-derived infectious agents, shifts in the distribution of fish increase the risk of introduction of non-endemic agents.

Climate change impacts on cultured species will likely drive farmers to re-evaluate species reared and reassess net-cage placement, further expanding development of offshore sites and/or relocating to higher latitudes. Relocation of cage-culture sites due to unfavourable conditions resulting from short- and long-term climate-related changes may result in different contact patterns with different stocks or species of wild hosts. Prior to relocation of cage-culture facilities, the identification of susceptible wild hosts and/or wild hosts capable of carrying infectious agents of concern should be assessed.

The consequence that climate change will have on the wild–farmed fish exchange of infectious agents is complex; in some cases there will be more exchange of agents with greater impact on one or both wild/farmed populations, while in others there may be no change or a decline in pathogen movement. The effects realistically can only be measured and assessed on a case-by-case basis at the local level. The consequences will likely change with the rate of climate change in the region, and it will be important that assessment and management are adaptive to changes.

9.8 Adaptive Management and EcoHealth to Reduce Management Risks Associated with Pathogen Exchange

To date the majority of the research and regulatory efforts surrounding pathogen exchange between cage-cultured and wild populations has focused on salmon cage culture, which occurs primarily in developed countries (e.g. Canada, Chile, Ireland, Norway, UK). In these countries, cage culture of salmon does not directly impact food security/availability and levels of socio-economic security are such that there is stronger public and government interest in understanding and mitigating possible risks that cage culture has on the environment and populations of wild fish. However, even with the significant effort to mitigate pathogen exchange between cage-cultured and wild salmon populations, there continues to be uncertainty with respect to the significance of wild–farmed pathogen exchange in structuring populations of wild salmon. With respect to Mediterranean waters, the risk of pathogen exchange between cage-cultured and wild fish is recognized, but not as thoroughly studied or regulated (Tičina *et al.*, 2020). Outside these areas, the risks of pathogen exchange between net-cage-cultured and wild fish have received almost no attention.

The Food and Agriculture Organization of the United Nations (FAO) estimates that world farmed fish production and trade are expected to increase in the next decade, with much of the growth occurring in Asia and Africa (FAO, 2020). In many of the regions in which aquaculture production is likely to increase there is a long history of livestock agriculture. These regions are often resource-poor and carry a disproportionately higher global burden of livestock diseases compared with more developed regions (Brooks-Pollock *et al.*, 2015). This is due (in part) to the lack of resources that results in inadequate animal health systems (people, institutions/organizations and resources) to deliver veterinary services to support not only farmers and livestock production, but also the collection of epidemiological and other data necessary for the establishment of health management practices, policies and regulations.

It is likely that similar situations will arise in aquaculture in these regions, especially with increases in scale of production. Currently even basic essential information including farm location, size and species is limited in many developing regions, hindering the ability to implement even basic health/biosecurity support and regulatory oversight (FAO, 2022).

This is compounded by the fact that wild aquatic animals and pathogens are transboundary, moving freely across geographical borders and zones. There are serious concerns regarding the impact of wild–farmed animal pathogen exchange, therefore international collaborations are essential for developing appropriate programmes to ensure a sustainable aquaculture sector within a thriving ecosystem. To assess risk factors contributing to population declines or extinction, we must move towards a systematic analysis of the relative effects of the widely proposed causal threats, namely habitat destruction, overexploitation, invasive species, pollution and infectious disease (Smith *et al.*, 2006).

Pathogen exchanges between cage-cultured and wild fish occur and they are bidirectional. This chapter has emphasized the high level of uncertainty in assessing the consequences of pathogen exchange on wild populations currently and into the future with the changes likely to occur due to climate change. The authors propose that the way forward is to better understand and manage wild–farmed pathogen exchange through adaptive management and EcoHealth. The high connectivity of the aquatic ecosystem suggests that management should be transboundary or even international.

9.8.1 Adaptive management – what is it?

Adaptive management is a 'decision process that, promotes flexible decision making that can be adjusted in the face of uncertainties as outcomes from management actions and other events become better understood' (Williams *et al.*, 2009; Tett, 2017).

Adaptive management is an iterative process that is designed to take such complexities and uncertainties into account, in part by formally managing programmes through a 'learning by doing' approach (Fig. 9.4). This approach focuses on defining the scope of the management problem, synthesizing existing knowledge about the system, monitoring and evaluation to provide information about changes that could increase an intervention's efficiency and effectiveness to achieve the management problem.

The adaptive approach deals with uncertainty through a structured improvement of relevant knowledge, while seeking to minimize risks associated with ongoing management which inevitably arise from imperfect information about system response (Ebi 2011; Keith *et al.*, 2011).

A key element of an effective adaptive management programme is the engagement and development of collaborations between stakeholders (those who will implement, monitor and/or be affected by plans), as well as managers/regulators and scientists who participate in problem assessment and the identification of knowledge gaps, while taking social and political preferences into consideration. In general, stakeholders – including those who make management decisions and those who are affected by these decisions – can make important contributions to understanding problems and identifying appropriate solutions. Wide stakeholder participation facilitates achieving consensus, cooperation, buy-in and ownership of the problem within and across the communities involved.

A key hurdle is the need to embrace uncertainty, so that management effort is spread over

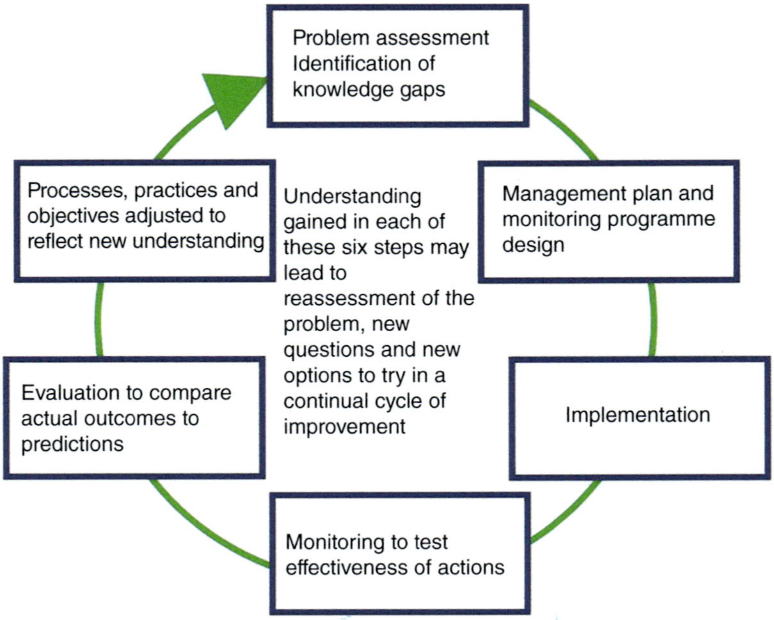

Fig. 9.4. Steps involved in adaptive management. During Step 1 (problem assessment, exploration and forecasting process), key gaps in understanding of the system (i.e. those that limit the ability to predict outcomes) are identified. Step 2 (design) involves designing a management plan and monitoring programme that will provide reliable feedback about the effectiveness of the chosen actions. It is useful to evaluate one or more proposed plans or designs on the basis of costs, risks, informativeness and ability to meet management objectives. In Step 3 (implementation), the plan is put into practice. In Step 4 (monitoring), indicators are monitored to determine how effective actions are in meeting management objectives and to test the hypothesized relationships that formed the basis for the forecasts. Step 5 (evaluation) involves comparing the actual outcomes to forecasts and interpreting the reasons underlying any differences. In Step 6 (adjustment), practices, objectives and the models used to make forecasts are adjusted to reflect new understanding. (Adapted from Nyberg, 1999.)

competing options in order to learn about them experimentally, rather than overinvesting in one option that is currently perceived as 'best practice' (Keith *et al.*, 2011). The challenge to researchers is to shift their focus from discovery to the science of implementation, while managers and policy makers must depart from their socio-political norms and institutional frameworks to embrace new thinking and effectively utilize the learning by doing through the adaptive management process.

Another essential element is clear and concise communication. Early stakeholder engagement and clear communication increase management effectiveness and the likelihood of solving the problem, as well as providing transparency and likely a higher level of acceptance in the process. As an example, an evaluation of an adaptive sea lice management programme in Norway (Traffic Light System) recommended improving communications to the lay person and increased involvement of stakeholders (e.g. farmers) in the evaluation and decision-making process (Eliasen *et al.*, 2021).

Adaptive management has been used widely in natural resource management (Keith *et al.*, 2011) and has been proposed but not yet widely adopted in aquaculture (Tett, 2017). With respect to reducing potential disease impacts of cage-cultured fish on wild fish, the adaptive management approach is a good option due to our limited understanding of and the high levels of uncertainty associated with this issue. It will also allow for the adaptation of management practices to overcome changes brought about by climate change.

9.8.2 EcoHealth

EcoHealth or Ecosystem Health is a fairly new discipline that focuses on research examining how changes in ecosystems can affect animal and human health (Binot and Morand, 2015; Lerner and Berg, 2017) and uses the main principles of adaptive management in managing health. EcoHealth integrates scientific understanding of ecological relationships within societal contexts, and focuses on the relationship between health, ecosystems and sustainable development where the latter is based on equality (Binot and Morand, 2015). Ultimately, the objective of EcoHealth research and practice is to develop environmentally sustainable, community-based interventions to improve the health of affected communities. It also promotes participatory action research that includes scholars in all relevant scientific disciplines, public health and healthcare professionals, community members and decision makers. EcoHealth emphasizes the involvement of communities as a key parameter for success before government's role.

Effectively managing the health risks of climate and other global environmental changes requires a broad understanding of the complex interactions existing in the ecosystem and the environment. Without this understanding, implementation of policies and programmes is likely not to be very effective. Interventions are more effective when they are designed to take equity into account, as socio-economic factors play a critical role in people's vulnerability and sensitivity to injury and disease by interacting with biological factors that mediate risk or affect the ability to adapt or respond to stressors.

9.9 Conclusion

Infectious agent exposure occurs between cage-cultured fish and wild populations. There are numerous transmission routes and factors that affect the outcome of these exposures within the individual host and at the population level. This chapter has discussed these factors in relation to global climate change, understanding potential consequences and how these may alter management decisions. This review is by no means prescriptive, but rather aimed to identify clear examples of how climate change may alter current interactions in the host–pathogen–environment triad and highlight the effects. The authors advocate for broader transboundary and multi-stakeholder involvement in aquatic animal health research and management with the concepts of adaptive management and EcoHealth as possible ways forward to maintain sustainability of net-cage aquaculture. Ultimately, from an understanding of the magnitude of interactions between cultured and wild fish, adaptive management practices can be implemented to mitigate the potential for pathogen exchange between wild and cultured fish populations.

References

Abdel-Latif, H.M., Dawood, M.A., Menanteau-Ledouble, S. and El-Matbouli, M. (2020) The nature and consequences of co-infections in tilapia: a review. *Journal of Fish Diseases* 43(6), 651–664.

Abram, Q.H., Dixon, B. and Katzenback, B.A. (2017) Impacts of low temperature on the teleost immune system. *Biology* 6, 39. https://doi.org/10.3390/biology6040039

Alcorn, S.W., Murray, A.L. and Pascho, R.J. (2002) Effects of rearing temperature on immune functions in sockeye salmon (*Oncorhynchus nerka*). *Fish & Shellfish Immunology* 12(4), 303–334.

Alfonso, S., Gesto, M. and Sadoul, B. (2021) Temperature increase and its effects on fish stress physiology in the context of global warming. *Journal of Fish Biology* 98(6), 1496–1508.

Adam, R., Alexander, J.D., Leenheer, P.D. and Bartholomew, J.L. (2015) Modeling the effects of climate change on disease severity: a case study of *Ceratonova* (syn *Ceratomyxa*) *shasta* in the Klamath River. In: Okamura, B., Gruhl, A. and Bartholomew, J.L. (eds) *Myxozoan Evolution, Ecology and Development*. Springer, Cham, Switzerland, pp. 363–378.

Anacleto, O., Cabaleiro, S., Villanueva, B., Saura, M., Houston, R.D., *et al.* (2019) Genetic differences in host infectivity affect disease spread and survival in epidemics. *Scientific Reports* 9(1), 4924.

Arechavala-Lopez, P., Sanchez-Jerez, P., Bayle-Sempere, J.T., Uglem, I. and Mladineo, I. (2013) Reared fish, farmed escapees and wild fish stocks – a triangle of pathogen transmission of concern to Mediterranean aquaculture management. *Aquaculture Environment Interactions* 3(2), 153–161.

Arechavala-Lopez, P., Toledo-Guedes, K., Izquierdo-Gomez, D., Šegvić-Bubić, T. and Sanchez-Jerez, P. (2018) Implications of sea bream and sea bass escapes for sustainable aquaculture management: a review of interactions, risks and consequences. *Reviews in Fisheries Science & Aquaculture* 26(2), 214–234.

Asche, F. and Smith, M.D. (2010) *Trade and Fisheries: Key Issues for the World Trade Organization*. WTO Staff Working Paper No. ERSD-2010-03. World Trade Organization, Geneva, Switzerland.

Bain, M.B., Cornwell, E.R., Hope, K.M., Eckerlin, G.E., Casey, R.N., *et al.* (2010) Distribution of an invasive aquatic pathogen (viral hemorrhagic septicemia virus) in the Great Lakes and its relationship to shipping. *PLoS ONE* 5(4), e10156.

Balfry, S.K., Albright, L.J. and Evelyn, T.P.T. (1996) Horizontal transfer of *Renibacterium salmoninarum* among farmed salmonids via the fecal–oral route. *Diseases of Aquatic Organisms* 25(1–2), 63–69.

Barrett, L.T., Swearer, S.E. and Dempster, T. (2019) Impacts of marine and freshwater aquaculture on wildlife: a global meta-analysis. *Reviews in Aquaculture* 11(4), 1022–1044.

Baumann, H., Talmage, S.C. and Gobler, C.J. (2012) Reduced early life growth and survival in a fish as a direct response to elevated CO_2 levels. *Nature Climate Change* 2(1), 38–41.

Bennett, R. (2012) Economic rationale for interventions to control livestock disease. *EuroChoices* 11(2), 5–11. https://doi.org/10.1111/j.1746-692X.2012.00227.x

Binning, S.A., Shaw, A.K. and Roche, D.G. (2017) Parasites and host performance: incorporating infection into our understanding of animal movement. *Integrative and Comparative Biology* 57(2), 267–280.

Binot, A. and Morand, S. (2015) Implementation of the One Health Strategy: lessons learnt from community-based natural resource programs for communities' empowerment and equity within an EcoHealth approach. In: Morand, S., Dujardin, J.-P., Lefait-Robin, R. and Apiwathnasorn, C. (eds) *Socio-Ecological Dimensions of Infectious Diseases in Southeast Asia*. Springer, Singapore, pp. 325–335.

Bishop, T.M., Smalls, A., Wooster, G.A. and Bowser, P.R. (2003) Aerobiological (airborne) dissemination of the fish pathogen, *Ichthyophthirius multifiliis* and the implications in fish health management. In: Lee, C.-S. and O'Bryen, P.J. (eds) *Biosecurity in Aquaculture Production Systems: Exclusion of Pathogens and Other Undesirables*. The World Aquaculture Society, Baton Rouge, Louisiana, pp. 51–64.

Bøhn, T., Nilsen, R., Gjelland, K.Ø., Biuw, M., Sandvik, A.D., *et al.* (2022) Salmon louse infestation levels on sea trout can be predicted from a hydrodynamic lice dispersal model. *Journal of Applied Ecology* 59(3), 704–714.

Boltana, S., Rey, S., Roher, N., Vargas, R., Huerta, M., *et al.* (2013) Behavioural fever is a synergic signal amplifying the innate immune response. *Proceedings of the Royal Society B: Biological Sciences* 280(1766), 20131381.

Boltana, S., Sanhueza, N., Donoso, A., Aguilar, A., Crespo, D., *et al.* (2018) The expression of TRPV channels, prostaglandin E_2 and pro-inflammatory cytokines during behavioural fever in fish. *Brain, Behavior, and Immunity* 71, 169–181.

Bondad-Reantaso, M.G., Fejzic, N., MacKinnon, B., Huchzermeyer, D., Seric-Haracic, S., *et al.* (2021) A 12-point checklist for surveillance of diseases of aquatic organisms: a novel approach to assist multidisciplinary teams in developing countries. *Reviews in Aquaculture* 13(3), 1469–1487. https://doi.org/10.1111/raq.12530

Bovo, G., Giorgetti, G., Jørgensen, P.E.V. and Olesen, N.J. (1987) Infectious haematopoietic necrosis: first detection in Italy. *Bulletin of the European Association of Fish Pathologists* 7(5), 124.

Brassard, P., Rau, M., and Curtis, M. (1982) Parasite-induced susceptibility to predation in diplostomiasis. *Parasitology* 85(3), 495–501. https://doi.org/10.1017/S0031182000056274

Brazenor, A.K. and Hutson, K.S. (2015) Effects of temperature and salinity on the life cycle of *Neobenedenia* sp. (Monogenea: Capsalidae) infecting farmed barramundi (*Lates calcarifer*). *Parasitology Research* 114(5), 1875–1886.

Bricknell, I. (2017) Types of pathogens in fish, waterborne diseases. In: Jeney, G. (ed.) *Fish Diseases: Prevention and Control Strategies*. Academic Press, London, pp. 53–80.

Brooks-Pollock, E., De Jong, M.C.M., Keeling, M.J., Klinkenberg, D. and Wood, J.L.N. (2015) Eight challenges in modelling infectious livestock diseases. *Epidemics* 10, 1–5. https://doi.org/10.1016/j.epidem.2014.08.005

Brudeseth, B.E., Wiulsrød, R., Fredriksen, B.N., Lindmo, K., Løkling, K.E., *et al.* (2013) Status and future perspectives of vaccines for industrialised fin-fish farming. *Fish & Shellfish Immunology* 35(6), 1759–1768.

Brugere, C., Onuigbo, D.M. and Morgan, K.L. (2017) People matter in animal disease surveillance: challenges and opportunities for the aquaculture sector. *Aquaculture* 467, 158–169. https://doi.org/10.1016/j.aquaculture.2016.04.012

Callaway, R., Shinn, A.P., Grenfell, S.E., Bron, J.E., Burnell, G., *et al.* (2012) Review of climate change impacts on marine aquaculture in the UK and Ireland. *Aquatic Conservation: Marine and Freshwater Ecosystems* 22(3), 389–421.

Carballo, J., Seoane, R.M. and Nieto, T.P. (2000) Adhesion of *Aeromonas salmonicida* to materials used in aquaculture. *Bulletin of the European Association of Fish Pathologists* 20(2), 77–82.

Cascarano, M.C., Stavrakidis-Zachou, O., Mladineo, I., Thompson, K.D., Papandroulakis, N. and Katharios, P. (2021) Mediterranean aquaculture in a changing climate: temperature effects on pathogens and diseases of three farmed fish species. *Pathogens* 10(9), 1205.

Chaikin, S., Dubiner, S. and Belmaker, J. (2022) Cold-water species deepen to escape warm water temperatures. *Global Ecology and Biogeography* 31(1), 75–88.

Chang, B.D., Page, F.H., Beattie, M.J. and Hill, B.W. (2011) Sea louse abundance on farmed salmon in the southwestern New Brunswick area of the Bay of Fundy. In: Jones, S. and Beamish, R. (eds) *Salmon Lice: An Integrated Approach to Understanding Parasite Abundance and Distribution*. Wiley, Chichester, UK, pp. 83–115.

Chang, B.D., Coombs, K.A. and Page, F.H. (2014) The development of the salmon aquaculture industry in southwestern New Brunswick, Bay of Fundy, including steps toward integrated coastal zone management. *Aquaculture Economics & Management* 18(1), 1–27. https://doi.org/10.1080/13657305.2014.855952

Chen, Z., Anttila, K., Wu, J., Whitney, C.K., Hinch, S.G. and Farrell, A.P. (2013) Optimum and maximum temperatures of sockeye salmon (*Oncorhynchus nerka*) populations hatched at different temperatures. *Canadian Journal of Zoology* 91, 265–274.

Cheung, W.W., Watson, R. and Pauly, D. (2013) Signature of ocean warming in global fisheries catch. *Nature* 497(7449), 365–368. https://doi.org/10.1038/nature12156

Chua, T.E. and Tech, E. (2002) Introduction and history of cage culture. In: Woo, P.T.K.., Bruno, D.W. and Lim, L.H.S. (eds) *Diseases and Disorders of Finfish in Cage Culture*, 1st edn. CAB International, Wallingford, UK, pp. 1–39.

Collins, C., Bresnan, E., Brown, L., Falconer, L., Guilder, J., *et al.* (2020) Impacts of climate change on aquaculture. *MCCIP Science Review* 2020, 482–520.

Coquet, L., Cosette, P., Quillet, L., Petit, F., Junter, G.A., and Jouenne, T. (2002) Occurrence and phenotypic characterization of *Yersinia ruckeri* strains with biofilm-forming capacity in a rainbow trout farm. *Applied and Environmental Microbiology* 68(2), 470–475.

Costello, C., Cao, L., Gelcich, S., Cisneros-Mata, M.A., Free, C.M., *et al.* (2020) The future of food from the sea. *Nature* 588(7836), 95–100. https://doi.org/10.1038/s41586-020-2616-y

Cribb, T.H., Adlard, R.D., Hayward, C.J., Bott, N.J., Ellis, D., *et al.* (2011) The life cycle of *Cardicola forsteri* (Trematoda: Aporocotylidae), a pathogen of ranched southern bluefin tuna, *Thunnus maccoyi*. *International Journal for Parasitology* 41(8), 861–870.

Doubleday, Z.A., Clarke, S.M., Li, X., Pecl, G.T., Ward, T.M., *et al.* (2013) Assessing the risk of climate change to aquaculture: a case study from south-east Australia. *Aquaculture Environment Interactions* 3(2), 163–175.

Douglas-Helders, M., Saksida, S., Raverty, S. and Nowak, B.F. (2001) Temperature as a risk factor for outbreaks of amoebic gill disease in farmed Atlantic salmon (*Salmo salar*). *Bulletin of the European Association of Fish Pathologists* 21(3), 114–116.

Ebi, K. (2011) Climate change and health risks: assessing and responding to them through adaptive management. *Health Affairs* 30(5), 924–930. https://doi.org/10.1377/hlthaff.2011.0071

Eliasen, K., Jackson, D., Koed, A., Revie, C., Swanson, H.A., *et al.* (2021) *An Evaluation of the Scientific Basis of the Traffic Light System for Norwegian Salmonid Aquaculture*. The Research Council of Norway, Lysaker, Norway.

Eldoy, S.H., Davidsen, J.G., Thorstad, E.B., Whoriskey, F., Aarestrup, K., *et al.* (2015) Marine migration and habitat use of anadromous brown trout (*Salmo trutta*). *Canadian Journal of Fisheries and Aquatic Sciences* 72(9), 1366–1378.

Elliott, D.G., Pascho, R.J. and Palmisano, A.N. (1995) Brood stock segregation for the control of bacterial kidney disease can affect mortality of progeny chinook salmon (*Oncorhynchus tshawytscha*) in seawater. *Aquaculture* 132(1–2), 133–144.

Encarnação, P. (2015) Functional feed additives in aquaculture feeds. In: Nates, S.F. (ed.) *Aquafeed Formulation*. Academic Press, Oxford, UK, pp. 217–237.

Engelhard, G.H. and Heino, M. (2006) Climate change and condition of herring (*Clupea harengus*) explain long-term trends in extent of skipped reproduction. *Oecologia* 149, 593–603.
Falconer, L., Hjøllo, S.S., Telfer, T.C., McAdam, B.J., Hermansen, Ø. and Ytteborg, E. (2020) The importance of calibrating climate change projections to local conditions at aquaculture sites. *Aquaculture* 514, 734487.
FAO (Food and Agriculture Organization of the United Nations) (2011) *Technical Guidelines on Aquaculture Certification*. FAO, Rome. Available at: https://www.fao.org/documents/card/fr/c/0ec15d4a-6295-51a5-af16-5c724c62de25/ (accessed 8 April 2022).
FAO (Food and Agriculture Organization of the United Nations) (2020) *The State of World Fisheries and Aquaculture 2020*. FAO, Rome. Available at: http://www.fao.org/publications/sofia/2020/en/ (accessed 25 April 2020).
FAO (Food and Agriculture Organization of the United Nations) (2022) Global Aquaculture Production 1950–2019 (database). Available at: https://www.fao.org/figis/servlet/TabSelector?tb_ds=Aquaculture&tb_mode=TABLE&tb_act=SELECT&tb_grp=ENVIRONMENT (accessed April 2022).
Fast, M.D. and Dalvin, S. (2020) Global climate change impacts on *Lepeophtheirus salmonis* infections in salmon. In: Woo, P.T.K., Leong, J. and Buchmann, K. (eds) *Climate Change and Infectious Fish Diseases*. CAB International, Wallingford, UK, pp. 471–498.
Fitridge, I., Dempster, T., Guenther, J. and de Nys, R. (2012) The impact and control of biofouling in marine aquaculture: a review. *Biofouling* 28(7), 649–669.
Flaten, A.C., Davidsen, J.G., Thorstad, E.B., Whoriskey, F., Ronning, L., *et al.* (2016) The first months at sea: marine migration and habitat use of sea trout *Salmo trutta* post-smolts. *Journal of Fish Biology* 89(3), 1624–1640.
Foreman, M.G.G., Stucchi, D.J., Garver, K.A., Tuele, D., Isaac, J., *et al.* (2012) A circulation model for the Discovery Islands, British Columbia. *Atmosphere–Ocean* 50(3), 301–316.
Foreman, M.G.G., Guo, M., Garver, K.A., Stucchi, D., Chandler, P., *et al.* (2015) Modelling infectious hematopoietic necrosis virus dispersion from marine salmon farms in the Discovery Islands, British Columbia, Canada. *PLoS ONE* 10(6), e0130951.
Frommel, A.Y., Maneja, R., Lowe, D., Malzahn, A.M., Geffen, A.J., *et al.* (2012) Severe tissue damage in Atlantic cod larvae under increasing ocean acidification. *Nature Climate Change* 2(1), 42–46.
Galappaththi, E.K., Susarla, V.B., Loutet, S.J., Ichien, S.T., Hyman, A.A. and Ford, J.D. (2022) Climate change adaptation in fisheries. *Fish and Fisheries* 23(1), 4–21.
Gardner, I., Saksida, S., Dixon, B., McKenzie, P. and Johnson, S. (2014) Pathogen exchange between wild and farmed finfish: evidence to assess pathogen source and factors associated with clinical disease occurrence. *Bulletin of the Aquaculture Association of Canada* 111(3), 4–35. Available at: http://aquacultureassociation.ca/wp-content/uploads/2017/01/AAC-Bulletin-111-3.pdf (accessed 18 October 2022).
Garver, K.A. and Hawley, L.M. (2020) *Characterization of Viral Haemorrhagic Septicaemia Virus (VHSV) to Inform Pathogen Transfer Risk Assessments in British Columbia*. DFO Canadian Science Advisory Secretariat Research Document, 2020/064. Fisheries and Oceans Canada, Ottawa.
Garver, K. and Wade, J. (2017) *Characterization of Infectious Hematopoietic Necrosis Virus (IHNV)*. DFO Canadian Science Advisory Secretariat Research Document, 2017/073. Fisheries and Oceans Canada, Ottawa.
Garver, K.A., Traxler, G.S., Hawley, L.M., Richard, J., Ross, J. and Lovy, J. (2013a) Molecular epidemiology of viral haemorrhagic septicaemia virus (VHSV) in British Columbia, Canada, reveals transmission from wild to farmed fish. *Diseases of Aquatic Organisms* 104, 93–104.
Garver, K.A., Mahony, A.A.M., Stucchi, D., Richard, J., Van Woensel, C. and Foreman, M. (2013b) Estimation of parameters influencing waterborne transmission of infectious hematopoietic necrosis virus (IHNV) in Atlantic salmon (*Salmo salar*). *PLoS ONE* 8(12), e82296.
Godwin, S.C., Fast, M.D., Kuparinen, A., Medcalf, K.E. and Hutchings, J.A. (2020) Increasing temperatures accentuate negative fitness consequences of a marine parasite. *Scientific Reports* 10, 18467. https://doi.org/10.1038/s41598-020-74948-3
Gorgoglione, B., Bailey, C. and Ferguson, J.A. (2020) Proliferative kidney disease in Alaskan salmonids with evidence that pathogenic myxozoans may be emerging north. *International Journal for Parasitology* 50(10–11), 797–807.
Graham, D.A., Cherry, K., Wilson, C.J. and Rowley, H.M. (2007) Susceptibility of salmonid alphavirus to a range of chemical disinfectants. *Journal of Fish Diseases* 30(5), 269–277.

Gregory, A., Munro, L.A., Wallace, I.S., Bain, N. and Raynard, R.S. (2007) Detection of infectious pancreatic necrosis virus (IPNV) from the environment in the vicinity of IPNV-infected Atlantic salmon farms in Scotland. *Journal of Fish Diseases* 30(10), 621–630.

Groner, M.L., McEwan, G.F., Rees, E.E., Gettinby, G. and Revie, C.W. (2016) Quantifying the influence of salinity and temperature on the population dynamics of a marine ectoparasite. *Canadian Journal of Fisheries and Aquatic Sciences* 73(8), 1281–1291.

Groves, L., Whyte, S.K., Purcell, S.L., Parrish, K., Perreira, J., *et al.* (2021) Infectious salmon anemia (ISAv) and climate change; the combined effects on Atlantic salmon's (*Salmo salar*) immunological response. *Presented at North American Comparative Immunology Workshop, 8–10 June 2021*.

Gustafson, L., Antognoli, M., Fica, M.L., Ibarra, R., Mancilla, J., *et al.* (2014) Risk factors perceived predictive of ISA spread in Chile: applications to decision support. *Preventive Veterinary Medicine* 117(1), 276–285.

Haesemeyer M. (2020) Thermoregulation in fish. *Molecular and Cellular Endocrinology* 518, 110986.

Hamed, S.B., Ranzani-Paiva, M.J.T., Tachibana, L., de Carla Dias, D., Ishikawa, C.M. and Esteban, M.A. (2018) Fish pathogen bacteria: adhesion, parameters influencing virulence and interaction with host cells. *Fish & Shellfish Immunology* 80, 550–562.

Hoai, T.D. (2020) Reproductive strategies of parasitic flatworms (Platyhelminthes, Monogenea): the impact on parasite management in aquaculture. *Aquaculture International* 28(1), 421–447.

Hammell, L., Stephen, C., Bricknell, I., Evensen, Ø. and Bustos, P. (2009) *Salmeon Aquaculture Dialogue Working Group Report on Salmon Disease*. Comealogue. Available at: https://c402277.ssl.cf1.rackcdn.com/publications/172/files/original/Salmon_Aquaculture_Dialogue_%E2%80%93_Working_Group_Report_on_Salmon_Disease_SalmonY.pdf?1344873463 (accessed 18 October 2022).

Harvey, A.C., Glover, K.A., Wennevik, V. and Skaala, O. (2020) Atlantic salmon and sea trout display synchronised smolt migration relative to linked environmental cues. *Scientific Reports* 10(1), 3529.

Håstein, T. and Lindstad, T. (1991) Diseases in wild and cultured salmon: possible interaction. *Aquaculture* 98(1–3), 277–288.

Hawley, L.M. and Garver, K.A. (2008) Stability of viral hemorrhagic septicemia virus (VHSV) in freshwater and seawater at various temperatures. *Diseases of Aquatic Organisms* 82, 171–178.

Henriksson, P.J., Rico, A., Troell, M., Klinger, D.H., Buschmann, A.H., *et al.* (2018) Unpacking factors influencing antimicrobial use in global aquaculture and their implication for management: a review from a systems perspective. *Sustainability Science* 13(4), 1105–1120.

Hershberger, P., Gregg, J., Grady, C., Collins, R. and Winton, J. (2010) Kinetics of viral shedding provide insights into the epidemiology of viral hemorrhagic septicemia in Pacific herring. *Marine Ecology Progress Series* 400, 187–193.

Hershberger, P.K., MacKenzie, A.H., Gregg, J.L., Wilmot, M.D., Powers, R.L. and Purcell, M.K. (2021) Long-term shedding from fully convalesced individuals indicates that Pacific herring are a reservoir for viral hemorrhagic septicemia virus. *Diseases of Aquatic Organisms* 144, 245–252. https://doi.org/10.3354/dao03595

Hick, P., Schipp, G., Bosmans, J., Humphrey, J. and Whittington, R. (2011) Recurrent outbreaks of viral nervous necrosis in intensively cultured barramundi (*Lates calcarifer*) due to horizontal transmission of betanodavirus and recommendations for disease control. *Aquaculture* 319(1–2), 41–52.

Hoinville, L.J., Alban, L., Drewe, J.A., Gibbens, J.C., Gustafson, L., *et al.* (2013) Proposed terms and concepts for describing and evaluating animal-health surveillance systems. *Preventive Veterinary Medicine* 112(1–2), 1–12. https://doi.org/10.1016/j.prevetmed.2013.06.006

Holmer, M. (2010) Environmental issues of fish farming in offshore waters: perspectives, concerns and research needs. *Aquaculture Environment Interactions* 1(1), 57–70.

Huntingford, F.A., Kadri, S. and Saraiva, J.L. (2023) Welfare of cage-cultured fish under climate change. In: Woo, P.T.K. and Subasinghe, R.P. (eds) *Climate Change on Diseases and Disorders of Finfish in Cage Culture*, 3rd edn. CAB International, Wallingford, UK, pp. 462–498.

Islam, M.J., Kunzmann, A. and Slater, M.J. (2022) Responses of aquaculture fish to climate change-induced extreme temperatures: a review. *Journal of the World Aquaculture Society* 53(2), 314–366.

Jackson, R. (2020) Eunice Foote, John Tyndall and a question of priority. *Notes and Records* 74(1), 105–118.

Jarp, J. and Karlsen, E. (1997) Infectious salmon anaemia (ISA) risk factors in sea-cultured Atlantic salmon *Salmo salar*. *Diseases of Aquatic Organisms* 28(2), 79–86.

Johansen, L.H., Jensen, I., Mikkelsen, H., Bjørn, P.A., Jansen, P.A. and Bergh, O. (2011) Disease interaction and pathogens exchange between wild and farmed fish populations with special reference to Norway. *Aquaculture* 315(3–4), 167–86. https://doi.org/10.1016/j.aquaculture.2011.02.014

Jones, D.T., Moffitt, C.M. and Peters, K.K. (2007) Temperature-mediated differences in bacterial kidney disease expression and survival in *Renibacterium salmoninarum*-challenged bull trout and other salmonids. *North American Journal of Fisheries Management* 27(2), 695–706.

Jones, S.R.M. (2017) *Characterization of* Piscirickettsia salmonis *and Salmonid Rickettsial Septicaemia to Inform Pathogen Transfer Risk Assessments in British Columbia*. DFO Canadian Science Advisory Secretariat Research Document, 2019/20. Fisheries and Oceans Canada, Ottawa.

Jones, S.R.M. and Price, D. (2022) Elevated seawater temperature and infection with *Neoparamoeba perurans* exacerbate complex gill disease in farmed Atlantic salmon (*Salmo salar*) in British Columbia, Canada. *Microorganisms* 10(5), 1039.

Jones, S.R.M., Bruno, D.W., Madsen, L. and Peeler, E.J. (2015) Disease management mitigates risk of pathogen transmission from maricultured salmonids. *Aquaculture Environment Interactions* 6(2), 119134. https://doi.org/10.3354/aei00121

Jun, L.J., Jeong, J.B., Kim, J.H., Nam, J.H., Shin, K.W., *et al.* (2009) Influence of temperature shifts on the onset and development of red sea bream iridoviral disease in rock bream *Oplegnathus fasciatus*. *Diseases of Aquatic Organisms* 84(3), 201–208.

Kaimuddin, A.H., Laë, R. and Tito De Morais, L. (2016) Fish species in a changing world: the route and timing of species migration between tropical and temperate ecosystems in Eastern Atlantic. *Frontiers in Marine Science* 3, 162.

Karvonen, A., Rintamäki, P., Jokela, J. and Valtonen, E.T. (2010) Increasing water temperature and disease risks in aquatic systems: climate change increases the risk of some, but not all, diseases. *International Journal for Parasitology* 40(13), 1483–1488.

Kawakami, H. and Nakajima, K. (2002) Cultured fish species affected by red sea bream iridoviral disease from 1996 to 2000. *Fish Pathology* 37(1), 45–47.

Keith, D.A., Martin, T.G., McDonald-Madden, E. and Walters, C. (2011) Uncertainty and adaptive management for biodiversity conservation. *Biological Conservation* 144(4), 1175–1178. https://doi.org/10.1016/j.biocon.2010.11.022

Kennedy, C.J. and Picard, C. (2012) Chronic low pH exposure affects the seawater readiness of juvenile Pacific sockeye salmon. *Fish Physiology and Biochemistry* 38, 1131–1143.

Kennedy, D.A., Kurath, G., Brito, I.L., Purcell, M.K., Read, A.F., *et al.* (2016) Potential drivers of virulence evolution in aquaculture. *Evolutionary Applications* 9(2), 344–354.

Kim, K.I., Kim, Y.C., Kwon, W.J., and Jeong, H.D. (2017) Evaluation of blue mussel *Mytilus edulis* as vector for viral hemorrhagic septicemia virus (VHSV). *Diseases of Aquatic Organisms* 126(3), 239–246.

Kitching, R.P., Hutber, A.M. and Thrusfield, M.V. (2005) A review of foot-and-mouth disease with special consideration for the clinical and epidemiological factors relevant to predictive modelling of the disease. *The Veterinary Journal* 169(2), 197–209.

Kitamura, S.I., Ko, J.Y., Lee, W.L., Kim, S.R., Song, J.Y., *et al.* (2007) Seasonal prevalence of lymphocystis disease virus and aquabirnavirus in Japanese flounder, *Paralichthys olivaceus* and blue mussel, *Mytilus galloprovincialis*. *Aquaculture* 266(1–4), 26–31.

Kocan, R.M. (2019) Transmission models for the fish pathogen *Ichthyophonus*: synthesis of field observations and empirical studies. *Canadian Journal of Fisheries and Aquatic Sciences* 76(4), 636–642.

Kocan, R.M., Hershberger, P.K. and Elder, N.E. (2001) Survival of the North American strain of viral hemorrhagic septicemia virus (VHSV) in filtered seawater and seawater containing ovarian fluid, crude oil and serum-enriched culture medium. *Diseases of Aquatic Organisms* 44, 75–78.

Krkošek, M. (2017) Population biology of infectious diseases shared by wild and farmed fish. *Canadian Journal of Fisheries and Aquatic Sciences* 74(4), 620–628.

Kunttu, H.M., Valtonen, E.T., Jokinen, E.I. and Suomalainen, L.R. (2009) Saprophytism of a fish pathogen as a transmission strategy. *Epidemics* 1(2), 96–100. https://doi.org/10.1016/j.epidem.2009.04.003

Kurath, G. and Winton, J. (2011) Complex dynamics at the interface between wild and domestic viruses of finfish. *Current Opinion in Virology* 1, 73–80.

Lafferty, K.D., Harvell, C.D., Conrad, J.M., Friedman, C.S., Kent, M.L., *et al.* (2015) Infectious diseases affect marine fisheries and aquaculture economics. *Annual Review of Marine Science* 7, 471–496. https://doi.org/10.1146/annurev-marine-010814-015646

Larsen, A.K., Nymo, I.H., Sørensen, K.K., Seppola, M., Rødven, R., *et al.* (2018) Concomitant temperature stress and immune activation may increase mortality despite efficient clearance of an intracellular bacterial infection in Atlantic cod. *Frontiers in Microbiology* 9, 2963. https://doi.org/10.3389/fmicb.2018.02963

Laurin, E., Jaramillo, D., Vanderstichel, R., Ferguson, H., Kaukinen, K.H., *et al.* (2019) Histopathological and novel high-throughput molecular monitoring data from farmed salmon (*Salmo salar* and *Oncorhynchus* spp.) in British Columbia, Canada, from 2011–2013. *Aquaculture* 499, 220–234. https://doi.org/10.1016/j.aquaculture.2018.08.072

Lauringson, M., Nousiainen, I., Kahar, S., Burimski, O., Gross, R., *et al.* (2021) Climate change-driven disease in sympatric hosts: temporal dynamics of parasite burden and proliferative kidney disease in wild brown trout and Atlantic salmon. *Journal of Fish Diseases* 44(6), 689–699. https://doi.org/10.1111/jfd.13330

Legrand, T.P., Wynne, J.W., Weyrich, L.S. and Oxley, A.P. (2020) A microbial sea of possibilities: current knowledge and prospects for an improved understanding of the fish microbiome. *Reviews in Aquaculture* 12(2), 1101–1134.

Leong, J., Evensen, O. and Woo, P.T.K. (2023) Infectious diseases of coldwater fish in marine and brackish waters. In: Woo, P.T.K. and Subasinghe, R.P. (eds) *Climate Change on Diseases and Disorders of Finfish in Cage Culture*, 3rd edn. CAB International, Wallingford, UK, pp. 34–75.

Lerner, H. and Berg, C. (2017) A comparison of three holistic approaches to health: One Health, EcoHealth, and Planetary Health. *Frontiers in Veterinary Science* 4, 163. https://doi.org/10.3389/fvets.2017.00163

Levipan, H.A., Tapia-Cammas, D., Molina, V., Irgang, R., Toranzo, A.E., *et al.* (2019) Biofilm development and cell viability: an undervalued mechanism in the persistence of the fish pathogen *Tenacibaculum maritimum*. *Aquaculture* 511, 734267.

Long, A., Richard, J., Hawley, L., LaPatra, S.E. and Garver, K.A. (2017) Transmission potential of infectious hematopoietic necrosis virus in APEX-IHN® vaccinated Atlantic salmon. *Diseases of Aquatic Organisms* 122, 213–221.

Lyngstad, T.M., Qviller, L., Sindre, H., Brun, E. and Kristoffersen, A.B. (2018) Risk factors associated with outbreaks of infectious salmon anemia (ISA) with unknown source of infection in Norway. *Frontiers in Veterinary Science* 5, 308.

Marcogliese, D.J. (2008) The impact of climate change on the parasites and infectious diseases of aquatic animals. *Revue Scientifique et Technique (International Office of Epizootics)* 27(2), 467–484.

Marcos-López, M., Gale, P., Oidtmann, B.C. and Peeler, E.J. (2010) Assessing the impact of climate change on disease emergence in freshwater fish in the United Kingdom. *Transboundary and Emerging Diseases* 57(5), 293–304.

Markussen, T., Jonassen, C.M., Numanovic, S., Braaen, S., Hjortaas, M., *et al.* (2008) Evolutionary mechanisms involved in the virulence of infectious salmon anaemia virus (ISAV), a piscine orthomyxovirus. *Virology* 374, 515–527. https://doi.org/10.1016/j.virol.2008.01.019

Mateus, A., Costa, R., Cardoso, J.C., Andree, K.B., Estévez, A., *et al.* (2017) Thermal imprinting modifies adult stress and innate immune responsiveness in the teleost sea bream. *Journal of Endocrinology* 233(3), 381–394.

Mather, C. and Fanning, L. (2019) Social licence and aquaculture: towards a research agenda. *Marine Policy* 99, 275–282. https://doi.org/10.1016/j.marpol.2018.10.049

Maulu, S., Hasimuna, O.J., Haambiya, L.H., Monde, C., Musuka, C.G., *et al.* (2021) Climate change effects on aquaculture production: sustainability implications, mitigation, and adaptations. *Frontiers in Sustainable Food Systems* 5, 609097. https://doi.org/10.3389/fsufs.2021.609097

McCallum, H., Harvell, D. and Dobson, A. (2003) Rates of spread of marine pathogens. *Ecology Letters* 6(12), 1062–1067.

McIlgorm, A., Hanna, S., Knapp, G., Le Floc'H, P., Millerd, F. and Pan, M. (2010) How will climate change alter fishery governance? Insights from seven international case studies. *Marine Policy* 34(1), 170–177.

McVicar, A.H. (1997) Disease and parasite implications of the coexistence of wild and cultured Atlantic salmon populations. *ICES Journal of Marine Science* 54, 1093–1103.

Mengual, I.L., Sanchez-Jerez, P. and Ballester-Berman, J.D. (2021) Offshore aquaculture as climate change adaptation in coastal areas: sea surface temperature trends in the western Mediterranean Sea. *Aquaculture Environment Interactions* 13, 515–526.

Mennerat, A., Nilsen, F., Ebert, D. and Skorping, A. (2010) Intensive farming: evolutionary implications for parasites and pathogens. *Evolutionary Biology* 37, 59–67.

Meyers, T.R. and Hickey, N. (2022) A perspective: molecular detections of new agents in finfish – interpreting biological significance for fish health management. *Journal of Aquatic Animal Health* 34(2), 47–47. https://doi.org/10.1002/aah.10155

Meyers, T.R., Bentz, C., Ferguson, J., Stewart, D., Starkey, N. and Burton, T. (2019) *Diseases of Wild and Cultured Fishes in Alaska*. Alaska Department of Fish and Game, Fish Pathology Laboratories, Anchorage/Juneau, Alaska. Available at: https://www.adfg.alaska.gov/static/species/disease/pdfs/fish_disease_book.pdf (accessed 15 June 2022).

Miller, G.M., Watson, S.-A., Donelson, J.M., McCormick, M.I. and Munday, P.L. (2012) Parental environment mediates impacts of increased carbon dioxide on a coral reef fish. *Nature Climate Change* 2(12), 858–861.
Mimeault, C., Wade, J., Foreman, M.G.G., Chandler, P.C., Aubry, P., *et al.* (2017) *Assessment of the Risks to Fraser River Sockeye Salmon due to Infectious Hematopoietic Necrosis Virus (IHNV) Transfer from Atlantic Salmon Farms in the Discovery Islands of British Columbia*. DFO Canadian Science Advisory Secretariat Research Document, 2017/075. Fisheries and Oceans Canada, Ottawa.
Mimeault, C., Nekouei, O., Garver, K.A., Jones, S.R.M., Johnson, S., *et al.* (2020) *Assessment of the Risk to Fraser River Sockeye Salmon due to* Moritella viscosa *Transfer from Atlantic Salmon Farms in the Discovery Islands Area, British Columbia*. DFO Canadian Science Advisory Secretariat Research Document, 2020/058. Fisheries and Oceans Canada, Ottawa.
Mjaaland, S., Hungnes, O., Teig, A., Dannevig, B.H., Thorud, K. and Rimstad, E. (2002) Polymorphism in the infectious salmon anemia virus hemagglutinin gene: importance and possible implications for evolution and ecology of infectious salmon anemia disease. *Virology* 304(2), 379–391.
Mo, T.A. (1994) Status of *Gyrodactylus salaris* problems and research in Norway. In: Lewis, J.W. (ed.) *Parasitic Diseases of Fish*. Samara Publishing, Tresaith, UK, pp. 43–48.
Molloy, S.D., Pietrak, M.R., Bricknell, I. and Bouchard, D.A. (2013) Experimental transmission of infectious pancreatic necrosis virus from the blue mussel, *Mytilus edulis*, to cohabitating Atlantic salmon (*Salmo salar*) smolts. *Applied and Environmental Microbiology* 79(19), 5882–5890.
Molloy, S.D., Pietrak, M.R., Bouchard, D.A. and Bricknell, I. (2014) The interaction of infectious salmon anaemia virus (ISAV) with the blue mussel, *Mytilus edulis*. *Aquaculture Research* 45(3), 509–518.
Mugwanya, M., Dawood, M.A., Kimera, F. and Sewilam, H. (2022) Anthropogenic temperature fluctuations and their effect on aquaculture: a comprehensive review. *Aquaculture and Fisheries* 7(3), 223–243.
Murray, A.G. (2009) Using simple models to review the application and implications of different approaches used to simulate transmission of pathogens among aquatic animals. *Preventive Veterinary Medicine* 88(3), 167–177.
Murray, A.G. and Salama, N.K. (2016) A simple model of the role of area management in the control of sea lice. *Ecological Modelling* 337, 39–47. https://doi.org/10.1016/j.ecolmodel.2016.06.007
Naylor, R.L., Hardy, R.W., Buschmann, A.H., Bush, S.R., Cao, L., *et al.* (2021) A 20-year retrospective review of global aquaculture. *Nature* 591(7851), 551–563.
Nekouei, O., Vanderstichel, R., Thakur, K., Arriagada, G., Patanasatienkul, T., *et al.* (2018) Association between sea lice (*Lepeophtheirus salmonis*) infestation on Atlantic salmon farms and wild Pacific salmon in Muchalat Inlet, Canada. *Scientific Reports* 8(1), 4023.
Nikoskelainen, S., Bylund, G. and Lilius, E.M. (2004) Effect of environmental temperature on rainbow trout (*Oncorhynchus mykiss*) innate immunity. *Developmental and Comparative Immunology* 28, 581–592.
Nowak, B.F. (2007) Parasitic diseases in marine cage culture – an example of experimental evolution of parasites? *International Journal for Parasitology* 37(6), 581–588.
Nyberg, B. (1999) *An Introductory Guide to Adaptive Management for Project Leaders and Participants*. Forest Practices Branch, BC Forest Service, Victoria, British Columbia, Canada. Available at: https://planet.uwc.ac.za/nisl/ESS/ESS121/Introductory-Guide-AM.pdf (accessed 2 June 2022).
Nylund, A., Devold, M., Plarre, H., Isdal, E. and Aarseth, M. (2003) Emergence and maintenance of infectious salmon anaemia virus (ISAV) in Europe: a new hypothesis. *Diseases of Aquatic Organisms* 56(1), 11–24.
Oidtmann, B., Dixon, P., Way, K., Joiner, C. and Bayley, A.E. (2018) Risk of waterborne virus spread – review of survival of relevant fish and crustacean viruses in the aquatic environment and implications for control measures. *Reviews in Aquaculture* 10(3), 641–669.
OIE (World Organisation for Animal Health) (2021) *Aquatic Animal Health Code*. OIE, Paris. Available at: https://www.oie.int/en/what-we-do/standards/codes-and-manuals/aquatic-code-online-access/?id=169&L=1&htmfile=chapitre_application_compartment.htm (accessed 8 April 2022).
Okocha, R.C., Olatoye, I.O. and Adedeji, O.B. (2018) Food safety impacts of antimicrobial use and their residues in aquaculture. *Public Health Reviews* 39(1), 1–22.
Oldham, T., Rodger, H. and Nowak, B.F. (2016) Incidence and distribution of amoebic gill disease (AGD) – an epidemiological review. *Aquaculture* 457, 35–42.
Olmos Soto, J., Paniagua-Michel, J.D.J., Lopez, L. and Ochoa, L. (2015) Functional feeds in aquaculture. In: Kim, S.-K. (ed.) *Springer Handbook of Marine Biotechnology*. Springer, Berlin/Heidelberg, Germany, pp. 1303–1319.
Overstreet, R.M. and Curran, S.S. (2004) Defeating diplostomoid dangers in USA catfish aquaculture. *Folia Parasitologica* 51, 153–165.

Oyinlola, M.A., Reygondeau, G., Wabnitz, C.C. and Cheung, W.W. (2020) Projecting global mariculture diversity under climate change. *Global Change Biology* 26(4), 2134–2148.

Peeler, E.J. and Taylor, N.G. (2011) The application of epidemiology in aquatic animal health – opportunities and challenges. *Veterinary Research* 42(1), 94.

Peeler, E.J., Oidtmann, B.C., Midtlyng, P.J., Miossec, L. and Gozlan, R.E. (2011) Non-native aquatic animals introductions have driven disease emergence in Europe. *Biological Invasions* 13, 1291–1303. https://doi.org/10.1007/s10530-010-9890-9

Pepi, M. and Focardi, S. (2021) Antibiotic-resistant bacteria in aquaculture and climate change: a challenge for health in the Mediterranean area. *International Journal of Environmental Research and Public Health* 18(11), 5723.

Pérez-Casanova, J.C., Rise, M.L., Dixon, B., Afonso, L.O.B., Hall, J.R., *et al.* (2008) The immune and stress responses of Atlantic cod to long-term increases in water temperature. *Fish & Shellfish Immunology* 24(5), 600–609.

Perreira, J., Purcell, S.L., Whyte, S.K. and Fast, M.D. (2019) The potential impact of climate change on bacterial kidney disease in Atlantic salmon (*Salmo salar*). Presented at *Ocean Frontiers Institute Researchers Workshop 2019, St John's, Newfoundland, Canada, 16–18 November 2019*.

Pert, C.C., Fryer, R.J., Cook, P., Kilburn, R., McBeath, S., *et al.* (2014) Using sentinel cages to estimate infestation pressure on salmonids from sea lice in Loch Shieldaig, Scotland. *Aquaculture Environment Interactions* 5(1), 49–59.

Phelps, N.B., Armién, A.G., Mor, S.K., Goyal, S.M., Warg, J.V., *et al.* (2012) Spring viremia of carp virus in Minnehaha Creek, Minnesota. *Journal of Aquatic Animal Health* 24(4), 232–237.

Polinski, M.P., Vendramin, N., Cuenca, A. and Garver, K.A. (2020) Piscine orthoreovirus: biology and distribution in farmed and wild fish. *Journal of Fish Diseases* 43(11), 1331–1352. https://doi.org/10.1111/jfd.13228

Pulkkinen, K., Suomalainen, L.R., Read, A.F., Ebert, D., Rintamäki, P. and Valtonen, E.T. (2010) Intensive fish farming and the evolution of pathogen virulence: the case of columnaris disease in Finland. *Proceedings of the Royal Society B: Biological Sciences* 277(1681), 593–600. https://doi.org/10.1098/rspb.2009.1659

Purbayu, M.A., Shavalier, M.A., Faisal, M. and Loch, T.P. (2021) Experimental evidence of epizootic epitheliotropic disease virus (salmonid herpesvirus-3, *Alloherpesviridae*) transmission via contaminated fomites and subsequent prevention using a disinfectant. *Pathogens* 10(6), 724.

Purcell, M.K., Hard, J.J., Neely, K.G., Park, L.K., Winton, J.R. and Elliott, D.G. (2014) Genetic variation in bacterial kidney disease (BKD) susceptibility in Lake Michigan chinook salmon and its progenitor population from the Puget Sound. *Journal of Aquatic Animal Health* 26(1), 9–18. https://doi.org/10.1080/08997659.2013.860061

Purcell, M.K., McKibben, C.L., Pearman-Gillman, S., Elliott, D.G. and Winton, J.R. (2016) Effects of temperature on *Renibacterium salmoninarum* infection and transmission potential in chinook salmon, *Oncorhynchus tshawytscha* (Walbaum). *Journal of Fish Diseases* 39(7), 787–798. https://doi.org/10.1111/jfd.12409

Ramsay, J.M., Watral, V., Schreck, C.B. and Kent, M.L. (2009) *Pseudoloma neurophilia* infections in zebrafish *Danio rerio*: effects of stress on survival, growth, and reproduction. *Diseases of Aquatic Organisms* 88(1), 69–84.

Rechisky, E.L., Porter, A.D., Johnston, S.D., Stevenson, C.F., Hinch, S.G., *et al.* (2021) Exposure time of wild, juvenile sockeye salmon to open-net-pen Atlantic salmon farms in British Columbia, Canada. *North American Journal of Fisheries Management* 41(3), 650–660. https://doi.org/10.1002/nafm.10574

Reid, G.K., Gurney-Smith, H.J., Marcogliese, D.J., Knowler, D., Benfey, T., *et al.* (2019) Climate change and aquaculture: considering biological response and resources. *Aquaculture Environment Interactions* 11, 569–602.

Reverter, M., Sarter, S., Caruso, D., Avarre, J.C., Combe, M., *et al.* (2020) Aquaculture at the crossroads of global warming and antimicrobial resistance. *Nature Communications* 11(1), 1870.

Rhyan, J.C. and Spraker, T.R. (2010) Emergence of diseases from wildlife reservoirs. *Veterinary Pathology* 47(1), 34–39. https://doi.org/10.1177/0300985809354466

Rico, A., Phu, T.M., Satapornvanit, K., Min, J., Shahabuddin, A.M., *et al.* (2013) Use of veterinary medicines, feed additives and probiotics in four major internationally traded aquaculture species farmed in Asia. *Aquaculture* 412, 231–243.

Rijkers, G.T., Frederix-Wolters, E.M. and van Muiswinkel, W.B. (1980) The immune system of cyprinid fish. Kinetics and temperature dependence of antibody-producing cells in carp (*Cyprinus carpio*). *Immunology* 41(1), 91–97.

Ritchie, R.J., McDonald, J.T., Glebe, B., Young-Lai, W., Johnsen, E. and Gagne, N. (2009) Comparative virulence of Infectious salmon anaemia virus isolates in Atlantic salmon, *Salmo salar* L. *Journal of Fish Diseases* 32(2), 157–171. https://doi.org/10.1111/j.1365-2761.2008.00973.x

Roberts-Thomson, A., Barnes, A., Fielder, D.S., Lester, R.J.G. and Adlard, R.D. (2006) Aerosol dispersal of the fish pathogen, *Amyloodinium ocellatum. Aquaculture* 257(1–4), 118–123.

Romalde, J.L., Magarinos, B., Nunez, S., Barja, J.L. and Toranzo, A.E. (1996) Host range susceptibility of *Enterococcus* sp. strains isolated from diseased turbot: possible routes of infection. *Applied and Environmental Microbiology* 62(2), 607–611.

Romero, J.F., Gardner, I.A., Saksida, S., McKenzie, P., Garver, K., *et al.* (2022) Simulated waterborne transmission of infectious hematopoietic necrosis virus among farmed salmon populations in British Columbia, Canada following a hypothetical virus incursion. *Aquaculture* 548, 737658.

Ros, A., Schmidt-Posthaus, H. and Brinker, A. (2022) Mitigating human impacts including climate change on proliferative kidney disease in salmonids of running waters. *Journal of Fish Diseases* 45(4), 497–521.

Rosa, R., Marques, A. and Nunes, M.L. (2012) Impact of climate change in Mediterranean aquaculture. *Reviews in Aquaculture* 4(3), 163–177.

Rosseland, B.O. and Skogheim, O.K. (1984) A comparative study on salmonid fish species in acid aluminum-rich water. II Physiological stress and mortality of one- and two-year-old fish. *Institution of Freshwater Research Drottningholm Report* 61, 186–194.

Rozas, M., Bohle, H., Grothusen, H. and Bustos, P. (2012) Epidemiology of amoebic gill disease (AGD) in Chilean salmon industry between 2007 and 2010. *Bulletin of the European Association of Fish Pathologists* 32(5), 181–188.

Saksida, S.M. (2006) Infectious haematopoietic necrosis epidemic (2001 to 2003) in farmed Atlantic salmon *Salmo salar* in British Columbia. *Diseases of Aquatic Organisms* 72(3), 213–223. https://doi.org/10.3354/dao072213

Saksida, S.M., Gardner, I. and Kent, M.L. (2014) Transmission of infectious agents between wild and farmed fish. In: Woo, P.T.K. and Bruno, D.W. (eds) *Diseases and Disorders of Finfish in Cage Culture*, 2nd edn. CAB International, Wallingford, UK, pp. 313–330.

Salama, N.K. and Rabe, B. (2013) Developing models for investigating the environmental transmission of disease-causing agents within open-cage salmon aquaculture. *Aquaculture Environment Interactions* 4(2), 91–115. https://doi.org/10.3354/aei00077

Samsing, F., Oppedal, F., Dalvin, S., Johnsen, I., Vågseth, T. and Dempster, T. (2018) Corrigendum: Salmon lice (*Lepeophtheirus salmonis*) development times, body size, and reproductive outputs follow universal models of temperature dependence. *Canadian Journal of Fisheries and Aquatic Sciences* 75(2), 329.

Semple, S.L., Vo, N.T.K., Li, A.R., Pham, P.H., Bols, N.C. and Dixon, B. (2017) Development and use of an Arctic charr cell line to study antiviral responses at extremely low temperatures. *Journal of Fish Diseases* 40(10), 1423–1439.

Shefat, S.H.T. (2018) Vaccines for use in finfish aquaculture. *Acta Scientific Pharmaceutical Sciences* 2(11), 15–19.

Sieracki, J.L., Bossenbroek, J.M. and Faisal, M. (2014) Modeling the secondary spread of viral hemorrhagic septicemia virus (VHSV) by commercial shipping in the Laurentian Great Lakes. *Biological Invasions* 16(5), 1043–1053.

Skar, C.K. and Mortensen, S. (2007) Fate of infectious salmon anaemia virus (ISAV) in experimentally challenged blue mussels *Mytilus edulis*. *Diseases of Aquatic Organisms* 74(1), 1–6.

Smith, K.F., Sax, D.F. and Lafferty, K.D. (2006) Evidence for the role of infectious disease in species extinction and endangerment. *Conservation Biology* 20(5), 1349–1357. https://doi.org/10.1111/j.1523-1739.2006.00524.x

Snow, M., Raynard, R.S., Murray, A.G., Bruno, D.W., King, J.A., *et al.* (2003) An evaluation of current diagnostic tests for the detection of infectious salmon anaemia virus (ISAV) following experimental water-borne infection of Atlantic salmon, *Salmo salar* L. *Journal of Fish Diseases* 26(3), 135–145.

Staurnes, M., Blix, P. and Reite, O.B. (1993) Effects of acid water and aluminum on parr–smolt transformation and seawater tolerance in Atlantic salmon, *Salmo salar*. *Canadian Journal of Fisheries and Aquatic Sciences* 50(9), 1816–1827.

St-Hilaire, S., Stephen, C., Kent, M. and Ribble, C. (2001) Sentinels in the bay: using farm fish to monitor for pathogens in the environment. *ICES Journal of Marine Science* 58(2), 369–373.

St-Hilaire, S., Ribble, C.S., Stephen, C., Anderson, E., Kurath, G. and Kent, M.L. (2002) Epidemiological investigation of infectious hematopoietic necrosis virus in salt water net-cage reared Atlantic salmon in British Columbia, Canada. *Aquaculture* 212(1–4), 49–67. https://doi.org/10.1016/S0044-8486(02)00201-6

St-Hilaire, S.N., Gomes, G., Pinheiro Marques, A.R., Chang, S.F. and Brun, E. (2023) Sporadic emerging infectious and non-infectious diseases and disorders. In: Woo, P.T.K. and Subasinghe, R.P. (eds) *Climate Change on Diseases and Disorders of Finfish in Cage Culture*, 3rd edn. CAB International, Wallingford, UK, pp. 315–350.

Subasinghe, R.P. and Shinn, A.P. (2023) Biosecurity: current and future strategies. In: Woo, P.T.K. and Subasinghe, R.P. (eds) *Climate Change on Diseases and Disorders of Finfish in Cage Culture*, 3rd edn. CAB International, Wallingford, UK, pp. 430–461.

Subasinghe, R.P., Delamare-Deboutteville, J., Mohan, C.V. and Phillips, M.J. (2019) Vulnerabilities in aquatic animal production. *Revue Scientifique et Technique (International Office of Epizootics)* 38, 423–436.

Sudirman, Halide, H., Jompa, J., Zulfikar, Iswahyudin and McKinnon, A.D. (2009) Wild fish associated with tropical sea cage aquaculture in South Sulawesi, Indonesia. *Aquaculture* 286(3), 233–239.

Sundberg, L.R., Ketola, T., Laanto, E., Kinnula, H., Bamford, J.K., *et al.* (2016) Intensive aquaculture selects for increased virulence and interference competition in bacteria. *Proceedings of the Royal Society B: Biological Sciences* 283(1826), 20153069. https://doi.org/10.1098/rspb.2015.3069

Tett, P. (2017) *Monitoring for Adaptive Management of Aquaculture, v2*. LMC Working Paper. Scottish Association for Marine Science, Oban, UK.

Teixeira Alves, M. and Taylor, N.G. (2020) Models suggest pathogen risks to wild fish can be mitigated by acquired immunity in freshwater aquaculture systems. *Scientific Reports* 10(1), 7513.

Tičina, V., Katavić, I. and Grubišić, L. (2020) Marine aquaculture impacts on marine biota in oligotrophic environments of the Mediterranean Sea – a review. *Frontiers in Marine Science* 7, 217.

Tobback, E., Decostere, A., Hermans, K., Haesebrouck, F. and Chiers, K. (2007) *Yersinia ruckeri* infections in salmonid fish. *Journal of Fish Diseases* 30(5), 257–268.

Tolo, I.E., Bajer, P.G., Wolf, T.M., Mor, S.K. and Phelps, N.B.D. (2022) Investigation of cyprinid herpesvirus 3 (CyHV-3) disease periods and factors influencing CyHV-3 transmission in a low stocking density infection trial. *Animals* 12(1), 2.

Traxler, G.S., Roome, J.R. and Kent, M.E. (1993) Transmission of infectious hematopoietic necrosis virus in seawater. *Diseases of Aquatic Organisms* 16(2), 111–114.

Troyer, R.M. and Kurath, G. (2003) Molecular epidemiology of infectious hematopoietic necrosis virus reveals complex virus traffic and evolution within southern Idaho aquaculture. *Diseases of Aquatic Organisms* 55(3), 175–185.

True, K., Bolick, A. and Foott, J. (2011) *Myxosporean Parasite* (Ceratomyxa shasta *and* Parvicapsula minibicornis) *Annual Prevalence of Infection in Klamath River Basin Juvenile Chinook Salmon, April–August 2010*. US Fish and Wildlife Service, California-Nevada Fish Health Center, Anderson, California.

Ugelvik, M.S., Skorping, A., Moberg, O. and Mennerat, A. (2017) Evolution of virulence under intensive farming: salmon lice increase skin lesions and reduce host growth in salmon farms. *Journal of Evolutionary Biology* 30(6), 1136–1142. https://doi.org/10.1111/jeb.13082

Vågsholm, I., Djupvik, H.O., Willumsen, F.V., Tveit, A.M. and Tangen, K. (1994) Infectious salmon anaemia (ISA) epidemiology in Norway. *Preventive Veterinary Medicine* 19(3–4), 277–290.

Vennerström, P., Maunula, L., Välimäki, E. and Virtala, A.M. (2020) Presence of viral haemorrhagic septicaemia virus (VHSV) in the environment of virus-contaminated fish farms and processing plants. *Diseases of Aquatic Organisms* 138, 145–154.

Vike, S., Nylund, S. and Nylund, A. (2009) ISA virus in Chile: evidence of vertical transmission. *Archives of Virology* 154(1), 1–8.

Vike, S., Oelckers, K., Duesund, H., Erga, S.R., Gonzalez, J., *et al.* (2014) Infectious salmon anemia (ISA) virus: infectivity in seawater under different physical conditions. *Journal of Aquatic Animal Health* 26(1), 33–42.

Vollset, K.W., Barlaup, B.T., Mahlum, S., Bjorn, P.A. and Skilbrei, O.T. (2016) Estimating the temporal overlap between post-smolt migration of Atlantic salmon and salmon lice infestation pressure from fish farms. *Aquaculture Environment Interactions* 8, 511–525.

Wade, J. and Weber, L. (2020) *Characterization of* Moritella viscosa *and Winter Ulcer to Inform Pathogen Transfer Risk Assessments in British Columbia*. DFO Canadian Science Advisory Secretariat Research Document, 2020/060. Fisheries and Oceans Canada, Ottawa.

Walker, P.J. and Winton, J.R. (2010) Emerging viral diseases of fish and shrimp. *Veterinary Research* 41(6), 51.

Wallace, I.S., Gregory, A., Murray, A.G., Munro, E.S. and Raynard, R.S. (2008) Distribution of infectious pancreatic necrosis virus (IPNV) in wild marine fish from Scottish waters with respect to clinically infected aquaculture sites producing Atlantic salmon, *Salmo salar* L. *Journal of Fish Diseases* 31(3), 177–186.

Wang, H.Y., Shen, S.F., Chen, Y.S., Kiang, Y.K. and Heino, M. (2020) Life histories determine divergent population trends for fishes under climate warming. *Nature Communications* 11(1), 4088.

Warheit, K. (2018) WDFW denies permit for company to place 800,000 Atlantic salmon into Puget Sound net pens. *Aquahoy*, 18 May 2018. Available at: https://aquahoy.com/wdfw-denies-permit-for-company-to-place-800-000-atlantic-salmon-into-puget-sound-net-pens/?amp=1 (accessed 26 May 2022).

Williams, B.K., Szaro, R.C. and Shapiro, C.D. (2009) *Adaptive Management: the US Department of the Interior Technical Guide*. US Department of the Interior, Washington, DC.

Williams, I.V. and Amend, D.F. (1976) A natural epizootic of infectious hematopoietic necrosis in fry of sockeye salmon (*Oncorhynchus nerka*) at Chilko Lake, British Columbia. *Journal of the Fisheries Research Board of Canada* 33(7), 1564–1567.

Wise, A.L., LaFrentz, B.R., Kelly, A.M., Khoo, L.H., Xu, T., *et al.* (2021) A review of bacterial co-infections in farmed catfish: components, diagnostics, and treatment directions. *Animals* 11(11), 3240.

Woo, P.T.K. (2003) *Cryptobia* (*Trypanoplasma*) *salmositica* and salmonid cryptobiosis. *Journal of Fish Diseases* 26(11–12), 627–646.

Wooster, G.A. and Bowser, P.R. (1996) The aerobiological pathway of a fish pathogen: survival and dissemination of *Aeromonas salmonicida* in aerosols and its implications in fish health management. *Journal of the World Aquaculture Society* 27(1), 7–14.

Xie, W.W., Li, Y.P., Bai, W.Y., Hou, J.L., Ma, T.F., *et al.* (2021) The source and transport of bioaerosols in the air: a review. *Frontiers of Environmental Science & Engineering* 15(3), 44.

10 Diseases and Disorders in Fish due to Harmful Algal Blooms

René S. Shahmohamadloo[1]*, Thijs Frenken[2], Seth M. Rudman[1], Pieter van West[3], Bastiaan W. Ibelings[4] and Vera L. Trainer[5]

[1]*Washington State University, Vancouver, Washington, USA;* [2]*HAS University of Applied Sciences, 's-Hertogenbosch, The Netherlands;* [3]*University of Aberdeen, Aberdeen, UK;* [4]*University of Geneva, Geneva, Switzerland;* [5]*National Oceanic and Atmospheric Administration, Seattle, Washington, USA*

10.1 Introduction

Sustaining clean and healthy waters for aquaculture and fisheries to meet the growing demand for aquatic foods is a great challenge of the 21st century (Brown *et al.*, 2020). Presently, approximately 820 million people (one in nine people in the world) are malnourished (FAO, IFAD, UNICEF, WFP and WHO, 2018) and the human population is projected to rise from 7.6 to 11.2 billion by 2100 (UN, 2017; Brown *et al.*, 2020). Aquaculture and fisheries have emerged as a sustainable protein source to improve future food security (FAO, 2018; Gephart *et al.*, 2020) providing more than half of the world's seafood (FAO, 2018; Lenzen *et al.*, 2021). This additional production will come from the expansion of aquaculture, both into marine and freshwater environments. However, this expansion is under threat by harmful algal blooms (HABs), which are increasing in frequency, severity and toxicity worldwide (Huisman *et al.*, 2018; IPCC, 2022; Parmesan *et al.*, 2022) and pose health risks to humans and wildlife (Carmichael, 2001; Shahmohamadloo *et al.*, 2022b, 2023). The total economic loss due to HABs in aquaculture is estimated at US$8 billion/year globally (Brown *et al.*, 2020), which represents 3.2% of the US$250 billion/year revenue (FAO, 2020a).

HABs are the excessive growth of phytoplankton (e.g. microalgae, cyanobacteria) or biomagnification of their toxins in marine, estuarine and freshwater systems (Huisman *et al.*, 2018). Several factors induce the formation of HABs. First, eutrophication from anthropogenic activity (e.g. agriculture, aquaculture, wastewater disposal) has dramatically increased nitrogen and phosphorus inputs into aquatic systems dating from the 1960s and remains a global management challenge (Schindler, 1974). Second, rising carbon dioxide (CO_2) concentrations in the atmosphere result in the acidification of marine and freshwater systems (Gobler, 2020). Third, rising temperature (IPCC, 2022) is a keystone parameter of climate change that influences phytoplankton growth rates, photosynthesis, stratification through the water column and nutrient uptake rates, as well as the seasonal window of growth and geographical distribution of HABs (Huisman *et al.*, 2018; Trainer *et al.*, 2020; Wells *et al.*, 2020). Fourth, key functional traits enable the competitive

*rene.shahmohamadloo@wsu.edu

 Climate Change on Diseases and Disorders of Finfish in Cage Culture, 3rd Edition (eds P.T.K. Woo and R.P. Subasinghe)
DOI: 10.1079/9781800621640.0010

advantage of some species of phytoplankton, including nitrogen-fixation abilities (Stal, 2009), CO_2-concentrating mechanisms (Verspagen *et al.*, 2014), buoyancy (Walsby, 1994), rapid generation times and short life cycles (Govaert *et al.*, 2021), and the ability to produce toxic secondary metabolites (Huisman *et al.*, 2018).

HABs remain a long-standing concern to aquaculture and fisheries, mainly because their toxins and other indirect effects can kill aquatic organisms (Díaz *et al.*, 2019; Brown *et al.*, 2020; Lenzen *et al.*, 2021; IPCC, 2022; Parmesan *et al.*, 2022). These toxins may also have sublethal effects on fish (e.g. non-lethal impacts on phenotype through the release of toxins into water after the lysis of a HAB event), of which associated acute and chronic impacts on fish are still largely unknown. Aquaculture is particularly threatened by HABs compared with wild capture fisheries (Trainer *et al.*, 2020) because movement of cultured fish is restricted and they cannot evade HABs (Lenzen *et al.*, 2021). Fish-farm operational procedure options for managing HABs are limited and additionally these options are expensive or sacrifice a major part of the yield (Shumway, 1990). For example, a HAB event in the Patagonian fjords of Chile in 2016 led to 40,000 tonnes of fish mortalities estimated at US$800 million in economic losses (Díaz *et al.*, 2019) and major social unrest (Trainer *et al.*, 2020). The numerous and widespread HAB-driven fish kills, like the event in Chile, demonstrate the need for new insights, management actions and policies that are informed by a mechanistic understanding of the adverse health effects in fish from exposure to HABs.

This chapter focuses on the sublethal (or non-lethal) impacts in fish from exposure to commonly occurring toxins produced by HABs. Although mass-mortality events such as the one described in Chile gained prominent attention worldwide, repeated exposure of animals to sublethal levels may become more common (Huisman *et al.*, 2018; Shahmohamadloo *et al.*, 2020a). Food recalls connected to tissue accumulation of HAB toxins in fish and poisoning cases in humans and animals are also rising (Svirčev *et al.*, 2019; IPCC, 2022; Parmesan *et al.*, 2022). Climate change is further projected to increase the mean number of days of a HAB event from 7 days presently to 16–23 days in 2050 and 18–39 days in 2090 (Chapra *et al.*, 2017), posing greater health and economic risks to aquaculture and fisheries. For these reasons, the Food and Agriculture Organization of the United Nations (FAO) stressed fish consumption as a primary route of exposure to HAB toxins in humans and called for a deeper understanding of the gene–environment interconnections in HAB species that continue to damage aquatic systems and the global blue economy (FAO, 2020b).

10.2 Diseases and Disorders

This section describes diseases and disorders in finfish resulting from sublethal (or non-lethal) exposure to marine and freshwater toxins produced by HABs, explaining the mechanisms of toxicity, non-clinical effects and clinical signs (with pathological lesions) on fish. The goal of this section is to provide an account of the detection, fate, occurrence and toxicity of HABs in fish, with particular attention on the acute (or short-term) and chronic (or long-term) effects on fish physiology and health.

10.2.1 Microcystins

Microcystins (MCs) are a class of cyclic peptides that are produced by several freshwater cyanobacteria including *Microcystis*, *Dolichospermum*, *Oscillatoria* and *Planktothrix* (Huisman *et al.*, 2018) (Fig. 10.1). More than 250 structural variants of MCs have been identified (Bouaïcha *et al.*, 2019) and have been detected in every continent worldwide (Harke *et al.*, 2016). Most MCs are hydrophilic, resistant to boiling and typically occur at high concentrations in fresh waters (WHO, 2020a), although their fate, occurrence and toxicity depend on a variety of factors including their molecular structure as well as pH, temperature, light intensity and nutrient concentrations (Wicks and Thiel, 1990; Pineda-Mendoza *et al.*, 2016; Puddick *et al.*, 2016). MCs are also seasonally present in temperate regions although year-round exposure is rare; however, greater risks of year-round exposure to MCs are possible in areas that have high seasonal temperatures favouring HAB persistence (WHO, 2020a).

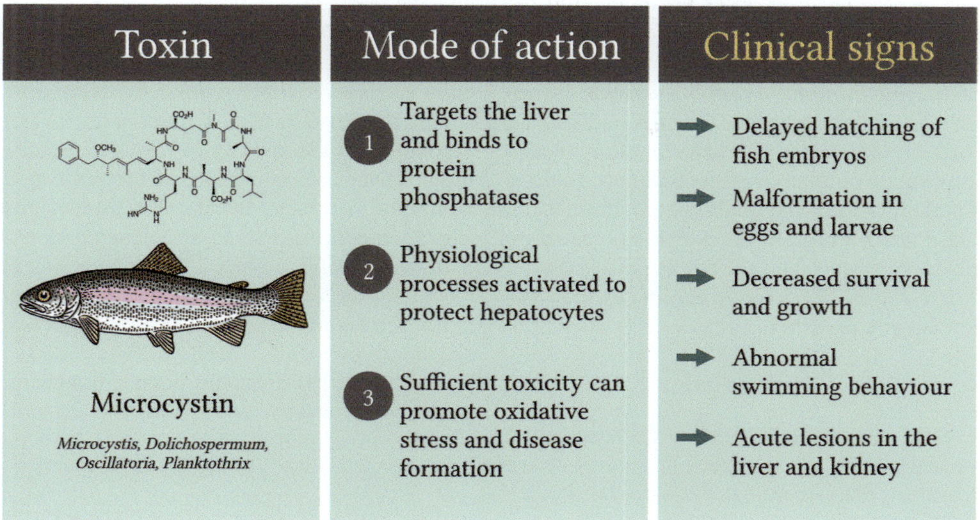

Fig. 10.1. Microcystin mode of action and clinical signs of toxicity in fish species.

Fish encounter MCs through direct contact with contaminated water, feeding, or by accumulation in aquatic food webs. The main route by which MCs are taken up by fish is thought to be through the gastrointestinal (GI) tract via dietary intake (Ibelings and Havens, 2008); however, it is also hypothesized that MCs can pass through fish gills when cells are lysing and releasing their toxins (Tencalla *et al.*, 1994; Xie *et al.*, 2005; Dyble *et al.*, 2011). Recent evidence suggests MCs inside healthy cyanobacterial cells can also affect fish populations, even in the early stages of a bloom's development when biomass is low and the bloom is not yet visible to humans (Shahmohamadloo *et al.*, 2021). Mortality in fish from MC exposure has been experimentally demonstrated by intraperitoneal injection and oral gavage (Tencalla *et al.*, 1994; Kotak *et al.*, 1996; Malbrouck and Kestemont, 2006). However, these routes of MC administration are not realistic and it is more common for fish to experience sublethal effects from balneation (Shahmohamadloo *et al.*, 2022a) and dietary intake (Ibelings and Havens, 2008).

Impact on fish production

The global occurrence of HABs has raised widespread concerns that MCs can have serious economic consequences to aquaculture and fisheries (Zimba *et al.*, 2001). For decades it has been postulated that MCs are one among several stress factors involved in fish kills during HAB events (Ibelings *et al.*, 2005). Harmful, sublethal impacts from MCs are evident in various aquatic organisms (Gene *et al.*, 2019; Shahmohamadloo *et al.*, 2020a,b, 2021, 2022a, 2023) and MC-producing HABs have occurred at record levels in some of the world's largest sources of fresh water (e.g. the Great Lakes, USA, see Michalak *et al.*, 2013; Hellweger *et al.*, 2022). Waterborne MCs can degrade within days to weeks (Edwards *et al.*, 2008), and climate change is projected to increase the mean number of days of a HAB event (Chapra *et al.*, 2017). Consequently, aquaculture and fisheries may be at greater health and economic risks since fish can be exposed to MCs for longer periods of time (Shahmohamadloo *et al.*, 2022a,b, 2023).

Mechanism of toxicity

MC toxicity in fish frequently starts in the liver through irreversible binding with high affinity to protein phosphatases (PP1, PP2A), which are connected to regulatory pathways that are responsible for cell replication, cytoskeletal structure, stress responses and DNA repair (Buratti *et al.*, 2017; WHO, 2020a). Several physiological processes are activated at the cellular level to protect hepatocytes from disease and death including detoxification as well as preventing cellular apoptosis, cellular proliferation, and possibly cancer (Pearson *et al.*, 2010). Depending on the length of exposure and severity of the HAB, MC toxicity can promote tumour formation, haemorrhage and organ failure (Chorus and Welker, 2021). It is important to note that MCs can also accumulate in other areas including the kidney (Kotak *et al.*, 1996; Shahmohamadloo *et al.*, 2021), which serves an important role of removing toxic compounds, and the edible muscle tissues (Xie *et al.*, 2005; Dyble *et al.*, 2011; Shahmohamadloo *et al.*, 2021, 2022a).

Clinical signs

In early life stages, MC exposure at sublethal concentrations can cause dose-dependent delays in hatching of fish embryos (Oberemm *et al.*, 1999), decreases in survival and growth rate (Oberemm *et al.*, 1999; Palikova *et al.*, 2003), abnormal swimming behaviour (Råbergh *et al.*, 1991), disturbances in reproductive success by causing malformations in eggs and larvae (Ernst *et al.*, 2001), mutagenic effects (Vasconcelos *et al.*, 2013), and histopathological effects including an enlarged and opaque yolk sac, small head, and curved body and tail (Oberemm *et al.*, 1999; Best *et al.*, 2002; Malbrouck and Kestemont, 2006). Liver disease signs of MC toxicity (or hepatotoxicity) typically involve acute lesions including necrosis, apoptosis and haemorrhage in juvenile and adult life stages (Tencalla *et al.*, 1994; Shahmohamadloo *et al.*, 2021; Shartau *et al.*, 2022). Lesions can be either predominantly peribiliary or perivenular (or around the vein). The character of the lesions may also vary depending on the severity of toxicity exposure, ranging from individualized hepatocytes to more basophilic shrunken hepatocytes. Kidney toxicity (or nephrotoxicity) also follows a similar pattern of histopathological alterations including dilatations of Bowman's capsule, vacuolization, necrosis and pyknosis of tubular cells, and oedema (Kotak *et al.*, 1996; Svirčev *et al.*, 2015; Wang *et al.*, 2019). Myopathy has also been found in muscle tissues that may be caused by an indirect downstream effect from MC toxicity in the liver and kidney (Shahmohamadloo *et al.*, 2021).

10.2.2 Nodularins

Nodularins (NODs) are hepatotoxic, cyclic, non-ribosomal pentapeptides that are structurally very similar to MCs and so far have been found mainly to be produced by *Nodularia* (Wiegand and Pflugmacher, 2005; Chorus and Welker, 2021) (Fig. 10.2). These toxins are relatively stable and occur in freshwater, brackish and marine systems (Codd *et al.*, 1999; Pearson *et al.*, 2010). NOD degradation is stimulated by

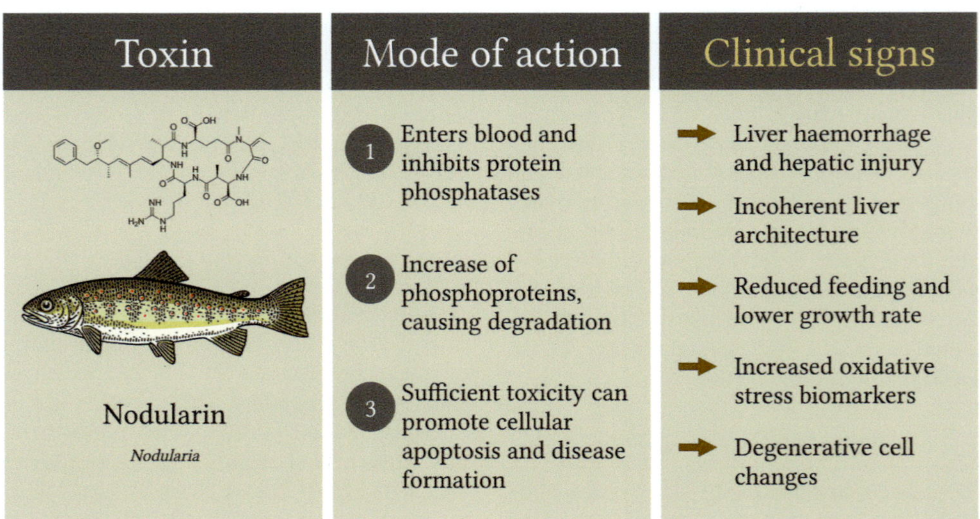

Fig. 10.2. Nodularin mode of action and clinical signs of toxicity in fish species.

ultraviolet (UV) radiation, microbial activity and by binding to copper sulfate (Heresztyn and Nicholson, 1997; Mazur-Marzec *et al.*, 2006; Edwards *et al.*, 2008; Toruńska *et al.*, 2008). NODs are primarily bound to proteins in viable cyanobacterial cells, and less than 20% is generally released into the surrounding water (Chorus and Welker, 2021). Both NODs from within cyanobacteria and in the surrounding water can bioaccumulate in fish (i.e. the gradual accumulation of the toxins in fish from consuming lower trophic-level organisms), thus posing a risk to humans from seafood consumption (Van Buynder *et al.*, 2001; Kankaanpää *et al.*, 2002; Chen *et al.*, 2013), although NODs are not classifiable as carcinogens due to a lack of exposure data in humans (Chen *et al.*, 2013).

Impact on fish production

Massive fish kills of greasy rockcod (*Epinephelus tauvina*), longfin eel (*Anguilla reinhardtii*), yellowfin bream (*Acanthopagrus australis*) and sea mullet (*Mugil cephalus*) have been linked to *Nodularia* blooms in Queensland, Australia (Stewart *et al.*, 2012). Sea mullet in particular contained high concentrations of NODs in livers, with high hepatic levels maintained in fish at 10 months after the HAB event. NODs were also detected in muscles, although concentrations were below human consumption guideline values for adults and children. However, no abnormal behaviours were observed in the sea mullets, raising concerns that NOD exposure can go undetected and toxin exposure via fish consumption can occur if fish are consumed whole.

Mechanism of toxicity

NODs enter the blood via bile acid carriers and are transported preferentially to hepatocytes (Van Apeldoorn *et al.*, 2007). Here, NODs inhibit protein phosphatase (PP1 and PP2A) activity that results in an increase of phosphoproteins, which causes cytoskeleton degradation, loss of cell junctions, disturbances of cell metabolism and cell-cycle control, and oxidative stress (Gulledge *et al.*, 2002; Pearson *et al.*, 2010). NODs are also considered a tumour promoter (Van Apeldoorn *et al.*, 2007). *In vitro* studies further revealed a dose-dependent apoptotic reaction of lymphocytes to NOD exposure, which can cause condensed cytoplasm, DNA fragmentation, and increased reactive oxygen species followed by programmed cell death (Sotton *et al.*, 2015). This suggests NOD exposure can rapidly and strongly affect mitochondrial-mediated pathways in fish cell apoptosis (Sotton *et al.*, 2015).

Clinical signs

In animals NODs can induce lethal liver haemorrhage, hepatic insufficiency, and oxidative stress in the tissues where they accumulate (Eriksson *et al.*, 1988; Van Apeldoorn *et al.*, 2007; Pearson *et al.*, 2010). Information on the effects of NODs on fish is still relatively scarce (Sotton *et al.*, 2015). NOD exposure has been shown to cause slightly incoherent liver architecture, degenerative cell changes and increased liver glutathione *S*-transferase (GST) activity (Vuorinen *et al.*, 2009). Comparing repeated exposure of fish to a single exposure has shown that NODs remain absent in the bile, indicating that they can be rapidly detoxified, and that by-products are quickly disintegrated and excreted (Vuorinen *et al.*, 2009). In early life stages, fish larvae exposed to NODs show reduced feeding and lower growth rates, likely as a result of the metabolic cost of detoxification (Karjalainen *et al.*, 2007). In adult life stages, dietary exposure to NODs does not appear to cause adverse effects on fish swimming activity and behaviour (Kankaanpää *et al.*, 2002); however, complete loss of liver architecture was observed after 1 to 2 days of oral exposure to NODs, although 4 to 8 days later there was partial recovery of hepatocytes (Kankaanpää *et al.*, 2002).

10.2.3 Cylindrospermopsins

Cylindrospermopsins (CYNs) are a class of cyclic guanidine alkaloids that are produced by several freshwater cyanobacteria including *Raphidiopsis* (formerly *Cylindrospermopsis*), *Aphanizomenon*, *Dolichospermum* and *Umezakia* (WHO, 2020b; Chorus and Welker, 2021) (Fig. 10.3). Four different structural variants of CYNs have been identified (Wimmer *et al.*, 2014) and have been detected in surface waters worldwide (Armah *et al.*, 2013), although the organisms producing cyanotoxins can vary with geography (WHO, 2020b). CYNs are chemically stable (Sotton

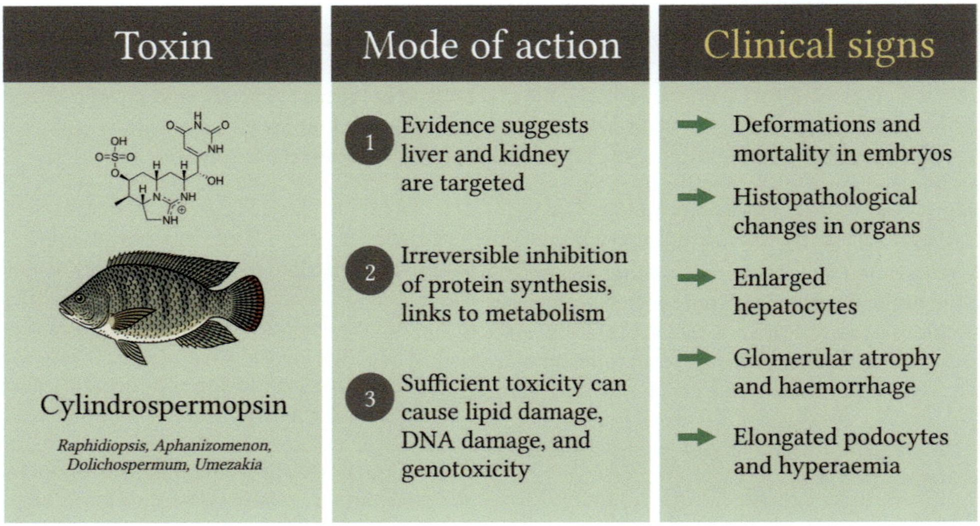

Fig. 10.3. Cylindrospermopsin mode of action and clinical signs of toxicity in fish species.

et al., 2015), hydrophilic, and resistant to boiling and variable pH (Chiswell *et al.*, 1999), although temperatures >50°C in combination with alkaline conditions can cause degradation (Chiswell *et al.*, 1999; Adamski *et al.*, 2016). CYNs typically occur at lower concentrations in fresh water because the cyanobacterial producers rarely form scums with high cell densities (WHO, 2020b; Chorus and Welker, 2021). CYNs also seem to occur more frequently in tropical and subtropical regions; communities that rely on local fish as a primary source of protein, in particular those who consume the entire fish, are at increased risk of exposure to CYNs given the mounting evidence of higher concentrations in fish liver and kidney (WHO, 2020b).

Fish encounter CYNs through direct contact with contaminated water, feeding, uptake through the gills or skin, or by accumulation in aquatic food webs (Guzmán-Guillén *et al.*, 2014; WHO, 2020b). Approximately 90% of CYNs in natural surface waters are released from cyanobacteria in the dissolved fraction (Rücker *et al.*, 2007) and are available for accumulation by fish via the intestine and gills (Guzmán-Guillén *et al.*, 2014).

Impact on fish production

Fish kills to cyprinids (Cyprinidae) have been linked to *Raphidiopsis* blooms recurring in a lake in Aleksandrovac, Serbia (Đorđević *et al.*, 2015). Fish kills occurred within 24 h when CYNs reached maximum concentrations of 24 μg/l in the lake, although it is postulated that other factors including uncharacterized toxic metabolites in *Raphidiopsis* may have also contributed to fish mortality (Svirčev *et al.*, 2016). Nevertheless, there are increasing concerns that CYN exposure to humans can occur through fish consumption.

Mechanism of toxicity

CYN toxicity in fish is rare and the mechanism has not been elucidated. Preliminary evidence suggests that CYNs target the liver (hepatotoxicity) and kidney (nephrotoxicity) and can reveal magnitude differences in mode of action depending on the length of exposure and concentration of the dose (Guzmán-Guillén *et al.*, 2013, 2014; Chorus and Welker, 2021). At low concentrations CYNs cause irreversible inhibition of protein synthesis (Terao *et al.*, 1994; Froscio *et al.*, 2003), whereas at higher concentrations CYNs interact with metabolites and mechanisms linked to cytochrome P450 which serve an important role in the detoxification of xenobiotics (Froscio *et al.*, 2003; Falconer and Humpage, 2006). A concentration-dependent increase in reactive oxygen species, lipid peroxidation and stress responses have also been observed from exposure to CYNs, causing damage to lipids, proteins and DNA (Gutiérrez-Praena *et al.*, 2011;

Liebel *et al.*, 2011; Guzmán-Guillén *et al.*, 2013). Evidence further suggests CYNs can cause genotoxicity that is linked to pronounced and prolonged oxidative stress (Guzmán-Guillén *et al.*, 2014). Cellular mechanisms to maintain cell viability and prevent DNA damage are activated to counteract these toxic effects (Liebel *et al.*, 2011). It is important to note that CYNs can also accumulate in other areas including the edible muscle tissues in various fish species (Berry *et al.*, 2012), although this work utilized the enzyme-linked immunoassay (ELISA) which can overestimate toxin concentrations and undermine the confidence in the data on toxin levels in seafood (Testai *et al.*, 2016).

Clinical signs

CYN exposure at sublethal concentrations can cause deformations and rapid mortality in fish embryos after injection of pure toxins (Berry *et al.*, 2009). However, it is suggested that CYNs cannot readily permeate cellular membranes of embryos, and companion studies exposing embryos to waterborne toxins demonstrated no developmental toxicity or mortality (Berry *et al.*, 2009; Sotton *et al.*, 2015). In juvenile and adult life stages, histopathological damages are dose-dependent and primarily occur in the liver and kidney. The liver can show enlarged hepatocytes with central nuclei and cytoplasmic vacuolization and hyalinization, increased hepatocyte nuclear diameter, steatosis and scarce cytoplasmic organelles (Gutiérrez-Praena *et al.*, 2012; Puerto *et al.*, 2012), and the kidney can show glomerular atrophy, dilations of Bowman's capsule, haemorrhage, elongated podocytes and hyperaemia (Gutiérrez-Praena *et al.*, 2012; Puerto *et al.*, 2012). Histopathological damages have also been shown in the heart, gills and intestines as well (Gutiérrez-Praena *et al.*, 2012; Puerto *et al.*, 2012). The clinical signs investigated here are all considered acute (short-term) exposure, and further work is needed to assess chronic (long-term) exposure from CYNs (Sotton *et al.*, 2015).

10.2.4 Anatoxin-a

Anatoxin-a (ATX) is a secondary amine alkaloid produced by a number of cyanobacteria, including *Dolichospermum*, *Aphanizomenon*, *Raphidiopsis*, *Cylindrospermum*, *Oscillatoria*, *Planktothrix* and *Phormidium* (Van Apeldoorn *et al.*, 2007; WHO, 2020c) (Fig. 10.4). It has a worldwide distribution and occurs in freshwater and brackish environments, as well as temperate, tropical and cold climatic regions (WHO, 2020c). ATX is relatively stable in low-light or acidic environments (Stevens and Krieger, 1991; Kaminski *et al.*,

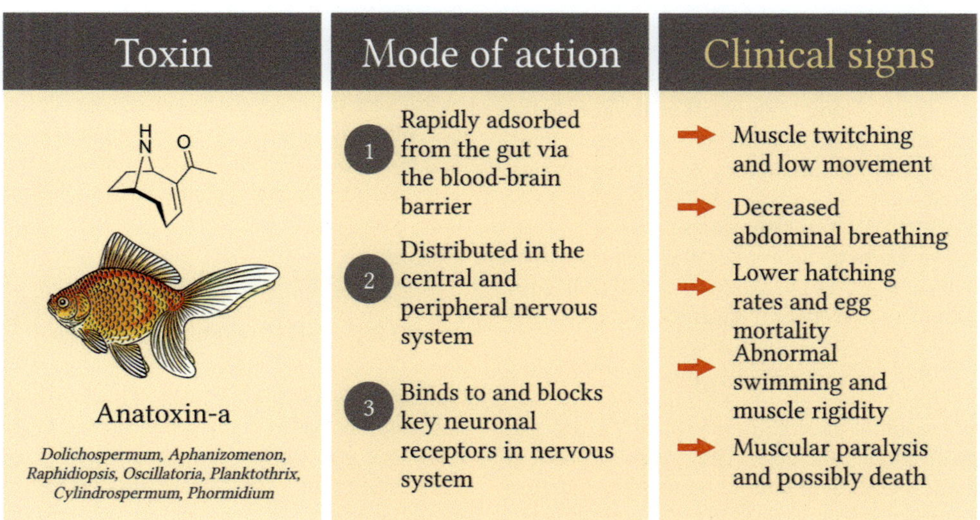

Fig. 10.4. Anatoxin-a mode of action and clinical signs of toxicity in fish species.

2013). However, ATX degradation is promoted by high pH, high temperature, increased light availability, UV-B irradiation and by bacterial activity (Stevens and Krieger, 1991; Rapala *et al.*, 1994; Van Apeldoorn *et al.*, 2007; Kaminski *et al.*, 2013). Under normal conditions, the half-life of ATX ranges roughly between 4 and 14 days (WHO, 2020c). However, during HAB conditions, when phytoplankton growth often causes increases in water pH, half-life may be as short as a few hours (Stevens and Krieger, 1991). ATX is highly soluble in water and is not susceptible to enzymatic hydrolysis (Van Apeldoorn *et al.*, 2007; Chorus and Welker, 2021). There is no clear evidence that ATX is released in large amounts from healthy cyanobacterial cells; exposure can thus be mainly expected during bloom lysis (Chorus and Welker, 2021). Limited data suggest that ATX bioaccumulation, and thus risk of human exposure via seafood consumption, is low (Testai *et al.*, 2016).

Impact on fish production

Although no known examples of socio-economic impacts on aquaculture and fisheries have been implicated with ATX, the risks of fish kills and animal poisonings are concerning in lakes that form shore scums with high concentrations of ATX (Viaggiu *et al.*, 2004).

Mechanism of toxicity

Acute toxicity studies in animals (i.e. mice, trout) show that ATX is rapidly adsorbed from the gut across the blood–brain barrier following oral exposure and is most likely distributed widely in the central and peripheral nervous systems, binding to receptors that play a key role in neuronal communication (Wonnacott and Gallagher, 2006; WHO, 2020c). It is likely that as a result ATX exposure leads to neuromuscular blocking (Carmichael *et al.*, 1975; Wonnacott and Gallagher, 2006; Van Apeldoorn *et al.*, 2007).

Clinical signs

As a result of neuromuscular blocking, depending on species and dose, ATX can cause death in animals within minutes to hours after exposure as a result of muscular paralysis and consecutive respiratory arrest (Carmichael *et al.*, 1975; Van Apeldoorn *et al.*, 2007; Chorus and Welker, 2021). Clinical signs of poisoning progress from muscle twitching, decreased movement, abdominal breathing and cyanosis, to convulsions and eventually death (Van Apeldoorn *et al.*, 2007). ATX had no acute toxic effect on zebrafish (*Danio rerio*) embryos, although at very high concentrations temporal changes in heart rate could be observed (Oberemm *et al.*, 1999). Similarly, exposure to pure ATX was almost harmless to fish in early stages of development, except that larval length was reduced at very high, but ecologically relevant, concentrations of the toxin (Osswald *et al.*, 2009). Interestingly, effects of cyanobacterial cell extracts containing ATX were more harmful than when ATX was administered as a pure toxin, but this may also be a result of other toxic substances or bacteria (Osswald *et al.*, 2009). The toxic effects of ATX depend on the life stage of fish. Cyanobacterial extracts containing ATX caused higher mortality of common carp (*Cyprinus carpio*) eggs and a lower hatching rate, and the larvae that hatched were smaller in size and had a higher incidence of skeletal malformations (Osswald *et al.*, 2009). In juvenile common carp, mortality occurred within 26–29 h, but morphological effects and lesions could not be identified (Osswald *et al.*, 2007). Juvenile common carp also showed rapid opercular movement and abnormal swimming compared with controls (Osswald *et al.*, 2007). Goldfish (*Carassius auratus*) given oral or intraperitoneal doses of cyanobacterial cell extracts containing ATX showed a latent period of 2 to 4 mins followed by muscle rigidity and death after 12 to 14 mins due to respiratory arrest (Carmichael *et al.*, 1975). However, goldfish were not affected when placed directly into algal culture, lyophilized culture or an aqueous medium containing the cell extracts, indicating that the toxin is not readily absorbed across the gill membranes (Carmichael *et al.*, 1975).

10.2.5 β-Methylamino-L-alanine

β-Methylamino-L-alanine (BMAA) is a polar and non-lipophilic neurotoxin produced by *Microcystis*, *Nostoc* and other cyanobacteria (Huisman *et al.*, 2018) (Fig. 10.5). There are also indications that it may be produced by diatoms and dinoflagellates (Metcalf *et al.*, 2021). BMAA has been reported in a variety of aquatic and

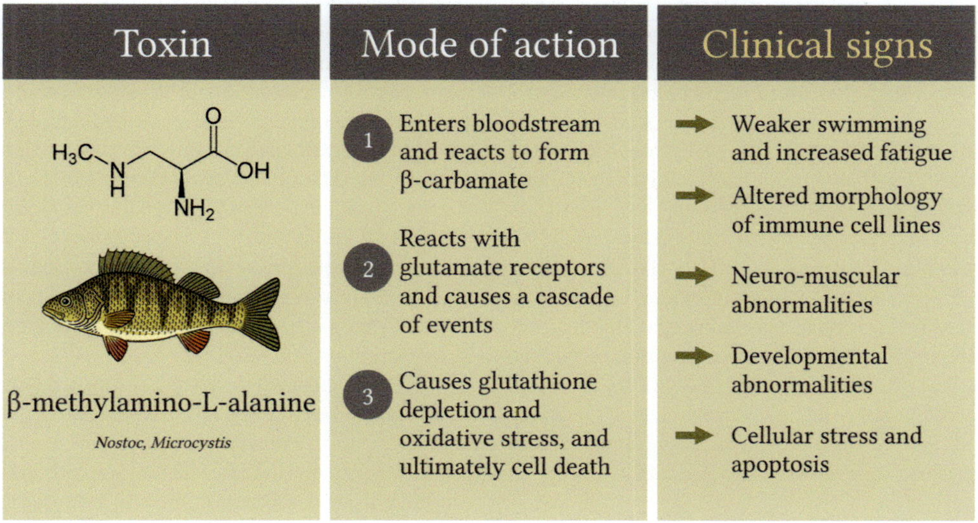

Fig. 10.5. β-Methylamino-L-alanine mode of action and clinical signs of toxicity in fish species.

terrestrial environments worldwide, suggesting that it is ubiquitous (Chiu *et al.*, 2011). In natural environments BMAA concentrations are generally low, but can vary by several orders of magnitude, and there are indications that it may bioaccumulate (Jonasson *et al.*, 2010; Lürling *et al.*, 2011; Metcalf *et al.*, 2021). Since it is also often found in seafood, this may be a pathway to human exposure (Jonasson *et al.*, 2010; Jiang *et al.*, 2014; Salomonsson *et al.*, 2015). Also inhalation of aerosolized BMAA is becoming of increasing concern (Metcalf *et al.*, 2021). However, methodological limitations, reporting and prolific analytical errors limit the conclusions that can be drawn from many existing BMAA studies (Faassen, 2014).

Since BMAA is hydrophilic, fish may be exposed to BMAA in the dissolved fraction, but also when feeding on phytoplankton or zooplankton containing protein-bound BMAA (Jonasson *et al.*, 2010; Lürling *et al.*, 2011; Lance *et al.*, 2018). Although BMAA has been found in fish, they seem to be considerably less contaminated compared with shellfish and aquatic invertebrates (Lance *et al.*, 2018; Metcalf *et al.*, 2021).

Impact on fish production

Although no known examples of socio-economic impacts on aquaculture and fisheries have been implicated with BMAA, exposure is expected to be minimal as BMAA's acute toxicity to fish is relatively low (Lance *et al.*, 2018).

Mechanism of toxicity

BMAA toxicity mechanisms are mainly based on animal models (i.e. rodents, birds and primates). When BMAA is consumed orally it enters the bloodstream and passes the blood–brain barrier where it reacts with bicarbonate (HCO_3^-) to form β-carbamate (Weiss and Choi, 1988; Duncan *et al.*, 1991). There, it can react with several glutamate receptors and cause a cascade of events beginning with: (i) changes in cellular ion concentrations; (ii) depolarization of cells; (iii) permeabilization of cell membranes; and eventually (iv) release of noradrenaline (Chiu *et al.*, 2011 and references therein). BMAA inhibits the cystine/glutamate antiporter (system Xc^-)-mediated cystine uptake, which in turn leads to glutathione depletion, increased oxidative stress and ultimately cell death (Liu *et al.*, 2009; Metcalf *et al.*, 2021). Additionally, BMAA disrupts calcium and mitochondrial homeostasis and can propagate neurotoxic effects between adjacent cells (Metcalf *et al.*, 2021).

Clinical signs

Acute toxic effects on fish are not described, and it is expected that BMAA mainly affects fish via prolonged or chronic exposure. Behavioural effects of BMAA on fish have been observed, in line with its neurotoxic potential (Purdie *et al.*, 2009). Zebrafish have shorter embryonic nerves, weaker swimming performance and increased fatigue (Powers *et al.*, 2017). Cytotoxic effects of BMAA on fish immune cell lines are known, leading to a reduction in their total count, altered morphology and decreased integrity (Sieroslawska and Rymuszka, 2019). BMAA exposure induces a range of neuromuscular and developmental abnormalities in zebrafish which can be directly related to disruptions to glutamatergic signalling pathways (Purdie *et al.*, 2009). Increased misfolding in proteins leads to protein aggregation, which may lead to cellular stress and increased apoptosis (Sieroslawska and Rymuszka, 2019).

10.2.6 Lipopolysaccharides

Lipopolysaccharides (LPSs) are large, complex molecules first discovered in membranes of Gram-negative bacteria. Also, many marine and freshwater cyanobacterial species are able to produce LPSs including *Anabaena*, *Microcystis*, *Planktothrix*, *Synechococcus*, *Agmenellum* and *Schizothrix* (Codd et al., 1999; Durai et al., 2015) (Fig. 10.6). Generally, LPSs consist of three structural components: (i) a glycan with an *O*-specific polysaccharide that is attached to (ii) a glycolipid anchor lipid A through (iii) a connecting polysaccharide core region (Caroff and Karibian, 2003). The function of LPSs is considered structural, thereby acting as a permeability barrier against antimicrobials but also as an active immune modulator (at low concentrations) inducing resistance to other invading microbes (Bertani and Ruiz, 2018). The structure of LPSs in water can be quite variable (i.e. it is not species and/or strain specific) but this also depends on abiotic factors such as temperature and osmolarity (Moosová *et al.*, 2019).

LPSs are normally excreted in low amounts when bacterial cells divide or are lysed (Caroff and Karibian, 2003). Phytoplanktivorous fish are expected to be exposed to LPSs while feeding, although no earlier studies can confirm this. As a result, fish may mainly encounter LPSs during the senescence of a HAB. Consequently, LPSs are degraded enzymatically by various organisms including mammals, molluscs, moulds and bacteria (Jamieson and Wardlaw, 1989). LPSs also easily form aggregates and complex molecules with a number of other natural products (Nowotny, 1969).

Impact on fish production

Bacterial disease is common and causes large economic losses in aquaculture (Toranzo *et al.*, 2005); however, it remains unclear to what

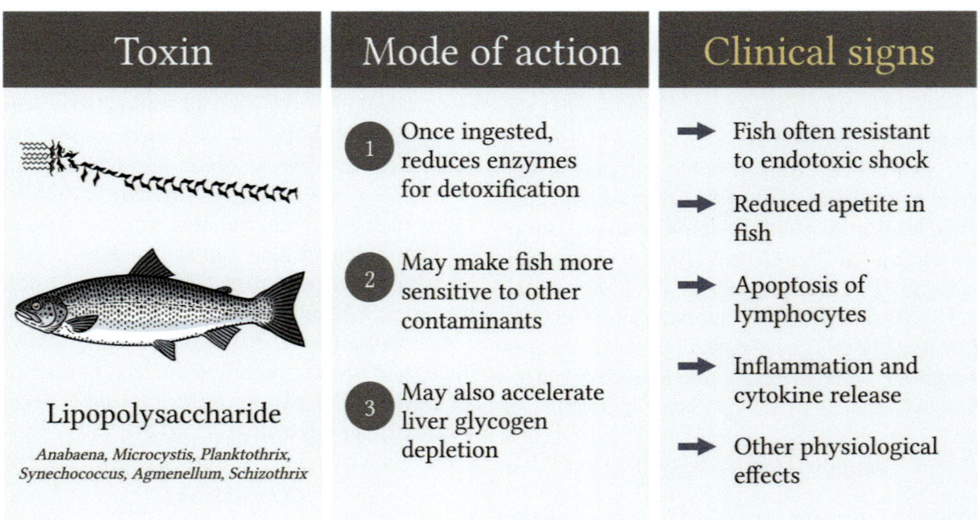

Fig. 10.6. Lipopolysaccharide mode of action and clinical signs of toxicity in fish species.

extent LPSs themselves cause direct fish mortality in aquaculture. Indirectly, there are indications that naturally occurring, low concentrations of LPSs may stimulate the immune response of fish, thus making them more resistant against bacterial infections and resulting in higher survival (Nya and Austin, 2010; Ispir and Dorucu, 2014). LPSs may also potentiate the toxic effects of heavy metals, representing a significant risk to organisms exposed to combinations of LPSs and metals in the environment (Notch *et al.*, 2011).

Mechanism of toxicity

No other natural product is known to elicit such a variety of reactions as endotoxins do when injected into the proper host (Nowotny, 1969). It is, however, well known that the lipid A part of the LPS structure is responsible for both the toxicity and the immune response of fish to LPS (Iliev *et al.*, 2005). At higher doses exposure may be lethal to animals. Fish are relatively resistant to LPSs compared with other animals (Wedemeyer *et al.*, 1969; Sepulcre *et al.*, 2009; Bi *et al.*, 2018). In zebrafish embryos, exposure to purified cyanobacterial LPS can significantly reduce the activity of microsomal and soluble GST, a group of enzymes that are important in detoxification (Best *et al.*, 2002; Jaja-Chimedza *et al.*, 2012). This reduction, however, only occurs in an *in vivo* experiment, whereas *in vitro* preparations of GST show no significant change in GST activity in response to LPS (Best *et al.*, 2002), which may suggest that LPS may modulate *de novo* synthesis of GST (Wang *et al.*, 2006). This reduced detoxification capacity induced by LPS exposure may make organisms more sensitive to co-exposure with other contaminants such as MCs. LPS exposure may also accelerate liver glycogen depletion in salmonids (Wedemeyer *et al.*, 1969).

Clinical signs

The properties of cyanobacterial LPSs are poorly characterized in comparison with those of other heterotrophic bacteria (Durai *et al.*, 2015). Adverse effects from LPS exposure in animals include pyrogenicity, hypotension, neutropenia, intravascular coagulation, hypoferraemia, leucocytosis, leucopenia, sepsis, abortion and shock (Swain *et al.*, 2008). Fish, however, are often resistant to endotoxic shock (Iliev *et al.*, 2005; Swain *et al.*, 2008). At high-dose exposures to fish no clinical signs (i.e. changes in body coloration, abnormalities, behavioural changes) were observed (Wedemeyer *et al.*, 1969; Nayak *et al.*, 2008). Other work indicates that LPSs may reduce the appetite of goldfish and may induce apoptosis of lymphocytes (Volkoff and Peter, 2004; Xiang *et al.*, 2008). Fish may also show pronounced inflammation, cytokine release and other physiological effects in response to LPSs (Swain *et al.*, 2008).

10.2.7 Saxitoxins

Saxitoxins (STXs), also known as paralytic shellfish poisons, are a class of natural alkaloids that are produced in both marine and freshwater systems (Fig. 10.7). In marine waters, STXs are produced by dinoflagellates including *Alexandrium*, *Gymnodinium* and *Pyrodinium*; in fresh water, STXs are produced by cyanobacteria including *Dolichospermum*, *Raphidiopsis*, *Cylindrospermum*, *Aphanizomenon*, *Scytonema*, *Lyngbya*, *Oxynema* and *Planktothrix* (WHO, 2020d; Chorus and Welker, 2021). More than 50 analogues of STXs have been identified (Wiese *et al.*, 2010) and detected in marine waters worldwide (Kleinteich *et al.*, 2013; Murray *et al.*, 2015; Chorus and Welker, 2021), although production of STXs depends on the HAB species (WHO, 2020d). Nearly all STXs are hydrophilic, except those produced by *Lyngbya* in freshwater systems (Chorus and Welker, 2021). Evidence suggests STXs persist in surface waters for 1–2 months (Batoréu *et al.*, 2005) and can remain stable at alkaline pH (>8.5) (Castro *et al.*, 2004) and high temperature (Jellett *et al.*, 1995). STX production is also influenced by several environmental factors including pH, temperature, light intensity, nutrient concentrations and high conductivity (Sivonen and Jones, 1999; Neilan *et al.*, 2008). STXs seem to occur more frequently in warm temperate regions (Laabir *et al.*, 2011; Murray *et al.*, 2015). Consequently, communities should take caution when consuming fish that may have encountered STX producers (Galvão *et al.*, 2009; de Morais Calado *et al.*, 2019; WHO, 2020d).

Fish encounter STXs through direct contact with contaminated water, feeding, uptake after the lysis of a HAB via epithelial absorption, or by accumulation in aquatic food webs (Lefebvre

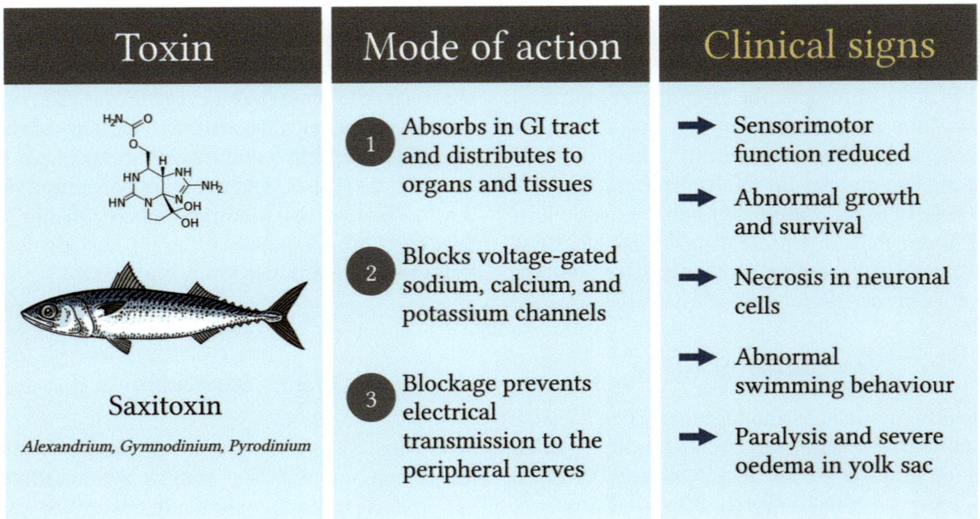

Fig. 10.7. Saxitoxin mode of action and clinical signs of toxicity in fish species.

et al., 2004; Galvão *et al.*, 2009). There is increasing interest to understand the sublethal effects in fish from exposure to STXs at naturally occurring concentrations, in particular for finfish populations that are endangered (Lefebvre *et al.*, 2004; Galvão *et al.*, 2009; Berry *et al.*, 2012; Fire *et al.*, 2012).

Impact on fish production

Fish kills are well documented and coincide with direct consumption of STX-producing dinoflagellates and cyanobacteria or by dietary intake of zooplankton that accumulated STXs (White, 1981; Fire *et al.*, 2012; Moustaka-Gouni *et al.*, 2016; Barrientos *et al.*, 2019). Recent examples include large-scale and multi-year sharp-nose puffer (*Canthigaster rostrata*) mortality events on the southern Caribbean coast of Costa Rica (Barrientos *et al.*, 2019).

Mechanism of toxicity

STX toxicity is well documented in humans from shellfish consumption. However, information for fish is scarce. As highly potent neurotoxins, STXs are readily absorbed in the GI tract and distributed to various organs and tissues including the central nervous system (Pearson *et al.*, 2010; WHO, 2020d; Chorus and Welker, 2021). Once inside the body, STXs are potent blockers of voltage-gated sodium channels in neuronal cells and calcium and potassium channel blockers in cardiac cells (Wang *et al.*, 2003; Su *et al.*, 2004; Testai *et al.*, 2016). Blockage of these channels prevents electrical transmission to the peripheral nerves including skeletal and cardiac muscles (Chorus and Welker, 2021). Neurological symptoms and mortality can occur, in some cases within minutes, depending on the length and severity of STX exposure (FAO, 2004). STXs can also produce free radicals in fish and induce cytotoxicity, genotoxicity and apoptosis in neuronal cells (Banerjee *et al.*, 2021). Generation of reactive oxygen species can further disrupt cellular antioxidants and cause lipid peroxidation and DNA damage in neuronal cells (da Silva *et al.*, 2014). In response, cellular detoxification mechanisms are activated in fish to chelate free radicals and prevent cellular damage (Banerjee *et al.*, 2021). STXs are further metabolized as glucuronides then rapidly excreted in the urine, suggesting glucuronidation as a metabolic pathway of detoxification in humans and animals (Munday *et al.*, 2013; Testai *et al.*, 2016; de Morais Calado *et al.*, 2019). However, glucuronidation can, in some instances, form more potent STX analogues. It is important to note that STXs can also accumulate in the liver of fish and induce oxidative stress and membrane damage (de Assis *et al.*, 2013). STXs can remain in fish muscles for 90 days after exposure, raising concerns for higher trophic-level species including humans (Galvão *et al.*, 2009; de Morais Calado *et al.*, 2019).

Clinical signs

In early life stages of fish, STX exposure at sublethal concentrations can cause reductions in sensorimotor function and paralysis by 96 h post-fertilization, severe oedema in the yolk sac, eye and pericardium, reduced yolk sac size, and abnormal growth and survival during larval development (Lefebvre *et al.*, 2004; Tian *et al.*, 2014). Morphological and sensorimotor effects in fish can be reversible if transferred into clean water (Lefebvre *et al.*, 2004). In juvenile and adult life stages, necrosis can form in neuronal cells from increased lipid peroxidation levels (da Silva *et al.*, 2014; Banerjee *et al.*, 2021). STX exposure can also alter locomotor activities (i.e. swimming behaviours) in fish at sublethal concentrations (Lopes *et al.*, 2017). Consequently, changes in behaviour can alter the reproductive fitness and predator–prey relationships in fish populations (Banerjee *et al.*, 2021).

10.2.8 Domoic acid

Domoic acid (DA) is a naturally occurring excitotoxin produced by diatoms *Nitzschia*, *Pseudo-nitzschia*, *Amphora*, and the red macro-alga *Chondria* in the bays and coastal areas of marine systems worldwide (Trainer *et al.*, 2012), although recently it has been detected in estuaries as well (Peacock *et al.*, 2018) (Fig. 10.8). Monitoring of DA has intensified globally since its discovery in 1987 after 145 people experienced amnesic shellfish poisoning in Prince Edward Island, Canada (Bates *et al.*, 1989). Since then, numerous incidences of extreme neurodegenerative disorders and lethality have been reported in birds and mammals (Scholin *et al.*, 2000; Bejarano *et al.*, 2008; Trainer *et al.*, 2012). DA is a cyclic amino acid with three carboxylic acid groups that give it high hydrophilicity and polarity (Quilliam *et al.*, 1989). It is known that DA acts as a glutamate agonist (Hampson and Manalo, 1998), mimicking glutamate, the principal neurotransmitter in the central nervous system that sends signals in the brain and throughout the nerves in the body (Landsberg, 2002). DA only has one major analogue and an epimer (epi-DA) that is of toxicological relevance (Ramsdell, 2007). It is also heat stable and not typically destroyed after cooking, but evidence suggests it is light-sensitive and can undergo epimerization with warming (Quilliam, 2003).

Fish encounter DA through direct contact with contaminated water, feeding, or by accumulation in aquatic food webs (Scholin *et al.*, 2000; Lefebvre *et al.*, 2012; Lewitus *et al.*, 2012). Although behavioural toxicity and mortality

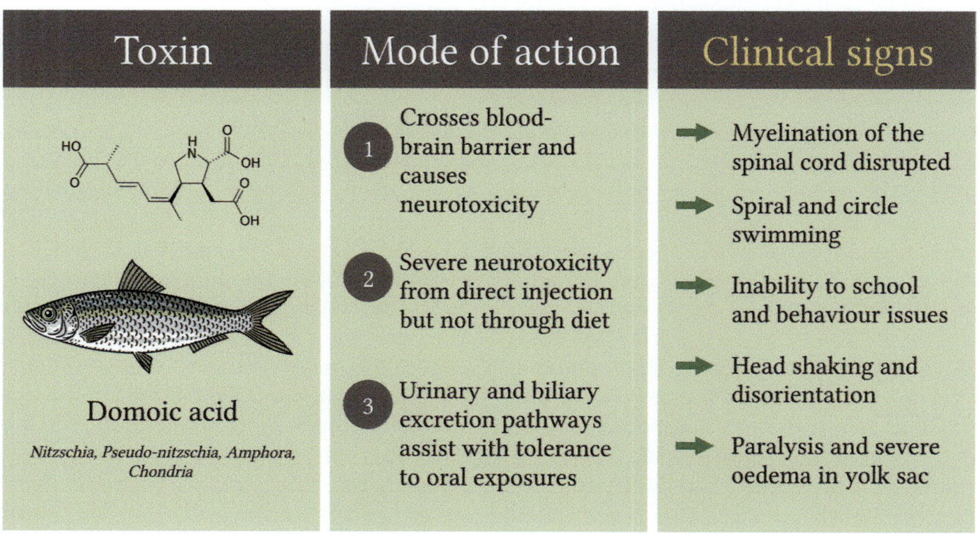

Fig. 10.8. Domoic acid mode of action and clinical signs of toxicity in fish species.

have been observed in seabirds and marine mammals, evidence suggests fish may be tolerant to DA under natural exposure conditions (Lefebvre *et al.*, 2012), resulting in high DA content, particularly in planktivorous fish such as sardines and anchovies (Trainer *et al.*, 2020). This information is important because under natural conditions DA-producing HABs may not necessarily cause diseases and disorders in fish. However, its sublethal accumulation at high concentrations in fish and subsequent consumption by higher trophic-level species raises concerns for its toxicity at higher levels of the food chain.

Impact on fish production

The global occurrence of HABs has raised widespread concerns that DA can have serious economic consequences to aquaculture and fisheries. DA can poison planktivorous finfish (e.g. anchovies and sardines) and serve as a vector for toxicity and mass mortality to higher trophic-level species including birds, sea mammals and humans (Scholin *et al.*, 2000; Lewitus *et al.*, 2012; Trainer *et al.*, 2012).

Mechanism of toxicity

DA toxicity in fish has been described in detail but its ecologically relevant route of exposure is debated. Toxicological studies administering sublethal concentrations of DA by direct injection resulted in severe neurotoxicity in various fish species (e.g. spinning, head shaking, disorientation, inability to school) and proof DA could cross the blood–brain barrier (Lefebvre *et al.*, 2001; Nogueira *et al.*, 2010; Panlilio *et al.*, 2020, 2021). This raised serious concerns because disoriented and intoxicated fish could be easily preyed upon by higher trophic species. However, other work indicated that direct injection is not an ecologically relevant route of exposure (Lefebvre *et al.*, 2012). In fact, a follow-up study found that dietary exposure to DA at naturally occurring concentrations is unlikely to cause behavioural defects in fish or significant impacts on fish populations (Lefebvre *et al.*, 2012). This conclusion was reached after recognizing DA shows remarkably similar lethal concentration values and impacts on the central nervous system in fish, birds and mammals that received the toxin by intracoelomic or intraperitoneal injection (Lefebvre *et al.*, 2001) yet starkly different responses after oral exposure (Lefebvre *et al.*, 2012). A similar study was performed in coho salmon to confirm whether a maximum ecologically relevant dose of DA would accumulate in relevant organs and tissues and cause behavioural changes (Lefebvre *et al.*, 2007). No behavioural symptoms were observed, and it was suggested that urinary and biliary excretion pathways assist with fish tolerance to DA oral exposures (Lefebvre *et al.*, 2007). These findings highlight the importance of the route of administration of DA in fish, which has implications for its absorption, distribution, metabolism and excretion pathways throughout the body.

Clinical signs

Although developmental defects (e.g. disruption in myelination of the spinal cord) and behavioural signs of excitotoxicity or neurotoxicity (e.g. spiral swimming, circle swimming, upside-down swimming, inability to school) have been found in fish that received direct injection of DA in laboratory studies (Lefebvre *et al.*, 2001; Nogueira *et al.*, 2010; Panlilio *et al.*, 2020, 2021), other work suggests oral exposure of DA in the field will not cause clinical signs in fish (Lefebvre *et al.*, 2012). Thus, it is difficult to describe clinical signs for DA since the relevant route of exposure from HABs must be considered.

10.2.9 Ciguatoxins

Ciguatoxins (CTXs) are a class of large polyether compounds that contain 13 to 14 fused rings giving them ladder-like structures (Nicolaou *et al.*, 2008; FAO and WHO, 2020) (Fig. 10.9). They are potent neurotoxins that are produced by the epiphytic benthic dinoflagellates *Gambierdiscus* and *Fukuyoa*, which have pantropical distribution (Yong *et al.*, 2018; FAO and WHO, 2020). CTXs cause the tropical disease ciguatera fish poisoning in humans, regarded as the most common fish poisoning resulting in >50,000 global cases annually (Traylor and Singhal, 2018). Ciguatera poisoning stems from small herbivorous reef fish grazing on toxic algae and detritus found on the dead corals, which are then preyed upon by larger carnivorous fish (Lehane, 2000;

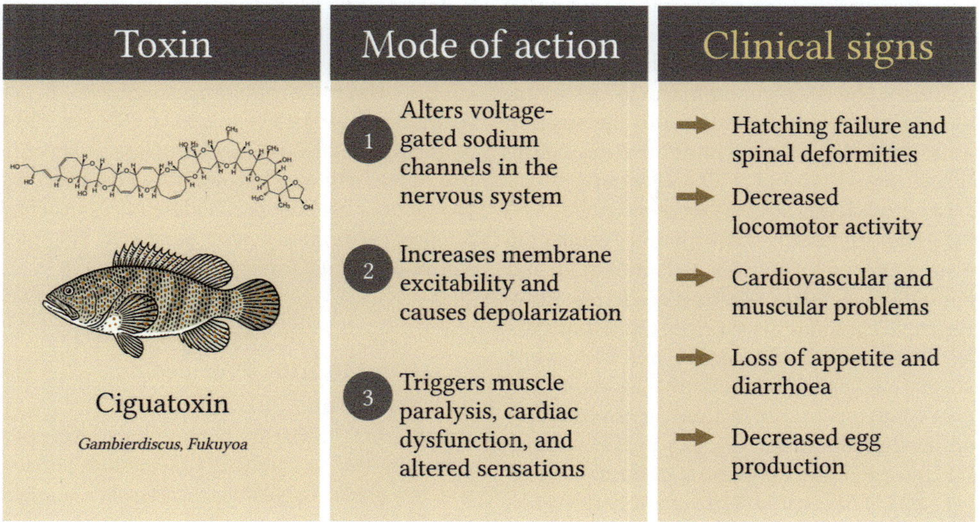

Fig. 10.9. Ciguatoxin mode of action and clinical signs of toxicity in fish species.

FAO, 2004; Ledreux *et al.*, 2014). From here CTXs can accumulate in these predatory reef fish (Lehane and Lewis, 2000) and biomagnify up the food chain, with levels reaching 50–100 times more concentrated in the viscera, liver and gonads (De Fouw *et al.*, 2001).

Ciguatera fish poisoning is of ongoing concern for aquaculture and fisheries because CTXs are odourless, tasteless, lipid-soluble, heat stable and resistant to mild pH fluctuations (Guzmán-Pérez and Park, 2000; FAO and WHO, 2020). CTXs are often present at very low concentrations in seafood (<ppb) and are not destroyed by cooking or freezing, making them difficult to detect in the absence of advanced detection methods (FAO and WHO, 2020). Intoxicated fish taste and smell normal (Lehane, 2000; FAO, 2004), which further complicates human perceptions of seafood safety and HABs toxicity. The impacts of global warming on sea-level rises, precipitation and nutrient inputs into aquatic systems support the growth and expansion of CTX-producing HABs, raising additional concerns for fish populations in marine waters (Gingold *et al.*, 2014; Yong *et al.*, 2018; FAO and WHO, 2020).

Impact on fish production

Over 425 fish species from the pantropics have been impacted by CTXs (Pérez-Arellano *et al.*, 2005). Coral reef fishes are an important seafood for the global market yet frequently contribute to the worldwide occurrence of ciguatera poisoning (FAO and WHO, 2020). Reef fish affected by CTXs include amberjack (*Seriola*), barracuda (Sphyraenidae), grouper (Serranidae), jack (Carangidae spp.), moray eel (Muraenidae spp.), parrotfish (Scaridae spp.), po'ou (Labridae spp.), roi (*Cephalopholis* spp.), snapper (Lutjanidae), surgeonfish (Lutjanidae spp.), trevally (*Caranx* spp.) and wrasse (Labridae spp.) (FDA, 2011).

Mechanism of toxicity

CTX toxicity is thoroughly documented in human poisoning cases from fish consumption. As highly potent neurotoxins, CTXs bind with high affinity and decrease the threshold for opening voltage-gated sodium channels in synapses of the nervous system (Bidard *et al.*, 1984). Open sodium channels increase membrane excitability and cause depolarization, which can trigger muscle paralysis, cardiac dysfunction, and altered sensations from heat and cold (FAO, 2004; Zimmermann *et al.*, 2013; Traylor and Singhal, 2018; FAO and WHO, 2020).

CTX toxicity in fish continues to be elucidated, in part because several congeners of CTXs exist that vary in oxygenation (i.e. oxocene and oxopene CTXs) and may metabolize differently depending on the trophic level of the species (Yogi *et al.*, 2011). For instance, in response to controlled dietary exposure to toxic *Gambierdiscus polynesiensis* cells, the planktivorous fish mullet (*M. cephalus*) showed rapid accumulation of CTXs in the GI tract and in the bloodstream, followed by rapid distribution into the somatic tissues, with the flesh and intestine carrying the highest proportion of CTXs (Ledreux *et al.*, 2014). High levels of CTXs were also measured in the gills, suggesting the respiratory route may be an important route for accumulation and elimination of CTXs (Ledreux *et al.*, 2014). However, rapid elimination of the oxocene congeners was observed in the blood, bile and liver, while oxopene congeners were retained (Ledreux *et al.*, 2014). Since it is known that carnivorous fish can accumulate oxopene CTXs in their tissues and cause ciguatera poisoning in humans (Yogi *et al.*, 2011), these findings suggest herbivorous fish metabolize oxopene CTXs in a time-dependent manner and these toxins will biomagnify through higher trophic-level species (Ledreux *et al.*, 2014).

Clinical signs

CTX exposure at sublethal concentrations can cause severe embryonic defects including hatching failure, spinal deformities and caudal fin malformation, haemorrhaging and discoloration of the gallbladder, immune dysfunction, decreased locomotor activity and altered muscle physiology (Colman *et al.*, 2004; Mak *et al.*, 2017; Yan *et al.*, 2017). CTXs can also cause cardiovascular, muscular and skeletal abnormalities and significantly reduce the hatching success of finfish (Edmunds *et al.*, 1999). In adults, CTX can cause abnormal behaviours including loss of appetite, diarrhoea, abnormal swimming, decreased egg production, gender-specific differences in reproductive performance and decreased hatching rate of offspring (Yan *et al.*, 2020).

10.2.10 Prymnesins

Prymnesins (PRMs) are potent phycotoxins produced by the haptophyte *Prymnesium parvum*, widely regarded as 'golden or golden-brown alga', which causes surface waters to appear golden and foamy (Manning and La Claire, 2010; Taylor *et al.*, 2021) (Fig. 10.10). This mixotrophic HAB-forming alga recurs in brackish estuarine waters and in mainland freshwater reservoirs across six continents in the northern and southern hemisphere (Guo *et al.*, 1996; Manning and La Claire, 2010; Taylor *et al.*, 2021). Three different PRMs have been identified (Manning and La Claire, 2010; Rasmussen *et al.*, 2016; Binzer *et al.*, 2019), and little is known about their molecular mechanisms of synthesis, mode of transport and biological relevance beyond the

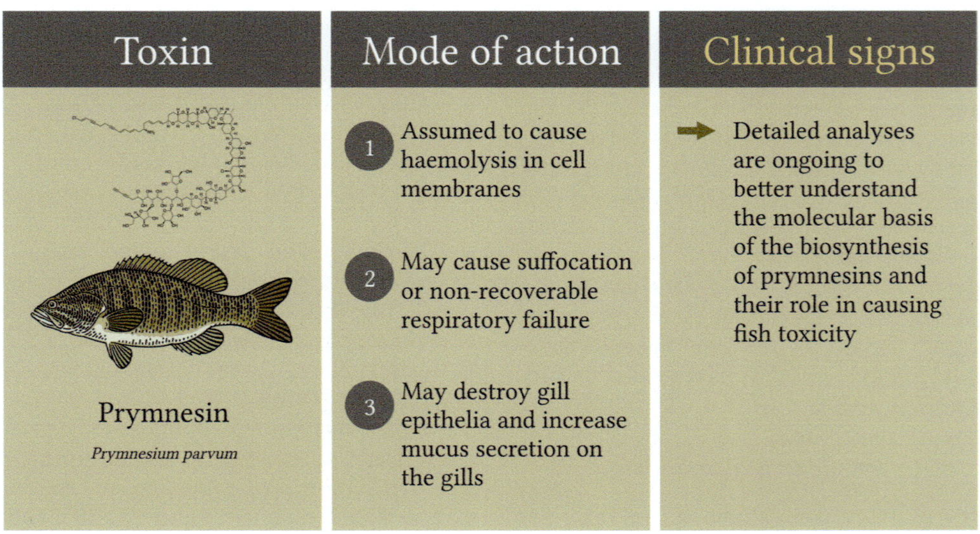

Fig. 10.10. Prymnesin mode of action and clinical signs of toxicity in fish species.

suggestion that they play an important role in the physiology of *P. parvum* (Binzer *et al.*, 2019; Medić *et al.*, 2022). PRMs are remarkably complex molecules with ladder-like, polycyclic ethers that make them potent polyketides with ichthyotic and haemolytic activities (Igarashi *et al.*, 1999). These toxins can have deleterious effects on co-occurring plankton by influencing trophic interactions and altering community structure (Tillmann, 2003; Blossom *et al.*, 2014). The production of PRMs is largely dictated by abiotic factors including moderate-to-low light, temperature ranging between 2 and 30°C, alkaline pH > 8.0, nutrient availability and low salinity (Guo *et al.*, 1996; Larsen *et al.*, 1998; Baker *et al.*, 2007; Manning and La Claire, 2010; Roelke *et al.*, 2016). Recent evidence suggests high irradiance can cause photodegradation of PRMs (Medić *et al.*, 2022), which corresponds with reduced toxicity to aquatic organisms including fish (Taylor *et al.*, 2021).

Fish encounter PRMs through direct contact with contaminated water by uptake through the gills (Baker *et al.*, 2007; Manning and La Claire, 2010; Taylor *et al.*, 2021). It is postulated that PRMs are excreted into surrounding waters during the senescent stages of a HAB and are positively correlated with acute toxicity in fish (Taylor *et al.*, 2020).

Impact on fish production

HABs of *P. parvum* continue to have devastating effects on aquaculture and fisheries around the world by causing fish kills (Brooks *et al.*, 2011; Roelke *et al.*, 2016) hypothesized to be caused by PRMs (Taylor *et al.*, 2021). A profiled example comes from Texas, USA, that reported a fish kill of over 34 million fish, valued at US$13 million and representing important smallmouth bass (*Micropterus dolomieu*), striped bass (*Morone saxatilis*), channel catfish (*Ictalurus punctatus*) and blue catfish (*Ictalurus furcatus*) sport fisheries (Southard *et al.*, 2010).

Mechanism of toxicity

Despite decades of research, it remains debated how *P. parvum* and PRMs affect fish. It is assumed that toxins released by *P. parvum* cause haemolysis (Kozakai *et al.*, 1982), destroy fish gill epithelia (Ulitzur and Shilo, 1966) and increase mucus secretion on the gills (Otterstrøm and Nielsen, 1939), acting as a barrier for oxygen transport (Bergsson *et al.*, 2019). It is believed massive fish kills result from suffocation, or non-recoverable respiratory failure, due to severe internal oxygen deficiency from exposure to *P. parvum* and its toxins (Ulitzur and Shilo, 1966; Svendsen *et al.*, 2018; Bergsson *et al.*, 2019; Medić *et al.*, 2022). Recent evidence of PRMs in the gill tissues of dead fish that encountered a *P. parvum* HAB suggests that PRMs caused toxicity (Wagstaff *et al.*, 2021).

Clinical signs

Detailed analyses of *P. parvum* are ongoing to better understand the molecular basis of the biosynthesis of PRMs and their role in causing fish toxicity. Clinical impacts from direct exposure to PRMs are at best hypothesized until analytical standards can be developed to explain their effects. Advancing our understanding of PRM toxicity in fish is of high interest given the massive socio-economic and ecological damage *P. parvum* HABs cause globally.

10.2.11 Blooms coinciding with fish kills but cannot be pinpointed to a toxin

Other HAB species with an uncharacterized toxin or combinations of toxins are associated with recurring fish kills worldwide. The mechanism of mortality may involve direct physical damage to fish gills. Alternatively, a suite of toxins or toxins acting in combination with environmental stressors, such as elevated temperature, low oxygen, high pH and ammonium, and co-occurrence with pathogens may play a role. Two commonly occurring species – the dinoflagellate, *Karenia* spp., and the raphidophyte, *Chattonella* spp. – have been selected and are discussed below to represent organisms that can cause fish mortalities through mechanisms that are not yet fully elucidated. Other fish-killing raphidophytes that are discussed include members of the genera *Pseudochattonella* (Eckford-Soper and Daugbjerg, 2016), *Chrysochromulina* (Simonsen and Moestrup, 1997), *Cochlodinium* (Tang and Gobler, 2009) and *Heterosigma* (Chang *et al.*, 1990).

Karenia spp.

Karenia is a genus of at least 12 species of marine dinoflagellates found in both oceanic and coastal waters worldwide (Glibert *et al.*, 2002; Haywood *et al.*, 2004; Brand *et al.*, 2012). This bloom-forming genus causes mortality of marine life (Oda, 1935; Gunter *et al.*, 1948) and huge financial losses for aquaculture (e.g. loss of US$330 million, Fujian Province, China, 2012) (Li *et al.*, 2017, 2019). *Karenia brevis* receives the most attention because it plagues coastal waters and releases toxins including brevetoxin, a potent ichthyotoxic neurotoxin that readily absorbs across the gill membranes in fish (Naar *et al.*, 2007) and shellfish (Landsberg *et al.*, 2009), the latter of which can cause acute neurotoxic shellfish poisoning in humans and mammals. However, the trophic transfer of brevetoxins makes it difficult to pinpoint this toxin as the cause of fish kills observed during *K. brevis* HABs. For instance, evidence suggests fish can safely accumulate brevetoxins by dietary transfer, yet when fish mortality from *K. brevis* occurs, brevetoxins are not detected in their tissues and viscera (Tester *et al.*, 2000; Naar *et al.*, 2007; Landsberg *et al.*, 2009).

The elusive toxicity of *Karenia* can also be due to *Karenia mikimotoi*, a species that does not produce brevetoxins, yet it is implicated in fish kills across Europe and Asia (Li *et al.*, 2019). Reason(s) for its toxicity are unclear because its HABs cause fish mortality well before the senescent stages, ruling out hypoxia as the culprit and supporting the possibility that healthy, intact cells are the cause (Li *et al.*, 2019). Some hypotheses for the toxicity of *Karenia* include: (i) toxins other than haemolysins and cytotoxins are involved which may break down rapidly once released by *K. mikimotoi*, therefore making it difficult to isolate and study them; (ii) these unknown toxins may reside predominantly near the cell membrane of *K. mikimotoi* at lethal concentrations; or (iii) the toxicity of *K. mikimotoi* might increase due to environmental or grazing pressures (Li *et al.*, 2017).

Histopathological changes from *Karenia* toxicity are scarce but have been observed in Atlantic salmon (*Salmo salar*) and rainbow trout (*Oncorhynchus mykiss*). Changes include gill disorders such as acute necrosis, sloughing of epithelial cells, severe oedematous separation of the epithelium from the lamellar branchial vessels, and swelling and pyknosis of branchial vessels (Mitchell and Rodger, 2007; Rodger *et al.*, 2011). Additional pathologies include pyknosis of the outer epithelium and sloughing of cells into the lumen of the intestine, and liver necrosis (Mitchell and Rodger, 2007).

Chattonella spp.

Chattonella is a genus of five species of marine raphidophytes found in tropical, subtropical and temperate regions worldwide (Edvardsen and Imai, 2006; Imai and Yamaguchi, 2012; Lum *et al.*, 2021). *Chattonella* spp. produce 'red tides' and perform diel vertical migration in coastal embayments between 10 and 20 m, which under favourable environmental conditions allows them to take up nutrients, expand and eventually cause massive fish kills (Watanabe *et al.*, 1995; Imai and Yamaguchi, 2012). However, mechanisms of toxicity remain unclear beyond the general understanding that suffocation is the ultimate cause of fish mortality.

Chattonella physically clog fish gills and cause mucus excretion (Matsusato and Kobayashi, 1974). Previous theories suggest gill damage is caused by polyunsaturated fatty acids (Shimada *et al.*, 1983) or brevetoxins (Endo *et al.*, 1992). Other evidence suggests the generation of reactive oxygen species (e.g. superoxides) and their synergistic role with free fatty acids could be responsible for gill tissue injury and mucus production causing fish mortalities (Marshall *et al.*, 2003; Shikata *et al.*, 2021). More recently, light was reported to be responsible for the haemolytic activity in *Chattonella*, demonstrating a significant relationship between haemolytic activity and chlorophyll c2 biosynthesis, suggesting that haemolytic toxins may be generated during electron/energy transfer through the chlorophyll c2 biosynthesis pathway (Wu *et al.*, 2021).

Histopathological changes from *Chattonella* toxicity are scarce but have been observed in northern bluefin tuna (García-Mendoza *et al.*, 2018). During a mass-mortality event in 2016 (from May to August), tuna were disoriented, gasping and swimming erratically prior to death that occurred hours after these symptoms manifested (García-Mendoza *et al.*, 2018). Histopathology in dead fish revealed abundant mucus and congestion in the gills, characterized by hyperplasia, fusion of gill filaments and lamellae, telangiectasia, oedemas, increased mucus cells and severe haemorrhage (García-Mendoza *et al.*, 2018).

10.3 Future Implications

10.3.1 Climate change

One obvious impact of climate change will be increasing lake and ocean temperatures which favour stratification and shallowing of mixed-layer depth in many locations, enhanced by greater precipitation and runoff in others (Hays *et al.*, 2005; IPCC, 2022). Increased stratification is expected to worsen the impacts of HABs in coastal seas as rapid depletion of surface nutrients may favour phytoplankton, including harmful algae, with unique nutrient acquisition strategies, including swimming towards nutrient-rich areas and mixotrophy (Smayda, 2010). Many dinoflagellate and raphidophyte HAB species have competitive physiological advantages for survival under stratified conditions, such as the ability to migrate vertically towards nutrients or sunlight required for their photosynthesis using flagella or gas vacuoles (Smayda and Reynolds, 2001). In particular, the raphidophyte HABs have caused massive economic damage to fish farms around the world, and it is believed that their competitive advantage and potentially also their impacts on fish farms will increase under warming conditions (Wells *et al.*, 2020).

Bloom-forming cyanobacteria are believed to benefit from anthropogenically driven changes in lakes (Hellweger *et al.*, 2022). However, cyanobacteria are a very diverse group, consisting of taxa that are morphologically, physiologically and ecologically radically different. Whereas *Microcystis* spp. may indeed benefit from climate-induced changes (Wilhelm *et al.*, 2020), the dominant cyanobacterium that forms blooms in alpine lakes, *Planktothrix rubescens*, actually does less well in warm summers because it is a cold-adapted species (Anneville *et al.*, 2015). 'Blooms like it hot' (Paerl and Huisman, 2008) has been interpreted to mean that cyanobacteria have higher optimal growth temperatures than their eukaryotic competitors, something that was, however, disproved in experiments (Lürling *et al.*, 2013). Still, cyanobacterial growth rate tends to increase with increasing temperature (Huisman *et al.*, 2018). Climate warming, furthermore, has already enhanced stability in the water column, and will select for buoyant cyanobacteria that can use intracellular gas vesicles to float (Walsby, 1994). In particular, colony-forming taxa like *Microcystis* translate their buoyancy into effective diurnal migration through the water column (Ibelings *et al.*, 1991). Additionally, in the absence of mixing, buoyant cyanobacteria float to the lake surface where they may produce thick scums, resulting in extremely high toxin concentrations (Chorus and Welker, 2021). Scum formation, however, should not be seen as adaptive, given the extreme conditions (e.g. high irradiance, strongly elevated temperatures, desiccation) leading to population losses in the scum. A second environmental driver is increasing atmospheric CO_2 concentrations; they are relevant as a result of a long-standing reputation that cyanobacteria do well at high pH and low CO_2, given their capacity to use HCO_3^- (Miller *et al.*, 1990). Again, studies show that the reality is more complex. Cyanobacteria possess a range of carbon-concentrating mechanisms (CCMs), some of which make species like *Microcystis* well adapted to increased CO_2 availability, with the additional consequence of changes in the genetic composition of HABs since different strains possess different CCMs (Sandrini *et al.*, 2016).

10.3.2 Socio-economic impacts

The largest known losses to the marine fish aquaculture industry worldwide have been due to the raphidophytes, including the genera *Chattonella*, *Pseudochattonella*, *Chrysochromulina*, *Heterosigma* and *Cochlodinium*. Together, Norway, Chile, Scotland and Canada generate more than 90% of the global farmed Atlantic salmon, and each of these countries has faced US$ millions in losses due to HABs, primarily because of raphidophyte blooms (Trainer, 2020). This section highlights the major economic losses to marine fish aquaculture worldwide due to HABs while discussing considerations for impacts of HABs to inland aquaculture.

A fish-kill event in Chile occurred in 2016 following a *Pseudochattonella* bloom, resulting in the mortality of more than 39 million salmon, approximately 15% of Chile's annual production, and an estimated US$800 million loss (Trainer *et al.*, 2020; Mardones *et al.*, 2021). In 2019, a bloom of *Chrysochromulina leadbeateri* in northern Norway killed ~8 million salmon with a direct value of ~US$93.5 million. A bloom of the raphidophyte, *Cochlodinium*, caused massive losses to aquaculture along the coast of south-east Korea in 1995, resulting in US$60 million loss, with losses occurring almost annually thereafter (Trainer, 2020). Similar impacts on fish aquaculture have been documented in other Asian countries such as Japan and China (Guo *et al.*, 2014; Itakura and Imai, 2014), where *Chattonella* causes fish mortality (e.g. 14.2 million fish worth US$90 million were lost in Harima-Nada Sea, Japan in 1972) (Imai and Yamaguchi, 2012). Mass mortalities of farmed fish include Atlantic salmon (*S. salar*), northern bluefin tuna (*Thunnus orientalis*), bluefin tuna (*Thunnus maccoyii*) and yellowtail (*Seriola quinqueradiata*) (Hallegraeff *et al.*, 1998; García-Mendoza *et al.*, 2018; Lum *et al.*, 2021). Recurring threats from the raphidophyte, *Heterosigma akashiwo*, caused extensive devastation (US$2 million to US$6 million per episode) to net-penned salmon in Washington State, USA (Trainer *et al.*, 2015). The total direct losses due to lost fish sales have been compounded by the loss of future sales, clean-up costs, losses of tax income and unemployment benefits. Total direct and indirect gross costs of up to US$300 million were estimated for the kill event in Norway (Marthinussen *et al.*, 2020). Other costs of HABs arise due to mitigation measures, including aeration, oxygenation, increased monitoring (including artificial intelligence methods), fish treatment, movement of fish to waters with reduced HAB concentrations, and clay dispersal. Consumer price of farmed fish can also decrease due to public perception of risk after a HAB event (Adams *et al.*, 2018).

The potential impacts of HABs to inland aquaculture are substantial (Brown *et al.*, 2020) but not yet fully described. Inland fisheries provide valuable sources of food for billions of people and jobs for millions of workers around the world (FAO, 2014). Approximately 80% of inland fisheries, including aquaculture operations, are found in the developing world (FAO, 2020a) and are important for reducing poverty in communities including minorities, rural impoverished people and women (Weeratunge *et al.*, 2004). However, as both the interest in land-based aquaculture and the occurrence of freshwater HABs are expected to increase around the world, due to anthropogenic activities and climate change, sources of water for inland aquaculture must be carefully considered. Cyanotoxins have been detected in freshwater fish worldwide but rarely at levels that would seriously impact the health of human consumers if guidelines for fish preparation, above all removal of viscera, are followed (Ibelings *et al.*, 2021). Because microbes or their toxins have been known to pass through some filtration systems used for aquaculture (King *et al.*, 2021), there is a likelihood that HAB toxins pose a threat to inland aquaculture operations through toxin incorporation into fish flesh (Hardy *et al.*, 2015) or via fish mortalities.

The global insurance industry has documented an increase over the last decade in HAB-related claims for aquaculture losses, constituting approximately 32% of all claims and totalling US$225 million in 2021 (G. Myer, AXA XL Global Commercial Insurance and Reinsurance, 2022, personal communication), in large part due to expansion of marine fish aquaculture into new locations, entry of new countries into the aquaculture industry and changing environmental conditions due to global warming that promote the occurrence of some HABs. Many insurers are not willing to support aquaculture operations unless plans for effective mitigation of HABs, government involvement in monitoring and warning of HABs, and development of new technologies for HAB prevention are demonstrated by the industry.

Mitigation measures should be planned by establishment of early warning systems that include autonomous underwater vehicle sampling (Free *et al.*, 2022) and other automated tools that can be used to identify HABs and their toxins, such as the environmental sample processor (ESP) and imaging flow cytobot (IFCB) (Jochens *et al.*, 2010; Anderson *et al.*, 2019). The cost–benefits of different mitigation or early warning measures should be assessed in each region as one approach may not be cost-effective in all locations and for all species (Wells *et al.*, 2020).

10.3.3 Management and monitoring

When monitoring of HABs, performing risk assessment and management, or deciding on the management and control of blooms, it is important to consider that HABs respond to different environmental drivers, with nutrient inputs a major contributing factor for freshwater HABs. The control of nutrient loading should be at the basis of freshwater HAB management (Ibelings *et al.*, 2016) with a focus at the level of the catchment. For instance, lakes with a watershed dominated by natural forest, rather than agriculture, rarely suffer from blooms of toxic cyanobacteria (Hamilton *et al.*, 2016). If nutrient control fails to result in preventing HAB formation, lake managers have several methods they can choose from to control the blooms, ranging from artificial lake mixing (Visser *et al.*, 2016) to biomanipulation (Triest *et al.*, 2016) or even use of hydrogen peroxide (Matthijs *et al.*, 2016). However, these measures can support but not replace control of nutrients, tackling both external and internal (resulting from the fish-farming operation itself) eutrophication. In order to prevent the impacts of nutrients on the development of freshwater HABs, there are initiatives exploring the use of artificial ponds using manufactured seawater and recirculating water systems (Timmons and Ebeling, 2013).

For monitoring and management of HABs affecting fish farms in natural lakes or ponds or associated with net-penned fish in nearshore marine systems, not all countries have policies in place to ensure regular monitoring using advanced automated technologies nor have access to the highly sophisticated analytical equipment that is needed, for instance, to do a full survey of the suite of toxins – and their multiple variants – present in HABs. Analysis of toxins in complex matrices of animal tissues is even more complicated (Anaraki *et al.*, 2020) and the scientific literature is full of studies that report unreliable data (see Testai *et al.*, 2016). For the majority of countries, it seems wise to base HABs monitoring on simpler, yet robust and reliable parameters based upon detection of cells, especially when analysis of toxins is not automated or when toxins produced by the HABs of concern have not been characterized. The second edition of the World Health Organization's (WHO) handbook *Toxic Cyanobacteria in Water* (Chorus and Welker, 2021) shows how this may be achieved for freshwater HABs. For instance, when alert levels for chlorophyll a or cyanobacterial biovolume are exceeded, it is likely that guideline values for cyanotoxins in drinking-water or food will be exceeded (see updated guideline values in Chorus and Welker, 2021), confirming that general chlorophyll- or cell-based monitoring can be effective. Even this being the case, the monitoring challenge is to provide sufficient temporal and spatial coverage. While these simpler, cell-based methods are broadly applicable in many parts of the world, progress is also being made using automated, high-frequency monitoring of lakes and oceans, even for phytoplankton (Marcé *et al.*, 2016; Wüest *et al.*, 2021), that allows data to be acquired and reported in real time. Likewise, algorithm development using the latest generation of satellites like Sentinel-2 is promising HAB detection with wide spatial coverage (Sòria-Perpinyà *et al.*, 2020). To safeguard consumers against toxin exposure, the FAO's Hazard Analysis and Critical Control Points (HCAPPs) for food or the WHO's Water Safety Plans (WSPs) for drinking-water provides optimal guidance. HACCPs and WSPs are tools to assess hazards and establish control systems that focus on prevention rather than relying mainly on end-product testing. They systematically assess hazards, risks and control measures at multiple stages (from catchment to consumer).

Similarly, comparable monitoring systems are being developed for live, *in situ* monitoring of HABs upstream of marine fish net-pen aquaculture systems. These use artificial intelligence tools such as the IFCB, an advanced flow cytometry system that is trained through classifier development to detect phytoplankton cells of concern (Jochens *et al.*, 2010; Anderson *et al.*, 2019). In Scotland, RS Aqua salmon farmers are developing early warning systems that use a suite of autonomous sensors developed by the manufacturer Innovasea and deployed upstream from the farm. These sensors detect environmental factors known to be indicative of HABs, such as chlorophyll and oxygen, or physical and chemical conditions potentially conducive to HAB development in the area, such as currents, turbidity and salinity. Data are sent via wireless networks to the cloud where they are analysed by algorithms to establish a HAB risk index which is then relayed to fish farmers in real time. This enables them to take action to minimize the impact of HABs. Another company, OTAQ, is developing an automated deep learning-based microscopic image analysis system called LPAS (live plankton and algae sensor), which will process live HAB images, in real time, and provide a digital output of the HAB species present in the farm and measure their abundance. Similarly, Grieg Seafoods in Canada, one of the world's largest salmon producers, is using a data collection system integrating collection of environmental data with phytoplankton data in conjunction with machine learning to provide early warning of HABs and prediction of risk. This information is used to determine when additional samples need to be collected or when mitigation strategies need to be triggered, such as platform diffusers used to upwell deep water and push surface water away (Brown, 2021). Fish aquaculture operations without proven monitoring and HABs mitigation strategies will have difficulty finding insurance, resulting in unsustainable losses (Trainer, 2020).

10.3.4 Pathogens

Pathogens infecting HAB species have been known for some time; however, only relatively recently have pathogens been acknowledged for their ecological roles in aquatic ecosystems and for the ecosystem services they provide to humans (Suttle, 2005; Frenken *et al.*, 2017; Paseka *et al.*, 2020). For instance, pathogens may be very effective top-down control agents of HABs (Wilhelm and Suttle, 1999; Frenken *et al.*, 2017) and as such may drive phytoplankton community succession, population subdivision, and even increases in within-species genetic diversity (Sønstebø and Rohrlack, 2011; Gsell *et al.*, 2013). Pathogens may also infect very selectively within host subpopulations and are able to adapt quickly to new or novel host strains (De Bruin *et al.*, 2008; Laundon *et al.*, 2021), thus showing potential for co-evolution.

Although HABs are often considered trophic dead ends, from a pathogen's perspective they are certainly not (Haraldsson *et al.*, 2018). Pathogens may kill a set of cells within a phytoplankton filament or colony, reducing filament length or colony size, and can thus make HABs more edible to herbivores (Frenken *et al.*, 2020; Park *et al.*, 2021). Possibly, subsequent shifts in HAB cell size distribution as a result of these pathogens (Šulčius *et al.*, 2017; Frenken *et al.*, 2020) may negate their physical impact on fish. Fungal parasites infecting inedible HABs nutritionally upgrade HABs into fungal zoospores that are edible to zooplankton (Frenken *et al.*, 2018; Gerphagnon *et al.*, 2019) and support high fitness levels (Kagami *et al.*, 2007). Similarly, viruses released after lysis of phytoplankton hosts may also serve as food to non-host organisms (Welsh *et al.*, 2020). Pathogens thus produce edible free-living stages that may fuel aquatic food webs and ultimately be beneficial to fish populations.

Many HABs can produce toxic metabolites, and the quantity and identity of toxin produced may be affected by pathogens (Rohrlack *et al.*, 2013; Šulčius *et al.*, 2018). When HABs are terminated by pathogens, toxins may be suddenly released into the surrounding water, as may have happened in Lake Erie, where viral infection of HABs contributed to the shutdown of Toledo's water supply when toxin concentrations were too high to be removed by conventional water treatment (Steffen *et al.*, 2017; McKindles *et al.*, 2020). Still, it is important to remember that although the maximum levels of MCs in Toledo's drinking-water (1.2 μg/l for a few days) exceeded the WHO's guideline value for lifetime exposure (1 μg/l), MCs remained well below the guideline value for short-term exposure (12 μg/l for up to 2 weeks). Consequently, these mass-release HAB events of toxins may also affect fish in aquaculture and should be considered when designing HAB control strategies to be used at finfish production sites. In fact, in the derivation of lifetime exposure guideline values, the WHO allocates 80% to drinking-water and 20% to food, although in certain situations this allocation may be incorrect (Ibelings and Chorus, 2007).

Parasites and bacteria may be used to biologically control HABs; however, testing and application have mostly been limited to laboratory studies (Sigee *et al.*, 1999; Pal *et al.*, 2020). Scaling up these approaches will pose a formidable challenge. Moreover, the potential for host–parasite antagonistic co-evolution presumably means that periods of host resistance will occur during which a selected parasite strain may not be effective until a further round of co-evolution restores that parasite (Brockhurst *et al.*, 2007).

10.3.5 The role of the microbiome

The holobiont concept posits that to understand the ecology and evolution of a particular species, its associated microbiota must also be studied (Zilber-Rosenberg and Rosenberg, 2008). While there has been debate about the holobiont concept (Doolittle and Booth, 2017), there is strong evidence across a range of taxa that the associated microbiome strongly influences host physiology (Bäckhed *et al.*, 2005), ecology (Rennison *et al.*, 2019) and evolution (Rudman *et al.*, 2019a; Lim and Bordenstein, 2020). There are two separate, but potentially critical, ways in which microbiomes may shape fish responses to HABs: (i) the effect of algae–microbe symbiosis on the formation and toxicity of blooms; and (ii) the effect of exposure to HABs on the fish microbiome.

The effect of algae–microbe symbiosis on the formation and toxicity of HABs

There is considerable evidence that the microbiome is a critical component in the growth and toxicity of *Microcystis*. Metagenomic sequencing of *Microcystis* colonies has been used to investigate the functional genomics of both cyanobacteria and associated microbes. These data suggest mutualistic interactions and functional complementation that influence a range of characteristics, including nitrogen cycling (Li *et al.*, 2018). Direct study of host-associated microbiomes of *Microcystis aeruginosa* has uncovered patterns of convergence in microbiome function associated with a trade-off between host fitness in low- and high-phosphorus conditions (Jackrel *et al.*, 2019), providing a potential mechanism by which *M. aeruginosa* maintains abundance across a range of phosphorus conditions. Microbiome composition can also influence the outcome of competitive dynamics between toxigenic strains of *Microcystis* and green algae, with microbiome presence an important component of *Microcystis* growth in establishing green algal cultures (Schmidt *et al.*, 2020). This suggests that host–microbiome associations may be a key part of the domination of cyanobacteria over green algae. Finally, a survey across 12 lakes demonstrated that the environmental microbiome in lakes where *M. aeruginosa* is present is remarkably consistent across populations spanning considerable geographical variation (Cook *et al.*, 2020). This pattern is similar to that observed in the genetic diversity of *M. aeruginosa*, suggesting that associations with components of the microbiome may not be geographically limited and could play out similarly across the globe.

The importance of microbiomes in the formation of blooms is best studied in *M. aeruginosa*, however work on other species of harmful algae has also demonstrated an important role of host–microbiome association in the formation of HABs. Both *Alexandrium fundyense* and *Dinophysis acuminata* are associated with unique prokaryotic and eukaryotic microbes that are likely to influence the formation and severity of HABs (Hattenrath-Lehmann and Gobler, 2017). The microbiome composition of *Alexandrium tamarense* and *Cochlodinium polykrikoides* shows substantial variation, even when held across generations in laboratory media, leading authors to suggest that the microbiome may be critical to formation of blooms (Shin *et al.*, 2018). Overall, it is clear that host–microbiome interactions are part of the ecology of many species of harmful algae. Moving forward, larger comparative data sets over space and time, in areas where HABs form and where they do not, are needed to determine whether there are particular host–microbiome associations that underlie the formation and severity of blooms in nature. This information could be critical to understanding why particular algae species become dominant, what causes temporal fluctuations in their abundance and the environmental mechanisms underpinning the production of toxins.

Effect of exposure to HABs on fish microbiomes

Several researchers have examined the effects of HAB exposure on fish microbiomes as a way of understanding the toxic effects of HABs on fish. One such study of the microbiome of Asian sea bass (*Lateolabrax maculatus*) exposed to *Microcystis* bloom and control conditions found no significant effects in microbiome composition or diversity but did uncover some family-level effects in bacterial abundance (Duan *et al.*, 2020). In contrast, zebrafish exposed for short durations to various concentrations of *M. aeruginosa* showed marked shifts in microbiome composition, including increases in the abundance of pathogenic members of the microbiome. The putative mechanism suggested by the authors was a host inflammatory response in the gut driven by *Microcystis* exposure that resulted in an increase in the proportion of pathogenic bacteria present (Qian *et al.*, 2019). Duperron *et al.* (2019) investigated the effects of pure MC and crude metabolite extracts from *M. aeruginosa* to determine whether exposure to either could drive shifts in microbiome composition in medaka (*Oryzias latipes*). They reported shifts associated only with exposures to extracts, suggesting that detrimental effects may come from secondary metabolites and not direct exposure to MCs. These represent some early examples investigating a previously unidentified way in which harmful algae can impact fish. To better integrate this work into a holistic understanding of the lethal and sublethal effects of HABs on fish health, future work that quantifies the effects of any observed shifts in microbiome composition on function and whole-organism growth or performance is key.

10.3.6 Leveraging -omics tools

Genomic tools have demonstrated utility in the study of toxicology, ecology and fisheries biology. There have been several well-cited reviews on the application of -omics data to understanding harmful algae (Anderson *et al.*, 2012; McLean, 2013). Hence, the focus in this section is to provide a brief description of some of the data types and the promise of some emerging applications.

Demonstrated utility for understanding HABs and their effects on fish health

Genomic tools have been useful in identifying the species and morphospecies responsible for HAB events (Pérez-Carrascal *et al.*, 2019) and in quantifying the prevalence of pathogens (McKindles *et al.*, 2021). Transcriptomics can provide links between ecology and physiology in harmful algae, including how gene regulation changes with environmental conditions (Harke and Gobler, 2015) and gene regulation related to toxin production (Zhang *et al.*, 2014). Proteomics approaches can provide new data types to document sublethal effects from exposure to harmful algae or toxins (Shahmohamadloo *et al.*, 2020b, 2022a). A meta-analysis of the proteomic effects from MC found 39 proteins that showed altered abundances in multiple toxicity studies including evidence that exposure may often induce oxidative stress (Welten *et al.*, 2020). Similar non-targeted proteomic approaches have been used to investigate the sublethal effects of several other harmful algae (Rodrigues *et al.*, 2016). Metabolomics has been demonstrated to have similar utility as a way of quantifying the effects of exposure on fish (Le Manach *et al.*, 2018). Determining the predictability of proteomic and metabolomic responses across disparate fish species and how strongly changes in the proteome relate to animal health and fitness are areas of future work.

Emerging technologies to enhance the study of HABs and their impact on fish health

There are numerous promising applications of -omics tools that are not yet widely used. CRISPR (clustered regularly interspaced short palindromic repeats)-based detection of algal toxins using SHERLOCK (sensitive high-efficiency reporter unLOCKing) can provide a remarkably simple and inexpensive way to test for algal toxins, and detection can greatly enhance the spatial and temporal resolution of HABs both in nature and aquaculture settings. Amplicon sequencing approaches that allow for the genotyping of thousands of individuals quickly and inexpensively (Meek and Larson, 2019) have great promise to improve understanding of the biological basis of HABs. Given the extensive intraspecific genetic variation in harmful algae (Guedes *et al.*, 2019; Dick *et al.*, 2021; Geffroy *et al.*, 2021), information on the spatial and temporal variation in strain presence can be critical to predict the location, severity and duration of blooms. Ionomics, the measurement of the total elemental composition of an organism (Salt *et al.*, 2008), has promising applications for understanding sublethal effects in fish (Jeyasingh *et al.*, 2017; Rudman *et al.*, 2019b) and the stoichiometric flows of nutrients that may sustain (Ipek and Jeyasingh, 2021) or reduce the severity of HABs.

10.3.7 Eco-evolutionary dynamics

The integration between ecology and evolutionary biology has grown considerably stronger in recent decades as empirical data demonstrating that evolution often occurs over contemporary timescales have mounted (Hairston *et al.*, 2005; Rudman *et al.*, 2022). In addition, it has become clear that rapid evolution can be a prominent driver of ecological patterns at the population, community and ecosystem levels (Yoshida *et al.*, 2003; Bassar *et al.*, 2010; Rudman and Schluter, 2016). Rapid evolution is likely pervasive in algal communities, where many species are both clonal and have short generation times. Hence, changes in the relative frequency of clones could profoundly shape the characteristics of algal populations over short timescales. Genetic variation within species (i.e. intraspecific variation) is likely also a factor in the response of fish to exposure to HABs, as a number of studies have demonstrated that genetic variation and adaptation can occur in response to toxic insults (Wu *et al.*, 1975; Di Giulio and Clark, 2015; Reid *et al.*, 2016; Oziolor *et al.*, 2019).

Rapid adaptation as a driver of HABs

Bloom-forming algae typically are rapidly reproducing species with large population sizes, hence the potential for fast evolution is considerable. This includes rapid evolution in toxin production and growth rate that can evolve within a single bloom. The extent of intraspecific genetic variation has been well studied in *Microcystis* where genomic sequencing and phylogenetic reconstruction have revealed some deeply divergent clades within this single species (Pérez-Carrascal *et al.*, 2019; Dick *et al.*, 2021). There are also notable examples of mutants that complicate some algal control methods, such as copper-tolerant *Microcystis* mutants (García-Villada *et al.*, 2004). Considerable intraspecific diversity is present in some other HAB-forming genera, including *Alexandrium*, which exhibits variation in genome size and copy number variation in genes related to toxin production (Geffroy *et al.*, 2021). A comparative analysis of interspecific and intraspecific variation across five strains each of *M. aeruginosa* and *Raphidiopsis raciborskii* showed a greater degree of intraspecific variation (Guedes *et al.*, 2019). Considerable differences in key functional traits between strains suggest a trade-off between suites of traits (Wilson *et al.*, 2006). This creates the potential for rapid clonal sorting that could have profound effects on the conditions over which HABs form, their severity and duration. Yet there has been little research documenting temporal clonal sorting and considerable work is needed to determine whether and how rapid evolution contributes to HABs (Dick *et al.*, 2021).

Intraspecific genetic variation and adaptation in response to HABs

When genetic variation is present and selection is strong, adaptation can occur quickly (Barrett and Schluter, 2008). As such, taxa that interact strongly with HABs may undergo rapid adaptation in response. *Daphnia* are noted for their ability to consume *Microcystis* and clones of *Daphnia* vary considerably in their ability to do so (Hairston *et al.*, 1999, 2001). The extent of adaptation in *Daphnia* to cyanobacterial HABs has been quantified (Sarnelle and Wilson, 2005; Chislock *et al.*, 2019) and *Daphnia*'s ability to remediate the severity of HABs due to rapid evolution has also been measured (Sarnelle, 2007; Chislock *et al.*, 2013). The role of intraspecific variation and rapid adaptation in response to other HAB-forming species is not well known, nor is the importance of intraspecific genetic variation in susceptibility to HAB toxins in fish responses. Future work aimed at filling in these gaps is crucial to understanding the full effects of HABs on co-occurring species and in understanding factors that control and limit HABs.

References

Adams, S.M., Shugart, L.R. and Hinton, D.E. (2018) Application of bioindicators in assessing the health of fish populations experiencing contaminant stress. In: McCarthy, J.F. and Shugart, L.R. (eds) *Biomarkers of Environmental Contamination*. CRC Press, Boca Raton, Florida, pp. 333–353.

Adamski, M., Żmudzki, P., Chrapusta, E., Bober, B., Kaminski, A., *et al.* (2016) Effect of pH and temperature on the stability of cylindrospermopsin. Characterization of decomposition products. *Algal Research* 15, 129–134. doi: 10.1016/j.algal.2016.02.020.

Anaraki, M.T., Shahmohamadloo, R.S., Sibley, P.K., MacPherson, K., Bhavsar, S.P., *et al.* (2020) Optimization of an MMPB Lemieux Oxidation method for the quantitative analysis of microcystins in fish tissue by LC-QTOF MS. *Science of the Total Environment* 737, 140209. doi: 10.1016/j.scitotenv.2020.140209.
Anderson, C.R., Berdalet, E., Kudela, R.M., Cusack, C.K., Silke, J., *et al.* (2019) Scaling up from regional case studies to a global harmful algal bloom observing system. *Frontiers in Marine Science* 6, 250. doi: 10.3389/fmars.2019.00250.
Anderson, D.M., Cembella, A.D. and Hallegraeff, G.M. (2012) Progress in understanding harmful algal blooms: paradigm shifts and new technologies for research, monitoring, and management. *Annual Review of Marine Science* 4, 143–176. doi: 10.1146/annurev-marine-120308-081121.
Anneville, O., Domaizon, I., Kerimoglu, O., Rimet, F. and Jacquet, S. (2015) Blue-green algae in a 'greenhouse century'? New insights from field data on climate change impacts on cyanobacteria abundance. *Ecosystems* 18(3), 441–458. doi: 10.1007/s10021-014-9837-6.
Armah, A., Hiskia, A., Kaloudis, T., Chernoff, N., Hill, D., *et al.* (2013) A review on cylindrospermopsin: the global occurrence, detection, toxicity and degradation of a potent cyanotoxin. *Environmental Science: Processes & Impacts* 15(11), 1979–2003. doi: 10.1039/C3EM00353A.
Bäckhed, F., Ley, R.E., Sonnenburg, J.L., Peterson, D.A. and Gordon, J.I. (2005) Host–bacterial mutualism in the human intestine. *Science* 307(5717), 1915–1920. doi: 10.1126/science.1104816.
Baker, J.W., Grover, J.P., Brooks, B.W., Ureña-Boeck, F., Roelke, D.L., *et al.* (2007) Growth and toxicity of *Prymnesium parvum* (Haptophyta) as a function of salinity, light, and temperature. *Journal of Phycology* 43(2), 219–227. doi: 10.1111/j.1529-8817.2007.00323.x.
Banerjee, S., Maity, S., Guchhait, R., Chatterjee, A., Biswas, C., *et al.* (2021) Toxic effects of cyanotoxins in teleost fish: a comprehensive review. *Aquatic Toxicology* 240, 105971. doi: 10.1016/j.aquatox.2021.105971.
Barrett, R.D. and Schluter, D. (2008) Adaptation from standing genetic variation. *Trends in Ecology & Evolution* 23(1), 38–44. doi: 10.1016/j.tree.2007.09.008.
Barrientos, R.G., Hernández-Mora, G., Alegre, F., Field, T., Flewelling, L., *et al.* (2019) Saxitoxin poisoning in green turtles (*Chelonia mydas*) linked to scavenging on mass mortality of Caribbean sharpnose puffer fish (*Canthigaster rostrata* – Tetraodontidae). *Frontiers in Veterinary Science* 6, 466. doi: 10.3389/fvets.2019.00466.
Bassar, R.D., Marshall, M.C., López-Sepulcre, A., Zandonà, E., Auer, S.K., *et al.* (2010) Local adaptation in Trinidadian guppies alters ecosystem processes. *Proceedings of the National Academy of Sciences USA* 107(8), 3616–3621. doi: 10.1073/pnas.0908023107.
Bates, S.S., Bird, C.J., Freitas, A.D., Foxall, R., Gilgan, M., *et al.* (1989) Pennate diatom *Nitzschia pungens* as the primary source of domoic acid, a toxin in shellfish from eastern Prince Edward Island, Canada. *Canadian Journal of Fisheries and Aquatic Sciences* 46(7), 1203–1215. doi: 10.1139/f89-156.
Batoréu, M.C.C., Dias, E., Pereira, P. and Franca, S. (2005) Risk of human exposure to paralytic toxins of algal origin. *Environmental Toxicology and Pharmacology* 19(3), 401–406. doi: 10.1016/j.etap.2004.12.002.
Bejarano, A.C., VanDola, F.M., Gulland, F.M., Rowles, T.K. and Schwacke, L.H. (2008) Production and toxicity of the marine biotoxin domoic acid and its effects on wildlife: a review. *Human and Ecological Risk Assessment: An International Journal* 14(3), 544–567. doi: 10.1080/10807030802074220.
Bergsson, H., Andersen, N.R., Svendsen, M.B.S., Hansen, P.J. and Steffensen, J.F. (2019) Respiratory physiology of European plaice (*Pleuronectes platessa*) exposed to *Prymnesium parvum*. *Fishes* 4(2), 32. doi: 10.3390/fishes4020032.
Berry, J.P., Gibbs, P.D., Schmale, M.C. and Saker, M.L. (2009) Toxicity of cylindrospermopsin, and other apparent metabolites from *Cylindrospermopsis raciborskii* and *Aphanizomenon ovalisporum*, to the zebrafish (*Danio rerio*) embryo. *Toxicon* 53(2), 289–299. doi: 10.1016/j.toxicon.2008.11.016.
Berry, J.P., Jaja-Chimedza, A., Dávalos-Lind, L. and Lind, O. (2012) Apparent bioaccumulation of cylindrospermopsin and paralytic shellfish toxins by finfish in Lake Catemaco (Veracruz, Mexico). *Food Additives & Contaminants: Part A* 29(2), 314–321. doi: 10.1080/19440049.2011.597785.
Bertani, B. and Ruiz, N. (2018) Function and biogenesis of lipopolysaccharides. *EcoSal Plus* 8(1). doi: 10.1128/ecosalplus.ESP-0001-2018.
Best, J.H., Pflugmacher, S., Wiegand, C., Eddy, F.B., Metcalf, J.S. and Codd, G.A. (2002) Effects of enteric bacterial and cyanobacterial lipopolysaccharides, and of microcystin-LR, on glutathione *S*-transferase activities in zebra fish (*Danio rerio*). *Aquatic Toxicology* 60(3–4), 223–231. doi: 10.1016/S0166-445X(02)00010-3.
Bi, D., Wang, Y., Gao, Y., Li, X., Chu, Q., *et al.* (2018) Recognition of lipopolysaccharide and activation of NF-κB by cytosolic sensor NOD1 in teleost fish. *Frontiers in Immunology* 9, 1413. doi: 10.3389/fimmu.2018.01413.

Bidard, J.N., Vijverberg, H.P., Frelin, C., Chungue, E., Legrand, A.M., *et al.* (1984) Ciguatoxin is a novel type of Na^+ channel toxin. *Journal of Biological Chemistry* 259(13), 8353–8357. doi: 10.1016/S0021-9258(17)39735-1.

Binzer, S.B., Svenssen, D.K., Daugbjerg, N., Alves-de-Souza, C., Pinto, E., *et al.* (2019) A-, B-and C-type prymnesins are clade specific compounds and chemotaxonomic markers in *Prymnesium parvum*. *Harmful Algae* 81, 10–17. doi: 10.1016/j.hal.2018.11.010.

Blossom, H.E., Rasmussen, S.A., Andersen, N.G., Larsen, T.O., Nielsen, K.F. and Hansen, P.J. (2014) *Prymnesium parvum* revisited: relationship between allelopathy, ichthyotoxicity, and chemical profiles in 5 strains. *Aquatic Toxicology* 157, 159–166. doi: 10.1016/j.aquatox.2014.10.006.

Bouaïcha, N., Miles, C.O., Beach, D.G., Labidi, Z., Djabri, A., *et al.* (2019) Structural diversity, characterization and toxicology of microcystins. *Toxins* 11(12), 714. doi: 10.3390/toxins11120714.

Brand, L.E., Campbell, L. and Bresnan, E. (2012) *Karenia*: the biology and ecology of a toxic genus. *Harmful Algae* 14, 156–178. doi: 10.1016/j.hal.2011.10.020.

Brockhurst, M.A., Morgan, A.D., Fenton, A. and Buckling, A. (2007) Experimental coevolution with bacteria and phage: the *Pseudomonas fluorescens*–Φ2 model system. *Infection, Genetics and Evolution* 7(4), 547–552. doi: 10.1016/j.meegid.2007.01.005.

Brooks, B.W., Grover, J.P. and Roelke, D.L. (2011) *Prymnesium parvum*: an emerging threat to inland waters. *Environmental Toxicology and Chemistry* 30(9), 1955–1964. doi: 10.1002/etc.613.

Brown, A.R., Lilley, M., Shutler, J., Lowe, C., Artioli, Y., *et al.* (2020) Assessing risks and mitigating impacts of harmful algal blooms on mariculture and marine fisheries. *Reviews in Aquaculture* 12(3), 1663–1688. doi: 10.1111/raq.12403.

Brown, J. (2021) Taking seafood farming to the next level. *Aquaculture North America*, 2 February 2021. Available at: https://www.aquaculturenorthamerica.com/taking-seafood-farming-to-the-next-level/ (accessed 27 March 2022).

Buratti, F.M., Manganelli, M., Vichi, S., Stefanelli, M., Scardala, S., *et al.* (2017) Cyanotoxins: producing organisms, occurrence, toxicity, mechanism of action and human health toxicological risk evaluation. *Archives of Toxicology* 91(3), 1049–1130. doi: 10.1007/s00204-016-1913-6.

Carmichael, W.W. (2001) Health effects of toxin-producing cyanobacteria: 'the cyanoHABs'. *Human and Ecological Risk Assessment: An International Journal* 7(5), 1393–1407. doi: 10.1080/20018091095087.

Carmichael, W.W., Biggs, D.F. and Gorham, P.R. (1975) Toxicology and pharmacological action of *Anabaena flos-aquae* toxin. *Science* 187(4176), 542–544. doi: 10.1126/science.803708.

Caroff, M. and Karibian, D. (2003) Structure of bacterial lipopolysaccharides. *Carbohydrate Research* 338(23), 2431–2447. doi: 10.1016/j.carres.2003.07.010.

Castro, D., Vera, D., Lagos, N., García, C. and Vásquez, M. (2004) The effect of temperature on growth and production of paralytic shellfish poisoning toxins by the cyanobacterium *Cylindrospermopsis raciborskii* C10. *Toxicon* 44(5), 483–489. doi: 10.1016/j.toxicon.2004.06.005.

Chang, F.H., Anderson, C. and Boustead, N.C. (1990) First record of a *Heterosigma* (Raphidophyceae) bloom with associated mortality of cage-reared salmon in Big Glory Bay, New Zealand. *New Zealand Journal of Marine and Freshwater Research* 24(4), 461–469. doi: 10.1080/00288330.1990.9516437.

Chapra, S.C., Boehlert, B., Fant, C., Bierman, V.J. Jr, Henderson, J., *et al.* (2017) Climate change impacts on harmful algal blooms in US freshwaters: a screening-level assessment. *Environmental Science & Technology* 51(16), 8933–8943. doi: 10.1021/acs.est.7b01498.

Chen, Y., Shen, D. and Fang, D. (2013) Nodularins in poisoning. *Clinica Chimica Acta* 425, 18–29. doi: 10.1016/j.cca.2013.07.005.

Chislock, M.F., Sarnelle, O., Jernigan, L.M. and Wilson, A.E. (2013) Do high concentrations of microcystin prevent *Daphnia* control of phytoplankton? *Water Research* 47(6), 1961–1970. doi: 10.1016/j.watres.2012.12.038.

Chislock, M.F., Kaul, R.B., Durham, K.A., Sarnelle, O. and Wilson, A.E. (2019) Eutrophication mediates rapid clonal evolution in *Daphnia pulicaria*. *Freshwater Biology* 64(7), 1275–1283. doi: 10.1111/fwb.13303.

Chiswell, R.K., Shaw, G.R., Eaglesham, G., Smith, M.J., Norris, R.L., *et al.* (1999) Stability of cylindrospermopsin, the toxin from the cyanobacterium, *Cylindrospermopsis raciborskii*: effect of pH, temperature, and sunlight on decomposition. *Environmental Toxicology* 14(1), 155–161. doi: 10.1002/(SICI)1522-7278(199902)14:1%3C155::AID-TOX20%3E3.0.CO;2-Z.

Chiu, A.S., Gehringer, M.M., Welch, J.H. and Neilan, B.A. (2011) Does α-amino-β-methylaminopropionic acid (BMAA) play a role in neurodegeneration? *International Journal of Environmental Research and Public Health* 8(9), 3728–3746. doi: 10.3390/ijerph8093728.

Chorus, I. and Welker, M. (2021) *Toxic Cyanobacteria in Water: A Guide to their Public Health Consequences, Monitoring and Management*, 2nd edn. CRC Press, Boca Raton, Florida, on behalf of World Health Organization, Geneva, Switzerland. doi: 10.1201/9781003081449.

Codd, G.A., Bell, S.G., Kaya, K., Ward, C.J., Beattie, K.A. and Metcalf, J.S. (1999) Cyanobacterial toxins, exposure routes and human health. *European Journal of Phycology* 34(4), 405–415. doi: 10.1080/09670269910001736462.

Colman, J.R., Dechraoui, M.Y.B., Dickey, R.W. and Ramsdell, J.S. (2004) Characterization of the developmental toxicity of Caribbean ciguatoxins in finfish embryos. *Toxicon* 44(1), 59–66. doi: 10.1016/j.toxicon.2004.04.007.

Cook, K.V., Li, C., Cai, H., Krumholz, L.R., Hambright, K.D., *et al.* (2020) The global *Microcystis* interactome. *Limnology and Oceanography* 65, S194–S207. doi: 10.1002/lno.11361.

da Silva, C.A., de Morais, E.C.P., Costa, M.D.M., Ribas, J.L.C., Guiloski, I.C., *et al.* (2014) Saxitoxins induce cytotoxicity, genotoxicity and oxidative stress in teleost neurons *in vitro*. *Toxicon* 86, 8–15. doi: 10.1016/j.toxicon.2014.04.016.

de Assis, H.C.S., da Silva, C.A., Oba, E.T., Pamplona, J.H., Mela, M., *et al.* (2013) Hematologic and hepatic responses of the freshwater fish *Hoplias malabaricus* after saxitoxin exposure. *Toxicon* 66, 25–30. doi: 10.1016/j.toxicon.2013.01.012.

De Bruin, A., Ibelings, B.W., Kagami, M., Mooij, W.M. and Van Donk, E. (2008) Adaptation of the fungal parasite *Zygorhizidium planktonicum* during 200 generations of growth on homogeneous and heterogeneous populations of its host, the diatom *Asterionella formosa*. *Journal of Eukaryotic Microbiology* 55(2), 69–74. doi: 10.1111/j.1550-7408.2008.00306.x.

De Fouw, J.C., Van Egmond, H.P. and Speijers, G.J.A. (2001) *Ciguatera Fish Poisoning: A Review*. RIVM Rapport 388802021. Rijksinstituut voor Volksgezondheid en Milieu RIVM, Bilthoven, The Netherlands.

de Morais Calado, S.L., Santos, G.S., Wojciechowski, J., de Magalhães, V.F. and de Assis, H.C.S. (2019) The accumulation dynamics, elimination and risk assessment of paralytic shellfish toxins in fish from a water supply reservoir. *Science of the Total Environment* 651, 3222–3229. doi: 10.1016/j.scitotenv.2018.10.046.

Díaz, P.A., Álvarez, G., Varela, D., Pérez-Santos, I., Díaz, M., *et al.* (2019) Impacts of harmful algal blooms on the aquaculture industry: Chile as a case study. *Perspectives in Phycology* 6(1–2), 39–50. doi: 10.1127/pip/2019/0081.

Dick, G.J., Duhaime, M.B., Evans, J.T., Errera, R.M., Godwin, C.M., *et al.* (2021) The genetic and ecophysiological diversity of *Microcystis*. *Environmental Microbiology* 23(12), 7278–7313. doi: 10.1111/1462-2920.15615.

Di Giulio, R.T. and Clark, B.W. (2015) The Elizabeth River story: a case study in evolutionary toxicology. *Journal of Toxicology and Environmental Health, Part B: Critical Reviews* 18(6), 259–298. doi: 10.1080/15320383.2015.1074841.

Doolittle, W.F. and Booth, A. (2017) It's the song, not the singer: an exploration of holobiosis and evolutionary theory. *Biology & Philosophy* 32(1), 5–24. doi: 10.1007/s10539-016-9542-2.

Đorđević, N.B., Simić, S.B. and Ćirić, A.R. (2015) First identification of the cylindrospermopsin (CYN)-producing cyanobacterium *Cylindrospermopsis raciborskii* (Woloszyńska) Seenayya & Subba Raju in Serbia. *Fresenius Environmental Bulletin* 24(11a), 3736–3742.

Duan, Y., Xiong, D., Li, Y., Dong, H., Wang, W. and Zhang, J. (2020) Changes in the gastrointestinal microbial community of *Lateolabrax maculatus* in a naturally occurring *Microcystis aeruginosa* bloom environment. *Aquaculture* 528, 735444. doi: 10.1016/j.aquaculture.2020.735444.

Duncan, M.W., Villacreses, N.E., Pearson, P.G., Wyatt, L., Rapoport, S.I., *et al.* (1991) 2-Amino-3-(methylamino)-propanoic acid (BMAA) pharmacokinetics and blood–brain barrier permeability in the rat. *Journal of Pharmacology and Experimental Therapeutics* 258(1), 27–35.

Duperron, S., Halary, S., Habiballah, M., Gallet, A., Huet, H., *et al.* (2019) Response of fish gut microbiota to toxin-containing cyanobacterial extracts: a microcosm study on the medaka (*Oryzias latipes*). *Environmental Science & Technology Letters* 6(6), 341–347. doi: 10.1021/acs.estlett.9b00297.

Durai, P., Batool, M. and Choi, S. (2015) Structure and effects of cyanobacterial lipopolysaccharides. *Marine Drugs* 13(7), 4217–4230. doi: 10.3390/md13074217.

Dyble, J., Gossiaux, D., Landrum, P., Kashian, D.R. and Pothoven, S. (2011). A kinetic study of accumulation and elimination of microcystin-LR in yellow perch (*Perca flavescens*) tissue and implications for human fish consumption. *Marine Drugs* 9(12), 2553–2571. doi: 10.3390/md9122553.

Eckford-Soper, L. and Daugbjerg, N. (2016) The ichthyotoxic genus *Pseudochattonella* (Dictyochophyceae): distribution, toxicity, enumeration, ecological impact, succession and life history – a review. *Harmful Algae* 58, 51–58. doi: 10.1016/j.hal.2016.08.002.

Edmunds, J.S.G., McCarthy, R.A. and Ramsdell, J.S. (1999) Ciguatoxin reduces larval survivability in finfish. *Toxicon* 37(12), 1827–1832. doi: 10.1016/S0041-0101(99)00119-1.

Edvardsen, B. and Imai, I. (2006) The ecology of harmful flagellates within Prymnesiophyceae and Raphidophyceae. In: Granéli, E. and Turner, J.T. (eds) *Ecology of Harmful Algae*. Springer, Berlin/Heidelberg, Germany, pp. 67–79. doi: 10.1007/978-3-540-32210-8_6.

Edwards, C., Graham, D., Fowler, N. and Lawton, L.A. (2008) Biodegradation of microcystins and nodularin in freshwaters. *Chemosphere* 73(8), 1315–1321. doi: 10.1016/j.chemosphere.2008.07.015.

Endo, M., Onoue, Y. and Kuroki, A. (1992) Neurotoxin-induced cardiac disorder and its role in the death of fish exposed to *Chattonella marina*. *Marine Biology* 112(3), 371–376. doi: 10.1007/BF00356281.

Eriksson, J.E., Meriluoto, J.A.O., Kujari, H.P., Österlund, K., Fagerlund, K. and Hällbom, L. (1988) Preliminary characterization of a toxin isolated from the cyanobacterium *Nodularia spumigena*. *Toxicon* 26(2), 161–166. doi: 10.1016/0041-0101(88)90168-7.

Ernst, B., Hitzfeld, B. and Dietrich, D. (2001) Presence of *Planktothrix* sp. and cyanobacterial toxins in Lake Ammersee, Germany and their impact on whitefish (*Coregonus lavaretus* L.). *Environmental Toxicology* 16(6), 483–488. doi: 10.1002/tox.10006.

Faassen, E.J. (2014) Presence of the neurotoxin BMAA in aquatic ecosystems: what do we really know? *Toxins* 6(3), 1109–1138. doi: 10.3390/toxins6031109.

Falconer, I.R. and Humpage, A.R. (2006) Cyanobacterial (blue-green algal) toxins in water supplies: cylindrospermopsins. *Environmental Toxicology* 21(4), 299–304. doi: 10.1002/tox.20194.

FAO (Food and Agriculture Organization of the United Nations) (2004) *Marine Biotoxins*. FAO Food and Nutrition Paper No. 80. FAO, Rome.

FAO (Food and Agriculture Organization of the United Nations) (2014) *The State of World Fisheries and Aquaculture 2014: Opportunities and Challenges*. FAO, Rome.

FAO (Food and Agriculture Organization of the United Nations) (2018) *The State of World Fisheries and Aquaculture 2018: Meeting the Sustainable Development Goals*. FAO, Rome. Available at: http://www.fao.org/3/i9540en/I9540EN.pdf (accessed 20 October 2022).

FAO (Food and Agriculture Organization of the United Nations) (2020a) *The State of World Fisheries and Aquaculture 2020: Sustainability in Action*. FAO, Rome. doi: 10.4060/ca9229en.

FAO (Food and Agriculture Organization of the United Nations) (2020b) *Climate Change: Unpacking the Burden on Food Safety*. Food Safety and Quality Series No. 8. FAO, Rome. doi: 10.4060/ca8185en.

FAO, IFAD, UNICEF, WFP and WHO (Food and Agriculture Organization of the United Nations, International Fund for Agricultural Development, United Nations Children's Fund, World Food Programme and World Health Organization) (2018) *The State of Food Security and Nutrition in the World 2018: Building Climate Resilience for Food Security and Nutrition*. FAO, Rome, Available at: http://www.fao.org/3/i9553en/i9553en.pdf (accessed 20 October 2022).

FAO and WHO (Food and Agriculture Organization of the United Nations and World Health Organization) (2020) *Report of the Expert Meeting on Ciguatera Poisoning, Rome, 19–23 November 2018*. Food Safety and Quality Series No. 9. FAO, Rome. doi: 10.4060/ca8817en.

FDA (Food and Drug Administration) (2011) *Fish and Fishery Products Hazards and Controls Guidance*. US Department of Health and Human Services, Food and Drug Administration, Center for Food Safety and Applied Nutrition, College Park, Maryland.

Fire, S.E., Pruden, J., Couture, D., Wang, Z., Bottein, M.Y.D., *et al*. (2012) Saxitoxin exposure in an endangered fish: association of a shortnose sturgeon mortality event with a harmful algal bloom. *Marine Ecology Progress Series* 460, 145–153. doi: 10.3354/meps09768.

Free, C.M., Moore, S.K. and Trainer, V.L. (2022) The value of monitoring in efficiently and adaptively managing biotoxin contamination in marine fisheries. *Harmful Algae* 114, 102226. doi: 10.1016/j.hal.2022.102226.

Frenken, T., Alacid, E., Berger, S.A., Bourne, E.C., Gerphagnon, M., *et al*. (2017) Integrating chytrid fungal parasites into plankton ecology: research gaps and needs. *Environmental Microbiology* 19(10), 3802–3822. doi: 10.1111/1462-2920.13827.

Frenken, T., Wierenga, J., van Donk, E., Declerck, S.A., de Senerpont Domis, L.N., *et al*. (2018) Fungal parasites of a toxic inedible cyanobacterium provide food to zooplankton. *Limnology and Oceanography* 63(6), 2384–2393. doi: 10.1002/lno.10945.

Frenken, T., Wolinska, J., Tao, Y., Rohrlack, T. and Agha, R. (2020) Infection of filamentous phytoplankton by fungal parasites enhances herbivory in pelagic food webs. *Limnology and Oceanography* 65(11), 2618–2626. doi: 10.1002/lno.11474.

Froscio, S.M., Humpage, A.R., Burcham, P.C. and Falconer, I.R. (2003) Cylindrospermopsin-induced protein synthesis inhibition and its dissociation from acute toxicity in mouse hepatocytes. *Environmental Toxicology* 18(4), 243–251. doi: 10.1002/tox.10121.

Galvão, J.A., Oetterer, M., do Carmo Bittencourt-Oliveira, M., Gouvêa-Barros, S., Hiller, S., *et al.* (2009) Saxitoxins accumulation by freshwater tilapia (*Oreochromis niloticus*) for human consumption. *Toxicon* 54(6), 891–894. doi: 10.1016/j.toxicon.2009.06.021.

García-Mendoza, E., Cáceres-Martínez, J., Rivas, D., Fimbres-Martinez, M., Sánchez-Bravo, Y., *et al.* (2018) Mass mortality of cultivated northern bluefin tuna *Thunnus thynnus orientalis* associated with *Chattonella* species in Baja California, Mexico. *Frontiers in Marine Science* 5, 454. doi: 10.3389/fmars.2018.00454.

García-Villada, L., Rico, M., Altamirano, M., Sánchez-Martín, L., López-Rodas, V. and Costas, E. (2004) Occurrence of copper resistant mutants in the toxic cyanobacteria *Microcystis aeruginosa*: characterisation and future implications in the use of copper sulphate as algaecide. *Water Research* 38(8), 2207–2213. doi: 10.1016/j.watres.2004.01.036.

Geffroy, S., Lechat, M.M., Le Gac, M., Rovillon, G.A., Marie, D., *et al.* (2021) From the *sxtA4* gene to saxitoxin production: what controls the variability among *Alexandrium minutum* and *Alexandrium pacificum* strains? *Frontiers in Microbiology* 12, 613199. doi: 10.3389/fmicb.2021.613199.

Gene, S.M., Shahmohamadloo, R.S., Ortiz, X. and Prosser, R.S. (2019) Effect of *Microcystis aeruginosa*-associated microcystin-LR on the survival of 2 life stages of freshwater mussel (*Lampsilis siliquoidea*). *Environmental Toxicology and Chemistry* 38(10), 2137–2144. doi: 10.1002/etc.4527.

Gephart, J.A., Golden, C.D., Asche, F., Belton, B., Brugere, C., *et al.* (2020) Scenarios for global aquaculture and its role in human nutrition. *Reviews in Fisheries Science & Aquaculture* 29(1), 122–138. doi: 10.1080/23308249.2020.1782342.

Gerphagnon, M., Agha, R., Martin-Creuzburg, D., Bec, A., Perriere, F., *et al.* (2019) Comparison of sterol and fatty acid profiles of chytrids and their hosts reveals trophic upgrading of nutritionally inadequate phytoplankton by fungal parasites. *Environmental Microbiology* 21(3), 949–958. doi: 10.1111/1462-2920.14489.

Gingold, D.B., Strickland, M.J. and Hess, J.J. (2014) Ciguatera fish poisoning and climate change: analysis of National Poison Center data in the United States, 2001–2011. *Environmental Health Perspectives* 122(6), 580–586. doi: 10.1289/ehp.1307196.

Glibert, P.M., Landsberg, J.H., Evans, J.J., Al-Sarawi, M.A., Faraj, M., *et al.* (2002) A fish kill of massive proportion in Kuwait Bay, Arabian Gulf, 2001: the roles of bacterial disease, harmful algae, and eutrophication. *Harmful Algae* 1(2), 215–231. doi: 10.1016/S1568-9883(02)00013-6.

Gobler, C.J. (2020) Climate change and harmful algal blooms: insights and perspective. *Harmful Algae* 91, 101731. doi: 10.1016/j.hal.2019.101731.

Govaert, L., De Meester, L., Spaak, P. and Hairston, N.G. Jr (2021) Eco-evolutionary dynamics in freshwater systems. In: Mehner, T. and Tockner, K. (eds) *Encyclopedia of Inland Waters*, Vol. 1, 2nd edn. Elsevier, Amsterdam, pp. 302–316. doi: 10.1016/B978-0-12-819166-8.00028-1.

Gsell, A.S., de Senerpont Domis, L.N., Verhoeven, K.J., Van Donk, E. and Ibelings, B.W. (2013) Chytrid epidemics may increase genetic diversity of a diatom spring-bloom. *The ISME Journal* 7(10), 2057–2059. doi: 10.1038/ismej.2013.73.

Guedes, I.A., Pacheco, A.B.F., Vilar, M.C., Mello, M.M., Marinho, M.M., *et al.* (2019) Intraspecific variability in response to phosphorus depleted conditions in the cyanobacteria *Microcystis aeruginosa* and *Raphidiopsis raciborskii*. *Harmful Algae* 86, 96–105. doi: 10.1016/j.hal.2019.03.006.

Gulledge, B.M., Aggen, J.B., Huang, H., Nairn, A.C. and Chamberlin, A.R. (2002) The microcystins and nodularins: cyclic polypeptide inhibitors of PP1 and PP2A. *Current Medicinal Chemistry* 9(22), 1991–2003. doi: 10.2174/0929867023368845.

Gunter, G., Williams, R.H., Davis, C.C. and Smith, F.W. (1948) Catastrophic mass mortality of marine animals and coincident phytoplankton bloom on the west coast of Florida, November 1946 to August 1947. *Ecological Monographs* 18(3), 310–324. doi: 10.2307/1948575.

Guo, H., Liu, X., Ding, D., Guan, C. and Yi, X. (2014) The economic cost of red tides in China from 2008–2012. In: Trainer, V.L. and Yoshida, T. (eds) *Proceedings of the Workshop on Economic Impacts of Harmful Algal Blooms on Fisheries and Aquaculture*. PICES Scientific Report No. 47. North Pacific Marine Science Organization (PICES), Sidney, British Columbia, Canada, pp. 27–34.

Guo, M., Harrison, P.J. and Taylor, F.J.R. (1996) Fish kills related to *Prymnesium parvum* N. Carter (Haptophyta) in the People's Republic of China. *Journal of Applied Phycology* 8(2), 111–117. doi: 10.1007/BF02186313.

Gutiérrez-Praena, D., Pichardo, S., Jos, Á. and Cameán, A.M. (2011) Toxicity and glutathione implication in the effects observed by exposure of the liver fish cell line PLHC-1 to pure cylindrospermopsin. *Ecotoxicology and Environmental Safety* 74(6), 1567–1572. doi: 10.1016/j.ecoenv.2011.04.030.

Gutierrez-Praena, D., Jos, A., Pichardo, S., Moyano, R., Blanco, A., *et al.* (2012) Time-dependent histopathological changes induced in Tilapia (*Oreochromis niloticus*) after acute exposure to pure cylindrospermopsin by oral and intraperitoneal route. *Ecotoxicology and Environmental Safety* 76(2), 102–113. doi: 10.1016/j.ecoenv.2011.10.008.

Guzmán-Guillén, R., Prieto, A.I., Vasconcelos, V.M. and Cameán, A.M. (2013) Cyanobacterium producing cylindrospermopsin cause oxidative stress at environmentally relevant concentrations in sub-chronically exposed tilapia (*Oreochromis niloticus*). *Chemosphere* 90(3), 1184–1194. doi: 10.1016/j.chemosphere.2012.09.027.

Guzmán-Guillén, R., Gutiérrez-Praena, D., De los Ángeles Risalde, M., Moyano, R., Prieto, A.I., *et al.* (2014) Immunohistochemical approach to study cylindrospermopsin distribution in tilapia (*Oreochromis niloticus*) under different exposure conditions. *Toxins* 6(1), 283–303. doi: 10.3390/toxins6010283.

Guzmán-Pérez, S.E. and Park, D.L. (2000) Ciguatera toxins: chemistry and detection. In: Botana, L.M. (ed.) *Seafood and Freshwater Toxins: Pharmacology, Physiology and Detection*. Marcel Dekker, New York, pp. 401–418.

Hairston, N.G., Lampert, W., Cáceres, C.E., Holtmeier, C.L., Weider, L.J., *et al.* (1999) Rapid evolution revealed by dormant eggs. *Nature* 401(6752), 446–446. doi: 10.1038/46731.

Hairston, N.G. Jr, Holtmeier, C.L., Lampert, W., Weider, L.J., Post, D.M., *et al.* (2001) Natural selection for grazer resistance to toxic cyanobacteria: evolution of phenotypic plasticity? *Evolution* 55(11), 2203–2214. doi: 10.1111/j.0014-3820.2001.tb00736.x.

Hairston, N.G. Jr, Ellner, S.P., Geber, M.A., Yoshida, T. and Fox, J.A. (2005) Rapid evolution and the convergence of ecological and evolutionary time. *Ecology Letters* 8(10), 1114–1127. doi: 10.1111/j.1461-0248.2005.00812.x.

Hallegraeff, G.M., Munday, B.L., Baden, D.G. and Whitney, P.L. (1998) *Chattonella marina* raphidophyte bloom associated with mortality of cultured bluefin tuna (*Thunnus maccoyii*) in South Australia. In: Reguera, B., Blanco, J., Fernandez, M.L. and Wyatt, T. (eds) *Harmful Algae, Proceedings of VIII Conference on Harmful Algae, Vigo, Spain, 25–29 June 1997*. Xunta de Galicia and International Oceanographic Commission of UNESCO, Spain, pp. 93–96.

Hamilton, D.P., Salmaso, N. and Paerl, H.W. (2016) Mitigating harmful cyanobacterial blooms: strategies for control of nitrogen and phosphorus loads. *Aquatic Ecology* 50(3), 351–366. doi: 10.1007/s10452-016-9594-z.

Hampson, D.R. and Manalo, J.L. (1998) The activation of glutamate receptors by kainic acid and domoic acid. *Natural Toxins* 6(3–4), 153–158. doi: 10.1002/(SICI)1522-7189(199805/08)6:3/4<153::AID-NT16>3.0.CO;2-1.

Haraldsson, M., Gerphagnon, M., Bazin, P., Colombet, J., Tecchio, S., *et al.* (2018) Microbial parasites make cyanobacteria blooms less of a trophic dead end than commonly assumed. *The ISME Journal* 12(4), 1008–1020. doi: 10.1038/s41396-018-0045-9.

Hardy, F.J., Johnson, A., Hamel, K. and Preece, E. (2015) Cyanotoxin bioaccumulation in freshwater fish, Washington State, USA. *Environmental Monitoring and Assessment* 187(11), 667. doi: 10.1007/s10661-015-4875-x.

Harke, M.J. and Gobler, C.J. (2015) Daily transcriptome changes reveal the role of nitrogen in controlling microcystin synthesis and nutrient transport in the toxic cyanobacterium, *Microcystis aeruginosa*. *BMC Genomics* 16(1), 1068. doi: 10.1186/s12864-015-2275-9.

Harke, M.J., Steffen, M.M., Gobler, C.J., Otten, T.G., Wilhelm, S.W., *et al.* (2016) A review of the global ecology, genomics, and biogeography of the toxic cyanobacterium, *Microcystis* spp. *Harmful Algae* 54, 4–20. doi: 10.1016/j.hal.2015.12.007.

Hattenrath-Lehmann, T.K. and Gobler, C.J. (2017) Identification of unique microbiomes associated with harmful algal blooms caused by *Alexandrium fundyense* and *Dinophysis acuminata*. *Harmful Algae* 68, 17–30. doi: 10.1016/j.hal.2017.07.003.

Hays, G.C., Richardson, A.J. and Robinson, C. (2005) Climate change and marine plankton. *Trends in Ecology & Evolution* 20(6), 337–344. doi: 10.1016/j.tree.2005.03.004.

Haywood, A.J., Steidinger, K.A., Truby, E.W., Bergquist, P.R., Bergquist, P.L., *et al.* (2004) Comparative morphology and molecular phylogenetic analysis of three new species of the genus *Karenia* (Dinophyceae) from New Zealand. *Journal of Phycology* 40(1), 165–179. doi: 10.1111/j.0022-3646.2004.02-149.x.

Hellweger, F.L., Martin, R.M., Eigemann, F., Smith, D.J., Dick, G.J. and Wilhelm, S.W. (2022) Models predict planned phosphorus load reduction will make Lake Erie more toxic. *Science* 376(6596), 1001–1005. doi: 10.1126/science.abm6791.

Heresztyn, T. and Nicholson, B.C. (1997) Nodularin concentrations in Lakes Alexandrina and Albert, South Australia, during a bloom of the cyanobacterium (blue-green alga) *Nodularia spumigena* and degradation of the toxin. *Environmental Toxicology and Water Quality* 12(4), 273–282. doi: 10.1002/(SICI)1098-2256(1997)12:4<273::AID-TOX1>3.0.CO;2-5.

Huisman, J., Codd, G.A., Paerl, H.W., Ibelings, B.W., Verspagen, J.M. and Visser, P.M. (2018) Cyanobacterial blooms. *Nature Reviews Microbiology* 16(8), 471–483. doi: 10.1038/s41579-018-0040-1.

Ibelings, B.W. and Chorus, I. (2007) Accumulation of cyanobacterial toxins in freshwater 'seafood' and its consequences for public health: a review. *Environmental Pollution* 150(1), 177–192. doi: 10.1016/J.ENVPOL.2007.04.012.

Ibelings, B.W. and Havens, K.E. (2008) Cyanobacterial toxins: a qualitative meta-analysis of concentrations, dosage and effects in freshwater, estuarine and marine biota. In: Hudnell, H.K. (ed.) *Cyanobacterial Harmful Algal Blooms: State of the Science and Research Needs*. Advances in Experimental Medicine and Biology, Vol. 619. Springer, New York, pp. 675–732. doi: 10.1007/978-0-387-75865-7_32.

Ibelings, B.W., Mur, L.R. and Walsby, A.E. (1991) Diurnal changes in buoyancy and vertical distribution in populations of *Microcystis* in two shallow lakes. *Journal of Plankton Research* 13(2), 419–436. doi: 10.1093/plankt/13.2.419.

Ibelings, B.W., Bruning, K., De Jonge, J., Wolfstein, K., Pires, L.M., *et al.* (2005) Distribution of microcystins in a lake foodweb: no evidence for biomagnification. *Microbial Ecology* 49(4), 487–500. doi: 10.1007/s00248-004-0014-x

Ibelings, B.W., Bormans, M., Fastner, J. and Visser, P.M. (2016) CYANOCOST special issue on cyanobacterial blooms: synopsis – a critical review of the management options for their prevention, control and mitigation. *Aquatic Ecology* 50(3), 595–605. doi: 10.1007/s10452-016-9596-x.

Ibelings, B.W., Foss, A. and Chorus, I. (2021) Exposure to cyanotoxins: understanding it and short-term interventions to prevent it. In: Chorus, I. and Welker, M. (2021) *Toxic Cyanobacteria in Water: A Guide to their Public Health Consequences, Monitoring and Management*, 2nd edn. CRC Press, Boca Raton, Florida, on behalf of World Health Organization, Geneva, Switzerland, pp. 368–384.

Igarashi, T., Satake, M. and Yasumoto, T. (1999) Structures and partial stereochemical assignments for prymnesin-1 and prymnesin-2: potent hemolytic and ichthyotoxic glycosides isolated from the red tide alga *Prymnesium parvum*. *Journal of the American Chemical Society* 121(37), 8499–8511. doi: 10.1021/ja991740e.

Iliev, D.B., Roach, J.C., Mackenzie, S., Planas, J.V. and Goetz, F.W. (2005) Endotoxin recognition: in fish or not in fish? *FEBS Letters* 579(29), 6519–6528. doi: 10.1016/j.febslet.2005.10.061.

Imai, I. and Yamaguchi, M. (2012) Life cycle, physiology, ecology and red tide occurrences of the fish-killing raphidophyte *Chattonella*. *Harmful Algae* 14, 46–70. doi: 10.1016/j.hal.2011.10.014.

IPCC (Intergovernmental Panel on Climate Change) (2022) *Climate Change 2022: Impacts, Adaptation, and Vulnerability. Contribution of Working Group II to the Sixth Assessment Report of the Intergovernmental Panel on Climate Change* (Pörtner, H.-O., Roberts, D.C., Tignor, M., Poloczanska, E.S., Mintenbeck, K., *et al.*, eds). Cambridge University Press, Cambridge, UK.

Ipek, Y. and Jeyasingh, P.D. (2021) Growth and ionomic responses of a freshwater cyanobacterium to supplies of nitrogen and iron. *Harmful Algae* 108, 102078. doi: 10.1016/j.hal.2021.102078.

Ispir, U. and Dorucu, M. (2014) Efficacy of lipopolysaccharide antigen of *Yersinia ruckeri* in rainbow trout by intraperitoneal and bath immersion administration. *Research in Veterinary Science* 97(2), 271–273. doi: 10.1016/j.rvsc.2014.07.020.

Itakura, S. and Imai, I. (2014) Economic impacts of harmful algal blooms on fisheries and aquaculture in western Japan – an overview of interannual variability and interspecies comparison. In: Trainer, V.L. and Yoshida, T. (eds) *Proceedings of the Workshop on Economic Impacts of Harmful Algal Blooms on Fisheries and Aquaculture*. PICES Scientific Report No. 47. North Pacific Marine Science Organization (PICES), Sidney, British Columbia, Canada, pp. 17–26.

Jackrel, S.L., White, J.D., Evans, J.T., Buffin, K., Hayden, K., *et al.* (2019) Genome evolution and host–microbiome shifts correspond with intraspecific niche divergence within harmful algal bloom-forming *Microcystis aeruginosa*. *Molecular Ecology* 28(17), 3994–4011. doi: 10.1111/mec.15198.

Jaja-Chimedza, A., Gantar, M., Mayer, G.D., Gibbs, P.D. and Berry, J.P. (2012) Effects of cyanobacterial lipopolysaccharides from *Microcystis* on glutathione-based detoxification pathways in the zebrafish (*Danio rerio*) embryo. *Toxins* 4(6), 390–404. doi: 10.3390/toxins4060390.

Jamieson, T. and Wardlaw, A.C. (1989) Degradation of bacterial lipopolysaccharides by digestive-gland extracts of marine bivalve molluscs. *Comparative Biochemistry and Physiology Part B: Comparative Biochemistry* 94(4), 837–843. doi: 10.1016/0305-0491(89)90174-0.

Jellett, J.F., Stewart, J.E. and Laycock, M.V. (1995) Toxicological evaluation of saxitoxin, neosaxitoxin, gonyautoxin II, gonyautoxin II plus III and decarbamoylsaxitoxin with the mouse neuroblastoma cell bioassay. *Toxicology In Vitro* 9(1), 57–65. doi: 10.1016/0887-2333(94)00194-Y.

Jeyasingh, P.D., Goos, J.M., Thompson, S.K., Godwin, C.M. and Cotner, J.B. (2017) Ecological stoichiometry beyond Redfield: an ionomic perspective on elemental homeostasis. *Frontiers in Microbiology* 8, 722. doi: 10.3389/fmicb.2017.00722.

Jiang, L., Kiselova, N., Rosén, J. and Ilag, L.L. (2014) Quantification of neurotoxin BMAA (β-*N*-methylamino-L-alanine) in seafood from Swedish markets. *Scientific Reports* 4(1), 6931. doi: 10.1038/srep06931.

Jochens, A.E., Malone, T.C., Stumpf, R.P., Hickey, B.M., Carter, M., *et al.* (2010) Integrated ocean observing system in support of forecasting harmful algal blooms. *Marine Technology Society Journal* 44(6), 99–121. doi: 10.4031/MTSJ.44.6.16.

Jonasson, S., Eriksson, J., Berntzon, L., Spáčil, Z., Ilag, L.L., *et al.* (2010) Transfer of a cyanobacterial neurotoxin within a temperate aquatic ecosystem suggests pathways for human exposure. *Proceedings of the National Academy of Sciences USA* 107(20), 9252–9257. doi: 10.1073/pnas.0914417107.

Kagami, M., von Elert, E., Ibelings, B.W., de Bruin, A. and Van Donk, E. (2007) The parasitic chytrid, *Zygorhizidium*, facilitates the growth of the cladoceran zooplankter, *Daphnia*, in cultures of the inedible alga, *Asterionella*. *Proceedings of the Royal Society B: Biological Sciences* 274(1617), 1561–1566. doi: 10.1098/rspb.2007.0425.

Kaminski, A., Bober, B., Lechowski, Z. and Bialczyk, J. (2013) Determination of anatoxin-a stability under certain abiotic factors. *Harmful Algae* 28, 83–87. doi: 10.1016/j.hal.2013.05.014.

Kankaanpää, H., Vuorinen, P.J., Sipiä, V. and Keinänen, M. (2002) Acute effects and bioaccumulation of nodularin in sea trout (*Salmo trutta* m. *trutta* L.) exposed orally to *Nodularia spumigena* under laboratory conditions. *Aquatic Toxicology* 61(3–4), 155–168. doi: 10.1016/S0166-445X(02)00054-1.

Karjalainen, M., Engström-Öst, J., Korpinen, S., Peltonen, H., Pääkkönen, J.P., *et al.* (2007) Ecosystem consequences of cyanobacteria in the northern Baltic Sea. *Ambio* 36(2), 195–202. doi: 10.1579/0044-7447(2007)36[195:ECOCIT]2.0.CO;2.

King, T.L., Nguyen, N., Doucette, G.J., Wang, Z., Bill, B.D., *et al.* (2021) Hiding in plain sight: shellfish-killing phytoplankton in Washington State. *Harmful Algae* 105, 102032. doi: 10.1016/j.hal.2021.102032.

Kleinteich, J., Wood, S.A., Puddick, J., Schleheck, D., Küpper, F.C. and Dietrich, D. (2013) Potent toxins in Arctic environments – presence of saxitoxins and an unusual microcystin variant in Arctic freshwater ecosystems. *Chemico-Biological Interactions* 206(2), 423–431. doi: 10.1016/j.cbi.2013.04.011.

Kotak, B.G., Semalulu, S., Fritz, D.L., Prepas, E.E., Hrudey, S.E. and Coppock, R.W. (1996) Hepatic and renal pathology of intraperitoneally administered microcystin-LR in rainbow trout (*Oncorhynchus mykiss*). *Toxicon* 34(5), 517–525. doi: 10.1016/0166-445X(94)90059-0.

Kozakai, H., Oshima, Y. and Yasumoto, T. (1982) Isolation and structural elucidation of hemolysin from the phytoflagellate *Prymnesium parvum*. *Agricultural and Biological Chemistry* 46(1), 233–236. doi: 10.1271/bbb1961.46.233.

Laabir, M., Jauzein, C., Genovesi, B., Masseret, E., Grzebyk, D., *et al.* (2011) Influence of temperature, salinity and irradiance on the growth and cell yield of the harmful red tide dinoflagellate *Alexandrium* catenella colonizing Mediterranean waters. *Journal of Plankton Research* 33(10), 1550–1563. doi: 10.1093/plankt/fbr050.

Lance, E., Arnich, N., Maignien, T. and Biré, R. (2018) Occurrence of β-*N*-methylamino-L-alanine (BMAA) and isomers in aquatic environments and aquatic food sources for humans. *Toxins* 10(2), 83. doi: 10.3390/toxins10020083.

Landsberg, J.H. (2002) The effects of harmful algal blooms on aquatic organisms. *Reviews in Fisheries Science* 10(2), 113–390. doi: 10.1080/20026491051695.

Landsberg, J.H., Flewelling, L.J. and Naar, J. (2009) *Karenia brevis* red tides, brevetoxins in the food web, and impacts on natural resources: decadal advancements. *Harmful Algae* 8(4), 598-607. doi: 10.1016/j.hal.2008.11.010.

Larsen, A., Bryant, S. and Båmstedt, U. (1998) Growth rate and toxicity of *Prymnesium parvum* and *Prymnesium patelliferum* (Haptophyta) in response to changes in salinity, light and temperature. *Sarsia* 83(5), 409–418. doi: 10.1080/00364827.1998.10413700.

Laundon, D., Mock, T., Wheeler, G. and Cunliffe, M. (2021) Healthy herds in the phytoplankton: the benefit of selective parasitism. *The ISME Journal* 15(7), 2163–2166. doi: 10.1038/s41396-021-00936-8.

Ledreux, A., Brand, H., Chinain, M., Bottein, M.Y.D. and Ramsdell, J.S. (2014) Dynamics of ciguatoxins from *Gambierdiscus polynesiensis* in the benthic herbivore *Mugil cephalus*: trophic transfer implications. *Harmful Algae* 39, 165–174. doi: 10.1016/j.hal.2014.07.009.

Lefebvre, K.A., Dovel, S.L. and Silver, M.W. (2001) Tissue distribution and neurotoxic effects of domoic acid in a prominent vector species, the northern anchovy *Engraulis mordax*. *Marine Biology* 138(4), 693–700. doi: 10.1007/s002270000509.

Lefebvre, K.A., Trainer, V.L. and Scholz, N.L. (2004) Morphological abnormalities and sensorimotor deficits in larval fish exposed to dissolved saxitoxin. *Aquatic Toxicology* 66(2), 159–170. doi: 10.1016/j.aquatox.2003.08.006.

Lefebvre, K.A., Noren, D.P., Schultz, I.R., Bogard, S.M., Wilson, J. and Eberhart, B.T.L. (2007) Uptake, tissue distribution and excretion of domoic acid after oral exposure in coho salmon (*Oncorhynchus kisutch*). *Aquatic Toxicology* 81(3), 266–274. doi: 10.1016/j.aquatox.2006.12.009.

Lefebvre, K.A., Frame, E.R. and Kendrick, P.S. (2012) Domoic acid and fish behavior: a review. *Harmful Algae* 13, 126–130. doi: 10.1016/j.hal.2011.09.011.

Lehane, L. (2000) Ciguatera update. *Medical Journal of Australia* 172(4), 176–179. doi: 10.5694/j.1326-5377.2000.tb125546.x.

Lehane, L. and Lewis, R.J. (2000) Ciguatera: recent advances but the risk remains. *International Journal of Food Microbiology* 61(2–3), 91–125. doi: 10.1016/S0168-1605(00)00382-2.

Le Manach, S., Sotton, B., Huet, H., Duval, C., Paris, A., *et al.* (2018) Physiological effects caused by microcystin-producing and non-microcystin producing *Microcystis aeruginosa* on medaka fish: a proteomic and metabolomic study on liver. *Environmental Pollution* 234, 523–537. doi: 10.1016/j.envpol.2017.11.011.

Lenzen, M., Li, M. and Murray, S.A. (2021) Impacts of harmful algal blooms on marine aquaculture in a low-carbon future. *Harmful Algae* 110, 102143. doi: 10.1016/j.hal.2021.102143.

Lewitus, A.J., Horner, R.A., Caron, D.A., Garcia-Mendoza, E., Hickey, B.M., *et al.* (2012) Harmful algal blooms along the North American west coast region: history, trends, causes, and impacts. *Harmful Algae* 19, 133–159. doi: 10.1016/j.hal.2012.06.009.

Li, Q., Lin, F., Yang, C., Wang, J., Lin, Y., *et al.* (2018) A large-scale comparative metagenomic study reveals the functional interactions in six bloom-forming *Microcystis*-epibiont communities. *Frontiers in Microbiology* 9, 746. doi: 10.3389/fmicb.2018.00746.

Li, X., Yan, T., Lin, J., Yu, R. and Zhou, M. (2017) Detrimental impacts of the dinoflagellate *Karenia mikimotoi* in Fujian coastal waters on typical marine organisms. *Harmful Algae* 61, 1–12. doi: 10.1016/j.hal.2016.11.011.

Li, X., Yan, T., Yu, R. and Zhou, M. (2019) A review of *Karenia mikimotoi*: bloom events, physiology, toxicity and toxic mechanism. *Harmful Algae* 90, 101702. doi: 10.1016/j.hal.2019.101702.

Liebel, S., Ribeiro, C.O., Silva, R.C., Ramsdorf, W.A., Cestari, M.M., *et al.* (2011) Cellular responses of *Prochilodus lineatus* hepatocytes after cylindrospermopsin exposure. *Toxicology In Vitro* 25(7), 1493–1500. doi: 10.1016/j.tiv.2011.05.010.

Lim, S.J. and Bordenstein, S.R. (2020) An introduction to phylosymbiosis. *Proceedings of the Royal Society B: Biological Sciences* 287(1922), 20192900. doi: 10.1098/rspb.2019.2900.

Liu, X., Rush, T., Zapata, J. and Lobner, D. (2009) β-*N*-Methylamino-L-alanine induces oxidative stress and glutamate release through action on system Xc^-. *Experimental Neurology* 217(2), 429–433. doi: 10.1016/j.expneurol.2009.04.002.

Lopes, K.C., Ferrão-Filho, A.D.S., Dos Santos, E.G., Cunha, R.A. and Santos, C.P. (2017) Effects of crude extracts of a saxitoxin-producer strain of the cyanobacterium *Cylindrospermopsis raciborskii* on the swimming behavior of wild and laboratory reared guppy *Poecilia vivipara*. *Toxicon* 129, 44–51. doi: 10.1016/j.toxicon.2017.02.002.

Lum, W.M., Benico, G., Doan-Nhu, H., Furio, E., Leaw, C.P., *et al.* (2021) The harmful raphidophyte *Chattonella* (Raphidophyceae) in Western Pacific: its red tides and associated fisheries damage over the past 50 years (1969–2019). *Harmful Algae* 107, 102070. doi: 10.1016/j.hal.2021.102070.

Lürling, M., Faassen, E.J. and Van Eenennaam, J.S. (2011) Effects of the cyanobacterial neurotoxin β-*N*-methylamino-L-alanine (BMAA) on the survival, mobility and reproduction of *Daphnia magna*. *Journal of Plankton Research* 33(2), 333–342. doi: 10.1093/plankt/fbq130.

Lürling, M., Eshetu, F., Faassen, E.J., Kosten, S. and Huszar, V.L. (2013) Comparison of cyanobacterial and green algal growth rates at different temperatures. *Freshwater Biology* 58(3), 552–559. doi: 10.1111/j.1365-2427.2012.02866.x.

Mak, Y.L., Li, J., Liu, C., Cheng, S.H., Lam, P.K.S., *et al.* (2017) Physiological and behavioural impacts of Pacific ciguatoxin-1 (P-CTX-1) on marine medaka (*Oryzias melastigma*). *Journal of Hazardous Materials* 321, 782–790. doi: 10.1016/j.jhazmat.2016.09.066.

Malbrouck, C. and Kestemont, P. (2006) Effects of microcystins on fish. *Environmental Toxicology and Chemistry* 25(1), 72–86. doi: 10.1897/05-029R.1.

Manning, S.R. and La Claire, J.W. (2010) Prymnesins: toxic metabolites of the golden alga, *Prymnesium parvum* Carter (Haptophyta). *Marine Drugs* 8(3), 678–704. doi: 10.3390/md8030678.

Marcé, R., George, G., Buscarinu, P., Deidda, M., Dunalska, J., *et al.* (2016) Automatic high frequency monitoring for improved lake and reservoir management. *Environmental Science & Technology* 50(20), 10780–10794. doi: 10.1021/acs.est.6b01604.

Mardones, J.I., Paredes, J., Godoy, M., Suarez, R., Norambuena, L., *et al.* (2021) Disentangling the environmental processes responsible for the world's largest farmed fish-killing harmful algal bloom: Chile (2016). *Science of the Total Environment* 766, 144383. doi: 10.1016/j.scitotenv.2020.144383.

Marshall, J.A., Nichols, P.D., Hamilton, B., Lewis, R.J. and Hallegraeff, G.M. (2003) Ichthyotoxicity of *Chattonella marina* (Raphidophyceae) to damselfish (*Acanthochromis polycanthus*): the synergistic role of reactive oxygen species and free fatty acids. *Harmful Algae* 2(4), 273–281. doi: 10.1016/S1568-9883(03)00046-5.

Marthinussen, A., Nystøyl, R., Storhaug, H.R., Valle, P.S. and Gaarder, M. (2020) *Økonomiske og samfunnsmessige konsekvenser av algeoppblomstringen i Nord-Norge*. Kontali Analyse AS, Kristiansund, Norway. Available at: https://www.kontali.no/uploads/6VY1FKI6/Sluttrapport_901574-Konsekvenser_av_algesituasjonen_i_nord.pdf (accessed 21 October 2022).

Matsusato, T. and Kobayashi, H. (1974) Studies on death of fish caused by red tide. *Bulletin of the Nansei Regional Fisheries Research Laboratory* 7, 43–67.

Matthijs, H.C., Jančula, D., Visser, P.M. and Maršálek, B. (2016) Existing and emerging cyanocidal compounds: new perspectives for cyanobacterial bloom mitigation. *Aquatic Ecology* 50(3), 443–460. doi: 10.1007/s10452-016-9577-0.

Mazur-Marzec, H., Meriluoto, J. and Pliński, M. (2006) The degradation of the cyanobacterial hepatotoxin nodularin (NOD) by UV radiation. *Chemosphere* 65(8), 1388–1395. doi: 10.1016/j.chemosphere.2006.03.072.

McKindles, K.M., Manes, M.A., DeMarco, J.R., McClure, A., McKay, R.M., *et al.* (2020) Dissolved microcystin release coincident with lysis of a bloom dominated by *Microcystis* spp. in western Lake Erie attributed to a novel cyanophage. *Applied and Environmental Microbiology* 86(22), e01397-20. doi: 10.1128/AEM.01397-20.

McKindles, K.M., Manes, M.A., McKay, R.M., Davis, T.W. and Bullerjahn, G.S. (2021) Environmental factors affecting chytrid (Chytridiomycota) infection rates on *Planktothrix agardhii*. *Journal of Plankton Research* 43(5), 658–672. doi: 10.1093/plankt/fbab058.

McLean, T.I. (2013) 'Eco-omics': a review of the application of genomics, transcriptomics, and proteomics for the study of the ecology of harmful algae. *Microbial Ecology* 65(4), 901–915. doi: 10.1007/s00248-013-0220-5.

Medić, N., Varga, E., Van de Waal, D.B., Larsen, T.O. and Hansen, P.J. (2022) The coupling between irradiance, growth, photosynthesis and prymnesin cell quota and production in two strains of the bloom-forming haptophyte, *Prymnesium parvum*. *Harmful Algae* 112, 102173. doi: 10.1016/j.hal.2022.102173.

Meek, M.H. and Larson, W.A. (2019) The future is now: amplicon sequencing and sequence capture usher in the conservation genomics era. *Molecular Ecology Resources* 19(4), 795–803. doi: 10.1111/1755-0998.12998.

Metcalf, J.S., Dunlop, R.A., Cox, P.A. and Banack, S.A. (2021) BMAA neurotoxicity. In: Kostrzewa, R.M. (ed.) *Handbook of Neurotoxicity*. Springer, Cham, Switzerland, pp. 1–16. doi: 10.1007/978-3-030-71519-9_225-1.

Michalak, A.M., Anderson, E.J., Beletsky, D., Boland, S., Bosch, N.S., *et al.* (2013) Record-setting algal bloom in Lake Erie caused by agricultural and meteorological trends consistent with expected future conditions. *Proceedings of the National Academy of Sciences USA* 110(16), 6448–6452. doi: 10.1073/pnas.1216006110.

Miller, A.G., Espie, G.S. and Canvin, D.T. (1990) Physiological aspects of CO_2 and HCO_3^- transport by cyanobacteria: a review. *Canadian Journal of Botany* 68(6), 1291–1302. doi: 10.1139/b90-165.

Mitchell, S. and Rodger, H. (2007) Pathology of wild and cultured fish affected by a *Karenia mikimotoi* bloom in Ireland, 2005. *Bulletin of the European Association of Fish Pathologists* 27(1), 39–42.

Moosová, Z., Šindlerová, L., Ambrůzová, B., Ambrožová, G., Vašíček, O., *et al.* (2019) Lipopolysaccharides from *Microcystis* cyanobacteria-dominated water bloom and from laboratory cultures trigger human immune innate response. *Toxins* 11(4), 218. doi: 10.3390/toxins11040218.

Moustaka-Gouni, M., Hiskia, A., Genitsaris, S., Katsiapi, M., Manolidi, K., *et al.* (2016) First report of *Aphanizomenon favaloroi* occurrence in Europe associated with saxitoxins and a massive fish kill in Lake Vistonis, *Greece. Marine and Freshwater Research* 68(4), 793–800. doi: 10.1071/MF16029.

Munday, R., Thomas, K., Gibbs, R., Murphy, C. and Quilliam, M.A. (2013) Acute toxicities of saxitoxin, neosaxitoxin, decarbamoyl saxitoxin and gonyautoxins 1&4 and 2&3 to mice by various routes of administration. *Toxicon* 76, 77–83. doi: 10.1016/j.toxicon.2013.09.013.

Murray, S., John, U. and Kremp, A. (2015) *Alexandrium* spp.: genetic and ecological factors influencing saxitoxin production and proliferation. In: Botana, L.M., Louzao, M.C. and Vilarino, N. (eds) *Climate Change and Marine and Freshwater Toxins*. De Gruyter, Berlin/Boston, Massachusetts, pp. 123–139. doi: 10.1515/9783110625738-004.

Naar, J.P., Flewelling, L.J., Lenzi, A., Abbott, J.P., Granholm, A., *et al.* (2007) Brevetoxins, like ciguatoxins, are potent ichthyotoxic neurotoxins that accumulate in fish. *Toxicon* 50(5), 707–723. doi: 10.1016/j.toxicon.2007.06.005.

Nayak, S.K., Swain, P., Nanda, P.K., Dash, S., Shukla, S., *et al.* (2008) Effect of endotoxin on the immunity of Indian major carp, *Labeo rohita. Fish & Shellfish Immunology* 24(4), 394–399. doi: 10.1016/j.fsi.2007.09.005.

Neilan, B.A., Pearson, L.A., Moffitt, M.C., Mihali, K.T., Kaebernick, M., *et al.* (2008) The genetics and genomics of cyanobacterial toxicity. In: Hudnell, H.K. (ed.) *Cyanobacterial Harmful Algal Blooms: State of the Science and Research Needs*. Advances in Experimental Medicine and Biology, Vol. 619. Springer, New York, pp. 417–452. doi: 10.1007/978-0-387-75865-7_17.

Nicolaou, K.C., Frederick, M.O. and Aversa, R.J. (2008) The continuing saga of the marine polyether biotoxins. *Angewandte Chemie International Edition*, 47(38), 7182–7225. doi: 10.1002/anie.200801696.

Nogueira, I., Lobo-da-Cunha, A., Afonso, A., Rivera, S., Azevedo, J., *et al.* (2010) Toxic effects of domoic acid in the seabream *Sparus aurata*. *Marine Drugs* 8(10), 2721–2732. doi: 10.3390/md8102721.

Notch, E.G., Miniutti, D.M., Berry, J.P. and Mayer, G.D. (2011) Cyanobacterial LPS potentiates cadmium toxicity in zebrafish (*Danio rerio*) embryos. *Environmental Toxicology* 26(5), 498–505. doi: 10.1002/tox.20578.

Nowotny, A. (1969) Molecular aspects of endotoxic reactions. *Bacteriological Reviews* 33(1), 72–98.

Nya, E.J. and Austin, B. (2010) Use of bacterial lipopolysaccharide (LPS) as an immunostimulant for the control of *Aeromonas hydrophila* infections in rainbow trout *Oncorhynchus mykiss* (Walbaum). *Journal of Applied Microbiology* 108(2), 686–694. doi: 10.1111/j.1365-2672.2009.04464.x.

Oberemm, A., Becker, J., Codd, G.A. and Steinberg, C. (1999) Effects of cyanobacterial toxins and aqueous crude extracts of cyanobacteria on the development of fish and amphibians. *Environmental Toxicology* 14(1), 77–88. doi: 10.1002/(SICI)1522-7278(199902)14:1<77::AID-TOX11>3.0.CO;2-F.

Oda, M. (1935) *Gymnodinium* mikimotoi Miyake et Kominami n. sp. (MS.) no akashiwo to ryusando no koka (The red tide of *Gymnodinium* mikimotoi Miyake et Kominami and the influence of copper sulfate on the red tide). *Dobutsugaku Zasshi (Zoological Magazine)* 47, 35–48 (in Japanese).

Osswald, J., Rellán, S., Carvalho, A.P., Gago, A. and Vasconcelos, V. (2007) Acute effects of an anatoxin-a producing cyanobacterium on juvenile fish – *Cyprinus carpio* L. *Toxicon* 49(5), 693–698. doi: 10.1016/j.toxicon.2006.11.010.

Osswald, J., Carvalho, A.P., Claro, J. and Vasconcelos, V. (2009) Effects of cyanobacterial extracts containing anatoxin-a and of pure anatoxin-a on early developmental stages of carp. *Ecotoxicology and Environmental Safety* 72(2), 473–478. doi: 10.1016/j.ecoenv.2008.05.011.

Otterstrøm, C.V. and Nielsen, S. (1939) Two cases of extensive mortality in fishes caused by the flagellate *Prymnesium parvum*. *Report of the Danish Biological Station* 44, 5–24.

Oziolor, E.M., Reid, N.M., Yair, S., Lee, K.M., Guberman VerPloeg, S., *et al.* (2019) Adaptive introgression enables evolutionary rescue from extreme environmental pollution. *Science* 364(6439), 455–457. doi: 10.1126/science.aav4155.

Paerl, H.W. and Huisman, J. (2008) Blooms like it hot. *Science* 320(5872), 57–58. doi: 10.1126/science.1155398.

Pal, M., Yesankar, P.J., Dwivedi, A. and Qureshi, A. (2020) Biotic control of harmful algal blooms (HABs): a brief review. *Journal of Environmental Management* 268, 110687. doi: 10.1016/j.jenvman.2020.110687.

Palikova, M., Navratil, S., Maršálek, B. and Bláha, L. (2003) Toxicity of crude extracts of cyanobacteria for embryos and larvae of carp (*Cyprinus carpio* L.). *Acta Veterinaria Brno* 72(3), 437–443. doi: 10.2754/avb200372030437.

Panlilio, J.M., Aluru, N. and Hahn, M.E. (2020) Developmental neurotoxicity of the harmful algal bloom toxin domoic acid: cellular and molecular mechanisms underlying altered behavior in the zebrafish model. *Environmental Health Perspectives* 128(11), 117002. doi: 10.1289/EHP6652.

Panlilio, J.M., Jones, I.T., Salanga, M.C., Aluru, N. and Hahn, M.E. (2021) Developmental exposure to domoic acid disrupts startle response behavior and circuitry in zebrafish. *Toxicological Sciences* 182(2), 310–326. doi: 10.1093/toxsci/kfab066.

Park, M.G., Kim, A., Jeon, B.S. and Kim, M. (2021) Parasite-mediated increase in prey edibility in the predator–prey interaction of marine planktonic protists. *Harmful Algae* 103, 101982. doi: 10.1016/j.hal.2021.101982.

Parmesan, C., Morecroft, M.D., Trisurat, Y., Adrian, R., Anshari, G.Z., *et al.* (2022) Terrestrial and freshwater ecosystems and their services. In: Pörtner, H.-O., Roberts, D.C., Tignor, M., Poloczanska, E.S., Mintenbeck, K., *et al.* (eds) *Climate Change 2022: Impacts, Adaptation, and Vulnerability. Contribution of Working Group II to the Sixth Assessment Report of the Intergovernmental Panel on Climate Change*. Cambridge University Press, Cambridge, UK, pp. 197–377.

Paseka, R.E., White, L.A., Van de Waal, D.B., Strauss, A.T., González, A.L., *et al.* (2020) Disease-mediated ecosystem services: pathogens, plants, and people. *Trends in Ecology & Evolution* 35(8), 731–743. doi: 10.1016/j.tree.2020.04.003.

Peacock, M.B., Gibble, C.M., Senn, D.B., Cloern, J.E. and Kudela, R.M. (2018) Blurred lines: multiple freshwater and marine algal toxins at the land–sea interface of San Francisco Bay, California. *Harmful Algae* 73, 138–147. doi: 10.1016/j.hal.2018.02.005.

Pearson, L., Mihali, T., Moffitt, M., Kellmann, R. and Neilan, B. (2010) On the chemistry, toxicology and genetics of the cyanobacterial toxins, microcystin, nodularin, saxitoxin and cylindrospermopsin. *Marine Drugs* 8(5), 1650–1680. doi: 10.3390/md8051650.

Pérez-Arellano, J.L., Luzardo, O.P., Brito, A.P., Cabrera, M.H., Zumbado, M., *et al.* (2005) Ciguatera fish poisoning, Canary Islands. *Emerging Infectious Diseases* 11(12), 1981. doi: 10.3201/eid1112.050393.

Pérez-Carrascal, O.M., Terrat, Y., Giani, A., Fortin, N., Greer, C.W., *et al.* (2019) Coherence of *Microcystis* species revealed through population genomics. *The ISME Journal* 13(12), 2887–2900. doi: 10.1038/s41396-019-0481-1.

Pineda-Mendoza, R.M., Zúñiga, G. and Martínez-Jerónimo, F. (2016) Microcystin production in *Microcystis aeruginosa*: effect of type of strain, environmental factors, nutrient concentrations, and N:P ratio on *mcyA* gene expression. *Aquatic Ecology* 50(1), 103–119. doi: 10.1007/s10452-015-9559-7.

Powers, S., Kwok, S., Lovejoy, E., Lavin, T. and Sher, R.B. (2017) Editor's highlight: embryonic exposure to the environmental neurotoxin BMAA negatively impacts early neuronal development and progression of neurodegeneration in the Sod1-G93R zebrafish model of amyotrophic lateral sclerosis. *Toxicological Sciences* 157(1), 129–140. doi: 10.1093/toxsci/kfx020.

Puddick, J., Prinsep, M.R., Wood, S.A., Cary, S.C. and Hamilton, D.P. (2016) Modulation of microcystin congener abundance following nitrogen depletion of a *Microcystis* batch culture. *Aquatic Ecology* 50(2), 235–246. doi: 10.1007/s10452-016-9571-6.

Puerto, M., Jos, A., Pichardo, S., Moyano, R., Blanco, A. and Cameán, A.M. (2012) Acute exposure to pure cylindrospermopsin results in oxidative stress and pathological alterations in tilapia (*Oreochromis niloticus*). *Environmental Toxicology* 29(4), 371–385. doi: 10.1002/tox.21764.

Purdie, E.L., Samsudin, S., Eddy, F.B. and Codd, G.A. (2009) Effects of the cyanobacterial neurotoxin β-*N*-methylamino-L-alanine on the early-life stage development of zebrafish (*Danio rerio*). *Aquatic Toxicology* 95(4), 279–284. doi: 10.1016/j.aquatox.2009.02.009.

Qian, H., Zhang, M., Liu, G., Lu, T., Sun, L. and Pan, X. (2019) Effects of different concentrations of *Microcystis aeruginosa* on the intestinal microbiota and immunity of zebrafish (*Danio rerio*). *Chemosphere* 214, 579–586. doi: 10.1016/j.chemosphere.2018.09.156.

Quilliam, M.A. (2003) Chemical methods for domoic acid, the amnesic shellfish poisoning (ASP) toxin. In: Hallegraeff, M., Anderson, D.M. and Cembella, A.D. (eds) *Manual on Harmful Marine Microalgae.* Monographs on Oceanography Methodology, Vol. 11. Intergovernmental Oceanographic Commission, UNESCO, Paris, pp. 247–266.

Quilliam, M.A., Thomson, B.A., Scott, G.J. and Siu, K.M. (1989) Ion-spray mass spectrometry of marine neurotoxins. *Rapid Communications in Mass Spectrometry* 3(5), 145–150. doi: 10.1002/rcm.1290030508.

Råbergh, C.M.I., Bylund, G. and Eriksson, J.E. (1991) Histopathological effects of microcystin-LR, a cyclic peptide toxin from the cyanobacterium (blue-green alga) *Microcystis aeruginosa* on common carp (*Cyprinus carpio* L.). *Aquatic Toxicology* 20(3), 131–145. doi: 10.1016/0166-445X(91)90012-X.

Ramsdell, J.S. (2007) The molecular and integrative basis to domoic acid toxicity. In: Botana, L.M. (ed.) *Phycotoxins: Chemistry and Biochemistry*. Blackwell Publishing, Ames, Iowa, pp. 223–250. doi: 10.1002/9780470277874.ch13.

Rapala, J., Lahti, K., Sivonen, K. and Niemelä, S.I. (1994) Biodegradability and adsorption on lake sediments of cyanobacterial hepatotoxins and anatoxin-a. *Letters in Applied Microbiology* 19(6), 423–428. doi: 10.1111/j.1472-765X.1994.tb00972.x.

Rasmussen, S.A., Meier, S., Andersen, N.G., Blossom, H.E., Duus, J.Ø., *et al.* (2016) Chemodiversity of ladder-frame prymnesin polyethers in *Prymnesium parvum*. *Journal of Natural Products* 79(9), 2250–2256. doi: 10.1021/acs.jnatprod.6b00345.

Reid, N.M., Proestou, D.A., Clark, B.W., Warren, W.C., Colbourne, J.K., *et al.* (2016) The genomic landscape of rapid repeated evolutionary adaptation to toxic pollution in wild fish. *Science* 354(6317), 1305–1308. doi: 10.1126/science.aah4993.

Rennison, D.J., Rudman, S.M. and Schluter, D. (2019) Parallel changes in gut microbiome composition and function during colonization, local adaptation and ecological speciation. *Proceedings of the Royal Society B: Biological Sciences* 286(1916), 20191911. doi: 10.1098/rspb.2019.1911.

Rodger, H.D., Henry, L. and Mitchell, S.O. (2011) Non-infectious gill disorders of marine salmonid fish. *Reviews in Fish Biology and Fisheries* 21(3), 423–440. doi: 10.1007/s11160-010-9182-6.

Rodrigues, P.M., Schrama, D., Campos, A., Osório, H. and Freitas, M. (2016) Applications of proteomics in aquaculture. In: Salekdeh, G. (ed.) *Agricultural Proteomics*, Vol. 1. Springer, Cham, Switzerland, pp. 175–209. doi: 10.1007/978-3-319-43275-5_10.

Roelke, D.L., Barkoh, A., Brooks, B.W., Grover, J.P., Hambright, K.D., *et al.* (2016) A chronicle of a killer alga in the west: ecology, assessment, and management of *Prymnesium parvum* blooms. *Hydrobiologia* 764(1), 29–50. doi: 10.1007/s10750-015-2273-6.

Rohrlack, T., Christiansen, G. and Kurmayer, R. (2013) Putative antiparasite defensive system involving ribosomal and nonribosomal oligopeptides in cyanobacteria of the genus *Planktothrix*. *Applied and Environmental Microbiology* 79(8), 2642–2647. doi: 10.1128/AEM.03499-12.

Rücker, J., Stüken, A., Nixdorf, B., Fastner, J., Chorus, I. and Wiedner, C. (2007) Concentrations of particulate and dissolved cylindrospermopsin in 21 Aphanizomenon-dominated temperate lakes. Toxicon 50(6), 800–809. doi: 10.1016/j.toxicon.2007.06.019.

Rudman, S.M. and Schluter, D. (2016) Ecological impacts of reverse speciation in threespine stickleback. *Current Biology* 26(4), 490–495. doi: 10.1016/j.cub.2016.01.004.

Rudman, S.M., Greenblum, S., Hughes, R.C., Rajpurohit, S., Kiratli, O., *et al.* (2019a) Microbiome composition shapes rapid genomic adaptation of *Drosophila melanogaster*. *Proceedings of the National Academy of Sciences USA* 116(40), 20025–20032. doi: 10.1073/pnas.1907787116.

Rudman, S.M., Goos, J.M., Burant, J.B., Brix, K.V., Gibbons, T.C., *et al.* (2019b) Ionome and elemental transport kinetics shaped by parallel evolution in threespine stickleback. *Ecology Letters* 22(4), 645–653. doi: 10.1111/ele.13225.

Rudman, S.M., Greenblum, S.I., Rajpurohit, S., Betancourt, N.J., Hanna, J., *et al.* (2022) Direct observation of adaptive tracking on ecological timescales in *Drosophila*. *Science* 375(6586), eabj7484. doi: 10.1126/science.abj7484.

Salomonsson, M.L., Fredriksson, E., Alfjorden, A., Hedeland, M. and Bondesson, U. (2015) Seafood sold in Sweden contains BMAA: a study of free and total concentrations with UHPLC–MS/MS and dansyl chloride derivatization. *Toxicology Reports* 2, 1473–1481. doi: 10.1016/j.toxrep.2015.11.002.

Salt, D.E., Baxter, I. and Lahner, B. (2008) Ionomics and the study of the plant ionome. *Annual Review of Plant Biology* 59, 709–733. doi: 10.1146/annurev.arplant.59.032607.092942.

Sandrini, G., Ji, X., Verspagen, J.M., Tann, R.P., Slot, P.C., *et al.* (2016) Rapid adaptation of harmful cyanobacteria to rising CO_2. *Proceedings of the National Academy of Sciences USA* 113(33), 9315–9320. doi: 10.1073/pnas.1602435113.

Sarnelle, O. (2007) Initial conditions mediate the interaction between *Daphnia* and bloom-forming cyanobacteria. *Limnology and Oceanography* 52(5), 2120–2127. doi: 10.4319/lo.2007.52.5.2120.

Sarnelle, O. and Wilson, A.E. (2005) Local adaptation of *Daphnia pulicaria* to toxic cyanobacteria. *Limnology and Oceanography* 50(5), 1565–1570. doi: 10.4319/lo.2005.50.5.1565.

Schindler, D.W. (1974) Eutrophication and recovery in experimental lakes: implications for lake management. *Science* 184(4139), 897–899. doi: 10.1126/science.184.4139.897.

Schmidt, K.C., Jackrel, S.L., Smith, D.J., Dick, G.J. and Denef, V.J. (2020) Genotype and host microbiome alter competitive interactions between *Microcystis aeruginosa* and *Chlorella sorokiniana*. *Harmful Algae* 99, 101939. doi: 10.1016/j.hal.2020.101939.

Scholin, C.A., Gulland, F., Doucette, G.J., Benson, S., Busman, M., *et al.* (2000) Mortality of sea lions along the central California coast linked to a toxic diatom bloom. *Nature* 403(6765), 80–84. doi: 10.1038/47481.

Sepulcre, M.P., Alcaraz-Pérez, F., López-Muñoz, A., Roca, F.J., Meseguer, J., *et al.* (2009) Evolution of lipopolysaccharide (LPS) recognition and signaling: fish TLR4 does not recognize LPS and negatively regulates NF-κB activation. *The Journal of Immunology* 182(4), 1836–1845. doi: 10.4049/jimmunol.0801755.

Shahmohamadloo, R.S., Poirier, D.G., Almirall, X.O., Bhavsar, S.P. and Sibley, P.K. (2020a) Assessing the toxicity of cell-bound microcystins on freshwater pelagic and benthic invertebrates. *Ecotoxicology and Environmental Safety* 188, 109945. doi: 10.1016/j.ecoenv.2019.109945.

Shahmohamadloo, R.S., Simmons, D.B. and Sibley, P.K. (2020b) Shotgun proteomics analysis reveals sub-lethal effects in *Daphnia magna* exposed to cell-bound microcystins produced by *Microcystis aeruginosa*. *Comparative Biochemistry and Physiology Part D: Genomics and Proteomics* 33, 100656. doi: 10.1016/j.cbd.2020.100656.

Shahmohamadloo, R.S., Ortiz Almirall, X., Simmons, D.B., Lumsden, J.S., Bhavsar, S.P., *et al.* (2021) Cyanotoxins within and outside of *Microcystis aeruginosa* cause adverse effects in rainbow trout (*Oncorhynchus mykiss*). *Environmental Science & Technology* 55(15), 10422–10431. doi: 10.1021/acs.est.1c01501.

Shahmohamadloo, R.S., Almirall, X.O., Simmons, D.B., Poirier, D.G., Bhavsar, S.P. and Sibley, P.K. (2022a) Fish tissue accumulation and proteomic response to microcystins is species-dependent. *Chemosphere* 287, 132028. doi: 10.1016/j.chemosphere.2021.132028.

Shahmohamadloo, R.S., Bhavsar, S.P., Almirall, X.O., Marklevitz, S.A., Rudman, S.M. and Sibley, P.K. (2022b) Low human health risks of algal toxins from consuming fish caught in Lake St. Clair. *bioRxiv*, doi: 10.1101/2022.09.08.507173.

Shahmohamadloo, R.S., Bhavsar, S.P., Ortiz Almirall, X., Rudman S.M. and Sibley, P.K. (2023) Lake Erie fish safe to eat yet afflicted by algal hepatotoxins. *Science of the Total Environment*, 861, 160474. doi: 10.1016/j.scitotenv.2022.160474.

Shartau, R.B., Snyman, H.N., Turcotte, L., McCarron, P., Bradshaw, J.C. and Johnson, S.C. (2022) Acute microcystin exposure induces reversible histopathological changes in chinook salmon (*Oncorhynchus tshawytscha*) and Atlantic salmon (*Salmo salar*). *Journal of Fish Diseases* 45(5), 729–742. doi: 10.1111/jfd.13599.

Shikata, T., Yuasa, K., Kitatsuji, S., Sakamoto, S., Akita, K., *et al.* (2021) Superoxide production by the red tide-producing *Chattonella* marina complex (Raphidophyceae) correlates with toxicity to aquacultured fishes. *Antioxidants* 10(10), 1635. doi: 10.3390/antiox10101635.

Shimada, M., Murakami, T.H., Imahayashi, T., Ozaki, H.S., Toyoshima, T. and Okaichi, T. (1983) Effects of sea bloom, *Chattonella antiqua*, on gill primary lamellae of the young yellowtail, *Seriola quinqueradiata*. *Acta Histochemica et Cytochemica* 16(3), 232–244. doi: 10.1267/ahc.16.232.

Shin, H., Lee, E., Shin, J., Ko, S.R., Oh, H.S., *et al.* (2018) Elucidation of the bacterial communities associated with the harmful microalgae *Alexandrium tamarense* and *Cochlodinium polykrikoides* using nanopore sequencing. *Scientific Reports* 8(1), 5323. doi: 10.1038/s41598-018-23634-6.

Shumway, S.E. (1990) A review of the effects of algal blooms on shellfish and aquaculture. *Journal of the World Aquaculture Society* 21(2), 65–104. doi: 10.1111/j.1749-7345.1990.tb00529.x.

Sieroslawska, A. and Rymuszka, A. (2019) Assessment of the cytotoxic impact of cyanotoxin beta-*N*-methylamino-L-alanine on a fish immune cell line. *Aquatic Toxicology* 212, 214–221. doi: 10.1016/j.aquatox.2019.05.012.

Sigee, D.C., Glenn, R., Andrews, M.J., Bellinger, E.G., Butler, R.D., *et al.* (1999) Biological control of cyanobacteria: principles and possibilities. In: Harper, D.M., Brierley, B., Ferguson, A.J.D. and Phillips, G. (eds) *The Ecological Bases for Lake and Reservoir Management.* Developments in Hydrobiology, Vol. 136. Springer, Dordrecht, The Netherlands, pp. 161–172. doi: 10.1007/978-94-017-3282-6_15.

Simonsen, S. and Moestrup, Ø. (1997) Toxicity tests in eight species of *Chrysochromulina* (Haptophyta). *Canadian Journal of Botany* 75(1), 129–136. doi: 10.1139/b97-015.

Sivonen, K. and Jones, G. (1999) Cyanobacterial toxins. In: Chorus, I. and Bartram, J. (eds) *Toxic Cyanobacteria in Water: A Guide to Their Public Health Consequence, Monitoring and Management*. E&FN Spon, London, pp. 55–124.

Smayda, T.J. (2010) Adaptations and selection of harmful and other dinoflagellate species in upwelling systems. 2. Motility and migratory behaviour. *Progress in Oceanography* 85(1–2), 71–91. doi: 10.1016/j.pocean.2010.02.005.

Smayda, T.J. and Reynolds, C.S. (2001) Community assembly in marine phytoplankton: application of recent models to harmful dinoflagellate blooms. *Journal of Plankton Research* 23(5), 447–461. doi: 10.1093/plankt/23.5.447.

Sønstebø, J.H. and Rohrlack, T. (2011) Possible implications of chytrid parasitism for population subdivision in freshwater cyanobacteria of the genus *Planktothrix*. *Applied and Environmental Microbiology* 77(4), 1344–1351. doi: 10.1128/AEM.02153-10.

Sòria-Perpinyà, X., Urrego, E.P., Pereira-Sandoval, M., Ruiz-Verdú, A., Soria, J.M., *et al.* (2020) Monitoring water transparency of a hypertrophic lake (the Albufera of València) using multitemporal Sentinel-2 satellite images. *Limnetica* 39(1), 373–386. doi: 10.23818/limn.39.24.
Sotton, B., Domaizon, I., Anneville, O., Cattanéo, F. and Guillard, J. (2015) Nodularin and cylindrospermopsin: a review of their effects on fish. *Reviews in Fish Biology and Fisheries* 25(1), 1–19. doi: 10.1007/s11160-014-9366-6.
Southard, G.M., Fries, L.T. and Barkoh, A. (2010) *Prymnesium parvum*: the Texas experience. *JAWRA Journal of the American Water Resources Association* 46(1), 14–23. doi: 10.1111/j.1752-1688.2009.00387.x.
Stal, L.J. (2009) Is the distribution of nitrogen-fixing cyanobacteria in the oceans related to temperature? *Environmental Microbiology* 11(7), 1632–1645. doi: 10.1111/j.1758-2229.2009.00016.x.
Steffen, M.M., Davis, T.W., McKay, R.M.L., Bullerjahn, G.S., Krausfeldt, L.E., *et al.* (2017) Ecophysiological examination of the Lake Erie *Microcystis* bloom in 2014: linkages between biology and the water supply shutdown of Toledo, OH. *Environmental Science & Technology* 51(12), 6745–6755. doi: 10.1021/acs.est.7b00856.
Stevens, D.K. and Krieger, R.I. (1991) Stability studies on the cyanobacterial nicotinic alkaloid snatoxin-A. *Toxicon* 29(2), 167–179. doi: 10.1016/0041-0101(91)90101-V.
Stewart, I., Eaglesham, G.K., McGregor, G.B., Chong, R., Seawright, A.A., *et al.* (2012) First report of a toxic *Nodularia spumigena* (Nostocales/Cyanobacteria) bloom in sub-tropical Australia. II. Bioaccumulation of nodularin in isolated populations of mullet (Mugilidae). *International Journal of Environmental Research and Public Health* 9(7), 2412–2443. doi: 10.3390/ijerph9072412.
Su, Z., Sheets, M., Ishida, H., Li, F. and Barry, W.H. (2004) Saxitoxin blocks L-type I_{Ca}. *Journal of Pharmacology and Experimental Therapeutics* 308(1), 324–329. doi: 10.1124/jpet.103.056564.
Šulčius, S., Slavuckytė, K. and Paškauskas, R. (2017) The predation paradox: synergistic and antagonistic interactions between grazing by crustacean predator and infection by cyanophages promotes bloom formation in filamentous cyanobacteria. *Limnology and Oceanography* 62(5), 2189–2199. doi: 10.1002/lno.10559.
Šulčius, S., Mazur-Marzec, H., Vitonytė, I., Kvederavičiūtė, K., Kuznecova, J., *et al.* (2018) Insights into cyanophage-mediated dynamics of nodularin and other non-ribosomal peptides in *Nodularia spumigena*. *Harmful Algae* 78, 69–74. doi: 10.1016/j.hal.2018.07.004.
Sullivan, T.J., Dhar, A.K., Cruz-Flores, R. and Bodnar, A.G. (2019) Rapid, CRISPR-based, field-deployable detection of white spot syndrome virus in shrimp. *Scientific Reports* 9(1), 19702. doi: 10.1038/s41598-019-56170-y.
Suttle, C.A. (2005) Viruses in the sea. *Nature* 437(7057), 356–361. doi: 10.1038/nature04160.
Svendsen, M.B.S., Andersen, N.R., Hansen, P.J. and Steffensen, J.F. (2018) Effects of harmful algal blooms on fish: insights from *Prymnesium parvum*. *Fishes* 3(1), 11. doi: 10.3390/fishes3010011.
Svirčev, Z., Lujić, J., Marinović, Z., Drobac, D., Tokodi, N., *et al.* (2015) Toxicopathology induced by microcystins and nodularin: a histopathological review. *Journal of Environmental Science and Health, Part C: Environmental Carcinogenesis and Ecotoxicology Reviews* 33(2), 125–167. doi: 10.1080/10590501.2015.1003000.
Svirčev, Z., Obradović, V., Codd, G.A., Marjanović, P., Spoof, L., *et al.* (2016) Massive fish mortality and *Cylindrospermopsis raciborskii* bloom in Aleksandrovac Lake. *Ecotoxicology* 25(7), 1353–1363. doi: 10.1007/s10646-016-1687-x.
Svirčev, Z., Lalić, D., Bojadžija Savić, G., Tokodi, N., Drobac Backović, D., *et al.* (2019) Global geographical and historical overview of cyanotoxin distribution and cyanobacterial poisonings. *Archives of Toxicology* 93(9), 2429–2481. doi: 10.1007/s00204-019-02524-4.
Swain, P., Nayak, S.K., Nanda, P.K. and Dash, S. (2008) Biological effects of bacterial lipopolysaccharide (endotoxin) in fish: a review. *Fish & Shellfish Immunology* 25(3), 191–201. doi: 10.1016/j.fsi.2008.04.009.
Tang, Y.Z. and Gobler, C.J. (2009) Characterization of the toxicity of *Cochlodinium polykrikoides* isolates from Northeast US estuaries to finfish and shellfish. *Harmful Algae* 8(3), 454–462. doi: 10.1016/j.hal.2008.10.001.
Taylor, R.B., Hill, B.N., Bobbitt, J.M., Hering, A.S., Brooks, B.W. and Chambliss, C.K. (2020) Suspect and non-target screening of acutely toxic *Prymnesium parvum*. *Science of the Total Environment* 715, 136835. doi: 10.1016/j.scitotenv.2020.136835.
Taylor, R.B., Hill, B.N., Langan, L.M., Chambliss, C.K. and Brooks, B.W. (2021) Sunlight concurrently reduces *Prymnesium parvum* elicited acute toxicity to fish and prymnesins. *Chemosphere* 263, 127927. doi: 10.1016/j.chemosphere.2020.127927.
Tencalla, F.G., Dietrich, D.R. and Schlatter, C. (1994) Toxicity of *Microcystis aeruginosa* peptide toxin to yearling rainbow trout (*Oncorhynchus mykiss*). *Aquatic Toxicology* 30(3), 215–224. doi: 10.1016/0166-445X(94)90059-0.

Terao, K., Ohmori, S., Igarashi, K., Ohtani, I., Watanabe, M.F., *et al.* (1994) Electron microscopic studies on experimental poisoning in mice induced by cylindrospermopsin isolated from blue-green alga *Umezakia natans*. *Toxicon* 32(7), 833–843. doi: 10.1016/0041-0101(94)90008-6.

Testai, E., Buratti, F.M., Funari, E., Manganelli, M., Vichi, S., *et al.* (2016) Review and analysis of occurrence, exposure and toxicity of cyanobacteria toxins in food. *EFSA Supporting Publications* 13(2), 998E. doi: 10.2903/sp.efsa.2016.EN-998.

Tester, P.A., Turner, J.T. and Shea, D. (2000) Vectorial transport of toxins from the dinoflagellate *Gymnodinium breve* through copepods to fish. *Journal of Plankton Research* 22(1), 47–62. doi: 10.1093/plankt/22.1.47.

Tian, L., Cheng, J., Chen, X., Cheng, S.H., Mak, Y.L., *et al.* (2014) Early developmental toxicity of saxitoxin on medaka (*Oryzias melastigma*) embryos. *Toxicon* 77, 16–25. doi: 10.1016/j.toxicon.2013.10.022.

Tillmann, U. (2003) Kill and eat your predator: a winning strategy of the planktonic flagellate *Prymnesium parvum*. *Aquatic Microbial Ecology* 32(1), 73–84. doi: 10.3354/ame032073.

Timmons, M.B. and Ebeling, J.M. (2013) *Recirculating Aquaculture*, 3rd edn. Ithaca Publishing Company, LLC, Ithaca, New York.

Toranzo, A.E., Magariños, B. and Romalde, J.L. (2005) A review of the main bacterial fish diseases in mariculture systems. *Aquaculture* 246(1–4), 37–61. doi: 10.1016/j.aquaculture.2005.01.002.

Toruńska, A., Bolałek, J., Pliński, M. and Mazur-Marzec, H. (2008) Biodegradation and sorption of nodularin (NOD) in fine-grained sediments. *Chemosphere* 70(11), 2039–2046. doi: 10.1016/j.chemosphere.2007.09.015.

Trainer, V.L. (ed.) (2020) *GlobalHAB: Evaluating, Reducing and Mitigating the Cost of Harmful Algal Blooms: A Compendium of Case Studies*. PICES Scientific Report No. 59. North Pacific Marine Science Organization (PICES), Sidney, British Columbia, Canada.

Trainer, V.L., Bates, S.S., Lundholm, N., Thessen, A.E., Cochlan, W.P., *et al.* (2012) *Pseudo-nitzschia* physiological ecology, phylogeny, toxicity, monitoring and impacts on ecosystem health. *Harmful Algae* 14, 271–300. doi: 10.1016/j.hal.2011.10.025.

Trainer, V.L., Moore, L., Eberhart, B.T.L., Bill, B.D., Cochlan, W.P., *et al.* (2015) Characterizing toxic activity from *Heterosigma akashiwo*: a tale of two assays. In: MacKenzie, A.L. (ed.) *Marine and Freshwater Harmful Algae: Proceedings of the 16th International Conference on Harmful Algae, 27–31 October 2014, Wellington, New Zealand*. Cawthron Institute, Nelson, New Zealand and the International Society for the Study of Harmful Algae (ISSHA), pp. 82–85.

Trainer, V.L., Moore, S.K., Hallegraeff, G., Kudela, R.M., Clement, A., *et al.* (2020) Pelagic harmful algal blooms and climate change: lessons from nature's experiments with extremes. *Harmful Algae* 91, 101591. doi: 10.1016/j.hal.2019.03.009.

Traylor, J. and Singhal, M. (2018) *Ciguatera Toxicity*. StatPearls Publishing, Treasure Island, Florida.

Triest, L., Stiers, I. and Van Onsem, S. (2016) Biomanipulation as a nature-based solution to reduce cyanobacterial blooms. *Aquatic Ecology* 50(3), 461–483. doi: 10.1007/s10452-015-9548-x.

Ulitzur, S. and Shilo, M. (1966) Mode of action of *Prymnesium parvum* ichthyotoxin. *The Journal of Protozoology* 13(2), 332–336. doi: 10.1111/j.1550-7408.1966.tb01915.x.

UN (United Nations) (2017) *World Population Prospects: The 2017 Revision*. UN, New York. Available at: https://www.un.org/development/desa/publications/world-population-prospects-the-2017-revision.html#:~:text=The%20current%20world%20population%20of,Nations%20report%20being%20launched%20today. (accessed 21 October 2022).

Van Apeldoorn, M.E., Van Egmond, H.P., Speijers, G.J. and Bakker, G.J. (2007) Toxins of cyanobacteria. *Molecular Nutrition & Food Research* 51(1), 7–60. doi: 10.1002/mnfr.200600185.

Van Buynder, P.G., Oughtred, T., Kirkby, B., Phillips, S., Eaglesham, G., *et al.* (2001) Nodularin uptake by seafood during a cyanobacterial bloom. *Environmental Toxicology* 16(6), 468–471. doi: 10.1002/tox.10004.

Vasconcelos, J.F., Barbosa, J.E.L., Lira, W. and Azevedo, S.M.F.O. (2013) Microcystin bioaccumulation can cause potential mutagenic effects in farm fish. *The Egyptian Journal of Aquatic Research* 39(3), 185–192. doi: 10.1016/j.ejar.2013.11.002.

Verspagen, J.M., Van de Waal, D.B., Finke, J.F., Visser, P.M., Van Donk, E. and Huisman, J. (2014) Rising CO_2 levels will intensify phytoplankton blooms in eutrophic and hypertrophic lakes. *PLoS ONE* 9(8), e104325. doi: 10.1371/journal.pone.0104325.

Viaggiu, E., Melchiorre, S., Volpi, F., Di Corcia, A., Mancini, R., *et al.* (2004) Anatoxin-a toxin in the cyanobacterium *Planktothrix rubescens* from a fishing pond in northern Italy. *Environmental Toxicology* 19(3), 191–197. doi: 10.1002/tox.20011.

Visser, P.M., Ibelings, B.W., Bormans, M. and Huisman, J. (2016) Artificial mixing to control cyanobacterial blooms: a review. *Aquatic Ecology* 50(3), 423–441. doi: 10.1007/s10452-015-9537-0.

Volkoff, H. and Peter, R.E. (2004) Effects of lipopolysaccharide treatment on feeding of goldfish: role of appetite-regulating peptides. *Brain Research* 998(2), 139–147. doi: 10.1016/j.brainres.2003.11.011.

Vuorinen, P.J., Sipiä, V.O., Karlsson, K., Keinänen, M., Furey, A., *et al.* (2009) Accumulation and effects of nodularin from a single and repeated oral doses of cyanobacterium *Nodularia spumigena* on flounder (*Platichthys flesus* L.). *Archives of Environmental Contamination and Toxicology* 57(1), 164–173. doi: 10.1007/s00244-008-9258-7.

Wagstaff, B.A., Pratscher, J., Rivera, P.P.L., Hems, E.S., Brooks, E., *et al.* (2021) Assessing the toxicity and mitigating the impact of harmful *Prymnesium* blooms in eutrophic waters of the Norfolk Broads. *Environmental Science & Technology* 55(24), 16538–16551. doi: 10.1021/acs.est.1c04742.

Walsby, A. (1994) Gas vesicles. *Microbiological Reviews* 58(1), 94–144. doi: 10.1128/mr.58.1.94-144.1994.

Wang, J., Salata, J.J. and Bennett, P.B. (2003) Saxitoxin is a gating modifier of HERG K^+ channels. *Journal of General Physiology*, 121(6), 583–598. doi: 10.1085/jgp.200308812.

Wang, L., Liang, X.F., Liao, W.Q., Lei, L.M. and Han, B.P. (2006) Structural and functional characterization of microcystin detoxification-related liver genes in a phytoplanktivorous fish, Nile tilapia (*Oreochromis niloticus*). *Comparative Biochemistry and Physiology Part C: Toxicology & Pharmacology* 144(3), 216–227. doi: 10.1016/j.cbpc.2006.08.009.

Wang, Z., Li, G., Wu, Q., Liu, C., Shen, J. and Yan, W. (2019) Microcystin-LR exposure induced nephrotoxicity by triggering apoptosis in female zebrafish. *Chemosphere* 214, 598–605. doi: 10.1016/j.chemosphere.2018.09.103.

Watanabe, M., Kohata, K., Kimura, T., Takamatsu, T., Yamaguchi, S.I. and Ioriya, T. (1995) Generation of a *Chattonella antiqua* bloom by imposing a shallow nutricline in a mesocosm. *Limnology and Oceanography* 40(8), 1447–1460. doi: 10.4319/lo.1995.40.8.1447.

Wedemeyer, G., Ross, A.J. and Smith, L. (1969) Some metabolic effects of bacterial endotoxins in salmonid fishes. *Journal of the Fisheries Research Board of Canada* 26(1), 115–122. doi: 10.1139/f69-010.

Weeratunge, N., Béné, C., Siriwardane, R., Charles, A., Johnson, D., *et al.* (2014) Small-scale fisheries through the wellbeing lens. *Fish and Fisheries* 15(2), 255–279. doi: 10.1111/faf.12016.

Weiss, J.H. and Choi, D.W. (1988) Beta-*N*-methylamino-L-alanine neurotoxicity: requirement for bicarbonate as a cofactor. *Science* 241(4868), 973–975. doi: 10.1126/science.3136549.

Wells, M.L., Karlson, B., Wulff, A., Kudela, R., Trick, C., *et al.* (2020) Future HAB science: directions and challenges in a changing climate. *Harmful Algae* 91, 101632. doi: 10.1016/j.hal.2019.101632.

Welsh, J.E., Steenhuis, P., de Moraes, K.R., van der Meer, J., Thieltges, D.W. and Brussaard, C.P. (2020) Marine virus predation by non-host organisms. *Scientific Reports* 10(1), 5221. doi: 10.1038/s41598-020-61691-y.

Welten, R.D., Meneely, J.P. and Elliott, C.T. (2020) A comparative review of the effect of microcystin-LR on the proteome. *Exposure and Health* 12(2), 111–129. doi: 10.1007/s12403-019-00303-1.

White, A.W. (1981) Marine zooplankton can accumulate and retain dinoflagellate toxins and cause fish kills. *Limnology and Oceanography* 26(1), 103–109. doi: 10.4319/lo.1981.26.1.0103.

Wicks, R.J. and Thiel, P.G. (1990) Environmental factors affecting the production of peptide toxins in floating scums of the cyanobacterium *Microcystis aeruginosa* in a hypertrophic African reservoir. *Environmental Science & Technology* 24(9), 1413–1418. doi: 10.1021/es00079a017.

Wiegand, C. and Pflugmacher, S. (2005) Ecotoxicological effects of selected cyanobacterial secondary metabolites a short review. *Toxicology and Applied Pharmacology* 203(3), 201–218. doi: 10.1016/j.taap.2004.11.002.

Wiese, M., D'Agostino, P.M., Mihali, T.K., Moffitt, M.C. and Neilan, B.A. (2010) Neurotoxic alkaloids: saxitoxin and its analogs. *Marine Drugs* 8(7), 2185–2211. doi: 10.3390/md8072185.

Wilhelm, S.W. and Suttle, C.A. (1999) Viruses and nutrient cycles in the sea: viruses play critical roles in the structure and function of aquatic food webs. *BioScience* 49(10), 781–788. doi: 10.2307/1313569.

Wilhelm, S.W., Bullerjahn, G.S. and McKay, R.M.L. (2020) The complicated and confusing ecology of *Microcystis* blooms. *mBio* 11(3), e00529-20. doi: 10.1128/mBio.00529-20.

Wilson, A.E., Wilson, W.A. and Hay, M.E. (2006) Intraspecific variation in growth and morphology of the bloom-forming cyanobacterium *Microcystis aeruginosa*. *Applied and Environmental Microbiology* 72(11), 7386–7389. doi: 10.1128/AEM.00834-06.

Wimmer, K.M., Strangman, W.K. and Wright, J.L. (2014) 7-Deoxy-desulfo-cylindrospermopsin and 7-deoxy-desulfo-12-acetylcylindrospermopsin: two new cylindrospermopsin analogs isolated from a Thai strain of *Cylindrospermopsis raciborskii*. *Harmful Algae* 37, 203–206. doi: 10.1016/j.hal.2014.06.006.

Wonnacott, S. and Gallagher, T. (2006) The chemistry and pharmacology of anatoxin-a and related homotropanes with respect to nicotinic acetylcholine receptors. *Marine Drugs* 4(3), 228–254. doi: 10.3390/md403228.

WHO (World Health Organization) (2020a) *Cyanobacterial Toxins: Microcystins. Background Document for Development of WHO Guidelines for Drinking-Water Quality and Guidelines for Safe Recreational Water Environments*. WHO/HEP/ECH/WSH/2020.6. WHO, Geneva, Switzerland.

WHO (World Health Organization) (2020b) *Cyanobacterial Toxins: Cylindrospermopsins. Background Document for Development of WHO Guidelines for Drinking-Water Quality and Guidelines for Safe Recreational Water Environments*. WHO/HEP/ECH/WSH/2020.4. WHO, Geneva, Switzerland.

WHO (World Health Organization) (2020c) *Cyanobacterial Toxins: Anatoxin-a and Analogues. Background Document for Development of WHO Guidelines for Drinking-Water Quality and Guidelines for Safe Recreational Water Environments*. WHO/HEP/ECH/WSH/2020.1. WHO, Geneva, Switzerland.

WHO (World Health Organization) (2020d) *Cyanobacterial Toxins: Saxitoxins. Background Document for Development of WHO Guidelines for Drinking-Water Quality and Guidelines for Safe Recreational Water Environments*. WHO/HEP/ECH/WSH/2020.8. WHO, Geneva, Switzerland.

Wu, L., Bradshaw, A.D. and Thurman, D.A. (1975) Potential for evolution of heavy metal tolerance in plants. III. The rapid evolution of copper tolerance in *Agrostis stolonifera*. *Heredity* 34(2), 165–187. doi: 10.1038/hdy.1975.21.

Wu, N., Tong, M., Gou, S., Zeng, W., Xu, Z. and Jiang, T. (2021) Hemolytic activity in relation to the photosynthetic system in *Chattonella marina* and *Chattonella ovata*. *Marine Drugs* 19(6), 336. doi: 10.3390/md19060336.

Wüest, A., Bouffard, D., Guillard, J., Ibelings, B.W., Lavanchy, S., *et al.* (2021) LéXPLORE: a floating laboratory on Lake Geneva offering unique lake research opportunities. *WIREs: Water* 8(5), e1544. doi: 10.1002/wat2.1544.

Xiang, L.X., Peng, B., Dong, W.R., Yang, Z.F. and Shao, J.Z. (2008) Lipopolysaccharide induces apoptosis in *Carassius auratus* lymphocytes, a possible role in pathogenesis of bacterial infection in fish. *Developmental & Comparative Immunology* 32(8), 992–1001. doi: 10.1016/j.dci.2008.01.009.

Xie, L., Xie, P., Guo, L., Li, L., Miyabara, Y. and Park, H.D. (2005) Organ distribution and bioaccumulation of microcystins in freshwater fish at different trophic levels from the eutrophic Lake Chaohu, China. *Environmental Toxicology* 20(3), 293–300. doi: 10.1002/tox.20120.

Yan, M., Leung, P.T., Ip, J.C., Cheng, J.P., Wu, J.J., *et al.* (2017) Developmental toxicity and molecular responses of marine medaka (*Oryzias melastigma*) embryos to ciguatoxin P-CTX-1 exposure. *Aquatic Toxicology* 185, 149–159. doi: 10.1016/j.aquatox.2017.02.006.

Yan, M., Mak, M.Y., Cheng, J., Li, J., Gu, J.R., *et al.* (2020) Effects of dietary exposure to ciguatoxin P-CTX-1 on the reproductive performance in marine medaka (*Oryzias melastigma*). *Marine Pollution Bulletin* 152, 110837. doi: 10.1016/j.marpolbul.2019.110837.

Yogi, K., Oshiro, N., Inafuku, Y., Hirama, M. and Yasumoto, T. (2011) Detailed LC-MS/MS analysis of ciguatoxins revealing distinct regional and species characteristics in fish and causative alga from the Pacific. *Analytical Chemistry* 83(23), 8886–8891. doi: 10.1021/ac200799j.

Yong, H.L., Mustapa, N.I., Lee, L.K., Lim, Z.F., Tan, T.H., *et al.* (2018) Habitat complexity affects benthic harmful dinoflagellate assemblages in the fringing reef of Rawa Island, Malaysia. *Harmful Algae* 78, 56–68. doi: 10.1016/j.hal.2018.07.009.

Yoshida, T., Jones, L.E., Ellner, S.P., Fussmann, G.F. and Hairston, N.G. (2003) Rapid evolution drives ecological dynamics in a predator–prey system. *Nature* 424(6946), 303–306. doi: 10.1038/nature01767.

Zhang, Y., Zhang, S.F., Lin, L. and Wang, D.Z. (2014) Comparative transcriptome analysis of a toxin-producing dinoflagellate *Alexandrium catenella* and its non-toxic mutant. *Marine Drugs* 12(11), 5698–5718. doi: 10.3390/md12115698.

Zilber-Rosenberg, I. and Rosenberg, E. (2008) Role of microorganisms in the evolution of animals and plants: the hologenome theory of evolution. *FEMS Microbiology Reviews* 32(5), 723–735. doi: 10.1111/j.1574-6976.2008.00123.x.

Zimba, P.V., Khoo, L., Gaunt, P.S., Brittain, S. and Carmichael, W.W. (2001) Confirmation of catfish, *Ictalurus punctatus* (Rafinesque), mortality from *Microcystis* toxins. *Journal of Fish Diseases* 24(1), 41–47. doi: 10.1046/j.1365-2761.2001.00273.x

Zimmermann, K., Deuis, J.R., Inserra, M.C., Collins, L.S., Namer, B., *et al.* (2013) Analgesic treatment of ciguatoxin-induced cold allodynia. *Pain* 154(10), 1999–2006. doi: 10.1016/j.pain.2013.06.015.

11 Biosecurity: Current and Future Strategies

Rohana P. Subasinghe[1]* and Andrew P. Shinn[2]
[1]FUTUREFISH Co. Ltd., London, UK; [2]INVE (Thailand), Nonthaburi, Thailand

11.1 Introduction

Aquaculture is the fastest-growing food production sector, and its contributions to food, nutrition security and poverty alleviation have been highlighted by Salin *et al.* (Chapter 1, this volume, 2023). It is also a highly diverse production sector. More than 500 species of fish are farmed in an array of environments (fresh water, brackish water and marine), production systems (e.g. ponds, cages, recirculating systems, integrated multitrophic systems, polyculture, nurseries, grow-out facilities and hatcheries), using three main management strategies (extensive, semi-intensive and intensive) under several different sizes of operation (from backyard and subsistence-level to small-, medium- and large-scale (and industrial) operations). Cage aquaculture (see Chapter 1, this volume) contributes significantly to global aquaculture production of several important freshwater and marine finfish species, bridging the supply and demand gap for aquatic foods in the coming decades.

Various diseases and disorders negatively impacting production of cultured finfish, especially in cage-culture systems, are described in this volume. In many instances the lack of efficient and effective biosecurity is the key factor underlying episodes of disease and subsequent production losses. During the last three decades, the aquaculture sector has experienced several re-emerging and newly emerging diseases at a rate of approximately one new disease every 3 to 5 years. While measures to prevent and reduce impacts of diseases that affect aquaculture have been launched by many stakeholders (e.g. government, producer, academic sectors), these efforts have not adequately addressed the root causes of disease problems challenging the industry (FAO, 2019).

The aquaculture value chain is complex with many segments and actors operating, and with each segment posing a source of risk. As a result, pathogen emergence, re-emergence and spread have, over the past few decades, been a continuous constraint to sustainable aquaculture worldwide.

11.2 Impacts of Diseases in Aquaculture

Disease is the most important threat and hurdle to sustainable aquaculture production. Impacts and losses due to disease are divided into the 'predictable' and the 'unpredictable'. While there

*rohana@futurefish.org

DOI: 10.1079/9781800621640.0011

are costs in treating and managing both, for predictable disease events there are also additional costs associated with prophylactic treatments and management (Shinn *et al.*, 2015). Although the World Organisation for Animal Health (OIE) lists major aquatic pathogens and diseases (OIE, 2021a) (Table 11.1), there are many other non-listed pathogens which cause serious losses to the aquaculture industry.

Disease events often pass undetected, undiagnosed, unreported and/or misdiagnosed, thereby hiding the true severity of the disease. If the actual cost pathogens pose to the industry can be estimated, then this information can be used to raise awareness with governments and to prioritize research and development programmes focusing on those pathogens. Loss estimates can further help investment in the aquaculture industry when used to calculate the risks for potential investors, as well as help farmers in their biosecurity and financial contingency plans.

At a farm level, the first critical step is in recognizing health issues. Unfortunately, many losses are just accepted as part of normal production. This fatalistic acceptance means that many diseases go unreported, hiding their true severity and threatening the advance of the aquaculture sector.

The impacts of complacency can also be seen in routine farm practices. While the execution of many procedures may be simple, it is human nature to look for shortcuts. This happens frequently in biosecurity. Measures are implemented but then are either not maintained or omitted in the absence of disease episodes, and consequentially the probability of a disease with high mortalities increases.

Limited knowledge or understanding about proper biosecurity planning combined with a strong desire for high crop production can, and has, resulted in catastrophic losses (see Fig. 11.1). For example, poor site design, high cage density with limited spacing between cages for water exchange and overstocking can result in poor flush rates, a rapid deterioration in water quality, a rise in nitrogenous wastes and can facilitate disease transmission. Overstocking results in additional stress on animals kept under culture, reducing their general health condition and increasing their predisposition to infections.

11.2.1 Economic impacts

Loss estimates for the top 100 diseases impacting aquaculture production have been compiled (Shinn *et al.*, 2018b) and include a wide range of viral, bacterial and parasitic pathogens (see Figs 11.2 and 11.3). Leading viral candidates include infectious haematopoietic necrosis virus (IHNV) (Saksida, 2006; Garver *et al.*, 2013), *Infectious pancreatic necrosis virus* (IPNV) (Christie, 1997; Munro and Midtlyng, 2011; Jensen and Kristoffersen, 2015), infectious salmon anaemia (ISA) (Rimstad *et al.*, 2011; Kibenge *et al.*, 2013), pancreas disease caused by salmonid alphavirus (Aunsmo *et al.*, 2010; Pettersen *et al.*, 2015), tilapia lake virus (TiLV) (Ferguson *et al.*, 2014; Dong *et al.*, 2017a,b; Surachetpong *et al.*, 2017) and viral haemorrhagic septicaemia (VHS) (Hill, 1992; Dale *et al.*, 2009; Kim *et al.*, 2009). For diseases like that caused by TiLV, an orthomyxo-like RNA virus that causes a syncytial hepatitis with a loss of condition, altered swimming behaviour and high rates of mortality, there is at the present time a general lack of data on losses, as production can readily compensate for losses.

Key bacterial pathogens include *Aeromonas hydrophila* (see Rasmussen-Ivey *et al.*, 2016), *Flavobacterium columnare* (see Shoemaker *et al.*, 2007, 2011), *Piscirickettsia salmonis* (see Berger, 2014) and *Vibrio anguillarum* (see Kitao *et al.*, 1983). Likewise, losses resulting from infections of *Streptococcus agalactiae* and *Streptococcus iniae* are substantial (Shoemaker *et al.*, 2001; Klesius *et al.*, 2008; Shinn *et al.*, 2018b). Shinn *et al.* (2018b) estimate from reports of *Streptococcus* spp. across Asia that a 7.5% mortality across the industry can be assumed, except for countries like Bangladesh and the Philippines which harvest fish at smaller size where a 3.75% rate of mortality is applied. Using a figure of 6,192,963 tonnes of 13 farmed species of tilapia (i.e. species and hybrids belonging to the genera *Coptodon*, *Oreochromis*, *Sarotherodon* and *Tilapia*) produced in 2019 (FishStatJ, 2021), then losses due to *Streptococcus* spp. are close to 440,860 tonnes and a harvest value of US$876.87 million by applying an average value of US$1,989 per tonne.

Argulus appears to be a leading cause of financial loss due to parasites. In the cyprinid industry, *Argulus*-associated losses are estimated at US$6.4 billion/year (i.e. 19.6% of the value of

Table 11.1. Key details regarding the OIE-notifiable diseases of fish. For each pathogen, comments on its transmission, persistence in the environment and if a vaccine is available are provided.

		Transmission					
Disease	Pathogen	Horizontal	Vertical	Temperature range (°C)	Survival	Vaccine available?	Reference
Epizootic haematopoietic necrosis	Virus	Yes	No data	11–20	Presumed that the virus can survive months to years on a fish farm and in sediment, plants and farm equipment. There are no data regarding the presence of the virus on eggs and larvae	No	Langdon (1989); OIE (2019a)
Epizootic ulcerative syndrome	Oomycete	Yes	No	Mostly 18–22	Encysted zoospores survive for at least 19 days. Fish eggs and larvae should be disinfected but there is no report of *Aphanomyces invadans* presence on them	Immunization of snakehead with a crude extract of *A. invadans* elicited a humoral immune response	Thompson *et al.* (1997); Lilley *et al.* (2001); OIE (2019b)
Gyrodactylus salaris	Monogenean	Yes	No	0–25	Detached flukes can survive for 132 h at 3°C to 24 h at 19°C. Eggs should be disinfected	No	Olstad *et al.* (2006); OIE (2019c)
Infectious salmon anaemia	Virus	Yes	No strong data	–	Over 14 days at 4°C to >10 days at 15°C. Eggs and embryo could be transmission risk	Yes, since 1999	Falk *et al.* (1997); OIE (2019d)
Infectious haematopoietic necrosis	Virus	Yes	Insufficient evidence	3–18	Virus decay rate is 4.11–4.37 days (8–12°C) in natural seawater and can be inactivated by suspended sediments and the microbial community within the water source. Survival is longer in fresh water than seawater. Transmission on inadequately disinfected eggs possible	Plasmid DNA vaccines containing the glycoprotein gene of the IHNV administered intramuscularly are efficacious	Lapatra (1998); Garver *et al.* (2005, 2013); Alonso and Leong (2013); Kell *et al.* (2014); OIE (2021b)

Koi herpesvirus disease	Virus	Yes	Egg-associated transmission cannot be ruled out	opt. 23–25 (>13–<29)	Studies in Israel found the virus survived >4 h and <21 h in water at 23–25°C. Studies in Japan found viral titres dropped within 3 days in environmental water and sediment at 15°C but remained infective for >7 days in filtered water. Virus DNA has been found in river water at 9–11°C, 4 months before an outbreak	A live attenuated virus has been used. Oral administration of a liposome-based vaccine containing inactivated virus has also proven effective	Perelberg *et al.* (2003); Haramoto *et al.* (2007); Ilouze *et al.* (2011); OIE (2019e)
Red sea bream iridoviral disease	Virus	Yes	Not been investigated	25+	Details regarding survival outside the host are unknown	An effective formalin-killed commercial vaccine is able for use in *Epinephelus coioides*, *Epinephelus malabaricus*, *Pagrus major*, *Pseudocaranx dentex* and *Seriola* spp.	OIE (2019f)
Salmonid alphavirus infection	Virus	Yes	No convincing evidence/ negligible	opt. 12–15	Salmonid alphavirus (SAV) subtype 3 (SAV3) shed by fish was completely inactivated after 3 weeks at 1°C and 4 weeks at 12°C. Graham *et al.* (2007) indicated the half-life of SAV is 2.5 days at 20°C increasing up to 61 days at 4°C (seawater, no organic loading), 2.5 days (20°C) to 7.7 days (4°C) in seawater with organic loading. In fresh water, SAV half-lives were 2.2 days (20°C) to 7.6 days (4°C) with organic loading, and 5.0 days (20°C) to 13.2 days (4°C) with no organic loading	DNA-based and virus-inactivated commercial vaccines available	Graham *et al.* (2007); OIE (2019g); Jarungsriapisit *et al.* (2020)

Continued

Table 11.1. Continued.

Disease	Pathogen	Transmission		Temperature range (°C)	Survival	Vaccine available?	Reference
		Horizontal	Vertical				
Spring viraemia of carp	Virus	Yes	Egg-associated transmission cannot be ruled out	>10–<22	5 weeks in river water at 10°C, >6 weeks in pond mud at 4°C, 4 days in pond mud at 10°C	A DNA vaccine that consists of plasmid DNA expression vectors given intramuscularly results in gene expression of immunogenic proteins for koi herpesvirus in fish muscle tissue	Ahne (1976); Zhou *et al.* (2014); OIE (2019h)
Viral haemorrhagic septicaemia	Virus	Yes	No	<28	Virus more stable in fresh water (28–35 days at 4°C; 1 year in filtered water at 4°C; 13 days at 15°C) than seawater (10–12 days at 4°C; 4 days at 15°C).	No commercial vaccine. Candidate vaccines, however, have included killed vaccines, attenuated live vaccines, recombinant vaccine in prokaryotic and eukaryotic expression systems, and DNA-based vaccines	Parry and Dixon (1997); Lorenzen and Lapatra (2005); Hawley and Garver (2008); OIE (2021c)

Fig. 11.1. The Nile tilapia, *Oreochromis niloticus*, cultured at this undisclosed Nile tilapia riverine site were reared in cages 12 units deep and in blocks which ran for several hundred metres without a break. The poor water exchange, water quality, stress and bacterial problems that were created led to a significant loss of stock. (Image by A. Shinn.)

the industry) (Shinn *et al.*, 2018b). Other key parasite-related losses include those due to amoebic gill disease (AGD) (Adams and Nowak, 2001; Munday *et al.*, 2001; Vass, 2013; Shinn *et al.*, 2015) and salmon lice (i.e. *Lepeophtheirus salmonis* and *Caligus* spp.), which remains the number one parasite problem in salmonid aquaculture with losses of >US$600 million/year (Shinn *et al.*, 2015). There are also notable losses due to the monogenetic skin flukes *Benedenia* and *Neobenedenia* which, notably, infect seriolids and can account for 22% of production costs, that is, approximately US$400 million/year (Whittington *et al.*, 2001; Ernst *et al.*, 2002). In fresh water, huge losses of small fish during hatchery–nursery production are due to the ciliate protozoans *Ichthyophthirius multifiliis* and *Trichodina* spp., which together account for an estimated loss of US$56.4 million/year assuming a 1% mortality (Shinn *et al.*, 2018b; Shinn *et al.*, Chapter 6, this volume, 2023).

Estimating annual disease losses is challenging, can be variable and is dependent on many biotic, abiotic and anthropogenic factors, so some caution should be exercised in the interpretation and application of these figures. By attempting to provide estimates of loss though, it is hoped that this will raise awareness within the industry, help lower the threshold of acceptable loss and drive further improvements in animal welfare, codes of practice, system management and animal health. With a greater transparency on losses associated with disease outbreaks the aquaculture sector could reduce instabilities, improve production and profitability, and achieve greater sustainability.

11.3 Minimizing Disease Impacts in Aquaculture

Aquaculture expansion and globalization of aquatic animal trade have caused unprecedented spread of diseases and pathogens. Several aquatic disease epidemics have substantially impacted

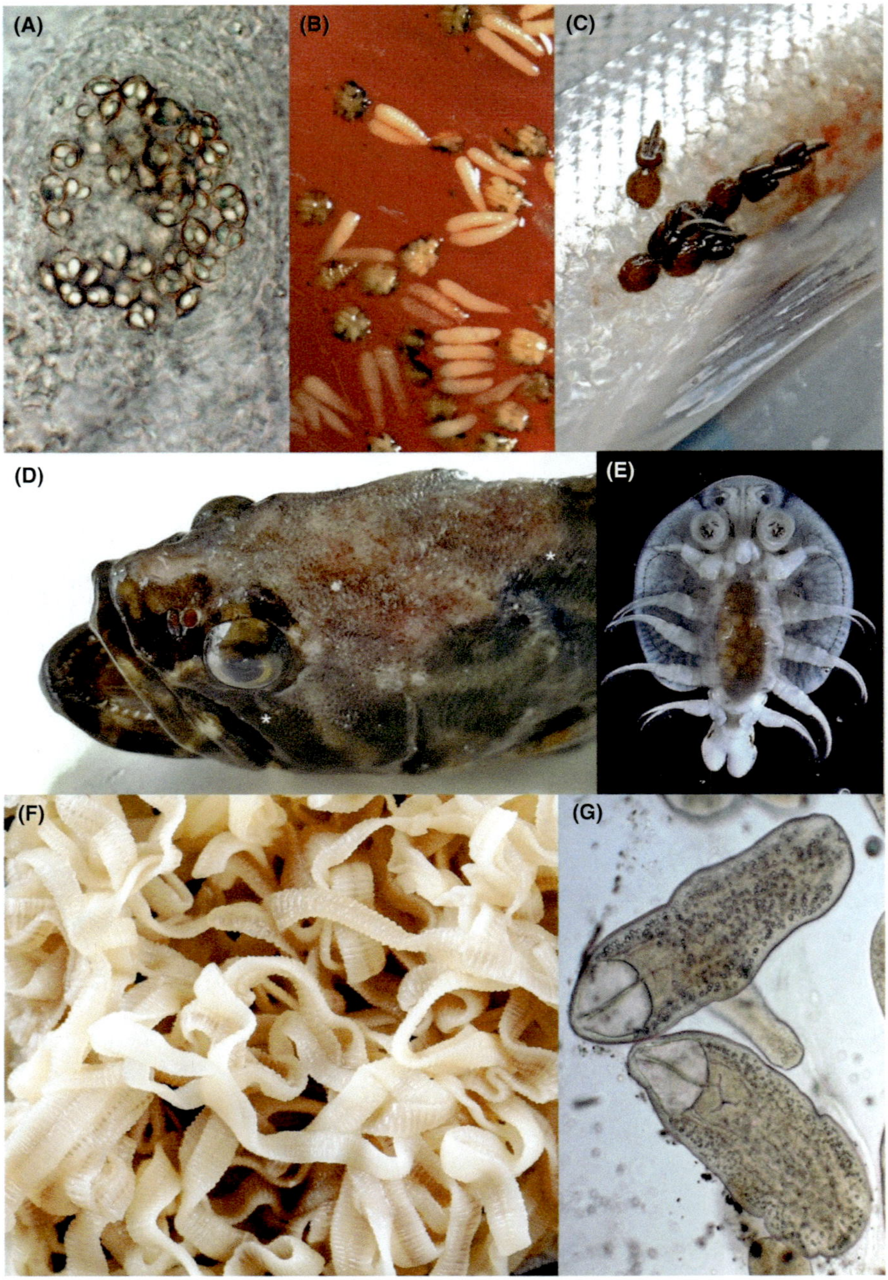

Fig. 11.2. Pathogens of cage-reared finfish. (A) *Myxobolus* sp. from Vietnamese striped catfish, *Pangasianodon hypophthalmus* (image by A. Shinn). (B) Adult female *Ergasilus sieboldi* on the gills of a rainbow trout, *Oncorhynchus mykiss* (image by A. Shinn). (C) Adult female *Lepeophtheirus salmonis* on Atlantic salmon, *Salmo salar* (image courtesy of D. Conway). (D) Feeding activity of *Caligus epidemicus* (asterisk) has denuded the epithelium of a brown-marbled grouper, *Epinephelus fuscoguttatus* (image by A. Shinn). (E) Gravid female *Argulus foliaceus* (image by A. Shinn). (F) Strobila of *Eubothrium crassum* (image courtesy of D. Conway). (G) Metacercariae of *Diplostomum spathaceum* released from the lens of rainbow trout (image by A. Shinn).

Fig. 11.3. Diseases and pathogens of cage-reared finfish. (A) Dual exophthalmia in a Nile tilapia, *Oreochromis niloticus*, infected with *Streptococcus agalactiae* (image by A. Shinn). (b) An *Aeromonas hydrophila* infection in Nile tilapia (image by A. Shinn). (C) Brown trout, *Salmo trutta*, infected with *Saprolegnia* sp. (image by A. Shinn). (D) *Apiosoma* sp. and *Gyrodactylus* sp. (asterisk) infection (image courtesy of G. Paladini). (E) *Ichthyobodo* (*Costia*) *necatrix* on the skin of a brown trout (image courtesy of G. Paladini).

regional, national and, in some cases, international industries and trade (Lafferty *et al.*, 2015). One recent example is the emergence and rapid spread of 'early mortality syndrome' or acute hepatopancreatic necrosis disease (AHPND) caused by pathogenic *Vibrio parahaemolyticus* carrying a *Pir*-toxin bearing plasmid in shrimp (De Schryver *et al.*, 2014; Shinn *et al.*, 2018a). This has severely impacted global shrimp production. Other examples include ISA outbreaks on salmon farms in Norway, Canada, the USA and Chile (Lafferty *et al.*, 2015) and the spread of epizootic ulcerative syndrome (EUS) in freshwater fish in almost all parts of the world that has caused heavy social and economic impacts worldwide (Kamilya and Baruah, 2014).

Risks to aquaculture, including cage culture, from climate change have been discussed widely (e.g. De Silva and Soto, 2009; Howes *et al.*, 2015; FAO, 2016). Temperature rises above and below the optimal range of tolerance can increase stress on fish due to suboptimal physiological conditions. Changes in water temperature may result in increased virulence of otherwise dormant pathogens, rampant growth of parasites, and shifts in, and/or extensions to, their distribution. It is also evident that the appearance of harmful algal blooms increases with increasing water temperatures.

Implementation of well-designed and practical biosecurity programmes, both at national and farm level, is therefore essential to reducing disease outbreaks in aquaculture. Climate change issues, challenges and opportunities must be carefully assessed, analysed and considered during designing biosecurity strategies and programmes. With such biosecurity programmes, when they are efficiently implemented, the chances of pathogen introduction, establishment and spread could be reduced. A consistent approach for the prevention, control and eradication of infectious and contagious diseases in aquaculture operations, through stringent biosecurity, is therefore necessary to improve aquaculture sustainability in the coming decades (Palić *et al.*, 2015).

11.4 Biosecurity Principles and Strategies

Biosecurity in aquaculture consists of practices that minimize the risk of pathogen transfer, establishment and spread. These include practices for reducing the stress to fish, thus making them less susceptible to pathogens and disease.

11.4.1 Prevention is better than cure

The implementation of biosecurity measures can help farms and production systems in limiting the entry of undesirable organisms. In a comprehensive assessment of sea lice infestations in the Atlantic salmon, *Salmo salar*, industry, Barrett *et al.* (2020) concluded that the effective use of barrier technologies such as skirts, snorkels or closed-containment systems, coupled with supplementary preventive methods, could make delousing treatments unnecessary at many sites, while high-risk locations might require additional management and regulation. They emphasized that preventive methods would be preferable to reactive delousing and moving towards a prevention-focused paradigm on Atlantic salmon farms might improve fish welfare and productivity, while avoiding significant environmental impacts.

Prevention can be achieved not only by using barriers to pathogens, but also employing other technologies such as genetics and vaccination as part of a comprehensive programme of integrated pathogen management. Enhancing host resistance to infectious disease has received increasing attention as a major goal of farm-animal breeding programmes. Selective breeding programmes improving genetic resistance to a range of fish pathogens have already resulted in moderate to high heritability for resistance to viral nervous necrosis (VNN) in European sea bass, *Dicentrarchus labrax* (see Palaiokostas *et al.*, 2018; Griot *et al.*, 2021a,b) and for resistance to pasteurellosis caused by *Photobacterium damselae* subsp. *piscicida* in gilthead sea bream, *Sparus aurata* (see Palaiokostas *et al.*, 2016; Aslam *et al.*, 2018). In discussing the opportunities for identifying TiLV resistance in Nile tilapia, *Oreochromis niloticus* Genetically Improved Farmed Tilapia (GIFT), Barría *et al.* (2021) reported that genetic markers from a quantitative trait locus (QTL) region have potential in marker-assisted selection to improve host resistance, providing a genetic solution to an infectious disease where few other control or mitigation options currently exist. Similarly, research into developing sea louse-resistant salmon has begun; heritable variation exists, and cumulative improvements are reducing susceptibility to sea lice in some salmon lineages (Barrett *et al.*, 2020).

Several effective vaccines against key bacterial and viral pathogens of fish have already been developed and applied in aquaculture (Bøgwald and Dalmo, 2019; Adams and Subasinghe, 2021) (see Table 11.1). Successful development of effective vaccines for other pathogens would also reduce the risk of diseases in the aquaculture industry.

11.4.2 On-farm versus off-farm biosecurity

Designing and implementing a good biosecurity plan at the farm level (on-farm biosecurity) is an important step towards protecting farm stocks from disease. New Zealand's Ministry for Primary Industries highlighted seven important steps to be included in daily farm operation process which will help in protecting farm stocks (NZMPI, 2016) (Fig. 11.4). They are:

- maintain stock health and welfare;
- be aware of the pathways of risk organisms into, within and off your farm;
- obtain pest- and disease-free stocks;
- keep things clean;
- check your farm;
- report anything unusual; and
- have a biosecurity management plan in place.

The following are considered important and essential for an effective on-farm biosecurity plan:

- Site design should minimize cross-contamination between different sections; permit the management/treatment of water flow through the site; and permit each production unit to be effectively isolated and disinfected on a routine basis (see Fig. 11.5).
- Use approved clean sources of seed and feed.
- Establish quarantine procedures for the receipt of new livestock.
- Request health certificates and conduct own health evaluations.
- Written standard operating procedures relating to biosecurity for each hatchery and farm procedure must be in place. There should, for example, also be protocols for routine cleaning and disinfection. All procedures should be communicated to all staff, with training and proficiency monitored where necessary, regularly updated and accompanied by a data log.
- Conduct risk assessments of all procedures.
- Identify control points where procedures could be implemented to prevent, reduce or remove the risks of disease.
- Biosecurity should be an overall management system. The entire production pipeline must be considered, including: (i) broodstock/juvenile source/quality; (ii) stocking densities; (iii) feed and feeding regimens; (iv) hatchery disinfection and management; (v) water system preparation; (vi) monitoring of water and soil parameters; (vii) disease surveillance; and (viii) training and record keeping.
- Have emergency plans and corrective action plans pre-prepared in advance of a disease outbreak. Each plan should detail the specific actions required for pathogen containment, control and eradication.
- Learn to recognize the signs of disease and/or deviations from normal.
- Conduct health surveillance on a regular basis to monitor the health of hatchery and farm stock and to control on-site sanitary status.
- Understand the economic benefits to 'closing the gate' on disease as opposed to the costs of corrective action.
- On discovering a biosecurity breach or disease outbreak, avoid procrastination and act. Avoid reckless action but act appropriately with veterinary support where required.
- Avoid the promiscuous, unsupervised use of medicants to control aquatic pathogens; seek proper veterinary advice instead.
- Conduct regular reviews of biosecurity on-site practices. Revise where necessary. Monitor. Avoid lapses in biosecurity rigour.

While an on-farm biosecurity plan helps in protecting farm stocks, it should be complemented by an off-farm biosecurity plan, which in actual terms should be a 'national biosecurity plan'. Such biosecurity plans covering political boundaries help reduce pathogen transfer risks and maintain safe transboundary movement of live animals and their products, provided that

Fig. 11.4. Main risk pathways for pests and diseases to spread on to, within and from a farm. (Adapted from NZMPI, 2016; inserts by A. Shinn.)

the plan is efficiently implemented. Since off-farm biosecurity plans address measures at the border level and beyond, they are bound by relevant international standards and treaties.

There are many examples of comprehensive biosecurity action plans, such as the Aquaculture Farm Biosecurity Plan: Generic guidelines and template (Sub-Committee on Aquatic Animal Health, 2016). Risk assessment is one such component. Risk analysis is an internationally accepted standard method for assessing whether trade in a particular commodity, such as a live aquatic animal or its product, poses a significant risk to human, animal or plant health; and, if so, what measures could be adopted to reduce that risk to an acceptable level. Several international factors have spurred the development of risk analysis. They include the liberalization of international trade through the General Agreement on Tariffs and Trade and the establishment of the World Trade Organization (WTO) and its Agreement on the Application of Sanitary and Phytosanitary Measures (SPS Agreement). WTO member countries are now required to use the risk analysis process to justify any restrictions on international trade beyond those specified by the Aquatic Animal Health Code (OIE, 2021a). These must be based on risks to human, animal or plant health.

The general framework for an import risk assessment (IRA) for live aquatic animals and their products is laid out in the OIE's Aquatic Animal Health Code (OIE, 2021a). The members of both the OIE and the WTO are obligated to follow OIE and WTO procedures. Although not

Fig. 11.5. Building biosecurity into the infrastructure of production sites. The design of a site should minimize cross-contamination between different sections; permit the management/treatment of water flow through the site; and permit each production unit to be effectively disinfected on a routine basis. (Image courtesy of Fish Vet Group Asia, Benchmark.)

obligatory, the International Council for the Exploration of the Sea (ICES) Code of Practice on the Introductions and Transfers of Marine Organisms (ICES, 2005, 2012) also has wide global acceptance and is considered the key framework for assessing proposals to introduce exotic species into new environments outside their native range or to transfer new strains of established species to countries where they are either native or previously introduced. Among other risk analysis frameworks, the ICES Code addresses the evaluation of potential genetic, ecology and pathogen risks associated with the transfer of aquatic organisms.

11.4.3 Emergency preparedness and response

On many occasions, unsuccessful preventive measures have led to the incursion of pathogens into production systems causing disease emergencies. Emergency preparedness has, therefore, been developed and implemented by some based on experience from unprepared, or incompletely planned, response to disease emergencies. In many instances, this approach has now proven to be economically sustainable; however, learning from these experiences is important. Pre-emergency and contingency investment in insurance, especially for high-value species, has become a welcome trend (Subasinghe *et al.*, 2021). Transparent reporting is of utmost importance. However, there is no longer a plausible excuse, with so many international examples as well as guidelines, audit systems and shared experiences, to avoid investment in prevention strategies and infrastructure, which can offset the much greater cost of repeated response/emergency approaches (Subasinghe *et al.*, 2021).

11.4.4 Disease and pathogen surveillance

Disease surveillance is a critical component of a good biosecurity plan. While some progress exists with aquatic disease surveillance, many barriers remain to the development of surveillance systems that support effective national and farm biosecurity. Although the awareness has been increased and efforts have been made, aquatic animal health services are inadequate in many geographies to design and apply/implement efficient surveillance strategies at national and/or regional levels.

Active surveillance is a systematic tool aimed at gathering data to demonstrate the presence or absence of diseases of national concern, to determine their distribution where present, and to detect emerging or exotic diseases that require investigation. Surveillance activities are often used to provide information for disease control programmes, or to support health certification requirements, import risk analysis and international trade. Delayed detection of disease outbreaks due to weak surveillance results in higher response costs to control outbreaks or, typically, disease establishment and rapid spread. Improving surveillance and biosecurity requires strengthened government services and regulation. A comprehensive account on the status and prospects of surveillance programmes in reducing the risks of disease in aquaculture is provided elsewhere (Subasinghe *et al*., 2021).

There are number of logistical, practical challenges and considerations to the collection and processing of samples from the field. In addition to the sampling methodologies that may be required in acquiring the target species/specimens and due to the physical attributes of the site and access to it, there may be additional environmental, social, ethical and governance-based issues to address. While many of the following comments have relevance to activities conducted on land, much of the following discussion focuses on the challenges of sampling finfish reared in a diverse range of systems and environments.

For a health surveillance programme to be effective there is a need to sample and conduct regular checks, but each sampling point should be comprehensive to provide a clear picture of the health status of the fish stock at that moment in time. At the same time, some caution should be exercised in taking accurate single point samples and in the consequential decisions that are made. Depending on the sensitivity and specificity of the diagnostic tests used, or in the suspicion of an early-stage infection where the probability of pathogen prevalence is expected to be low, a larger sample size may be required. This is particularly important in health certification prior to the movements of stock or before the stocking of new systems to ensure stocks and their culture systems are free of infection. A robust programme of on-farm health surveillance is also centred in good sample sizes as a means of the early detection of pathogen threats but also allowing for the implementation of appropriate interventions in a timely manner, as opposed to responding to a crisis where the management options may become limited.

11.4.5 Safe movement of live aquatic animals

Application of good biosecurity at all levels will reduce the risk of pathogen transfer and establishment. As aquaculture has become a global industry requiring the transboundary movements of live animals and animal products, risk analysis for aquatic animal pathogens has become a major component of global strategies aimed at providing appropriate health management protocols and biosecurity measures that protect national biological, social and economic resources. At the same time, these support economically and environmentally sustainable aquaculture development, while also facilitating trade. Although the focus of this book and the current chapter is to discuss pathogen-related issues in aquaculture, especially for cage aquaculture, a complete risk assessment is also necessary to guide acceptable levels of risk of a movement, covering assessment of ecological and genetic risks of such a movement. One recent example is the comprehensive transfer guidelines and risk management strategy developed by WorldFish facilitating the safe transfer of GIFT from Malaysia to Nigeria (Amarasinghe, 2021; Arthur, 2021; Bartley, 2021; WorldFish, 2021).

Subasinghe *et al.* (2021) discuss the non-conventional ways of dispersal of Harmful Aquatic Organisms and Pathogens (HAOP)[1] between ecosystems. HAOP may settle and reproduce beyond control to become pests in areas outside their original geographical distribution. The successful transfers are exacerbated by the changes in natural habitats (i.e. as a consequence of global warming, physical barrier/habitat destructions), facilitating and increasing species' direct transfer across natural boundaries. Unlike pollution,[2] HAOP often exhibit robust biological traits and can reproduce over time which makes their eradication almost impossible. This makes HAOP a challenging hazard to manage. Bio-invasions are seriously impacting aquatic ecosystems that are used by multiple industries, including aquaculture and fisheries. Therefore, science-based policies to protect marine ecosystems and the communities living on them must consider the specificity related to the risk of transfer and spread of HAOP.

11.5 Applying Biosecurity in Aquaculture

Applying practical and efficient biosecurity principles through well-designed strategies in aquaculture requires an in-depth understanding of the production system, including input supplies, service provisions, husbandry management practices and the strength of the human capital. While good biosecurity in aquaculture production requires a holistic approach, especially in cage aquaculture, there are a few major aspects requiring strong emphasis.

11.5.1 Clean seed

Years of experience have now convinced the aquaculture industry and community that the use of clean and healthy seed should be given high priority in biosecurity for preventing disease outbreaks and subsequent losses (Subasinghe *et al.*, 2021). Clean seed requires a strategy for excluding pathogens. This specific pathogen-free (SPF) strategy used in aquaculture began with Pacific whiteleg shrimp, *Penaeus vannamei*. Over the years, SPF shrimp jumped into industrial-scale commercial operations taking the lead within the aquaculture industry and allowing the exponential growth of *P. vannamei* in Asia. The SPF strategy is now also applied in the salmon industry and is increasingly being embraced by other aquaculture stakeholders. Demonstrated benefits of applying SPF principles and technology in shrimp aquaculture for producing healthy seed have prompted research into 'SPF Fish' by several companies and research groups. Invention of healthy 'SPF Fish', targeting what is called 'people's fish' (i.e. carps, tilapias and species of catfish, etc.), accessible and affordable to smallholders, will offer significant opportunities and benefits in the future for improving biosecurity in global aquaculture (Subasinghe *et al.*, subm.).

The use of genetic technologies in biosecurity goes beyond SPF seed. A recent breakthrough showed promise in combating TiLV through selective breeding of fish with resistance genes. In 2020, researchers from the University of Edinburgh's Roslin Institute and WorldFish analysed the genes of 1821 GIFT fish, which were tagged and placed in a pond that had an outbreak of TiLV. The variation in TiLV resistance was found to be independent of genetic variation in growth, meaning that any future breeding programmes for GIFT that produce fish resistant to TiLV will not adversely affect the growth of the fish and will benefit farmers' yields. Use of TiLV-resistant seed for tilapia in cage culture would be highly desirable and effective as producing fish in open waters poses serious biosecurity challenges.

11.5.2 Clean feed

The biosecurity of aquatic diets lies in the quality (food safety) of the ingredients used and in the way in which ingredients and diets are stored. Mycotoxins are common contaminants of plant ingredients, produced by toxigenic fungi belonging to *Alternaria*, *Aspergillus*, *Fusarium* and *Penicillium*, among others, and those most encountered are as follows. Aflatoxin, particularly AFB1 produced by *Aspergillus flavus*, is the most prevalent and toxic. A dose of 20 ppb in the diet presented to *Oncorhynchus mykiss* resulted in 56% of the stock developing hepatocellular carcinomas after 8 months and 83% after 12 months (Schoenhard *et al.*, 1981). In Nile tilapia, *O. niloticus*, doses of 100 ppb in the diet resulted in lower weight gain, food conversion ratio (FCR) and haematological profiles, liver histopathology, elevated liver enzymes and a lower resistance to *A. hydrophila* infection (Mahfouz and Sherif, 2015). Cyclopiazonic acid (CPA), a mycotoxin produced by *A. flavus*, can depress weight gain in channel catfish, *Ictalurus punctatus* (Jantrarotai and Lovell, 1990). Deoxynivalenol

(DON) produced by *Fusarium graminearum*, added to diets at 0.3–2.6 ppm and presented to *O. mykiss* for 8 weeks, resulted in significant decreases in feed intake, weight gain, growth rate and FCR (Hooft *et al.*, 2011). Atlantic salmon, *S. salar*, given 0.5–6 ppm DON in the diet for up to 8 weeks showed inhibited protein synthesis and growth. There was also a statistically significant dose-related reduction in antibody levels following vaccination against *Aeromonas salmonicida* (see Bernhoft *et al.*, 2018). Fumonisin produced by *Fusarium verticillioides*, and especially FB1, is among the most toxic of mycotoxins produced. Doses of >80 ppm result in increased mortality (The Fish Site, 2013). Diets containing >20 ppm moniliformin (MON) from *Fusarium proliferatum* result in reduced weight gain in *I. punctatus* (Yildirim *et al.*, 2000). Ochratoxin A (OTA) is produced by species belonging to the genera *Aspergillus* and *Penicillium*. *I. punctatus* fed 4 ppm OTA in the diet had kidney damage and lower weight gain (Manning *et al.*, 2003). T-2 toxin, a *Fusarium* mycotoxin, depressed weight gain and lowered haematocrits at low doses of 0.625 ppm, while >1 ppm cause high rates of mortality (Manning *et al.*, 2005). *I. punctatus* fed either 4 ppm OTA or 1–2 ppm T-2 for 6 weeks and then subsequently challenged with *Edwardsiella ictaluri* on day 21 had statistically significantly ($P < 0.05$) higher mortality than control groups (Manning *et al.*, 2005). Zearalenone (ZEN), a *Fusarium* mycotoxin, accumulates in the ovaries with impacts on reproduction in zebrafish, *Danio rerio* (see Schwartz *et al.*, 2010); however, the impacts on reproduction in other species like *O. mykiss* are unknown (Woźny *et al.*, 2013).

To address the biosecurity of aquatic diets, binders are added to prevent them from being absorbed. Hydrated sodium calcium aluminosilicate (HSCAS) clays work well in binding aflatoxins but less so for the other mycotoxins (Zhao *et al.*, 2010). Modified fractions of *Saccharomyces cerevisiae* are also reported to work well on a wider range of mycotoxins (Zhao *et al.*, 2010).

11.5.3 Vaccines

Vaccines are recognized as important tools for the prevention and control of diseases in fish. The number of fish vaccines commercially available has grown in recent years but there are still numerous key diseases where no vaccines are available or cases where existing vaccines do not perform well (see Table 11.1). Many vaccines are efficiently used in cage aquaculture, mostly among salmonids.

Commercial fish vaccines are available for a wide range of fish species (see Evensen, 2016), including Atlantic salmon (*S. salar*), rainbow trout (*O. mykiss*), sea bass (*D. labrax*) and sea bream (*S. aurata*), tilapia (*O. niloticus*/*Oreochromis mossambicus*), amberjack (*Seriola dumerili*) and yellowtail (*Seriola quinqueradiata*) in Japan, catfish (*I. punctatus*) and Vietnamese catfish (*Pangasionodon hypophthalmus*). Many scientists and vaccinologists have made significant progress towards producing a commercial vaccine against TiLV and an affordable and accessible vaccine will be available shortly.

Effective improvements on current vaccines are still required for some of the traditional inactivated vaccines regarding efficacy (identification and optimization of antigen components), choice of adjuvant and route of administration. Development of successful vaccines against intracellular bacterial pathogens and viruses may require the use of live attenuated vaccines and application as oral vaccines, although there are safety concerns with live vaccines. Vaccines against parasites and fungi are also in development and these may need to rely on recombinant or DNA vaccine technology. There is also still a requirement for basic information on pathogenesis, immune response and identification of potentially protective antigens for such parasites and fungi (Adams and Subasinghe, 2021).

11.5.4 Diagnostics

As finfish aquaculture continues to expand, there is a commensurate need for improvements to aquatic biosecurity practices and in the management of aquatic animal health and the support that can be offered at the individual farm level. This can be done partly by improving the capabilities of individual farmers to carry out health checks in 'real time' (i.e. within a time frame that allows effective decision making on therapeutic or preventive actions). Point-of-care diagnostics (POCDs) are tests that are designed to be used on site to provide rapid results without the need for dedicated laboratory facilities. Examples of POCDs include water quality test kits, pH strips, bacteriology testing, lateral flow immunoassay, loop-mediated isothermal amplification (LAMP), sentinel water quality probes (for the measurement of potentially deleterious nitrogenous wastes and oxygen) and temperature probes, etc. Even basic POCDs can facilitate decision making on the health status of animals without the delays associated with conventional laboratory testing. POCDs can be used in investigation of disease outbreaks; in passive and active health surveillance for pathogen screening; as an early warning system to prevent disease outbreaks; and in certification for animal movements.

To ensure the accuracy and consistency of results, individual POCDs need to be validated and approved by the relevant national competent authorities. As each POCD may have different sensitivity and specificity for specific diseases, their application and the interpretation of results should also be considered.

POCDs should be portable, easy to operate ('user-friendly'), and the cost of each test should be low. Obviously, devices should be sufficiently accurate, specific and sensitive to facilitate early-stage detection of an infection and improve decision making to avoid costs associated with advanced infections. POCDs can play an important role in remote settings and in resource-limited farms where access to disease diagnostic services is limited and where there are practical considerations regarding what diagnostics can be conducted 'pond-side' or in basic farm laboratories. POCDs can also facilitate the correct application of treatment regimens (drug, dose and time), thereby reducing the indiscriminate (and ineffective) use of drugs and chemicals.

Results of POCDs are interpreted in context when making a diagnosis. Test results are one element in making a disease diagnosis and should not be considered as definitive or taken in isolation. The application of POCDs in human medicine is often at the level of the individual patient, whereas in aquaculture they are usually applied at a population level. Thus, the final diagnosis should be made by a health professional based on the POCD results and clinical signs, if available.

While POCDs can be effectively applied at farm level to improve diagnostic capability, they should be considered within the context of health management at farm level. It is important to realize the critical roles played by veterinarians and health diagnosticians in developing and managing local, regional and national animal healthcare strategies and the role of POCDs within this framework.

POCDs should also be viewed within the complete framework of diagnostic methods used in health management. In addition to POCDs, it is important to consider physical and behavioural assessment of individuals and populations, microscopic examination of tissues and body fluids, and environmental parameters (e.g. water quality parameters including harmful phytoplankton/zooplankton) when making a presumptive diagnosis.

As aquaculture continues to intensify and develop, there will also be an increased risk of infection from existing and exotic pathogens. While many advanced aquaculture production industries move towards greater degrees of precision farming with the employment of a suite of tools that integrate artificial intelligence (AI), image analysis (IA) and big data-based technologies to support biosecurity and aquatic health management, these developments are not, currently, a global requirement across finfish aquaculture. For most producers, represented by small- to medium-sized farms, there is a greater need to have increased accessibility to POCDs and veterinary support in empowering farmers to make real-time management decisions. In support of this, national governments might consider setting standards for aquaculture diagnostics including what testing should be conducted, the approval of standard methods/POCD devices, stating who is authorized to do the testing and setting standards for training of animal health workers. Commercial development of solutions should be encouraged within a health management framework to reduce the burden on government services, drive innovation and reduce costs for farmers. There needs to be a parallel development of the reporting system to capture and process disease diagnostic data and communicate them back to the farming community to provide timely warnings of potential disease threats and enable national health agencies to target their activities (e.g. in disease containment or in the deployment of support services, etc.) more effectively. This reporting system must incorporate disease diagnostic data from public and private sources such as aquaculture veterinarians and trained health diagnosticians. It should also include a concerted communications strategy from national health authorities to farmers stressing the importance of approved diagnostic tools for the surveillance/testing of the major pathogens of concern.

For POCD devices, improvements in operational sensitivity, accuracy, analysis run times and unit cost of pathogen testing will ultimately determine their rate of uptake, use and benefit to farmers. There will be a need for multiplexing – multiple pathogen screening capability – and for the adoption of sensitive environmental DNA (eDNA)-like based technologies to assess the health of the entire population or systems, rather than a small number of specimens subsampled from it. It is expected that the next decade will see further developments in virtual veterinary diagnostics supported by smartphone with mobile healthcare packages or software applications to: capture data, images and video that feed into centralized health support laboratories to allow the health, welfare and behaviour of stocks to be assessed; provide early warning of disease threats relevant to discrete water catchments/regions; use data that feed into trained neural network/AI-based decision-making systems; permit tracking and tracing of diseases; and aid in the provision of online health training. As with POCDs themselves, it is important that such approaches are validated and accredited.

11.6 Biosecurity Governance

Biosecurity governance is complex but necessary to ensure the efficient implementation of biosecurity plans and strategies. The global nature of aquaculture has created national, regional and/or international frameworks comprising treaties, agreements, declarations, guidelines and policies. However, the binding (obligatory) and non-binding (voluntary) nature of these instruments has reduced the efficacy of biosecurity implementation and management worldwide. Interestingly, most international and regional biosecurity frameworks (57%) do not require compliance, and all others (43%) require compliance only when ratified by a nation state (Campbell *et al.*, 2020).

National legislation and regulatory frameworks are critically important for sound biosecurity. There should be incentives for good compliance while disincentives for bad compliance. National legislation must project the interest of all actors along the aquaculture value chain. However, legislation is of no value if it is not adequately enforced.

Administration of shared water bodies under different national jurisdictions, which may challenge biosecurity and disease control, requires strong coordination and collaboration among states, countries and authorities. Harmonized legislation and enforcement procedures are equally important. In this regard, various frameworks covering groups of countries based on political and/or geographical coverages exist (APEC/FAO/NACA/SEMARNAP, 2001; OIE, 2021a). However, the effectiveness of such arrangements in protecting geographies from incursion and spread of pathogens remains questionable.

11.7 Progressive Management Pathway for Improving Aquaculture Biosecurity

While many efforts have been exerted by national competent authorities, industry and academe, as well as regional and international entities and development institutions, successful biosecurity practices have not been properly implemented in many areas and so actions have been reactive or *ad hoc*, which is significantly more costly than investment in preventive measures (Subasinghe *et al.*, 2021).

The Progressive Management Pathway for Improving Aquaculture Biosecurity (PMP/AB) was developed by the Food and Agriculture Organization of the United Nations (FAO) and partners as a 'paradigm shift' after analysing the pathways and factors to disease emergence and seeing a need for strategic planning to further guide and support countries towards achieving sustainable aquaculture biosecurity and health management systems. The PMP/AB is an extension of FAO's 'Progressive Control Pathways' (PCP) approach, which has been adopted internationally to assist countries develop risk mitigation strategies that reduce or prevent losses from major livestock diseases. However, whereas most

PCPs focus on control of specific diseases, the PMP/AB focuses on diseases faced by aquaculture at the commodity and enterprise level. The PMP/AB uses a comprehensive, holistic approach to improving aquaculture biosecurity and supporting sustainable development.

The PMP/AB is progressive, collaborative and risk based (see Fig. 11.6). The four stages of PMP/AB involve strong public and private stakeholder input to promote the application of risk management at the sector level as part of a national approach. Countries decide the appropriate entry point, how far and how fast to progress to the next stage. Due to the wide variation of species farmed in each country, aquaculture sectors may advance independently, at different speeds or with different goals, but a common requisite is strong cooperation between government and industry, such as a public–private sector partnership (PPP). This is necessary to ensure clarity on roles and responsibilities, identify key gaps requiring improved capacity and infrastructure, and increase awareness of the cost/benefits of biosecurity along the whole value chain. Risk analysis is a key aspect of all stages of the PMP/AB. Risk hotspots (critical control points) are identified for biosecurity investment (training, diagnostic capacity, etc.). All this feeds into development of a national strategy on aquatic animal health (NSAAH) or national aquaculture biosecurity strategy, which sets the foundation for ongoing review and updating as the industry develops.

The benefits of the PMP/AB can be summarized as follows:

- It builds on management capacity using bottom-up and top-down approaches, is evidence-based and supported by transparent and ongoing review. The co-management approach ensures problems are clearly defined and management solutions have common understanding and buy-in.
- The PMP/AB provides a degree of consistency between participating countries or regions that is essential for reducing risks from trade, as well as for addressing biosecurity-related trade challenges.

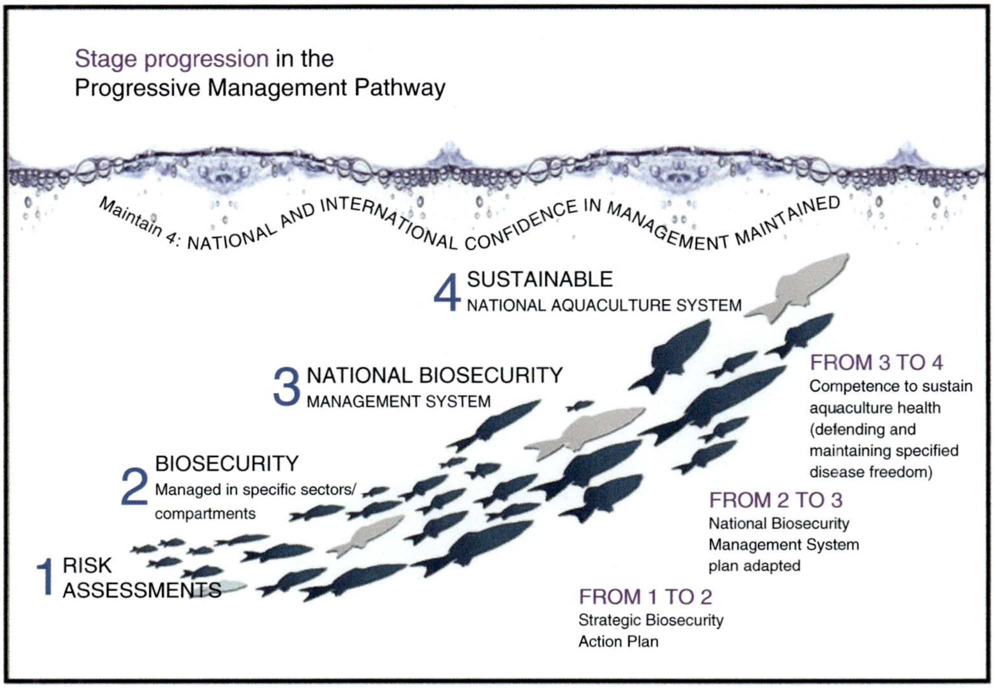

Fig. 11.6. Proposed FAO implementation of aquaculture biosecurity, using the Progressive Management Pathway for Improving Aquaculture Biosecurity (PMP/AB). (Adapted from FAO, 2019.)

- The PMP/AB provides a framework that is adaptable and can respond to changes in aquaculture production scope and objectives (small to large scale; local to international industries), as well as to environmental and anthropological changes that impact aquaculture production.

11.8 Climate Change Impacts on Diseases in Aquaculture

Climate change effects on aquaculture production and their sustainability implications, mitigation and adaptations have recently been reviewed by Maulu *et al.* (2021) and a general account on the impacts on cage aquaculture is provided in Chapter 1 (this volume). Most of the climate changes occurring in the world, both in the north and in the south, that appear to impact the aquatic environment and its life, affect aquatic production and aquaculture. However, hard evidence on the scale of climate change impacts on aquaculture is yet to be obtained and their socio-economic and environmental implications are yet to be understood.

The health of fish is governed by the quality of their culture conditions, including temperature, oxygen solubility, salinity, content of dissolved nitrogenous waste products, etc. (Mydlarz *et al.*, 2006), and adverse environmental conditions or rapid changes can negatively impact welfare. Complex interrelationships exist between many chemical and physical parameters in the aquatic environment where changes in one parameter can affect the toxicity of others, with consequential impacts to fish health.

Marine cage sites are frequently positioned in 'shallow' coastal waters and estuaries and are susceptible to changes in water temperature, storm intensity or frequency, and sea level. The biosecurity of exposed sites can be threatened by wave action and is susceptible to storm damage. Wave energy is greatest at the surface and as waves approach the coast, the velocity of the waves decreases but their height increases. The severity and frequency of storms have been predicted to increase (Frost *et al.*, 2012), resulting in potential structural damage to fish enclosures with consequential breaches to biosecurity via either the loss of stock or through the introduction of disease agents, vectors and predators with the movement of flood waters.

For pen sites situated in rivers and estuaries, there are concerns regarding the frequency of flood events, their magnitude, the changes to hydrology and the consequential biosecurity impacts associated with runoff and the increased transport of pathogens, sediments, pollutants and nutrients (Callaway *et al.*, 2012).

According to Collins *et al.* (2020), diseases in aquaculture are likely to be affected by a changing temperature regime, but in a largely unpredictable manner. However, what is certain is that when fish are exposed to thermal stress, they become more susceptible to diseases and that warmer conditions may result in the establishment of exotic diseases. Since fish become more vulnerable to diseases and pathogens through both direct and indirect climate impacts, especially through thermal stress (Chiaramonte *et al.*, 2016), warmwater disease outbreaks are therefore predicted to occur more frequently in addition to the possibility of discovering new ones under a changing climate (Sae-Lim *et al.*, 2017).

Logically, at least up to certain limit, rising temperature will accelerate the replication rate, virulence, life-cycle longevity and transmission of pathogens in fish, promoting the emergence of diseases in production systems. In salmon cage culture, warmwater pathogens such as sea lice will remain a challenge and further warming is likely to worsen the infections in cold temperate conditions, requiring more treatments thus more expense (Collins *et al.*, 2020). Undoubtedly, increased disease incidents in aquaculture production systems will lead to reduced profits, consequently affecting the socio-economic viability of aquaculture. On the contrary, rising temperature may positively impact coldwater diseases, such as vibriosis and ulcers that currently affect Atlantic salmon and might gradually disappear due to the emerging unfavourable conditions (Sae-Lim *et al.*, 2017), thereby favouring aquaculture production.

After water, oxygen is the first limiting component of the aquatic environment. Oxygen solubility decreases with increasing salinity and temperature; the solubility of oxygen in seawater is approximately 20% lower than in fresh water. Pörtner and Knust (2007) warn that the thermal limitation of oxygen will be among the main physiological effects of climate change.

This has been examined by Daufresne *et al.* (2009), Baudron *et al.* (2014) and van Rijn *et al.* (2017), among others, who highlight the 5–20% decline in fish body size over recent decades in response to changes in global temperatures. For some aquaculture fish species, rises in temperature may push them beyond their optimal or tolerant culture ranges (see Table 1 in Jahangiri *et al.*, 2021) and may have impacts on egg equality and production (King and Pankhurst, 2004), their capacity to tolerate environmental stresses and/or their susceptibility to infections. The immune function of some farmed fish species may increase within their optimal temperature range (Bly *et al.*, 1997; Bowden *et al.*, 2007), but this may be compromised by the increased stress associated with lower dissolved oxygen concentrations in warmer waters. Increased water temperatures may also result in the increased loading of pathogens within sites and extend periods of exposure, potential susceptibility and infection (Harvell *et al.*, 2002).

It is well known that harmful algal blooms occur due to change in climatic conditions. They are a serious threat to the environmental sustainability of aquaculture production. For example, some flagellates and dinoflagellates are potentially toxic or nuisance species which could be responsible for stress or mortality among finfish and shellfish (Gubbins *et al.*, 2013). Trainer *et al.* (2019) reported an unprecedented loss of fish, the largest ever recorded worldwide, in Chilean aquaculture due to the expansion of *Pseudochattonella* and *Alexandrium* species whose outbreaks were associated with climate-induced changes in water column stratification. Some studies have also confirmed pathologies such as inflammation, atrophy and necrosis in several organs of bivalve molluscs resulting from harmful algal blooms (Haberkorn *et al.*, 2010; Basti *et al.*, 2011; Hégaret *et al.*, 2012). Although there is limited information on the mechanisms through which climate change will affect toxic substances in aquaculture, Farrell *et al.* (2015) reported that temperature variation can affect the metabolism of most widespread harmful algae.

While bad weather events and damage to culture infrastructures account for significant percentages of non-pathogen-associated stock losses, predators also account for a high proportion of losses. Predators include birds (EIFAC, 1988; Kumar *et al.*, 2021; Radwan, 2022), fish (Bevan *et al.*, 2002), sea lions and seals (Tillapaugh *et al.*, 1993; Nash *et al.*, 2000; Sepúlveda and Oliva, 2005; Vilata *et al.*, 2009; The Fish Site, 2021), otters and mink (Bruun-Schmidt *et al.*, 2000; Bevan *et al.*, 2002), sharks (Angel and Edelist, 2013; Ceyhan *et al.*, 2020), etc. Predators can damage pen enclosures and can impact production in several ways: (i) directly from the resulting fish mortalities; (ii) by inflicting damage of sufficient severity that fish are subsequently lost from production; (iii) as a result of an attack leading to non-lethal wounding, transferring pathogens or creating wounds in the fish that are subject to secondary infection; or (iv) by stressing the fish, reducing production performance (Bevan *et al.*, 2002). Predators can be excluded by culturing stock within enclosed systems (e.g. greenhouses or enclosed netted enclosures using a small size mesh to keep aerial predators out), by using anti-predator nets around submerged pen enclosures or by adopting overhead wires to keep birds out. For larger sites or where the former is not possible, then predators can be deterred by changing the design of the culture site (e.g. using ponds with an increased water depth, steep banks, removing structures that birds can perch upon) or by using lights, scarecrows, sonic propane cannons, alarms, water spray devices, reflectors, silhouettes of predatory birds, or having dogs and personnel on site (Bevan *et al.*, 2002). Constant sound-emitting acoustic deterrent devices (ADDs) have now been discontinued due to the potential disturbance that they can cause to other marine mammals. Instead, these are being replaced by the next generation of AI devices employing night and thermal-imaging cameras – acoustic startle response (ASR) and electric startle response (ESR) devices which recognize approaching predators and then emit a randomized frequency of pulses to deter them (Fish Farming Expert, 2021). On the Isle of Lewis, Western Isles, Scotland, fibreglass sound-emitting orcas have been used to scare away seals where historically seal attacks have accounted for a 1% mortality of fish (Fish Farming Expert, 2019).

11.9 Future Strategies for Improving Aquaculture Biosecurity

Sievers *et al.* (2022) reviewed the biological challenges and opportunities of submerged cage aquaculture of marine fish. They concluded that submerged aquaculture holds the promise of providing relief from periods of less-than-optimal environmental conditions, reducing fish interactions with harmful organisms and unlocking new production areas devoid of conflict with other coastal users. They also noted that not all fish species will be similarly suited to submerged culture, and a suite of key challenges and bottlenecks stands in the way of commercial production of several species. If submerged culture is to mature and fulfil its promise, research to empirically document production and environmental benefits as well as issues surrounding fish welfare throughout the production cycle needs to lead the way.

The importance of POCDs was mentioned earlier in this chapter. While POCDs could help smallholders with limited budgets, for industry stakeholders with larger biosecurity budgets, a suite of advanced technologies is employed across the production pipeline to monitor and safeguard the health of stock and culture systems. These include, as examples, systems/sensors for:

1. Fish recognition/detection (e.g. You Only Look Once (YOLO) real-time object detection models, see Xu and Matzner, 2018; Stavelin *et al.*, 2021).
2. Calculating stock numbers, biomass, growth performance, and for grading (Lopes *et al.*, 2017; Zhao *et al.*, 2021).
3. Estimating appetite, detecting hunger levels and ensuring diet delivery in all weathers (e.g. Anderson, 2015; Måløy *et al.*, 2019; The Fish Site, 2022) as well as reducing feed waste (e.g. Observe Technologies, 2022).
4. Behavioural analysis of pen-reared stocks using recurrent neural networks (e.g. Zhao *et al.*, 2018; Romero-Ferrero *et al.*, 2019).
5. Health status assessments; for example, of health and survival (Chang *et al.*, 2021), of salmon skin health (Sveen *et al.*, 2021), and for screening for sea lice (e.g. counting, see Moore, 2020) and their removal (e.g. Stingray Marine Solutions AS, 2020).
6. Predicting disease outbreaks (e.g. of sea lice using AquaCloud, see NCE Seafood Innovation, 2017), including phytoplankton and algal blooms (Zhang *et al.*, 2016; Holmyard, 2019).
7. Monitoring water quality through the use of, for example, AI, satellite remote-sensing technologies, sensor-equipped drones and sentinel probes to collect water quality data (Ubina and Cheng, 2022).
8. Improving farm efficiency, such as of aerators, pump flow, etc. (Lee, 2000; Manoharan *et al.*, 2020).
9. Monitoring environmental impacts (e.g. in assessing bacterial and ciliate populations around cages) through the analysis of eDNA barcodes (e.g. Frühe *et al.*, 2020).
10. Predator recognition (Fish Farming Expert, 2021).
11. Fish breeding programmes using AI algorithms to increase the prediction accuracy for disease resistance (tolerance or resistance) (Bargelloni *et al.*, 2021; Vu *et al.*, 2022).

In diagnostics, advanced tools include the development of lab-on-a-chip/microarray diagnostics for a suite of aquatic pathogens, including MinION-like technologies (Gallagher *et al.*, 2018; Delamare-Deboutteville *et al.*, 2021), eDNA approaches (Bastos Gomes *et al.*, 2017) and microfluidic-based technology using layered paper to control fluid flow (see Martinez *et al.*, 2010), among others. MinION, for example, allows for the multiplex, rapid, high-throughput sequencing of samples but its utility requires a parallel development in the rapid analysis of bioinformatic data.

Precision tools alone are not sufficient to make an accurate diagnosis; trained human resources are equally important. Access to qualified aquatic veterinary healthcare provision in many countries is a challenge; the number of providers may be limited or expertise covering the full breadth of species and health issues encountered within the country may not exist. In such cases, the use of POCDs can provide significant benefits in healthcare of aquaculture stocks. However, there are also significant challenges, not least in demonstrating to farmers the economic benefits of investing in POCDs and in training of farmers, technicians and local health practitioners on their proper use and application.

11.10 New and Next-Generation Farm Systems

In addition to the implementation of intelligent technologies within culture systems to directly address disease (i.e. systems for the early detection of disease conditions or to remove pathogens in waters where they are currently found), many of the challenges from disease in traditional flow-through systems have led to the design of systems to exclude pathogens. These include the redesign of existing sites to closed-water systems and managing water quality on site, thereby negating the need to abstract water from common water pools where the disease threats are potentially higher and unknown. Such systems may include the use of recirculatory aquaculture systems (RAS) or the move to land-based farming systems (e.g. there were >75 land salmon farms in 2020; Cox, 2020). At the same time, some enterprises have looked to offshore farming to avoid some of the challenges associated with coastal aquaculture (lower water quality; higher prevalence of certain pathogens like sea lice; availability of new site permits; etc.). Among the most notable are SalMar's 250,000 m^3 volume Ocean Farm 1 (SalMar, 2021) (see Fig. 11.7) and Nordlaks' 385 m long Havfarm 1, which has the capacity to produce 10,000 tonnes of salmon (Soltveit, 2020). The design of several closed-containment production units also looks to address problems of sea lice infestation, predator attacks and escapees. The use of net snorkels and skirts ('anti-sea lice shields') made of a fluid-permeable polyester material around salmon cages has reduced sea lice infestations by 80% (Dickie, 2018; Stien *et al.*, 2018). Nekkar's 'Starfish', a closed cage system (Nekkar, 2022), Hauge's 1850 m^3 capacity 'Egget' (Hauge Aqua Solutions AS, 2016) and FishGLOBE's 3500 m^3 enclosed salmon culture units protect stock from sea lice. Plans to increase the size of the FishGLOBE units to hold 30,000 m^3 of salmon are currently being explored through industrial partnership (Salmon Business, 2022).

11.11 Closing Remarks

If aquaculture is to help meet the demands for animal proteins of a growing population in the face of climate change, then substantial gains in production can be made by first reducing the magnitude of current losses attributable to disease due to lapses in biosecurity and, second, by adapting culture systems which meet the necessary biological demands for each species. This

Fig. 11.7. Owned and operated by SalMar, Ocean Farm 1 is the first salmon farm designed and built for exposed operation, with a novel design that combines solutions from aquaculture and the offshore industry. (https://www.dnv.com/news/ocean-farm-1-receives-first-ever-offshore-fish-farming-class-certificate-from-dnv-gl-165626, accessed 12 January 2023.)

chapter highlights the importance of health surveillance in the early detection and treatment of disease. While high levels of site biosecurity may require high levels of investment, if not appropriately maintained, when there are breaches of biosecurity in these sites, the magnitude of stock and economic losses can be large (i.e. due to the size of culture units, stocking densities used, etc.). Some aquaculture stakeholders are moving towards greater degrees of precision farming using, for example, a suite of AI-based tools as part of their biosecurity protocols for monitoring stocks, their behaviours and health status. The remainder, the most substantial part of global aquaculture production, is represented by farming conducted in lower investment systems. As such, the probability of disease events in these systems is higher but the magnitude of economic loss is typically lower. While the onus of implementing robust biosecurity practices is frequently heaped upon the grow-out farmers, biosecurity is a responsibility throughout the entire production chain, which needs to operate within regulatory frameworks, enforced by law, to support sustainable production. The FAO (2018) recently stated that 'a paradigm shift is needed in dealing with aquaculture biosecurity risks'. A subsequent review of biosecurity across the specific segments of the aquaculture value chain conducted by Subasinghe *et al.* (2021) calls attention to the need for efficient and effective biosecurity strategies and protocols at all levels in the next decade. Emphases should be on the need for: (i) emergency preparedness and response to disease outbreaks; (ii) diagnostics; (iii) microbial management at production level; (iv) disease and pathogen surveillance; (v) welfare; (vi) regulations and legislation; (vii) fish trade; (viii) research and technology development; (ix) antimicrobial resistance; (x) the identification of non-conventional means of pathogen transfer; and (xi) in training and access to health support networks.

Notes

[1] Wording established by Article 1 of the International Convention for the Control and Management of Ships' Ballast Water and Sediments, 2004 (also known as the Ballast Water Management Convention or BWM Convention).

[2] The United Nations Convention on the Law of the Sea distinguishes 'pollution of marine environment' ('[...] the introduction by man, directly or indirectly, of substances or energy into the marine environment [...]'; Article 1, #4) from biohazards (Article 196).

References

Adams, A. and Subasinghe, R. (2021) Fish vaccines. In: Metwalli, S., Viljoen, G. and El Idrisi, A. (eds) *Veterinary Vaccines: Principles and Applications*, 1st edn. Wiley-Blackwell, Chichester, UK, pp. 113–118.

Adams, M.B. and Nowak, B.F. (2001) Distribution and structure of lesions in the gills of Atlantic salmon, *Salmo salar* L., affected with amoebic gill disease. *Journal of Fish Diseases* 24, 535–542.

Ahne, W. (1976) Untersuchungen über die Stabilität des karpfenpathogenen Virusstammes 10/3. *Fisch und Umwelt* 2, 121–127.

Alonso, M. and Leong, J. (2013) Licensed DNA vaccines against infectious hematopoietic necrosis virus (IHNV). *Recent Patents on DNA & Gene Sequences* 7, 62–65.

Amarasinghe, U.S. (2021) *GIFT Transfer Risk Management: Ecology. Ecology Risk Analysis and Recommended Risk Management Plan for the Transfer of GIFT (Oreochromis niloticus) from Malaysia to Nigeria*. Program Report No. 2021-13. WorldFish, Penang, Malaysia.

Anderson, J. (2015) eFishery fish farm feeder dispenses feed based on fish hunger levels. *New Atlas*, 20 October 2015. Available at: https://newatlas.com/efishery-fish-farm-smart-feeder/39940/ (accessed 20 January 2022).

Angel, D.L. and Edelist, D. (2013) Sustainable development of marine aquaculture off the-coast and offshore – a review of environmental and ecosystem issues and future needs in tropical zones. In: Lovatelli, A., Aguilar-Manjarrez, J. and Soto, D. (eds) *Expanding Mariculture Farther Offshore: Technical, Environmental, Spatial and Governance Challenges. FAO Technical Workshop, 22–25 March 2010,*

Orbetello, Italy. FAO Fisheries and Aquaculture Proceedings No. 24. Food and Agriculture Organization of the United Nations, Rome, pp. 173–200.

APEC/FAO/NACA/SEMARNAP (2001) *Trans-Boundary Aquatic Animal Pathogen Transfer and the Development of Harmonized Standards on Aquaculture Health Management (FWG/03/2000). Report of the Joint APEC/FAO/NACA/SEMARNAP Workshop, Puerto Vallarta, Jalisco, Mexico, 24–28th July 2000*. Network of Aquaculture Centres in Asia-Pacific, Bangkok.

Arthur, J.R. (2021) *GIFT Transfer Risk Management: Pathogen. Pathogen Risk Analysis and Recommended Risk Management Plan for Transferring GIFT* (Oreochromis niloticus) *from Malaysia to Nigeria*. Program Report No. 2021-17. WorldFish, Penang, Malaysia.

Aslam, M.L., Carraro, R., Bestin, A., Cariou, S., Sonesson, A.K., *et al.* (2018) Genetics of resistance to photobacteriosis in gilthead sea bream (*Sparus aurata*) using 2b-RAD sequencing. *BMC Genetics* 19, 43.

Aunsmo, A., Valle, P.S., Sandberg, M., Midtlyng, P.J. and Bruheim, T. (2010) Stochastic modelling of direct costs of pancreas disease (PD) in Norwegian farmed Atlantic salmon (*Salmo salar* L.). *Preventive Veterinary Medicine* 93, 233–241.

Bargelloni, L., Tassiello, O., Babbucci, M., Ferraresso, S., Franch, R., *et al.* (2021) Data imputation and machine learning improve association analysis and genomic prediction for resistance to fish photobacteriosis in the gilthead sea bream. *Aquaculture Reports* 20, 100661.

Barrett, L.T., Oppedal, F., Robinson, N. and Dempster, T. (2020) Prevention not cure: a review of methods to avoid sea lice infestations in salmon aquaculture. *Reviews in Aquaculture* 12(4), 2527–2543.

Barría, A., Trịnh, T.Q., Mahmuddin, M., Peñaloza, C., Papadopoulou, A., *et al.* (2021) A major quantitative trait locus affecting resistance to tilapia lake virus in farmed Nile tilapia (*Oreochromis niloticus*). *Heredity* 127, 334–343.

Bartley, D.M. (2021) *GIFT Transfer Risk Management: Genetics. Genetic Risk Analysis and Recommended Risk Management Plan for the Transfer of GIFT* (Oreochromis niloticus) *from Malaysia to Nigeria*. Program Report No. 2021-12. WorldFish, Penang, Malaysia.

Basti, L., Endo, M. and Segawa, S. (2011) Physiological, pathological, and defense alterations in Manila clams (short-neck clams), *Ruditapes philippinarum*, induced by *Heterocapsa circularisquama*. *Journal of Shellfish Research* 30, 829–844.

Bastos Gomes, G., Hutson, K.S., Domingos, J.A., Chung, C., Hayward, S., *et al.* (2017) Use of environmental DNA (eDNA) and water quality data to predict protozoan parasites outbreaks in fish farms. *Aquaculture* 479, 467–473.

Baudron, A.R., Needle, C.L., Rijnsdorp, A.D. and Marshall, C.T. (2014) Warming temperatures and smaller body sizes: synchronous changes in growth of North Sea fishes. *Global Change Biology* 20, 1023–1031.

Berger, E.K. (2014) *Piscirickettsia salmonis*; characterization and infection in the zebrafish model. MSc thesis, University of Oslo, Oslo.

Bernhoft, A., Høgåsen, H.R., Rosenlund, G., Moldal, T., Grove, S., *et al.* (2018) Effects of dietary deoxynivalenol or ochratoxin A on performance and selected health indices in Atlantic salmon (*Salmo salar*). *Food and Chemical Toxicology* 121, 374–386.

Bevan, D.J., Chandroo, K.P. and Moccia, R.D. (2002) Predator control in commercial aquaculture in Canada. *AEC Order No.* 02-001. Available at: https://animalbiosciences.uoguelph.ca/aquacentre/files/misc-factsheets/Predator%20Control%20in%20Commercial%20Aquaculture%20in%20Canada.pdf (accessed 19 February 2022).

Bly, J.E., Quiniou, S.M. and Clem, L.W. (1997) Environmental effects on fish immune mechanisms. In: Gudding, R., Lillehaug, A., Midtlyng, P.J. and Brown, F. (eds) *Fish Vaccinology*. Development of Biological Standards, Vol. 90. Karger, Basel, Switzerland, pp. 33–43.

Bøgwald, J. and Dalmo, R.A. (2019) Review on immersion vaccines for fish: an update 2019. *Microorganisms* 7, 627.

Bowden, T.J., Thompson, K.D., Morgan, A.L., Gratacap, R.M.L. and Nikoskelainen, S. (2007) Seasonal variation and the immune response: a fish perspective. *Fish & Shellfish Immunology* 22, 695–706.

Bruun-Schmidt, I.O., Madsen, A.B., Asferg, T. and Larsen, O.N. (2000) Oddere (*Lutra lutra*) og dambrug in Danmark [Otters (*Lutra lutra*) and fish farms in Denmark]. *Flora og Fauna* 106(3–4), 79–89.

Callaway, R., Shinn, A.P., Bron, J.E., Cook, E.J., Crumlish, M., *et al.* (2012) Climate change and aquaculture in the UK. *Aquatic Conservation: Marine and Freshwater Ecosystems* 22, 389–421.

Campbell, I., Kambey, C.S.B., Mateo, J.P., Rusekwa, S.B., Hurtado, A.Q., *et al.* (2020) Biosecurity policy and legislation for the global seaweed aquaculture industry. *Journal of Applied Phycology* 32, 2133–2146.

Ceyhan, T., Akyol, O., Demir-Sağlam, Y., Sen, H., Özgül, A. and Düzbastilar, F.O. (2020) The gathering of predators around the sea-cage fish farms in the Aegean Sea. *Cahiers de Biologie Marine* 61, 361–368.

Chang, C.C., Wang, J.H., Wu, J.L., Hsieh, Y.-Z., Wu, T.-D., *et al.* (2021) Applying artificial intelligence (AI) techniques to implement a practical smart cage aquaculture management system. *Journal of Medical and Biological Engineering* 41, 652–658.

Chiaramonte, L., Munson, D. and Trushenski, J. (2016) Climate change and considerations for fish health and fish health professionals. *Fisheries* 41(7), 396–399.

Christie, K.E. (1997) Immunization with viral antigens: infectious pancreatic necrosis. *Developments in Biological Standardization* 90, 191–199.

Collins, C., Bresnan, E., Brown, L., Falconer, L., Guilder, J., *et al.* (2020) Impacts of climate change on aquaculture. *MCCIP Science Review* 2020, 482–520.

Cox, S. (2020) The rise of the land salmon. *The Narwhal*, 19 December 2020. Available at: https://thenarwhal.ca/rise-of-land-salmon-farming/ (accessed 26 March 2022).

Dale, O.B., Ørpetveit, I., Lyngstad, T.M., Kahns, S., Skall, H.F., *et al.* (2009) Outbreak of viral haemorrhagic septicaemia (VHS) in seawater-farmed rainbow trout in Norway caused by VHS virus genotype III. *Diseases of Aquatic Organisms* 85, 93–103.

Daufresne, M., Lengfellner, K. and Sommer, U. (2009) Global warming benefits the small in aquatic ecosystems. *Proceedings of the National Academy of Sciences USA* 106, 12788–12793.

Delamare-Deboutteville, J., Taengphu, S., Gan, H.M., Kayansamruaj, P., Debnath, P.P., *et al.* (2021) Rapid genotyping of tilapia lake virus (TiLV) using Nanopore sequencing. *Journal of Fish Diseases* 44(10), 1491–1502.

De Schryver, P., Defoirdt, T. and Sorgeloos, P. (2014) Early mortality syndrome outbreaks: a microbial management issue in shrimp farming? *PLoS Pathogens* 10(4), e1003919.

De Silva, S.S. and Soto, D. (2009) Climate change and aquaculture: potential impacts, adaptation and mitigation. In: Cochran, K., De Young, C., Soto, D. and Bahri, T. (eds) *Climate Change Implications for Fisheries and Aquaculture: Overview of Current Scientific Knowledge*. FAO Fisheries and Aquaculture Technical Paper No. 530. Food and Agriculture Organization of the United Nations, Rome, pp. 151–212.

Dickie, M. (2018) Scotland's salmon farmers tackle sea lice scourge. *Financial Times*, 2 April 2018. Available at: https://www.ft.com/content/ef70820a-3341-11e8-b5bf-23cb17fd1498 (accessed 26 March 2022).

Dong, H.T., Ataguba, G.A., Khunrae, P., Tattanarojpong, T. and Senapin, S. (2017a) Evidence of TiLV infection in tilapia hatcheries in Thailand from 2012 to 2017 reveals probable global spread of the disease. *Aquaculture* 479, 579–583.

Dong, H.T., Siriroob, S., Meemetta, W., Santimanawong, W., Gangnonngiw, W., *et al.* (2017b) Emergence of tilapia lake virus in Thailand and an alternative semi-nested RT-PCR for detection. *Aquaculture* 476, 111–118.

EIFAC (European Inland Fisheries Advisory Commission) (1988) *Report of the EIFAC Working Party on Prevention and Control of Bird Predation in Aquaculture and Fisheries Operations*. EIFAC Technical Paper No. 51. Food and Agriculture Organization of the United Nations, Rome.

Ernst, I., Whittington, I., Corneillie, S. and Talbot, C. (2002) Monogenean parasites in sea-cage aquaculture. *Austasia Aquaculture* (February/March), 46–48.

Evensen, Ø. (2016) Development of fish vaccines: focusing on methods in fish vaccines. In: Adams, A. (ed.) *Birkhäuser Advances in Infectious Diseases*. Springer, Basel, Switzerland, pp. 53–74.

FAO (Food and Agriculture Organization of the United Nations) (2016) *The State of World Fisheries and Aquaculture 2016*. FAO, Rome.

FAO (Food and Agriculture Organization of the United Nations) (2018) *The State of World Fisheries and Aquaculture 2018: Meeting the Sustainable Development Goals*. FAO, Rome, pp. 224. Available at: https://www.fao.org/3/i9540en/i9540en.pdf (accessed 3 November 2022).

FAO (Food and Agriculture Organization of the United Nations) (2019) *Report of the FAO/MSU/WB First Multi-Stakeholder Consultation on a Progressive Management Pathway to Improve Aquaculture Biosecurity (PMP/AB), Washington, DC, USA, April 10–12, 2018*. Fisheries and Aquaculture Report No. 1254. FAO, Rome.

Falk, K., Namork, E., Rimstad, E., Mjaaland, S. and Dannevig, B.H. (1997) Characterization of infectious salmon anemia virus, an orthomyxo-like virus isolated from Atlantic salmon (*Salmo salar* L). *Journal of Virology* 71, 9016–9023.

Farrell, H., Seebacher, F., O'Connor, W., Zammit, A., Harwood, D.T. and Murray, S. (2015) Warmer temperature acclimation impacts metabolism of paralytic shellfish toxins from *Alexandrium minutum* in commercial oysters. *Global Change Biology* 21, 3402–3413.

Ferguson, H.W., Kabuusu, R., Beltran, S., Reyes, E., Lince, J.A. and del Pozo, J. (2014) Syncytial hepatitis of farmed tilapia, *Oreochromis niloticus* (L.): a case report. *Journal of Fish Diseases* 37(6), 583–589.

Fish Farming Expert (2019) Shock and orca: Mowi uses fake whale to scare seals. *Fish Farming Expert*, 3 June 2019. Available at: https://www.fishfarmingexpert.com/article/shock-and-orca-mowi-uses-fake-whale-to-scare-seals/ (accessed 28 March 2022).

Fish Farming Expert (2021) Harm-free seal scarer puts Ace Aquatec in contention for second GSA award. *Fish Farming Expert*, 27 October 2021. Available at: https://www.fishfarmingexpert.com/article/ace-aquatec-in-the-running-for-second-global-aquaculture-innovation-award/ (accessed 28 March 2021).

FishStatJ (2021) FAO global fishery and aquaculture production statistics. v2021.1.2. In: Berger, T., Sibeni, F. and Calderini, F. (eds) *FishStatJ, A Tool for Fishery Statistics Analysis. Release 4.01.06*. Fisheries Division (NFI), Statistics and Information Branch (NFIS), Food and Agriculture Organization of the United Nations, Rome.

Frost, M., Baxter, J.M., Buckley, P., Cox, M., Dye, S. and Withers Harvey, N. (2012) Impacts of climate change on fish, fisheries and aquaculture. *Aquatic Conservation: Marine and Freshwater Ecosystems* 22, 331–336.

Frühe, L., Cordier, T., Dully, V., Breiner, H.W., Lentendu, G., *et al.* (2020) Supervised machine learning is superior to indicator value inference in monitoring the environmental impacts of salmon aquaculture using eDNA metabarcodes. *Molecular Ecology* 30, 2988–3006.

Gallagher, M.D., Matejusova, I., Nguyen, L., Ruane, N.M., Falk, K. and Macqueen, D.J. (2018) Nanopore sequencing for rapid diagnostics of salmonid RNA viruses. *Scientific Reports* 8, 16307.

Garver, K.A., Conway, C.M., Elliott, D.G. and Kurath, G. (2005) Analysis of DNA-vaccinated fish reveals viral antigen in muscle, kidney and thymus, and transient histopathologic changes. *Marine Biotechnology* 7, 540–553.

Garver, K.A., Mahony-Grant, A.M., Stucchi, D.J., Richard, J., Van Woensel, C. and Foreman, M.G.G. (2013) Estimation of parameters influencing waterborne transmission of infectious hematopoietic necrosis virus (IHNV) in Atlantic salmon (*Salmo salar*). *PLoS ONE* 8(12), e82296.

Graham, D.A., Staples, V., Wilson, C.J., Jewhurst, H., Cherry, K., *et al.* (2007) Biophysical properties of salmonid alphaviruses: influences of temperature and pH on virus survival. *Journal of Fish Diseases* 30, 533–543.

Griot, R., Allal, F., Phocas, F., Brard-Fudulea, S., Morvezen, R., *et al.* (2021a) Genome-wide association studies for resistance to viral nervous necrosis in three populations of European sea bass (*Dicentrarchus labrax*) using a novel 57k SNP array DlabChip. *Aquaculture* 530, 735930.

Griot, R., Allal, F., Phocas, F., Brard-Fudulea, S., Morvezen, R., *et al.* (2021b) Optimization of genomic selection to improve disease resistance in two marine fishes, the European sea bass (*Dicentrarchus labrax*) and the gilthead sea bream (*Sparus aurata*). *Frontiers in Genetics* 12, 665920.

Gubbins, M., Bricknell, I. and Service, M. (2013) Impacts of climate change on aquaculture. *MCCIP Science Review* 2013, 318–327.

Haberkorn, H., Lambert, C., Le Goïc, N., Guéguen, M., Moal, J., *et al.* (2010) Effects of *Alexandrium minutum* exposure upon physiological and hematological variables of diploid and triploid oysters, *Crassostrea gigas*. *Aquatic Toxicology* 97, 96–108.

Haramoto, E., Kitajima, M., Katayama, H. and Ohgaki, S. (2007) Detection of koi herpesvirus DNA in river water in Japan. *Journal of Fish Diseases* 30, 59–61.

Harvell, C.D., Mitchell, C.E., Ward, J.R., Altizer, S., Dobson, A.P., *et al.* (2002) Climate warming and disease risks for terrestrial and marine biota. *Science* 296, 2158–2162.

Hauge Aqua Solutions AS (2016) Addressing key challenges through new innovative technology. Available at: https://www.haugeaqua.com/Technology/ (accessed 20 January 2022).

Hawley, L.M. and Garver, K.A. (2008) Stability of viral hemorrhagic septicemia virus (VHSV) in freshwater and seawater at various temperatures. *Diseases of Aquatic Organisms* 82, 171–178.

Hégaret, H., Brokordt, K.B., Gaymer, C.F., Lohrmann, K.B., García, C. and Varela, D. (2012) Effects of the toxic dinoflagellate *Alexandrium catenella* on histopathogical and escape responses of the Northern scallop *Argopecten purpuratus*. *Harmful Algae* 18, 74–83.

Hill, B. (1992) Impact of viral diseases on salmonid fish in Europe. In: Kimura, T. (ed.) *Proceedings of the OJI International Symposium on Salmonid Diseases*. Hokkaido University Press, Sapporo, Japan, pp. 48–59.

Holmyard, N. (2019) Killers at sea: harmful algal blooms and their impact on aquaculture. *Global Seafood Alliance*, 15 July 2019. Available at: https://www.globalseafood.org/advocate/killers-at-sea-harmful-algal-blooms-and-their-impact-on-aquaculture/ (accessed 4 February 2022).

Hooft, J.M., Elmor, A.E.H.I., Encarnação, P. and Bureau, D.P. (2011) Rainbow trout (*Oncorhynchus mykiss*) is extremely sensitive to the feed-borne *Fusarium* mycotoxin deoxynivalenol (DON). *Aquaculture* 311(1–4), 224–232.

Howes, E.L., Joos, F., Eakin, M. and Gattuso, J.-P. (2015) *The Oceans 2015 Initiative, Part I: An Updated Synthesis of the Observed and Projected Impacts of Climate Change on Physical and Biological Processes in the Oceans*. Study No. 02/15. IDDRI, Paris.

ICES (International Council for the Exploration of the Sea) (2005) *ICES Code of Practice on the Introductions and Transfers of Marine Organisms*. ICES, Copenhagen.

ICES (International Council for the Exploration of the Sea) (2012) Appendices to the ICES Code of Practice (CoP) on the Introductions and Transfers of Marine Organisms (2005), Appendix C: Quarantine. ICES, Copenhagen, pp. 263–265.

Ilouze, M., Davidovich, M., Diamant, A., Kotler, M. and Dishon, A. (2011) The outbreak of carp disease caused by CyHV-3 as a model for new emerging viral diseases in aquaculture: a review. *Ecological Research* 26, 885–892.

Jahangiri, L., Shinn, A.P., Pratoomyot, J. and Bastos Gomes, G. (2021) Unveiling associations between ciliate parasites and bacterial microbiomes under warm-water fish farm conditions – a review. *Reviews in Aquaculture* 13, 1097–1118.

Jantrarotai, W. and Lovell, R.T. (1990) Acute and subchronic toxicity of cyclopiazonic acid to channel catfish. *Journal of Aquatic Animal Health* 2(4), 255–260.

Jarungsriapisit, J., Nuñez-Ortiz, N., Nordbø, J., Moore, L.J., Mæhle, S. and Patel, S. (2020) The effect of temperature on the survival of salmonid alphavirus analysed using *in vitro* and *in vivo* methods. *Aquaculture* 516, 734647.

Jensen, B.B. and Kristoffersen, A.B. (2015) Risk factors for outbreaks of infectious pancreatic necrosis (IPN) and associated mortality in Norwegian salmonid farming. *Diseases of Aquatic Organisms* 114, 177–187.

Kamilya, D. and Baruah, A. (2014) Epizootic ulcerative syndrome (EUS) in fish: history and current status of understanding. *Reviews in Fish Biology and Fisheries* 24(1), 369–380.

Kell, A.M., Wargo, A.R. and Kurath, G. (2014) Viral fitness does not correlate with three genotype displacement events involving infectious hematopoietic necrosis virus. *Virology* 464, 146–155.

Kibenge, F.S.B., Godoy, M.G. and Kibenge, M.J.T. (2013) Worldwide update on infectious salmon anaemia (ISA): efforts to successfully manage ISA. Presented at *GOAL Conference, Paris, 7–10 October 2013*. Available at: https://www.aquaculturealliance.org/wp-content/uploads/2015/04/goal13-kibenge.pdf (accessed 3 February 2022).

Kim, W.-S., Kim, S.-R., Kim, D., Kim, J.-O., Park, M.-A., *et al.* (2009) An outbreak of VHSV (viral hemorrhagic septicemia virus) infection in farmed olive flounder *Paralichthys olivaceus* in Korea. *Aquaculture* 296, 165–168.

King, H.R. and Pankhurst, N.W. (2004) Effect of maintenance at elevated temperatures on ovulation and luteinizing hormone releasing hormone analogue (LHRHa) responsiveness of female Atlantic salmon (*Salmo salar*) in Tasmania. *Aquaculture* 233, 583–597.

Kitao, T., Aoki, T., Fukudome, M., Kawano, K., Wada, Y.O. and Mizuno, Y. (1983) Serotyping of *Vibrio anguillarum* isolated from diseased freshwater fish in Japan. *Journal of Fish Diseases* 6, 175–181.

Klesius, P.H., Shoemaker, C.A. and Evans, J.J. (2008) *Streptococcus*: a worldwide fish health problem. In: *Proceedings of the 8th International Symposium on Tilapia in Aquaculture, Cairo, Egypt, 12–14 October 2008*. The Central Laboratory for Aquaculture Research. Co-Sponsored by The American Tilapia Association (ATA) and Aqua/Fish Collaborative Research Support Program (A/F CRSP), pp. 83–107. Available at: https://ag.arizona.edu/azaqua/ista/ISTA8/ISTA8.htm (accessed 3 February 2022).

Kumar, G., Hegde, S., Wise, D. and Mischke, C. (2021) Economic losses of catfish to avian predation: a case report. *North American Journal of Aquaculture* 83, 127–137.

Lafferty, K.D., Harvell, C.D., Conrad, J.M., Friedman, C.S., Kent, M.L., *et al.* (2015) Infectious diseases affect marine fisheries and aquaculture economics. *Annual Review of Marine Science* 7(1), 471–496.

Langdon, J.S. (1989) Experimental transmission and pathogenicity of epizootic haematopoietic necrosis virus (EHNV) in redfin perch, *Perca fluviatilis* L., and 11 other teleosts. *Journal of Fish Diseases* 12, 295–310.

Lapatra, S.E. (1998) Factors affecting pathogenicity of infectious hematopoietic necrosis virus (IHNV) for salmonid fish. *Journal of Aquatic Animal Health* 10(2), 121–131.

Lee, P.G. (2000) Process control and artificial intelligence software for aquaculture. *Aquacultural Engineering* 23, 13–36.

Lilley, J.H., Petchinda, T. and Panyawachira, V. (2001) *Aphanomyces invadans* zoospore physiology: 4. *In vitro* viability of cysts. *The AAHRI Newsletter* 10, 1–4.

Lopes, F., Silva, H., Almeida, J.M., Pinho, C. and Silva, E. (2017) Fish farming autonomous calibration system. *OCEANS 2017 – Aberdeen* 2017, 1–6. https://doi.org/10.1109/OCEANSE.2017.8084565.

Lorenzen, N. and Lapatra, S.E. (2005) DNA vaccines for aquacultured fish. *Revue Scientifique et Technique (Office International des Épizooties)* 24, 201–213.

Mahfouz, M.E. and Sherif, A.H. (2015) A multiparameter investigation into adverse effects of aflatoxin on *Oreochromis niloticus* health status. *The Journal of Basic & Applied Zoology* 71, 48–59.

Måløy, H., Aamodt, A. and Misimi, E. (2019) A spatio-temporal recurrent network for salmon feeding action recognition from underwater videos in aquaculture. *Computers and Electronics in Agriculture* 167, 105087.

Manning, B.B., Ulloa, R.M., Li, M.H., Robinson, E.H. and Rottinghaus, G.E. (2003) Ochratoxin A fed to channel catfish (*Ictalurus punctatus*) causes reduced growth and lesions of hepatopancreatic tissue. *Aquaculture* 219(1–4), 739–750.

Manning, B.B., Terhune, J.S., Li, M.H., Robinson, E.H., Wise, D.J. and Rottinghaus, G.E. (2005) Exposure to feedborne mycotoxins T-2 toxin or ochratoxin A causes increased mortality of channel catfish challenged with *Edwardsiella ictaluri*. *Journal of Aquatic Animal Health* 17(2), 147–152.

Manoharan, H., Teekaraman, Y., Kshirsagar, P.R., Sundaramurthy, S. and Manoharan, A. (2020) Examining the effect of aquaculture using sensor-based technology with machine learning algorithm. *Aquaculture Research* 51, 4748–4758.

Martinez, A.W., Phillips, S.T., Whitesides, G.M. and Carrilho, E. (2010) Diagnostics for the developing world: microfluidic paper-based analytical devices. *Analytical Chemistry* 82(1), 3–10.

Maulu, S., Hasimuna, O.J., Haambiya, L.H., Monde, C., Musuka, C.G., *et al.* (2021) Climate change effects on aquaculture production: sustainability implications, mitigation, and adaptations. *Frontiers in Sustainable Food Systems* 5, 609097.

Moore, G. (2020) Automatic lice counting on trial at Scottish farms. *Fish Farming Expert*, 25 November 2020. Available at: https://www.fishfarmingexpert.com/aquabyte-automatic-lice-counting-bryton-shang/automatic-lice-counting-on-trial-at-scottish-farms/1164435 (accessed 31 October 2022).

Munday, B.L., Zilberg, D. and Findlay, V. (2001) Gill disease of marine fish caused by infection with *Neoparamoeba pemaquidensis*. *Journal of Fish Diseases* 24, 497–507.

Munro, E.S. and Midtlyng, P.J. (2011) Infectious pancreatic necrosis and associated birnaviruses. In: Woo, P.T.K. and Bruno, D.W. (eds) *Fish Diseases and Disorders*. Vol. 3. *Viral, Bacterial and Fungal Infections*, 2nd edn. CAB International, Wallingford, UK, pp. 1–65.

Mydlarz, L.D., Jones, L.E. and Harvell, C.D. (2006) Innate immunity, environmental drivers and disease ecology of marine and freshwater invertebrates. *Annual Review of Ecology, Evolution, and Systematics* 37, 251–288.

Nash, C.E., Iwamoto, R.N. and Mahnken, C.V.W. (2000) Aquaculture risk management and marine mammal interactions in the Pacific Northwest. *Aquaculture* 183, 307–323.

NCE Seafood Innovation (2017) AquaCloud – The use of artificial intelligence in sea lice management. *News*, 17 May 2017. Available at: https://seafoodinnovation.no/2017/05/17/aquacloud_the_use_of_artificial_intelligence_in_sea_lice_management/ (accessed 4 February 2022).

Nekkar (2022) Closed cage for fish farming. Manage your environment. Available at: https://nekkar.com/aquaculture (accessed 28 March 2022).

NZMPI (New Zealand Ministry for Primary Industries) (2016) *Aquaculture Biosecurity Handbook: New Zealand's Commercial and Non-commercial Aquaculture to Minimise On-farm Biosecurity Risk*. Ministry for Primary Industries, Wellington.

Observe Technologies (2022) Artificial intelligence for aquaculture. Available at: https://observe.tech/ (accessed 4 February 2022).

OIE (World Organisation for Animal Health) (2019a) *Chapter 2.3.1. Infection with epizootic haematopoietic necrosis virus*. Manual of Diagnostic Tests for Aquatic Animals, 14/11/2019. Available at: https://www.woah.org/fileadmin/Home/eng/Health_standards/aahm/current/chapitre_ehn.pdf (accessed 3 February 2022).

OIE (World Organisation for Animal Health) (2019b) Chapter 2.3.2. Infection with *Aphanomyces invadans* (epizootic ulcerative syndrome). Manual of Diagnostic Tests for Aquatic Animals, 14/11/2019. Available at: https://www.woah.org/fileadmin/Home/eng/Health_standards/aahm/current/chapitre_aphanomyces_invadans.pdf (accessed 3 February 2022).

OIE (World Organisation for Animal Health) (2019c) Chapter 2.3.3. Infection with *Gyrodactylus salaris*. Manual of Diagnostic Tests for Aquatic Animals, 14/11/2019. Available at: https://www.woah.org/fileadmin/Home/eng/Health_standards/aahm/current/chapitre_gyrodactylus_salaris.pdf (accessed 3 February 2022).

OIE (World Organisation for Animal Health) (2019d) Chapter 2.3.5. Infection with HPR-deleted or HPR0 infectious salmon anaemia virus. Manual of Diagnostic Tests for Aquatic Animals, 14/11/2019. Available at: https://www.woah.org/fileadmin/Home/eng/Health_standards/aahm/current/chapitre_isav.pdf (accessed 3 February 2022).

OIE (World Organisation for Animal Health) (2019e) Chapter 2.3.7. Infection with koi herpesvirus. Manual of Diagnostic Tests for Aquatic Animals, 14/11/2019. Available at: https://www.woah.org/fileadmin/Home/eng/Health_standards/aahm/current/chapitre_koi_herpesvirus.pdf (accessed 3 February 2022).

OIE (World Organisation for Animal Health) (2019f) Chapter 2.3.8. Red sea bream iridoviral disease. Manual of Diagnostic Tests for Aquatic Animals, 14/11/2019. Available at: https://www.woah.org/fileadmin/Home/eng/Health_standards/aahm/current/chapitre_rsbid.pdf (accessed 3 February 2022).

OIE (World Organisation for Animal Health) (2019g) Chapter 2.3.6. Infection with salmonid alphavirus. Manual of Diagnostic Tests for Aquatic Animals, 14/11/2019. Available at: https://www.woah.org/fileadmin/Home/eng/Health_standards/aahm/current/chapitre_salmonid_alphavirus.pdf (accessed 3 February 2022).

OIE (World Organisation for Animal Health) (2019h) Chapter 2.3.9. Spring viraemia of carp. Manual of Diagnostic Tests for Aquatic Animals, 14/11/2019. Available at: https://www.woah.org/fileadmin/Home/eng/Health_standards/aahm/current/chapitre_svc.pdf (accessed 3 February 2022).

OIE (World Organisation for Animal Health) (2021a) Aquatic Animal Health Code. Available at: https://www.oie.int/en/what-we-do/standards/codes-and-manuals/aquatic-code-online-access/ (accessed 29 January 2022).

OIE (World Organisation for Animal Health) (2021b) Chapter 2.3.5. Infection with infectious hematopoietic necrosis virus. Manual of Diagnostic Tests for Aquatic Animals 2021. Available at: https://www.woah.org/fileadmin/Home/eng/Health_standards/aahm/current/2.3.05_IHN.pdf (accessed 3 February 2022).

OIE (World Organisation for Animal Health) (2021c) Chapter 2.3.10. Infection with viral haemorrhagic septicaemia virus. Manual of Diagnostic Tests for Aquatic Animals 2021. Available at: https://www.woah.org/fileadmin/Home/fr/Health_standards/aahm/current/2.3.10_VHS.pdf (accessed 3 February 2022).

Olstad, K., Cable, J., Robertsen, G. and Bakke, T.A. (2006) Unpredicted transmission strategy of *Gyrodactylus salaris* (Monogenea: Gyrodactylidae): survival and infectivity of parasites on dead hosts. *Parasitology* 133, 33–41.

Palaiokostas, C., Ferraresso, S., Franch, R., Houston, R.D. and Bargelloni, L. (2016) Genomic prediction of resistance to pasteurellosis in gilthead sea bream (*Sparus aurata*) using 2b-RAD sequencing. *G3: Genes, Genomes, Genetics* 6(11), 3693–3700.

Palaiokostas, C., Cariou, S., Bestin, A., Bruant, J.-S., Haffray, P., *et al.* (2018) Genome-wide association and genomic prediction of resistance to viral nervous necrosis in European sea bass (*Dicentrarchus labrax*) using RAD sequencing. *Genetics Selection Evolution* 50, 30.

Palić, D., Scarfe, A.D. and Walster, C.I. (2015) A standardized approach for meeting national and international aquaculture biosecurity requirements for preventing, controlling, and eradicating infectious diseases. *Journal of Applied Aquaculture* 27, 185–219.

Parry, L. and Dixon, P.F. (1997) Stability of nine viral haemorrhagic septicaemia virus (VHSV) isolates in seawater. *Bulletin of the European Association of Fish Pathologists* 17, 31–36.

Perelberg, A., Smirnov, M., Hutoran, M., Diamant, A., Bejerano, Y. and Kotler, M. (2003) Epidemiological description of a new viral disease afflicting cultured *Cyprinus carpio* in Israel. *Israeli Journal of Aquaculture – Bamidgeh* 55, 5–12.

Pettersen, J.M., Rich, K.M., Bang Jensen, B. and Aunsmo, A. (2015) The economic benefits of disease triggered early harvest: a case study of pancreas disease in farmed Atlantic salmon from Norway. *Preventive Veterinary Medicine* 121(3–4), 314–324.

Pörtner, H.O. and Knust, R. (2007) Climate change affects marine fishes through the oxygen limitation of thermal tolerance. *Science* 315, 95–97.

Radwan, M. (2022) Vital economic threat of predatory birds and parasites to cultivated fishes in Egypt. *Aquaculture* 548, 737666.

Rasmussen-Ivey, C.R., Hossain, M.J., Odom, S.E., Terhune, J.S., Hemstreet, W.G., *et al.* (2016) Classification of a hypervirulent *Aeromonas hydrophila* pathotype responsible for epidemic outbreaks in warm-water fishes. *Frontiers in Microbiology* 7, 1615.

Rimstad, E., Dale, O.B., Dannevig, B.H. and Falk, K. (2011) Infectious salmon anaemia. In: Woo, P.T.K. and Bruno, D. (eds) *Fish Diseases and Disorders*. Vol. 3. *Viral, Bacterial and Fungal Infections*, 2nd edn. CAB International, Wallingford, UK, pp. 143–165.
Romero-Ferrero, F., Bergomi, M.G., Hinz, R.C., Heras, F.J.H. and de Polavieja, G.G. (2019) idtracker.ai: tracking all individuals in small or large collectives of unmarked animals. *Nature Methods* 16, 179–182.
Sae-Lim, P., Kause, A., Mulder, H.A. and Olesen, I. (2017) Breeding and genetics symposium: climate change and selective breeding in aquaculture. *Journal of Animal Science* 95, 1801–1812.
Saksida, S.M. (2006) Infectious haematopoietic necrosis epidemic (2001 to 2003) in farmed Atlantic salmon *Salmo salar* in British Columbia. *Diseases of Aquatic Organisms* 72, 213–223.
Salin, K.R., Subasinghe, R.P., Senarathna, D. and Shinn, A.P. (2023) Cage culture of finfish: its importance, distributions and future modifications in ongoing climate change. In: Woo, P.T.K. and Subasinghe, R.P. (eds) *Climate Change on Diseases and Disorders of Finfish in Cage Culture*, 3rd edn. CAB International, Wallingford, UK, pp. 1–33.
SalMar (2021) Offshore fish farming: a new era in fish farming is on its way. Available at: https://www.salmar.no/en/offshore-fish-farming-a-new-era/ (accessed 28 March 2022).
Salmon Business (2022) Grieg and FishGLOBE to develop 30,000m³ closed-containment fish farms together. *Salmon Business*, 10 February 2022. Available at: https://salmonbusiness.com/grieg-seafood-and-fishglobe-enter-partnership-to-develop-30000m%C2%B3-closed-containment-fish-farms/ (accessed 28 March 2022).
Schoenhard, G.L., Hendricks, J.D., Nixon, J.E., Lee, D.J., Wales, J.H., *et al*. (1981) Aflatoxicol-induced hepatocellular carcinoma in rainbow trout (*Salmo gairdneri*) and the synergistic effects of cyclopropenoid fatty acids. *Cancer Research* 41(3), 1011–1014.
Schwartz, P., Thorpe, K.L., Bucheli, T.D., Wettstein, F.E. and Burkhardt-Holm, P. (2010) Short-term exposure to the environmentally relevant estrogenic mycotoxin zearalenone impairs reproduction in fish. *Science of the Total Environment* 409, 326–333.
Sepúlveda, M. and Oliva, D. (2005) Interactions between South American sea lions *Otaria flavescens* (Shaw) and salmon farms in southern Chile. *Aquaculture Research* 36, 1062–1068.
Shinn, A.P., Pratoomyot, J., Bron, J.E., Paladini, G., Brooker, E.E. and Brooker, A.J. (2015) Economic costs of protistan and metazoan parasites to global mariculture. *Parasitology* 142, 196–270.
Shinn, A.P., Pratoomyot, J., Griffiths, D., Jiravanichpaisal, J. and Briggs, M. (2018a) Asian shrimp production and the economic costs of disease. *Asian Fisheries Science* 31S, 29–58.
Shinn, A.P., Pratoomyot, J., Metselaar, M. and Bastos Gomes, G. (2018b) Diseases in aquaculture – counting the costs of the top 100. In: Binder, E.M. (ed.) *Proceedings of the Biomin World Nutrition Forum. SCOPE: Scientific Challenges and Opportunities in the Protein Economy, Cape Town, South Africa, 3–5th October 2018*. Erber AG, Getzersdorf, Austria, pp. 227–262.
Shinn, A.P., Dong, H.T., Vinh, N.T., Wongwaradechkul, R. and Lio-Po, G.D. (2023) Infectious diseases of warmwater fish in fresh water. In: Woo, P.T.K. and Subasinghe, R.P. (eds) *Climate Change on Diseases and Disorders of Finfish in Cage Culture*, 3rd edn. CAB International, Wallingford, UK, pp. 202–277.
Shoemaker, C.A., Klesius, P.H. and Evans, J.J. (2001) Prevalence of *Streptococcus iniae* in tilapia, hybrid striped bass, and channel catfish on commercial fish farms in the United States. *American Journal of Veterinary Research* 62, 174–177.
Shoemaker, C.A., Klesius, P.H. and Evans, J.J. (2007) Immunization of eyed channel catfish, *Ictalurus punctatus*, eggs with monovalent *Flavobacterium columnare* vaccine and bivalent *F. columnare* and *Edwardsiella ictaluri* vaccine. *Vaccine* 25, 1126–1131.
Shoemaker, C.A., Klesius, P.H., Drennan, J.D. and Evans, J. (2011) Efficacy of a modified live *Flavobacterium columnare* vaccine in fish. *Fish & Shellfish Immunology* 30, 304–308.
Sievers, M., Korsøen, Ø., Warren-Myers, F., Oppedal, F., Macaulay, G., *et al*. (2022) Submerged cage aquaculture of marine fish: a review of the biological challenges and opportunities. *Reviews in Aquaculture* 14, 106–119.
Soltveit, T. (2020) In the shadow of the giant: watch a video of the Havfarm. *Fish Farming Expert*, 26 August 2020. Available at: https://www.fishfarmingexpert.com/article/in-the-shadow-of-the-giant-watch-a-video-of-the-havfarm/ (accessed 20 January 2022).
Stavelin, H., Rasheed, A., San, O. and Hestnes, A.J. (2021) Applying object detection to marine data and exploring explainability of a fully convolutional neural network using principal component analysis. *Ecological Informatics* 62, 101269.
Stien, L.H., Bendiksen Lind, M., Oppedal, F., Wright, D.W. and Seternes, T. (2018) Skirts on salmon production cages reduced salmon lice infestations without affecting fish welfare. *Aquaculture* 490, 281–287.

Stingray Marine Solutions AS (2020) The Stingray system. Available at: https://www.stingray.no/delousing-with-laser/?lang=en (accessed 20 January 2022).

Subasinghe, R., Alday-Sanz, V., Bondad-Reantaso, M., Huang, J., Shinn, A. and Sorgeloos, P. (2021) Biosecurity: reducing the burden of disease. *Presented at Global Conference on Aquaculture + 20: Aquaculture for Food and Sustainable Development, Shanghai, People's Republic of China, 22–25 September 2021*. Organized by Food and Agriculture Organization of the United Nations, Rome.

Sub-Committee on Aquatic Animal Health (2016) Aquaculture Farm Biosecurity Plan: generic guidelines and template. Australian Government, Department of Agriculture, Fisheries and Forestry. Available at: https://www.agriculture.gov.au/agriculture-land/fisheries/aquaculture/farm-biosecurity-plan (accessed 23 October 2022).

Surachetpong, W., Janetanakit, T., Nonthabenjawan, N., Tattiyapong, P., Sirikanchana, K. and Amonsin, A. (2017) Outbreaks of tilapia lake virus infection, Thailand, 2015–2016. *Emerging Infectious Diseases* 23, 1031–1033.

Sveen, L., Timmerhaus, G., Johansen, L.H. and Ytteborg, E. (2021) Deep neural network analysis – a paradigm shift for histological examination of health and welfare of farmed fish. *Aquaculture* 532, 736024.

The Fish Site (2013) Mycotoxins in aquaculture feeds. *The Fish Site*, 4 March 2013. Available at: https://thefishsite.com/articles/mycotoxins-in-aquaculture-feeds (accessed 30 March 2022).

The Fish Site (2021) The predator that's killing 500,000 Scottish farmed salmon a year. *The Fish Site*, 14 January 2021. Available at: https://thefishsite.com/articles/the-predator-thats-killing-500-000-scottish-farmed-salmon-a-year (accessed 19 February 2022).

The Fish Site (2022) Umitron teams up with leading Japanese sushi firm. *The Fish Site*, 17 January 2022. Available at: https://thefishsite.com/articles/umitron-teams-up-with-leading-japanese-sushi-firm-kawa-sushi (accessed 4 February 2022).

Thompson, K.D., Lilley, J.H., Chinabut, S. and Adams, A. (1997) The antibody response of snakehead, *Channa striata* Bloch, to *Aphanomyces invadans*. *Fish & Shellfish Immunology* 7, 349–353.

Tillapaugh, D., Brenton, C. and Harrower, B. (1993) *Predation on Salmon Farms in British Columbia: The Impacts of Harbour Seals* (Phoca vitulina). *The Results of a 1991 Survey*. Ministry of Agriculture, Food & Fisheries, Abbotsford, British Columbia, Canada.

Trainer, V.L., Moore, S.K., Hallegraeff, G., Kudela, R.M., Clement, A., *et al*. (2019) Pelagic harmful algal blooms and climate change: lessons from nature's experiments with extremes. *Harmful Algae* 91, 101591.

Ubina, N.A. and Cheng, S.C. (2022) A review of unmanned system technologies with its application to aquaculture farm monitoring and management. *Drones* 6, 12.

van Rijn, I., Buba, Y., DeLong, J., Kiflawi, M. and Belmaker, J. (2017) Large but uneven reduction in fish size across species in relation to changing sea temperatures. *Global Change Biology* 23, 3667–3674.

Vass, S. (2013) Gill disease to cost salmon farmers £30 m. *The Herald Scotland*, 19 January 2013. Available at: http://www.heraldscotland.com/business/markets-economy/gill-disease-to-cost-salmon-farmers-30m.19956340 (accessed 15 January 2022).

Vilata, J., Oliva, D. and Sepúlveda, M. (2009) The predation of farmed salmon by South American sea lions (*Otaria flavescens*) in southern Chile. *ICES Journal of Marine Science* 67(3), 475–482.

Vu, N.T., Phuc, T.H., Oanh, K.T.P., Sang, N.V., Trang, T.T. and Nguyen, N.H. (2022) Accuracies of genomic predictions for disease resistance of striped catfish to *Edwardsiella ictaluri* using artificial intelligence algorithms. *G3: Genes, Genomes, Genetics* 12(1), jkab361.

Whittington, I.D., Corneillie, S., Talbot, C., Morgan, J.A.T. and Adlard, R.D. (2001) Infections of *Seriola quinqueradiata* Temminck & Schlegel and *S. dumerili* (Risso) in Japan by *Benedenia seriolae* (Monogenea) confirmed by morphology and 28S ribosomal DNA analysis. *Journal of Fish Diseases* 24, 421–425.

WorldFish (2021) *Guidelines for Managing the Risks of Introductions and Transfers of Genetically Improved Farmed Tilapia (GIFT)*. Guideline No. 2021-18. WorldFish, Penang, Malaysia.

Woźny, M., Obremski, K., Jakimiuk, E., Gusiatin, M. and Brzuzan, P. (2013) Zearalenone contamination in rainbow trout farms in north-eastern Poland. *Aquaculture* 416–417, 209–211.

Xu, W. and Matzner, S. (2018) Underwater fish detection using deep learning for water power applications. *2018 International Conference on Computational Science and Computational Intelligence (CSCI)* 2018, 313–318.

Yildirim, M., Manning, B.B., Lovell, R.T., Grizzle, J.M. and Rottinghaus, G.E. (2000) Toxicity of moniliformin and fumonisin B1 fed singly and in combination in diets for young channel catfish *Ictalurus punctatus*. *Journal of the World Aquaculture Society* 31, 599–608.

Zhang, F., Wang, Y., Cao, M., Sun, X., Du, Z., *et al.* (2016) Deep-learning-based approach for prediction of algal blooms. *Sustainability* 8(10), 1060.

Zhao, J., Shirley, R.B., Dibner, J.D., Uraizee, F., Officer, M., *et al.* (2010) Comparison of hydrated sodium calcium aluminosilicate and yeast cell wall on counteracting aflatoxicosis in broiler chicks. *Poultry Science* 89(10), 2147–2156.

Zhao, J., Bao, W., Zhang, F., Zhu, S., Liu, Y., *et al.* (2018) Modified motion influence map and recurrent neural network-based monitoring of the local unusual behaviors for fish school in intensive aquaculture. *Aquaculture* 493, 165–175.

Zhao, S., Zhang, S., Liu, J., Wang, H., Zhu, J., *et al.* (2021) Application of machine learning in intelligent fish aquaculture: a review. *Aquaculture* 540, 736724.

Zhou, J.X., Wang, H., Li, X.W., Zhu, X., Lu, W.L. and Zhang, D.M. (2014) Construction of KHV-CJ ORF25 DNA vaccine and immune challenge test. *Journal of Fish Diseases* 37(4), 319–325.

12 Welfare of Cage-Cultured Fish under Climate Change

Felicity A. Huntingford[1]*, Sunil Kadri[2] and João L. Saraiva[3]
[1]*University of Glasgow, Glasgow, UK;* [2]*University of Stirling, Stirling, UK;* [3]*FishEthoGroup Association, Olhão, Portugal and Centre of Marine Sciences (CCMAR), Faro, Portugal*

12.1 Introduction

12.1.1 Increasing importance of the welfare of cultured fish

The welfare of cultured fish, including those reared in cages, is of increasing concern for the general public. Consequently, there is strong interest in fish welfare among researchers, as reflected in a steady increase in publications on fish welfare, with citations in Web of Science increasing from 34 in the five-year period 2001–2006 to 167 in 2017–2021. There have been numerous conferences and workshops on this topic, such as the Second Aquatic Animal Welfare Conference (2020), the Second Symposium on Welfare in Aquaculture (2020) and the Conferencia Internacional 'Bienestar Animal en Salmónidos' (2021), run by non-governmental organizations, academic institutions and the aquaculture industry, respectively. Public concern about the welfare of farmed fish has been responsible for enhanced control and regulation at national and international/European level (e.g. by the World Organisation for Animal Health (OIE) and the Norwegian Food Safety Authority), as well as for welfare assurance schemes (e.g. RSPCA Assured, Friend of the Sea, CertifiedHumane.org and the Global Animal Partnership).

12.1.2 Why welfare is important

Concern for the welfare of cultured fish arises in part from ethical considerations based, for example, on the moral importance of caring for captive animals (Bovenkerk and Meijboom, 2020). Over and above this, good welfare potentially benefits all parties in fish culture. From an economic perspective, up to a point and in many cases, good production and good welfare go hand in hand, with poor welfare often impacting production-related traits (Saraiva *et al.*, 2019). In addition, in some markets (e.g. Europe), there are premiums to be charged for fish cultured under welfare assurance schemes (see above and Noble *et al.*, 2018). Additionally, ensuring good welfare in farmed fish is an important aspect of job satisfaction for fish-farm workers (Medaas *et al.*, 2021; Turnbull, 2022). As a final point in this context, poor welfare often increases the adverse impacts of aquaculture on the environment, such as when appetite is suppressed in stressed fish which increases feed wastage into and pollution of the water surrounding farm cages.

*Felicity.Huntingford@glasgow.ac.uk

DOI: 10.1079/9781800621640.0012

12.1.3 Topics in this chapter

Before discussing how climate change might affect the welfare of cage-farmed fish, considerable background information is needed on what welfare is and how it is challenged and protected in cage farming as currently practised. This chapter starts by stressing the great diversity of fish, both between and within species, concentrating on a few traits that have potential implications for how fish respond to environmental challenges (Section 12.2). Section 12.3 presents some basic but important fish welfare science. This is the third edition of *Diseases and Disorders of Finfish in Cage Culture*, speaking to a huge amount of research aimed at understanding and preventing disease and other sources of ill health. This is entirely appropriate from a welfare perspective because, generally and under most conditions, good health is a necessary foundation for good welfare. However, there are complexities in the relationship between health and welfare and these are covered in Section 12.4. Many features of cage culture potentially compromise welfare, and these are discussed, together with some strategies for mitigating such effects, in Sections 12.5 to 12.8. Section 12.9 looks to the future, trying to predict what cage farming of fish for food will look like from a welfare perspective in a decade. This includes consideration of the impacts of accelerating climate change on the welfare of fish farmed in cages and some strategies for mitigating these effects. Finally, Section 12.10 presents some concluding comments.

12.2 Fish Diversity

12.2.1 Differences between species

The term 'finfish' ('fish' in what follows) refers to a great variety of animals that have little in common, except they are vertebrates that live in water, have fins and mostly respire using gills (Fig. 12.1). More than 34,000 species of fish are recognized; less than 1000 are cartilaginous fish (elasmobranchs), the remainder being bony fish. Few if any elasmobranchs are cultured, except perhaps for captive breeding programmes. Among the bony fish, with the exception of sturgeon (which are usually cultured in tanks), most cage-farmed species belong to the *c.*30,000 species of teleosts. An increasing number of ornamental teleosts are cultured to protect natural populations (Pouil *et al.*, 2020), but these too are mostly reared in tanks and ponds, except for a long-established cage culture of various carp species in China (Chen *et al.*, 2007). These groups are not considered further in this chapter, which concentrates on teleost fish that are reared in cages for food.

The great diversity within the teleosts in physiology, behaviour and ecology influences how they cope with environmental challenges (Fernö *et al.*, 2020). Some of this diversity arises because fish belong to different evolutionary lineages (Fig. 12.1), each with particular inherited traits. For example, the majority of teleost fish control buoyancy by means of the swim bladder, derived originally from an outgrowth of the gut. In some lineages (physoclistous fish, including perch-like fish such as bass), there is no remaining connection with the gut and the swim bladder is filled by gas secreted from the blood. In other lineages (physostomous fish such as carp, salmon and eels), the swim bladder remains connected to the gut and is filled by gulping air at the water surface. So, while physostomous fish need to make regular trips to the water surface to top up their swim bladders, physoclistous fish are free of this constraint. Other important traits in teleost fish have diverged through evolved adaptation to different lifestyles and environments. For example, gill area is larger in active fast-swimming fish compared with more sluggish species, reflecting their greater metabolic requirements (Bigman *et al.*, 2018). Knife fish (*Brachyhypopomus* spp.) living in environments subjected to periodic hypoxia have strikingly larger gill areas than do closely related species from more oxygen-rich sites (Crampton *et al.*, 2008).

12.2.2 Differences within species

There are also striking differences within a given species, associated with life history stage and among populations adapted to different habitats. For example, Atlantic salmon (*Salmo salar*) in the juvenile, freshwater stage that live in fast running water have less gas in their swim bladder than do those in stiller water, negative buoyancy giving better control of movement in currents

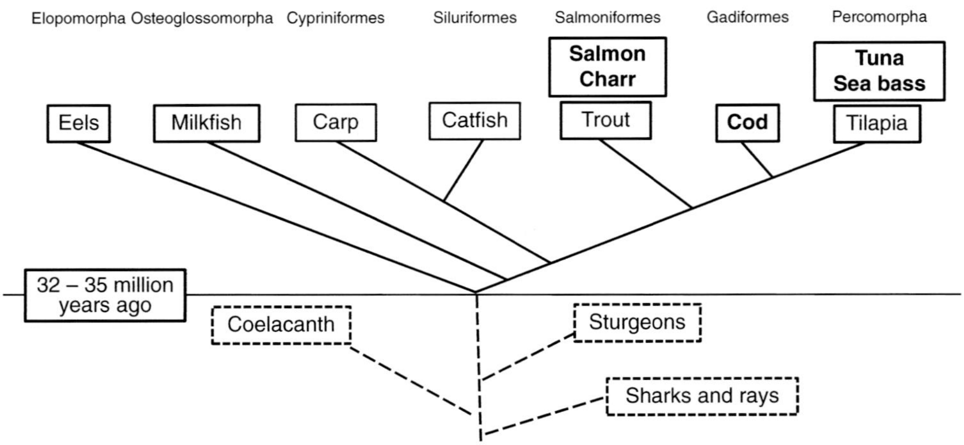

Fig. 12.1. A simplified phylogeny of living jawed finfish, with some farmed species indicated (in solid boxes). Species that are sometimes farmed in cages are indicated in plain font; those that are usually farmed in cages are in bold font. Taxonomic groups within the teleosts (above the line) are indicated along the top. (Drawn by the chapter authors.)

(Saunders, 1965). In the cichlid species *Pseudocrenilabrus multicolor victoriae*, fish from hypoxic swamps have 29% larger gill areas than do those living in well-oxygenated lakes and rivers (Chapman *et al.*, 2000). Some of these differences are inherited, others being the result of developmental plasticity (Crispo and Chapman, 2010).

Gill area also varies among individuals of the same population. Some of this variation is associated with marked and consistent individual differences in reaction to a variety of environmental challenges, often referred to as 'stress coping styles'. Thus, individuals of many fish species lie at different points along a so-called 'proactive–reactive axis'. Those at the proactive end of the spectrum are aggressive and risk-taking (usually summarized by the term 'bold' in the behavioural literature), respond actively when challenged, tend to form and stick to routines, and show a predominantly adrenaline-based physiological stress response. By contrast, reactive fish are usually non-aggressive, avoid risk (often described as 'timid'), respond to challenge by freezing or hiding, are behaviourally flexible and show a predominantly cortisol-based stress response (Koolhaas *et al.*, 1999; Castanheira *et al.*, 2017; Johansen *et al.*, 2020). In several cases, proactive fish have a higher metabolic rate than do their reactive counterparts – for example, common carp, *Cyprinus carpio* (Huntingford *et al.*, 2010); Arctic charr, *Salvelinus alpinus* (Magnhagen *et al.*, 2018); and rainbow trout, *Oncorhynchus mykiss* (Skov *et al.*, 2019) – and in carp this is associated with a larger gill surface area (Jenjan *et al.*, 2013). In nature proactive and reactive animals flourish in different environments, proactive individuals being favoured when resources are abundant and predictable, population densities are high and predation is low, while reactive animals do best in the opposite conditions. In fish this has clear implications for how well they adapt to conditions in cage aquaculture, and this is discussed further below (Section 12.6.4).

12.3 Some Basic Fish Welfare Science

12.3.1 What is welfare?

The word 'welfare' is so commonly used that there are bound to be different interpretations, meaning different things to a fish farmer, a biologist, a policy maker or a consumer, for example. Only by clearly defining the term can it be properly assessed, monitored and incorporated into laws and regulations. A simple definition proposed by Broom (1991), that welfare is the state of the animal as it copes with the environment, is widely accepted. This has the following important implications: (i) welfare is a characteristic of an animal, not something that is given to it by carers; (ii) welfare varies along a continuum, from negative to positive; (iii) welfare can be measured independently of ethical considerations; (iv) direct measurements of the state of the animal must also be used to assess its welfare, over and above knowledge of its biology; (v) such direct measures of how well an animal copes with the environment give essential information about its welfare; and (vi) coping mechanisms may vary among different species.

12.3.2 Conceptualizing and protecting welfare

Three distinct approaches are used when addressing animal welfare (Fraser *et al.*, 1999, 2009). According to a *feelings-based approach*, to experience good welfare the animal should be free from negative experiences and have access to positive ones. According to a *function-based approach*, welfare is good if the animal can adapt to its environment, with all its biological functions working effectively. According to a *nature-based approach*, each species has an inherent biological nature and the ability to express this is essential for good welfare. Applying each of these approaches separately has led to important improvements in animal welfare. However, as suffering, health problems and impairment of natural behaviour often accompany each other, an integrated, multidisciplinary approach helps to promote the objective measurements of welfare (Saraiva *et al.*, 2018). All three of these approaches can be seen in the concept of Five Freedoms (Box 12.1), first presented in the UK Brambell Report into husbandry of livestock (1965) and revised by the Farm Welfare Council of the UK in 1979 into its present form (McCulloch, 2013). Current considerations of welfare in farmed animals draw heavily on this concept of five freedoms, which forms the basis of recommendations and legislations worldwide and is still extensively employed for academic, educational and veterinary purposes, with great practical utility (McCulloch, 2013). It paved the way for animals to be considered by European law as sentient beings in the Lisbon Treaty of 2007. The approach is open to criticism; for example, it implies incorrectly that captive animals are passive, rather than engaging actively with their environment (Ohl and van der Staay, 2012).

Box 12.1. The Five Freedoms, as presented in the Brambell Report (1965) for terrestrial farm animals.

Good welfare requires:

1. Freedom from hunger and thirst, with ready access to fresh water and a diet to maintain full health and vigour.
2. Freedom from discomfort, with an appropriate environment, including shelter and a comfortable resting area.
3. Freedom from pain, injury or disease, with prevention or rapid diagnosis and treatment.
4. Freedom to express normal behaviour, having sufficient space, proper facilities and (where appropriate) the company of the animal's own kind.
5. Freedom from fear and distress, with ensured conditions and treatment that avoid mental suffering.

In addition, understandably at the time, the emphasis was very much on protecting animals from negative experiences, epitomizing the view that 'free from harm equals good'. The idea that animals can experience positive states that culture conditions should accommodate has historically been under-represented in current frameworks and, to an extent, in current welfare research. A more dynamic perspective on welfare that accommodates some of these complications is based on the concept of allostasis (Korte *et al.*, 2007), according to which, up to a point, challenges in the environment are beneficial to an animal because they provide stimulation. Too much stimulation represents an allostatic load that the animal fails to cope with, but too little stimulation may also be harmful. This view indicates that proper environmental stimulation may promote what is often called 'positive welfare' and also provides a theoretical basis for seeking to improve fish welfare by making changes in cage design that provide stimulation (see 'Environmental enrichment' in Section 12.8.1).

12.3.3 Fish sentience and positive welfare

The feelings-based approach towards fish welfare (Section 12.3.2) obviously depends on these animals having feelings, or at least being capable of some sort of affective appraisal of their surroundings and experiences. The term most often used to describe such an ability is 'sentience', the capacity to experience subjective, perceptual experiences such as pleasure, joy, fear or pain (Dawkins, 1998; Appleby and Sandøe, 2002; Brown, 2015). This apparently simple requirement carries important ethical consequences; most would agree that if an animal is sentient, it can probably suffer and therefore warrants protection (Brown, 2015).

The extent to which fish (and other non-mammalian animals) are sentient has been the motive of heated debates (see e.g. Learmouth, 2020; Browning and Birch, 2022). This debate should be appraised in the context of the extensive and growing body of evidence demonstrating complex cognitive capabilities of fish generally (e.g. Brown, 2015; Luchiari *et al.*, 2021). Understanding the subjective experiences of any non-human animal is complex, but there are sources of powerful, objective evidence on this issue, based on detailed behavioural, physiological and neurobiological studies.

Much of the debate over fish sentience has been focused on whether they experience pain (see e.g. Sneddon *et al.*, 2018 and subsequent commentary thread). There is abundant and accumulating evidence that fish can feel pain (defined by the International Association for the Study of Pain as 'an unpleasant sensory and emotional experience associated with actual or potential tissue damage'). Box 12.2 lists some general multidisciplinary criteria for considering that a given category of animal has the capacity to experience the emotion of pain, rather than simply responding to nociception. These have all been demonstrated for at least one species of fish, as they have in mammals and birds (Sneddon *et al.*, 2014).

Less is known about the capacity of fish to experience positive emotions. However, there is accumulating evidence of some commonality in neuroendocrine and cognitive correlates of positive affective states in fish and mammals. Thus, fish have neural structures and systems which are similar to those that modulate the feeling of well-being and positive reinforcement in mammalian brains (see Fife-Cook and Franks, 2019). In this context, a direct test using fish was recently made of an influential model of emotions in humans. This model (summarized in Fig. 12.2A) recognizes four core emotional states in humans, based on valence (whether the experience is positive or negative) and salience or arousal (how strong it is) (Mendl *et al.*, 2010). When the model was tested on sea bream (results summarized in Fig. 12.2B), four comparable states (positive as well as negative) were identified, associated with comparable behavioural, physiological and neuromolecular responses that are unique to each valence–arousal quadrant (Fig. 12.2B). This led the authors to conclude that fish show 'emotion-like' states (Cerqueira *et al.*, 2017).

While there are many knowledge gaps regarding the extent and nature of mental states in fish (and there will surely be differences between species), the evidence for sentience in fish is very strong and certainly sufficient to warrant giving fish the benefit of doubt in this context. In fact, in most of what follows, the welfare of caged fish is well covered by the least controversial, functional approach.

Box 12.2. Some criteria for determining whether a particular category of animal has the capacity to feel pain, specifying features that suggest the emotion of pain rather than a much simpler (avoidance) response to nociception. Note that no single criterion is sufficient on its own. (From information in Sneddon *et al.*, 2014.)

The nature of the direct responses to nociceptive stimulation:

- involves central processing in brain areas that regulate motivated behaviour;
- is sensitive to opioid modulation;
- includes physiological responses;
- is more complex than simply avoidance;
- is accompanied by protective behaviour, such as wound guarding; and
- is reduced by analgesia.

Indicative of longer-term motivational changes and rapid learning:

- response to noxious stimulus suppresses other motivations;
- reduced attention is paid to other important stimuli;
- repeat noxious stimulation is avoided (conditioned place avoidance and avoidance learning);
- costs are accepted in order to avoid noxious stimulus; and
- self-administration of analgesia occurs, with cost for access accepted.

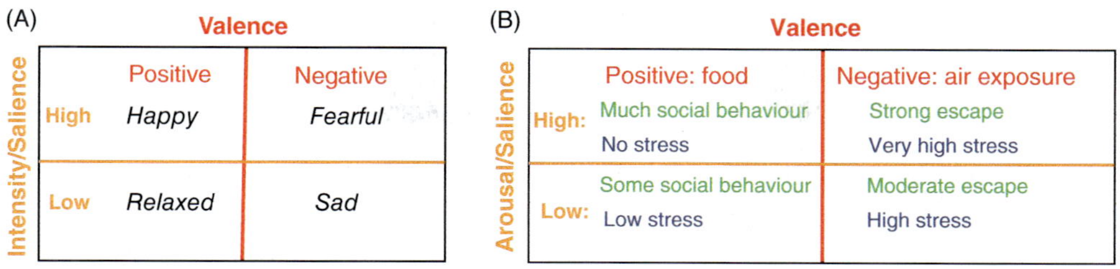

Fig. 12.2. (A) Schematic representation of a two-dimensional model of core emotions in humans, determined by different combinations of valence and salience, as described in Mendl *et al.* (2010). (B) Summary of the results of a direct test of this model in sea bream (Cerqueira *et al.*, 2017). Different groups of fish were trained with either food or brief exposure in air (manipulating valence). Arousal was manipulated by making the experience either predictable (always associated with a light stimulus) or unpredictable. The figure summarizes behaviour and stress levels in response to the light alone at the end of training. Each quadrant is characterized by a unique combination of behaviour, stress physiology and activation in specific parts of the forebrain comparable to that seen in humans and other mammals. (Drawn by the chapter authors.)

The view that animals experience mental and physical states exceeding what is strictly necessary for short-term survival (performing certain actions for their own sakes, rather than simply seeking resources such as food or safety) is often referred to as positive welfare. This has been discussed by many authors (e.g. Lawrence *et al.*, 2019) and is considered further in Section 12.6.2. The application of positive welfare to fish was proposed by Fife-Cook and Franks (2019), based on the assumption that fish (as other vertebrates) are naturally equipped to seek positive affective states that can be identified from knowledge of their biology and behaviour. Suggested examples of behaviours that fish may perform for their own sake include preferential attachment (Heathcote *et al.*, 2017), social buffering of fear (Faustino *et al.*, 2017), social motivation (Galhardo *et al.*, 2011; Maia and Volpato, 2017) and voluntary exploration of a novel area (Graham *et al.*, 2018).

12.4 The Relationship Between Health and Welfare

12.4.1 Direct adverse effects of disease on welfare

To protect the well-being of farmed fish, much effort has gone into developing systems for improving fish health, directed at pathogen-induced diseases and other threats to health, such as malnutrition and injury. This is entirely appropriate because ill health is usually a powerful cause and indicator of poor welfare. This is not just because disease generates poor health and mortality, but also because survivors may experience welfare problems after the disease has passed. Chapters in this volume speak to the success of such a health-based approach to fish welfare, but there are some interesting complications.

12.4.2 Stress, disease and disease treatment

Poor general welfare often makes fish more susceptible to disease

There is a well-documented relationship between good welfare and good immune function, making a direct link between welfare and disease. Thus, improved welfare benefits humans as well as fish by promoting healthy, disease-free food production (Saraiva *et al.*, 2022). For example, in Atlantic salmon mortality rates increase in the weeks following transfer to sea cages, a stressful procedure that potentially compromises welfare in a number of ways (Stien *et al.*, 2016).

Disease treatments may also compromise welfare

Effects of stress on immune function can potentially generate a vicious circle whereby stressful disease treatments generate poor welfare, which may render the fish more susceptible to other pathogens (or malnutrition, if appetite is impaired). For example, commercial delousing treatments using hot water and mechanical removal cause damage and stress, which impair welfare (e.g. Sommerset *et al.*, 2021). Vaccine delivery against infectious diseases has several adverse consequences for welfare, including the stress of capture, handling and skin puncture (mitigated by light anaesthesia) and side effects such as abdominal adhesions (Fraser *et al.*, 2014).

12.4.3 The non-equivalence of health and welfare

Good health does not necessarily equate to good welfare

As emphasized by current interest in positive welfare in fish (Section 12.3.3), the absence of disease and injury, even when combined with good nutritional status, does not necessarily ensure good welfare, although it is clearly an appropriate aspiration for those who care for cultured fish. A healthy, well-fed fish may be free of negative experiences induced by disease and unsatisfied hunger, for example, but its welfare may yet be compromised by lack of access to positive experiences, such as social companionship and the opportunity for exploration (Saraiva *et al.*, 2018).

Poor health does not always equate to poor welfare

Although ill health and injury in cultured fish probably mean these animals are experiencing poor welfare, a functional perspective suggests some interesting reasons why this may not always be the case. In nature, animals balance the costs of potentially dangerous behaviour against the fitness benefits that such behaviour may confer. For example, when fighting for food or mates, many animals accept significant costs (reduced energy reserves, injury and impaired immunocompetence, all indicators of poor welfare, see Section 12.6.3) when fighting leads to better growth and more offspring (enhanced Darwinian fitness). This is called 'adaptive self-expenditure' (Barnard and Hurst, 1996; Barnard, 2007) and must involve motivational systems that prompt fish to fight in the face of its immediate negative consequences. That such systems exist is indicated by the fact that, for example, a perceived strong social threat reduces pain responsiveness in male mice, *Mus musculus* (Langford *et al.*, 2011); by in a sense anticipating injury, such pain suppression can be seen as preparing for adaptive self-expenditure. When fitness is likely to be enhanced by fighting, Barnard and Hurst (1996) argue that there is no reason to suppose that an animal suffers from any costs incurred in a fight, as long as it is free to decide whether or not to fight 'on its own adaptive terms'. Conversely, if fitness is likely to be reduced by fighting, there is no reason to suppose that a fish suffers if it does not fight. In recognizing that the consequences of adaptive responses to challenge do not necessarily indicate poor welfare, this approach is similar to the view that, within limits, experiencing an allostatic load may be good for welfare (Section 12.3.2).

This may well seem an academic point in cage farming where most fish are reared for food and killed before breeding. However, even farmed Atlantic salmon are not fully domesticated (Saraiva *et al.*, 2018) and they bring with them evolved mechanisms whereby they are motivated to fight over resources when the benefits of getting the resource outweigh the possible costs of fighting (see Damsgård and Huntingford, 2012). Where food is predictable in space and time (as it can be on fish farms), the mechanisms for adaptive self-expenditure may click in and fish may choose to fight for food; any injuries gained while competing for food in this way could represent adaptive self-expenditure and are not necessarily indicative of poor welfare. This is the case even though on farms the resulting improved growth will never actually be translated into improved fitness. There are, however, many other good reasons for preventing fighting and injury among fish held in farm cages.

12.5 Promoting the Welfare of Fish in Cages: Challenges

Farmed fish production relies on at least 466 species, with a notable increase in recent years. Although the 20 most common species account for over 80% of total fish production, this diversity makes accommodating their (species specific) welfare needs extremely challenging. The contrast with land animal production is striking, as fewer than 30 species are generally farmed for food on land (FAO Domestic Animal Diversity Information System (DAD-IS), http://www.fao.org/dad-is/data/en/, accessed 17 August 2022). In terms of numbers of animals, the portal www.fishcount.org.uk (accessed 16 August 2022) estimated the total number of farmed fish in 2017 to be between 51,107 and 167,476 million animals, representing a 4–6% increase since 2015.

An additional complexity is domestication, a human-induced process that gradually alters a cultured organism through both inadvertent selection (usually for tameness) and targeted selective breeding (often for important production traits, such as milk yield in cows) (Price, 2001). While domestication of land animals has been taking place for thousands of years, in fish the inadvertent domestication and targeted selection are very recent processes (Teletchea and Fontaine, 2012; Saraiva *et al.*, 2018). Selection programmes are largely focused on production-related traits such as faster growth or disease resistance. Selection for fast growth potentially creates a number of problems for fish welfare. For example, because the capacity for fast growth is a polygenically inherited trait, there may be carry-over effects of selection to other inherited traits that are important for welfare (Pasquet, 2018). Fast-growing fish are often bolder and more aggressive (see Section 12.2.2), which can cause welfare problems for less aggressive fish, especially if fast growth in some fish increases size variation in a cage population. Box 12.3 presents some additional adverse consequences of fast growth for welfare. So, while the absence of real domestication means that, unlike animals farmed on land, cultured fish are not tamer and less easily stressed, targeted selection for fast growth may itself compromise the welfare of selected populations.

12.6 Promoting the Welfare of Fish in Cages: Assessing Welfare

The task of ensuring that fish farmed in cages enjoy good welfare involves developing tools for monitoring welfare in cage conditions, identifying husbandry procedures that compromise good welfare and developing systems for mitigating such adverse effects. There have been many reviews about assessing and protecting welfare (e.g. Kristiansen *et al.*, 2020; Saraiva *et al.*, 2022); the following subsections point out some key issues and discuss their relevance to fish in cages.

12.6.1 Frameworks for assessing fish welfare

Much research effort has gone into developing schemes for assessing the welfare of key species of cage-farmed fish (Stien *et al.*, 2020). Although, as pointed out above, such schemes must be specific for both species and stage, welfare protection schemes for different species and stages are similar in broad principles. These are illustrated here by reference to Atlantic salmon, because they make up a very large proportion of cage-farmed fish (*c*.2.7 million tonnes were produced globally in 2020) (FAO, 2020) and because there has been a strong focus on the assessment of welfare in this species, integrated under the heading of FISHWELL (Noble *et al.*, 2018). Similar systems have been developed for Mediterranean aquaculture species such as European sea bass, *Dicentrarchus labrax* (Yildiz *et al.*, 2021), and gilthead sea bream, *Sparus aurata* (Roque *et al.*, 2020), as well as for Nile tilapia, *Oreochromis niloticus* (Pedrazzani *et al.*, 2020; Tschirren *et al.*, 2021).

12.6.2 Welfare needs

To assess fish welfare on farms, it is important to identify what are called the 'welfare needs' of the species concerned, or the conditions that must be provided to ensure good welfare. In the case of Atlantic salmon (Noble *et al.*, 2018) these needs include access to resources such as food and an appropriate environment that allows for respiration, temperature control and osmotic balance, as well as good water quality. Access to appropriate resources and environmental conditions potentially allows health-related needs to be met, which include protection from disease and/or injury.

All the different features of the cage environment that are necessary for welfare potentially

Box 12.3. Some additional welfare costs of fast growth.

Fast growth:

- often causes early reproduction – more fast-growing cage-farmed fish mature during on-growing, compromising welfare (Noble *et al.*, 2018) (Section 12.6.3 and Fig. 12.3);
- increases energy expenditure (e.g. Pedersen, 1997);
- can involve impairments in muscle structure (Nebo *et al.*, 2013);
- and skeletal deformities (e.g. in fast-growing triploid salmon, Gaffney *et al.*, 2018);
- and cataracts (e.g. Ersdal *et al.*, 2001; Hamre *et al.*, 2022);
- and impaired hearing (Reimer *et al.*, 2017);
- and reduced longevity (Metcalfe and Monaghan, 2003).

Fig. 12.3. Maturing caged salmon (photographed from above) that have left the school and swim against the current direction. This is an example of a natural element of a species' behavioural repertoire being used as an early warning sign for welfare problems. (Photograph reproduced with kind permission from Jan Erik Fosseidengen, Institute of Marine Research, Norway.)

map on to feelings, such that meeting each need is expected to activate brain systems that cause the animals to feel 'good', whereas failure to meet them will activate systems that cause animals to feel 'bad'. There is much still to be learned about sentience in fish and the circumstances that create positive and negative emotions (Section 12.3.3). However, schemes such as those described above (Section 12.6.1) provide a comprehensive and practical approach, given current understanding.

12.6.3 Operational welfare indicators

Input and outcome indicators

Having identified a set of welfare needs for fish of a given species and stage, reliable indicators are required of the extent to which these needs are met in any given case. The subset of indicators that can be applied relatively easily and accurately on working farms, so-called 'operational welfare indicators' (OWIs), come in two broad categories. *Environment-based* (or input) indicators are centred on the resources and the environment that fish experience. These include water temperature, salinity, dissolved oxygen and carbon dioxide, pH and alkalinity, total nitrogen, turbidity and current speed. They also include aspects of the husbandry regime, such as amount of food delivered, cage lighting and stocking density. Using such OWIs involves measuring the status of each relevant variable and ensuring it falls within pre-identified required limits for the well-being of the species and life history stage. Attention has focused on physical and nutritional needs, but fish also have behavioural needs, in the sense of being strongly driven to perform specific actions because these are rewarding in themselves, independent of any direct physiological benefits they may bring. Recognizing this emphasizes the importance of ensuring that caged fish can experience positive welfare (Fife-Cook and Franks, 2019).

Animal-based (or outcome) indicators use attributes of the fish themselves. Table 12.1 lists the main outcome OWIs for Atlantic salmon from the FISHWELL scheme (Noble *et al.*, 2018), with brief explanations of what they indicate about welfare and some complexities in their interpretation. Some of these indicators reflect the status of the whole group (e.g. percentage mortality rates). Others are collected on individual fish (e.g. body weight) and may be collected via observation of live fish in the cage or from fish killed for sampling. Such individual scores are usually combined to assess group averages and variability.

As Table 12.1 shows, interpreting any single outcome OWI is not necessarily straightforward; for example, since fish often experience periods of natural anorexia, as in the case of overwintering juvenile Atlantic salmon (Metcalfe and Thorpe, 1992), low appetite, reduced condition factor and resulting slow growth do not inevitably equate to poor welfare. Furthermore, collecting representative samples for groups of fish held in cages is a major challenge for all these outcome OWIs. However, a portfolio of such indicators offers a multifaceted picture of the welfare of the fish population in a given cage.

Behavioural welfare indicators

A nature-based approach to welfare (Section 12.3.2) suggests that performance of natural behaviour is the 'gold standard' for assessing welfare in farmed fish, although there are complexities with this approach (Saraiva *et al.*, 2019). Over and above this, several of the outcome OWIs in Table 12.1 concern how fish behave, either individually or in schools; these have a special role in monitoring welfare, often acting as an early warning of welfare problems (Martins *et al.*, 2012). Behavioural observations as fish come near the surface to feed or from video footage from underwater cameras can provide valuable qualitative information about their welfare (Føre *et al.*, 2018; Noble *et al.*, 2018). For example, in spite of husbandry systems designed to delay breeding, some Atlantic salmon reach sexual maturity in production cages, with adverse effects on their welfare. Maturing fish leave the school and swim into the cage wall facing the current, which is atypical of immature fish, but natural in mature wild fish migrating towards the coast (Noble *et al.*, 2018) (Fig. 12.3); in cages, this provides a behavioural indicator of reproductive status and hence poor welfare. Quantitative changes in behaviour are also informative (e.g. Begout *et al.*, 2012), although often difficult and/or time-consuming to collect. Newly developed computer vision systems for accurate quantification of behaviour are increasingly used in this context (Føre *et al.*, 2018) (see Section 12.9.2). Examples of how behavioural change can be used to assess the welfare of fish in cage conditions are presented in Fig. 12.4.

Schooling behaviour can provide a sensitive indicator of inappropriate cage conditions. For example, in undisturbed schooling species, individual fish swim at broadly similar distances from their nearest neighbours and head in a similar direction: they are 'polarized'. In nature, loss of such a regular, polarized spatial arrangement within a group of fishes occurs in response to a variety of stressors, such as the presence of a predator. Polarized schooling (Fig. 12.4A) is reasonably well-maintained during transfer of Atlantic salmon smolts from freshwater farms to sea cages by well-boat; this suggests that fish welfare is reasonable, confirmed by a low incidence of erratic behaviour (Fig. 12.4A) and a small, transient increase in cortisol levels (Nomura *et al.*, 2009). Other features of swimming provide information about welfare. For example, tail beat frequency falls in response to oxygen supersaturation in Atlantic salmon (Espmark

Table 12.1. Overview of some animal-based/outcome operational welfare indicators (OWIs) for Atlantic salmon, including the rationale for the use of each indicator and some complexities experienced when applying them. (Compiled from information provided in Noble *et al*., 2018.)

	OWI	Rationale and some complexities of application and interpretation
Group level	Mortality rate	*Rationale*: Cumulative mortality gives retrospective information about welfare over time. Daily mortality potentially reflects current welfare *Complexity*: Cause of death is needed for corrective action
	Surface activity	*Rationale*: Jumping at the surface happens when salmon need to fill their swim bladder. A poorly filled swim bladder restricts behavioural control *Complexity*: Surface activity may also be caused by hunger or presence of underwater predators
	Appetite	*Rationale*: Stress suppresses appetite, so reduced feeding may imply poor welfare *Complexity*: Diverse causes of appetite loss complicate diagnosis of welfare problems
	Growth	*Rationale*: Poor growth indicates unmet nutritional needs *Complexity*: Many other relevant factors influence growth rate
	Disease prevalence	*Rationale*: Good health contributes to welfare. Stress increases susceptibility, so disease may flag up other welfare issues *Complexity*: Detection of disease necessarily gives retrospective information on welfare
Individual level	Operculum beat	*Rationale*: Ventilation rate increases with hypoxia and stress, so can indicate welfare problems. Positive experiences may increase ventilation rate *Complexity*: Various adverse events can increase ventilation rate, making interpretation difficult
	Gill status	*Rationale*: Poor gill condition is an immediate welfare problem and an indicator of adverse holding conditions *Complexity*: Definitive gill evaluation requires histological examination
	Various condition indices	*Rationale*: Reduced body condition is associated with inadequate feed intake and can indicate other welfare problems *Complexity*: Cause of condition loss needs to be determined for interpretation and remedial action
	Status of fins (skin, eyes and opercula)	*Rationale*: Damage to fins and body is a direct threat to welfare and also makes fish vulnerable to infection *Complexity*: There are many causes of damage that need to be diagnosed for remediation
	Plasma cortisol	*Rationale*: Secretion of cortisol is part of physiological stress, so consistently high levels potentially indicate chronic stress *Complexity*: Standard methods for measuring cortisol are invasive and stressful. One-off cortisol measures are hard to interpret. Cortisol secretion is part of an adaptive response to challenge and may indicate coping rather than poor welfare (see Section 12.3.2)

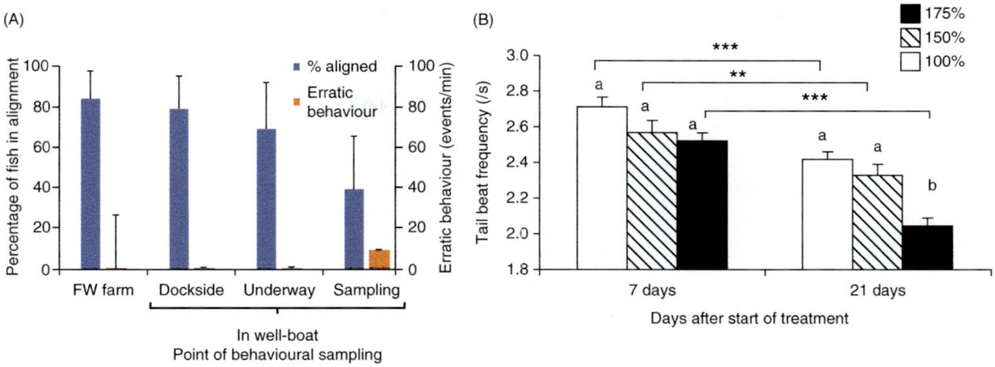

Fig. 12.4. Examples of the use of quantitative behavioural OWIs for Atlantic salmon. (A) An index of polarized swimming in Atlantic salmon smolts at various points during transfer from fresh water (FW) to marine cages (including after routine sampling during transit) for one particular smolt production unit, together with the incidence of erratic behaviour (darting or turning suddenly). Mean values with their standard deviations represented by vertical bars. (Figure reproduced with slight modifications with permission from Nomura *et al.*, 2009). (B) Tail beat frequency in parr at 7 and 21 days after exposure to different degrees of oxygen supersaturation. Mean values with their standard errors represented by vertical bars. Different lower-case letters indicate differences between dissolved oxygen levels within the same day. Asterisks indicate differences within oxygen groups across days: ** $P < 0.01$; *** $P < 0.001$. (Figure reproduced with slight modifications with permission from Espmark and Baeverfjord, 2009.)

and Baeverfjord, 2009) (Fig. 12.4B). Atlantic salmon held in submerged cages swim in a distinctive head-up posture, a natural reaction to a natural challenge – the need to get to the surface to fill the swim bladder – that serves as a warning of compromised welfare (Korsøen *et al.*, 2009).

12.6.4 Variability in welfare needs and indicators

Although the broad principles of welfare assessment and protection are equivalent across species, it would clearly be inappropriate to extrapolate specific needs and indicators from one species (or life history stage) to another. Considering the various welfare needs identified by Noble *et al.* (2018) and others, all species of cage-farmed fish need appropriate resources, but, taking food as an example, what nutritional form this takes and how it is best presented vary. For example, bottom-feeding fish such as Atlantic halibut (*Hippoglossus hippoglossus*) feed and grow better when fed sinking rather than floating pellets (Kristiansen and Fernö, 2007). Good feed management also depends on matching feed delivery to natural appetite rhythms, which often vary between and within species. For example, tambaqui (*Colossoma macropomum*), an Amazonian species that is intensively farmed in Brazil, shows nocturnal feeding behaviour with 84.98% of daily food intake occurring at night, and grows better when fed during the night (Reisa *et al.*, 2019).

Although all cage-farmed fish species require an appropriate environment for good welfare, just what that comprises is strongly dictated by the natural environment to which they are adapted. Optimal water quality obviously varies greatly between, for example, freshwater and marine species and between tropical and temperate species. Likewise, while ensuring good health is necessary for all cage-farmed species, as the other chapters in this volume illustrate, the precise spectrum of diseases from which they need to be protected is variable across these same dimensions. Ensuring that cage-farmed fish of a particular species are held in conditions that allow them to meet their behavioural needs is a complex issue, but flatfish such as the olive flounder (*Paralichthys olivaceus*) rest more and use less energy when provided with a substrate for burying (Dou *et al.*, 2000), as do Atlantic cod (*Gadus morhua*) reared with structures among which to hide (Zimmerman *et al.*, 2012).

Taking a few outcome OWIs, how well the relationship between weight and length embodied in the commonly used condition factor ($K = 100 \times$ weight (g) $\times$ [length (cm)]$^{-3}$) indicates healthy plumpness depends on species and life history stage (Noble *et al.*, 2018). In terms of behavioural indicators, increased surface activity in physostomous fish often reflects a need to fill the swim bladder, but in physoclistous fish it indicates other strong responses, such as hunger or avoidance of adverse conditions in deeper water. Specific responses to adverse conditions also vary among species; for example, whereas Atlantic cod freeze in response to hypoxia, thereby saving oxygen (Herbert and Steffensen, 2005), brook charr (*Salvelinus fontinalis*) respond by swimming faster, a form of escape (Tang and Boisclaire, 1995). Swimming speed decreases in relation to stocking density in European sea bass (Santos *et al.*, 2010), but increases in Atlantic halibut (Kristiansen *et al.*, 2004).

Many cage-cultured fish have different stress coping styles (Section 12.2.2) and these have implications for assessing and protecting welfare (Huntingford and Adams, 2005; Castanheira *et al.*, 2017; Johansen *et al.*, 2020). For example, behavioural indicators of alarm are different in proactive fish, which respond to perceived danger by active escape, and reactive fish, which respond by freezing. A different pattern of physiological stress response means, for example, that low cortisol levels in proactive fish do not necessarily indicate an absence of stress. Stress coping style also impacts on the kinds of welfare challenges that cage-farmed fish face; Table 12.2 gives a broad overview of the various ways in which protecting welfare can potentially be influenced by stress coping style.

In the light of such extensive between-species variation in welfare needs and welfare indicators, clearly a great deal of information is needed to ensure good welfare for all species of fish that are farmed in cages. For many, such information is lacking (Franks *et al.*, 2021) and a considerable research effort is needed (and to an extent is underway) to make good this gap.

12.6.5 Special features of cage farming in relation to fish welfare

Input OWIs

Cages are often much larger than tanks and raceways, which allows fish much greater freedom of movement (Noble *et al.*, 2018). They are also open structures, with renewal of cage water and removal of waste being determined by local currents and tidal flows. Up to a point, this

Table 12.2. Some potential implications of stress coping style for protecting welfare in (cage) farmed fish. Note that in most species there is a continuum of stress coping style ranging from strongly proactive to strongly reactive.

In general:	Potential implications for protecting welfare
Proactive fish are bold; reactive fish are timid	Reactive fish are more sensitive to disturbance, including stronger suppression of feeding
Proactive fish are more aggressive	Aggressive, proactive fish are more likely to be injured with associated risk of infection
Proactive fish follow routines; reactive fish are flexible	Predictable, regular husbandry practices may favour proactive fish. Greater flexibility may make reactive fish more responsive to environmental enrichment
Proactive fish have higher metabolic rates and larger gill areas	Higher metabolic rate increases food requirements and makes hypoxia more of a threat. Larger gill area makes proactive fish more vulnerable to poor water quality
Proactive tilapia gravitate towards higher temperatures	Optimal temperature range will vary with stress coping style, although there are probably sufficient temperature gradients within cages to accommodate this
Proactive and reactive fish show different patterns of susceptibility to disease and other pathologies	Susceptibility to a different spectrum of health conditions and disease suggests that veterinary regimes need to be vigilant for different conditions

makes it easier to ensure good water quality without the requirement for additional technology (Jobling, 2010). Set against this, compared for example with tank-based systems, tight control of ambient water conditions is much more challenging and very much at the mercy of weather conditions. Consequently, the environment within a given cage is non-uniform in space and time. Within limits, this is potentially advantageous from a welfare perspective, allowing fish to exercise choice with respect to currents and temperature and representing a form of environmental enrichment (Section 12.8.1). For example, Atlantic salmon in sea cages adjust their behaviour during periods of high tidal flow by switching from circular schooling to holding station against the current (Johansson *et al.*, 2014) and take advantage of reduced swimming velocities in the centre of the circular school (Gansel *et al.*, 2014). Above certain limits, however, lack of tight control over environmental conditions represents a challenge when it comes to using input indicators to ensure fish welfare. As several chapters in this volume show, unlike fish farmed in tanks, cage-farmed fish are also more exposed to pathogens and other harmful organisms.

Outcome OWIs

Observation during routine husbandry is easier for fish in cages compared with those in deeper natural ponds (Jobling, 2010). However, from time to time, adverse weather conditions will inevitably hinder monitoring outcome OWIs due to interrupted access and impaired visibility, for example resulting from eutrophication and algal blooms (Noble *et al.*, 2018; Shahmohamadloo *et al.*, Chapter 10, this volume, 2023). A number of additional practical challenges arise when assessing the welfare of fish farmed in cages, given the large body of water that cages enclose (much of it out of sight) and the large numbers of individuals involved, especially in intensive cage aquaculture (Føre *et al.*, 2018; Franks *et al.*, 2021); for example, in European Atlantic salmon farms, a single cage may be stocked initially with more than 200,000 fish. Many outcome OWIs give information at the group level and even OWIs collected from individual fish are usually averaged in some way. Combining these points (large volumes of water, sometimes with poor visibility, very large population sizes and essentially group-level OWIs), there is a real concern that there may be a significant number of undetected individual fish experiencing poor welfare, even though group-level OWIs may indicate an acceptable level of welfare (Franks *et al.*, 2021). Considerable effort has gone into developing systems for monitoring fish (often individual fish, although not usually individually identified) throughout large cages. These include increasingly sophisticated echo-sounding and hydroacoustic techniques and underwater cameras, often combined with computer-based systems for real-time analysis (Section 12.9.2).

12.6.6 The importance of what farmers know

The expertise and experience of farm workers play a key role in ensuring the welfare of fish in cages. In a recent commentary, an experienced fish veterinarian who has worked in many parts of the world, wrote: 'In all that time, meeting and talking with many fish farmers, I have never encountered one who intentionally mistreated their animals. Most expressed considerable concern for their animals' (Turnbull, 2022). Similarly, Medaas *et al.* (2021) wrote of comments made by Norwegian fish-farm workers attending courses on fish welfare: '... we were struck by the profound sense of occupational and personal or moral responsibility for fish welfare that mobilized these accounts ...'.

In a recent study using qualitative behavioural assessment (QBA), a technique developed for probing the mental states and welfare of terrestrial animals, experienced fish farmers were shown video footage of Atlantic salmon collected in a range of situations and asked to produce a list of words describing what their behaviour was expressing (Jarvis *et al.*, 2022). The final agreed list (Fig. 12.5) included terms such 'agitated', 'relaxed', 'tense' and 'content'. It is notable that these farmers were comfortable using such emotional terms to describe what they saw and that they agreed among themselves about the appropriate descriptors. A separate group of (inexperienced) observers was then asked to apply these descriptors to the behaviour of salmon in a separate set of video clips. Principal

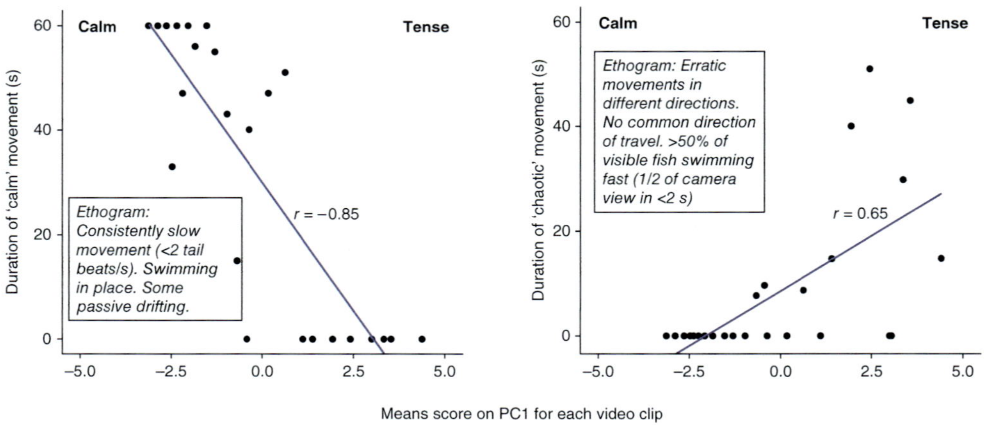

Fig. 12.5. Some results from a quantitative behavioural assessment of farmed salmon, which generated a final agreed list of descriptors from experienced farmers as follows: 'inquisitive'; 'unsure'; 'agitated'; 'relaxed'; 'flighty'; 'listless'; 'startled'; 'tense'; 'crowded'; 'calm'; 'aggressive'; 'fearful'; 'tranquil'; 'irritated'; 'skittish'; 'mellow'; 'anxious'; 'energetic'; 'stressed'; 'content'. The first axis identified by principal components analysis accounted for 56% of the total variability, opposing terms such as 'anxious' and 'skittish' (counting positively for this component) to those such as 'calm', 'mellow' and 'relaxed' (counting negatively). The figure shows scatterplots of mean scores for PC1 (tense/calm) for each video clip versus ethological quantification of movement patterns for the same clips (ethological definition inserted in italics). (Adapted from Jarvis *et al.*, *Frontiers in Veterinary Science*, 2022. Free to reproduce under Attribution 4.0 International, CC BY 4.0.)

components analysis was used to analyse associations between the various descriptors for each clip. One important dimension opposed descriptors such as 'anxious' and 'skittish' to those such as 'mellow' and 'relaxed'; very high intra- and inter-observer reliabilities showed that observers were consistent in applying the defined terms. There was a strong and significant relationship between this QBA-derived score and classical ethological measures extracted from the same video clips (Fig. 12.5), demonstrating that the intuitions of experienced farmers strongly predict the results of detailed (and very time-consuming) ethological assessment.

12.7 Promoting the Welfare of Fish in Cages: Practices that Compromise Fish Welfare

12.7.1 General farm environment and management

There have been many reviews about farm management and husbandry practices that potentially compromise the welfare of farmed fish, including those cultured in cages (for recent examples see van de Vis *et al.*, 2020; Saraiva *et al.*, 2022). These involve both general aspects of the farm environment and specific husbandry practices, both of which are covered briefly here, emphasizing issues that are particularly relevant to the welfare of fish farmed in cages. Many features of the general cage environment can impair fish welfare. Failure to ensure good water quality is obviously important, but detrimental conditions also include the following:

Space limitations and stocking density

In nature many species of fish show sustained swimming at species-specific optimal speeds, and this is also shown in farmed fish circling within sufficiently large cages (Oppedal *et al.*, 2011). Fish in cages located in sites with strong currents or tidal flows may experience restricted space due to cage deformation (Noble *et al.*, 2018). Given the known benefits of sustained swimming (McKenzie *et al.*, 2021), cages that are too small or inappropriately shaped to allow sustained swimming may compromise fish welfare. High stocking density can result in space limitation and impaired water quality, but inappropriate stocking densities also impact on welfare through the occurrence of aggressive behaviour, which is widespread among fish in nature and on farms (Huntingford, 2020). Welfare is often better at intermediate densities, being compromised, for example, by fighting among fish at low stocking densities (when the benefits of preferential access to good feeding outweigh the costs of fighting, see Section 12.4.3) and by poor water quality at high densities (e.g. for Atlantic salmon, see Turnbull *et al.*, 2005; Adams *et al.*, 2007).

Inappropriate feed management

Clearly, failure to provide a diet that meets the nutritional needs of the fish concerned in a form that is readily detectable and sufficiently attractive to elicit feeding will result in malnutrition and impaired welfare. Issues that are particularly relevant to cage culture include ensuring that delivered food items remain in the cage for a sufficient time for the fish to eat them and that the fish can detect the food. Larval cod use visual cues to detect prey and feed most efficiently at high light intensities (Puvanendran and Brown, 2002) and in water inoculated with phytoplankton ('green water'), which enhances contrast of prey against their background (Muller-Feuga, 2000). In addition, failure to provide the right amount of food at the right time will potentially lead to malnutrition, while failure to ensure that all fish in a cage can feed means that some fish will be inadequately nourished. Arguably, this is a particular problem for fish farmed in cages rather than in ponds, due to restricted possibility for the fish to benefit from natural prey (Jobling, 2010), although coho salmon (*Oncorhynchus kisutch*) farmed in Chile feed extensively on small fish that swim through their cages (S. Kadri, Puerto Varas, 2022, personal communication).

Exposure to harmful organisms (predators and pathogens)

Compared with fish in tanks, raceways or recirculating systems, cage-farmed fish are more exposed to a variety of harmful organisms and usually lack opportunities to hide or escape which makes them more vulnerable to natural predators (Jobling, 2010), including many birds and mammals (Huntingford, 2020). Jellyfish blooms pose an increasing threat to the welfare of cage-farmed fish (Wahli *et al.*, Chapter 4, this volume, 2023), small species being able to enter cages directly and larger species entering in still-dangerous pieces after disintegrating against the cage walls. Salmon in cages can escape from individual jellyfish, but not when these arrive in swarms, which can cause great damage (Noble *et al.*, 2018). Health and welfare of many important cage-farmed species, including Atlantic salmon, European sea bass and gilthead sea bream, are impaired by exposure to various micro-hydrozoan species, which damage gill tissue, impairing function and increasing risk of infection (Bosch-Belmar *et al.*, 2017; Clinton *et al.*, 2021).

Absence of structures necessary for positive welfare

In general, it is clear that an impoverished, broadly uniform physical environment has negative effects on welfare in several species (Arechavala-Lopez *et al.*, 2021). The absence of an appropriate substrate compromises welfare for cage-reared flatfish (e.g. Dou *et al.*, 2000). Compared with fish provided with hiding places, juvenile Atlantic cod reared in bare tanks show impaired learning, less resting, and abnormal swimming patterns and shoaling (Salvanes *et al.*, 2007). In some senses, it is more difficult to provide appropriate structures in a cage compared with a tank, but, as above (Section 12.6.5), the environment to which caged fish are exposed is non-uniform in space and time in terms of temperature, light levels and currents. Within limits, this could be advantageous from a welfare perspective, allowing fish to exercise choice with respect to currents and temperature.

12.7.2 Specific husbandry practices

A number of specific husbandry practices impact negatively on fish welfare (Noble *et al.*, 2018) (Table 12.3).

While such adverse effects of specific husbandry practices are common to all aquaculture systems, there are some special issues for cage farming. For example, the continual replacement of water and removal of waste through the

Table 12.3. Common activities on cage fish farms that induce stress in fish, their potential negative impact on fish welfare and the husbandry procedures during which they occur, with special reference to Atlantic salmon. (Compiled from information in Noble *et al.*, 2018.)

Activity	Potential adverse effects	Used in the following husbandry practices
Objects/divers in cages	Disturbance Stress	System sanitization Cage monitoring
Anaesthesia	Stress Possible overdose Finite recovery time	Removing fish for sampling Disease prevention and treatment
Feed withdrawal	Short term (7–10 days) Long term (>10 days)	Before transport Before handling Before slaughter In response to low oxygen levels In response to disease, because of low appetite, etc.
Crowding Pumping With/without handling	Injury (pain) Poor water quality	Without handling: In-farm transfer between cages With handling: Grading Disease prevention and treatment Off-farm transport Emergency euthanasia Harvest
Well-boat transport	Poor water quality Noise and vibration 'Jostling' Injury	Stocking cages Preharvest transfer for off-farm harvesting
Holding in temporary cages or tanks	Poor water quality Injury	Disease prevention and treatment Off-farm transport Emergency euthanasia Holding station pre-slaughter
Exposure to therapeutic treatments: Injection Vaccination Chemical bath	Irritation of body surfaces Pain Local injury Local inflammation Abdominal adhesions Poor water quality	Disease prevention and treatment
Stunning Slaughter	Severe stress Pain	Harvest Emergency euthanasia

cage walls means that system sanitization can be carried out less frequently. Because cages are cheaper to install and run than tanks for example, extremely high stocking densities are not necessary for economic profitability (S. Kadri, Puerto Varas, 2022, personal communication). However, the sheer size of many cages, while allowing greater freedom of movement, means that even at moderate stocking densities the number of fish per cage can be very large. Inevitably, therefore, procedures take longer, increasing stress and, for all procedures requiring crowding and pumping, the risk of injury. Both the large number of individuals involved and the depth of many modern cages mean that exhaustive monitoring is challenging. In addition, because fish in cages are more exposed to disease, there is a greater need for therapeutic treatments than for fish held in tanks and raceways. Since on-farm slaughter is rare in cage culture, the need to transfer to onshore slaughter sites is special to cage farming.

12.8 Promoting the Welfare of Fish in Cages: Mitigating Welfare Problems

The sustainable production of large numbers of high-quality fish is the main goal of modern aquaculture and ensuring good welfare in its broadest sense is critical for reaching this goal. There has therefore been an extensive research and development effort on how to rear fish in a way that fosters their welfare. Once again, it is clearly impossible and inappropriate to summarize the results in this chapter; instead, some key issues are discussed briefly, with special reference to the culture of fish in cages.

12.8.1 General features of the cage environment

Ensuring that the fish have an appropriate environment for their various welfare needs is achieved by careful positioning of cages in sites where conditions are favourable for the species concerned, designing suitable cages, monitoring relevant environment indicators and setting up remedial procedures when poor welfare is indicated. Monitoring and ameliorating the various aspects of water quality are obviously critical (Noble *et al.*, 2018), but the focus here is on mitigation of the specific adverse effects highlighted above (Section 12.7.1).

Mitigating adverse effects of space limitation and high fish densities

Even where cages are large and stocking density is strongly regulated, fish in cages may sometimes experience very high densities. For example, Atlantic salmon in sea cages fed at the water surface often congregate at high densities before and during feed delivery. Natural response to spatial cues such as temperature and light gradients can also result in aggregations that are sufficiently high to cause collisions and local oxygen depletion (Johansson *et al.*, 2006). This problem can be addressed by means of strategically placed underwater lights; the natural response of swimming towards these results in more even use of the available space and lower fish densities (Juell *et al.*, 2003) (Fig. 12.6).

When fish do aggregate, collisions can be avoided, and several benefits gained (McKenzie *et al.*, 2021), if they swim steadily as a polarized school. Atlantic salmon in sea cages show such polarized swimming in response to natural currents, especially at high-energy sites (Johansson *et al.*, 2014). In tanks, generating appropriate currents by directed water flow is relatively easy, but this is not an option in cages. An alternative approach is to exploit the natural response that

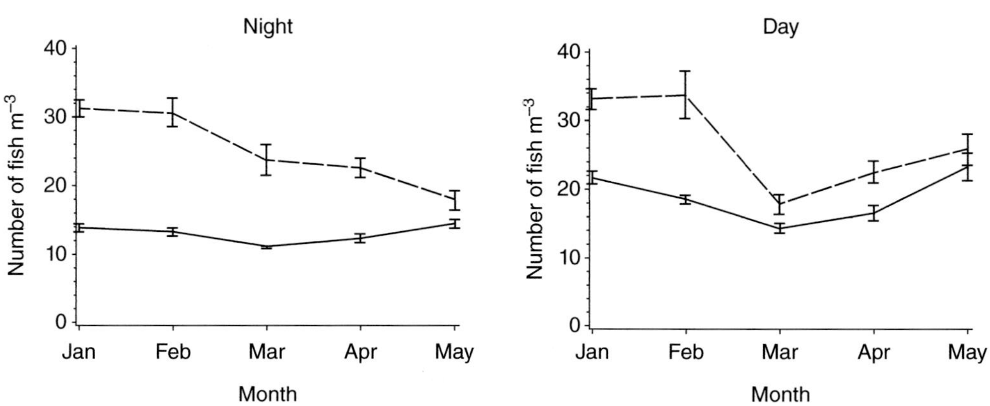

Fig. 12.6. Density (number of fish per m^3) in Atlantic salmon by night and by day in 20 m deep production cages illuminated from the water surface (dashed lines) or by submerged lights (solid lines), in different months. Effects of light position, time of day and month analysed using non-parametric Wilcoxon rank sum tests, with a significant effect of light position ($P < 0.001$), except in May during the day. (Reproduced with slight modifications from Juell *et al.*, 2003.)

stimulates fish of many species to follow moving stimuli (Herbert, 2013). For example, Atlantic salmon post-smolts in tanks show sustained polarized swimming in response to a central column of apparently moving lights and this is associated with reduced stress (Herbert *et al.*, 2011). Attempts to deploy this same system in sea cages were successful to the extent that the light system promoted polarized swimming in favourable conditions, but were unsuccessful, in that it did not work reliably in all conditions, providing an example of the challenge of deploying complex welfare mitigation systems in sea cages (S. Kadri, Puerto Varas, 2022, personal communication).

Ensuring effective feed management

There is a huge literature on fish nutrition and feed management in aquaculture (e.g. Jobling *et al.*, 2012a,b; Raubenheimer *et al.*, 2012; Noble *et al.*, 2018). For well-researched species at least, this helps to ensure that fish in cages receive a nutritionally adequate diet delivered in ways that avoid both underfeeding and overfeeding. Much of this knowledge is encapsulated in feed tables that incorporate important fish-related variables such as species, life history stage and size, as well as environmental variables, such as daily and seasonal changes in light and temperature regimes. These can be used to inform effective hand feeding or programmed into computer-controlled feeders. For smaller rearing systems, several commercially available feeders have been developed that automatically match delivery to appetite, such that fish are always fed when hungry, but not otherwise. Use of such feeders promotes fast, uniform growth and reduces fighting, stress and injury in several farmed species (Attia *et al.*, 2012). Such 'smart' feeders have dropped out of practical use on production farms as cages have got larger (S. Kadri, Puerto Varas, 2022, personal communication), although a next generation for use in large cages is in development (Zhao *et al.*, 2019) (see Section 12.9.2). Currently, manual delivery of feed is common; properly deployed by experienced farmers, this involves continuing delivery until all fish are satiated, identified by direct observation of surface activity and by video footage from above the water surface and from underwater feeding cameras. Feed delivery is adjusted in real time to environmental conditions such as currents and cage depth.

Exposure to harmful organisms

Concentrating here on how cage-farmed fish can be protected from harmful free-living organisms, several different cage designs and a variety of in-cage equipment have been developed to protect Atlantic salmon from the infective, larval stages of various species of sea lice, which congregate in well-illuminated surface water. Protective cage designs include skirted cages and semi-enclosed containment systems that physically separate lice and salmon; submerged cages that prevent the fish from moving near to the water surface; and submerged cages with snorkels that allow fish access to the water surface to fill their swim bladders within a protected area (Noble *et al.*, 2018). Several 'in cage' solutions make use of the natural behaviour of the fish concerned. Drawing on their natural response to light, salmon for example can be kept away from the water surface by submerged lights (Frenzl *et al.*, 2014).

On coral reefs, many species of fish regularly visit special sites (cleaning stations) where they find blue-streak cleaner wrasse (*Labroides dimidiatus*) that remove and eat ectoparasites. This symbiosis involves a complex set of behavioural capacities in the so-called 'client fish', including learning the location of cleaning stations belonging to good cleaners (e.g. Soares, 2017). Such behaviour provided the rationale for farming Atlantic salmon together with temperate fish species such as lumpfish (*Cyclopterus lumpus*) and corkwing wrasse (*Symphodus melops*), both of which ingest ectoparasites removed from other fish. This strategy can be successful in reducing numbers of sea lice (e.g. Powell *et al.*, 2018), but has run into problems, such as depletion of natural populations of cleaners, importing of diseases and concern about welfare of the wrasse (e.g. Geitung *et al.*, 2020). Salmon held in semi-closed containment systems are physically protected from swarms of jellyfish, although welfare may be indirectly compromised, for example, if pump intake screens become clogged by jellyfish (Noble *et al.*, 2018). Plumes of bubbles created by pumping compressed air through underwater porous tubes (bubble curtains) create vertical currents that can deflect larger jellyfish from fish farm cages, but this is dependent on depth and local wave regimes (Haberlin *et al.*, 2021).

Environmental enrichment

Recognizing the importance of giving captive animals access to positive experiences, as well as protecting them against negative ones, there has been a considerable research focus on designing enriched captive environments that offer fish access to such experiences. Benefits arising from such environmental enrichment have been well documented in terrestrial farm animals and also, increasingly, in fish (Arechavala-Lopez *et al.*, 2021). Successful environmental enrichment for fish includes physical enrichment (e.g. adding structures to the water column and a variety of substrates), sensorial enrichment (providing fish with more complex sensory environments) and occupational enrichment (introducing a variety of challenges to the rearing environment, such as the need to locate hidden food). As an example of the benefits of simple structural enrichments, provision of vertical plant-fibre ropes in otherwise bare cages increased the use made by sea bream of the inner part of their cages, resulting in less fin erosion (Fig. 12.7) and reduced aggression (Arechavala-Lopez *et al.*, 2019). There are also on-farm reports of Atlantic salmon repeatedly moving in shoals through artificial kelp (added to cages for the benefit of cleaner fish) and swimming hard to remain within a bubble stream from an in-cage aeration system (S. Kadri, Puerto Varas, 2022, personal communication).

As an example of sensorial enrichment, swimming became more regular and growth rates increased in koi carp exposed to recordings of certain types of music, as opposed to control conditions or exposure to recordings of urban noise (Kusku *et al.*, 2019). In this case, a simple enhancement of the auditory environment produced non-trivial improvements in welfare, differences probably being the result of the range and patterns of sound frequencies concerned.

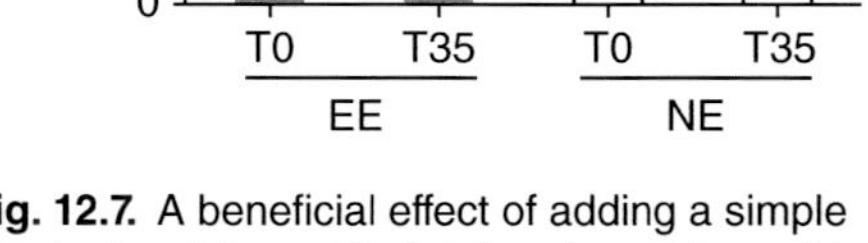

Fig. 12.7. A beneficial effect of adding a simple physical enrichment to fish farming systems. Mean index values of pectoral fin damage (on a scale from 0 for perfect fin to 4 for short and dysfunctional fins), with standard deviations represented by vertical bars, in juvenile sea bream reared in cages enriched with vertical plant-fibre ropes (EE) compared with controls reared in bare cages (NE), at the start of the study (T0) and after 35 days (T35). Significant differences among treatments and times are marked with dissimilar lower-case letters ($P < 0.05$). (Reproduced with slight modifications from Arechavala-Lopez *et al.*, 2019.)

Giving fish the opportunity to look after themselves

A special case of environmental enrichment, with potential benefits for avoiding infection or mitigating disease, involves tapping into natural behaviours that allow wild fish to protect themselves from infection and take care of their body surfaces. For example, sea trout can learn to avoid visual cues associated with previous encounters with infective trematode eye fluke larvae (Klemme and Karvonen, 2016). To the extent that it is successful, use of cleaner fish to remove sea lice provides an example of exploiting natural behaviour for mitigating the effects of harmful organisms.

In nature, fish often control their physiological status by making use of natural temperature gradients to improve the efficiency of various important functions such as upstream migration (e.g. Tanaka *et al.*, 2000). Common carp raise their body temperature above ambient by periodically basking in sunspots, which increases growth rates (Nordahl *et al.*, 2018). Given the opportunity, fish also move into warmer water in response to an infection, showing behavioural, as opposed to physiological, fever (Reynolds *et al.*, 1976). For example, when offered a choice of water temperatures, infected zebrafish (*Danio rerio*) and common carp both show behavioural fever by choosing water at higher temperature. Their increased body temperature stimulates a strong and specific immune response, which lowers the incidence of serious clinical signs and mortality (Boltano *et al.*, 2013; Rakus *et al.*, 2017). Nile tilapia also show behavioural fever in response to an infection (Cerqueira *et al.*, 2016) and research is underway to explore how this may protect them against disease when reared in ponds (S. Rey Planellas, Stirling, 2022, personal communication).

12.8.2 Specific husbandry practices

In general, selective breeding for low stress responsiveness has been successful in the common carp (Tanck *et al.*, 2002) and rainbow trout (Pottinger and Carrick, 1999), offering the possibility of farming fish that are relatively undisturbed by common husbandry practices. Given the existence of stress coping styles in many farmed species (see Section 12.5), one complication here is that selection for reduced stress responsiveness may increase aggressiveness. For specific husbandry practices that challenge the welfare of farmed fish, Noble *et al.* (2018) describe meticulous regimes for minimizing any adverse effects on welfare. These include careful advanced preparation, continuous monitoring and adjustment during the procedures and appropriate aftercare protocols.

In addition, there has been a continuing programme of research and development aimed at finding less-invasive alternatives to established methods. This has been particularly the case for slaughter (Noble *et al.*, 2018; van de Vis *et al.*, 2020), where several technologies are now available for rapid and humane killing of fish. Developments aimed at reducing the need for capturing and handling fish during grading and sampling include the replacement of conventional grading by hand and in air by passive or self-grading, whereby fish swim voluntarily through an underwater grid, sometimes in response to simple directional stimuli such as light cues, thereby avoiding undue crowding, pumping and handling (e.g. for larval pikeperch, *Sander lucioperca*, see Tielmann *et al.*, 2016). Commercial self-grading systems are commonly used in salmon aquaculture in the UK and Norway. For example, the Flexi-Panel system (gradingsystems.com, accessed 28 February 2022) uses an adjustable lightweight flexible grid and, according to the promotional material, allows quick and accurate grading of fish of a wide range of sizes, with reduced stress, less damage and removing the need for feed withdrawal which is normally done prior to grading.

Non-invasive remote sampling systems for estimating fish size include rectangular frames that scan passing fish optically and calculate size and condition factor (e.g. the VAKI Biomass Daily system, MSD Animal Health Inc.), stereo cameras that capture paired images of passing fish from which their size can be estimated (Lines *et al.*, 2001) (e.g. OptoScale AS), and split-beam and multibeam sonar systems (Soliveres *et al.*, 2017) (e.g. Biometrics AS). Various camera systems, including hyperspectral cameras (Tillett *et al.*, 1999), can be used to monitor skin condition remotely in individual fish, identifying scratches, wounds and the presence of ectoparasites. Behavioural indicators of stress, such as changes in swimming dynamics and loss of polarized schooling (see Section 12.6.3), can be extracted from split-beam and multibeam sonar and from submerged cameras (see Føre *et al.*, 2018). Cortisol concentration in fish scales can potentially be used to identify chronic stress, for example in European sea bass (Samaras *et al.*, 2021). In some species, skin pigmentation patterns correlate with plasma cortisol levels and could be used as remote indicators of physiological stress and stress coping strategy (e.g. rainbow trout, Kittilsen *et al.*, 2009; Arctic charr, Magnhagen *et al.*, 2018). The effectiveness of such remote monitoring systems is being continually enhanced by technological developments in data collection and, increasingly, by the use of computer vision and various other methodologies for handling and interpreting large data sets (Section 12.9.2).

12.9 Looking to the Future

12.9.1 Cage aquaculture: what is likely to change?

The need for worldwide, large-scale aquaculture production is not going to go away, nor is legitimate public concern for the welfare of farmed fish. In the medium term at least, these potentially conflicting pressures are likely to increase and interact, helping to shape the future of aquaculture. There will be increased pressure to produce sufficient fish to feed a growing human population and, with declining wild fish stocks, much of this will inevitably come from aquaculture. To meet this need, a number of developments are likely; those that relate particularly to cage culture include the following.

Intensification and diversification

In the interest of increased production, greater intensification (up to a point) is likely in some cage-farming sectors (see Huntingford *et al.*, 2018). This may involve larger offshore sea cages, to reduce environmental impact and avoid conflict with other interests, which bring their own welfare problems (Hvas *et al.*, 2021). In other cage-farming sectors, there is likely to be further diversification in the species farmed, possibly with emphasis on omnivorous and herbivorous species instead of piscivores. Further diversification will call for an expanded research effort aimed at identifying the welfare needs of these new farmed species.

Disease prevention and health

Intensification of cage farming will surely lead to new disease threats, requiring new measures for treatment, prevention and management of infectious diseases (see Leong *et al.*, Chapter 2, this volume, 2023; Cain and Polinski, Chapter 3, this volume, 2023; Chapter 4, this volume; Parker-Graham *et al.*, Chapter 5, this volume, 2023; Shinn *et al.*, Chapter 6, this volume, 2023; Soares *et al.*, Chapter 7, this volume, 2023; St-Hilaire *et al.*, Chapter 8, this volume, 2023; Saksida *et al.*, Chapter 9, this volume, 2023). In intensive fish farming, veterinary care has for many years focused upon disease management, but in spite of improved treatments and diagnostic tools, overall mortalities on fish farms have not been reduced in the past two decades (Atlantic salmon: Norwegian Ministry of Fisheries, https://www.fiskeridir.no/English/Aquaculture/Statistics/Atlantic-salmon-and-rainbow-trout, accessed 31 October 2022; sea bass and sea bream: N. Steiropoulos, Glasgow, 2013, personal communication). Furthermore, few veterinary courses include more than a passing reference to fish and aquaculture. As aquaculture grows rapidly, there will be an increased need for veterinary graduates who have an interest in and capacity for caring for both the health and the welfare of large fish populations, encompassing both prevention of poor welfare and provision of positive welfare experiences. The industry needs to move towards preventive fish health management, with veterinarians responsible for welfare management having a '360-degree view' of everything going on in the farms and participating as members of production teams aimed at minimizing losses due to disease. This will involve close monitoring of husbandry practices, water quality, equipment, biosecurity and other factors pertinent to fish welfare.

Molecular technologies

There is likely to be an expansion in the use of molecular tools in aquaculture generally. These will surely be used increasingly in selective breeding programmes, with high-throughput genotyping, for example, having generated genome-wide markers for the main cage-farmed species, allowing breeding values for specific traits to be estimated from genotypes. This has already informed focused selective breeding programmes for important production traits, including growth rates, age of maturation, stress reactivity and resistance to various diseases. The technologies required for such analyses are species specific, but progress in this area is rapid and commercial resources will become available for more species (Boudry *et al.*, 2021). The potential for using genome editing technologies to develop genetically improved lines has also been demonstrated for several species of farmed fish, suppressed reproduction and increased disease resistance are among a number of regularly targeted traits (Yang *et al.*, 2021). Among cage-farmed species, Atlantic salmon and Nile tilapia genetically edited for fast growth are already commercially available (Okoli *et al.*, 2022). As with any husbandry procedures that generate fast growth, gains from such selective breeding programmes potentially come at a cost to fish welfare (Section 12.2.2). Such welfare costs are likely to be complex and need to be kept firmly on the research agenda.

12.9.2 Fish welfare science: what is likely to change

Ensuring positive welfare and appropriately enriched environments

Research developments driven by public concern for the welfare of farmed fish are likely to include increasing understanding of sentience in fish and how this impacts their welfare in culture systems. There is likely to be continued pressure to promote positive welfare and therefore to develop enrichment interventions that are feasible and economic in large cages full of fish.

Protecting individual welfare

Future concern for the welfare of cage-farmed fish is likely to include recognition that individuals within a given species show different stress responses and so of the need to ensure that the cage environment allows fish with different stress coping styles to flourish. One possible strategy is to make use of the many biosensors that have been developed as research tools for studying fish welfare at an individual level. In theory, it might be possible to fit every fish in a cage with such smart tags to allow an accurate assessment of their welfare (Brijs *et al.*, 2021; although see Macaulay *et al.*, 2021). This might just be a realistic economic prospect for cages lightly stocked with high-value fish, such as tuna and broodstock of certain species. There would be a problem of how to retrieve fish whose tag signals poor welfare, although with careful training individual salmon can be separated from small groups and sequestered (Lines and Frost, 1997). A possible compromise strategy would be to fit a subset of a population of fish farmed for food at high densities in cages to act as sentinels, including both proactive and reactive fish. However, this too would be problematic, since all fish would have to be scanned at slaughter and their tags removed. Arguably, the most realistic strategy for protecting the welfare for fish across the range of stress coping styles is to develop cage systems in which their distinct welfare needs can be met; this might involve ensuring access to a range of temperatures and currents, a degree of cover and well-distributed and abundant feed, possibly using the new generation of smart feeders.

Non-invasive monitoring of OWIs and smart analysis of big data

Aquaculture has access to an increasingly sophisticated array of sensors for remote monitoring of the aquatic environment, fish health and welfare, with equally sophisticated tools for analysing the resulting very large data sets. These may well change the way fish welfare is assessed and protected, at least in some intensive cage aquaculture sectors.

Information on the environment within specific cages (and hence on input OWIs) is increasingly collected automatically by *in situ* sensors recording variables such as temperature, current velocity and dissolved oxygen (Noble *et al.*, 2018). Modern cage aquaculture also makes increasing use of public data relative to a range of relevant environmental variables. The huge data sets from these two sources require a wide range of analytical and modelling tools to extract and interpret data relevant to, in this case, fish welfare (e.g. Yang *et al.*, 2021). Such digital technologies can also give advanced warning of conditions outside cages that threaten welfare (e.g. jellyfish or algae blooms), allowing protective measures such as adding skirts or deploying bubble curtains to be implemented (S. Kadri, Puerto Varas, 2022, personal communication).

Concerning outcome OWIs, non-invasive fish monitoring will surely have an increasing role in assessing the welfare of cage-farmed fish, although the most sophisticated applications so far have been deployed in tanks or circulating systems, which are less-challenging environments for this purpose. Monitoring tools include a variety of automatic sensors (mainly using visual and acoustic signals), with associated high-powered analytical technologies to determine fish size, numbers and biomass, thereby reducing the need to capture and handle fish (Li *et al.*, 2020a). A range of optical and acoustic systems are available for automated quantification of fish behaviour in cages and increasingly sophisticated machine vision is being used to analyse the signals they generate. Resulting behavioural information includes individual swimming speed and direction, tail beat frequency and opercular beat and, for groups of fish, the structure, dynamics and dispersal of schools and even the trajectory of individual fish within schools (An *et al.*, 2021). Such behavioural information is potentially of great importance in assessing fish welfare.

Data on appetite and feeding in cage-farmed fish are important for both production and welfare. A wide range of vision-based and acoustic technologies are available for collecting such information, as are tools for analysing and interpreting their output, mostly in tanks but some also in cages (Li *et al.*, 2020b). For example, the characteristic sounds made by fish when eating can be used to detect feeding in turbot (*Scophthalmus maximus*) (Lagardère and Mallekh, 2000). In silver perch (*Bidyanus bidyanus*) held in tanks, artificial intelligence methodologies allow the incidence of food-searching behaviour to be predicted with a very high degree of accuracy from local dissolved oxygen levels (Wu *et al.*, 2015). Even in cages, dissolved oxygen changes locally as a consequence of feeding activity. In grass carp (*Ctenopharyngodon idellus*) in ponds, a similar analytical approach generated a predictor of appetite using dissolved oxygen and temperature that was sufficiently accurate to control an effective on-demand feeding system (Zhao *et al.*, 2019).

Precision fish farming

The availability of multiple sensors to monitor environment and fish in cages, together with the power of modern computational and information systems to interpret the resulting data and to guide and implement husbandry decisions, map on to the main components of what is called 'precision fish farming' (Føre *et al.*, 2018; O'Donncha and Grant, 2020). This is a concept in development, aimed at improved accuracy and precision in future aquaculture operations, protecting fish welfare in the process, while reducing dependence on manual labour and human intervention. The concept may become a central element of environmentally sustainable and welfare-friendly fish culture. It is summarized very simply in Fig. 12.8 with reference to feed management, for which the four components of precision fish farming are currently developed, to an extent at least. As Føre *et al.* (2018) point out, the key logical steps in precision fish farming encompass and extend those that have always informed management in well-run fish farms.

The (convincing) arguments for the future importance of precision fish farming (Føre *et al.*, 2018; O'Donncha and Grant, 2020) are particularly relevant to cage farming. They include the large numbers of fish increasingly reared in single production units, lack of visibility of most of the population and exposed farms being periodically cut off by bad weather. In the short term, it is hard to envisage a total hands-off (and eyes-off) approach to all aspects of cage management. This is partly because the huge numbers of fish in cage culture means that failure of a fully automated system could be disastrous and partly because the accumulated knowledge and

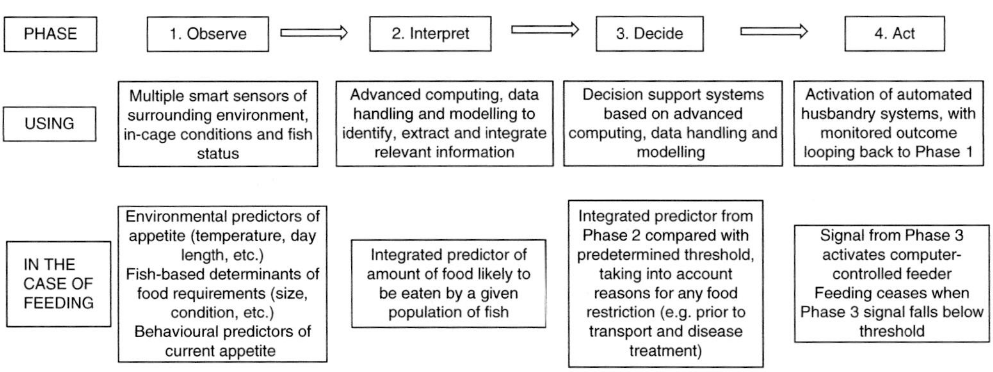

Fig. 12.8. Representation of the four phases of precision fish farming identified by Føre *et al.* (2018), illustrated with reference to feeding fish to appetite. Where appropriate, precision fish farming would replace monitoring by farm staff; research-based and experience-based decision making; and manual implementation of husbandry decisions, as in current practice. (Drawn by the chapter authors.)

wisdom of real fish farmers will be hard to replace completely, even with cleverly designed programs and powerful computers. All the same, given the increasing challenges of monitoring and protecting welfare in large cage populations, components of precision fish farming techniques are likely to be increasingly integrated into intensive cage aquaculture.

12.9.3 The added challenge of climate change for the welfare of cage-farmed fish

The dimensions of climate change and the vulnerability of cage culture

Some degree of accelerating climate change is inevitable, posing a serious threat to many human activities. Salin *et al.* (Chapter 1, this volume, 2023) describe the main changes that are already having an impact on cage culture; these vary with different cage-culture systems, but all pose many challenges for production and welfare. For convenience of reference here, Box 12.4 summarizes the main interrelated climatic changes that are expected and already underway, distinguishing between gradual, long-term trends and periodic extreme environmental events.

Also as discussed in Chapter 1 (this volume), compared with terrestrial farmed animals, finfish are particularly vulnerable to the various effects of climate change, being ectotherms and highly exposed to the aquatic environment through their gills. Any effects of climate change will be exacerbated in cage farming because the culture space is very much exposed to the surrounding environment, so fish are potentially 'at the mercy of what the environment throws at the farm' (S. Kadri, Puerto Varas, 2022, personal communication). On the other hand, the infrastructure for cage farming is perhaps easier and cheaper to relocate than are ponds and some complex land-based systems.

Box 12.4. Expected dimensions of climate change, separating gradual, chronic changes from acute weather events.

Gradual changes

General aquatic environment:

- stronger waves;
- changing currents;
- altered stratification;
- reduced ice cover;
- rising sea levels/coastal erosion; and
- higher solar radiation.

Temperature regimes:

- higher average temperatures in most cases; and
- changed seasonality.

Water quality:

- lower dissolved oxygen;
- higher carbon dioxide;
- lower pH; and
- increased turbidity.

Extreme weather events

- Heatwaves.
- Sudden drops in water temperature (especially in freshwater systems).
- Extreme precipitation (surge-based flooding and droughts).
- Storms and high winds.
- Rapid changes in salinity.
- Acute increases in turbidity.

Climate change and fish physiology

There have been many studies of the effects of climate changes on biological functions in fishes, most in their natural habitat, but with a good number considering farmed fish (e.g. Reid *et al.*, 2019; Cubillo *et al.*, 2021; Maulu *et al.*, 2021). Box 12.5 summarizes some probable effects of climate change on different physiological systems in fish, concentrating on temperature, just one of the 'lethal three' dimensions of rising temperature, falling oxygen levels and increasing acidity. This summary can only be indicative, because there are many complicating factors, including species- and location-specific interactions among these three dimensions and additional effects of other variables such as increasing pollution levels. Clearly, many of these effects will negatively impact fish welfare, requiring as it does that they can adapt effectively to their environment, are in good health and with all their biological functions working effectively (Section 12.3.1).

How fish adapt to environmental change

Within a given generation (in other words, ignoring long-term, evolutionary responses to climate

Box 12.5. Some interrelated effects of increased temperature on key physiological systems in fish. (Compiled from information in Whitney *et al.*, 2016; Teal *et al.*, 2018; Reid *et al.*, 2019; Islam *et al.*, 2021.)

Neuroendocrinology

Exposure to higher temperatures initiates neuroendocrine responses in the central nervous system, triggering release of corticosteroids and catecholamines. Chronically elevated plasma cortisol levels are costly, diverting energy from growth and reproduction and may result in morbidity and mortality.

Circulation

Both acute and chronic high temperature reduce cardiac function due to altered ionic balance and decrease blood flow to brain, liver, kidney and intestine, compromising biological functioning generally.

Respiratory metabolism

Standard metabolic rate (SMR), the minimum oxygen uptake needed to sustain life, increases rapidly with temperature, while maximum metabolic rate (MMR), the maximum rate of oxygen uptake (e.g. during intense exercise), increases to a plateau and then falls. Therefore, aerobic scope (MMR – SMR), the energetic potential of an organism to accomplish all of its tasks, eventually falls, constraining various functions in fish even at non-lethal temperatures.

Digestion

Increased metabolic demands may increase food requirements. High temperature may also influence feed conversion efficiency and the digestibility of specific nutritional categories, potentially leading to malnutrition.

Iono-regulation

Temperature change results in net ion loss in freshwater fish; the opposite occurs in seawater fish. The resulting ionic imbalances impair synaptic transmission in the central nervous system, with negative effects for general physiological functioning, amplified by rising salinities during droughts. Thus, maintaining the hydromineral balance becomes an increasing metabolic drain.

Immunology

Temperatures above the species' optimum have numerous adverse effects on immune function due to stress and increased maintenance costs are limited as energy is diverted to homeostasis. Impaired immune function at higher temperatures reduces the ability of fish to fight pathogens.

Reproduction

Deviations from optimal temperatures have numerous effects on sex ratios and the timing of and investment in reproduction in fishes, with implications for reproductive success in breeding fish. Early maturation in fish farmed for food has negative effects on welfare.

change), fish in nature show flexible adjustments to a changing environment through behavioural and physiological responses, for example by increasing food intake to cover the increased cost of many biological processes at higher temperatures. They may also show longer-term organizational effects; for example, exposure of parents or fry to moderate hypoxia can generate hypoxia resistance in later life (Naya-Català *et al.*, 2021). Up to a point, fish held in cages are also able to make such adjustments, but these all come with a cost such that, for example, eventually growth falls with increasing water temperature even though fish eat more (Holt and Jørgensen, 2015). Wild fish can also adjust to climate change by local tracking of favourable conditions; again, up to a point, fish in cages can also make such adaptive adjustments, as when Atlantic salmon move within sea cages to track favourable temperature and light conditions (Oppedal *et al.*, 2011). Eventually, however, long-term alterations in habitat due to climate change mean that currently suitable locations for a given species cease to provide the necessary environmental conditions for production, health and welfare, while currently unsuitable locations may become favourable. In nature, wild fish can and do make large-scale movements in response to such changes, which has already resulted in well-documented range changes. Since farmed fish cannot migrate, they can only arrive in newly favourable conditions through human intervention.

Production and welfare in farmed fish in the face of gradual climate change

The discussion here focuses on the direct effects of climate change on fish welfare, rather than indirect effects such as supply chain fragility (covered fully in Chapter 1, this volume), although these too will have consequences for welfare. Not all predicted effects of climate change on cage culture are negative from the perspective of production (see Chapter 1, this volume). For example, in some species on some time scales and in some climate change scenarios, faster growth is predicted at higher water temperatures. Increased growth rates are sometimes seen as a benefit to fish as well as to farmers, but the various costs of fast growth (Section 12.2.2) may well make the integrated consequences of fast growth detrimental to individual welfare.

In most cases, however, both direct experience and predictive modelling demonstrate adverse effects of climate changes on production. Thus, for a variety of cage-farming systems, Chapter 1 (this volume) documents reduced growth rates, poor energy reserves, increased problems with disease and parasites, and higher mortality rates. Such effects clearly indicate impaired welfare, in its most robust, functional sense. In brief and in general, it will be increasingly difficult to meet the identified input OWIs for many cage-farmed fish species at current farm sites. Modelling the future geography of aquaculture in general in relation to the known environmental requirements for 85 common mariculture species predicts a 10–40% decline in species suitability for tropical and subtropical zones and a 40% increase at higher latitudes by the mid-21st century, given no mitigation to climate change in place (Oyinlola *et al.*, 2020).

There are some less obvious effects of gradual climate change that have special relevance to welfare. One very direct negative effect of climate change on the welfare of farmed fish relates to the effectiveness of electrical stunning devices. Higher temperatures may lead to high localized evaporation and therefore an increase in salinity. As water conductivity is highly dependent on dissolved ions, adjustments to wet stunning devices should necessarily be made taking account of the local salinity, species-specific parameters, desired stunning level (and duration) and power requirements of such devices (Lines and Kestin, 2004; Hjelmstedt *et al.*, 2022). Slightly less direct but important additional effects include disruption of sensory systems and the interacting effects of high temperatures and low oxygen levels on metabolic scope for action. In both cases, fish welfare is likely to be impaired well before more obvious indicators of poor welfare, such as slow growth, disease and morbidity, become evident.

Several aspects of climate change impair the proper functioning of sensory systems, interfering with normal behaviour and thus compromising welfare and its assessment. Adverse effects mediated indirectly by changes in the immediate cage environment include increased turbidity, with consequent problems for visual detection of food by some species and for monitoring behavioural outcome OWIs. In addition, many direct effects of climate change on the detection and processing of sensory information have been reported. For example, otolith size and shape are altered in juvenile emerald rockcod (*Trematomus bernacchii*) at predicted temperatures, with likely adverse effects on hearing, orientation and control of movement (Reimer *et al.*, 2017; Naslund *et al.* 2021).

Effects of climate change on sensory processes have been particularly well studied with reference to effects of ocean acidity on olfaction (e.g. Porteus *et al.*, 2021). The sense of smell is vital for fish, with olfactory cues playing important roles in feeding, predator avoidance, social interactions and many aspects of reproduction (Arechavala-Lopez *et al.*, 2021). In spite of controversy over particular studies, reduced olfactory sensitivity to many ecologically relevant odorants (including amino acids, bile acids, bile and alarm cues) has been demonstrated in fish held in seawater with near future PCO_2 levels, with both detection and processing of important olfactory cues being compromised (Velez *et al.*, 2019). Such effects will impair feeding responses, for example, with adverse consequences for production and welfare. It is possible that future feed formulation will have to rely on more potent chemicals to increase attractability and palatability (Arechavala-Lopez *et al.*, 2021).

As outlined in Box 12.5, aerobic scope (the difference between maximal and standard metabolic rate) in any given set of conditions defines the energy available for aerobic activities (e.g. locomotion, growth, digestion, fighting diseases). Aerobic scope is reduced by temperatures that are non-lethal but outside the species' optimum and is an early predictor of wild fish moving to cooler waters (Teal *et al.*, 2018). In farmed fish, impaired functioning and poor welfare may therefore occur well before visible effects on growth and survival are evident. The relationship between aerobic scope and temperature is strongly influenced by other environmental variables including, unsurprisingly, dissolved oxygen levels. Aerobic scope is compromised particularly strongly by a combination of high temperatures and low dissolved oxygen. A study of wild grey mullet (*Mugil cephalus*) moving between a bay and a shallow lagoon (in which water temperature and oxygen levels vary independently, neither being limiting at any point across the annual cycle) found that fish moved between the two sites following the combination of temperature and oxygen that maximized aerobic scope (previously determined by laboratory experiments) (Cucco *et al.*, 2012). Thus, wild fish have the capacity to track optional combinations of, in this case, temperature and oxygen. If farmed fish species have equivalent behavioural mechanisms, as is very likely, being farmed in suboptimal conditions caused by climate change will have negative effects on welfare. These will arise both because farmed fish do not have access to conditions that maximize aerobic scope (compromising their ability to grow and fight off pathogens, for example) and because the drive to track conditions that optimize functioning might represent a genuine behavioural need that cannot be met.

In both these cases (impaired sensory ability and reduced aerobic scope) negative consequences for welfare may seem trivial in relation to other disastrous effects of climate change. This is not the case, firstly because behavioural consequences of impaired olfaction or reduced aerobic scope are potential early warning signals that things are going wrong. In addition, as the case of mullet tracking optimal conditions also shows, given the necessary conditions, fish can potentially help themselves to cope with some aspects of climate change, which can only be good for aquaculture.

Welfare in farmed fish in the face of acute weather events arising from climate change

In addition to impacts of gradual climate change on the welfare of cage-farmed fish, acute climate events such as droughts, storms, flooding and heatwaves pose their own serious threats. During storms and floods, cage-farmed fish may experience currents that are too strong for them to swim against, resulting in injuries or mortality when pinned to cage walls. Such effects will be exacerbated if farmers have no access to the cages during storms and so are unable to assess and protect the welfare of their stock. In addition, there are concerns about the welfare of individual farmed fish that escape as a consequence of storm damage, setting aside the possible adverse effects on wild fish populations. The combined effects of temperature stress and poor water quality during heatwaves have multiple, sometimes fatal, negative effects on physiological functioning in farmed fish and hence on their health and welfare (e.g. Wade *et al.*, 2019 for Atlantic salmon). Other acute events caused by climate change that compromise the welfare of caged fish include harmful algal blooms (Chapter 10, this volume) and jellyfish swarms (Section 12.7.1 and Chapter 4, this volume). These may require emergency harvesting (S. Kadri, Puerto Varas, 2022, personal communication), which has its own implications for welfare (Section 12.7.2).

Possible mitigation of the adverse effects of climate change on cage-farmed fish

In spite of the complex consequences of climate change for production and welfare in cage-farmed fish, a number of strategies are available for minimizing adverse effects, while capitalizing on new possibilities. It is important that the need for environmental sustainability and fish welfare remain firmly embedded in such systems, in spite of the pressing need for sustained production and food security (see Chapter 1, this volume). The concern about and expertise in fish welfare demonstrated by fish farmers (Section 12.6.6) reinforces the view expressed in Chapter 1 (this volume) that experienced farmers will be key to developing and deploying risk management strategies, certainly at a local level.

In terms of gradual climate shifts, where well-established cage-farmed species are concerned, mitigation strategies to protect welfare (see Chapter 1, this volume) include regular upgrading of cage design. For example, cages can be made stronger to withstand storms. Submerged cages can be used to protect fish from the most extreme effects of climate change; when fitted with air reservoirs these can be used for physostomous fish with open swim bladders as well as for those with closed swim bladders (Sievers *et al.*, 2022). Targeted selection for strains that are more robust in terms of resistance to heat, to stress generally and to current and new diseases is also possible (e.g. Sae-Lim *et al.*, 2017). Relocation of farm sites from newly unfavourable to newly favourable locations will be essential; since cages and infrastructure can be moved, direct physical relocation is worth consideration, as well as construction from scratch of new cage farms. Both these strategies will require forward planning by the industry, policy makers and regulators, with an orderly adjustment of cage-farming effort (as has been necessary for terrestrial crop farming) to minimize adverse effects on welfare. Monitoring systems need to become more adept at using available data to forecast global and local conditions in the aquatic environment and to model how long a given site or region is likely to remain suitable for a specific fish species.

An additional mitigation strategy is a shift towards farming species such as tilapia and milkfish (Chanidae) that are evolutionarily adapted to higher temperatures and/or are more tolerant of environmental fluctuations. As discussed in Chapter 1 (this volume), all these approaches need to be backed up by intensive research into tolerance and welfare in the face of the various environmental stressors that cage-farmed fish will encounter under climate change. Such forward planning and mitigation strategies will be essential if serious impacts on fish welfare, as well as on production, are to be avoided.

In terms of acute effects of climate change, a number of measures to secure farm sites during acute weather events are possible, for example: moving cages nearer to riverbanks in anticipation of floods (Chapter 1, this volume); reducing stocking densities in anticipation of droughts; withholding food and provision of aeration at the height of heatwaves (Wade *et al.*, 2019); and protecting cages from jellyfish blooms using bubble curtains (Haberlin *et al.*, 2021; S. Kadri, Puerto Varas, 2022, personal communication). Avoiding farming during seasons in which extreme events are likely is an additional, more drastic, possibility (Chapter 1, this volume). In all such cases, there is a pressing need for advanced warning based on monitoring systems that take advantage of new sensing technology and big data approaches. Properly deployed, these should enable the aquaculture industry to develop site-specific modelling and early warning of impending acute harmful events, allowing mitigation strategies to be implemented effectively.

12.10 Conclusions

Welfare is an important but complex phenomenon with several unresolved controversial issues concerning the welfare of fish in particular. Some of these are likely to be resolved to an extent through an intensive programme of multidisciplinary scientific research, for example, into the full extent and nature of sentience in fish. Current understanding in this context is already sufficient to indicate that positive experiences are an important component of fish welfare and, hence, that environmental enrichment is necessary for protecting the welfare of farmed fish. Other controversial issues, such as the importance of naturalness for welfare, will require a more philosophical approach. Even with these incompletely resolved issues, a great deal has been achieved using the (arguably most pragmatic) functional approach by which good welfare is indicated if the fish concerned can adapt effectively to its environment, is in good health and with all its biological functions, including activities that feel good, working effectively.

This chapter has summarized the huge amount of research, industry-level developments and hands-on effort by fish farmers that have gone into promoting the welfare of farmed fish and discussed issues relating particularly to cage-farmed fish. Compared with fish in tanks, for example, some issues are positive; caged fish have more space and access to greater fine-scale environmental variability. Others are negative; there is less possibility for control over the cage environment and more environmental pollution. Monitoring fish is particularly challenging, especially when many fish are held in large cages. The stress induced by capturing fish for welfare sampling in air and the challenge of monitoring the welfare of fish across the whole volume of a cage have stimulated an extraordinary development of systems for smart sensing of important variables, using the full power of modern computation and information management. These are likely to become increasingly powerful and increasingly well integrated into smart husbandry systems that approach the concept of precision fish farming. Such systems potentially offer some solutions to the problem of ensuring good welfare for individuals with different stress coping styles.

Climate change is underway, and a suite of gradual environmental changes is already having an impact, being set to alter profoundly the geography of cage farming, for example. Protection of fish welfare makes it imperative that such changes are anticipated, and that coordinated action is taken before it becomes impossible to ensure that cage-farmed fish have the environmental conditions they need for good welfare. We cannot simply leave things as they are and wait for fish to become ill and die, before taking action.

Climate change is also likely to influence the range of species and strains of fish farmed in cages, in favour of those that are more tolerant of high temperatures and of environmental fluctuations. The welfare imperative here is to ensure that enough is known about the biology of the species and strains concerned to allow their welfare needs to be met, before intensive farming of such fish gets under way and lessons have to be learned by trial and error.

Cage-farmed fish will also experience an increasing number of acute events such as storms, algal blooms and heatwaves, which can only have adverse effects on their welfare. Little can be done to protect fish against such events, except making good use of available big data to predict these accurately and to have systems ready and in place to protect them. All these responses will be challenging, but the fact that production and economic goals will be pulling in the same direction as the demands of protecting welfare is one reason for optimism.

References

Adams, C.E., Turnbull, J.F., Bell, A., Bron, J.E. and Huntingford, F.A. (2007) Multiple determinants of welfare in farmed fish: stocking density, disturbance, and aggression in Atlantic salmon (*Salmo salar*). *Canadian Journal of Fisheries and Aquatic Sciences* 64, 336–344.

An, D., Huang, J. and Wei, Y. (2021) A survey of fish behaviour quantification indexes and methods in aquaculture. *Reviews in Aquaculture* 13, 2169–2189.

Appleby, M. and Sandoe, P. (2002) Philosophical debate on the nature of well-being: implications for animal welfare. *Animal Welfare,* 11, 383–394.

Arechavala-Lopez, C., Diaz-Gil, J.L. Saraiva, D., Moranta, M.F., Castanheira, S., *et al.* (2019) Effects of structural environmental enrichment on welfare of juvenile seabream (*Sparus aurata*). *Aquaculture Reports* 15, 100224.

Arechavala-Lopez, P., Cabrere-Alvarez, M.J., Maia, C.M. and Saraiva, J.L. (2021) Environmental enrichment in fish aquaculture: a review of fundamental and practical aspects. *Reviews in Aquaculture* 2021, 1–25.

Attia, J., Millot, S., Di-Poi, C., Begout, M.-L., Noble, C., *et al.* (2012) Demand feeding and welfare in farmed fish. *Fish Physiology and Biochemistry* 38, 107–118.

Barnard, C. (2007) Ethical regulation and animal science: why behaviour is special. *Animal Behaviour* 74, 5–13.

Barnard, C.J. and Hurst, J.L. (1996) Welfare by design: the natural selection of welfare criteria. *Animal Welfare,* 5, 35–37.

Begout, M.L., Kadri, S., Huntingford, F. and Damsgård, B. (2012) Tools for studying the behaviour of farmed fish. In: Huntingford, F., Jobling, M. and Kadri, S. (eds) *Aquaculture and Behavior*. Wiley-Blackwell, Oxford, UK, pp. 65–86.

Bigman, J.S., Pardo, S.A., Prinzing, T.S., Dando, M., Wegner, N.C., *et al.* (2018) Ecological lifestyles and the scaling of shark gill surface area. *Journal of Morphology* 279, 1716–1724.
Boltano, S., Rey, S.P., Roher, N., Vargas, R., Huerta, M., *et al.* (2013) Behavioural fever is a synergistic signal amplifying the innate immune response. *Proceedings of the Royal Society B: Biological Sciences* 280, 20131381.
Bosch-Belmar, M., Azzurro, E., Puli, K., Milisenda, G., Fuentes, V., *et al.* (2017) Jellyfish blooms perception in Mediterranean finfish aquaculture. *Marine Policy* 76, 1–7.
Boudry, P., Allal, F., Aslam, M.L., Bargelloni, L., Bean, T.P., *et al.* (2021) Current status and potential of genomic selection to improve selective breeding in the main aquaculture species of International Council for the Exploration of the Sea (ICES) member countries. *Aquaculture Reports* 20, 100700.
Bovenkerk, B. and Meijboom, F.L.B. (2013) Fish welfare in aquaculture: explicating the chain of interactions between science and ethics. *Journal of Agricultural and Environmental Ethics* 26, 41–61.
Brambell, F.W.R. (1965) *Report of the Technical Committee to the Enquiry into the Welfare of Animals kept under Intensive Livestock Systems*. Her Majesty's Stationery Office, London.
Brijs, J., Føre, M., Gräns, A., Clark, T.D., Axelsson, M., *et al.* (2021) Biosensing technologies in aquaculture: how remote monitoring can bring us closer to our farm animals. *Philosophical Transactions of the Royal Society B: Biological Sciences* 376, 20200218.
Broom, D.M. (1991) Animal welfare: concepts and measurement. *Journal of Animal Science* 69, 4167–4175.
Brown, C. (2015) Fish intelligence, sentience and ethics. *Animal Cognition* 18, 1–17.
Browning, H. and Birch, J. (2022) Animal sentience. *Philosophy Compass* 17, e12822.
Cain, K.D. and Polinski, M.P. (2023) Infectious diseases of coldwater fish in fresh water. In: Woo, P.T.K. and Subasinghe, R.P. (eds) *Climate Change on Diseases and Disorders of Finfish in Cage Culture*, 3rd edn. CAB International, Wallingford, UK, pp. 76–124.
Castanheira, M.F., Conceição, L.E.C., Millot, S., Rey, S., Bégout, M.-L., *et al.* (2017) Coping styles in farmed fish: consequences for aquaculture. *Reviews in Aquaculture* 9, 23–41.
Cerqueira, M., Rey, S., Featherstone, Z., Crumlish, M. and MacKenzie, S. (2016) Thermal preference predicts animal personality in Nile tilapia *Oreochromis niloticus*. *Journal of Animal Ecology* 85, 1389-1400.
Cerqueira, M., Millot, S., Castanheira, M.F., Félix, A.S., Silva, T., *et al.* (2017) Cognitive appraisal of environmental stimuli induces emotion-like states in fish. *Scientific Reports* 7, 13181.
Chapman, L.J., Galis, F. and Shinn, J. (2000) Phenotypic plasticity and the possible role of genetic assimilation: hypoxia-induced trade-offs in the morphological traits of an African cichlid. *Ecology Letters* 3, 387–393.
Chen, J., Guang, C., Xu, H., Chen, Z., Xu, P., *et al.* (2007) A review of cage and open aquaculture: China. In: Halwart, M., Soto, D. and Arthur, J.R. (eds) *Cage Aquaculture – Regional Review and Global Overview*. FAO Fisheries Technical Paper No. 498. Food and Agriculture Organization of the United Nations, Rome, pp. 53–68.
Clinton, M., Ferrier, D.E.K., Martin, S.A.M. and Brierley, A.S. (2021) Impacts of jellyfish on marine cage aquaculture: an overview of existing knowledge and the challenges to finfish health. *ICES Journal of Marine Science* 78, 1557–1573. doi: 10.1093/icesjms/fsaa254.
Crampton, W.G.R., Chapman, L.J. and Bell, J. (2008) Interspecific variation in gill size is correlated to ambient dissolved oxygen in the Amazonian electric fish *Brachyhypopomus* (Gymnotiformes: Hypopomidae). *Environmental Biology of Fishes* 83, 223–235.
Crispo, E. and Chapman, L.J. (2010) Geographic variation in phenotypic plasticity in response to dissolved oxygen in an African cichlid fish. *Journal of Evolutionary Biology* 23(10), 2091–2103.
Cubillo, A.M., Ferreira, J.G., Lencart-Silva, J., Taylor, N.G.H., Kennerley, A., *et al.* (2021) Direct effects of climate change on productivity of European aquaculture. *Aquaculture International* 29, 1561–1590.
Cucco, A., Sinerchia, M., Lefrancois, C., Magni, P., Ghezzoc, M., *et al.* (2012) A metabolic scope-based model of fish response to environmental changes. *Ecological Modelling* 237–238, 132–141.
Damsgård, B. and Huntingford. F. (2012) Fighting and aggression. In: Huntingford, F., Jobling, M. and Kadri, S. (eds) *Aquaculture and Behavior*. Wiley-Blackwell, Oxford, UK, pp. 248–284.
Dawkins, M.S. (1998) Evolution and animal welfare. *Quarterly Review of Biology* 73(3), 305–328.
Dou, S., Seikai, T. and Tsukamoto, K. (2000) Cannibalism in Japanese flounder juveniles, *Paralichthys olivaceus*, reared under controlled conditions. *Aquaculture* 182, 149–159.
Ersdal, C., Midtlyng, P.J. and Jarp, J. (2001) An epidemiological study of cataracts in seawater farmed Atlantic salmon *Salmo salar*. *Diseases of Aquatic Organisms* 45, 229–236.
Espmark, A.M. and Baeverfjord, G. (2009) Effects of hyperoxia on behavioural and physiological variables in farmed Atlantic salmon (*Salmo salar*) parr. *Aquaculture* 17, 341–355.

FAO (Food and Agriculture Organisation of the United Nations) (2020) *The State of World Fisheries and Aquaculture 2020: Sustainability in Action*. FAO, Rome.

Faustino, A., Tacão-Monteiro, A. and Oliveira, R. (2017) Mechanisms of social buffering of fear in zebrafish. *Scientific Reports* 7, 44329.

Fernö, A., Bjelland, O. and Kristiansen, T.S. (2020) The diverse world of fishes. In: Kristiansen, T.S., Fernö, A., Pavlidis, M.A. and van de Vis, H. (eds) *The Welfare of Fish*. Springer, Cham, Switzerland, pp. 43–72.

Fife-Cook, I. and Franks, B. (2019) Positive welfare for fishes: rationale and areas for future study. *Fishes* 4, 31. doi: 10.3390/fishes4020031.

Føre, M., Frank, K., Norton, T., Svendsen, E., Alfredsen, J.A., *et al.* (2018) Precision fish farming: a new framework to improve production in aquaculture. *Biosystems Engineering* 173, 176–193.

Franks, B., Ewell, C. and Jacquet, J. (2021) Animal welfare risks of global aquaculture. *Science Advances* 7(14), eabg0677.

Fraser, D. (1999) Animal ethics and animal welfare science: bridging the two cultures. *Applied Animal Behavioural Science* 65, 171–189.

Fraser, D. (2009) Assessing animal welfare: different philosophies, different scientific approaches. *ZooBiology* 28, 507–518.

Fraser, T.W.K., Hansen, T., Mayer, I., Skjæraasen, J.E., Glover, K.A., *et al.* (2014) The effect of triploidy on vaccine side-effects in Atlantic salmon. *Aquaculture* 433, 481–490.

Frenzl, B., Stien, L.H., Cockerill, D., Oppedal, F., Richards, R.H., *et al.* (2014) Manipulation of farmed Atlantic salmon swimming behaviour through the adjustment of lighting and feeding regimes as a tool for salmon lice control. *Aquaculture* 424–425, 183–188.

Gaffney, L.P., Watson, B.M., Oakes, J.D. and Devlin, R.H. (2018) Internal morphological effects of growth hormone transgenesis in coho salmon. *Transactions of the American Fisheries Society* 147, 1199–1206.

Galhardo, L., Almeida, O. and Oliveira, R.F. (2011) Measuring motivation in a cichlid fish: an adaptation of the push-door paradigm. *Applied Animal Behaviour Science* 130, 60–70.

Gansel, L.C., Rackebrandt, S., Oppedal, F. and McClimans, T.A. (2014) Flow fields inside stocked fish cages and the near environment. *Journal of Offshore Mechanics and Arctic Engineering* 136, 031201.

Geitung, L., Wright, D.W., Oppedal, F., Stien, L.H., Vagseth, T., *et al.* (2020) Cleaner fish growth, welfare and survival in Atlantic salmon sea cages during an autumn–winter production. *Aquaculture* 528, 735623.

Graham, C., von Keyserling, M.A.G. and Franks, B. (2018) Free-choice exploration increases affiliative behaviour in zebrafish. *Applied Animal Behaviour Science* 203, 103–110.

Haberlin, D., McAllen, R. and Doyle, T.K. (2021) Field and flume tank experiments investigating the efficacy of a bubble curtain to keep harmful jellyfish out of finfish pens. *Aquaculture* 531, 735915.

Hamre, K., Micallef, G., Hillestad, M., Johansen, J., Remo, S., *et al.* (2022) Changes in daylength and temperature from April until August for Atlantic salmon (*Salmo salar*) reared in sea cages, increase growth rate, and may cause consumption of anti-oxidants, onset of cataracts and increased oxidation of fillet astaxanthin. *Aquaculture* 551, 737950.

Heathcote, R.J.P., Darden, S.K., Franks, D.W., Ramnarine, I.W. and Croft, D.P. (2017) Fear of predation drives stable and differentiated social relationships in guppies. *Scientific Reports* 7, 41679.

Herbert, N.A. (2013) Practical aspects of induced exercise in finfish aquaculture. In: Palstra, A.P. and Planas, J.V. (eds) *Swimming Physiology of Fish*. Springer, Berlin, pp. 377–405.

Herbert, N.A. and Steffensen, J.F. (2005) The response of Atlantic cod, *Gadus morhua*, to progressive hypoxia: fish swimming speed and physiological stress. *Marine Biology* 147, 1403–1412.

Herbert, N.A., Kadri, S. and Huntingford, F.A. (2011) A moving light stimulus elicits a sustained swimming response in farmed Atlantic salmon, *Salmo salar* L. *Fish Physiology and Biochemistry* 37, 317–325.

Hjelmstedt, P., Sundell, E., Brij, J., Berg, C., Sandblom, E., *et al.* (2022) Assessing the effectiveness of percussive and electrical stunning in rainbow trout: does an epileptic-like seizure imply brain failure? *Aquaculture* 552, 738012.

Holt, R.E. and Jørgensen, C. (2015) Climate change in fish: effects of respiratory constraints on optimal life history and behaviour. *Biology Letters* 11, 20141032.

Huntingford, F.A. (2020) Fish behaviour: determinants and implications for welfare. In: Kristiansen, T.S., Fernö, A., Pavlidis, M.A. and van de Vis, H. (eds) *The Welfare of Fish*. Springer, Cham, Switzerland, pp. 73–110.

Huntingford, F.A. and Adams, C.E. (2005) Behavioural syndromes in farmed fish: implications for production and welfare. *Behaviour* 142, 1213–1227.

Huntingford, F.A., Andrew, G., MacKenzie, S., Morera, D., Coyle, S.M., *et al.* (2010) Coping strategies in a strongly schooling fish, the common carp, *Cyprinus carpio*. *Journal of Fish Biology* 76, 1576–1591.

Huntingford, F.A., Turnbull, J. and Kadri, S. (2018) Methods to increase fish production: welfare and sustainability implications. In: Grandin, T. and Whiting, M. (eds) *Are We Pushing Animals to Their Biological Limits? Welfare and Ethical Considerations*. CAB International, Wallingford, UK, pp. 89–121.

Hvas, M., Folkedal, O. and Oppedal, F. (2021) Fish welfare in offshore salmon aquaculture. *Reviews in Aquaculture* 13, 836–852.

Islam, M.J., Kunzmann, A. and Slater, M.J. (2021) Responses of aquaculture fish to climate change induced extreme temperatures: a review. *Journal of the World Aquaculture Society* 53, 314–366.

Jarvis, S., Ellis, M.A., Turnbull, J., Rey, S.P. and Wemelsfelder, F. (2022) Qualitative behavioural assessment (QBA) in juvenile farmed Atlantic Salmon (*Salmo salar*): potential for on-farm welfare assessment. *Frontiers in Veterinary Science* 8, 702783.

Jenjan, H., Mesquita, F., Huntingford, F.A. and Adams, C. (2013) Respiratory function in common carp with different stress coping styles: a hidden cost of personality traits? *Animal Behaviour* 85(6), 1245–1249.

Jobling, M. (2010) Fish culture: the rearing environment. In: Le Francois, N., Jobling, M., Carter, C. and Blier, P. (eds) *Finfish Aquaculture Diversification*. CAB International, Wallingford, UK, pp. 33–60.

Jobling, M., Alanärä, A., Kadri, S. and Huntingford, F. (2012a) Feeding biology and foraging. In: Huntingford, F., Jobling, M. and Kadri, S. (eds) *Aquaculture and Behavior*. Wiley-Blackwell, Oxford, UK, pp. 121–149.

Jobling, M., Alanärä, A., Noble, C., Sánchez-Vázquez, J., Kadri, S., *et al.* (2012b) Appetite and feed intake. In: Huntingford, F., Jobling, M. and Kadri, S. (eds) *Aquaculture and Behavior*. Wiley-Blackwell, Oxford, UK, pp. 183–219.

Johansen, I.B., Höglund, E. and Øverli, Ø. (2020) Individual variations and coping style. In: Kristiansen, T.S., Fernö, A., Pavlidis, M.A. and van de Vis, H. (eds) *The Welfare of Fish*. Springer, Cham, Switzerland, pp. 283–302.

Johansson, D., Ruohonen, K., Kiessling, A., Oppedal, F., Stiansen, J.-E., *et al.* (2006) Effect of environmental factors on swimming depth preferences of Atlantic salmon (*Salmo salar*) and temporal and spatial variations in oxygen levels in sea cages at a fjord site. *Aquaculture* 254, 594–605.

Johansson, D., Laursen, F., Fernö, A., Fosseidengen, J.E., Klebert, P., *et al.* (2014) The interaction between water currents and salmon swimming behaviour in sea cages. *PLoS ONE* 9, e97635.

Juell, J.-E., Oppedal, F., Boxaspen, K. and Taranger, G.L. (2003) Submerged light increases swimming depth and reduces fish density of Atlantic salmon *Salmo salar* in production cages. *Aquaculture Research* 34, 469–477.

Kittilsen, S., Schjolden, J., Beitnes-Johansen, I., Shaw, J.C., Pottinger, T.G., *et al.* (2009) Melanin-based skin spots reflect stress responsiveness in salmonid fish. *Hormones and Behavior* 56, 292–298.

Klemme, I. and Karvonen, A. (2016) Learned parasite avoidance is driven by host personality and resistance to infection in a fish–trematode interaction. *Proceedings of the Royal Society B: Biological Sciences* 283, 20161148.

Korte, S.M., Olivier, B. and Koolhaas, J.M. (2007) A new animal welfare concept based on allostasis. *Physiology and Behavior* 92, 422–428.

Koolhaas, J., Korte, S.M., De Boer, S.F., Van der Vegt, B.J., Van Reenen, C.G., *et al.* (1999) Coping styles in animals: current status in behavior and stress-physiology. *Neuroscience and Biobehavioral Reviews* 23, 925–935.

Korsøen, Ø.J., Dempster. T., Fjelldal, P.G., Oppedal, F. and Kristiansen, T.S. (2009) Long-term culture of Atlantic salmon (*Salmo salar* L.) in submerged cages during winter affects behaviour, growth and condition. *Aquaculture* 296, 373–381.

Kristiansen, T.S. and Fernö, A. (2007) Individual behaviour and growth of halibut (*Hippoglossus hippoglossus* L.) fed sinking and floating feed: evidence of different coping styles. *Applied Animal Behaviour Science* 104, 236–250.

Kristiansen, T.S., Fernö, A., Holm, J.C., Privitera, L., Bakke, S., *et al.* (2004) Swimming behaviour as an indicator of low growth rate and impaired welfare in Atlantic halibut (*Hippoglossus hippoglossus* L.) reared at three stocking densities. *Aquaculture* 230, 137–151.

Kristiansen, T.S., Fernö, A., Pavlidis, M.A. and van de Vis, H. (2020) *The Welfare of Fish*. Springer, Cham, Switzerland.

Kusku, H., Ergun, S., Yilmaz, S., Guroy, B. and Yigit, M. (2019) Impacts of urban noise and musical stimuli on growth performance and feed utilization of koi fish (*Cyprinus carpio*) in recirculating water conditions. *Turkish Journal of Fisheries and Aquatic Sciences* 19, 513–523.

Lagardère, J.P. and Mallekh, R. (2000) Feeding sounds of turbot (*Scophthalmus maximus*) and their potential use in the control of food supply in aquaculture I. Spectrum analysis of the feeding sounds. *Aquaculture* 189, 251–258.

Langford, D.J., Tuttle, A.H., Briscoe, C., Harvey-Lewis, C., Baran, I., *et al.* (2011) Varying perceived social threat modulates pain behavior in male mice. *The Journal of Pain* 12, 125–132.

Lawrence, A.B., Vigors, B. and Sandøe, P. (2019) What is so positive about positive animal welfare? – A critical review of the literature. *Animals* 9(10), 783.

Learmouth, M.J. (2020) The matter of non-avian reptile sentience, and why it 'matters' to them: a conceptual, ethical and scientific review. *Animals* 10, 901. doi: 10.3390/ani10050901.

Leong, J., Evensen, O. and Woo, P.T.K. (2023) Infectious diseases of coldwater fish in marine and brackish waters. In: Woo, P.T.K. and Subasinghe, R.P. (eds) *Climate Change on Diseases and Disorders of Finfish in Cage Culture*, 3rd edn. CAB International, Wallingford, UK, pp. 34–75.

Li, D., Hao, Y. and Duan, Y. (2020a) Nonintrusive methods for biomass estimation in aquaculture with emphasis on fish: a review. *Reviews in Aquaculture* 12, 1390–1411.

Li, D., Wang, Z., Wu, S., Miao, Z., Du, L. *et al.* (2020b) Automatic recognition methods of fish feeding behavior in aquaculture: a review. *Aquaculture* 528, 735508.

Lines, J.A. and Frost, A.R. (1997) Selective attraction of salmon. *Aquaculture Engineering* 16, 261–273.

Lines, J.A. and Kestin, S. (2004) Electrical stunning of fish: the relationship between the electric field strength and water conductivity. *Aquaculture* 241, 219–234.

Lines, J.A., Tillett, R.D., Ross, L.G., Chan, D., Hockaday, S., *et al.* (2001) An automatic image-based system for estimating the mass of free-swimming fish. *Computers and Electronics in Agriculture* 31, 151–168.

Luchiari, A.C., Málaga-Trillo, E., Tran, S. and Gerlai, R. (2021) Editorial: Zebrafish cognition and behavior. *Frontiers in Behavioral Neuroscience* 15, 659501. doi: 10.3389/fnbeh.2021.659501.

Macaulay, G., Warren-Myers, F., Barrett, L.T., Oppedal, F., Føre, M., *et al.* (2021) Tag use to monitor fish behaviour in aquaculture: a review of benefits, problems and solutions. *Reviews in Aquaculture* 13, 1565–1582.

Magnhagen, C., Backström, T., Nilsson, J. and Brännäs, E. (2018) Oxygen consumption and swimming performance in Arctic charr with different pigmentation patterns. *Applied Animal Behaviour Science* 202, 119–124.

Maia, C.M and Volpato, G.L. (2017) Preference index supported by motivation tests in Nile tilapia. *PLoS ONE* 12(4), e0175821.

Martins, C.I.M., Galhardo, L., Noble, C., Damsgård, B., Spedicato, M.T., *et al.* (2012) Behavioural indicators of welfare in farmed fish. *Fish Physiology and Biochemistry* 38, 17–41.

Maulu, S., Hasimuna, O.J., Haambiya, L.H., Monde, C., Musuka, C.G., *et al.* (2021) Climate change effects on aquaculture production: sustainability implications, mitigation, and adaptations. *Frontiers in Sustainable Food Systems* 5, 609097.

McCulloch, S.P. (2013) A critique of FAWC's Five Freedoms as a framework for the analysis of animal welfare. *Journal of Agricultural and Environmental Ethics* 26, 959–975.

McKenzie, D.J., Palstra, A.P., Planas, J., MacKenzie, S., Begout, M.-L., *et al.* (2021) Aerobic swimming in intensive finfish aquaculture: applications for production, mitigation and selection. *Reviews in Aquaculture* 13, 138–155.

Medaas, C., Lien, M.E., Gismervik, K., Kristiansen, T., Osmundsen, T., *et al.* (2021) Minding the gaps in fish welfare: the untapped potential of fish farm workers. *Journal of Agricultural and Environmental Ethics* 34, 29.

Mendl, M., Burman, O.H.P. and Paul, E.S. (2010) An integrative and functional framework for the study of animal emotion and mood. *Proceedings of the Royal Society B: Biological Sciences* 277, 2895–2904.

Metcalfe, N.B. and Monaghan, P. (2003) Growth versus lifespan: perspectives from evolutionary ecology. *Experimental Gerontology* 38, 935–940.

Metcalfe, N.B. and Thorpe, J.E. (1992) Anorexia and defended energy levels in juvenile Atlantic salmon. *Journal of Animal Ecology* 61, 175–181.

Muller-Feuga, A. (2000) The role of microalgae in aquaculture: situation and trends. *Journal of Applied Phycology* 12, 527–534.

Naslund, A.W., Davis, B.E., Hobbs, J.A., Fangue, N.A. and Todgham, A.E. (2021) Warming, not CO_2-acidified seawater, alters otolith development of juvenile Antarctic emerald rockcod (*Trematomus bernacchii*). *Polar Biology* 44, 1917–1923.

Naya-Català, F., Simó-Mirabet, P., Calduch-Giner, J. and Pérez-Sánchez, J. (2021) Transcriptomic profiling of Gh/Igf system reveals a prompted tissue-specific differentiation and novel hypoxia responsive genes in gilthead sea bream. *Scientific Reports* 11, 16466.

Nebo, C., Portella, M.C., Carani, F.R., de Almeida, F.L.A., Padovani, C.R., *et al.* (2013) Short periods of fasting followed by refeeding change the expression of muscle growth-related genes in juvenile Nile tilapia (*Oreochromis niloticus*). *Comparative Biochemistry and Physiology Part B: Biochemistry and Molecular Biology* 164, 268–274.

Noble, C., Gismervik, K., Iversen, M.H., Kolarevic, J., Nilsson, J., *et al.* (eds) (2018) *Welfare Indicators for Farmed Atlantic Salmon: Tools for Assessing Fish Welfare*. Nofima, Tromsø, Norway.

Nordahl, O., Tibblin, P., Koch-Schmidt, P., Berggren, H., Larsson, P., *et al.* (2018) Sun-basking fish benefit from body temperatures that are higher than ambient water. *Proceedings of the Royal Society B: Biological Sciences* 285, 20180639.

Nomura, M., Sloman, K.A., von Keyserlingk, M.A.G. and Farrell, A.P. (2009) Physiology and behaviour of Atlantic salmon (*Salmo salar*) smolts during commercial land and sea transport *Physiology and Behavior* 96, 233–243.

O'Donncha, F. and Grant, J. (2019) Precision aquaculture. *IEEE Internet of Things Magazine* 2(4), 26–30.

Ohl, F. and van der Staay, F.J (2012) Animal welfare: at the interface between science and society. *The Veterinary Journal* 192, 13–19.

Okoli, A.S., Blix, T., Myhr, A.I., Xu, W. and Xu, X. (2022) Sustainable use of CRISPR/Cas in fish aquaculture: the biosafety perspective. *Transgenic Research* 31, 1–21.

Oppedal, F., Dempster, T. and Stien, L.H. (2011) Environmental drivers of Atlantic salmon behaviour in sea-cages: a review. *Aquaculture* 311, 1–18.

Oyinlola, M.M., Reygondeau, G., Wabnitz, C.C.C. and Cheung, W.W.L. (2020) Projecting global mariculture diversity under climate change. *Global Climate Change* 26, 2134–2145.

Parker-Graham, C.A., Heckman, T.I., Griffin, M.J. and Soto, E. (2023) Infectious diseases of warmwater fish in marine and brackish waters. In: Woo, P.T.K. and Subasinghe, R.P. (eds) *Climate Change on Diseases and Disorders of Finfish in Cage Culture*, 3rd edn. CAB International, Wallingford, UK, pp. 163–201.

Pasquet, A. (2018) Effects of domestication on fish behaviour. In: Teletchea, F. (ed.) *Animal Domestication*. IntechOpen. doi: 10.5772/intechopen.78752.

Pedersen, B.H. (1997) The cost of growth in young fish, a review of new hypotheses. *Aquaculture* 155, 259–269.

Pedrazzani, A.S., Quintiliano, M.H., Bolfe, F., de Oliveira Sans, E.C. and Molento, M.C.F. (2020) Tilapia on-farm welfare assessment protocol for semi-intensive production systems. *Frontiers in Veterinary Science* 7, 606388.

Porteus, C.S., Roggatz, C.C., Velez, Z., Hardege, J.D. and Hubbard, P.C. (2021) Acidification can directly affect olfaction in marine organisms. *Journal of Experimental Biology* 224, jeb237941.

Pottinger, T.G. and Carrick, T.R. (1999) Modification of the plasma cortisol response to stress in rainbow trout by selective breeding. *General and Comparative Endocrinology* 116, 122–132.

Pouil, S., Tlusty, M.F., Rhyne, A.L. and Metian, M. (2020) Aquaculture of marine ornamental fish: overview of the production trends and the role of academia in research progress. *Reviews in Aquaculture* 12, 1217–1230.

Powell, A., Treasurer, J.W., Pooley, C.L., Keay, A.J., Lloyd, R., *et al.* (2018) Use of lumpfish for sea-lice control in salmon farming: challenges and opportunities. *Reviews in Aquaculture* 10, 683–702.

Price, E.O. (2001) *Animal Domestication and Behavior*. CAB International, Wallingford, UK.

Puvanendran, V. and Brown, J.A. (2002) Foraging, growth and survival of Atlantic cod larvae reared in different light intensities and photoperiods. *Aquaculture* 214, 131–151.

Rakus, K., Ronsmans, M., Forlenza, M., Michiels, T., Wiegertjes, G.F., *et al.* (2017) Conserved fever pathways across vertebrates: a herpesvirus expressed decoy TNF-α receptor delays behavioral fever in fish. *Cell Host & Microbe* 211, 244–253.

Raubenheimer, D., Simpson, S., Sánchez-Vázquez, J., Huntingford, F., Kadri, S., *et al.* (2012) Nutrition and diet choice. In: Huntingford, F., Jobling, M. and Kadri, S. (eds) *Aquaculture and Behavior*. Wiley-Blackwell, Oxford, UK, pp. 150–182.

Reid, G.K., Gurney-Smith, H.J., Marcogliese, D.J., Knowler, D., Benfey, T., *et al.* (2019) Climate change and aquaculture: considering biological response and resources. *Aquaculture Environment Interactions* 11, 569–602.

Reimer, T., Dempster, T., Wargelius, A., Fjelldal, P.G., Hansen, T., *et al.* (2017) Rapid growth causes abnormal vaterite formation in farmed fish otoliths, *Journal of Experimental Biology* 220, 2965–2969.

Reisa, Y.d.S., Lenon, J., Leitea, R., Andrerson, C., de Almeida, L., *et al.* (2019) New insights into tambaqui (*Colossoma macropomum*) feeding behavior and digestive physiology by the self-feeding approach: effects on growth, dial patterns of food digestibility, amylase activity and gastrointestinal transit time. *Aquaculture* 498, 116–122.

Reynolds, W.W., Casterlin, M.E. and Covert, J.B. (1976) Behavioural fever in teleost fishes. *Nature* 259, 41–42.

Roque, A., Castanheira, M.F., Toffan, A., Arechavala-Lopez, P., Brun, E., *et al.* (2020) Report on Fish Welfare and List of Operational Welfare Indicators in Sea Bream. Deliverable 4.5 of the Horizon 2020 project MedAID (GA number 727315), published in the project web site on 17.02.2020. Available at: http://www.medaid-h2020.eu/index.php/deliverables/ (accessed 15 August 2022).

Sae-Lim, P., Kause, A., Mulder, H.A. and Olesen, I. (2017) Breeding and Genetics Symposium: climate change and selective breeding in aquaculture. *Journal of Animal Science* 95, 1801–1812.

Saksida, S.M., Fast, M.D., Garver, K.A. and Johnson, S.C. (2023) Movement of infectious agents between wild and farmed fish. In: Woo, P.T.K. and Subasinghe, R.P. (eds) *Climate Change on Diseases and Disorders of Finfish in Cage Culture*, 3rd edn. CAB International, Wallingford, UK, pp. 351–386.

Salin, K.R., Subasinghe, R.P., Senarathna, D. and Shinn, A.P. (2023) Cage culture of finfish: its importance, distributions and future modifications in ongoing climate change. In: Woo, P.T.K. and Subasinghe, R.P. (eds) *Climate Change on Diseases and Disorders of Finfish in Cage Culture*, 3rd edn. CAB International, Wallingford, UK, pp. 1–33.

Salvanes, A.G.V., Moberg, O. and Braithwaite, V.A. (2007) Effects of early experience on group behaviour in fish. *Animal Behaviour* 74, 805–811.

Samaras, A., Dimitroglou, A., Kollias, S., Skouradakis, G., Papadakis, I.E., *et al.* (2021) Cortisol concentration in scales is a valid indicator for the assessment of chronic stress in European sea bass, *Dicentrarchus labrax* L. *Aquaculture* 545, 737257.

Santos, G.A., Schrama, J.W., Mamauag, R.E.P., Rombout, J.H.W.M. and Verreth, J.A.J. (2010) Chronic stress impairs performance, energy metabolism and welfare indicators in European seabass (*Dicentrarchus labrax*): the combined effects of fish crowding and water quality deterioration. *Aquaculture* 299, 73–80.

Saraiva, J.L., Castanheira, M.F., Arechavala-López, P., Volstorf, J. and Heinzpeter, B. (2018) Domestication and welfare in farmed fish. In: Teletchea, F. (ed.) *Animal Domestication*. INTechOpen. doi: 10.5772/intechopen.77251.

Saraiva, J.L., Arechavala-Lopez, P., Castanheira, M.F., Volstorf, J. and Heinzpeter Studer, B. (2019) A global assessment of welfare in farmed fishes: the FishEthoBase. *Fishes* 4(2), 30.

Saraiva, J.L., Arechavala-Lopez, P. and Sneddon, L. (2022) Farming fish. In: Knight, A., Phillips, C. and Sparks, P. (eds) *Routledge Handbook of Animal Welfare*. Routledge, Abingdon, UK, pp. 115–127.

Saunders, R.L. (1965) Adjustment of buoyancy in young Atlantic salmon and brook trout by changes in swim bladder volume. *Journal of the Fisheries Board of Canada* 22, 335–352.

Shahmohamadloo, R.S., Frenken, T., Rudman, S.M., van West, P., Ibelings, B.W. and Trainer, V.L. (2023) Diseases and disorders in fish due to harmful algal blooms. In: Woo, P.T.K. and Subasinghe, R.P. (eds) *Climate Change on Diseases and Disorders of Finfish in Cage Culture*, 3rd edn. CAB International, Wallingford, UK, pp. 387–429.

Shinn, A.P., Dong, H.T., Vinh, N.T., Wongwaradechkul, R. and Lio-Po, G.D. (2023) Infectious diseases of warmwater fish in fresh water. In: Woo, P.T.K. and Subasinghe, R.P. (eds) *Climate Change on Diseases and Disorders of Finfish in Cage Culture*, 3rd edn. CAB International, Wallingford, UK, pp. 202–277.

Sievers, M., Korsøen, O., Warren-Myers, F., Oppedal, F., Macaulay, G., *et al.* (2022) Submerged cage aquaculture of marine fish: a review of the biological challenges and opportunities. *Reviews in Aquaculture* 14, 106–119.

Skov, P.V., Gregersen, K.J.D., Gesto, M. and Jokumsen, A. (2019) Proactive coping style in early emerging rainbow trout carries a metabolic cost with no apparent return. *Comparative Biochemistry and Physiology Part A: Molecular & Integrative Physiology* 231, 104–110.

Sneddon, L.U., Elwood, R.W., Adamo, S.A. and Leach, M.C. (2014) Defining and assessing animal pain. *Animal Behaviour* 97, 201–212.

Sneddon, L.U., Wolfenden, D.C.C., Leach, M.C., Valentim, A.M., Steenbergen, P.J., *et al.* (2018) Ample evidence for fish sentience and pain. *Animal Sentience* 21, 17.

Soares, F., Costas, B., Lourenço-Marques, C., Fernández, I. and Gavaia, P. (2023) Non-infectious disorders of warmwater fish. In: Woo, P.T.K. and Subasinghe, R.P. (eds) *Climate Change on Diseases and Disorders of Finfish in Cage Culture*, 3rd edn. CAB International, Wallingford, UK, pp. 278–314.

Soares, M.C. (2017) Neurobiology of mutualistic behaviour: the cleaner fish swims into the spotlight. *Frontiers in Behavioral Neuroscience* 11, 191. doi: 10.3389/fnbeh.2017.00191.

Soliveres, E., Poveda, P., Estruch, V.D., Pérez-Arjona, I., Puig, V., *et al.* (2017) Monitoring fish weight using pulse-echo waveform metrics. *Aquacultural Engineering* 77, 125–131.

Sommerset, I., Bang Jensen, B., Bornø, B., Haukaas, A. and Brun, E. (eds) (2021) *The Health Situation in Norwegian Aquaculture 2020. Report 41a/2021.* Norwegian Veterinary Institute, Ås, Norway.

St-Hilaire, S.N., Gomes, G., Pinheiro Marques, A.R., Chang, S.F. and Brun, E. (2023) Sporadic emerging infectious and non-infectious diseases and disorders. In: Woo, P.T.K. and Subasinghe, R.P. (eds) *Climate Change on Diseases and Disorders of Finfish in Cage Culture*, 3rd edn. CAB International, Wallingford, UK, pp. 315–350.

Stien, L.H., Bracke, M., Noble, C. and Kristiansen, T.S. (2020) Assessing fish welfare in aquaculture. In: Kristiansen, T.S., Fernö, A., Pavlidis, M.A. and van de Vis, H. (eds) *The Welfare of Fish*. Springer, Cham, Switzerland, pp. 303–321.

Tanaka, H., Takagi, Y. and Naito, Y. (2000) Behavioural thermoregulation of chum salmon during homing migration in coastal waters. *Journal of Experimental Biology* 203, 1825–1833.

Tanck, M.W.T., Claes, T., Bovenhuis, H. and Komen, J. (2002) Exploring the genetic background of stress using isogenic progenies of common carp selected for high or low stress-related cortisol response. *Aquaculture* 204, 419–434.

Tang, M. and Boisclair, D. (1995) Relationship between respiration rate of juvenile brook trout (*Salvelinus fontinalis*), water temperature, and swimming characteristics. *Canadian Journal of Fisheries and Aquatic Science* 52, 2138–2145.

Teal, L.R., Marras, S., Peck, M.A., and Domenici, P. (2018) Physiology-based modelling approaches to characterize fish habitat suitability: their usefulness and limitations. *Estuarine, Coastal and Shelf Science* 201, 56–63.

Teletchea, F. and Fontaine, P. (2012) Levels of domestication in fish: Implications for the sustainable future of aquaculture. *Fish and Fisheries* 15(2), 181–195.

Tielmann, M., Schulz, C. and Meyer, S. (2016) Self-grading of larval pike-perch (*Sander lucioperca*), triggered by positive phototaxis. *Aquaculture Engineering* 72–73, 13–19.

Tillett, R.D., Bull, C.R. and Lines, J.A. (1999) An optical method for the detection of sea lice, *Lepeophtheirus salmonis*. *Aquacultural Engineering* 21, 33–48.

Tschirren, L., Bachmann, D., Güler, A.C., Blaser, O., Rhyner, N., *et al.* (2021) MyFishCheck: a model to assess fish welfare in aquaculture. *Animals* 11(1), 145.

Turnbull, J.F. (2022) The complex influences on how we care for farmed fish. *Frontiers in Veterinary Science* 8, 765797.

Turnbull, J.F., Bell, A., Adams, C., Bron, J. and Huntingford, F. (2005) Stocking density and welfare of cage farmed Atlantic salmon: application of a multivariate analysis. *Aquaculture* 243, 121–132.

van de Vis, H., Kolarevic, J., Stien, L.H., Kristiansen, T.S., Gerritzen, M., *et al.* (2020) Welfare of farmed fish in different production systems and operations. In: Kristiansen, T.S., Fernö, A., Pavlidis, M.A. and van de Vis, H. (eds) *The Welfare of Fish*. Springer, Cham, Switzerland, pp. 323–332.

Velez, Z., Roggatz, C.C., Benoit, D.M., Hardege, J.D. and Hubbard, P.C. (2019) Short- and medium-term exposure to ocean acidification reduces olfactory sensitivity in gilthead seabream. *Frontiers in Physiology* 10, 731.

Wade, N.M., Clark, T.D., Maynard, B.T., Atherton, S., Wilkinson, R.J., *et al.* (2019) Effects of an unprecedented summer heatwave on the growth performance, flesh colour and plasma biochemistry of marine cage-farmed Atlantic salmon (*Salmo salar*). *Journal of Thermal Biology* 80, 64–74.

Wahli, T., Schmidt-Posthaus, H. and Segner, H. (2023) Non-infectious disorders of coldwater fish. In: Woo, P.T.K. and Subasinghe, R.P. (eds) *Climate Change on Diseases and Disorders of Finfish in Cage Culture*, 3rd edn. CAB International, Wallingford, UK, pp. 125–162.

Whitney, J.E., Al-Chokhachy, R., Bunnell, D.B., Caldwell, C.A., Cooke, S.J., *et al.* (2016) Physiological basis of climate change impacts on North American inland fishes. *Fisheries* 41, 332–345.

Wu, T., Huang, Y. and Chen, J. (2015) Development of an adaptive neural-based fuzzy inference system for feeding decision-making assessment in silver perch (*Bidyanus bidyanus*) culture. *Aquaculture Engineering* 66, 41–51.

Yang, X., Zhang, S., Liu, J., Gao, Q., Dong, S., *et al.* (2021) Deep learning for smart fish farming: applications, opportunities and challenges. *Reviews in Aquaculture* 13, 66–90.

Yildiz, H.Y., Chatzifotis, S., Anastasiadis, P., Parisi, G. and Papandroulakis, N. (2021) Testing of the Salmon Welfare Index Model (SWIM 1.0) as a computational welfare assessment for sea-caged European sea bass. *Italian Journal of Animal Science* 20, 1423–1430.

Zhao, S., Ding, W., Zhao, S. and Gu, J. (2019) Adaptive neural fuzzy inference system for feeding decision-making of grass carp (*Ctenopharyngodon idellus*) in outdoor intensive culturing ponds. *Aquaculture* 498, 28–36.

Zimmermann, E.W., Purchase, C.F. and Fleming, I.A. (2012) Reducing the incidence of net cage biting and the expression of escape-related behaviors in Atlantic cod (*Gadus morhua*) with feeding and cage enrichment. *Applied Animal Behaviour Science* 141, 71–78.

Index

Note: Page numbers in **bold** type refer to **figures**; page numbers in *italic* type refer to *tables*